T0343468

MILK PROTEINS: FROM EXPRESSION TO FOOD

Food Science and Technology
International Series

A complete list of books in this series appears at the end of this volume.

Milk Proteins: from Expression to Food

Edited by
Abby Thompson, Mike Boland and Harjinder Singh
Massey University
New Zealand

AMSTERDAM • BOSTON • HEIDELBERG • LONDON • NEW YORK • OXFORD
PARIS • SAN DIEGO • SAN FRANCISCO • SINGAPORE • SYDNEY • TOKYO
Academic Press is an imprint of Elsevier

Academic Press is an imprint of Elsevier
525 B Street, Suite 1900, San Diego, CA 92101-4495, USA
30 Corporate Drive, Suite 400, Burlington, MA 01803, USA
32, Jamestown Road, London NW1 7BY, UK
360 Park Avenue South, New York, NY 10010-1710, USA

Library of Congress Cataloging-in-Publication Data
A catalog record for this book is available from the Library of Congress

British Library Cataloguing in Publication Data
A catalogue record for this book is available from the British Library

ISBN: 978-0-12-374039-7

For information on all Academic Press publications
visit our website at www.elsevierdirect.com

Typeset by Charon Tec Ltd., A Macmillan Company
www.macmillansolutions.com

Printed and bound by CPI Group (UK) Ltd, Croydon, CR0 4YY

Transferred to Digital Print 2012

Contents

Contributors

Skelte G. Anema (Ch 8) Fonterra Co-operative group Ltd, Private Bag 11 029, Palmerston North, New Zealand

Aurelie Auguste (Ch 2) CRC for Innovative Dairy Products, Department of Zoology, University of Melbourne, Victoria, 3010, Australia.

Mike Boland (Ch 10, 18) Riddet Institute, Massey University, Private Bag 11 222, Palmerston North, New Zealand

Amelia J. Brennan (Ch 2) CRC for Innovative Dairy Products, Department of Zoology, University of Melbourne, Victoria, 3010, Australia.

Therese Considine (Ch 13) Fonterra Co-operative Group Ltd, Private Bag 11 029, Palmerston North, New Zealand

Lawrence K. Creamer (Ch 6, 7) Riddet Institute, Massey University, Private Bag 11 222, Palmerston North, New Zealand

Matthew Digby (Ch 2, 3) Department of Zoology and CRC for Innovative Dairy Products, University of Melbourne, Victoria 3010, Australia.

Mary Ann Drake (Ch 15) Department of Food Science, Southeast Dairy Foods Research Center, North Carolina State University, Raleigh, NC 27695, USA

Patrick B. Edwards (Ch 6) IFS, Massey University, Private Bag 11 222, Palmerston North, New Zealand

Mary Familari (Ch 2) CRC for Innovative Dairy Products, Department of Zoology, University of Melbourne, Victoria, 3010, Australia.

John Flanagan (Ch 13) Riddet Center, Massey University, Palmerston North, New Zealand, now at Glanbia Nutritionals, Glanbia Innovation Center, Kilkenny, Ireland.

Pat F. Fox (Ch 1) Department of Food and Nutritional Sciences, University College, Cork, Ireland

Kelvin K.T. Goh (Ch 12) Institute of Food, Nutrition and Human Health, Massey University, Private Bag 11222, Palmerston North, New Zealand

W. James Harper (Ch 14) Ohio State University, Dept. of Food Science & Technology, 110 Parker Fd., 2015 Fyffe Road, Columbus, OH 43210

Kerianne Higgs (Ch 10) Fonterra Co-operative Group Ltd, Private Bag 11 029, Palmerston North, New Zealand

David S. Horne (Ch 5) Formerly Hannah Research Institute, Ayr, KA6 5HL, Scotland, UK

John W. Holland (Ch 4) Institute for Molecular Bioscience, The University of Queensland, Brisbane 4072, Australia

Geoffrey B. Jameson (Ch 6) IFS, Massey University, Private Bag 11 222, Palmerston North, New Zealand

Elie Khalil (Ch 2) CRC for Innovative Dairy Products, Department of Zoology, University of Melbourne, Victoria, 3010, Australia.

Joly Kwek (Ch 2) CRC for Innovative Dairy Products, Department of Zoology, University of Melbourne, Victoria, 3010, Australia.

Christophe Lefevre (Ch 2, 3) CRC for Innovative Dairy Products, Department of Zoology, University of Melbourne, Victoria, 3010, Australia. Victorian Bioinformatics Consortium, Monash University, Clayton, Victoria, 3800, Australia.

John S. Lucey (Ch 16) Department of Food Science, University of Wisconsin-Madison, 1605 Linden Drive, Madison, WI 53706, USA

Sonia Mailer (Ch 2) CRC for Innovative Dairy Products, Department of Zoology, University of Melbourne, Victoria, 3010, Australia.

K. Menzies (Ch 3) Department of Zoology and CRC for Innovative Dairy Products, University of Melbourne, Victoria 3010, Australia.

R. E. Miracle (Ch 15) Department of Food Science, Southeast Dairy Foods Research Center, North Carolina State University, Raleigh, NC 27695, USA

Paul J. Moughan (Ch 17) Riddet Institute, Massey University, Private Bag 11 222, Palmerston North, New Zealand

Kevin R. Nicholas (Ch 2, 3) CRC for Innovative Dairy Products, Department of Zoology, University of Melbourne, Victoria, 3010, Australia

Hasmukh A. Patel (Ch 7) Fonterra Co-operative group Ltd, Private Bag 11 029, Palmerston North, New Zealand

Anwesha Sarkar (Ch 12) Riddet Institute, Massey University, Private Bag 11 222, Palmerston North, New Zealand

Pierre Schuck (Ch 9) UMR1253, Science et Technologie du Lait et de l'OEuf, Agrocampus Rennes INRA, F-35000 Rennes, France

Julie A. Sharp (Ch 2, 3) CRC for Innovative Dairy Products, Department of Zoology, University of Melbourne, Victoria, 3010, Australia.

P. A. Sheehy (Ch 3) Center for Advanced Technologies in Animal Genetics and Reproduction (REPROGEN), Faculty of Veterinary Science, The University of Sydney, PMB 3 PO, Camden, NSW 2570, Australia.

Harjinder Singh (Ch 11, 12) Riddet Institute, Massey University, Private Bag 11 222, Palmerston North, New Zealand

Denijal Topcic (Ch 2) CRC for Innovative Dairy Products, Department of Zoology, University of Melbourne, Victoria, 3010, Australia.

P. Williamson (Ch 3) Center for Advanced Technologies in Animal Genetics and Reproduction (REPROGEN), Faculty of Veterinary Science, The University of Sydney, PMB 3 PO, Camden, NSW 2570, Australia.

J. M. Wright (Ch 15) Department of Food Science, Southeast Dairy Foods Research Center, North Carolina State University, Raleigh, NC 27695, USA

P. C. Wynn (Ch 3) EH Graham Centre for Agricultural Innovation (Charles Sturt University and NSW Department of Primary Industries) Locked Bag 588, Wagga Wagga, NSW 2678, Australia.

Aiqian Ye (Ch 11) Riddet Institute, Massey University, Private Bag 11 222, Palmerston North, New Zealand.

Preface

Proteins are vital ingredients for the food industry because they provide all the essential amino acids needed for human health combined with a wide range of dynamic functional properties, such as the capacity to form network structures and stabilize emulsions and foams. The proteins of milk have excellent functional properties and nutritional value, and some have distinctive physiological properties, which are widely exploited in the food industry. Milk proteins have been the subject of intensive research during the last 50 years in an effort to unravel their molecular structures and interactions, relationship between structure and functional attributes, interactions of proteins during processing and, more recently, their physiological functions.

Recent studies on the interactions of milk proteins in complex food systems are leading to a new understanding of the nature of these interactions and their impact on food quality. The knowledge has resulted in the development of several specialized milk protein ingredients tailored to meet specific needs of the food industry. Currently, there is a growing demand by the food industry for milk protein ingredients for specialist high-value applications such as functional foods. In the future, application of novel analytical approaches (genomics, proteomics, nanotechnology) to milk proteins and food materials will provide further understanding of molecular structures and interactions to enable the dairy industry to produce highly functional and healthy protein ingredients for specific applications.

Several books have been published about milk and milk proteins—so why another one? Most of the earlier books have addressed different specialist aspects of dairy science and technology. The primary theme of this book is to present a view along the dairy food chain—starting at the cow (and its mammalian relatives) and finishing with nutritional aspects affecting the consumer, dipping into important current research topics along the way. The molecular structures and interactions of milk proteins under various processing environments are covered most prominently. More importantly, the book also contains a considerable amount of material from dairy industry-based or industry-funded research. Thus, it provides fresh perspectives on milk proteins, from an advanced dairy industry point of view.

The editors particularly thank Fonterra Cooperative Limited for making available time for staff members to contribute their chapters, and for making available hitherto unpublished material. This book is designed to provide an update and call for attention, for industry and academic researchers alike, to important and relevant milk protein science in areas that have the potential to advance the dairy industry.

The overall theme covered in this book was piloted at a meeting organized by the Riddet Center and Fonterra Cooperative Limited in February 2006, with invited presentations from a number of experts in the relevant fields from Australasia, the USA and the UK. This meeting was particularly successful, with a large number of international delegates attending from a broad range of disciplines. This confirmed the growing interest of milk protein scientists in looking beyond the boundaries of their immediate topic area to gain an understanding of how the whole food chain fits together. Such an understanding can help elucidate mechanisms and processes, identify novel research opportunities, and provide additional applications for new developments.

This book includes chapters covering many of the topics addressed at the meeting, as well as some new subjects that we felt were important in order to provide a more complete picture of the journey from expression to food. We would like to thank both the contributors who have been involved from the meeting in 2006 and those who have come on board more recently.

We have chosen to start the book with a comprehensive overview of the biology and chemistry of milk, to set the stage and give a broad underpinning of the later chapters for readers not familiar with this field. Attention is then turned to the biology, and particularly the molecular biology, of lactation, looking first at some "extreme" mammals—the tammar wallaby, which can express two different milk compositions at the same time, and the fur seal, which produces an extremely concentrated milk—to give an idea of the range of biology of milk production. The book thereafter focuses on bovine milk, with mention of the milk from other domestic species as appropriate. This starts with an update on the genomics of bovine milk proteins, and is followed by an overview on post-translational modifications, which completes our view of the biology of milk protein production.

The structural chemistry of milk proteins, including the latest model of casein micelle and molecular structures of whey proteins, is covered in detail. The behavior of milk proteins under a variety of processing regimes, including ultra high pressure, functional systems, drying and storage of powders, is dealt with in a series of chapters. These chapters address our current state of knowledge about existing and emerging processes for the production of milk protein-based food ingredients.

Attention is then turned to the behaviors of milk proteins in real and model food systems, and finally to consumer aspects—the sensory and nutritional/functional food aspects of milk proteins. A wrap-up chapter gives a view on likely issues of future importance for milk proteins, including the emerging area of nutrigenomics.

As with any volume written by a large number of contributors, this contains a variety of styles of presentation. We have made no attempt to homogenize the authors' styles, but have provided guidance on chapter content to make for best possible continuity.

A volume of this kind always requires a large amount of work by a large number of people. We would like particularly to thank all the contributing authors for their efforts and their expeditious preparation of manuscripts that allowed for the timely publication of this book. We are pleased to acknowledge Claire Woodhall for assisting with the technical editing, and the staff at Elsevier for producing this book.

Abby Thompson
Mike Boland
Harjinder Singh

Milk: an overview

1

Pat F. Fox

Abstract

Milk is the characterizing secretion of mammals, of which there are about 4500 species, produced to meet the complete nutritional requirements of the neonate of the species, as well as some defensive and other physiological requirements. The milks of all species are basically similar but there are very significant species-specific differences. In addition to supplying all the nutritional requirements of the neonate, many of the micro-constituents of milk, such as oligosaccharides, immunoglobulins, metal-binding proteins and enzymes, serve protective roles. Milk is an aqueous solution of lactose, inorganic and organic salts, and numerous compounds at trace levels (milk serum), in which are dispersed colloidal particles of three size ranges: whey proteins dissolved at the molecular level, the caseins dispersed as large (50–500 nm) colloidal aggregates (micelles), and lipids emulsified as large (1–20 μm) globules.

The colloidal stability of milk, especially of the casein micelles, is very important from the nutritional and technological viewpoints. The micelles are destabilized and aggregate or gel following limited proteolysis or acidification to ≈pH 4.6. *In vivo*, aggregation occurs in the stomach of the neonate, thereby slowing transit and improving digestibility. Technologically, destabilization of the micelles can be undesirable or can be exploited in the production of cheese and fermented milk products.

Man has used milk in his diet for about 8000 years and a major industry has developed around the processing of milk of a few species for human foods, especially milk from cattle, buffalo, sheep and goats. Milk processing, which exploits certain physico-chemical properties

Milk Proteins: From Expression to Food
ISBN: 978-0-12-374039-7

of milk, is practiced worldwide, especially in Europe and North America. Milk is a very flexible raw material, from which a very wide range of different products is produced, including about 1400 varieties of cheese.

In this introductory chapter, the chemical and biological characteristics of milk constituents and the physico-chemical and technological properties of milk will be summarized.

Introduction

Milk is a fluid secreted by the female of all mammalian species, of which there are about 4500 extant species (about 80% of mammalian species are extinct), primarily to meet the complete nutritional requirements of the neonate. The principal requirements are for energy (supplied by lipids and lactose and, when in excess, by proteins), essential amino acids and amino groups for the biosynthesis of non-essential amino acids (supplied by proteins), essential fatty acids, vitamins, inorganic elements and water. Because the nutritional requirements of the neonate depend on its maturity at birth, its growth rate and its energy requirements, which depend mainly on environmental temperature, the gross composition of milk shows large inter-species differences, which reflect these requirements (see Fox and McSweeney, 1998). Milk also serves a number of physiological functions, which are performed mainly by proteins and peptides, including immunoglobulins, enzymes, enzyme inhibitors, growth factors, hormones and antibacterial agents.

Of the 4500 species of mammal, the milk of only about 180 species has been analyzed, and of these the data for only about 50 species are considered to be reliable (i.e. a sufficient number of samples, samples taken properly, representative sampling, adequate coverage of the lactation period). Milk from the commercially important species—cow, goat, sheep, buffalo, yak, horse and pig—are quite well characterized; human milk is also well characterized, as is that of experimental laboratory animals, especially rats and mice. Reviews on non-bovine milks include: general (Evans, 1959; Jenness and Sloan, 1970), buffalo (Laxminarayana and Dastur, 1968), goat (Parkash and Jenness, 1968; Haenlein, 1980; Jenness, 1980), sheep (Bencini and Pulina, 1997), sheep and goats (International Dairy Federation, 1996; Jandal, 1996; Park *et al.*, 2007), camel (Rao *et al.*, 1970; Farah, 1993), horse (Doreau and Boulot, 1989; Solaroli *et al.*, 1993; Doreau, 1994; Park *et al.*, 2006), human (Atkinson and Lonnerdal, 1989; Jensen, 1989, 1995) and sow (Verstegen *et al.*, 1998). The *Handbook* edited by Park and Haenlein (2006) is a particularly useful source of information on the milk of non-bovine mammals; it includes chapters on goat, sheep, buffalo, mare, camel, yak, reindeer, sow, llama, minor species (moose, musk ox, caribou, alpaca, ass, elk, seal, sea lion and polar bear) and human. Inter-species comparisons of more specific aspects are cited in the appropriate sections.

The milks of certain domesticated animals, and dairy products therefrom, are major components of the human diet in many parts of the world. Domesticated goats, sheep and cattle have been used for milk production since about 8000 BC. Recorded milk production today is about 600×10^6 tonnes per annum, about 85% of which is bovine, 11% is buffalo and about 2% each is ovine and caprine, with small amounts produced from horses, donkeys, camels, yaks and reindeer. Milk and dairy products are consumed throughout the world but are particularly important in Europe, the USA, Canada, Argentina, India, Australia and New Zealand. The contribution of milk and dairy products to dietary intake varies widely for different regions of the world, e.g. the kilocalories per day supplied by milk range from 12 in China to 436 in Ireland; in the UK, milk and dairy products supply ≈30% of dietary protein consumed by young children, ≈27% of dietary lipids and ≈65% of calcium (Barker, 2003).

The chemistry and the physico-chemical properties of milk have been studied for about 200 years and are now understood in considerable detail, and described in a voluminous literature. The objectives of this chapter are to provide a summary and overview of the evolution of mammals and lactation and of the principal constituents of milk, especially proteins, which are the subject of this book. Where possible, inter-species comparisons of milk and its constituents are made. Numerous textbooks and review articles are cited and may be consulted for the primary literature.

Evolution of mammals and lactation

The secretion of milk is one of the characterizing features of mammals, which evolved from egg-laying, pre-mammalian reptiles, synapsids and cynodonts. Cynodonts are believed to be the ancestors of all mammals and evolved ≈200 million years ago (at the end of the Triassic Period). The word "mammal" is derived from "*mamma*," which is Latin for breast. Initially, mammals were small shrew-like creatures but they have evolved and diversified to occupy all niches on land, sea and air. They range in size from a few grams (pigmy shrew) to 200 tonnes (blue whale). Their dominance occurred especially after the extinction of the dinosaurs, 60–70 million years ago, at the interface between the Cretaceous and Tertiary Periods (C/T interface). Mammals have been successful: the young of most species are born alive (viviparous) and all are supplied with a specially designed food—milk—for the critical period after birth. No other class of animal is so pampered (for an interesting discussion on this point, see Peaker, 2002). Not surprisingly, the evolution of mammals is a popular subject; reviews include: Crampton and Jenkins (1973), Kemp (1982), Lillegraven *et al.* (1987), Benton (1999), Easteal (1999), Forsyth (2003), Lillegraven (2004), and Springer *et al.* (2004).

Mammals are distinguished from other classes of animals by four criteria:

- they secrete milk to nourish their young;
- they are endothermic, i.e. they can control their body temperature;
- they grow body hair or wool for insulation—even aquatic mammals have some hair; and
- they have different types of teeth (flat incisors, conical canines and multi-cusped molars), which allow them to masticate different types of food.

The class *Mammalia* contains two subclasses—*Prototheria* and *Theria* (young born alive).

Prototheria. These egg-laying mammals, known as monotremes because they have only one opening for the elimination of waste, mating and egg laying, were the first mammals, only five species of which survive, i.e. the duck-billed platypus and four species of echidna (also called spiny anteater), which are found only in Australia and New Guinea. Presumably, there were other species of monotreme, which have become extinct.

Theria. About 90 million years ago, the *Theria* split into two infra-classes, *Metatheria* and *Eutheria.* However, the fossil of a eutherian mammal, believed to be ≈125 million years old, was discovered recently in north-eastern China; it was named *Eomonia scansoria*, meaning "earliest eutherian mammal with specialized features for climbing" (Ji *et al.*, 2002).

Metatheria. Usually called marsupials, of which there are about 330 species. The young are born alive (*viviparous*) but very immature and develop in an abdominal pouch (*marse* = pouch, purse). Marsupials survive mainly in Australia and surrounding islands (>200 species), with several species in South America and one species, the Virginia opossum, in North America; there are none in Europe, Asia or Africa.

Eutheria. The fetus develops in utero, where it receives nourishment from the maternal blood via a highly specialized organ called the placenta (these are called placental mammals); ≈95% of all mammals are eutherians.

Classification of mammals

There are ≈4500 species of mammal, which are classified into 20 orders (see MacDonald, 2004). It is estimated that only about 20% of the species that have evolved over the last 200 million years are still extant. The classification and nomenclature of mammals commenced with the work of Carolus Linnaeus in 1758 and was based initially on morphological characteristics (see MacDonald, 2004). More soundly based classification is now possible, based on DNA sequences (Madsen *et al.*, 2001; Murphy *et al.*, 2001; Springer *et al.*, 2004) and on the primary sequence of certain proteins, e.g. growth hormone and prolactin (Forsyth and Wallis, 2002). It should be possible to classify mammals based on the structure of milk proteins, especially caseins, which are fast-mutating proteins; some preliminary work has been reported (Goldman, 2002; Rijnkels, 2002; Simpson and Nicholas, 2002). However, the sequences of milk proteins from a sufficient number of species have not been determined to make a comprehensive classification scheme possible, but considerable progress on the structure of milk proteins is being made (see Martin *et al.*, 2003) and is discussed further later.

A minor whey protein—whey acidic protein (WAP)—has already been useful for tracing the relationships between mammalian families (Hajjoubi *et al.*, 2006). To date, WAP has been found only in the milk of platypus, echidna, tammar wallaby, opossum, mouse, rat, rabbit, camel and pig. In humans and ruminants, the gene for WAP has been frame shifted and occurs as a pseudogene that is not transcribed. The distribution of WAP suggests that the loss of a functional WAP gene occurred after the divergence of the pig and ruminant lines but before the *Bovidae* diverged from the other ruminants. Analysis of the milks from a wider range of species for WAP should be interesting.

Classification and phylogenetic relationships of the principal dairying species

All of the principal, and many of the minor, dairying species belong to the family *Bovidae*, a member of the order *Artiodactyla* (even-toed ungalates [hoofed] mammals, i.e. cloven hoofed). A few minor dairying species (horse and ass) are members of *Perissodactyla* (odd-toed ungalates). The *Bovidae* evolved ~18 million years ago; the earliest fossil attributed to the *Bovidae* is *Eotragus*, found in 18-million-year-old deposits in Pakistan. The order has three sub-orders: *Ruminantia* (ruminants; to which all major dairying species belong), *Sunia* (pigs and related species) and *Tylopoda* (camels, llama, alpaca and guanaco).

The *Ruminantia* are classified into six families: *Tragulidae* (chrevrotains), *Moschidae* (musk deer), *Antilocapridae* (pronghorns), *Giraffidae* (giraffes and okapi), *Cervidae* (deer; 43 species in 16 genera) and *Bovidae* (137–38 species in 46–7 genera).

The *Bovidae* are divided into six subfamilies, of which the *Bovinae* is the most important. The *Bovinae* are divided into three tribes, of which the *Bovini* are the most important from our viewpoint. The *Bovini* are classified into five genera: *Bubalus* (water buffalo), *Bos* (cattle), *Pseudoryx*, *Syncerus* (African [Cape] buffalo) and *Bison* (American and European; the latter is also called wisent).

There are seven or eight species of *Bos*: *B. primigenus* (aurochs—the ancestor of domestic cattle; they are extinct; the last animal was killed in Poland in 1627), *B. javanicus* (banteng), *B. gaurus* (gaur), *B. frontalis* (gayol), *B. mutus* (yak), *B. sauvali* (kouprey), *B. taurus* (European cattle) and *B. indicus* (Indian, humped zebu cattle). (*B. taurus* and *B. indicus* may be sub-species rather than species.) The phylogenetic relationships of the *Bovini* have been studied by molecular biology techniques (see Finlay, 2005).

Today, there are about 1.3×10^9 cattle worldwide, of which there are two species—*B. taurus*, of European origin, and *B. indicus*,which originated in India. *B. indicus* (zebu) cattle also dominate in Africa but apparently African zebu cattle have some *B. taurus* genes, probably as a result of cross-breeding many centuries ago. Zebu are less efficient producers of milk or meat than *B. taurus* but are more resistant to heat stress and various diseases, and therefore dominate in tropical regions.

Since cattle were domesticated ~8000 years ago, they have been bred selectively, especially during the past 200 years, i.e. since Herd books have been kept. These breeding practices have selected for various characteristics, e.g. health, fertility, docility, milk or meat production, or both. Today, there are about 800 breeds of cattle, including dairy, beef or dual-purpose breeds. There are ~200 million dairy cows of many, mainly local, breeds; Holstein-Friesian is the principal breed of dairy cow, representing ~35% of the total (~70 million cows). Other important international dairy breeds are Brown Swiss (~4 million), Jersey (~2 million), Ayrshire, Guernsey and Red Dane.

There are about 160 million buffalo worldwide, of which there are two types—river and swamp—found mainly in South-East Asia, India and Egypt, with smaller numbers in Bulgaria, Italy, Brazil and Australia. Depending largely on the region, buffalo are used for milk, meat or work, or combinations of these. There are no breeds of buffalo, within the meaning applied to cattle; rather, they are named after the area from which they come.

Evolution of the mammary gland

Evolution of the mammary gland is believed to have commenced with the synapsids, ~300 million years ago. These reptiles laid membrane-shelled eggs, which lost water rapidly through evaporation and were kept moist by an aqueous oily secretion from sebaceous apocrine glands on the breast/abdomen of the mother; these were somewhat like the blood patches on the breast of birds. The secretions are believed to have contained a range of bactericidal substances, e.g. oligosaccharides, lysozyme, lactoferrin, transferrin, immunoglobulins and peroxidases, which protected the egg against microbial infection. Presumably, the secretions were licked by the neonate from the mother's abdomen and served as a source of nutrients.

Considering their importance in the evolution of mammals, including man, the evolution of the mammary gland and the origin of lactation have attracted considerable attention. Reviews on the subject include Pond (1977), Hayssen and Blackburn (1985), Blackburn *et al.*, (1989), Blackburn (1991, 1993), Hayssen (1993) and Oftedal (2000a, 2000b).

Structure of the mammary gland

The microscopic structure of the mammary gland of all species—monotremes, marsupials and eutherians—is basically similar. The structure of the bovine gland from the cellular level to the organ level has been described by Fox and McSweeney (1998). The cells (mammocytes), the structure of which is basically similar to that of all animal cells, are arranged as a monolayer in a pear-shaped organelle, called an alveolus. The alveoli are connected via a system of ducts to a cistern, where the milk is stored until it is expressed from the gland, usually through a teat which is sealed by a sphincter muscle. There is little *de novo* synthesis in the mammary gland; rather, the major constituents in milk are synthesized from molecules imported from the blood through the basal cell membrane. Within the mammocyte, mainly in the rough endoplasmic reticulum (ER), these molecules are polymerized to lactose, lipids or proteins. The mammocytes are provided with a good blood supply through an extensive system of capillaries and are surrounded by contractile mycoepithelial cells which, under the control of the hormones oxytocin and prolactin, contract and express milk from the alveoli through the ducts, and eventually from the gland. The hormonal control of mammary growth and function has been described by Forsyth (1986).

Although, at the microscopic level, the mammary gland is essentially similar across species, the number and the appearance of the gland are characteristic of the species. Monotremes have many glands on the abdomen; the glands do not end in a teat and the milk is licked from the abdomen by the young. Marsupials have two or four glands, which end in a teat, within the pouch. On entering the pouch, the young attaches to a teat and remains attached during the period it spends permanently in the pouch. During this period, an older offspring may use another gland during its visits to the pouch. The two glands secrete milk of very different composition, designed for the specific requirements of the neonate, and the composition changes markedly when the offspring leaves the pouch intermittently. The mammary glands of eutherians are located externally to the body cavity and end in a teat; their number varies from two (human, goat, sheep, horse, elephant, etc.), four (cattle), 14 (pig) to 24 (some insectivores). The glands are separate anatomically.

The external location of the mammary gland facilitates study of the biosynthesis of milk constituents by isotope dilution techniques, arterio-venus concentration differences, perfusion of the severed gland, tissue slices and cell homogenates.

Utilization of milk

As young mammals are born at different stages of maturity and with different nutritional requirements, the milk of each species is designed to meet the requirements of the neonate of that species, i.e. it is species specific. Milk is intended to be consumed unchanged by the young suckling its mother. However, man has consumed the milk of other species for at least 8000 years. Several species have been used for milk production but today cattle, especially *Bos taurus*, is the principal dairying species, accounting for ~85% of total milk production. The other important dairying species are buffalo (*Bubalus bubalus*) (11%), goats and sheep (2% each); other species are significant in certain regions, or for certain purposes, e.g. camel, yak, reindeer, horse and donkey.

Milk is often described as the most "nearly perfect" food; although this is true only for the young of the producing or closely related species, the milk of all species is a nutrient-rich and well-balanced food (Kon, 1959; du Puis, 2002; Patton, 2004). Many of the minor constituents of milk have biological properties, which are described in the appropriate section (see also Korhonen, 2006); these minor constituents have been attracting considerable attention recently.

However, milk is very susceptible to the growth of micro-organisms, which will cause spoilage if the milk is stored. To counteract this, man has developed a range of products that are more stable than milk; some of these date from 4000 BC and have evolved desirable epicurean characteristics, in addition to their nutritional value. Today, several thousand food products are produced from milk; these fall into the following principal groups: liquid/beverage milk (40%), cheese (35%), milk powders (15%), concentrated milks (2%), fermented milk products (2%), butter (30%; some of which is produced from cream/fat obtained as a by-product in the manufacture of other products), ice cream, infant formula, creams, protein-rich products and lactose. Some of these groups are very diverse, e.g. 1400 varieties of cheese have been listed.

Composition of milk

Milk is a very complex fluid containing several hundred molecular species (several thousand if all triglycerides are counted individually). The principal constituents are water, lipids, sugar (lactose) and proteins. In addition, there are numerous minor constituents, mostly at trace levels, e.g. minerals, vitamins, hormones, enzymes and miscellaneous compounds. The chemistry of these compounds is generally similar across species but in many cases their structure differs in detail, reflecting evolutionary changes. The concentration of the principal constituents varies widely among species: lipids, 2–55%; proteins, 1–20%; lactose, 0–10%, reflecting mainly the energy requirements (lipids and lactose) and growth rate (mainly proteins) of the neonate. The concentrations of the minor constituents also vary widely.

Within any species, the composition of milk varies among individual animals, between breeds, with the stage of lactation, feed and health of the animal and with many other factors. The fat content of bovine milk shows large inter-breed differences and within any breed there is a wide range of fat and protein content for individual animals; similar differences occur in the milk of the sheep, goat and buffalo.

Reflecting mainly the nutritional and physiological requirements of the neonate, the composition of milk, and even the profile of constituents therein, changes markedly during lactation. The changes are most marked during the first few days post-partum, especially in the immunoglobulin fraction of proteins. For marsupials, the milk changes from a high-carbohydrate (mainly oligosaccharides) secretion to a high-fat secretion when the neonate begins to leave the pouch, a time that corresponds roughly to the birth of eutherians. The composition of milk remains relatively constant during mid-lactation but changes considerably in late lactation, reflecting the involution of the mammary gland tissue and the greater influx of blood constituents.

Milk constituents

In the following sections, the chemistry of milk carbohydrates, lipids, proteins, salts and some minor constituents are described; where possible, inter-species comparisons are made.

Carbohydrates

Lactose

The principal carbohydrate in the milk of most species is the reducing disaccharide lactose, which is composed of galactose and glucose linked by a β1–4 glycosidic bond. Its concentration varies from 0 to ~10% (Fox and McSweeney, 1998) and milk is the only known source of lactose. Research on lactose commenced with the work of Carl Scheele in about 1780; its chemistry and its important physico-chemical properties have been described very thoroughly. The very extensive literature has been reviewed by Whittier (1925, 1944), Weisberg (1954), Zadow (1984, 1992), Fox (1985, 1997), Fox and McSweeney (1998) and McSweeney and Fox (2008).

Lactose is synthesized in the epithelial mammary cells from two molecules of glucose absorbed from the blood. One molecule of glucose is phosphorylated and converted (epimerized) to galactose-P via the Leloir pathway, which is widespread in animal tissues and bacterial cells. Galactose-P is condensed with a second molecule of glucose through the action of a unique two-component enzyme—lactose synthetase. One component is UDP-galactosyl transferase (EC 2.4.1.22), which transfers galactose from UDP-galactose to any of several acceptor molecules in the biosynthesis of glycoproteins and glycolipids. The specificity of the transferase is controlled and modified by a-lactalbumin (α-La), one of the principal milk proteins, which reduces the Michaelis constant (K_M) for glucose 1000-fold and, in its presence, most of the galactose is transferred to glucose, with the synthesis of lactose. There is a positive correlation between the concentrations of lactose and α-La in milk; the milk of the Californian sea lion or the hooded seal, which contains no lactose, also lacks α-La.

The synthesis of lactose draws water osmotically into the Golgi vesicles and hence affects the volume of milk and the concentration of casein, which is packaged in the Golgi vesicles. There is an inverse correlation between the concentrations of lactose and casein in milk (Jenness and Holt, 1987).

Lactose serves as a ready source of energy for the neonate, providing 30% of the calories in bovine milk and acting as an alternative to the more energy-dense lipids. Milks with a high lactose concentration tend to have lower levels of lipids (Jenness and Sloan, 1970).

Lactose is also responsible for about 50% of the osmotic pressure of milk, which is isotonic with blood and hence essentially constant. For milk with a low level of lactose, the concentration of inorganic salts is high to maintain the osmotic pressure at the desired level; there is an inverse relationship between the concentrations of lactose and salts (ash) in milk (Jenness and Sloan, 1970).

During mastitis or in late lactation, the integrity of the mammocyte cell membranes is damaged and there is an influx of blood constituents into milk; the osmotic pressure increases and, to adjust this, the concentration of lactose is reduced. This relationship is expressed as the Koesler Number (% chloride × 100 ÷ % lactose), which is normally <2; a value >3 is considered to be abnormal. Today, the Koesler Number is rarely used as a diagnostic indicator of mastitis. The electrical conductivity of milk is commonly used for this purpose, as it depends mainly on the milk salts and can be measured in-line during milking.

Why milk contains lactose rather than some other sugar(s) is not clear. The presence of a disaccharide rather than a monosaccharide can be explained on the basis that twice as much (mass) disaccharide as monosaccharide can be accommodated for any particular incremental increase in osmotic pressure, which is fixed. Maltose, which consists of two molecules of glucose, would seem to be the obvious choice of disaccharide. As energy is expended in converting glucose to galactose, some benefit must accrue from this conversion; a possible benefit is that galactose, or a derivative thereof, occurs in some physiologically important lipids and proteins, and a galactose-containing sugar in milk provides the neonate with a ready supply of this important monosaccharide.

The properties of lactose are generally similar to those of other sugars, but lactose differs in some technologically important respects. Some important characteristics of lactose are as follows:

- Lactose is a reducing sugar, i.e. it has a free, or potentially free, carbonyl group (an aldehyde group in the case of lactose).
- Like other reducing sugars, lactose exists partially as an open-chain form with an aldehyde group, which can form a hemi-acetal and thus a ring structure. Formation of a hemi-acetal creates a new chiral center (asymmetric carbon),

which may exist as two isomers (enantiomorphs), α or β. By alternately opening and forming the ring structure, the molecule can interchange between the α and β isomers, a process referred to as mutarotation.

- The α and β isomers of lactose have very different properties, the most important of which are specific rotation, $[\alpha]^{20}_{D}$ (+89° and +35° for α and β respectively), and solubility (70 and 500 g/L for α and β respectively).
- Like all reducing sugars, lactose can participate in the Maillard (non-enzymatic) browning reaction, resulting in the production of (off-)flavor compounds and brown polymers. The Maillard reaction contributes positively to the flavor and color of many foods, e.g. the crust of bread, toast and deep-fried products, but the effects in dairy products are negative and must be avoided.
- Redox titration using alkaline $CuSO_4$ (Fehling's solution) or chloramine-T is the principal standard method for the quantitative determination of lactose. It may also be determined by polarimetry, spectrophotometrically after reaction with phenol or anthrone in strongly acid solution, enzymatically or by high performance liquid chromatography.
- Among sugars, lactose, especially the α enantiomorph, has low solubility in water; however, when in solution, it is difficult to crystallize, which may cause problems in lactose-rich dairy products, e.g. skimmed milk powder and whey powder, unless precautions are taken to induce and control crystallization.
- α and β lactose are soluble in water to the extent of about 70 and 500 g/L respectively at 20°C; at equilibrium, the ratio of α to β is about 1:2, giving a total solubility of about 180 g/L at 20°C. The solubility of α lactose is more temperature dependent than that of the β isomer and α lactose is the more soluble at >94°C. Hence, α lactose is the form of lactose that crystallizes at <94°C and is the usual commercial form of lactose; β lactose may be prepared by crystallization at >94°C.
- α lactose crystallizes as a monohydrate whereas β lactose forms anhydrous crystals; thus, the yield of α lactose is 5% higher than that of β lactose.
- When milk or whey is spray dried, any lactose that has not been pre-crystallized forms an amorphous glass. The amorphous glass is stable if the moisture content of the powder is maintained low but, if the moisture content increases to >6%, the lactose crystallizes as the α hydrate, the crystals of which form interlocking masses and clumps which may render the powder unusable if this is very extensive, i.e. inadequately crystallized powder is hygroscopic. The problem can be avoided by adequate crystallization of lactose before drying or by using effective packaging.
- Interestingly, crystalline lactose has very low hygroscopicity and is used in icing sugar blends.
- Among sugars, lactose has a low level of sweetness; it is only about 16% as sweet as sucrose at 1% in solution and hence has limited value as a sweetening agent, the principal application of sugars in foods. However, it is a useful bulking agent when excessive sweetness is undesirable.
- Lactose is important in the manufacture of fermented dairy products, where it serves as a carbon source for lactic acid bacteria, which produce lactic acid.

Modification of the concentration of lactose in milk through genetic engineering

There has been considerable interest in modifying the lactose content of milk by genetic engineering. As the concentration of lactose is controlled by the concentration of α-La in the secretory cells, the approach to changing the concentration of lactose involves altering the level of α-La. There is interest in reducing the level of lactose for a number of reasons, not least because lactose is the least valuable constituent in milk but it costs energy on the part of the animal to synthesis it; therefore, it would be economically advantageous to reduce the lactose content of milk.

As lactose effectively controls the water content of milk, and most dairy processes require the removal of water, it would be advantageous to reduce the amount of water in milk by reducing the level of lactose. However, if the level of lactose is reduced too much, the viscosity of the milk will be too high for easy expression from the mammary gland; the viscosity of mouse milk engineered to contain no lactose was so high that the pups were unable to suckle and died. Obviously, this problem could be overcome by reducing the level of lactose rather than eliminating it. Alternatively, it may be possible to modify the milk secretory mechanism to produce a more useful, or at least a less problematic, sugar than lactose, e.g. glucose, maltose or lactulose (which is a laxative and a prebiotic). It might be possible to increase the concentration of salts in milk.

As discussed below, most adult humans are unable to digest lactose. If the problems arising from high viscosity were resolved, lactose-free or lactose-reduced milk would be nutritionally desirable. The possibility of engineering the mammary cell to secrete β-galactosidase into milk and to hydrolyze lactose in situ has been suggested.

In contrast, there are potentially cases where it would be advantageous to increase the lactose content of milk. The economic benefits of increasing the milk output of sows by increasing its lactose content have been discussed by Wheeler (2003).

Nutritional problems associated with lactose

Mammals cannot absorb disaccharides from the small intestine, where they are hydrolyzed to monosaccharides which are absorbed. Lactose is hydrolyzed by β-galactosidase (lactase) which is secreted by cells in the brush border of the small intestine. The young of most mammalian species secrete an adequate level of β-galactosidase but, as the animal ages, the secretion of β-galactosidase declines and eventually becomes inadequate to hydrolyze undigested lactose, which enters the large intestine into which it draws water, causing diarrhea, and is metabolized by bacteria with the production of gas, which causes cramps and flatulence. In humans, this may occur at 8–10 years of age. These problems cause many individuals to exclude milk from their diet. The problems may be avoided by pre-hydrolyzing the lactose using exogenous β-galactosidase (see Mahoney, 1997). The frequency and the intensity of lactose intolerance/malabsorption vary widely among populations from ≈100% in South-East Asia to ≈5% in north-west Europe (Mustapha *et al.*, 1997; Ingram and Swallow, 2008).

Production and utilization of lactose

Previously, whey from cheese or casein production was considered to be a waste material that was fed to farm animals, irrigated on land or disposed of into sewers.

Environmental and economic considerations now dictate that whey be used more efficiently. The principal product lines produced from whey are various whey powders, whey protein products produced by membrane technology and lactose and its derivatives.

Lactose is prepared commercially by crystallization from concentrated whey or ultrafiltrate. The crystals are usually recovered by centrifugation; this process is essentially similar to that used for sucrose or other sugars. About 400 000 tonnes of crystalline lactose is produced annually, compared with ≈100 million tonnes of sucrose.

Because of its relatively low sweetness and low solubility, the applications of lactose are much more limited than those of sucrose or glucose. Its principal application is in the production of "humanized" infant formulas based on cow's milk (human milk contains ≈7% lactose in comparison with ≈4.6% in bovine milk). The lactose used may be a purified crystalline product or in the form of demineralized whey (for physiological reasons, it is necessary to reduce the concentration of inorganic salts in whey).

Lactose has a number of low-volume, special applications in the food industry, e.g. as a free-flowing or agglomerating agent, to accentuate/enhance the flavor of some foods, to improve the functionality of shortenings, and as a diluent for pigments, flavors or enzymes. It is widely used in the tableting of drugs in the pharmaceutical industry, where low hygroscopicity is a critical property.

Lactose can be converted to several more valuable food-grade derivatives, of which the most significant are: glucose-galactose syrups (≈ three times as sweet as lactose; produced by hydrolysis by β-galactosidase), lactulose (galactose-fructose; a prebiotic and a laxative), lactitol (the alcohol of lactose), lactobionic acid (a sweet-tasting acid, which is a very rare property), tagatose, oligosaccharides (prebiotics) and fermentation products (ethanol, and lactic, acetic and propionic acids).

Oligosaccharides

In addition to lactose, a large number of other free saccharides have been found in milk, the concentration, proportions and types of which show large inter-species differences. Oligosaccharides are the most common form, but small amounts of monosaccharides are also present and some milk proteins, especially κ-casein, are glycosylated; there are low levels of highly glycosylated glycoproteins, especially mucins, and glycolipids in the milk fat globule membrane (MFGM).

Almost all of the oligosaccharides have lactose at the reducing end, contain three to eight monosaccharides, may be linear or branched and contain either or both of two unusual monosaccharides—fucose (a 6-deoxyhexose) and *N*-acetylneuraminic acid. Fucose occurs quite widely in the tissues of mammals and other animals, where it serves an array of functions (Becker and Lowe, 2003). Its significance in the oligosaccharides in milk is not clear; perhaps it is to supply the neonate with preformed fucose as the concentration of oligosaccharides is higher in colostrum than in milk. General reviews on the oligosaccharides in milk include Newburg and Newbauer (1995) and Urashima *et al.* (2001, 2007).

The oligosaccharides are synthesized in the mammary gland, catalyzed by special transferases that transfer galactosyl, sialyl, *N*-acetylglucosaminyl or fucosyl residues from nucleotide sugars to the core structures. These transferases are not affected by α-La and are probably similar to the transferases that catalyze the glycosylation of lipids and proteins.

The milk of all species examined contains oligosaccharides but the concentration varies markedly. The highest levels are in the milk of monotremes, marsupials, marine mammals, humans, elephants and bears. With the exception of humans, the milk of these species contains little or no lactose and oligosaccharides are the principal carbohydrates. The milk of the echidna contains mainly the trisaccharide fucosyllactose whereas that of the platypus contains mainly the tetrasaccharide difucosyllactose. Among marsupials, the best studied is the tammar wallaby; presumably, its lactation pattern and its milk composition are typical of marsupials. A low level of lactose is produced at the start of lactation but, about 7 days after birth, a second galactosyltransferase appears and tri- to pentasaccharides are produced, which by ≈180 days are the principal saccharides; during this period, the saccharide content is high (≈50% of total solids) and the level of lipids is low (≈15% of total solids). At about 180 days, the carbohydrates decrease to a very low level and are mainly monosaccharides, whereas the level of lipids increases to ≥60% of total solids.

Human milk contains ≈130 oligosaccharides, at a total concentration of ≈15 g/L; these are considered to be important for neonatal brain development. Bear milk contains very little lactose but a high level of total sugars (probably mainly oligosaccharides)—1.7 and 28.6 g/kg respectively (Oftedal *et al.*, 1993). Elephant milk contains ≈50 and 12 g/kg of lactose and oligosaccharides respectively a few days postpartum but, as lactation progresses, the concentration of lactose decreases whereas that of oligosaccharides increases, e.g. 12 and 18 g/kg respectively at 47 days (Osthoff *et al.*, 2005). The milk of seals contains both lactose and oligosaccharides but the milks of the Californian sea lion, Northern fur seal and Australian fur seal contain neither, probably because they contain no α-La (Urashima *et al.*, 2001).

Bovine, ovine, caprine and equine milks contain relatively low levels of oligosaccharides, which have been characterized (see Urashima *et al.*, 2001). Caprine milk contains about 10 times as much oligosaccharides as bovine milk and ovine milk, and a process for their isolation by nanofiltration has been reported (Martinez-Ferez *et al.*, 2006). Possible methods for producing oligosaccharides similar to human milk oligosaccharides, by fermentation or by transgenic animals or by recovering oligosaccharides from cow's milk whey or ultrafiltration permeate, have been discussed by Mehra and Kelly (2006).

As discussed earlier, oligosaccharides with bactericidal properties were probably the saccharides present in the mammary secretions of early mammals, and the high level of oligosaccharides in the milk of monotremes and marsupials conforms with their secretion early in evolution. Messer and Urashima (2002) proposed that the primitive mammary glands of the first common ancestor of mammals produced lysozyme (a predecessor of α-La) and a number of glycosyltransferases but little or no α-La, which resulted in the production of a low level of lactose that was utilized in the synthesis of oligosaccharides and did not accumulate. Initially, the oligosaccharides served mainly as bactericidal agents but later became a source of energy for the neonate and both of these functions persist for monotremes, marsupials and some eutherians, e.g. bears, elephants and marine mammals.

However, most eutherians evolved to secrete predominantly lactose as an energy source, due to the synthesis of an increased level of α-La, whereas oligosaccharides continued to play a bactericidal role. Human milk, which contains high levels of both lactose and oligosaccharides, seems to be anomalous. Work on the oligosaccharides of a wider range of species is needed to explain this situation.

The significance of oligosaccharides is not clear but the following aspects may be important. Firstly, for any particular level of energy, they have a smaller impact on osmotic pressure than smaller saccharides. Secondly, they are not hydrolyzed by β-galactosidase and neither fucosidase nor neuraminidase is secreted in the intestine; hence the oligosaccharides are not hydrolyzed and absorbed in the gastrointestinal tract and function as soluble fiber and prebiotics that affect the microflora of the large intestine. Thirdly, it is claimed that they prevent the adhesion of pathogenic bacteria in the intestine. And finally, galactose and especially *N*-acetylneuraminic acid are important for the synthesis of glycolipids and glycoproteins, which are important for brain development; hence, it has been suggested that the oligosaccharides are important for brain development (see Kunz and Rudloff, 2006).

Lipids

Lipids (commonly called oils or fats, which are liquid or solid respectively at ambient temperature) are those constituents of tissues, biological fluids or foods that are soluble in an apolar solvent, e.g. diethyl ether, chloroform or carbon tetrachloride. Historically, the fat of milk was regarded as its most valuable constituent and, until recently, milk was valued largely or totally on its fat content. This was due at least partially to the development in 1890 of rather simple methods for quantifying the fat content of milk by S. M. Babcock and N. Gerber, long before comparable methods for proteins became available. Milk lipids are very complex chemically and exist as a unique emulsion. Milk lipids have been thoroughly studied and characterized (see Fox, 1983, 1995; Fox and McSweeney, 1998, 2006; and references therein).

The level of fat in milk shows very large inter-species differences, ranging from $\approx$2% to $>$50% (see Fox and McSweeney, 1998). Lipids are 2.5 times more energy dense than lactose, so when a highly caloric milk is required (e.g. by animals in a cold environment, such as marine mammals or hibernating bears) this is achieved by increasing the fat content of the milk.

Lipids are commonly divided into three classes.

- Neutral lipids: these are esters of glycerol and one, two or three fatty acids for mono-, di- and triglycerides respectively. Neutral lipids are by far the dominant class of lipids in all foods and tissues, representing 98.5% of total milk lipids.
- Polar lipids (a complex mixture of fatty acid esters of glycerol or sphingosine): many contain phosphoric acid, a nitrogen-containing compound (choline, ethanolamine or serine) or a sugar/oligosaccharide. Although present at low levels ($\sim$1% of total milk lipids), the polar lipids play critical roles in milk and dairy products. They are very good natural emulsifiers and are concentrated in the MFGM, which maintains the milk lipids as discrete globules and ensures their physical and biochemical stability.

- Miscellaneous lipids: a heterogeneous group of compounds that are unrelated chemically to each other or to neutral or polar lipids. This group includes cholesterol, carotenoids and the fat-soluble vitamins, A, D, E and K. The cartenoids are important for two reasons: firstly, they are natural pigments (yellow, orange, red) and are responsible for the color of butter and cheese; some consumers prefer highly colored cheese, which is obtained by adding a carotenoid-containing extract from annatto beans; secondly, some carotenoids are converted to vitamin A in the liver.

Fatty acids

Fatty acids are carboxylic acids with the general formula R-COOH, where the alkyl group (R) is a hydrocarbon chain containing from 3 to 25 carbons (total number of carbons, from 4 to 26), which may be saturated or unsaturated (1 to 6 double bonds), and is usually straight (normal), with small amounts of branched chain, hydroxy and keto (oxo) acids. The vast majority of fatty acids have an even number of carbon atoms because they are synthesized from, and elongated by adding, a two-carbon compound, acetyl CoA, on each cycle of the multi-enzyme fatty acid synthetase (FAS). Although the hydroxy fatty acids are present at low levels, they are important in milk fat because they are converted on heating to lactones that give a desirable flavor to milk fat, which is considered to be the premium cooking fat. Although keto acids are also minor components, they are important flavor precursors because they are converted to highly flavored methyl ketones.

The melting point of fatty acids increases progressively with molecular weight (MW), whereas solubility in water decreases. The melting point decreases with the introduction of double bonds and, for unsaturated fatty acids, the melting point of the *cis* isomer is lower than that of the *trans* isomer.

Milk lipids are chemically similar to all other lipids but contain a very wide range of fatty acids (up to 400 fatty acids have been reported in milk lipids, although most of these are present at trace levels). The milk lipids of ruminants are unique in that they are the only natural lipids that contain butyric (butanoic) acid ($C_{4:0}$); they also contain substantial amounts of medium-chain fatty acids (hexanoic [$C_{6:0}$], octanoic [$C_{8:0}$] and decanoic [$C_{10:0}$]), the only other sources of which are coconut oil and palm kernel oil. The short- and medium-chain fats are water soluble and volatile, and have a strong aroma and taste.

The fatty acids in milk fat are obtained from three sources.

- Butanoic acid is produced by reducing β-hydroxybutanoic acid, which is synthesized by bacteria in the rumen; ruminant milk fat is the only natural lipid that contains this fatty acid.
- All hexanoic ($C_{6:0}$) to tetradecanoic ($C_{14:0}$) acids and 50% of hexadecanoic ($C_{16:0}$) acid are synthesized in the mammary gland from acetyl CoA ($CH_3COSCoA$); these fatty acids are released from the FAS by chain-length-specific thioesterases, the relative activities of which are responsible for inter-species differences in the proportions of medium-chain fatty acids. Decanoic

acid ($C_{10:0}$) and dodecanoic acid ($C_{12:0}$) are major fatty acids in the milk fats of elephant, horse, donkey, zebra, tapir, rhinoceros, rabbit and hare but these fats contain very little or no butanoic acid (Glass *et al.*, 1967; Glass and Jenness, 1971; Christie, 1995; Osthoff *et al.*, 2005). These species are non-ruminant herbivores with a large caecum, a feature that presumably is somehow responsible for the high levels of $C_{10:0}$ and $C_{12:0}$; some of the above species also practice coprophagy.
- All octadecanoic ($C_{18:0}$) acid and 50% of hexadecanoic ($C_{16:0}$) acid are obtained from dietary lipids.

The unsaturated fatty acids are synthesized as follows:

- $C_{18:1}$ is produced from $C_{18:0}$ in the liver by Δ-9 desaturase.
- $C_{18:2}$ is obtained from the diet, i.e. it is an essential fatty acid.
- The other unsaturated fatty acids are produced from $C_{18:2}$ by further desaturation and/or elongation.

Ruminant milk fats contain low levels of polyunsaturated fatty acids (PUFAs) because PUFAs in the diet are hydrogenated by bacteria in the rumen. Biohydrogenation can be prevented by encapsulating dietary PUFAs or PUFA-rich sources in cross-linked (by HCHO) protein or cross-linked crushed oilseeds. PUFA-enriched milk has improved spreadability of butter and perceived improved nutritional qualities.

Incomplete biohydrogenation by the rumen bacterium *Butyrivibrio fibrisolvens* results in the formation of conjugated linoleic acid (CLA; also called rumenic acid), which has potent anti-carcinogenic properties. Eight isomers of CLA are possible but *cis*-9, *trans*-11 is the most biologically active and is the predominant isomer found in bovine milk. The formation of CLA and its nutritional benefits have been the subject of considerable research during the past 15 years; this research has been reviewed by Bauman and Lock (2006) and Parodi (2006).

Distribution of fatty acids in triglycerides

As well as the constituent fatty acids, the position of the fatty acids in triglycerides affects their melting point and rheological properties. For these reasons, and to completely characterize the structure of triglycerides, the positions of the fatty acids in milk triglycerides have been determined. An index of this can be obtained by determining the acyl carbon number (ACN) of triglycerides, i.e. the sum of the number of carbons in the three component fatty acids, which can be done by gas chromatography. Probably the first study on this aspect was by Breckenridge and Kuksis (1967), in which the ACNs of the milk triglycerides from seven species were reported. More recent work has been reviewed by Christie (1995) and MacGibbon and Taylor (2006).

The complete structure of triglycerides can be determined by stereospecific analysis, the results of which for milk fat have been described by Christie (1995) and MacGibbon and Taylor (2006). The most notable feature is the almost exclusive esterification of the short-chain fatty acids—$C_{4:0}$ and $C_{6:0}$—at the Sn3 position.

Degradation of lipids

Food lipids are susceptible to two forms of deterioration: lipid oxidation, leading to oxidative rancidity, and the hydrolysis of lipids by lipases (lipolysis), leading to hydrolytic rancidity.

Lipid oxidation involves a very complex set of chemical reactions, which have been well characterized; the literature has been comprehensively reviewed by T. Richardson and M. Korycka-Dahl in Fox (1983), by T. P. O'Connor and N. M. O'Brien in Fox (1983), and in Fox and McSweeney (2006).

Milk contains an indigenous lipoprotein lipase (LPL), which is normally inactive because it is separated from the triglyceride substrates by the MFGM. However, if the membrane is damaged, lipolysis and hydrolytic rancidity ensue rapidly. When milk lipids are hydrolyzed by milk LPL, the short- and medium-chain fatty acids are preferentially released. These fatty acids are major contributors to flavor, which may be desirable or undesirable, depending on the product. Hydrolytic rancidity caused by milk LPL is potentially a very serious problem in raw milk and in some dairy products. Lipolysis in milk has been reviewed comprehensively by H. C. Deeth and C. H. FitzGerald in Fox (1983, 1995) and in Fox and McSweeney (2006).

A low level of lipolysis is desirable in all types of cheese, especially in blue cheeses, in which the principal lipases are those secreted by the blue mould *Penicillium roqueforti*. The free fatty acids are converted to alk-2-ones, the principal flavor compounds in blue cheeses. The characteristic piquant flavor of some cheeses, e.g. Pecorino Romano, is due to short- and medium-chain fatty acids, which are released mainly by an added lipase—pregastric esterase. Other derivatives of fatty acids are alk-2-ols (secondary alcohols), lactones, esters and thio-esters; these are important flavor compounds in cheese.

Milk lipids as an emulsion

Lipids are insoluble in water or aqueous systems. When mixed, a lipid and water (or aqueous solvent) form distinct layers and a force, interfacial tension (γ), exists between the layers. Lipids can be dispersed in water by vigorous agitation (homogenization) but, when agitation ceases, the droplets of lipid coalesce quickly into a single mass (i.e. phase separation). This is driven by the need to reduce the interfacial area and γ to a minimum. If γ is reduced, the droplets of lipid will remain discrete, although they will rise to the surface (i.e. cream) because of the lower density of lipids (0.9) compared with water (1.0). Interfacial tension can be reduced by using a surface-active agent (emulsifier, detergent). Natural emulsifiers include proteins, phospholipids and mono- and diglycerides; there is a wide range of synthetic emulsifiers.

In milk, the lipids are dispersed in the milk serum (specific gravity, 1.036) as globules with a diameter in the range from <1 to $\approx 20\,\mu m$ (mean 3–4 μm). The fatty acids (from the sources described above) and monoglycerides (from blood lipids) are synthesized to triglycerides in the rough endoplasmic reticulum (RER) at the basal end of the epithelial cells. The triglycerides form into globules within the RER and are released into the cell cytoplasm. The globules are stabilized by a complex layer of proteins and phospholipids, known as the MFGM. The inner layer of the MFGM is acquired within the epithelial cell as the fat globules, after release from the RER, move towards the apical membrane. The outer layer of the MFGM is the apical membrane of the secretory cell through which the lipid globules are pushed as they are expressed from the cell. Because the stability of the milk emulsion is critical in most dairy products, the structure and the stability of the MFGM have been the subject of research for more than 100 years. The MFGM is composed mainly of phospholipids and proteins.

Sodium dodecyl sulfate polyacrylamide gel electrophoresis (SDS-PAGE) indicates that there are eight main proteins in the MFGM, which have been isolated and characterized (see Mather, 2000). However, SDS-PAGE followed by micro-capillary high performance liquid chromatography-mass spectrometry (HPLC-MS) reveals 120 proteins, of which 23% are involved in protein trafficking, 23% in cell signalling, 21% in unknown functions, 11 in fat trafficking/metabolism, 9% in transport and 7% in protein synthesis/folding, 4% are immune proteins and 2% are contaminating skim milk proteins (Reinhardt and Lippolis, 2006). Many of the 70 indigenous enzymes in milk are concentrated in the MFGM. The very extensive literature on the MFGM has been the subject of many reviews, including Keenan and Mather (2006), who present an up-to-date model of the MFGM.

Some of the MFGM is shed during the aging of milk, specifically if agitated, and forms vesicles (sometimes called microsomes) in the skimmed milk. The MFGM may be damaged by agitation, homogenization, whipping or freezing and may lead to hydrolytic rancidity and non-globular fat which may cause cream plug, oiling-off in coffee and tea, and poor wettability of milk powder (for a review, see Evers, 2004). The MFGM is stripped from the fat globules by extensive agitation (usually of cream), a process referred to as churning; the free fat coalesces and is kneaded (worked) to give a water-in-oil emulsion—butter. The MFGM partitions into the aqueous phase, referred to as buttermilk. The phospholipids in buttermilk give it good emulsifying properties and there is commercial interest in using it as a food ingredient (Singh, 2006). Some of the polar lipids in the MFGM are reported to have desirable nutritional properties (see Ward *et al.*, 2006), but there are conflicting results (see Riccio, 2004).

Presumably, the fat globules in the milk of all species are stabilized by a membrane similar to that in bovine milk but there is very little information on the MFGM in non-bovine milks. Buchheim *et al.* (1989) and Welsch *et al.* (1990) studied the glycoproteins in the MFGM of human, rhesus monkey, chimpanzee, dog, sheep, goat, cow, gray seal, camel and alpaca by SDS-PAGE with Periodic Acid Schiff (PAS) staining, Western blotting and lectin biochemistry. Large intra- and inter-species differences were found; very highly glycosylated proteins were found in the MFGM of primates, horse, donkey, camel and dog (see Patton, 1994, 2004, and references therein). Long (0.5–1 μm) filamentous structures extend from the surface of the fat globules in equine and human milks; the filaments are composed of mucins (highly glycosylated proteins) which dissociate rapidly from the surface of globules in the bovine milk serum on cooling; they are also lost on heating human milk, e.g. at 80°C for 10 min (see Patton, 2004, and references therein).

The filaments facilitate the adherence of fat globules to the intestinal epithelium and probably improve the digestion of fat. The mucins prevent bacterial adhesion and may protect mammary tissue against tumors (mammary tumors are very rare in the cow). Why the filaments on bovine milk fat globules are lost much more easily than those in equine milk and human milk is not known; work in this area is warranted. Proteomic methodology is being applied to study the human MFGM and membranes of the mammary epithelial cells (for references, see Reinhardt and Lippolis, 2006).

The fat globules in bovine milk form a cream layer because of the difference in specific gravity between the fat and aqueous phases, but the cream layer is dispersed readily by gentle agitation. The rate of creaming can be calculated from Stokes' equation:

$$v = 2r^2(\rho 1 - \rho 2)g/9\eta$$

where v is the velocity of creaming, r is the radius of the fat globules, $\rho 1$ and $\rho 2$ are the specific gravities of the continuous and dispersed phases respectively, g is acceleration due to gravity and η is the viscosity of the continuous phase.

Based on the typical values for r, $\rho 1$, $\rho 2$ and η for milk, one would expect a cream layer to form in milk in about 60 h, but in fact a cream layer forms in about 30 min. The faster than expected rate of creaming is due to the aggregation of fat globules, aided by an immunoglobulin-M-type protein, called cryoglobulin, because it precipitates on to the fat globules when the milk is cooled. The clusters of globules then behave as a unit with a much larger radius. Creaming can be prevented by homogenizing the milk, which reduces the size of the globules and denatures the cryoglobulins. The fat globules in buffalo, ovine, caprine, equine and camel milks do not agglutinate, because these milks lack cryoglobulins.

Previously, the creaming of milk was a very important attribute and was a popular research topic; it has been the subject of several reviews, most recently by Huppertz and Kelly (2006). Traditionally, the fat was removed from milk by natural (gravity) creaming. Gravity creaming is still used to standardize the fat content for some cheese varieties, e.g. Parmigiano Reggiano, but the removal of fat from milk is now usually accomplished by centrifugal separation, in which g is replaced by ω^2R, where ω is the centrifugal velocity in radians per second and R is the radius of the centrifuge bowl. Centrifugal separation is very efficient, essentially instantaneous and continuous.

Proteins

The properties of milk and most dairy products are affected more by the proteins they contain than by any other constituent. The milk proteins also have many unique properties; because of this and their technological importance, the milk proteins have been studied extensively and are probably the best characterized food protein system.

Research on milk proteins dates from the early nineteenth century. Pioneering work was reported by J. Berzelius in 1814, by H. Schubler in 1818 on the physico-chemical status of milk proteins, and by H. Braconnot in 1830 who published the first paper in which the word casein was used. A method for the preparation of protein from milk by acid precipitation was described in 1938 by J. G. Mulder, who coined the term "protein." The acid-precipitated protein was referred to as casein (some early authors called acid-precipitated milk protein caseinogen, which is converted by rennin to casein, that coagulates in the presence of Ca^{2+}; this situation is analogous to the conversion of fibrinogen in blood by thrombin to fibrin, which coagulates in the presence of Ca^{2+}); about 70 years ago, the term "casein" was universally adopted as the English word for the pH-4.6-insoluble protein in milk. The method for acid (isoelectric) precipitation of casein was refined by Hammarsten (1883) and, consequently, isoelectric casein is frequently referred to as casein *nach* Hammarsten. An improved method for the isolation of casein was published in 1918 by L. L. van Slyke and J. C. Baker.

The liquid whey remaining after isoelectric precipitation of casein from skim or whole milk is a dilute solution of proteins (whey or serum proteins; ≈0.7% in bovine milk), lactose, organic and inorganic salts, vitamins and several constituents at trace levels. By salting-out with $MgSO_4$, the whey proteins were fractionated by J. Sebelein, in 1885, into soluble (albumin) and insoluble (globulin) fractions. According to McMeekin (1970), in 1889 A.Wichmann crystallized a protein from the albumin fraction of whey by addition of $(NH_4)_2SO_4$ and acidification, a technique used to crystallize blood serum albumin and ovalbumin. Using the techniques available 100 years ago, the whey proteins were found to be generally similar to the corresponding fractions of blood proteins and were considered to have passed directly from blood to milk; consequently, the whey proteins attracted little research effort until the 1930s.

In addition to the caseins and whey proteins, milk contains two other groups of proteinaceous materials—proteose peptones (PPs) and non-protein nitrogen (NPN)—which were recognized in 1938 by S. J. Rowland who observed that, after heating milk fat at 95°C for 10 min, the whey proteins co-precipitated with the caseins on acidification to pH 4.6. When the pH-4.6-soluble fraction of heated milk was made to 12% trichloroacetic acid (TCA), some nitrogenous compounds precipitated and were designated "proteose peptone"; nitrogenous compounds that remained soluble in 12% TCA were designated NPN. A modified version of S. J. Rowland's scheme is now used to quantify the principal nitrogenous groups in milk.

Thus, by 1938, the complexity of the milk protein system had been described (i.e. caseins, lactalbumin (now known to consist mainly of α-lactalbumin, β-lactoglobulin and blood serum albumin), lactoglobulin, PPs and NPN, which represent approximately 78, 12, 5, 2 and 3% respectively of the nitrogen in bovine milk). However, knowledge of the milk protein system was rudimentary and vague at this stage.

Knowledge on the chemistry of milk proteins has advanced steadily during the twentieth century, as can be followed through the progression of textbooks and reviews on dairy chemistry (see Fox, 1992, 2003; Fox and MacSweeney, 2003).

Preparation of casein and whey proteins

The protein fractions may be prepared from whole or skimmed milk, but the latter is usually used because the fat is occluded in isoelectric casein and interferes with further characterization of the proteins. The fat is easily removed from milk by centrifugation (e.g. $3000 \times g$ for 30 min) and any remaining fat may be removed by washing the precipitated protein with ether. Isoelectric precipitation is the most widely used method for separating the casein and non-casein fractions of milk protein but several other techniques are used in certain situations (see Fox, 2003).

- Isoelectric precipitation at ≈pH 4.6 at 20°C: the precipitate is recovered by filtration or low-speed centrifugation. Essentially similar methods are used to prepare casein on a laboratory scale or an industrial scale.
- Ultracentrifugation: in milk, the casein exists as large micelles that may be sedimented by centrifugation at 100 000 × *g* for 1 h; the whey proteins are not sedimentable. The casein pellet can be redispersed in a suitable buffer as micelles with properties similar to those of natural micelles.
- Salting-out methods: casein can be precipitated by any of several salts, usually by $(NH_4)_2SO_4$ at 260 g/L or saturated NaCl. The immunoglobulins co-precipitate with the caseins.
- Ultrafiltration and microfiltration: all the milk proteins are retained by small-pore, semi-permeable membranes and are separated from lactose and soluble salts. This process, ultrafiltration, is used widely for the industrial-scale production of whey protein concentrates (WPCs) and to a lesser extent for the production of total milk protein. Intermediate-pore membranes are used to separate casein micelles from whey proteins. In microfiltration, using large-pore membranes (0.4 μm), both the caseins and the whey proteins are permeable, but >99.9% of bacteria and other large particles are retained; microfiltration is used for the production of extended-shelf-life beverage milk or cheese milk or to remove lipoprotein particles from whey to improve the functionality of WPC.
- Gel filtration: it is possible to separate the caseins from the whey proteins by permeation chromatography but this method is not used industrially and is rarely used on a laboratory scale.
- Precipitation by ethanol: the caseins are precipitated from milk by ≈40% ethanol, whereas the whey proteins remain soluble; however, precipitation by ethanol is rarely used, on either a laboratory scale or an industrial scale, for the precipitation of casein.
- Cryoprecipitation: caseins, in a micellar form, may be destabilized and precipitated by freezing milk or, preferably, concentrated milk, at about −10°C. Precipitation is caused by a decrease in pH and an increase in Ca^{2+} concentration; the precipitated micelles may be redispersed as micelles by heating to about 55°C. Alternatively, the cryoprecipitated casein may be recovered, washed and dried; it has many interesting properties for food applications, but it is not produced commercially.
- Rennet coagulation: the casein micelles are destabilized by specific, limited proteolysis and coagulate in the presence of Ca^{2+}. The properties of rennet-coagulated casein are very different from those of isoelectric casein; rennet-coagulated casein is very suitable for certain food applications, e.g. cheese analogs.
- Caseinates: isoelectric casein is insoluble in water, but may be converted to water-soluble caseinates by dispersion in water and adjusting the pH to ≈6.7 with alkali, usually NaOH to yield sodium caseinate. KOH, NH_4OH and $Ca(OH)_2$ give the corresponding caseinates, which may be freeze dried or spray dried.

Comparison of key properties of casein and whey proteins

- Solubility at pH 4.6. The caseins are, by definition, insoluble at pH 4.6, whereas the whey proteins are soluble under the ionic conditions of milk. The isoelectric precipitation of casein is exploited in the production of caseins and caseinates, fermented milk products and acid-coagulated cheeses.
- Coagulability following limited proteolysis. The caseins are coagulable following specific, limited proteolysis, whereas the whey proteins are not. This property of the caseins is exploited in the production of rennet-coagulated cheese (≈75% of all cheese) and rennet casein.
- Heat stability. The caseins are very heat stable. Milk at pH 6.7 may be heated at 100°C for 24h without coagulation and withstands heating at 140°C for up to 20–25min; aqueous solutions of sodium caseinate may be heated at 140°C for several hours without apparent changes. The heat stability of the whey proteins is typical of globular proteins; they are denatured completely on heating at 90°C for 10min. The remarkably high heat stability of the caseins, which is probably due to their lack of typical stable secondary and tertiary structures, permits the production of heat-sterilized dairy products with relatively small physical changes.
- Amino acid composition. The caseins contain high levels of proline (17% of all residues in β-casein), which explains their lack of α- and β-structures. The caseins are phosphorylated, whereas the principal whey proteins are not. Whole isoelectric casein contains approximately 0.8% phosphorus, but the degree of phosphorylation varies among the individual caseins. The phosphate is attached to the polypeptides as phosphomonoesters of serine: the presence of phosphate groups has major significance for the properties of the caseins, e.g., (i) molecular charge and related properties and heat stability; and (ii) metal binding which affects their physico-chemical, functional and nutritional properties. Metal binding by casein is regarded as a biological function because it enables a high concentration of calcium phosphate to be carried in milk in a soluble form (to supply the requirements of the neonate); otherwise, calcium phosphate would precipitate in, and block, the ducts of the mammary gland, leading to the death of the gland and perhaps of the animal.
- The caseins are low in sulfur (0.8%), whereas the whey proteins are relatively rich (1.7%). The sulfur in casein is mainly in methionine, with little cystine or cysteine; the principal caseins are devoid of the latter two amino acids. The whey proteins are relatively rich in cysteine and/or cystine, which have major effects on the physico-chemical properties of these proteins and of milk.
- Site of biosynthesis. The caseins are synthesized in the mammary gland and are unique to this organ. Presumably, they are synthesized to meet the amino acid requirements of the neonate and as carriers of important metals required by the neonate. The principal whey proteins are also synthesized in, and are unique to, the mammary gland, but several minor proteins in milk are derived from blood, either by selective transport or due to leakage. Most of the whey proteins have a biological function.
- Physical state in milk. The whey proteins exist in milk as monomers or as small quaternary structures whereas the caseins exist as large aggregates, known

as micelles, with a mass of $\approx 10^8$ Da and containing about 5000 molecules. The white color of milk is due largely to the scattering of light by the casein micelles. The structure, properties and stability of the casein micelles are of major significance for the technological properties of milk and have been the subject of intensive research (see below).

Heterogeneity and fractionation of casein

O. Hammersten believed that isoelectric casein was a homogeneous protein but, during the early years of the twentieth century, evidence was presented by T. B. Osborne and A. J. Wakeman, and especially by K. Linderstrøm-Lang and collaborators, that it might be heterogeneous (see McMeekin, 1970). By extraction with ethanol-HCl mixtures, K. Linderstrøm-Lang and S. Kodoma obtained three major casein fractions, which contained about 1.0, 0.6 or 0.1% phosphorus, and several minor fractions. The heterogeneity of casein was confirmed by analytical ultracentrifugation and free-boundary electrophoresis by K. O. Pedersen and O. Mellander respectively (see McMeekin, 1970). Electrophoresis resolved isoelectric casein into three proteins, which were named α, β and γ in order of decreasing electrophoretic mobility and represented about 75, 22 and 3% of whole casein respectively.

Following the demonstration of its heterogeneity, several attempts were made to isolate the individual caseins. The first reasonably successful method was developed in 1944 by Warner, who exploited differences in the solubilities of α- and β-caseins at pH 4.4 and 2°C. A much more satisfactory fractionation method was developed in 1952 by N. J. Hipp and co-workers based on the differential solubility of α-, β- and γ-caseins in urea solutions at pH 4.9. This method was widely used for many years until the widespread application of ion-exchange chromatography.

In 1956, α-casein was resolved by D. F. Waugh and P. H. von Hippel into calcium-sensitive and calcium-insensitive proteins which were called α_s- and κ-caseins respectively. κ-Casein, which represents $\approx$12% of total casein, is responsible for the formation and stabilization of casein micelles, and affects many technologically important properties of the milk protein system. Numerous chemical methods were soon developed for the isolation of κ-casein (see Fox, 2003). α_s-Casein, prepared by the method of D. F. Waugh and P. H. von Hippel, contains two proteins, now called α_{s1}- and α_{s2}-caseins.

Chemical methods for fractionation of the caseins have now been largely superseded by ion-exchange chromatography, which gives superior results when urea and a reducing agent are used (see Strange *et al.*, 1992; Imafidon *et al.*, 1997).

Application of gel electrophoresis to the study of milk proteins

Zone electrophoresis on a solid medium, paper or cellulose acetate was introduced in the 1940s. This technique gave good results with many protein systems, but the caseins, because of a very strong tendency to associate hydrophobically, were resolved poorly on these media. Electrophoresis in starch gels (SGE) using discontinuous buffer systems was introduced to general protein chemistry by M. D. Poulik in 1957 and was applied to the study of the caseins by R. G. Wake and R. L. Baldwin in 1961. The resolving power of SGE was far superior to that of any of its predecessors. When urea (7 M) and a reducing agent, usually 2-mercaptoethanol, were incorporated into the starch gel, isoelectric casein was resolved into about 20 bands, most of which were due to the microheterogeneity of one or more of the caseins.

Electrophoresis on polyacrylamide disk gels (PAGE) was introduced by L. Ornstein in 1964 and was applied to the study of the caseins by R. F. Peterson in 1966. PAGE and SGE give similar results, but PAGE is far easier to use and has become the standard electrophoretic method for the analysis of caseins (and most other protein systems). Gel electrophoretic methods for the analysis of milk proteins have been reviewed by Swaisgood (1975), Strange *et al.* (1992) and Tremblay *et al.* (2003).

SDS-PAGE, which resolves proteins mainly on the basis of molecular mass, is very effective for most proteins but, because the masses of the four caseins are quite similar, SDS-PAGE is not very effective. β-Casein, which has very high surface hydrophobicity, binds a disproportionately high amount of SDS and, consequently, has a higher electrophoretic mobility than α_{s1}-casein, although it is a larger molecule. SDS-PAGE is very effective for the resolution of whey proteins and is the method of choice.

Microheterogeneity of the caseins

α_{s1}-, α_{s2}-, β- and κ-caseins represent approximately 38, 10, 35 and 12% respectively of whole bovine casein. However, SGE or PAGE indicates much greater heterogeneity due to small differences in one or more of the caseins, referred to as microheterogeneity, which arises from five factors.

Variability in the degree of phosphorylation

All the caseins are phosphorylated but to a variable degree (α_{s1}-, 8 or 9P; α_{s2}-, 10, 11, 12 or 13P; β-, 4 or 5P; κ-, 1 or 2P per molecule). The number of phosphate residues is indicated thus: α_{s1}-CN 8P, β-CN 5P etc.

Genetic polymorphism

In 1955, R. Aschaffenburg and J. Drewry discovered that β-lactoglobulin exists in two forms (variants, polymorphs) now called A and B, which differ by only two amino acids. The variant in the milk is genetically controlled and the phenomenon is called genetic polymorphism. It was soon shown that all milk proteins exhibit genetic polymorphism and at least 45 polymorphs have been detected by PAGE, which differentiates on the basis of charge and therefore only polymorphs that differ in charge have been detected. It is very likely that only a small proportion of the genetic polymorphs of milk proteins have been detected. The potential of peptide mapping of enzymatic hydrolysates by HPLC-MS has been assessed. The genetic polymorph(s) present is indicated by a Latin letter as follows: β-CN **A** 5P, α_{s1}-CN **B** 9P, κ-CN **A** 1P etc. Genetic polymorphism also occurs in the milk of sheep, goat, buffalo, pig and horse, and probably of all species.

Technologically important properties of milk, e.g. rennetability, heat stability, yield and proportions of milk proteins, are affected by the genetic polymorphs of the milk proteins present, and work in this area is being expanded and refined. The extensive literature on the genetic polymorphism of milk proteins has been the subject of several reviews, including Ng-Kwai-Hang and Grosclaude (2003).

Disulfide bonding

α_{s1}- and β-caseins lack cysteine and cystine, but both α_{s2}- and κ-caseins contain two ½ cystine residues which occur as intermolecular disulfide bonds. α_{s2}-Casein exists as a disulfide-linked dimer, whereas up to 10 κ-casein molecules may be linked by disulfide bonds. Inclusion of a reducing agent (usually mercaptoethanol) in SGE or PAGE gels is required for good resolution of κ-casein; in its absence, α_{s2}-casein appears as a dimer (originally called α_{s5}-casein).

Variations in the degree of glycosylation

κ-Casein is the only glycosylated casein; it contains galactose, *N*-acetylgalactosamine and *N*-acetylneuraminic (sialic) acid, which occur as tri- or tetrasaccharides, the number of which varies from 0 to 4 per molecule of protein (i.e. a total of 9 variants).

Hydrolysis of the caseins by plasmin

Milk contains several indigenous proteinases, the principal of which is plasmin, a trypsin-like, serine-type proteinase from blood; it is highly specific for peptide bonds with a lysine or arginine at the P1 position. The preferred casein substrates are β- and α_{s2}-caseins; α_{s1}-casein is also hydrolyzed, but κ-casein is very resistant, as are the whey proteins. All the caseins contain several lysine and arginine residues, but only a few bonds are hydrolyzed rapidly. β-Casein is hydrolyzed rapidly at the bonds Lys_{28}–Lys_{29}, Lys_{105}–His_{106} and Lys_{107}–Glu_{108}. The resulting C-terminal peptides are the γ-caseins (γ^1: β-CNf29–209; γ^2: β-CNf106–209; γ^3: β-CNf108–209), whereas the N-terminal peptides are proteose peptones 5, 8_{slow} and 8_{fast}. The γ-caseins, which represent $\approx$3% of total casein, are evident in PAGE gels. Other plasmin-produced peptides are probably present but either are too small to be readily detectable by PAGE or their concentrations are very low relative to those of the principal caseins.

Although α_{s2}-casein in solution is also quite susceptible to plasmin, α_{s2}-casein-derived peptides have not been identified in milk. α_{s1}-Casein in solution is also hydrolyzed readily by plasmin; members of a minor casein fraction, λ-casein, are N-terminal fragments of α_{s1}-casein produced by plasmin (O'Flaherty, 1997).

Molecular properties of the milk proteins

The principal, and many of the minor, milk proteins have been very well characterized. The principal properties of the six milk-specific proteins are summarized in Table 1.1. A number of features warrant comment.

The six principal lactoproteins are small molecules, a feature that contributes to their stability. The primary structures of the principal lactoproteins of several species are known, as are the substitutions in the principal genetic variants.

The whey proteins are highly structured, but the four caseins lack stable secondary structures; classical physical measurements indicate that the caseins are unstructured,

Table 1.1 Properties of the principal lactoproteins

Properties	Caseins				β-Lg	α-La
	α_{s1}	α_{s2}-	β-	κ-		
MW	23 612	25 228	23 980	19 005	18 362	14 174
Residues	199	207	209	169	162	123
Conc in milk (g/L)	12–15	3–4	9–11	2–4	3.0	0.7
Phosphate residues	8–9	10–13	4–5	1–2	0	0
½ Cystine	0	2	0	2	5	8
Sugars	0	0	0	Yes	0	0
Prolyl residue per molecule	17	10	35	20	8	2
A_{280}, 1% 1 cm	10.1	11.1	4.6	9.6	9.4	20.1
Secondary structure	Low	Low	Low	Low	High	High
$H\phi_{ave}$	4.89	4.64	5.58	5.12	5.03	4.68
pI	4.96	5.27	5.20	5.54	5.2	4.2–4.5
Partial specific volume (ml/g)	0.728	0.720	0.741	0.734	0.751	0.735

but theoretical considerations indicate that, rather than being unstructured, the caseins are very flexible molecules and have been referred to as rheomorphic (Holt and Sawyer, 1993; see also Horne, 2002; Farrell *et al.*, 2006a). The inability of the caseins to form stable structures is due mainly to their high content of the structure-breaking amino acid proline; β-casein is particularly rich in proline, with 35 of the 209 residues. The open, flexible structure of the caseins renders them very susceptible to proteolysis, which facilitates their natural function as a source of amino acids.

In contrast, the native whey proteins, especially β-lactoglobulin, are quite resistant to proteolysis, and at least some are excreted in the feces of infants. This feature is important because most of the whey proteins play a non-nutritional function in the intestine and, therefore, resistance to proteolysis is important.

The caseins are generally regarded as very hydrophobic proteins but, with the exception of β-casein, they are not exceptionally hydrophobic. Because of their lack of stable secondary and tertiary structures, most of their hydrophobic residues are exposed and, consequently, they have a high surface hydrophobicity.

One of the more notable features of the amino acid sequence of the caseins is that the hydrophobic and hydrophilic residues are not distributed uniformly, thereby giving the caseins a distinctly amphipatic structure. This feature, coupled with their open flexible structure, gives the caseins good surface activity, and good foaming and emulsifying properties, making casein the functional protein of choice for many applications. Because of their hydrophobic sequences, the caseins have a propensity to yield bitter hydrolysates.

Also because of their open structure, the caseins have a high specific volume and, consequently, form highly viscous solutions, which is a disadvantage in the production of caseinates. Because of its high viscosity, it is not possible to spray dry sodium caseinate solutions containing >20% protein, thereby increasing the cost of drying and resulting in low-bulk-density powders.

The lack of stable tertiary structures means that the caseins are not denaturable *stricto senso* and, consequently, are extremely heat stable; sodium caseinate, at pH 7, can withstand heating at 140°C for several hours without visible change. This very high heat stability makes it possible to produce heat-sterilized dairy products with very little change in physical appearance; other major food systems undergo major physical and sensoric changes on severe heating.

The caseins have a very strong tendency to associate, due mainly to hydrophobic bonding. Even in sodium caseinate, the most soluble form of casein, the molecules form aggregates of 250–500 kDa, i.e. containing 10–20 molecules. This strong tendency to associate makes it difficult to fractionate the caseins, for which a dissociating agent, e.g. urea or SDS, is required. On the other hand, a tendency to associate is important for some functional applications and in the formation and stabilization of casein micelles. In contrast, the whey proteins are molecularly dispersed in solution.

Because of their high content of phosphate groups, which occur in clusters, α_{s1}-, α_{s2}- and β-caseins have a strong tendency to bind metal ions, which in the case of milk are mainly Ca^{2+} ions. This property has many major consequences; the most important from a technological viewpoint is that these three proteins, which represent approximately 85% of total casein, are insoluble at calcium concentrations $>\approx 6$ mM at temperatures >20°C. As bovine milk contains ≈30 mM calcium, one would expect that the caseins would precipitate under the conditions prevailing in milk. However, κ-casein, which contains only one organic phosphate group, binds calcium weakly and is soluble at all calcium concentrations found in dairy products.

Furthermore, when mixed with the calcium-sensitive caseins, κ-casein can stabilize and protect ≈10 times its mass of the former by forming large colloidal particles called casein micelles. The micelles act as carriers of inorganic elements, especially calcium and phosphorus, but also magnesium and zinc, and are, therefore, very important from a nutritional viewpoint. Through the formation of micelles, it is possible to solubilize much higher levels of calcium and phosphate than would otherwise be possible.

Nomenclature of milk proteins

During the period of greatest activity on the fractionation of casein (1950–1970), several casein (and whey protein) fractions were prepared that either were similar to proteins already isolated and named, or were artifacts of the isolation procedure. In order to standardize the nomenclature of the milk proteins, the American Dairy Science Association established a Nomenclature Committee in 1955, which has published seven reports, the most recent of which is Farrell *et al.* (2004). In addition to standardizing the nomenclature of the milk proteins, the characteristics of the principal milk proteins are summarized in these articles.

Whey proteins

About 20% of the total proteins of bovine milk are whey (serum) proteins. The total whey protein fraction is prepared by any of the methods described for the preparation of casein, i.e. the proteins that are soluble at pH 4.6 or in saturated NaCl or, after rennet-induced coagulation of the caseins, are permeable on microfiltration, or not sedimented by ultracentrifugation.

The proteins prepared by these methods differ somewhat: acid whey contains PPs; immunoglobulins are co-precipitated with the caseins by saturated NaCl; rennet whey contains the macropeptides, produced from κ-casein by rennet, plus small amounts of casein; and small casein micelles remain in the ultracentrifugal serum.

On a commercial scale, whey-protein-rich products are prepared by:

- ultrafiltration/diafiltration of casein or rennet whey to remove various amounts of lactose, and spray drying to produce whey protein concentrates (WPCs, 30–85% protein);
- ion-exchange chromatography and spray drying to yield whey protein isolate (WPI), containing ≈95% protein;
- demineralization by electrodialysis or ion exchange, thermal evaporation of water and crystallization of lactose; and
- thermal denaturation, removal of precipitated protein by filtration/centrifugation and spray drying, to yield lactalbumin, which has very low solubility and poor functionality.

Fractionation of whey proteins

It was recognized early that acid whey contains two well-defined groups of proteins: lactalbumins, which are soluble in 50% saturated $(NH_4)_2SO_4$ or saturated $MgSO_4$, and lactoglobulins, which are salted out under these conditions. The lactoglobulin fraction contains mainly immunoglobulins. The lactalbumin fraction contains two principal proteins, β-lactoglobulin and α-lactalbumin, and several minor proteins, including blood serum albumin and lactoferrin, which have been isolated by various procedures and crystallized (see Imafidon *et al.*, 1997; Fox, 2003).

There is considerable interest in the production of the major and many minor whey proteins on a commercial scale for nutritional, nutraceutical or functional applications. Several methods have been developed for the industrial-scale production of several whey proteins (see Mulvihill and Ennis, 2003).

Major characteristics of whey proteins

β-Lactoglobulin

β-Lactoglobulin (β-Lg) represents ≈50% of the whey proteins, ≈12% of the total protein, in bovine milk. It is a typical globular protein and has been characterized very well. The extensive literature has been reviewed by, among others, Sawyer (2003).

β-Lg is the principal whey protein in the milk of the cow, buffalo, sheep and goat, although there are slight inter-species differences. Initially, it was considered that β-Lg occurs only in the milk of ruminants, but it is now known that a similar protein occurs in the milk of many other species, including the sow, mare, kangaroo, dolphin and manatee. However, β-Lg does not occur in the milk of human, rat, mouse, guinea pig, camel, llama or alpaca, in which α-La is the principal whey protein.

Bovine β-Lg consists of 162 residues per monomer, with a MW of ≈18 kDa; its amino acid sequence, and that of several other species, has been established. Its isoelectric point is ≈pH 5.2. It contains two intramolecular disulfide bonds and one mole of cysteine per monomer. The cysteine is especially important because it reacts, following thermal denaturation, with the intermolecular disulfide of κ-casein and significantly affects the rennet coagulation and heat stability of milk. It is also responsible for the cooked flavor of heated milk. Some β-Lgs (e.g. porcine) lack a sulfydryl group. Ten genetic variants, A–J, of bovine β-Lg have been identified, the most common being A and B. Genetic polymorphism also occurs in β-Lg of other species.

β-Lg is a highly structured protein: in the pH range 2–6, 10–15% of the molecule exists as α-helices, 43% as β-sheets and 47% as unordered structures, including β-turns; the β-sheets occur in a β-barrel-type calyx. The molecule has a very compact globular structure; each monomer exists almost as a sphere, about 3.6 nm in diameter. β-Lg exists as a dimer, MW ≈36 kDa, in the pH range 5.5–7.5, as a monomer at <pH 3.5 and >pH 7.5, and as a tetramer (MW ≈144 kDa) in the pH range 3.5–5.5. Porcine and other β-Lgs that lack a free thiol do not form dimers, probably not due directly to the absence of a thiol group.

β-Lg is very resistant to proteolysis in its native state; this feature suggests that its primary function is not nutritional. It may have either or both of two biological roles:

- It binds retinol (vitamin A) in a hydrophobic pocket, protects it from oxidation and transports it through the stomach to the small intestine where the retinol is transferred to a retinol-binding protein, which has a similar structure to β-Lg. It is not clear how retinol is transferred from the core of the fat globules, where it occurs in milk, to β-Lg and why some species lack this protein. β-Lg can bind many hydrophobic molecules and hence its ability to bind retinol may be incidental. β-Lg is a member of the lipocalin family, all of which have binding properties (Akerstrom *et al.*, 2000).
- Through its ability to bind fatty acids, β-Lg stimulates lipase activity, which may be its most important physiological function.

β-Lg is the most allergenic protein in bovine milk for human infants and there is interest in producing whey protein products free of β-Lg for use in infant formulas. β-Lg has very good thermo-gelling properties and determines the gelation of WPCs.

α-Lactalbumin

About 20% of the protein of bovine whey (3.5% of total milk protein) is α-La, which is the principal protein in human milk. It is a small protein containing 123 amino acid residues, with a mass of ≈14 kDa, and has been well characterized; the literature has been reviewed by, among others, McKenzie and White (1991) and Brew (2003).

α-La contains four tryptophan residues per mole, giving it a specific absorbance at 280 nm of 20. It contains four intramolecular disulfide bonds per mole but no cysteine, phosphate or carbohydrate. Its isoionic point is ≈pH 4.8. The milk of

Bos taurus breeds contains only one genetic variant of α-La, B, but zebu cattle produce two variants, A and B. α-La has been isolated from the milk of the cow, sheep, goat, sow, human, buffalo, rat, guinea pig, horse and many other species; there are minor inter-species differences in the composition and properties.

The primary structure of α-La is homologous with lysozyme; out of a total of 123 residues in α-La, 54 are identical to corresponding residues in chicken egg white lysozyme and 23 others are structurally similar. α-La is a compact, highly structured globular protein. Because of difficulties in preparing good crystals, its tertiary structure has not been determined but is similar to that of lysozyme, on which a model of the structure of α-La is based (see McKenzie and White, 1991). In evolutionary terms, lysozyme is a very ancient protein; it is believed that α-La evolved from it through gene duplication (see Nitta and Sugai, 1989).

As discussed earlier, α-La is a component of lactose synthetase, the enzyme that catalyzes the final step in the biosynthesis of lactose. There is a direct correlation between the concentrations of α-La and lactose in milk. The milk of some marine mammals contains very little or no α-La.

α-La is a metalloprotein containing one Ca^{2+} per mole in a pocket containing four Asp residues. The calcium-containing protein is the most heat stable of the principal whey proteins, or, more correctly, the protein renatures following heat denaturation, which occurs at a relatively low temperature, as indicated by differential scanning calorimetry. When the pH is reduced to $<\approx$pH 5, the Asp residues become protonated and lose their ability to bind Ca^{2+}. The metal-free protein is denatured at quite a low temperature and does not renature on cooling; this characteristic has been exploited to isolate α-La from whey. α-La has poor thermo-gelling properties. Most lysozymes do not bind Ca^{2+} but equine lysozyme is an exception (Nitta *et al.*, 1987) and seems to be an intermediate in the evolution of lysozyme to α-La (see Nitta and Sugai, 1989).

α-La is synthesized in the mammary gland, but a very low level is transferred, probably via leaky mammocyte junctions, into blood serum, in which the concentration of α-La increases during pregnancy or following administration of steroid hormones to male or female animals (Akers, 2000). The concentration of α-La in blood serum is a reliable, non-invasive indicator of mammary gland development and of the potential of an animal for milk production.

A high MW form of α-La, isolated recently from human acid-precipitated casein, has anti-carcinogenic activity; it was named HAMLET (human α-La made lethal to tumour cells). Bovine α-La can be converted to a form with similar activity, called BAMLET; it is the molten globular state formed from apo-α-La and *cis*Δ9-octadecenoic acid (see Chatterton *et al.*, 2006).

Blood serum albumin

Normal bovine milk contains 0.1–0.4 g/L of blood serum albumin (BSA; 0.3–1.0% of total nitrogen), presumably as a result of leakage from blood; it has no known biological function in milk. BSA has been studied extensively; for reviews, see Fox (2003). Because of its low concentration in milk, BSA has little effect on the physicochemical properties of WPC and WPI.

Immunoglobulins

Mature bovine milk contains 0.6–1 g immunoglobulins (Igs)/L (≈3% of total nitrogen), but colostrum contains ≈ 10% (w/v) Ig, the level of which decreases rapidly post-partum. IgG_1 is the principal Ig in bovine, caprine or ovine milk, with lesser amounts of IgG_2, IgA and IgM; IgA is the principal Ig in human milk. The cow, sheep and goat do not transfer Ig to the fetus in utero and the neonate is born without Ig in its blood; consequently, it is very susceptible to bacterial infection with a very high risk of mortality.

The young of these species can absorb Ig from the intestine for several days after birth and thereby acquire passive immunity until they synthesize their own Ig, within a few weeks of birth. The human mother transfers Ig in utero and the offspring are born with a broad spectrum of antibodies. Although the human baby cannot absorb Ig from the intestine, the ingestion of colostrum is still very important because its Igs prevent intestinal infection. Some species, e.g. the horse, transfer Ig both in utero and via colostrum (see Hurley, 2003).

It has been suggested that the neonate secretes chymosin rather than pepsin because the former is weakly proteolytic and does not inactivate Ig. Colostrum contains α_2-macroglobulin which may inhibit proteinases in the gastrointestinal tract and protect Ig.

The modern dairy cow produces colostrum far in excess of the requirements of its calf; surplus colostrum is available for the recovery of Ig and other nutraceuticals (Pakkanen and Aalto, 1997). There is considerable interest in hyperimmunizing cows against certain human pathogens, e.g. rota virus, for the production of antibody-rich milk for human consumption, especially by infants; the Ig could be isolated from the milk and presented as a "pharmaceutical" or consumed directly in the milk.

Whey acidic protein

WAP was identified first in the mouse milk and has since been found also in rat, rabbit, camel, wallaby, opossum, echidna and platypus milk. As the milks of all of these species lack β-Lg, it was thought that these proteins were mutually exclusive. However, porcine milk, which contains β-Lg, was recently found to contain WAP also. The MW of WAP is 14–30 kDa (the variation may be due to differences in glycosylation) and it contains two (in eutherians) or three (in monotremes and marsupials) four-disulfide domains. As human milk lacks β-Lg, it might be expected to contain WAP but there are no reports to this effect. In humans and ruminants, the WAP gene is frame shifted and is a pseudogene. WAP functions as a proteinase inhibitor, is involved in terminal differentiation in the mammary gland and has antibacterial activity (for reviews, see Simpson and Nicholas, 2002; Hajjoubi *et al.*, 2006).

Proteose peptone 3

The PP fraction of milk protein is a very complex mixture of peptides, most of which are produced by the action of indigenous plasmin (see above) but some are indigenous to milk. The fraction has been only partially characterized; the current status has been described by Fox (2003). The PP fraction is of little or no technological significance.

Bovine proteose peptone 3 (PP3) is a heat-stable phosphoglycoprotein that was first identified in the PP (heat-stable, acid-soluble) fraction of milk. Unlike the other peptides in this fraction, PP3 is an indigenous milk protein, synthesized in the mammary gland. Bovine PP3 consists of 135 amino acid residues, with five phosphorylation and three glycosylation sites. When isolated from milk, the PP3 fraction contains at least three components of MW ≈ 28, 18 and 11 kDa; the largest of these is PP3, and the smaller components are fragments thereof generated by plasmin (see Girardet and Linden, 1996). PP3 is present mainly in acid whey but some is present in the MFGM. Girardet and Linden (1996) proposed changing the name to lactophorin or lactoglycophorin; it has also been referred to as the hydophobic fraction of PP.

Because of its strong surfactant properties (Campagna et al., 1998), PP3 can prevent contact between milk lipase and its substrates, thus preventing spontaneous lipolysis. Although its amino acid composition suggests that PP3 is not a hydrophobic protein, it behaves hydrophobically, possibly because of the formation of an amphiphilic α-helix, one side of which contains hydrophilic residues whereas the other side is hydrophobic. The biological role of PP3 is unknown.

Non-protein nitrogen

The NPN fraction of milk contains those nitrogenous compounds that are soluble in 12% TCA; it represents ≈5% of total nitrogen (≈300 mg/L). The principal components are urea, creatine, uric acid and amino acids. Human milk contains a high level of taurine which can be converted to cysteine and may be nutritionally important for infants. Urea, the concentration of which varies considerably, has a significant effect on the heat stability of milk.

Minor proteins

Milk contains several proteins at very low or trace levels, many of which are biologically active (see Schrezenmeir *et al.*, 2000); some are regarded as highly significant and have attracted considerable attention as nutraceuticals. When ways of increasing the value of milk proteins are discussed, the focus is usually on these minor proteins but they are, in fact, of little economic value to the overall dairy industry. They are found mainly in the whey but some are also located in the fat globule membrane. Reviews on the minor proteins include Fox and Flynn (1992), Fox and Kelly (2003), and Haggarty (2003).

Metal-binding proteins

Milk contains several metal-binding proteins: the caseins (Ca, Mg, Zn), α-La (Ca), xanthine oxidase (Mo, Fe), alkaline phosphatase (Zn, Mg), lactoperoxidase (Fe), catalase (Fe), ceruloplasmin (Cu), glutathione peroxidase (Se), lactoferrin (Fe) and transferrin (Fe).

Lactoferrin (Lf), a non-haem iron-binding glycoprotein, is a member of a family of iron-binding proteins, which includes transferrin and ovotransferrin (conalbumin) (see Lonnerdal, 2003). It is present in several body fluids, including saliva, tears, sweat and semen. Lf has several potential biological functions: it improves the bioavailability of iron, is bacteriostatic (by sequestering iron and making it unavailable to intestinal bacteria) and has antioxidant, antiviral, anti-inflammatory, immunomodulatory and anti-carcinogenic activity.

Human milk contains a very high level of Lf ($\approx$20% of total nitrogen) and therefore there is interest in fortifying bovine-milk-based infant formulas with Lf. The pI of Lf is $\approx$9.0, i.e. it is cationic at the pH of milk, whereas most milk proteins are anionic, and it can be isolated on an industrial scale by adsorption on a cation-exchange resin. Hydrolysis of Lf by pepsin yields peptides called lactoferricins, which are more bacteriostatic than Lf and their activity is independent of iron status. Milk also contains a low level of serum transferrin.

Milk contains a copper-binding glycoprotein—ceruloplasmin—also known as feroxidase (EC 1.16.3.1). Ceruloplasmin is an α_2-globulin with a MW of $\approx$ 126 kDa; it binds six atoms of copper per molecule and may play a role in delivering essential copper to the neonate.

Glutathione peroxidase (GTPase) is a selenium-containing protein. It has been reported that milk contains GTPase and that it binds 30% of the total selenium in milk (see Fox and Kelly, 2006b). GTPase has no known enzymatic function in milk; the activity attributed to GTPase in milk may be due to sulfydryl oxidase.

β-Microglobulin

β_2-Microglobulin, initially called lactollin, was first isolated from bovine acid-precipitated casein in 1963 by M. L, Groves. Lactollin, reported to have a MW of 43 kDa, is a tetramer of β_2-microglobulin, which consists of 98 amino acids, with a calculated MW of 11 636 Da. β_2-Microglobulin is a component of the immune system and is probably produced by proteolysis of a larger protein, mainly within the mammary gland; it has no known significance in milk.

Osteopontin

Osteopontin (OPN) is a highly phosphorylated acidic glycoprotein, consisting of 261 amino acid residues with a calculated MW of 29 283 Da (total MW of the glycoprotein, $\approx$60 000 Da). OPN has 50 potential calcium-binding sites, about half of which are saturated under normal physiological concentrations of calcium and magnesium.

OPN occurs in bone (it is one of the major non-collagenous proteins in bone), in many other normal and malignant tissues, and in milk and urine; it can bind to many cell types. It is believed to have a diverse range of functions (Bayless *et al.*, 1997) but its role in milk is not clear. A rapid method for the isolation of OPN from milk was reported by Azuma *et al.* (2006) who showed that it binds Lf, lactoperoxidase and Igs, and may serve as a carrier for these proteins. An acidic whey protein fraction that contains OPN reduces bone loss in ovariectomized rats (Kruger *et al.*, 2006).

Vitamin-binding proteins

Milk contains binding proteins for at least the following vitamins: retinol (vitamin A, i.e. β-Lg), biotin, folic acid and cobalamine (vitamin B_{12}). The precise role of these proteins is not clear but they probably improve the absorption of vitamins from the intestine or act as antibacterial agents by rendering vitamins unavailable to bacteria. The concentration of these proteins varies during lactation but the influence of other factors such as individuality, breed and nutritional status is not known. The activity of these proteins is reduced or destroyed on heating at temperatures somewhat higher than high temperature, short time (HTST) pasteurization.

Angiogenins

Angiogenins induce the growth of new blood vessels, i.e. angiogenesis. They have high sequence homology with members of the RNase A superfamily of proteins and have RNase activity. Two angiogenins (ANG-1 and ANG-2) have been identified in bovine milk and blood serum; both strongly promote the growth of new blood vessels in a chicken membrane assay. The function(s) of the angiogenins in milk is unknown. They may be part of a repair system to protect either the mammary gland or the intestine of the neonate and/or part of the host-defence system.

Kininogen

Two kininogens have been identified in bovine milk, a high (88–129 kDa, depending on the level of glycosylation) and a low (16–17 kDa) MW form. Bradykinin, a biologically active peptide containing nine amino acids, which is released from the high MW kininogen by the action of the enzyme kallikrein, has been detected in the mammary gland, and is secreted into milk, from which it has been isolated. Plasma kininogen is an inhibitor of thiol proteinases and has an important role in blood coagulation. Bradykinin affects smooth muscle contraction and induces hypertension. The biological significance of bradykinin and kininogen in milk is unknown.

Glycoproteins

Many of the minor proteins discussed above are glycoproteins; in addition, several other minor glycoproteins have been found in milk and especially in colostrum, the function of which have not been elucidated. One of the high MW glycoproteins in bovine milk is prosaposin, a neurotrophic factor that plays an important role in the development, repair and maintenance of nervous tissue. It is a precursor of saposins A, B, C and D, which have not been detected in milk.

The physiological role of prosaposin in milk is not known, although saposin C, released by digestion, could be important for the growth and development of the young.

Proteins in the MFGM

About 1% of the total protein in milk is in the MFGM. Most of the proteins are present at trace levels, including many of the indigenous enzymes in milk. The principal proteins in the MFGM include mucin, adipophilin, butyrophilin and xanthine oxidase (see the section on milk lipids).

Growth factors

Milk contains many peptide hormones, including epidermal growth factor, insulin, insulin-like growth factors 1 and 2, three human growth factors ($\alpha 1$, $\alpha 2$ and β), two mammary-derived growth factors (I and II), colony-stimulating factor, nerve growth factor, platelet-derived growth factor and bombasin. It is not clear whether these factors play a role in the development of the neonate or in the development and functioning of the mammary gland, or both (see Fox and Flynn, 1992; Gauthier *et al.*, 2006; Baumrucker, 2007).

Indigenous milk enzymes

Milk contains about 70 indigenous enzymes, which are minor but very important members of the milk protein system (see Fox and Kelly, 2006a, 2006b). The enzymes originate from the secretory cells or the blood; many are concentrated in the MFGM and originate in the Golgi membranes of the cell or the cell cytoplasm, some of which becomes entrapped as crescents inside the encircling membrane during exocytosis. Plasmin and LPL are associated with the casein micelles and several enzymes are present in the milk serum; many of the latter are derived from the MFGM, which is shed as the milk ages.

The indigenous enzymes are significant for several reasons:

- deterioration of product quality (plasmin, LPL, acid phosphatase, xanthine oxidase);
- bactericidal agents (lactoperoxidase and lysozyme);
- indices of the thermal history of milk (alkaline phosphatase, γ-glutamyltransferase or lactoperoxidase);
- indices of mastitic infection (catalase, acid phosphatase and especially *N*-acetylglucosaminidase).

The concentrations/activities of the indigenous enzymes in milk show greater inter-species differences than any other constituent, e.g.,

- 3000 times more lysozyme in equine milk and human milk than in bovine milk;
- lactoperoxidase is a major enzyme in bovine milk but is absent from human milk;
- human milk and the milks of a few other species contain bile-salts-stimulated lipase but the milks of most species lack this enzyme;
- the principal lipase in milk is LPL, of which there is <500 times as much in guinea pig milk as in rat milk; and
- bovine milk has a high level of xanthine oxidoreductase (XOR) activity but all other milks that have been studied have low XOR activity because the protein lacks molybdenum (XOR plays a major role in the excretion of fat globules from the mammocyte but in this function it does not act as a non-enzyme).

The reason(s) for these inter-species differences is/are not known but some of them may be significant.

Biologically active cryptic peptides

One of the most exciting recent developments in milk proteins is the discovery that all milk proteins contain sequences that have biological/physiological activities when released by proteolysis. The best studied are phosphopeptides, angiotensin-converting-enzyme inhibitory peptides, platelet-modifying peptides, opiate peptides, immuno-modulating peptides and the caseinomacropeptides, which have many biological properties (see Fox and McSweeney, 2003; Korhonen, 2006; Korhonen and Pihlanto, 2006).

Casein micelles

It has been known since the work of H. Schuler in 1818 that the casein in milk exists as large particles, now called casein micelles. The stability of the micelles is critically important for many of the technologically important properties of milk and consequently has been the focus of much research, especially during the past 50 years. Early views and research on casein micelles have been reviewed by Fox and Brodkorb (2008). Although views on the detailed structure of the casein micelle are divided, there is widespread, or unanimous, agreement on their general structure and properties.

Electron microscopy shows that casein micelles are spheres with a diameter in the range 50–500 nm (average ≈120 nm) and a mass ranging from 10^6 to 3×10^9 Da (average ≈10^8 Da). There are numerous small micelles, but these represent only a small proportion of the mass. There are 10^{14}–10^{16} micelles/mL of milk, and they are roughly two micelle diameters (≈250 nm) apart. The dry matter of the micelles is ≈94% protein and 6% low-molecular mass species, mainly consisting of calcium phosphate with some magnesium and citrate and trace amounts of other species, referred to collectively as colloidal calcium phosphate (CCP). The micelles bind ≈2.0 g H_2O/g protein. They scatter light, and the white color of milk is due largely to light scattering by the casein micelles; the white color is lost if the micelles are disrupted, by dissolving CCP with citrate, EDTA or oxalate, by increasing pH, or by urea (>5 M) or ethanol (≈35% at 70°C).

Stability of casein micelles

The micelles are quite stable to the principal processes to which milk is normally subjected. They are very stable at high temperatures, and withstand heating at 140°C for 15–20 min at pH 6.7. Coagulation is caused by heat-induced changes, e.g. a decrease in pH due to the pyrolysis of lactose to acids, dephosphorylation of casein, cleavage of the carbohydrate-rich moiety of κ-casein, denaturation of the whey proteins and their precipitation on the casein micelles, and precipitation of soluble calcium phosphate on the micelles (see O'Connell and Fox, 2003).

The micelles are stable to compaction (e.g. they can be sedimented by ultracentrifugation and redispersed by mild agitation), to commercial homogenization and to Ca^{2+} concentrations up to at least 200 mM at temperatures up to 50°C. The effects of high pressure (up to 800 MPa) on the casein micelles in bovine, ovine, caprine and buffalo milks have been studied; the size of the micelles increases up to 200–300 MPa but decreases at higher pressure (see Huppertz *et al.*, 2006).

As the pH of milk is reduced, CCP dissolves and is fully soluble at = pH 4.9. Acidification of cold (4°C) milk to pH 4.6, followed by dialysis against bulk milk is a convenient technique for altering the CCP content of milk. If acidified cold milk is readjusted to pH 6.7, the micelles reform, provided that the pH had not been reduced below 5.2. This result seems to suggest that most of the CCP can be dissolved without destroying the structure of the micelles.

Some proteinases, especially chymosin, catalyze a very specific hydrolysis of κ-casein, as a result of which the casein coagulates in the presence of Ca^{2+} or other divalent ions. This is the key step in the manufacture of most cheese varieties. The proteinase preparations used for cheesemaking are called rennets.

At room temperature, the casein micelles are destabilized by ≈40% ethanol at pH 6.7 or by lower concentrations if the pH is reduced. However, if the system is heated to ≥70°C, the precipitate redissolves and the system becomes translucent. When the system is recooled, the white appearance of milk is restored and a gel is formed if the ethanol-milk mixture is held at 4°C, especially if concentrated milk is used. If the ethanol is removed by evaporation, very large aggregates (average diameter ≈3000 nm) are formed. The dissociating effect of ethanol is promoted by increasing the pH (35% ethanol causes dissociation at 20°C and pH 7.3) or adding NaCl. Methanol and acetone have an effect similar to ethanol, but propanol causes dissociation at ≈25°C. The mechanism by which ethanol dissociates casein micelles has not been established, but it is not due to the solution of CCP, which is unchanged. The micelles are also reversibly dissociated by urea (5 M), SDS or raising the pH to >9. Under these conditions, the CCP is not dissolved.

The micelles are destabilized by freezing (cryodestabilization) due to a decrease in pH and an increase in the Ca^{2+} concentration in the unfrozen phase of milk; concentrated milk is very susceptible to cryodestabilization. Cryodestabilized casein can be dispersed by warming the thawed milk to 55°C to give particles with micelle-like properties.

Micelle structure

There has been speculation since the beginning of the twentieth century on how the casein particles (micelles) are stabilized (see Fox and Brodkorb, 2008), but no significant progress was possible until the isolation and characterization of κ-casein by D. F. Waugh and P. H. von Hippel in 1956. The first attempt to describe the structure of the casein micelle was made by Waugh in 1958 and, since then, numerous models have been made and refined. Progress has been reviewed regularly (see Fox, 2003); recent reviews include de Kruif and Holt (2003), Horne (2002, 2003, 2006) and Farrell *et al.* (2006b).

The principal features that must be met by any micelle model are the following: κ-casein, which represents ≈15% of total casein, must be so located as to be able to stabilize the calcium-sensitive α_{s1}-, α_{s2}- and β-caseins, which represent approximately 85% of total casein; chymosin and other rennets, which are relatively large molecules (MW ≈35 kDa), very rapidly and specifically hydrolyze most of the κ-casein; when heated in the presence of whey proteins, κ-casein and β-Lg (MW ≈ 36 kDa) interact to form a disulfide-linked complex, which modifies the rennet and heat coagulation properties of the micelles.

The arrangement that would best explain these features is a surface layer of κ-casein surrounding the calcium-sensitive caseins, somewhat analogous to a lipid emulsion in which the triglycerides are surrounded by a thin layer of emulsifier. Most models of the casein micelle propose a surface location for κ-casein but some early models envisaged κ-casein serving as nodes in the interior of the micelle (see Fox, 2003).

Removal of CCP causes disintegration of the micelles into particles of MW ≈ 10^6 Da, suggesting that the casein molecules are held together in the micelles by CCP. The properties of the CCP-free system are very different from those of normal milk (e.g. it is precipitated by relatively low levels of Ca^{2+}, it is more stable to heat-induced coagulation and it is not coagulable by rennets). Many of these properties can be restored, at least partially, by an increased concentration of calcium. However, CCP is not the only integrating factor, as indicated by the dissociating effect of urea, SDS, ethanol or alkaline pH. At low temperatures, casein, especially β-casein, dissociates from the micelles.

There has been strong support for the view, first proposed by C. V. Morr in 1967, that the micelles are composed of sub-micelles (≈10^6 Da and 10–15 nm in diameter) linked together by CCP, giving a micelle with an open, porous structure. On removing CCP (by acidification/dialysis, EDTA, citrate or oxalate), the micelle disintegrates. Disintegration may also be achieved by treatment with urea, SDS, 35% ethanol at 70°C or pH > 9. These reagents do not solubilize CCP, suggesting that hydrophobic interactions and hydrogen bonds contribute to micelle structure. Much of the evidence for a sub-micellar structure relies on electron microscopy studies that appear to show variations in electron density, a raspberry-like structure, which was interpreted as indicating sub-micelles.

Views on the proposed structure of the sub-micelles have evolved over the years (see Fox, 2003; Fox and Kelly, 2003). Proposals have included a rosette-type structure similar to that of a classical soap micelle, in which the polar regions of α_{s1}-, α_{s2}- and β-caseins are orientated towards the outside of the sub-micelle to reduce electrostatic repulsion between neighboring charged groups; each sub-micelle is considered to be surrounded by a layer (coat) of κ-casein, thus providing a κ-casein coat for the entire micelle.

Several authors have suggested that the sub-micelles are not covered completely by κ-casein and that there are κ-casein-rich, hydrophilic regions and κ-casein-deficient, hydrophobic regions on the surface of each sub-micelle. The latter aggregate via the hydrophobic patches such that the entire micelle has a κ-casein-rich surface layer, but with some of the other caseins on the surface also. In a popular version of this model, it is proposed that the hydrophilic C-terminal region of κ-casein protrudes from the surface, forming a layer 5–10 nm thick and giving the micelles a hairy appearance. This hairy layer, functioning as an ionic brush, is responsible for micelle stability through major contributions to zeta potential (−20 mV) and steric stabilization. If the hairy layer is removed through specific hydrolysis of κ-casein or collapsed (e.g. by ethanol), the colloidal stability of the micelles is destroyed, and they aggregate.

A further variant of the sub-unit model envisages two main types of sub-units—one consisting of α_{s1}-, α_{s2}- and β-caseins, which are present in the core of the micelle; the other, consisting of α_{s1}- and α_{s2}- and κ-caseins, forms a surface layer. It has also been proposed that β-casein associates to form thread-like structures with which α_{s1}- and α_{s2}-caseins associate hydrophobically to form the core of the micelle or sub-micelles, which are surrounded by a layer of κ-casein; CCP cements neighboring sub-micelles within the micelle.

Although the sub-micelle model of the casein micelle explains many of the principal features of, and physico-chemical reactions undergone by, the micelles and has been supported widely, it has never enjoyed unanimous support. Indeed new electron microscopy techniques have cast doubts on the authenticity of sub-micelles. Using cryopreparation electron microscopy with stereo-imaging, McMahon and McManus (1998) found no evidence to support the sub-micellar model; they concluded that, if the micelles do consist of sub-micelles, these must be smaller than 2 nm or less densely packed than previously presumed. Like other forms of electron microscopy, field emission scanning electron microscopy showed that casein micelles have an irregular surface, but Dalgleish *et al.* (2004) concluded that the caseins form tubular structures rather than spherical sub-micelles; in principle, this model seems basically similar to earlier sub-unit models.

Three alternative models have been proposed recently. Visser (1992) proposed that the micelles are spherical conglomerates of randomly aggregated casein molecules held together by amorphous calcium phosphate and hydrophobic bonds, with a surface layer of κ-casein. Holt (1992) considered the casein micelle to be a tangled web of flexible casein molecules forming a gel-like structure in which micro-granules of CCP are an integral feature and from the surface of which the C-terminal region of κ-casein extends, forming a hairy layer. In what he referred to as the dual-binding model, Horne (2002, 2003, 2006) described how casein molecules interact hydrophobically and through calcium phosphate nanoclusters to form micelles. These three models retain the key features of the sub-micellar model, i.e. the cementing role of CCP and the predominantly surface location and micelle-stabilizing role of κ-casein, and differ from it mainly with respect to the internal structure of the micelle.

Inter-species comparison of milk proteins

The milks of the species for which data are available show considerable differences in protein content, from ≈1 to ≈20%. The protein content reflects the growth rate of the neonate of the species, i.e. its requirements for essential amino acids. The milks of all species for which data are available contain two groups of protein—caseins and whey proteins—but the ratio of these varies widely. Both groups show genus- and even species-specific characteristics, which presumably reflect some unique nutritional or physiological requirements of the neonate of the species.

Interestingly, and perhaps significantly, of the milks that have been characterized, human and bovine milks are more or less at opposite ends of the spectrum. Among the general inter-species comparisons of milk proteins are Woodward (1976), Jenness (1973, 1979, 1982), Ginger and Grigor (1999) and Martin *et al.* (2003); reviews on milk proteins of individual species include: buffalo (Addeo *et al.*, 1977), goat (Trujillo *et al.*, 1977, 2000), sheep (Amigo *et al.*, 2000), camel (Ochirkhuyag *et al.*, 1997; Kappler *et al.*, 1998), yak (Ochirkhuyag *et al.*, 1997), horse (Ochirkhuyag *et al.*, 2000; Park *et al.*, 2006) and sow (Gallagher *et al.*, 1997).

There is considerably more and better information on the inter-species comparison of individual milk proteins than on overall milk composition, probably because only one sample of milk from one animal is sufficient to yield a particular protein for characterization. The two principal milk-specific whey proteins, α-La and β-Lg, from quite a wide range of species have been characterized and, in general, show a high degree of homology (see Brew, 2003; Sawyer, 2003). The caseins show much greater inter-species diversity, especially in the α-casein fraction; all species that have been studied contain a protein with an electrophoretic mobility similar to that of bovine β-casein, but the β-caseins that have been sequenced show a low level of homology (Martin *et al.*, 2003).

Sheep's milk is used mainly for cheese production, with small amounts used for the production of fermented milks; hence the coagulation and gel-forming properties of ovine milk are particularly important. The α_{s1}-casein of ovine milk is very heterogeneous—to date, 10 genetic variants have been identified; not only do the properties of the variants differ, but also the concentration of α_{s1}-casein varies from 0 to 26% of total casein and, consequently, the total protein content varies considerably and this, in turn, has major effects on the rennet-induced coagulation properties of ovine milk and on the yield and quality of cheese produced therefrom (Amigo *et al.*, 2000; Clark and Sherbon, 2000a, 2000b).

Human β-casein occurs in multi-phosphorylated form (0–5 mol phosphorus per mol protein; see Atkinson and Lonnerdal, 1989), as does mare's β-casein (Ochirkhuyag *et al.*, 2000). Considering the critical role played by κ-casein, it would be expected that all casein systems contain this protein, but Ochirkhuyag *et al.* (2000) failed to identify κ-casein in mare's milk and suggested that the micelle-stabilizing role was played by β-casein with zero or a low level of phosphorylation; more recent work has shown that equine milk contains a low level of κ-casein (Iametti *et al.*, 2001; Egito *et al.*, 2002). Human κ-casein is very highly glycosylated, containing 40–60% carbohydrate (compared with $\approx$ 10% in bovine κ-casein). The α_s-casein fraction differs markedly between species; human milk probably lacks an α_s-casein whereas the α-casein fractions in horse and donkey milks are very heterogeneous. The caseins of only about 10 species have been studied in some detail.

There are very considerable inter-species differences in the minor proteins of milk. The milks of those species that have been studied in sufficient depth contain approximately the same profile of minor proteins, but there are very marked quantitative differences. Most of the minor proteins in milk have some biochemical or physiological function, and the quantitative inter-species differences presumably reflect the requirements of the neonate of the species. The greatest inter-species differences, in some cases 4000-fold, seem to occur in the indigenous enzymes (Fox and Kelly, 2006a, 2006b).

In the milk of all species, the caseins exist as micelles (at least the milks appear white), but the properties of the micelles in the milk of only a few non-bovine species have been studied: caprine (Ono and Creamer, 1986; Ono *et al.*, 1989), ovine (Ono *et al.*. 1989), buffalo (Patel and Mistry, 1997), camel (Attia *et al.*, 2000), mare (Welsch *et al.*, 1988; Ono *et al.*, 1989) and human (Sood *et al.*, 1997, 2002). The appearance and the size of casein micelles in the milk of 19 species—guinea pig, rat, nutria (coypu), dog, cat, grey seal, rabbit, donkey, horse, alpaca, dromedary camel, cow, red deer, sheep, pig, water buffalo, goat, porpoise and man—were studied by Buchheim *et al.* (1989); the structures of all micelles appeared to be similar on electron microscopy but there were large inter-species differences in size—human micelles were smallest (64 nm) whereas those of the alpaca, goat, camel and donkey were very large (300–500 nm).

Salts

When milk is heated at 500°C for ≈5 h, an ash derived mainly from the inorganic salts of milk, and representing ≈0.7% w/w of the milk, remains. However, the elements in the ash are changed from their original forms to oxides or carbonates; the ash contains phosphorus and sulfur derived from caseins, lipids, sugar phosphates or high-energy phosphates. The organic salts, the most important of which is citrate, are oxidized and lost during ashing; some volatile metals, e.g. sodium, are partially lost. Thus, ash does not accurately represent the salts of milk.

However, the principal inorganic and organic ions in milk can be determined directly by potentiometric, spectrophotometric or other methods. The typical concentrations of the principal elements—the macro-elements— are shown in Table 1.2; considerable variability occurs, due, in part, to poor analytical methods and/or to samples from cows in very early or late lactation or suffering from mastitis. Milk also contains 20–25 elements at very low or trace levels. These micro-elements are very important from a nutritional viewpoint: many, e.g. Zn, Fe, Mo, Cu, Ca, Se and Mg, are present in enzymes, and many are concentrated in the MFGM; some micro-elements, e.g. Fe and Cu, are very potent lipid pro-oxidants. Although the salts are relatively minor constituents of milk, they are critically important for many technological and nutritional properties of milk.

Some of the salts in milk are fully soluble but others, especially calcium phosphate, exceed their solubility under the conditions in milk and occur partly in the colloidal state, associated with the casein micelles; these salts are referred to as CCP, although some magnesium, citrate and traces of other elements are also present in the micelles. As discussed earlier, CCP plays a critical role in the structure and stability of the casein micelles. The typical distribution of the principal organic and inorganic ions between the soluble and colloidal phases is summarized in Table 1.2. The actual form of the principal species can be determined or calculated after making certain assumptions; typical values are shown in Table 1.2.

The solubility and the ionization status of many of the principal ionic species are interrelated, especially H^+, Ca^{2+}, PO_4^{3-} and citrate^{3-}. These relationships have major effects on the stability of the caseinate system and consequently on the processing properties of milk. The status of various species in milk can be modified by adding certain salts to milk, e.g. the Ca^{2+} concentration is reduced by adding PO_4^{3-} or citrate^{3-}; addition of $CaCl_2$ affects the distribution and ionization status of calcium and phosphate, and the pH of milk.

The precise nature and the structure of CCP are uncertain. It is associated with the caseins, probably via the casein phosphate residues; it probably exists as nanocrystals that include phosphate residues of casein. The simplest stoichiometry is $Ca_3(PO_4)_2$ but spectroscopic data suggest that $CaHPO_4$ is the most likely form.

The distribution of species between the soluble and colloidal phases is strongly affected by pH and temperature. As the pH is reduced, CCP dissolves and is

Table 1.2 Concentration and distribution of the principal milk salts

Species	Concentration (mg/l)	%	Soluble form	Colloidal
Sodium	500	92	Completely ionized	8
Potassium	1450	92	Completely ionized	8
Chloride	1200	100	Completely ionized	–
Sulfate	100	100	Completely ionized	–
Phosphate	750	43	10% bound to Ca and Mg 54% $H_2PO_4^-$ 36% HPO_4^{2-}	57
Citrate	1750	94	85% bound to Ca and Mg (undissociated) 15% $Citr^{3-}$	6
Calcium	1200	34	35% Ca^{2+} 55% bound to citrate 10% bound to phosphate	66
Magnesium	130	67	Probably similar to calcium	33

completely soluble at <≈pH 4.9; the reverse occurs when the pH is increased. The solubility of calcium phosphate decreases as the temperature is increased and soluble calcium phosphate is transferred to the colloidal phase, with the release of H^+ and a decrease in pH:

$$CaHPO_4/Ca(H_2PO_4)_2 \leftrightarrow Ca_3(PO_4)_2 + 3H^+$$

These changes are quite substantial and are at least partially reversible on cooling.

As milk is supersaturated with calcium phosphate, concentration of milk by evaporation of water increases the degree of supersaturation and the transfer of soluble calcium phosphate to the colloidal state, with the concomitant release of H^+. Dilution has the opposite effect.

Milk salts equilibria are also shifted on freezing; as pure water freezes, the concentrations of solutes in unfrozen liquid are increased. Soluble calcium phosphate precipitates as $Ca_3(PO_4)_2$, releasing H^+ (the pH of the unfrozen liquid may decrease to 5.8). The crystallization of lactose as a monohydrate increases the degree of supersaturation by reducing the amount of solvent water.

There are substantial changes in the concentrations of the macro-elements in milk during lactation, especially at the beginning and end of lactation and during mastitic infection. Changes in the concentration of some of the salts in milk, especially calcium phosphate and citrate, have major effects on the physico-chemical properties of the casein system and on the processability of milk, especially rennet coagulability and related properties and heat stability.

Reviews on the chemistry of milk salts include Pyne (1962), Holt (1985, 1997, 2003) and Lucey and Horne (2008). Reviews on the nutritional significance of milk salts include Flynn and Power (1985), Flynn and Cashman (1997), Cashman (2003, 2006) and Hunt and Nielsen (2008).

Vitamins

Milk contains all the vitamins in sufficient quantities to enable normal maintenance and growth of the neonate. Cow's milk is a very significant source of vitamins, especially biotin (B_7), riboflavin (B_2) and cobalamine (B_{12}), in the human diet. For general information on the vitamins and for specific aspects in relation to milk and dairy products, including stability during processing and storage, the reader is referred to a set of articles in Roginski *et al.* (2003) and in McSweeney and Fox (2007).

In addition to their nutritional significance, four vitamins are significant for other reasons: vitamin A (retinol) and carotenoids are responsible for the yellow-orange color of fat-containing products made from cow's milk; vitamin E (tocopherols) is a potent antioxidant; vitamin C (ascorbic acid) is an antioxidant or pro-oxidant, depending on its concentration; vitamin B_2 (riboflavin), which is greenish-yellow, is responsible for the color of whey or ultrafiltration permeate, co-crystallizes with lactose and is responsible for its yellowish color, which may be removed by recrystallization or bleached by oxidation. It also acts as a photocatalyst in the development of light-oxidized flavor in milk, which is due to the oxidation of methionine.

Water

Water is the principal constituent in the milk of most species. In addition to meeting the requirement of the neonate for water, the water in milk serves as a solvent for milk salts, lactose and proteins, and affects their properties and stability. It controls the rate of many reactions, e.g. Maillard browning, lipid oxidation, enzyme activity and microbial growth, thus affecting the stability of milk and milk products. The preservation of milk products is effected by reducing the water activity by dehydration or by adding sugar or NaCl; the stability and the quality of most dairy products are strongly affected by their water content and small differences can cause major instability problems.

The chemistry of water and its significance in foods in general and in dairy products are not discussed. The reader is referred to Duckworth (1975), Rockland and Beuchat (1987), Fennema (1996), Roos (1997), Franks (2000), Chieh (2006) and Simatos *et al.* (2008).

Conclusions

Milk is a very complex fluid. It contains several hundred molecular species, mostly at trace levels. Most of the micro-constituents are derived from blood or mammary tissue but most of the macro-constituents are synthesized in the mammary gland and are milk specific. The constituents of milk may be in true aqueous solution (e.g. lactose and most inorganic salts), or as a colloidal solution (proteins, which may be present as individual molecules or as large aggregates of several thousand molecules, called micelles) or as an emulsion (lipids). The macro-constituents can be fractionated readily and are used widely as food ingredients. The natural function of milk is to supply the neonate with its complete nutritional requirements for a period (sometimes several months) after birth and with many physiologically important molecules, including carrier proteins, protective proteins and hormones.

The properties of milk lipids and proteins may be readily modified by biological, biochemical, chemical or physical means and thus converted into novel dairy products.

In this overview, the chemical and physico-chemical properties of milk sugar (lactose), lipids, proteins and inorganic salts were discussed. There is a vast primary literature and a range of specialist textbooks on the technology used to convert milk into a range of food products but these aspects were not considered herein.

References

Addeo, F., Mercier, J-C. and Ribadeau-Dumas, B. (1977). The caseins of buffalo milk. *Journal of Dairy Research*, **44**, 455–68.

Akers, R. M. (2000). Selection for milk protein production from a lactation biology viewpoint. *Journal of Dairy Science*, **83**, 1151–58.

Akerstrom, B., Flower, D. R. and Salier, J. P. (2000). Lipocalins 2000. *Biochimica et Biophysica Acta*, **1482**, 1–356.

Amigo, L., Recio, I. and Ramos, M. (2000). Genetic polymorphism of ovine proteins: its influence on technological properties of milk – a review. *International Dairy Journal*, **10**, 135–49.

Atkinson, S. A. and Lonnerdal, B. (1989). *Protein and Non-protein Nitrogen in Human Milk*. Boca Raton, Florida: CRC Press.

Attia, H., Kherouatou, N., Nasri, M. and Khorchani, T. (2000). Characterization of the dromedary milk casein micelle and study of its changes during acidification. *Lait*, **80**, 503–15.

Azuma, N., Maeta, A., Fukuchi, K. and Kanno, C. (2006). A rapid method for purifying osteopontin from bovine milk and interaction between osteopontin and other milk proteins. *International Dairy Journal*, **16**, 370–78.

Barker, M. E. (2003). Contribution of dairy foods to nutrient intake. In *Encyclopedia of Dairy Sciences* (H. Roginski, J. W. Fuquay and P. F. Fox, eds) pp. 2133–37. London: Academic Press.

Bauman, D. E. and Lock, A. L. (2006). Conjugated linoleic acid: biosynthesis and nutritional significance. In *Advanced Dairy Chemistry,* Volume 2, *Lipids*, 3rd edn (P. F. Fox and P. L. H. McSweeney, eds) pp. 93–136. New York: Springer.

Baumrucker, C. R. (2008). Hormones and biologically active compounds in milk. In *Advanced Dairy Chemistry,* Volume 3, *Lactose, Water, Salts and Vitamins*, 3rd edn (P. L. H. McSweeney and P. F. Fox, eds), in press. New York: Springer.

Bayless, K. J., Davis, G. E. and Meininger, G. A. (1997). Isolation and biological properties of osteopontin from bovine milk. *Protein Expression and Purification*, **9**, 309–14.

Becker, D. J. and Lowe, J. R. (2003). Fucose: biosynthesis and biological functions in mammals. *Glycobiology*, **13**, 41R–53R.

Bencini, R. and Pulina, G. (1997). The quality of sheep milk. *Australian Journal of Experimental Agriculture*, **37**, 485–504.

Benton, M. J. (1999). Early origins of modern birds and mammals: molecules vs morphology. *BioEssays*, **21**, 1043–51.
Blackburn, D. G. (1991). Evolutionary origins of the mammary gland. *Mammal Review*, **21**, 81–96.
Blackburn, D. G. (1993). Lactation: historical patterns and potential for manipulation. *Journal of Dairy Science*, **76**, 3195–212.
Blackburn, D. G., Hayssen, V. and Murphy, C. J. (1989). The origin of lactation and the evolution of milk: a review with new hypothesis. *Mammal Review*, **19**, 1–26.
Breckenridge, W. C. and Kuksis, A. (1967). Molecular weight distribution of milk fat triglycerides from seven species. *Lipid Research*, **8**, 473–78.
Brew, K. (2003). α-Lactalbumin. In *Advanced Dairy Chemistry*, Volume 1, Part A, Proteins, 3rd edn (P. F. Fox and P. L. H. McSweeney, eds) pp. 387–419. New York: Kluwer Academic/Plenum Publishers.
Buchheim, W., Lund, S. and Scholtissek, J. (1989). Vergleichende Untersuchungen zur Struktur und Grosse von Caseinmicellen in der Milch verschiedener Species. *Kieler Milchwirtschaftliche Forschungsberichte*, **41**, 253–65.
Campagna, S., Vitoux, B., Humbert, G., Girardet, J-M., Linden, G., Haertle, T. and Gaillard, J-L. (1998). Conformational studies of a synthetic peptide from the putative lipid-binding domain of bovine milk component PP3. *Journal of Dairy Science*, **81**, 3139–48.
Cashman, K. (2003). Minerals in dairy products: macroelements, nutritional significance; trace elements, nutritional significance. In *Encyclopedia of Dairy Sciences* (H. Roginski, J. W. Fuquay and P. F. Fox, eds) pp. 2051–65. London: Academic Press.
Cashman, K. (2006). Milk minerals (including trace elements) and bone health. *International Dairy Journal*, **16**, 1389–98.
Chatterton, D. E. W., Smithers, G., Roupas, P. and Brodkorb, A. (2006). Bioactivity of β-lactoglobulin and α-lactalbumin – technological implications and processing: review. *International Dairy Journal*, **16**, 1229–40.
Chieh, C. (2006). Water chemistry and biochemistry. In *Food Biochemistry and Food Processing* (Y. H. Hui, ed.) pp. 103–33. Oxford: Blackwell Publishing.
Christie, W. W. (1995). Composition and structure of milk lipids. In *Advanced Dairy Chemistry, Volume 2, Lipids*, 2nd edn (P. F. Fox, ed.) pp. 1–36. London: Chapman and Hall.
Clark, S. and Sherbon, J. W. (2000a). Alpha$_{s1}$-casein, milk composition and coagulation properties of goat milk. *Small Ruminant Research*, **38**, 123–34.
Clark, S. and Sherbon, J. W. (2000b). Genetic variants of alpha$_{s1}$-casein in goat milk: breed distribution and associations with milk composition and coagulation properties. *Small Ruminant Research*, **38**, 135–43.
Crampton, A. W. and Jenkins, F. A. (1973). Mammals from reptiles: a review of mammalian origins. *Annual Review of Earth and Planetary Science*, **1**, 131–55.
Dalgleish, D. G., Spagnuolo, P. A. and Goff, H. D. (2004). A possible structure of the casein micelle based on high-resolution field-emission scanning electron microscopy. *International Dairy Journal*, **14**, 1025–31.
de Kruif, C. G. and Holt, C. (2003). Casein micelle structure, functions and interactions. In *Advanced Dairy Chemistry,* Volume 1, *Proteins*, 3rd edn (P. F. Fox and P. L. H. McSweeney, eds) pp. 233–76. New York: Kluwer Academic/Plenum Publishers.

Doreau, M. (1994). Le lait de jument et sa production: particularites et facteurs de variation. *Lait*, **74**, 401–18.
Doreau, M. and Boulot, S. (1989). Recent knowledge on mare milk production: a review. *Livestock Production Science*, **22**, 213–35.
du Puis, E. M. (2002). *Nature's Perfect Food: How Milk Became America's Drink.* New York: New York University Press.
Duckworth, R. B. (1975). *Water Relations in Foods.* London: Academic Press.
Easteal, S. (1999). Molecular evidence for the early divergence of placental mammals. *BioEssays*, **21**, 1052–58.
Egito, A. S., Miclo, L., Lopez, C., Adam, A., Girardet, J-M. and Gillard, J-L. (2002). Separation and characterization of mares' milk α_{s1}-, β- , κ-caseins, γ-casein-like, and proteose peptone component 5-like peptides. *Journal of Dairy Science*, **85**, 697–706.
Evans, D. E. (1959). Milk composition of mammals whose milk is not normally used for human consumption. *Dairy Science Abstracts*, **21**, 277–88.
Evers, J. (2004). The milkfat globule membrane – compositional and structural changes post secretion by the mammary secretory cell. *International Dairy Journal*, **14**, 661–74.
Farah, Z. (1993). Composition and characteristics of camel milk. *Journal of Dairy Research*, **60**, 603–26.
Farrell, H. M. Jr., Jimenez-Flores, R., Bleck, G. T., Brown, E. M., Butler, J. E., Creamer, L. K., Hicks, C. L., Holler, C. M., Ng-Kwai-Huang, K. F. and Swaisgood, H. E. (2004). Nomenclature of the proteins of cows' milk – sixth revision. *Journal of Dairy Science*, **87**, 1641–47.
Farrell, H. M. Jr., Qi, P. X. and Uversky, V. N. (2006a). New views on protein structure: application to the caseins: protein structure and functionality. In *Advances in Biopolymers: Molecules, Clusters, Networks, and Interactions* (M. L. Fishman, P. X. Qi and L. Whisker, eds), ACS Symposium Series 935 pp. 1–18. Washington, DC: American Chemical Society.
Farrell, H. M. Jr., Malin, E. L., Browne, E. M. and Qi, P. X. (2006b). Casein micelle structure: what can be learned from milk synthesis and structural biology?. *Current Opinion in Colloid and Interface Science*, **11**, 135–47.
Fennema, O. R. (1996). Water and ice. In *Food Chemistry*, 3rd edn (O. R. Fennema, ed.) pp. 17–94. New York: Marcel Dekker.
Finlay, E. (2005). *Bovini mitrochondrial DNA: demographic history and recombination.* PhD Thesis. Dublin University, Dublin, Ireland.
Flynn, A. and Cashman, K. (1997). Nutritional aspects of minerals in bovine and human milks. In *Advanced Dairy Chemistry,* Volume 3, *Lactose, Water, Salts and Vitamins*, 2nd edn (P. F. Fox, ed.) pp. 257–302. London: Chapman & Hall.
Flynn, A. and Power, P. (1985). Nutritional aspects of minerals in bovine and human milks. In *Developments in Dairy Chemistry,* Volume 3, *Lactose, and Minor Constituents* (P. F. Fox, ed.) pp. 183–215. London: Elsevier Applied Science.
Forsyth, I. A. (1986). Variation among species in the endocrine control of mammary growth and function: the roles of prolactin, growth hormone and placental lactogen. *Journal of Dairy Science*, **69**, 886–903.
Forsyth, I. A. (2003). Mammals. In *Encyclopedia of Dairy Sciences* (H. Roginski, J. W. Fuquay and P. F. Fox, eds) pp. 1672–80. London: Academic Press.

Forsyth, I. A. and Wallis, I. A. (2002). Growth hormone and prolactin – molecular and functional evolution. *Journal of Mammary Gland Biology and Neoplasia*, **7**, 291–312.

Fox, P. F. (1983). *Developments in Dairy Chemistry -2- Lipids.* London: Applied Science Publishers.

Fox, P. F. (1985). *Developments in Dairy Chemistry,* Volume 3, *Lactose and Minor Constituents.* London: Elsevier Applied Science.

Fox, P. F. (1992). *Advanced Dairy Chemistry,* Volume 1, *Proteins*, 2nd edn.. London: Elsevier Applied Science.

Fox, P. F. (1995). *Advanced Dairy Chemistry,* Volume 2, *Lipids*, 2nd edn.. London: Chapman & Hall.

Fox, P. F. (1997). *Advanced Dairy Chemistry,* Volume 3: Lactose, Water, Salts and Vitamins, 2nd edn.. London: Chapman & Hall.

Fox, P. F. (2003). Milk proteins: general and historical aspects. In *Advanced Dairy Chemistry,* Volume 1, *Proteins*, 3rd edn (P. F. Fox and P. L. H. McSweeney, eds) pp. 1–48. New York: Kluwer Academic/Plenum Publishers.

Fox, P. F. and Brodkorb, A. (2008). The casein micelle: historical aspects, current concepts and significance. *International Dairy Journal*, **18**, 677–684.

Fox, P. F. and Flynn, A. (1992). Biological properties of milk proteins. In *Advanced Dairy Chemistry,* Volume 1, *Proteins*, 2nd edn (P. F. Fox, ed.) pp. 255–84. London: Elsevier Applied Science.

Fox, P. F. and Kelly, A. L. (2003). Developments in the chemistry and technology of milk proteins. 2. Minor milk proteins. *Food Australia*, **55**, 231–34.

Fox, P. F. and Kelly, A. L. (2006a). Indigenous enzymes in milk: overview and historical aspects – Part 1. *International Dairy Journal*, **16**, 500–16.

Fox, P. F. and Kelly, A. L. (2006b). Indigenous enzymes in milk: overview and historical aspects – Part 2. *International Dairy Journal*, **16**, 517–32.

Fox, P. F. and McSweeney, P. L. H. (1998). *Dairy Chemistry and Biochemistry.* London: Chapman & Hall. , [reprinted by Kluwer Academic/Plenum Publishers, New York, 2000].

Fox, P. F. and McSweeney, P. L. H. (2003). *Advanced Dairy Chemistry,* Volume 1, *Proteins*, 3rd edn. New York: Kluver Academic-Plenum Plublishers.

Fox, P. F. and McSweeney, P. L. H. (2006). *Advanced Dairy Chemistry,* Volume 2, *Lipids*, 3rd edn. New York: Springer Verlag.

Franks, F. (2000). *Water – A Matrix of Life*, 2nd edn. Cambridge: Royal Society of Chemistry.

Gallagher, D. P., Cotter, P. F. and Mulvihill, D. M. (1997). Porcine milk proteins: a review. *International Dairy Journal*, **7**, 99–118.

Gauthier, S. F., Pouliot, Y. and Maubois, J-L. (2006). Growth factors in bovine milk and colostrum: composition, extraction and biological activities. *Lait*, **86**, 99–125.

Ginger, M. R. and Grigor, M. R. (1999). Comparative aspects of milk caseins. *Comparative Biochemistry and Physiology B*, **124**, 133–45.

Girardet, J-M. and Linden, G. (1996). PP3 component of bovine milk: a phosphorylated glycoprotein. *Journal of Dairy Research*, **63**, 333–50.

Glass, R. L. and Jenness, R. (1971). Comparative biochemical studies of milk. VI. Constituent fatty acids of milk fats of additional species. *Comparative Biochemistry and Physiology*, **38B**, 353–59.

Glass, R. L., Troolin, H. A. and Jenness, R. (1967). Comparative biochemical studies of milk. IV. Constituent fatty acids of milk fats. *Comparative Biochemistry and Physiology*, **22**, 415–25.

Goldman, A. S. (2002). Evolution of mammary gland defence system. *Journal of Mammary Gland Biology and Neoplasia*, **7**, 277–89.

Haenlein, G. F. W. (1980). Status of world literature on dairy goats: introductory remarks. *Journal of Dairy Science*, **63**, 1591–99.

Haggarty, N. W. (2003). Milk proteins: minor proteins, bovine serum albumin and vitamin-binding proteins. In *Encyclopedia of Dairy Sciences* (H. Roginski, W. Fuquay and P. F. Fox, eds) pp. 1939–46. London: Academic Press.

Hajjoubi, S., Rival-Gervier, S., Hayes, H., Floriot, S., Eggen, A., Pivini, F., Chardon, P., Houdebine, L-M. and Thepot, D. (2006). Ruminants genome no longer contains whey acidic protein gene but only a pseudogene. *Gene*, **370**, 104–12.

Hammarsten, O. (1883). Zur Frage, ob das Casein ein einheitleicher Stoff sei. *Z. Physiol Chemie*, **7**, 227–73.

Hayssen, V. (1993). Empirical and theoretical constraints on the evolution of lactation. *Journal of Dairy Science*, **76**, 3213–33.

Hayssen, V. and Blackburn, D. G. (1985). α-Lactalbumin and the origins of lactation. *Evolution*, **39**, 1147–49.

Holt, C. (1985). The milk salts: their secretion. concentration and physical chemistry. In *Developments in Dairy Chemistry,* Volume 3, *Lactose and Minor Constituents* (P. F. Fox, ed.) pp. 143–81. London: Elsevier Applied Science.

Holt, C. (1992). Structure and stability of bovine casein micelles. *Advances in Protein Chemistry*, **43**, 63–151.

Holt, C. (1997). The milk salts and their interaction with casein. In *Advanced Dairy Chemistry,* Volume 3, *Lactose, Water, Salts and Vitamins*, 2nd edn (P. F. Fox, ed.) pp. 233–56. London: Chapman & Hall.

Holt, C. (2003). The milk salts and their interaction with casein. In *Encyclopedia of Dairy Sciences* (H. Roginski, J. W. Fuquay and P. F. Fox, eds) pp. 2007–15. London: Academic Press.

Holt, C. and Sawyer, L. (1993). Caseins as rheomorphic proteins: interpretation of primary and secondary structures of α_{s1}-, β- and κ-caseins. *J. Chem. Soc. Faraday Trans.* **89**, 2683–92.

Horne, D. S. (2002). Casein structure, self-association and gelation. *Current Opinion in Colloid and Interface Science*, **7**, 456–61.

Horne, D. S. (2003). Caseins, micellar structure. In *Encyclopedia of Dairy Sciences* (H. Roginski, J. W. Fuquay and P. F. Fox, eds) pp. 1902–9. London: Academic Press.

Horne, D. S. (2006). Casein micelle structure: models and muddles. *Current Opinion in Colloid and Interface Science*, **11**, 148–53.

Hunt, C. and Nielsen, F. H. (2007). Milk salts: nutritional aspects. In *Advanced Dairy Chemistry,* Volume 3, *Lactose, Water, Salts and Vitamins*, 3rd edn (P. L. H. McSweeney and P. F. Fox, eds), in press. New York: Springer.

Huppertz, T. and Kelly, A. L. (2006). Physical chemistry of milk fat globules. In *Advanced Dairy Chemistry,* Volume 3, *Lipids*, 3rd edn (P. F. Fox and P. L. H. McSweeney, eds) pp. 173–212. New York: Springer.

Huppertz, T., Kelly, A. L. and Fox, P. F. (2006). High pressure-induced changes in ovine milk. 2. Effects on ovine casein micelles and whey proteins. *Milchwissenschaft*, **61**, 394–97.

Hurley, W. L. (2003). Immunoglobulins in mammary secretions. In *Advanced Dairy Chemistry,* Volume 1, *Proteins*, 3rd edn (P. F. Fox and P. L. H. McSweeney, eds) pp. 421–47. New York: Kluwer Academic/Plenum Publishers.

Iametti, S., Tedeschi, G., Oungre, E. and Bonomi, F. (2001). Primary structure of κ-casein isolated from mares' milk. *Journal of Dairy Research*, **68**, 53–61.

Imafidon, G. F., Farkye, N. Y. and Spanier, A. M. (1997). Isolation, purification and alteration of some functional groups of major milk proteins: a review. *CRC Critical Reviews in Food Science and Nutrition*, **37**, 663–89.

Ingram, C. J. E. and Swallow, D. (2007). Lactose malabsorption. In *Advanced Dairy Chemistry,* Volume 3, *Lactose, Water Walts and Vitamins*, 3rd edn (P. L. H. McSweeney and P. F. Fox, eds), in press. New York: Springer.

International Dairy Federation (1996). *Production and Utilization of Ewe and Goat Milk.* Proceedings of CIRVAL Seminar, Crete, October 19–21, 1995.

Jandal, J. M. (1996). Comparative aspects of goat and sheep milk. *Small Ruminant Research*, **22**, 177–85.

Jenness, R. (1973). Caseins and caseinate micelles of various species. *Netherlands Milk and Dairy Journal*, **27**, 251–57.

Jenness, R. (1979). Comparative aspects of milk proteins. *Journal of Dairy Research*, **46**, 197–210.

Jenness, R. (1980). Composition and characteristics of goat milk: review 1968–1979. *Journal of Dairy Science*, **63**, 1605–30.

Jenness, R. (1982). Inter-species comparison of milk proteins. In *Developments in Dairy Chemistry,* Volume 1, *Proteins* (P. F. Fox, ed.) pp. 87–114. London: Applied Science Publishers.

Jenness, R. and Holt, C. (1987). Casein and lactose concentrations in milk of 31 species are negatively correlated. *Experientia*, **43**, 1015–18.

Jenness, R. and Sloan, R. E. (1970). The composition of milk of various species: a review. *Dairy Science Abstracts*, **32**, 599–612.

Jensen, R. G. (1989). *The Lipids of Human Milk.* Boca Raton, Florida: CRC Press.

Jensen, R. G. (1995). *Handbook of Milk Composition.* San Diego, California: Academic Press.

Ji, Q., Luo, Z-X., Wible, J. R., Zhang, J. F. and Georgi, J. A. (2002). The earliest known eutherian mammal. *Nature*, **416**, 816–22.

Kappeler, S., Farah, Z. and Puhan, Z. (1998). Sequence analysis of *Camelus dromedarius* milk caseins. *Journal of Dairy Research*, **65**, 209–22.

Keenan, T. W. and Mather, I. H. (2006). Intracellular origin of milk fat globules and the nature of the milk fat globule membrane. In *Advanced Dairy Chemistry,* Volume 2, *Lipids*, 3rd edn (P. F. Fox and P. L. H. McSweeney, eds) pp. 137–71. New York: Springer.

Kemp, T. S. (1982). *Mammal-like Reptiles and the Origin of Mammals.* London: Academic Press.

Kon, S. K. (1959). *Milk and its Products in Human Nutrition.* Rome: Food and Agriculture Organisation of the United Nations.

Korhonen, H. (2006). Technological and health aspects of bioactive components of milk. *International Dairy Journal*, **16**, 1227–428.

Korhonen, H. and Pihlanto, A. (2006). Bioactive peptides: production and functionality. *International Dairy Journal*, **16**, 945–69.

Kruger, M. C., Poulsen, R. C., Schollum, L., Haggarty, N., Ram, S. and Palmano, K. (2006). A comparison between acidic and basic protein fractions from whey or milk for reduction of bone loss in the ovariectomized rat. *International Dairy Journal*, **16**, 1149–56.

Kunz, C. and Rudloff, S. (2006). Health promoting aspects of milk oligosaccharides: a review. *International Dairy Journal*, **16**, 1341–46.

Laxminarayana, H. and Dastur, N. N. (1968). Buffalo milk and milk products, Parts I & II. *Dairy Science Abstracts*, **30**, 177–86. , 231–241

Lillegraven, J. A. (2004). Polarities of mammalian evolution seen through homologs of the inner cell mass. *Journal of Mammalian Evolution*, **11**, 143–202.

Lillegraven, J. A., Thompson, S. D., McNab, B. K. and Patton, J. L. (1987). The origin of eutherian mammals. *Biological Journal of the Linnean Society*, **32**, 281–336.

Lonnerdal, B. (2003). Lactoferrin. In *Advanced Dairy Chemistry,* Volume 1, *Proteins*, 3rd edn (P. F. Fox and P. L. H. McSweeney, eds) pp. 449–66. New York: Kluwer Academic/Plenum Publishers.

Lucey, J. A. and Horne, D. S. (2008). Milk salts: technological significance. In *Advanced Dairy Chemistry,* Volume 3, *Lactose, Water, Salts and Vitamins*, 3rd edn (P. L. H. McSweeney and P. F. Fox, eds), in press. New York: Springer.

MacDonald, D. (2004). *The New Encyclopedia of Mammals.* Oxford: Oxford University Press.

MacGibbon, A. K. H. and Taylor, M. W. M. (2006). Composition and structure of bovine milk lipids. In *Advanced Dairy Chemistry,* Volume 2, *Lipids*, 3rd edn (P. F. Fox and P. L. H. McSweeney, eds) pp. 1–42. New York: Springer.

Madsen, O., Scally, M., Douady, C. J., Kao, D. J., De Bry, R. W., Adkins, R., Amrine, H. M., Stanhope, M. J., de Jong, W. W. and Springer, M. S. (2001). Parallel adaptive radiations in two major clades of placental mammals. *Nature*, **409**, 610–14.

Mahoney, R. R. (1997). Lactose: enzymic modification. In *Advanced Dairy Chemistry, Volume 3, Lactose, Water, Salts and Vitamins*, 2nd edn (P. F. Fox, ed.) pp. 77–125. London: Chapman & Hall.

Martin, P., Ferranti, P., Leroux, C. and Addeo, F. (2003). Non-bovine caseins: quantitative variability and molecular diversity. In *Advanced Dairy Chemistry,* Volume 1, *Proteins*, 3rd edn (P. F. Fox and P. L. H. McSweeney, eds) pp. 277–317. New York: Kluwer Academic/Plenum Publishers.

Martinez-Ferez, A., Rudloff, S., Gaudix, A., Henkel, C. A., Pohlentz, G., Boza, J. J., Gaudix, E. M. and Kunz, C. (2006). Goats milk as a natural source of lactose-derived oligosaccharides: isolation by membrane technology. *International Dairy Journal*, **16**, 173–81.

Mather, I. H. (2000). A review and proposed nomenclature for major milk proteins of the milk fat globule membrane. *Journal of Dairy Science*, **83**, 203–47.

McKenzie, H. A. and White, F. H. Jr. (1991). Lysozyme and α-lactalbumin: structure, function, and interrelationships. *Advances in Protein Chemistry*, **41**, 173–315.

McMahon, D. J. and McManus, W. R. (1998). Rethinking casein micelle structure using electron microscopy. *Journal of Dairy Science*, **81**, 2985–93.

McMeekin, T. L. (1970). Milk proteins in retrospect. In *Milk Proteins: Chemistry and Molecular Biology,* Volume I (H. A. McKenzie, ed.) pp. 3–15. New York: Academic Press.

McSweeney, P. L. H. and Fox, P. F. (2007). *Advanced Dairy Chemistry,* Voulme 3, *Lactose, Water, Salts and Vitamins*, 3rd edn. New York: Springer.

Mehra, R. and Kelly, P. (2006). Milk oligosaccharides: structural and technological aspects. *International Dairy Journal*, **16**, 1334–40.

Messer, M. and Urashima, T. (2002). Evolution of milk oligosaccharides and lactose. *Trends in Glycoscience and Glycotechnology*, **14**, 153–76.

Mulvihill, D. M. and Ennis, M. P. (2003). Functional milk proteins: production and utilization. In *Advanced Dairy Chemistry,* Volume 1, *Proteins*, 3rd edn (P. F. Fox and P. L. H. McSweeney, eds) pp. 1175–228. New York: Kluwer Academic/Plenum Publishers.

Murphy, W. J., Eziak, E., Johnson, W. E., Zhang, Y. P., Ryder, O. A. and O'Brien, S. J. (2001). Molecular phylogenetics and the origin of mammals. *Nature*, **409**, 416–18.

Mustapha, A., Hertzler, S. R. and Savaiano, D. A. (1997). Lactose: nutritional significance. In *Advanced Dairy Chemistry,* Volume 3, *Lactose, Water, Salts and Vitamins*, 2nd edn (P. F. Fox, ed.) pp. 127–54. London: Chapman & Hall.

Newburg, D. S. Newbauer, S. H. (1995). Carbohydrates in milk: analysis, quantities and significance, in: *Handbook of Milk Composition*, (R. G. Jensen, ed.) pp. 273–349. San Diego, California: Academic Press.

Ng-Kwai-Hang, K. F. and Grosclaude, F. (2003). Genetic polymorphism of milk proteins. In *Advanced Dairy Chemistry,* Volume 1, *Proteins*, 3rd edn (P. F. Fox and P. L. H. McSweeney, eds) pp. 739–816. New York: Kluwer Academic/Plenum Publishers.

Nitta, K. and Sugai, S. (1989). The evolution of lysozyme and α-lactalbumin. *European Journal of Biochemistry*, **182**, 111–18.

Nitta, K., Tsuge, H., Sugai, S. and Shimazaki, K. (1987). The calcium-binding properties of equine lysozyme. *FEBS Letters*, **223**, 405–08.

O'Connell, J. E. and Fox, P. F. (2003). Heat-induced coagulation of milk. In *Advanced Dairy Chemistry,* Volume 1, *Proteins*, 3rd edn (P. F. Fox and P. L. H. McSweeney, eds) pp. 879–945. New York: Kluwer Academic/Plenum Publishers.

O'Flaherty, F. (1997). *Characterization of some minor caseins and proteose peptones of bovine milk.* MSc Thesis. National University of Ireland, Cork, Ireland.

Ochirkhuyag, B., Chobert, J-M., Dalgalarrondo, M., Choiset, Y. and Haertle, T. (1997). Characterization of caseins from Mongolian yak, khainak, and bactrian camel. *Lait*, **77**, 601–13.

Ochirkhuyag, B., Chorbet, J-M., Dalgalarrondo, M. and Haertle, T. (2000). Characterization of mare caseins. Identification of α_{s1}- and α_{s2}-caseins. *Lait*, **80**, 223–35.

Oftedal, O. T. (2002a). The mammary gland and its origin during synapsid evolution. *Journal of Mammary Gland Biology and Neoplasia*, **7**, 225–52.

Oftedal, O. T. (2002b). The origin of lactation as a water source for parchment shelled eggs. *Journal of Mammary Gland Biology and Neoplasia*, **7**, 253–66.

Oftedal, O. T., Alt, G. L., Widdowson, E. N. and Jakubasz, M. R. (1993). Nutrition and growth of suckling black bears (*Urus americanus*) during their mother's winter fast. *British Journal of Nutrition*, **70**, 59–79.

Ono, T. and Creamer, L. K. (1986). Structure of goat casein micelles. *New Zealand Journal of Dairy Science and Technology*, **21**, 57–64.

Ono, T., Kohno, H., Odagiri, S. and Takagi, T. (1989). Subunit components of casein micelles from bovine, ovine, caprine and equine milks. *Journal of Dairy Research*, **56**, 61–68.

Osthoff, G., de Waal, H. O., Hugo, A., de Wit, M. and Botes, P. (2005). Milk composition of a free-ranging African elephant (*Loxodonta africana*) cow during early lactation. *Comparative Biochemistry and Physiology*, **141**, 223–29.

Pakkanen, R. and Aalto, J. (1997). Growth factors and antimicrobial factors in bovine colostrums. *International Dairy Journal*, **7**, 285–97.

Park, Y. W. and Haenlein, G. F. W. (2006). *Handbook of Milk of Non-bovine Mammals.* Oxford: Blackwell Publishing.

Park, Y. W., Zhang, H., Zhang, B. and Zhang, L. (2006). Mare milk. In *Handbook of Milk of Non-bovine Mammals* (Y. W. Park and G. F. W. Haenlein, eds) pp. 275–96. Oxford: Blackwell Publishing.

Park, Y. W., Juarez, M., Ramos, M. and Haenlein, G. F. W. (2007). Physico-chemical characteristics of goat and sheep milk. *Small Ruminant Research*, **68**, 88–113.

Parkash, S. and Jenness, R. (1968). The composition and characteristics of goat's milk: a review. *Dairy Science Abstracts*, **30**, 67–87.

Parodi, P. (2006). Nutritional significance of milk lipids. In *Advanced Dairy Chemistry,* Volume 2, *Lipids*, 3rd edn (P. F. Fox and P. L. H. McSweeney, eds) pp. 601–39. New York: Springer.

Patel, R. S. and Mistry, V. V. (1997). Physicochemical and structural properties of ultrafiltered buffalo milk and milk powder. *Journal of Dairy Science*, **80**, 812–17.

Patton, S. (1999). Some practical implications of the milk mucins. *Journal of Dairy Science*, **82**, 1115–17.

Patton, S. (2004). *Milk: Its Remarkable Contribution to Human Health and Well-being.* New Brunswick, New Jersey: Transaction Publishers.

Peaker, M. (2002). The mammary gland in mammalian evolution: a brief commentary on some of the concepts. *Journal of Mammary Gland Biology and Neoplasia*, **7**, 347–53.

Pond, C. M. (1977). The significance of lactation in the evolution of mammals. *Evolution*, **31**, 177–99.

Pyne, G. T. (1962). Reviews on the progress of dairy science. Section C. Dairy chemistry. Some aspects of the physical chemistry of the salts of milk. *Journal of Dairy Research*, **29**, 101–30.

Rao, M. B., Gupta, R. C. and Dastur, N. N. (1970). Camels' milk and milk products. *Indian Journal of Dairy Science*, **23**, 71–78.

Reinhardt, T. A. and Lippolis, J. D. (2006). Bovine milk fat globule membrane proteome. *Journal of Dairy Research*, **73**, 406–16.

Riccio, P. (2004). The proteins of the milk fat globule membrane in the balance. *Trends in Food Science and Technology*, **15**, 458–61.

Rijnkels, M. (2002). Comparison of the casein gene loci. *Journal of Mammary Gland Biology and Neoplasia*, **7**, 327–45.

Rockland, L. B. and Beuchat, L. R. (1987). *Water Activity: Theory and Application to Food.* New York: Marcel Dekker.

Roginski, H., Fuquay, J. W. and Fox, P. F. (2003). *Encyclopedia of Dairy Sciences.* London: Academic Press.

Roos, Y. (1997). Water in milk products. In *Advanced Dairy Chemistry,* Volume 3, *Lactose, Water, Salts and Vitamins*, 2nd edn, pp. 306–46 London: Chapman & Hall.

Sawyer, L. (2003). β-Lactoglobulin. In *Advanced Dairy Chemistry,* Volume 1, *Part A, Proteins*, 3rd edn (P. F. Fox and P. L. H. McSweeney, eds) pp. 319–86. New York: Kluwer Academic/Plenum Publishers.

Schrezenmeir, J., Korhonen, H., Williams, C. M., Gill, H. S. and Shah, N. P. (2000). Beneficial natural bioactive substances in milk and colostrums. *British Journal of Nutrition*, **84**(Suppl 1), S1–S166.

Simatos, D., Champion, D., Lorient, D., Loupiac, C. and Roudaut, G. (2008). Water in dairy products. In *Advanced Dairy Chemistry,* Volume 3, *Lactose, Water, Salts and Vitamins*, 3rd edn (P. F. Fox and P. L. H. McSweeney, eds), in press. New York: Springer.

Simpson, K. J. and Nicholas, K. R. (2002). Comparative biology of whey proteins. *Journal of Mammary Gland Biology and Neoplasia*, **7**, 313–26.

Singh, H. (2006). The milk fat globule membrane. A biophysical system for food applications. *Current Opinion in Colloid & Interface Science*, **11**, 154–63.

Solaroli, G., Pagliarini, E. and Peri, C. (1993). Composition and quality of mare's milk. *Italian Journal of Food Science*, **V**, 3–10.

Sood, S. M., Herbert, P. J. and Slattery, C. W. (1997). Structural studies on casein micelles of human milk: dissociation of β-casein of different phosphorylation levels induced by cooling and ethylenediaminetetraacetate. *Journal of Dairy Science*, **80**, 628–33.

Sood, S. M., Erickson, G. and Slattery, C. W. (2002). Formation of reconstituted casein micelles with human β-casein and bovine κ-casein. *Journal of Dairy Science*, **85**, 472–77.

Springer, M. S., Stanhope, S. J., Madsen, O. and de Jong, W. W. (2004). Molecules consolidate the placental mammal tree. *Trends in Ecology and Evolution*, **19**, 430–38.

Strange, E. D., Malin, E. L., van Hekken, D. L. and Basch, J. J. (1992). Chromatographic and electrophoretic methods for analysis of milk proteins. *Journal of Chromatography*, **624**, 81–102.

Swaisgood, H. E. (1975). *Methods of Gel Electrophoresis of Milk Proteins*, pp. 1–33, Champaign, Illinois: American Dairy Science Association.

Tremblay, L., Laporte, M. F., Leonil, J., Dupont, D. and Paquin, P. (2003). Quantitation of proteins in milk and milk products. In *Advanced Dairy Chemistry,* Volume 1, *Proteins*, 3rd edn (P. F. Fox and P. L. H. McSweeney, eds) pp. 49–139. New York: Kluwer Academic/Plenum Publishers.

Trujillo, A. J., Guamis, B. and Carretero, C. (1997). Las proteinas mayoritarias de la leche de cabra. *Alimentaria*, **285**, 19–28.

Trujillo, A. J., Casals, I. and Guamis, B. (2000). Analysis of major caprine milk proteins by reverse-phase high-performance liquid chromatography and electrospray ionization-mass spectrometry. *Journal of Dairy Science*, **83**, 11–19.

Urashima, T., Saito, T., Nakarmura, T. and Messer, M. (2001). Oligosaccharides of milk and colostrums in non-humam mammals. *Glycoconjugate Journal*, **18**, 357–71.

Urashima, T., Kitaoka, M., Asakuma, S. and Messer, M. (2007). Indigenous oligosaccharides in milk. In *Advanced Dairy Chemistry,* Volume 3, *Lactose, Water, Salts and Vitamins*, 3rd edn (P. L. H. McSweeney and P. F. Fox, eds), in press. New York: Springer.

Verstegen, M. W. A., Moughan, P. J. and Schrama, J. W. (1998). *The Lactating Sow.* Wageningen: Wageningen Pers.

Visser, H. (1992). A new casein micelle model and its consequences for pH and temperature effects on the properties of milk. In *Protein Interactions* (H. Visser, ed.) pp. 135–65. Weinheim: VCH.

Ward, R. E., German, J. B. and Corredig, M. (2006). Composition, applications, fractionation, technological and nutritional significance of milk fat globule material. In *Advanced Dairy Chemistry,* Volume 2, *Lipids*, 3rd edn (P. F. Fox and P. L. H. McSweeney, eds) pp. 213–44. New York: Springer.

Weisberg, S. M. (1954). Recent progress in the manufacture and use of lactose: a review. *Journal of Dairy Science*, **37**, 1106–15.

Welsch, U., Buchheim, W., Schumacher, U., Schinko, I. and Patton, S. (1988). Structural, histochemical and biochemical observations on horse milk-fat-globule membranes and casein micelles. *Histochemistry*, **88**, 357–65.

Welsch, U., Schumaker, U., Buchheim, W., Schinko, I., Jenness, R. and Patton, S. (1990). Histochemical and biochemical observations on milk-fat globule membranes from several species. *Acta Histochemica Supplement*, **40**, 59–64.

Wheeler, M. B. (2003). Production of transgenic livestock: promise fulfilled. *Journal of Animal Science*, **81**(Suppl 3), 32–37.

Whittier, E. O. (1925). Lactose: a review. *Chemical Reviews*, **2**, 85–125.

Whittier, E. O. (1944). Lactose and its utilization: a review. *Journal of Dairy Science*, **27**, 505–37.

Woodward, D. R. (1976). The chemistry of mammalian caseins: a review. *Dairy Science Abstracts*, **38**, 137–50.

Zadow, J. G. (1984). Lactose: properties and uses. *Journal of Dairy Science*, **67**, 2654–79.

Zadow, J. G. (1992). *Lactose and Whey Processing.* London: Elsevier Applied Science.

2

The comparative genomics of tammar wallaby and Cape fur seal lactation models to examine function of milk proteins

Julie A. Sharp, Matthew Digby, Christophe Lefevre, Sonia Mailer, Elie Khalil, Denijal Topcic, Aurelie Auguste, Joly Kwek, Amelia J. Brennan, Mary Familari and Kevin R. Nicholas

Abstract

The composition of milk includes nutritional and developmental factors that are crucial both to the function of the mammary gland and to the growth and physiological development of the suckled young. This chapter examines the option of exploiting the comparative biology of species with extreme adaptation to lactation to identify milk protein bioactive components with these functions that are likely to have commercial potential either as a nutraceutical in functional foods or as pharmaceuticals. Increasingly, we find that these molecules are more readily identified in animal models with very different reproductive strategies than by screening

Milk Proteins: From Expression to Food
ISBN: 978-0-12-374039-7

laboratory and livestock species that give birth to a developed neonate and subsequently provide milk with unchanging composition during lactation.

Two animal models with very different lactation cycles are the tammar wallaby (*Macropus eugenii*) and the Cape fur seal (*Arctocephalus pusillus pusillus*). The tammar wallaby has adopted a reproductive strategy that includes a short gestation (26.5 days), birth of an immature young and a relatively long lactation (300 days). The composition of the milk changes progressively during the lactation cycle and this is controlled by the mother and not by the sucking pattern of the young. The tammar wallaby can practice concurrent asynchronous lactation; the mother provides a concentrated milk for an older animal that is out of the pouch, and a dilute milk from an adjacent mammary gland for a newborn pouch young, demonstrating that the mammary gland is controlled locally. These changes in milk composition control the development of the young and therefore provide new opportunities to identify proteins regulating specific developmental processes in the pouch young.

The second study species, the Cape fur seal, has a lactation that is characterized by a repeated cycle of long at-sea foraging trips (up to 23 days) alternating with short suckling periods of 2–3 days on-shore. Lactation almost ceases while the seal is off-shore but the mammary gland does not progress to apoptosis and involution. It is likely that specific milk proteins have a role both to reduce apoptosis in the mammary gland during foraging and to meet the challenges of growth and fasting physiology of the pup on-shore. Technology platforms using genomics, proteomics and bioinformatics have been used to exploit these models to identify milk bioactive components. In addition, the availability of sequenced marsupial, dog and bovine genomes permits rapid transfer of information to the cow to provide outcomes for the dairy industry.

Introduction

The composition of milk includes all the factors required to provide appropriate nutrition for growth of the neonate. However, it is now clear that milk also comprises bioactive molecules that play a central role in regulating developmental processes in the young and providing a protective function for both the suckled young and the mammary gland during the time of milk production (Trott *et al.*, 2003; Waite *et al.*, 2005). Identifying these bioactive components and their physiological function in eutherians can be difficult and requires extensive screening of milk components to meet the demand for foods that improve well-being and options for the prevention and treatment of disease.

In addition, human targets for milk bioactive components are focused to specific and changing health needs during the progression from infant to elderly (Figure 2.1). For example, the need to support growth and development in early life differs from the maintenance of muscle, eyes and brain function in the elderly. Health issues such as obesity, diabetes, hypertension and cancer increase with an aging population. Therefore, functional foods that both improve general nutrition and well-being and target prevention and treatment of health problems are very attractive. New animal models with unique reproductive strategies are now becoming increasingly relevant to search for these factors. This chapter focuses on two animal models that provide new insights into how the lactation cycle is regulated and how, when combined with technology platforms that include genomics, proteomics and bioinformatics, it can be exploited to

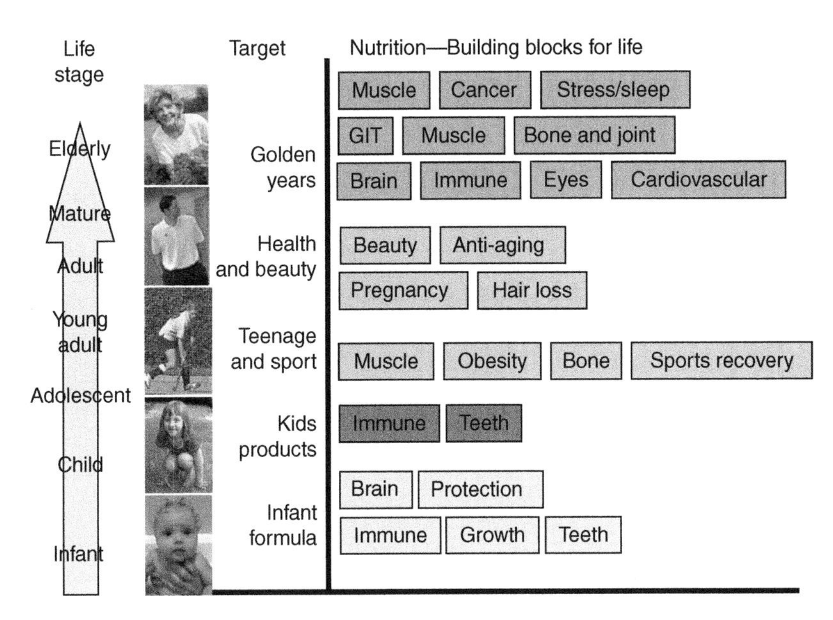

Figure 2.1 Nutraceuticals for the lifecycle. Potential life-stage-specific targets for milk bioactive components address the specific health needs associated with the progression from infant to elderly.

identify milk proteins with bioactivity and subsequent application as nutraceuticals and pharmaceuticals.

The lactation cycle is common to all mammals, although marsupials and some pinnipeds have evolved a reproductive strategy that is very different from that of most eutherians. These animal models have an extreme adaptation to lactation and have increased our understanding of the function of milk proteins by revealing mechanisms that are present but not readily apparent in many eutherian mammals (Brennan *et al.*, 2007). Eutherians have a long gestation relative to their lactation period, and the composition of the milk does not change substantially.

In contrast, reproduction in marsupials such as the tammar wallaby (*Macropus eugenii*) is characterized by a short gestation followed by a long lactation, and all the major milk constituents change progressively during lactation (Nicholas, 1988a, 1988b; Tyndale-Biscoe and Janssens, 1988). The tammar wallaby neonate is altricial and remains attached to the teat for the first 100 days of lactation, and may be considered as a fetus maintained in the pouch as opposed to the uterus. The conversion of milk to body mass is very similar to the conversion of precursors to body mass observed in the eutherian fetus in utero (Tyndale-Biscoe and Janssens, 1988). Therefore, examination of the milk provides a unique opportunity to identify specific molecules that play a primary role in regulating the development of the young.

The lactation cycle of the Cape fur seal (*Arctocephalus pusillus pusillus*) comprises maternal foraging and infant nursing that are spatially and temporally separate (Bonner, 1984). Lactating mothers suckle offspring over a period of 4–12 months and females alternate between short periods on-shore suckling their young and longer periods of up to 3–4 weeks foraging at sea (Gentry and Holt, 1986). Cape fur seals provide new opportunities to examine a potential role for milk in protecting the mammary gland from cell death and infection while the lactating animal is foraging, and to examine a function for milk in the growth and protection of the pup that remains on-shore.

The tammar wallaby (*Macropus eugenii*)

Milk protein gene expression; naturally occurring gene knockouts used to assess protein function

Lactation in the tammar wallaby has been divided into phases that are defined by the composition of the milk and the apparent sucking pattern of the young (Figure 2.2; Nicholas *et al.*, 1995, 1997). Phase 1 comprises a 26.5-day pregnancy followed by

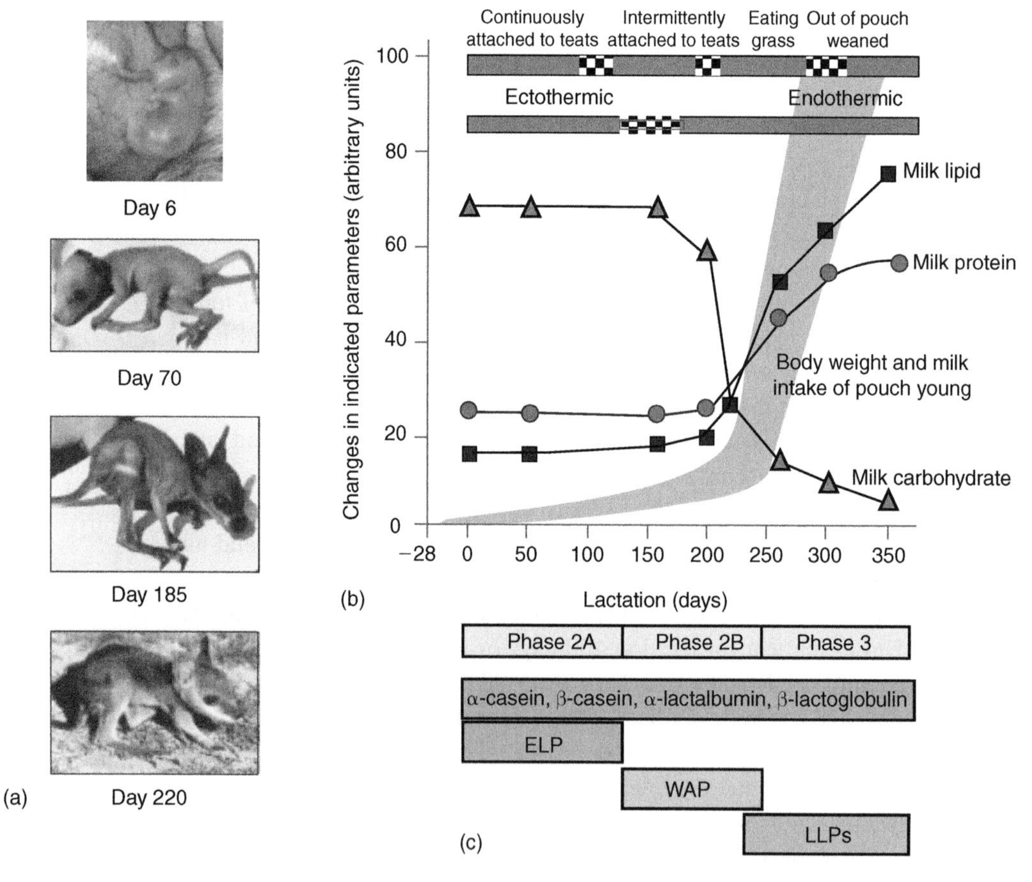

Figure 2.2 The lactation cycle of the tammar wallaby. (a) Development of the pouch young from day 6 to day 220 of age. (b) The lactation cycle in the tammar wallaby has been divided into four phases characterized by changes in milk composition and the sucking pattern of the pouch young. (c) Expression of the major milk protein genes during the lactation cycle. The α-casein, β-casein, α-lactalbumin and β-lactoglobulin genes are induced at parturition and expression remains elevated for the entire lactation. The genes encoding ELP (early lactation protein), WAP (whey acidic protein) and the LLPs (late lactation proteins A and B) are expressed only for specific phases of the lactation cycle.

parturition, and the subsequent 200 days of Phase 2 are characterized by lactogenesis and the secretion of small volumes of dilute milk high in complex carbohydrate and low in fat and protein. The pouch young remains attached to the teat for approximately the first 100 days (Phase 2A), after which it relinquishes permanent attachment to the teat and sucks less frequently while remaining in the pouch (Phase 2B). The onset of Phase 3 of lactation (200–330 days) is characterized by temporary exit from the pouch by the young, a large increase in milk production and a change in the composition of the milk to include elevated levels of protein and lipid, and low levels of carbohydrate.

There is a progressive increase in protein concentration (Nicholas, 1988a, 1988b) and protein production during lactation, and the composition of the proteins changes considerably (Nicholas, 1988a, 1988b; Nicholas *et al.*, 1997; Simpson *et al.*, 2000; Trott *et al.*, 2002). Interestingly, macropodids such as the tammar wallaby can practice concurrent asynchronous lactation; the mother provides a concentrated milk high in protein and fat for an older animal that is out of the pouch, and a dilute milk low in fat and protein but high in complex carbohydrate from an adjacent mammary gland for a newborn pouch young (Nicholas, 1988a, 1988b; Figure 2.3). This phenomenon shows that the mammary gland is controlled locally and it is likely that specific milk factors contribute to this process (Trott *et al.*, 2003).

To provide more detailed information on changes in milk protein during lactation in the tammar wallaby, we recently developed a custom tammar wallaby mammary expressed sequence tag (EST), array printed with 10 000 cDNAs representing genes

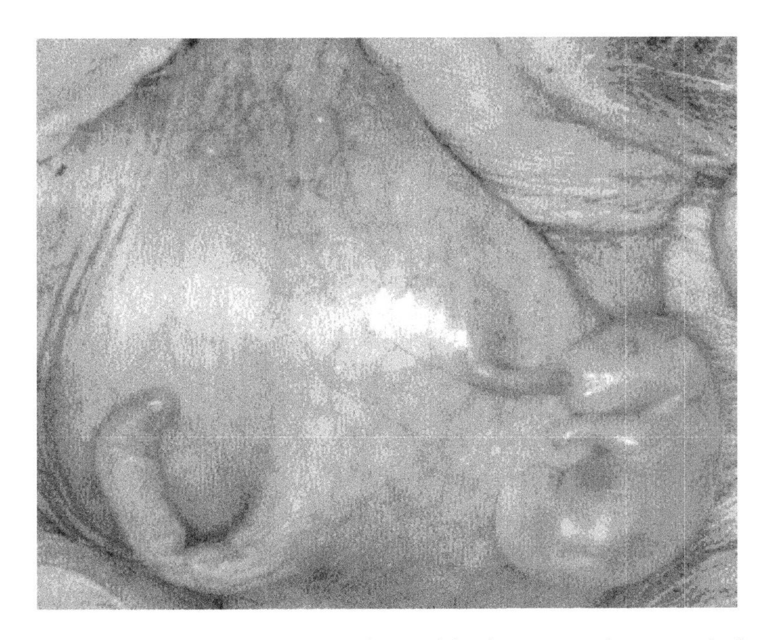

Figure 2.3 Concurrent asynchronous lactation. The pouch has been retracted to expose the four mammary glands. A 6-day-old young is attached to a teat from a mammary gland secreting Phase 2A milk. An older animal at approximately 275 days of age has vacated the pouch and sucks from the elongated teat, which provides Phase 3 milk from the enlarged mammary gland. The remaining two teats are from quiescent mammary glands.

expressed across the lactation cycle to transcript profile the mammary gland at all the major stages of pregnancy and lactation. A comprehensive set of cDNA libraries was produced from mammary tissues collected throughout the lactation cycle of the tammar wallaby and a total of 14 837 ESTs were produced by cDNA sequencing. A database was established to provide sequence analysis and sequence assembly, protein and peptide prediction and identification of motifs correlated with bioactivity. The microarray was used to transcript profile approximately 5000 genes; Figure 2.4 shows the expression profile of the major milk protein genes during lactation. These microarray data confirm earlier published results using Northern blot analysis examining

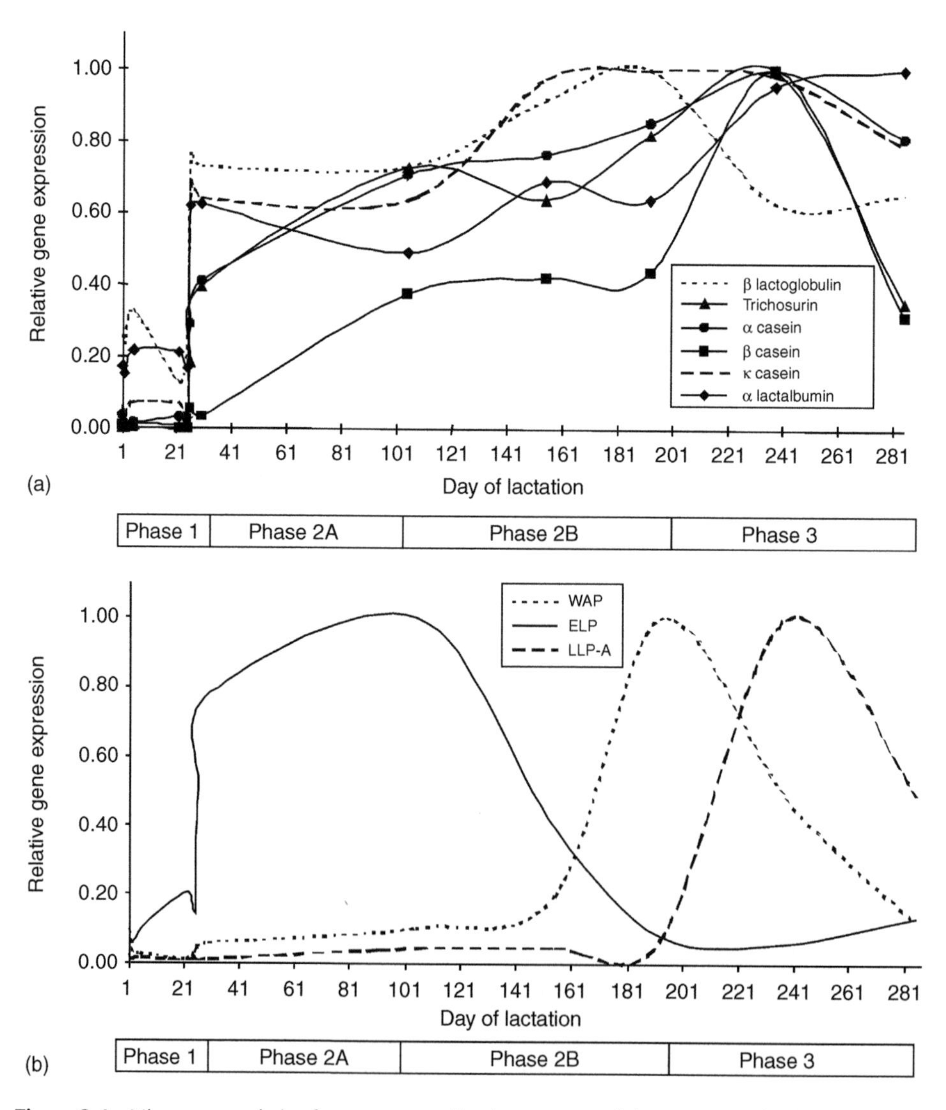

Figure 2.4 Microarray analysis of genes expressed in the tammar wallaby mammary gland during lactation. The phases of lactation are described in Figure 2.2. (a) The major milk protein genes expressed throughout lactation. (b) The major milk proteins expressed at specific phases of the lactation cycle.

the timing and level of expression of these genes (Bird *et al.*, 1994; Simpson *et al.*, 2000; Trott *et al.*, 2002) at each major phase of the lactation cycle.

There are two temporally different patterns of milk protein gene expression during the lactation cycle: one group of genes is induced to high levels around parturition and is expressed throughout lactation; a second group of genes is expressed highly only during specific phases of lactation (Simpson and Nicholas, 2002). For example, the genes for the whey proteins β-lactoglobulin, trichosurin and α-lactalbumin, and the α-casein and β-casein genes, are induced co-ordinately around the time of parturition, and are expressed for the duration of lactation (Figure 2.4a). In contrast, the early lactation protein (ELP) gene is expressed at very high levels in Phase 2A, the whey acidic protein (WAP) gene is expressed most highly in Phase 2B and a gene coding for an outlier lipocalin protein, referred to as late lactation protein A (LLP-A), is highly expressed in Phase 3 (Figure 2.4b).

A second LLP gene referred to as LLP-B is not on the microarray but is first expressed after 200 days of lactation (Trott *et al.*, 2002). Whereas some milk proteins may be specific to the tammar wallaby, in most cases, orthologues of the proteins are found in other species (Simpson and Nicholas, 2002). Therefore, the tammar wallaby is now emerging as a valuable model to provide a "temporal gene knockout" that allows for a more accurate assessment of the role of the gene product in regulating either a specific stage of development of the young or mammary function during each phase of the lactation cycle.

Changes in milk composition regulate growth of the tammar wallaby pouch young

A study by Trott *et al.* (2003) examined the hypothesis that the sucking pattern of the pouch young controls the pattern of milk secretion in the tammar wallaby. To test this hypothesis, groups of 60-day-old pouch young were fostered at 2-weekly intervals on to one group of host mothers so that a constant sucking stimulus on the mammary gland was maintained for 56 days. This allowed the lactational stage to progress ahead of the age of the young.

Analysis of the milk in fostered and control groups showed that the timing of changes in the concentration of protein and carbohydrate was unaffected by altering the sucking pattern. However, the rate of growth and development of the foster pouch young was significantly increased relative to that of the control pouch young. This probably resulted from the foster pouch young ingesting more milk of higher energy content and milk with a composition that was approximately 50 days more advanced than the milk normally consumed at their age (Trott *et al.*, 2003). It was concluded from these studies that the lactating tammar wallaby regulates both milk composition and the rate of milk production, and that these determine the rates of pouch young growth and development, irrespective of the age of the pouch young. Therefore, the tammar wallaby allows the correlation of milk with a specific composition, particularly the milk proteins, with defined developmental stages of the suckled young.

More recent experiments (Waite *et al.*, 2005) have extended the studies of Trott *et al.* (2003) and have examined a role for milk in specific changes in the development

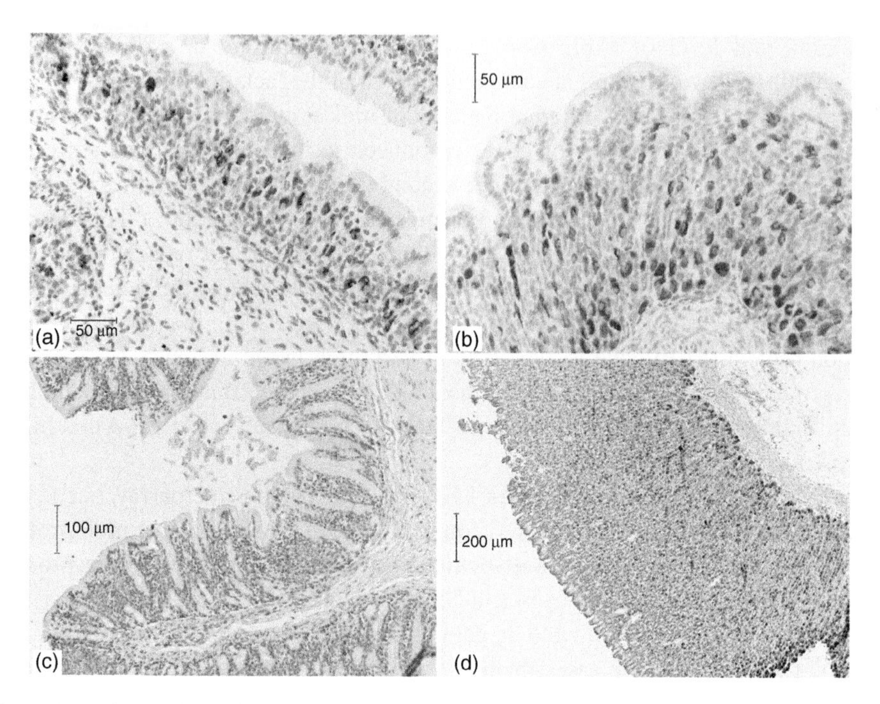

Figure 2.5 Distribution of parietal cells in the forestomach (a, c) and hindstomach (b, d) of the tammar wallaby pouch young. The parietal cells are identified by immunohistochemistry (brown staining H^+/K^+-ATPase pump) during development of the stomach in pouch young at day 168 (a, b) and day 260 (c, d).

of the gut of the tammar wallaby pouch young. Parietal cells positive for the alpha sub-unit of the H^+/K^+-ATPase pump were present in both the forestomach and the hindstomach of pouch young at 168 days of age (Figure 2.5). However, at 260 days of age, parietal cells were seen in sections from the hindstomach but were no longer detectable in the forestomach. It is likely that this change in the forestomach is essential to establish the appropriate microflora as the pouch young emerges from the pouch and starts to eat herbage.

A similar cross-fostering approach to that used by Trott *et al.* (2003) accelerated the growth rate of the fostered young but did not accelerate development of the stomach. However, these experiments need to be repeated by fostering 170-day-old pouch young to mothers at day 200 of lactation and then examining the stomach 30 days later. It is conceivable that the milk secreted after day 200 of lactation either contains a factor(s) that leads to a decrease in parietal cells or has lost a factor that is required to maintain parietal cells in the forestomach.

Identification of milk bioactive components in tammar wallaby milk using a functional genomics platform

It is likely that milk bioactive components have a variety of roles in regulating both mammary gland function and development of the suckled young. Depending on their function, some bioactive components will be required continuously whereas

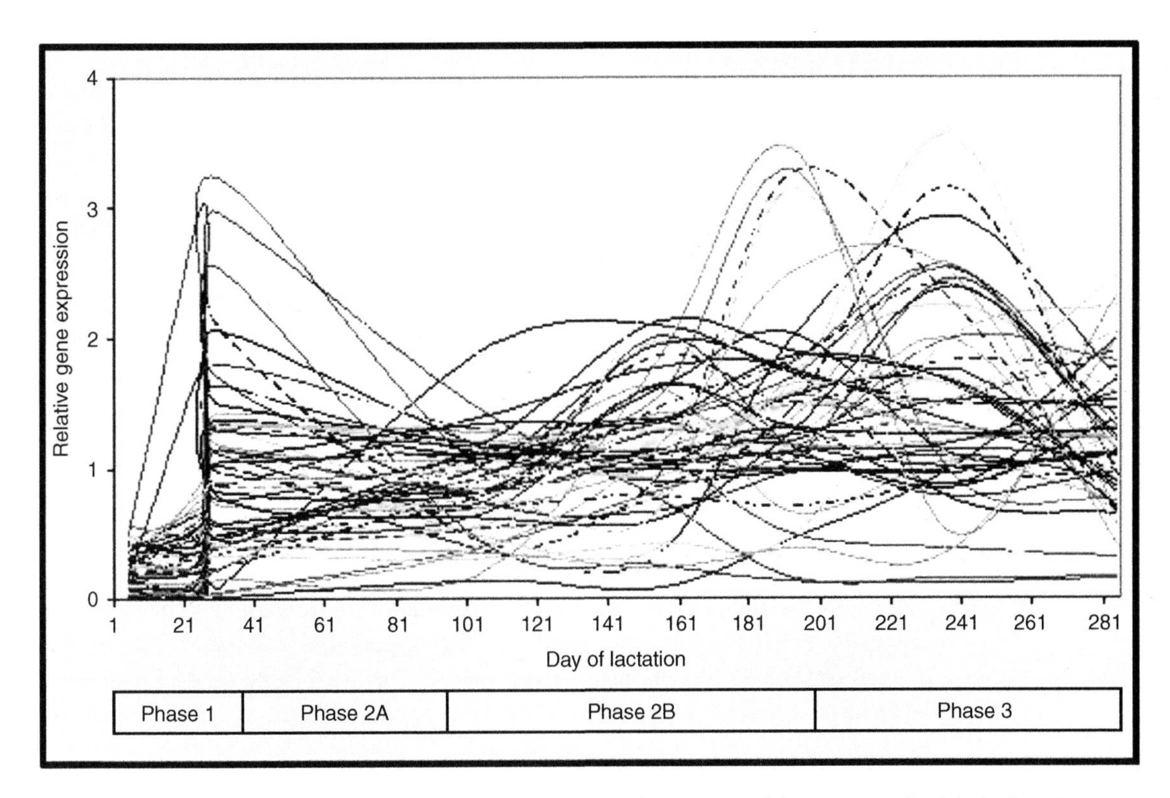

Figure 2.6 Microarray analysis of genes coding for secreted proteins in the tammar wallaby mammary gland during lactation. The phases of lactation are described in Figure 2.2. A total of 75 gene transcripts are shown (see also Plate 2.6).

others will have a role at specific stages of lactation and development of the young. Therefore, it is likely that their pattern of secretion can be matched to the developmental profile of the suckled young and may follow similar trends to the major milk proteins described in Figure 2.4.

Generally, two approaches can be used to identify bioactive proteins secreted in milk. The first is a proteomics approach that fractionates milk using a variety of separation technologies and examines fractions for bioactivity. A second approach uses a genomics platform to identify the differentially expressed genes that code for a secreted protein. Adopting the latter approach, the tammar wallaby EST database and the microarray data were used to predict the cDNAs coding for proteins with a signal peptide which represented differentially expressed genes when assessed by microarray analysis (M. Digby, C. Lefevre and K. R. Nicholas, unpublished data). This approach identified 75 genes, excluding all the identified major milk proteins, with the expression profile shown in Figure 2.6.

It is clear that the pattern of expression of these genes is variable and that many proteins are secreted into the milk at specific stages of the lactation cycle. Individual proteins encoded by the cDNA were expressed *in vitro* and the corresponding proteins were examined for activity in cell-based assays for immune modulation, inflammatory responses, growth and differentiative effects. At least 30 of these proteins had at least one of these bioactivities. In addition, some of the proteins showed a capacity to alter the development of three-dimensional structures of tammar wallaby mammary epithelial cells in a mammosphere culture model, raising the possibility that these proteins may also contribute to growth and function of the mammary gland during lactation.

A causal relationship between these bioactive components and the development of the pouch young remains to be established. However, bovine orthologues for each of the new tammar wallaby bioactive components have been identified and subsequent research will focus on identifying these factors in bovine milk, determining if they can be recovered from waste streams during milk processing, and assessing market application and potential.

Fractionation of milk and analysis of bioactive proteins

The fractionation of milk from tammar wallabies at the major phases of lactation provides an opportunity to identify fractions with bioactivity and to follow changes in activity in specific fractions across the duration of lactation. Temporal changes in activity would probably suggest an association with particular developmental process(es) in the developing young, although a potential autocrine effect on mammary gland function could also be considered. The microarray used to identify differentially expressed genes, described previously, was limited to approximately 5000 genes and it is likely that the number of differentially expressed genes coding for secreted bioactive components will increase when a microarray with whole genome coverage becomes available.

The genome of another marsupial, *Monodelphis domestica*, has been sequenced to eight-fold coverage and the genome of the tammar wallaby has been sequenced to two-fold coverage; this information will probably lead to a marsupial whole transcriptome microarray. The disadvantage of analyzing fractionated whey is that often the sample assayed comprises a mixture of proteins, requiring additional work to identify the specific protein(s) associated with the bioactivity. However, this approach potentially allows identification of proteins and peptides that are not currently represented on the tammar wallaby EST microarray and which may have either undergone post-translational processes or resulted from alternative splicing of genes.

Whey samples collected from tammar wallabies in Phases 2B and 3 were fractionated by reverse phase high performance liquid chromatography (HPLC) as shown in Figure 2.7. Proteins and peptides bound to the column were eluted with a linear acetonitrile gradient and the 60 fractions were analyzed in cell-based assays for potential to stimulate growth and differentiation of cells (ERK activity), pro- and anti-apoptotic assays, and assays for stimulation and inhibition of immune response. Analysis of ERK showed activity in specific fractions (Figure 2.7) in whey from tammar wallabies at Phase 2B. However, activity was evident not only in the same fractions in whey from tammar wallabies in Phase 3 of lactation, but also in an additional set of whey fractions. A specific role for this factor/s is not yet apparent, but it has considerable potential because it is correlated with a dramatic increase in milk production and growth of the young as they emerge from the pouch and begin to eat herbage and consume milk.

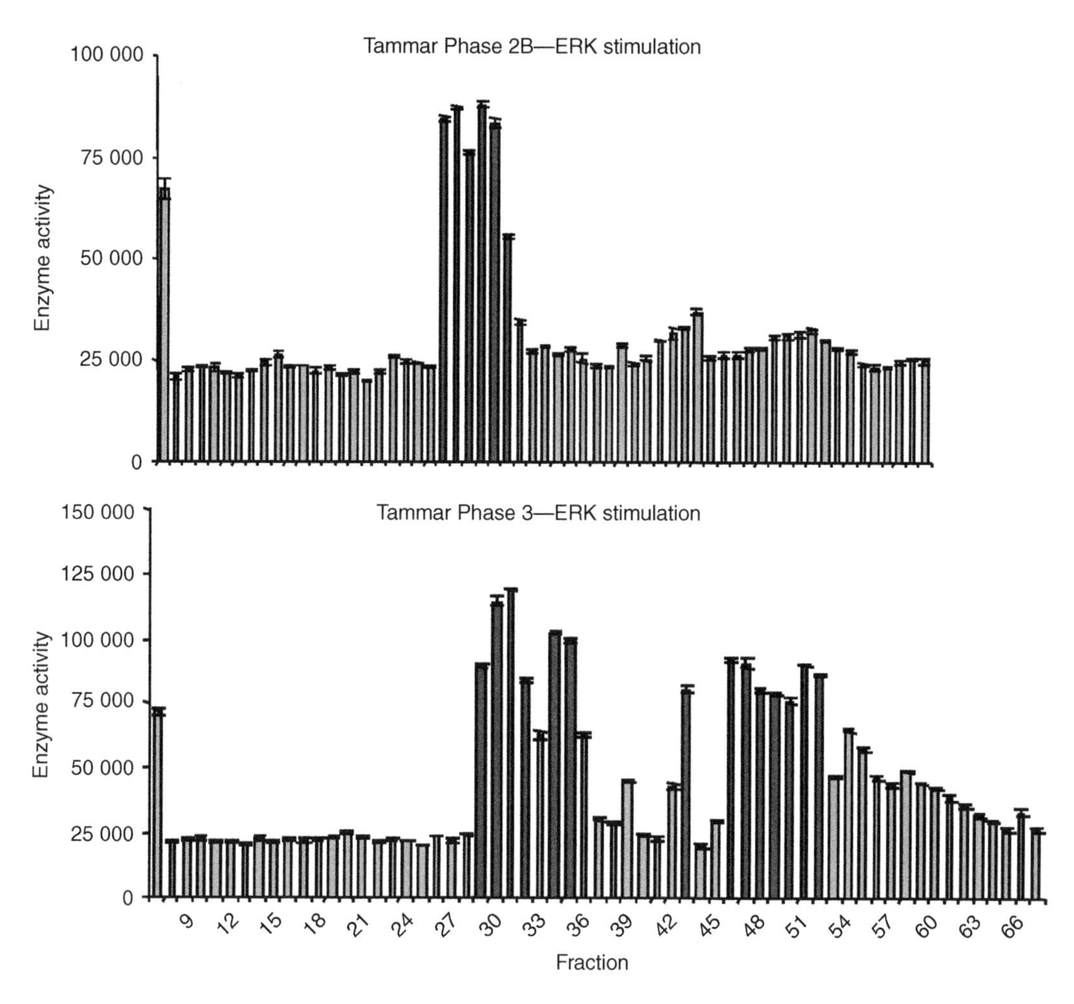

Figure 2.7 Fractionation and analysis of tammar wallaby whey. Whey from milk collected in Phase 2B and Phase 3 of lactation was fractionated by reverse phase HPLC. Proteins and peptides bound to the column were eluted with a linear acetonitrile gradient and the 60 fractions were analyzed for potential to stimulate ERK (growth and differentiation of cells) in a cell-based assay.

WAP; a potential role in development of the young, and local regulation of mammary development in the tammar wallaby

Although increasing attention is directed to identifying the minor protein components of milk that have bioactivity, it is surprising how little is known about the function of the major whey proteins in most species. Identifying the primary function for milk proteins such as β-lactoglobulin and WAP has proved to be challenging. WAP is a major whey protein and is secreted throughout lactation in many eutherian species including rat, mouse, rabbit, camel and pig (Simpson *et al.*, 2000; Simpson and Nicholas, 2002). WAP is secreted in the milk of all marsupials studied to date including the tammar wallaby (Simpson *et al.*, 2000), the red kangaroo (Nicholas *et al.*, 2001), the brush-tailed opossum (Demmer *et al.*, 2001) and the fat-tailed dunnart

(De Leo *et al.*, 2006), and WAP mRNA has been detected in the lactating mammary gland of a stripe-faced dunnart and a short-tailed opossum (D. Topcic and K. R. Nicholas, unpublished data).

However, the secretion of the protein is specific to a phase of lactation in all the marsupials mentioned above and therefore the tammar wallaby provides a naturally occurring gene knockout model to assess the role of this protein in regulating specific stages of development of the young, the mammary gland or potentially both. The amino acid sequence of WAP from the two monotremes platypus and echidna has been reported (Simpson *et al.*, 2000; Sharp *et al.*, 2007), but the pattern of secretion in these species is yet to be established.

Alignment of the amino acid sequences of WAPs from marsupial, monotreme and eutherian species shows limited sequence identity (Simpson *et al.*, 2000; Sharp *et al.*, 2007). However, these proteins are recognized by the presence of the WAP motif (KXGXCP) and a domain structure known as the four-disulfide-core (4-DSC) domain, which consists of eight cysteine residues (Ranganathan *et al.*, 1999). The WAPs secreted in the milk of eutherians have two 4-DSC domains (Simpson *et al.*, 2000), whereas the WAPs in all marsupials and monotremes studied to date have three domains (Simpson *et al.*, 2000; Demmer *et al.*, 2001; De Leo et al., 2006; Sharp *et al.*, 2007). The five exons of the marsupial and monotreme WAP genes contrast with the four exons only of the eutherian WAP genes (De Leo *et al.*, 2006; Sharp *et al.*, 2007). The significance of the loss of the third domain in eutherians remains unclear, particularly as convincing evidence for biological function of milk WAP remains to be established.

Recent studies have shown that mouse WAP added to a culture medium of mouse HC-11 cells (a mouse mammary epithelial cell line) is anti-proliferative, and may act by an autocrine or paracrine mechanism (Nukumi *et al.*, 2004). This is consistent with reports showing that over-expression of WAP in transgenic mice inhibits development of the mammary gland and the secretion of milk (Burdon *et al.*, 1991). Studies in our laboratory have used *in vitro* models to examine the effect of tammar wallaby WAP on proliferation of an epithelial-enriched population of tammar wallaby mammary cells. In contrast to the inhibitory action of mouse WAP on proliferation of HC-11 cells, results show that tammar wallaby WAP added to culture medium stimulates proliferation of mammary epithelial cells and increases expression of the cell cycle gene cyclin D1 (Figure 2.8).

Earlier studies have shown that DNA synthesis in mammary tissue is higher in Phase 2 than in Phase 3 (Nicholas, 1988a), which is consistent with a potential role of tammar wallaby WAP in development of the mammary gland. This is further supported by our studies showing that tammar wallaby WAP stimulates DNA synthesis in primary cultures of mammary epithelial cells (D. Topcic, unpublished data). Interestingly, studies using a WAP gene knockout mouse have shown that mammary development is normal. However, the pups had limited development at the later stages of lactation (Triplett *et al.*, 2005).

As tammar wallaby WAP is a major milk protein that is secreted during the middle third of lactation when the pouch young's diet comprises only milk, it could be speculated that this protein plays a specific role in the development of both the mammary

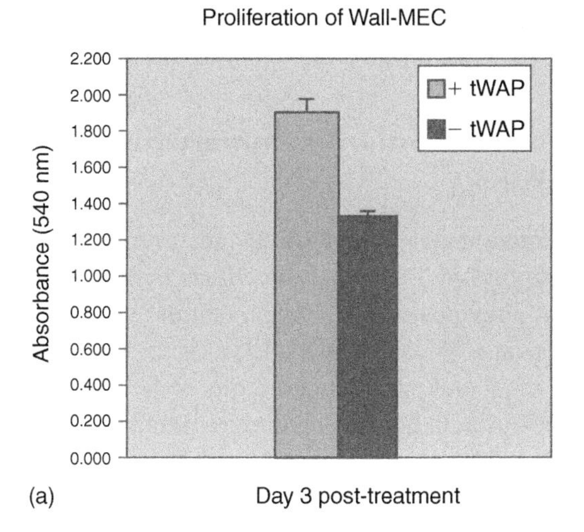

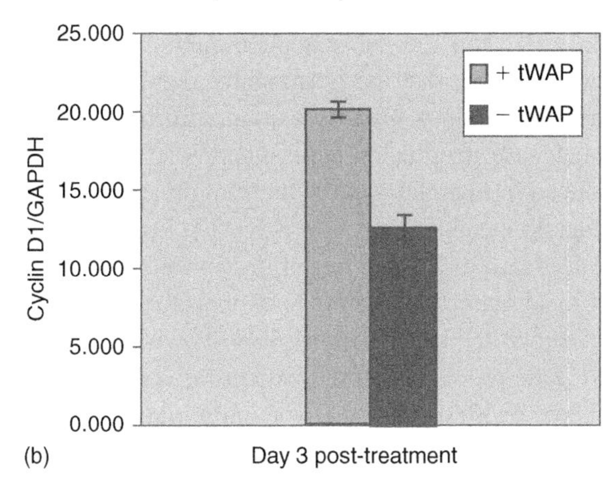

Figure 2.8 (a) Proliferation of wallaby mammary epithelial cells (Wall-MEC) cultured in the presence and absence of tammar wallaby WAP (tWAP). The results shown are after day 3 of treatment with (+) and without (−) tWAP. The vertical bars indicate SEM ($n = 3$), $P < 0.05$. (b) Expression of the cyclin D1 gene in Wall-MEC grown in the presence and absence of tammar wallaby WAP (day 3 post-treatment). Expression of the cyclin D1 and GAPDH genes was determined by RT-PCR analyses. The relative expression levels of the cyclin D1 gene in Wall-MEC was quantified by National Institutes of Health image software and was presented as a proportion of the housekeeping gene GAPDH. Reprinted from Brennan *et al.* (2007). URL http://jds.fass.org/, © American Dairy Science Association.

gland and the suckled young at this time. The apparent milk WAP genes in human, cow, ewe and goat have accumulated mutations, making them non-functional pseudogenes (see Hajjoubi *et al.*, 2006). It is likely that a putative function for WAP in mammary development is predominantly found only in marsupials and monotremes, and that the activity has been lost in human, cow, ewe and goat because of a loss of evolutionary pressure on this protein that relates to changes in the reproductive strategy of eutherians. Therefore, the marsupial may be a more appropriate model to explore the potential of other major milk proteins such as β-lactoglobulin which, like WAP, is not expressed in all eutherians. Resolution of the causes and effects of these comparative differences will provide valuable insights into the roles these molecules perform.

A role for milk in the control of mammary function in the tammar wallaby

The expression of milk protein genes is regulated concurrently by systemic endocrine factors, by paracrine factors such as the extracellular matrix and by autocrine factors secreted in the milk. Previous studies using a tammar wallaby mammary explant culture model (Nicholas and Tyndale-Biscoe, 1985) have shown that different combinations of insulin, cortisol and prolactin are required for expression of the α- and β-casein genes, and whey protein genes including α-lactalbumin and β-lactoglobulin (Simpson and Nicholas, 2002).

Tammar wallaby mammary explants from late pregnant tammar wallabies can be induced with insulin, cortisol, prolactin, thyroid hormone and oestrogen to express the WAP gene (Simpson et al., 2000). Therefore, the inhibition normally observed *in vivo* during Phase 2A, and the subsequent induction of WAP gene expression around 100 days post-partum, may be hormonally regulated. In addition, the LLP genes can be down-regulated in mammary explants from Phase 3 tammar wallabies and then restimulated with insulin, cortisol and prolactin, but expression of these genes cannot be induced in mammary explants from pregnant tammar wallabies with any hormone combination tested (Trott *et al.*, 2002). Either the appropriate hormonal milieu was not used and additional hormones are required, or the tissue requires additional serum or local mammary factors to express these genes.

This conclusion is supported by an earlier study showing that constructs with up to 1.8 kb of the LLP-A gene promoter did not express a reporter gene after transfection into CHO cells incubated with insulin, cortisol and prolactin, whereas control experiments showed that a rat β-casein gene construct was hormone responsive. In addition, the same construct was not expressed in lactating transgenic mice (Trott *et al.*, 2002).

Recent studies in our laboratory have demonstrated that constructs comprising short-tailed opossum LLP-A and tammar wallaby LLP-B promoters (up to 5 kb of DNA) with a reporter gene were not transcriptionally active following transfection into HC-11 cells, regardless of the hormonal combination in the culture. However, unlike the LLP gene promoters, constructs with various WAP promoter fragments from tammar wallaby, short-tailed opossum and stripe-faced dunnart showed increased transcriptional activity when prolactin was added to the medium (D. Topcic and K. R. Nicholas, unpublished data). This suggests that the mechanisms controlling the expression of these milk protein genes in marsupial species are likely to be different from those in eutherians.

There is increasing evidence to suggest that milk plays an important role in regulating mammary epithelial function and survival, and this is particularly evident during involution (Brennan *et al.*, 2007). Apoptosis was induced preferentially in the sealed teats of lactating mice (Li *et al.*, 1997; Marti *et al.*, 1997), while the litter suckled successfully on the remaining teats, which indicates that cell death is stimulated by an intra-mammary mechanism that is sensitive to milk accumulation (Quarrie *et al.*,

1995). A protein known as the feedback inhibitor of lactation (FIL) has been suggested as a candidate and is secreted in the milk of the tammar wallaby (Hendry *et al.* 1998) and other species. It acts specifically through interaction with the apical surface of the mammary epithelial cell to reduce secretion (Wilde *et al.*, 1995).

More recent studies using the tammar wallaby mammary explant culture model (Nicholas and Tyndale-Biscoe, 1985) to examine the process of involution have confirmed the likely role of milk, and particularly putative autocrine factors, for controlling mammary function during involution (Brennan *et al.*, 2007). Mammary explants from pregnant tammar wallabies were cultured for 3 days with insulin, cortisol and prolactin to induce milk protein gene expression. To mimic involution, all hormones were subsequently removed from the culture medium for 10 days to down-regulate expression of the milk protein genes (Figure 2.9). Surprisingly, the explants retained the same level of response during a subsequent challenge with lactogenic hormones.

Previous studies have shown that there is limited secretion of milk proteins from tammar wallaby mammary explants, but it is unlikely that milk constituents accumulated to elevated concentrations (Nicholas and Tyndale-Biscoe, 1985). The maintenance of epithelial cell viability and hormone responsiveness in explants cultured

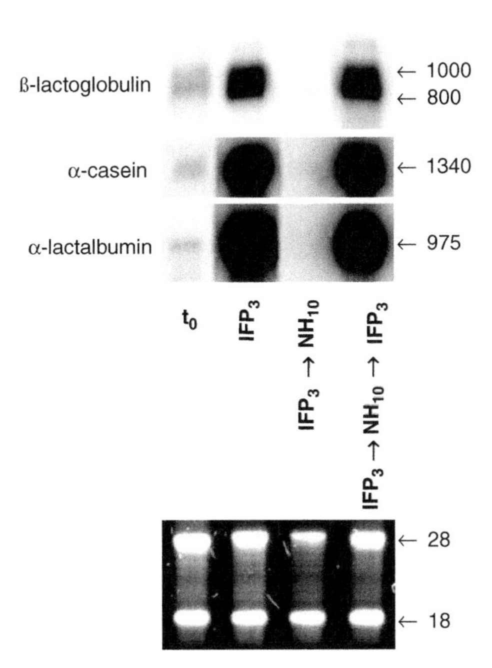

Figure 2.9 Northern blot analysis of β-lactoglobulin, α-casein and α-lactalbumin gene expression in tammar wallaby mammary explants. Gene expression is shown in: mammary tissue from day 24 pregnant tammar wallabies prior to culture (t_0); explants cultured in media with insulin, cortisol and prolactin (IFP) for 3 days; explants subsequently cultured in the absence of hormones (NH) for 10 days; and following the re-introduction of IFP for 3 days. The length of culture in days is shown by the subscript. Total RNA (10 μg, lower panel) was assayed by Northern blot analysis using [a-^{32}P]dCTP -labelled cDNA probes for the β-lactoglobulin, α-casein and α-lactalbumin genes (upper panels). Arrows indicate transcript size in nucleotides and RNA ribosomal bands.

in the absence of hormones is consistent with a more active mechanism, such as the accumulation of local factors in the milk being the primary stimulus for apoptosis of mammary epithelial cells in the tammar wallaby mammary gland. This model permits the uncoupling of hormone and milk-regulated involution and, clearly, a primary outcome of these studies is evidence for the extraordinary capacity for the survival and maintenance of hormone responsiveness by tammar wallaby mammary epithelial cells cultured in a chemically defined medium with no exogenous hormones and growth factors.

Assuming that bovine mammary epithelial cells show similar characteristics to tammar wallaby mammary epithelial cells, it is likely that milk components will play a major role in the shape of the lactation curve in dairy cattle following peak lactation at approximately 20 weeks of milk production. The lactation cycle in dairy cattle includes a period of increasing milk yield in early lactation followed by a steadily declining yield for the remainder of lactation. The amount of milk produced during lactation is determined by the peak yield and the persistency of lactation (see McFadden, 1997). Milk production is largely a function of the number and the activity of secretory cells in the udder, which decline significantly between the time of peak yield and late lactation. Therefore, it follows that approaches to address the decline in milk yield and lactational persistency after peak lactation must involve changes to the frequency of apoptosis in mammary secretory cells.

Endocrine treatment of cattle with bovine somatotrophin increases milk yield but, in many cases, persistency is not altered. Furthermore, although moderate heritability of persistency suggests that selection for this trait is possible, it is achieved at the cost of milk yield. More recently, increased frequency of milking has been shown to increase milk yield, suggesting that the mammary gland has a local intrinsic resistance to regression. Identification of the milk components that impact significantly on mammary cell fate may provide new approaches for strategies to improve lactational persistency.

The Cape fur seal (*Arctocephalus pusillus pusillus*)

The Cape fur seal suckles its pup for up to 10 months (Gentry and Holt, 1986) and lactation comprises alternate periods of several days on-shore suckling the young and extended periods at sea (Figure 2.10; Bonner, 1984; Oftedal *et al.*, 1987; Trillmich, 1996). Foraging trips are variable but can extend to 23 days (Gamel *et al.*, 2005). Nursing periods are generally 1–3 days and depletion of the mother's body reserves necessitates a return to the ocean to forage.

A study by Cane *et al.* (2005) showed that the mean protein contents of the milks of the Antarctic fur seal (*Arctocephalus gazella*) and the Australian fur seal (*Arctocephalus pusillus doriferus*) were 10.6 and 10.9% respectively, and did not change significantly across the lactation cycle. Electrophoresis of skim milk of the Australian fur seal showed 12 separate bands, representing five caseins and seven whey proteins. The majority of the whey protein was β-lactoglobulin, with possibly

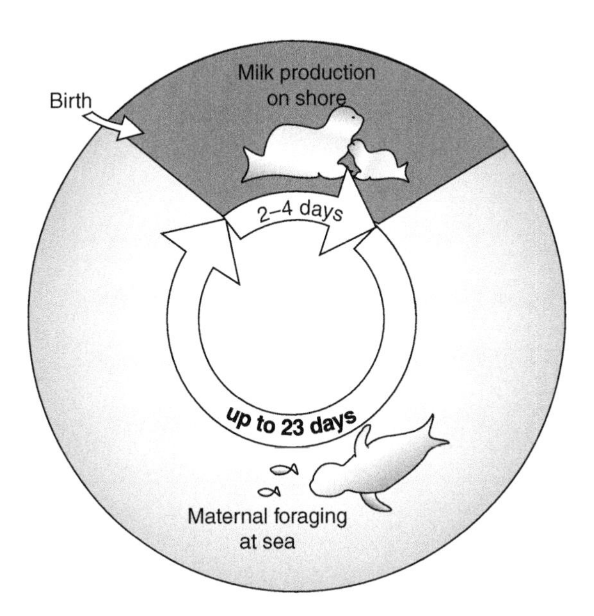

Figure 2.10 The "foraging" lactation strategy of the Cape fur seal. The pregnant female arrives at shore to give birth and remains with the pup for approximately 1 week. For the remainder of lactation (10–12 months), females alternate trips to sea with short trips on-shore to suckle their pup.

three different isoforms present. However, there was no significant change in the composition of milk proteins during lactation.

More recent studies have shown that milk protein gene expression also decreases during foraging trips as the β-casein, α_{s2}-casein and β-lactoglobulin genes are all down-regulated in the mammary gland of the foraging Cape fur seal (*Arctocephalus pusillus pusillus*) (Cane, 2005; Sharp et al., 2006). During time at sea, the lactating mammary gland does not progress to involution but produces less milk compared with that of the on-shore lactating female. For example, milk production in Antarctic fur seals *(A. gazella)* has been shown to continue while the female is foraging at sea but at only 19% of the rate of production on land (Arnould and Boyd, 1995).

Milk protein gene expression in the lactating Cape fur seal during suckling and foraging

During lactation on-shore, the mammary alveoli of the Cape fur seal are engorged with milk containing a large amount of lipid (Figures 2.11a, 2.11b; Sharp *et al.*, 2006). In contrast, during the mother's extended foraging trip, the alveoli appear less distended, epithelial cells surrounding the alveoli appear columnar and the lipid component within the milk is decreased. High sequence conservation between the Cape fur seal and the dog, 95% similarity at the DNA level, permits a significant detection rate of measurable hybridization signals between seal cDNA and the Affymetrix canine microarray (Sharp *et al.*, 2006).

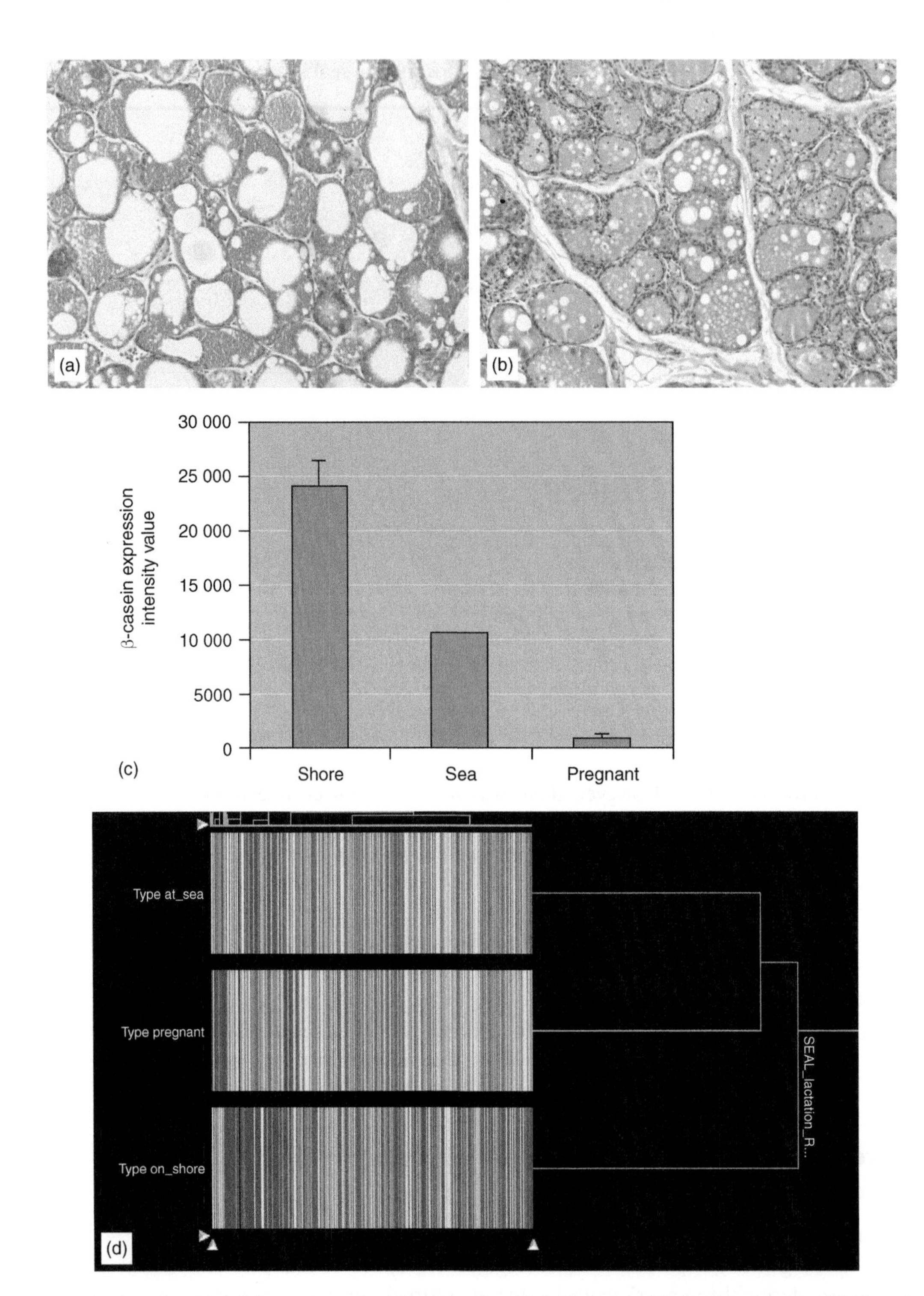

Figure 2.11 Histological sections of the mammary gland from Cape fur seals: (a) lactating while nursing on-shore and (b) lactating while foraging at sea. Sections are stained with hematoxylin and eosin. Magnification x 100. (c) Milk protein gene expression. β-Casein expression during Cape fur seal lactation cycle. Analysis of expression using canine Affymetrix chips hybridized to cDNA probes generated from RNA from pregnant (placental gestation and non-lactating, $n = 2$), lactating on-shore ($n = 2$) and lactating at sea ($n = 1$) (animals in embryonic diapause) Cape fur seals. (d) Cluster analysis of gene expression profiles from the Cape fur seal mammary gland during different stages of lactation. A total of 1020 Cape fur seal mammary messenger RNA (mRNA) transcripts were identified with expression levels above an intensity of 250 in any sample type. Hierarchical clustering was conducted using Euclidean distance. Pregnant and on-shore lactating data represent an average of two animals. Off-shore data represent a single sample. Reprinted from Current Topics in Developmental Biology, **72**, G. P. Schatten, snr ed., J. A. Sharp, K. N. Cane, C. Lefevre, J. P. Y. Arnold, and K. R. Nicholas, *Fur Seal Adaptations to Lactation: Insights into Mammary Gland Function*, pp. 276–308, New York: Academic Press. Copyright 2006, with permission from Elsevier (see also Plate 2.11).

Expression of the β-casein gene is barely detectable in the mammary gland of pregnant seals whereas the level of expression is significantly elevated in the mammary gland of seals lactating on-shore (Figure 2.11c). However, expression of this gene is also reduced during the foraging trip. Cluster analysis of gene expression profiles in mammary tissue from pregnant, on-shore lactating and off-shore lactating seals revealed that the overall expression profile of the lactating mammary gland of the foraging Cape fur seal is more closely related to the profile of pregnant non-lactating animals (placental gestation) than to the profile obtained from on-shore lactating animals (Figure 2.11d). This result suggests that the interruption of lactation in foraging animals involves a major reprogramming of mammary gland gene expression.

In most mammals during natural weaning, as the alveoli fill with milk due to cessation of suckling, there is a decline in milk protein gene expression in the mammary epithelial cells, and the epithelium regresses and enters involution (Li *et al.*, 1997). This process is characterized by apoptotic cell loss and mammary gland remodelling (Strange *et al.*, 1992; Lund *et al.*, 1996; Metcalfe *et al.*, 1999). Apoptosis associated with involution in the mammary gland of the foraging seal has been analyzed and found to be barely detectable; even after extended periods when there is no sucking stimulus, the gland does not regress (J. A. Sharp *et al.*, unpublished data).

The process by which the lactating seal reduces milk production and avoids entering apoptosis while foraging is unknown. However, it is clear that transcript profiling of the mammary gland of the foraging seal has shown that the acute immune response is associated with involution, presumably to protect the mammary gland from infection during the extended foraging trip while the milk remains in the gland (Sharp *et al.*, 2007). Gene expression profiles associated with survival and preservation of tissue architecture were also either maintained or up-regulated, preventing degradation of mammary tissue. These global gene expression data suggested that the immune and acute phase responses observed in the mouse at involution (Clarkson *et al.*, 2004), and mimicked in the foraging Cape fur seal mammary gland, are independent of the apoptotic phase of involution.

A consequence of reduced nursing is that putative factors responsible for regulating apoptosis are retained in the mammary gland. As discussed previously, it has been shown that the sealing of a single mammary teat of a mouse induces an accumulation of milk, resulting in changes in mammary gene expression and apoptosis within the sealed gland but not of the remaining glands of the same animal (Li *et al.*, 1997; Marti *et al.*, 1997). Studies in lactating animals from a variety of species indicate that regulation of milk secretion may involve an inhibitory factor (Knight *et al.*, 1994; Wilde *et al.*, 1995; Peaker *et al.*, 1998).

As mentioned previously, experiments have identified a small whey protein, termed feedback inhibitor of lactation (FIL) (Wilde *et al.*, 1995), that is synthesized by the secretory epithelial cells of the mammary gland and is secreted into the alveolar lumen along with other milk constituents. It has been proposed that FIL acts on the synthesis and secretory pathway by binding a putative receptor on the apical surface of the epithelial cells (Rennison *et al.*, 1993; Blatchford *et al.*, 1998). Furthermore, FIL may block translation of milk protein transcripts (Rennison *et al.*, 1993) and inhibit secretion of milk constituents.

Preliminary data (Cane, 2005) have demonstrated a FIL-like activity in fractionated seal milk but the level of inhibitory activity measured was similar to that reported for other species (Blatchford *et al.*, 1998). In addition, the activity did not differ in milk collected from seals either arriving on-shore after foraging at sea or after they had been on-shore suckling their pups for 1–2 days. Recently, we showed that mammary-epithelial-cell-enriched fractions from a pregnant Cape fur seal formed three-dimensional structures (mammospheres) in culture (Sharp *et al.*, 2007). The seal cells secreted their own extracellular matrix and formed mammospheres that responded to insulin, cortisol and prolactin in the medium to express milk protein genes. This model will provide new opportunities to assess the role of milk factors on mammary function in the Cape fur seal.

The role of α-lactalbumin in lactose synthesis is well established. However, studies have suggested that this protein may also be implicated in the process of involution (Hakansson *et al.*, 1995, 1999; Baltzer *et al.*, 2004). Milk from otariid pinnipeds contains little or no lactose (Urashima *et al.*, 2001) and α-lactalbumin protein has not been detected in otariid milk (Cane *et al.*, 2005) suggesting that this protein is probably absent in these species. It is interesting to speculate that the absence of biologically active α-lactalbumin in milk may be consistent with the absence of apoptosis in the mammary gland of lactating seals during foraging and that loss of this protein has provided an opportunity to alter the lactational strategy of the otarrid family of seals.

This unique capacity to inhibit apoptosis associated with involution, to maintain mammary epithelial cells in a viable and hormone-responsive state, and to stimulate mammary gland growth to increase milk production during the lactation cycle provides new opportunities to identify the mechanism controlling these events. An understanding of the genetic and/or local factors in the residual milk regulating this process may have application for improving lactational persistency in the dairy cow, either by improved breeding programs or by manipulation of milk factors.

A genomics platform in the Cape fur seal to identify milk bioactive components

The capacity of the Cape fur seal to uncouple involution from the cessation of sucking while foraging at sea is one example of how evolution has adapted this mammal's lactation cycle. It is likely that the mammary gland also makes other temporal changes such as preventing local infection and reducing inflammatory responses while at sea, and preparing to secrete a milk that delivers protection from infection and the appropriate growth factors to stimulate growth of the gut in the fasting pup that remains on-shore. The reprogramming of this gland offers new opportunities to identify mechanisms and specific mammary factors that are important for the gland to adapt to the stress and challenges during a long foraging trip in the ocean.

To assist in the identification of these molecules, a cDNA library produced from the mammary gland of a lactating off-shore Cape fur seal has been sequenced. Analysis of 10 000 EST sequences, representing approximately 4000 gene transcripts expressed in the lactating off-shore Cape fur seal mammary gland, allows for identification of putative secreted proteins and potentially improves annotation of sequences detected by Affymetrix analysis (J. A. Sharp, C. Lefevre and K. R. Nicholas, unpublished data). One interesting observation is the high abundance of lysozyme ESTs in the library, indicating that secretion of this protein in the milk may have a role in protecting the mammary gland of the mother and the gut of the pup from bacterial infection. Further analysis of the EST library has now identified the genes for more than 30 new putative bioactive components and a bovine orthologue for each protein has been identified. These proteins are now being characterized and examined for commercial potential.

Conclusions

This chapter has described two animals with extreme adaptation to lactation and has explored approaches to exploit their unique reproductive strategies to identify new proteins in milk with bioactivity that have the potential to regulate mammary function and growth of the suckled young. In contrast, the lactation cycle in most eutherian mammals is characterized by the initiation of lactation around parturition and the production of milk in which the individual components vary little during the entire period of lactation. For example, in the dairy cow, evolution and selection pressure have led to a lactation cycle that does not include significant changes in the secretory pattern of the milk proteins. Therefore, identification of new protein bioactive molecules in the dairy cow requires a broad screening program and will probably identify molecules in milk that are secreted at low and unchanged levels during lactation.

In contrast, in the tammar wallaby and the Cape fur seal, the temporal pattern of secretion of a milk bioactive component that is correlated with either a specific developmental phase of the suckled young, or altered mammary function, is more likely to indicate a cause-and-effect relationship, and these factors are more likely to have increased commercial value. The increasing availability of sequenced genomes in public databases has underpinned increased interest in comparative genomics and bioinformatics, and our results using the tammar wallaby and Cape fur seal models have led to the identification of many bovine and human orthologues of proteins with bioactivity. Conceptually, it is very likely that the relevance and use of species with an extreme adaptation to reproduction or environmental pressure will become increasingly attractive for improved understanding of the genetic regulation of physiological processes.

References

Arnould, J. P. and Boyd, L. L. (1995). Temporal patterns of milk production in Atlantic fur seals (*Arctocephalus gazella*). *Journal of Zoology (London)*, **237**, 1–12.

Baltzer, A., Svanborg, C. and Jaggi, R. (2004). Apoptotic cell death in the lactating mammary gland is enhanced by a folding variant of alpha-lactalbumin. *Cellular and Molecular Life Sciences*, **61**, 1221–28.

Bird, P., Hendry, K. A. K., Shaw, D., Wilde, C. and Nicholas, K. R. (1994). Progressive changes in milk protein gene expression and prolactin binding during lactation in the tammar wallaby (*Macropus eugenii*). *Journal of Molecular Endocrinology*, **13**, 117–25.

Blatchford, D. R., Hendry, K. A. and Wilde, C. J. (1998). Autocrine regulation of protein secretion in mouse mammary epithelial cells. *Biochemical and Biophysical Research Communications*, **248**, 761–66.

Bonner, W. N. (1984). Lactation strategies of pinnipeds: problems for a marine mammalian group. *Symposia of the Zoological Society of London*, **51**, 253–72.

Brennan, A. J., Sharp, J. A., Lefevre, C., Topcic, D., Auguste, A., Digby, M. and Nicholas, K. R. (2007). The tammar wallaby and fur seal; models to examine local control of lactation. *Journal of Dairy Science*, **90**(Suppl 1), E66–E75.

Burdon, T., Wall, R. J., Shamay, A., Smith, G. H. and Hennighausen, L. (1991). Overexpression of an endogenous milk protein gene in transgenic mice is associated with impaired mammary alveolar development and a milchlos phenotype. *Mechanisms of Development*, **36**, 67–74.

Cane, K. N. (2005). *The physiological and molecular regulation of lactation in fur seals*. PhD Thesis. Department of Zoology, University of Melbourne, Melbourne, Australia.

Cane, K., Arnould, J. P. and Nicholas, K. R. (2005). Characterisation of proteins in the milk of fur seals. *Comparative Biochemistry and Physiology. Part B,*, **141**, 111–20.

Clarkson, R. W., Wayland, M. T., Lee, J., Freeman, T. and Watson, C. J. (2004). Gene expression profiling of mammary gland development reveals putative roles for death receptors and immune mediators in post-lactational regression. *Breast Cancer Research* **6**(2), 92–109.

De Leo, A., Lefevre, C., Topcic, D., Pharo, E., Cheng, J. F., Frappell, P., Westerman, M., Graves, J. and Nicholas, K. R. (2006). Isolation and characterization of two whey protein genes in the Australian dasyurid marsupial the stripe faced dunnart (Sminthopsis macroura). *Cytogenetic and Genome Research*, **115**, 62–69.

Demmer, J., Stasiuk, S. J., Grigor, M. R., Simpson, K. J. and Nicholas, K. R. (2001). Differential expression of the whey acidic protein during lactation in the brushtail possum (Trichosurus vulpecula). *Biochimica et Biophysica Acta*, **1522**, 187–94.

Gamel, C. M., Davis, R. W., David, J. H., Meÿer, M. A. and Brandon, E. (2005). Reproductive energetics and female attendance patterns of Cape fur seals (Arctocephalus pusillus pusillus) during early lactation. *American Midland Naturalist*, **153**, 152–70.

Gentry, R. L. and Holt, J. R. (1986). Attendance behaviour of northern fur seals. In *Fur Seals: Maternal Strategies on Land and at Sea* (R. L. Gentry and R. L. Kooyman, eds) pp. 41–60. Princeton, New Jersey: Princeton University Press.

Hajjoubi, S., Rival-Gervier, S., Hayes, H., Floriot, H., Eggan, S., Piumi, F., Chardon, P., Houdebine, L. and Thepot, D. (2006). Ruminants genome no longer contains whey acidic protein gene but only a pseudogene. *Gene*, **370**, 104–12.

Hakansson, A., Zhivotovsky, B., Orrenius, S., Sabharwal, H. and Svanborg, C. (1995). Apoptosis induced by a human milk protein. *Proceedings of the National Academy of Sciences of the USA*, **92**, 8064–68.

Hakansson, A., Andreasson, J., Zhivotovsky, B., Karpman, D., Orrenius, S. and Svanborg, C. (1999). Multimeric alpha-lactalbumin from human milk induces apoptosis through a direct effect on cell nuclei. *Experimental Cell Research*, **246**, 451–60.

Hendry, K. A., Simpson, K. J., Nicholas, K. R. and Wilde, C. J. (1998). Autocrine inhibition of milk secretion in the lactating tammar wallaby (*Macropus eugenii*). *Journal of Molecular Endocrinology*, **21**, 169–77.

Knight, C. H., Hirst, D. and Dewhurst, R. J. (1994). Milk accumulation and distribution in the bovine udder during the interval between milkings. *Journal of Dairy Research*, **61**, 167–77.

Li, M., Liu, X., Robinson, G., Bar-Peled, U., Wagner, K. U., Young, W. S., Hennighausen, L. and Furth, P. A. (1997). Mammary-derived signals activate programmed cell death during the first stage of mammary gland involution. *Proceedings of the National Academy of Sciences of the USA*, **94**, 3425–30.

Lund, L. R., Romer, J., Thomasset, N., Solberg, H., Pyke, C., Bissell, M. J., Dano, K. and Werb, Z. (1996). Two distinct phases of apoptosis in mammary gland involution: proteinase-independent and -dependent pathways. *Development*, **122**, 181–93.

Marti, A., Feng, Z., Altermatt, H. J. and Jaggi, R. (1997). Milk accumulation triggers apoptosis of mammary epithelial cells. *European Journal of Cell Biology*, **73**, 158–65.

McFadden, T. B. (1997). Prospects for improving lactational persistency. In *Milk Composition, Production and Biotechnology* (R. A. Welch, D. J. W. Burns, S. R. Davis, A. I. Popay and C. G. Prosser, eds) pp. 319–39. Wallingford: CAB International.

Metcalfe, A. D., Gilmore, A., Klinowska, T., Oliver, J., Valentijn, A. J., Brown, R., Ross, A., MacGregor, G., Hickman, J. A. and Streuli, C. H. (1999). Developmental regulation of Bcl-2 family protein expression in the involuting mammary gland. *Journal of Cell Science*, **112**, 1771–83.

Nicholas, K. R. (1988a). Control of milk protein synthesis in the marsupial *Macropus eugenii*: a model system to study prolactin-dependent development. In *The Developing Marsupial: Model for Biomedical Research* (C. H. Tyndale-Biscoe and P. A. Janssens, eds) pp. 68–85. Heidelberg: Springer-Verlag.

Nicholas, K. R. (1988b). Asynchronous dual lactation in a marsupial, the tammar wallaby (*M. eugenii*). *Biochemical and Biophysical Research Communications*, **154**, 529–36.

Nicholas, K. R. and Tyndale-Biscoe, C. H. (1985). Prolactin-dependent accumulation of alpha-lactalbumin in mammary gland explants from the pregnant tammar wallaby (*Macropus eugenii*). *Journal of Endocrinology*, **106**, 337–42.

Nicholas, K. R., Wilde, C. J., Bird, P. H., Hendry, K. A. K. K., Tregenza, K. and Warner, B. (1995). Asynchronous concurrent secretion of milk proteins in the tammar wallaby (*Macropus eugenii*). In *Intercellular Signalling in the Mammary Gland* (C. J. Wilde, ed.) pp. 153–70. New York: Plenum Press.

Nicholas, K. R., Simpson, K., Wilson, M., Trott, J. and Shaw, D. (1997). The tammar wallaby: a model to study putative autocrine-induced changes in milk composition. *Journal of Mammary Gland Biology and Neoplasia*, **7**, 313–25.

Nicholas, K. R., Fisher, J. A., Muths, E., Trott, J., Janssens, P. A., Reich, C. and Shaw, D. C. (2001). Secretion of whey acidic protein and cystatin is down regulated at mid-lactation in the red kangaroo (*Macropus rufus*). *Comparative and Biochemical Physiology. Part A, Molecular and Interactive Physiology*, **129**, 851–58.

Nukumi, N., Ikeda, K., Osawa, M., Iwamori, T., Naito, K. and Tojo, H. (2004). Regulatory function of whey acidic protein in the proliferation of mouse mammary epithelial cells in vivo and in vitro. *Developmental Biology*, **274**, 31–44.

Oftedal, O. T., Boness, D. J. and Tedmam, R. A. (1987). The behaviour, physiology, and anatomy of lactation in the Pinnipedia. *Current Mammalogy*, **1**, 175–245.

Peaker, M., Wilde, C. J. and Knight, C. H. (1998). Local control of the mammary gland. *Biochemical Society Symposia*, **63**, 71–79.

Quarrie, L. H., Addey, C. V. and Wilde, C. J. (1995). Apoptosis in lactating and involuting mouse mammary tissue demonstrated by nick-end DNA labelling. *Cell Tissue Research*, **281**, 413–19.

Ranganathan, S., Simpson, K. J., Shaw, D. C. and Nicholas, K. R. (1999). The whey acidic protein family: a new signature motif and three-dimensional structure by comparative modeling. *Journal of Molecular Graphics and Modelling*, **17**, 106–13.

Rennison, M. E., Kerr, M., Addey, C. V., Handel, S. E., Turner, M. D., Wilde, C. J. and Burgoyne, R. D. (1993). Inhibition of constitutive protein secretion from lactating mouse mammary epithelial cells by FIL (feedback inhibitor of lactation), a secreted milk protein. *Journal of Cell Science*, **106**, 641–48.

Sharp, J. A., Cane, K. N., Lefevre, C., Arnould, J. P. and Nicholas, K. R. (2006). Fur seal adaptations to lactation: insights into mammary gland function. *Current Topics in Developmental Biology*, **72**, 275–308.

Sharp, J. A., Lefevre, C. and Nicholas, K. R. (2007). Molecular evolution of monotreme and marsupial whey acidic protein genes. *Evolution and Development*, **9**(4), 379–92.

Simpson, K. J. and Nicholas, K. R. (2002). The comparative biology of whey proteins. *Journal of Mammary Gland Biology and Neoplasia*, **7**, 313–25.

Simpson, K. J., Ranganathan, S., Fisher, J. A., Janssens, P. A., Shaw, D. C. and Nicholas, K. R. (2000). The gene for a novel member of the whey acidic protein family encodes three four-disulfide core domains and is asynchronously expressed during lactation. *Journal of Biological Chemistry*, **275**, 23074–81.

Strange, R., Li, F., Saurer, S., Burkhardt, A. and Friis, R. R. (1992). Apoptotic cell death and tissue remodelling during mouse mammary gland involution. *Development*, **115**, 49–58.

Trillmich, F. (ed.) (1996). *Parental Investment in Pinnipeds*. San Diego, California: Academic Press.

Triplett, A. A., Sakamoto, K., Matulka, L. A., Shen, L., Smith, G. H. and Wagner, K. U. (2005). Expression of the whey acidic protein (WAP) is necessary for adequate nourishment of the offspring but not functional differentiation of mammary epithelial cells. *Genesis*, **43**, 1–11.

Trott, J., Wilson, M., Hovey, R., Shaw, D. C. and Nicholas, K. R. (2002). Expression of novel lipocalin-like milk protein gene is developmentally-regulated during lactation in the tammar wallaby, *Macropus eugenii*. *Gene*, **283**, 287–97.

Trott, J. F., Simpson, K. J., Moyle, R. L. C., Hearn, C. M., Shaw, G., Nicholas, K. R. and Renfree, M. B. (2003). Maternal regulation of milk composition, milk production, and pouch young development during lactation in the tammar wallaby (*Macropus eugenii*). *Biology of Reproduction*, **68**, 929–36.

Tyndale-Biscoe, C. H. and Janssens, P. A. (1988). *The Developing Marsupial: Model for Biomedical Research*. Heidelberg: Springer-Verlag.

Urashima, T., Arita, M., Yoshida, M., Nakamura, T., Arai, I., Saito, T., Arnould, J. P., Kovacs, K. M. and Lydersen, C. (2001). Chemical characterisation of the oligosaccharides in hooded seal (*Cystophora cristata*) and Australian fur seal (*Arctocephalus *pusillus doriferus*) milk. *Comparative and Biochemical Physiology. Part B, Biochemistry and Molecular Biology*, **128**, 307–23.

Waite, R., Giraud, A., Old, J., Howlett, M., Shaw, G., Nicholas, K. R. and Familari, M. (2005). Cross-fostering in *Macropus eugenii* leads to increased weight but not accelerated gastrointestinal maturation. *Journal of Experimental Zoology. Part A, Comparative Experimental Biology*, **303**, 331–44.

Wilde, C. J., Addey, C. V. P., Boddy-Finch, L. and Peaker, M. (1995). In *Intercellular Signalling in the Mammary Gland.* (C. J. Wilde, ed.) pp. 227–37New York: Plenum Press.

3

Significance, origin and function of bovine milk proteins: the biological implications of manipulation or modification

P. A. Sheehy, P. Williamson, J. A. Sharp, K. Menzies, C. Lefevre, M. Digby, K. R. Nicholas and P. C. Wynn

Abstract

The rapid advances in technology for both evaluating and understanding the structure of animal genomes and their functional significance have presented opportunities for scientists to more clearly understand the complexity of bovine milk proteins and the control of their expression. Used in conjunction with proteomic databases, we can start to expand our knowledge of how milk proteins are processed into peptides, which represent much of the biological activity

Milk Proteins: From Expression to Food
ISBN: 978-0-12-374039-7

residing within colostrum and milk. The challenge then remains to translate this information into products that form the basis of a functional foods industry, helping to underpin the commercial viability of the dairy industry. In this chapter, we present a review of the current status of bovine milk genomics and functional genomics, and describe the roles, characteristics and key bioactivities of the major bovine milk proteins and their encrypted peptides. The application of these analytical tools to the full spectrum of lactation strategies adopted by eutherians, marsupials and monotremes to improve our understanding of the milk proteome is discussed.

Introduction

As well as being the source of nutrition for the neonates of mammals, milk has long been considered to be a healthy source of dietary proteins and minerals for human consumption, and a global dairy industry has evolved to cater to that need. With competition from other foods and beverages, there has been an impetus for diversification of dairy-based foods, manipulation of milk composition to remove saturated fats, and fortification of milk with fatty acids, minerals and vitamins.

The next evolution of dairy products is to characterize, accentuate and refine naturally occurring biological activities to develop dairy-based functional foods or "nutraceuticals." Although a significant amount of information in relation to the biological activities present in the major bovine milk proteins is already available, the advances in genomic technologies and the utilization of comparative genomics approaches provide opportunities for further discovery and characterization of the bioactivities in bovine milk.

Milk genomics: a contemporary approach to milk composition

Trends in the consumer market are driving an increased focus on the health and dietary functions of food products, and there is an emerging market for milk-derived products. The major focus in dairy is on milk proteins, which are rich in bioactive properties and may therefore be particularly valuable to dairy processors and suppliers providing specialist milk to specification. This area of product development is poised to undergo a significant expansion based on the bovine genome sequencing project.

Milk has been a subject for nutrition and dairy science research over many years, but especially during the post Second World War period, when it received significant support from the British government for its role as a major source of public nutrition. Increasingly, milk is the source of animal protein of choice worldwide because of its ease of handling. Industrial processing of milk into dairy products has been based largely on the insights into milk chemistry that arose from seminal studies conducted during this period. These studies were concerned primarily with nutritional and processing properties of major milk proteins. Recent advances in bovine genomics now extend the capacity to analyze milk at a more detailed level, to provide a basis to understand the properties of milk proteins and, along with parallel advances in humans and other animals, to open a window on the evolutionary origins and specific benefits of milk. Milk genomics is the study of milk using post-genomic approaches.

Advances in bovine genome science

Genomics is concerned with the study of an organism's gene sequence, gene expression and gene function. Genome science has developed rapidly in the past decade as a result of a vast expansion of genome sequence data, initially arising from the human genome sequencing project, and the concomitant development of applied information technology (bioinformatics). In December 2003, only months after the completion of the human genome sequencing project, an international consortium formally declared the initiation of the bovine genome sequencing project. The human genome sequencing project resulted in the establishment of significant infrastructure, expertise and technological advances. Consequently, sequencing of the bovine genome progressed quickly and was completed in 2005, well ahead of the original schedule.

DNA sequence

Bovine genetics, with a focus on the identification of quantitative trait loci (QTL), has been a very active area of study for many years (Khatkar *et al.*, 2004). Using classical genetic and molecular genetic approaches, over 1000 bovine genes had been mapped by the mid 1990s. A phase then followed that was based on candidate gene mapping, and creation and sequencing of bovine expressed sequence tag (EST) libraries. The Institute for Genome Research assembled a bovine gene index that contained information on more than 100 000 sequences. The bovine sequencing strategy employed for the bovine genome sequencing project was a whole genome shotgun approach. An individual animal, a Hereford cow, was chosen as the source of DNA. The total sequencing effort provided approximately 7x coverage of the genome, representing about 2.7 billion base pairs (reviewed in Tellam, 2007). The project complements and in many cases supersedes existing resources for bovine genome science, and provides a platform for developing bovine-specific post-genome research tools and applied industry applications.

Genome map/structure

The bovine genome comprises 29 autosomes, plus the X and Y chromosomes. An international bovine bacterial artificial chromosome (BAC) map was produced to assist in assembling the genome, complemented by an integrated mapping strategy using radiation hybrid and existing physical mapping data. A virtual assembly was released in 2005 and the first de novo assembly was released in 2006 (http://genomes.tamu.edu/bovine/). The current genome assembly is version btau_3.1 with a more accurate version 4 due to be released in 2008.

Comparative genetic maps have revealed significant similarity (synteny) between bovine genomes and the genomes of other species. This can be viewed most effectively using the OXGRID display at (http://oxgrid.angis.org.au/). The synteny between bovine and other species is striking; for example, the mean length of syntenic segments between bovine and human is approximately twice as long as the average length of conserved syntenic segments between human and mouse.

Polymorphism

Understanding genome polymorphism within the bovine is a primary goal of the genome sequencing project, and has implications for variants of milk proteins that affect processing characteristics and in some cases consumer health. The sequencing project itself resulted in the discovery of a large number of single nucleotide polymorphisms (SNPs). This effort has been extended to resequence six breeds—Holstein, Angus, Jersey, Brahman, Limousine and Norwegian Red—at low coverage, with the aim of identifying genome-wide SNPs (Tellam, 2007). In addition, mining of existing ESTs has contributed a putative 17 344 SNPs (Hawken *et al.*, 2004).

Initial analysis of dairy herds using a subset of SNPs revealed that the predominant cattle breed used internationally for milk production, Holstein Friesian, has low genetic diversity and may be considered to be a single population unit (Zenger *et al.*, 2007). The average minor allele frequency for SNPs was 0.29 based on 9195 SNP genotypes in 1000 bulls (Khatkar *et al.*, 2007). Further, using the same genotyping data, Khatkar and colleagues in the Australian Cooperative Research Center for Innovative Dairy Products determined that the extent of linkage disequilibrium in the Holstein Friesian breed is significantly greater than that reported for humans. They went on to construct the first bovine haplotype map and identified 727 blocks (covering just 2.18% of the genome) with a median length of 2.9 kb.

Functional genomics

In addition to increased understanding of dairy herd genetics, the bovine genome sequencing project has led to improved annotation of bovine genes, which underlies comprehensive studies of bovine gene expression. This was enabled by the release of a comprehensive bovine DNA chip. Based on oligonucleotide microarrays and developed using the Affymetrix platform (Dalma-Weiszhausz *et al.*, 2006), the chip contains ≈22 000 probesets, representing ≈19 000 genes. More recently, a long-oligonucleotide spotted array has been developed by an international consortium of researchers. These tools are now employed for the study of mammary gland gene expression, and consequently lactation functional genomics data are emerging.

Studies reporting lactation profiles using microarrays constructed from a bovine cDNA library or ESTs have already been published (Suchyta *et al.*, 2003). We completed a study of the bovine lactation cycle using the first-release comprehensive bovine Affymetrix DNA chip. This study compared gene expression during pregnancy, lactation and involution phases, and identified highly expressed and differentially expressed genes (unpublished data). As a result, a library of almost 4000 genes that were differentially expressed between phases was created. Utilizing these data, further analysis identified a large group of predicted cow's milk polypeptides. These peptides are the subject of ongoing validation and development as dairy products.

A complementary approach to gene expression studies of the mammary gland is provided by milk proteomics (Smolenski *et al.*, 2007). Cow's milk comprises approximately 3.5% protein, 80% of which are caseins; the remainder are predominantly the major whey proteins, but there are also a large number of low-abundant proteins, some of which have significant bioactivity. This area has also benefited markedly from the bovine genome sequencing project. It is now possible to identify variants of both major and low-abundant proteins in milk based on bovine protein sequences. Together with rapid advances in mass spectrometry, particularly improvements in sensitivity and quantification, and complemented by gene expression data, the protein composition of milk can now be dissected with great accuracy.

Comparative milk genomics

The bovine genome sequencing initiative was not the only project to follow the human genome sequencing effort. As this initial project abated, international consortia formed to sequence the genomes of other organisms, including the chicken, dog, opossum, chimpanzee, macaque and platypus. Currently, there is a major project underway, referred to as the Mammalian Genome Project, to sequence 24 animal genomes. Funded by the National Institutes of Health (NIH), diverse species were included to maximize the branch length of the evolutionary tree. Control of lactation and milk composition vary considerably across different species. The Mammalian Genome Project will contribute enormously to the study of the evolution of milk proteins, and will complement a growing body of work on whole genome expression studies in mammary tissue, now mostly in cattle, humans and mice.

Aligned with the goal to develop a deep understanding of milk proteins, we have developed a comparative approach to assess the genomic basis of diversity in milk composition between species. Experimental models have been chosen on the basis of reproductive/lactation strategy, and have been cross-referenced to the dairy cow. The study is based on gene expression profiles and integration with genomic mapping information and dairy QTL analysis. Within this framework, a specific methodology has been applied to identify those genes that may code for milk proteins that are either inherently active in milk or could be developed as dairy products.

A recent example of how comparative genomics may be exploited to alter milk protein composition is the use of functional genomics to exploit animal models with extreme adaptation to lactation, as an alternative approach to identifying key genes that specifically regulate protein synthesis in the mammary gland. For example, milk protein is the only component of milk that is synthesized de novo in the mammary gland of the Cape fur seal. The protein content of this seal milk is more than double that of bovine milk (Cane *et al.*, 2005). The second model, the tammar wallaby, increases both the milk protein concentration and the total production of milk protein in the latter half of lactation (Nicholas *et al.*, 1997). Changes in global gene expression in mammary tissues from the tammar wallaby at early and late lactation using a cDNA array (see Chapter 2), and in pregnant and lactating seals and dairy cows using an Affymetrix microarray, indicate that folate metabolism, and particularly the role of the folate receptor, may be a crucial regulatory point of milk protein synthesis in mammary epithelial cells.

Although this analysis is limited because of redundancy on the arrays (number of probes and number of genes are different), non-symmetric mapping between species, dependence upon threshold for mapping and sequence quality and reliance on UniGene gene assemblies, feeding supplement experiments in the dairy cow have indicated a possible role for folate in manipulating milk proteins.

The importance of the folate receptor 1 (or α) for cellular uptake of folate was established by the analysis of renal folate handling in mice with targeted gene knockouts of folate-binding proteins 1 and 2 (folbp1 and folbp2, equivalent to human and cow folrα and folrβ) (Birn, 2006). Molecular studies in human, monkey and mouse cell lines have shown that the folate receptor α mediates cellular uptake of folate and that folate receptor populations are regulated at multiple levels. Folate may either have a direct effect on the mammary gland or may reduce the competition for precursors between gluconeogenesis and methylneogenesis and increase metabolic efficiency of the mammary gland to influence milk protein production. Furthermore, the folate receptor population may play a crucial role in the capacity of mammary epithelial cells to respond and utilize circulating serum folate.

The folate requirements of lactating dairy cows have been extensively reviewed by Girard *et al.,* (2005). Lactation increases the demands both for methylated compounds (synthesis of milk choline, creatine, creatinine and carnitine) and for methionine to support milk protein synthesis (Xue and Snoswell, 1985; Girard and Matte, 2005). The high demand for folate during lactation is demonstrated by the observation that total serum folates decrease by 40% across the lactation period in dairy cows (Girard *et al.*, 1989).

Folate supplement experiments in dairy cows have shown a positive response to milk production and milk protein yields (Girarde and Matte, 1998; Girard *et al.*, 2005; Graulet *et al.*, 2007). This positive response was not consistent amongst all experimental cows in the first two studies, and appeared to be dependent on cow lactation status (premiparous were non-responsive, multiparous were responsive), stage of lactation and serum vitamin B_{12} status.

The more recent study by Graulet *et al.* (2007), including vitamin B_{12} supplementation, suggested that the increases in milk and milk protein yields as a result of folic acid supplements were not dependent on the supply of vitamin B_{12} and were probably closely related to the role of folic acid in the DNA cycle. However, this does not diminish the importance of the role that folate metabolism plays in milk protein synthesis. This is an example of how the combined approach of comparative genomics and bioinformatics, plus species with extreme adaptation to lactation, may be utilized to identify key genes and cellular processes involved specifically in modifying milk composition and milk protein production.

Origins of milk proteins

The study of milk protein genes and encoded milk proteins of domesticated mammals and undomesticated mammals alike has made these data relevant to dairy science.

The data highlight adaptive features and the diversity of milk protein genes over evolutionary time. Evolutionary evidence suggests that complex lactation preceded divergence of the mammalian lineages. Monotreme lactation appears to be the most ancestral form of lactation whereas marsupials and eutherians developed divergent strategies. The study of the evolution of milk proteins assists in determining the significance of maintaining certain proteins while changing or losing others. These differences may reflect divergence of lactation strategies and may provide clues as to why some proteins have become non-essential and are in the process of being lost by some eutherians.

Bioinformatic predictions of genes encoding secreted proteins up-regulated during lactation in the mammary gland suggest that as many as 300 different proteins may be found in milk. However, caseins are the major proteins in milk, representing about 80% of total milk protein content. Expression of a number of casein genes has been confirmed in monotremes (platypus and echidna) and marsupials.

Three evolutionarily related calcium-sensing casein-like genes and the functionally related κ-casein were identified in monotremes. As in other mammals, a number of casein splice variants could be identified. Interestingly, all these casein genes cluster tightly within a 100 kb region of the platypus genome, a physical linkage that has been observed in the casein locus of all mammalian genomes examined so far (Rijnkels, 2002). The respective organization of the casein locus is also highly conserved in all mammalian lineages. Casein genes occur in the order α_{s1}-, β-, α_{s2}- and κ-, where the β-casein gene is encoded in the opposite direction from the other genes. The close proximity of α_{s1}- and β-casein genes in an inverted tail—tail orientation and the relative orientation of the more distant κ-casein genes are similar in all mammalian genome sequences available so far. This observation suggests that the synchronized tissue-specific expression of casein genes in the lactating mammary gland may be controlled, in part, by a common mechanism at this locus. However, genome sequence analysis has also revealed variation in the α_{s2}-casein gene content of mammalian lineages.

Marsupials possess only one copy of α-casein corresponding to α_{s1}-casein (Lefevre *et al.*, 2007). In the monotreme lineage, a recent duplication or a gene conversion event involving β-casein occurred to produce an α-like casein gene at a genomic location similar to that of the eutherian α_{s2}-casein. A similar but more ancestral duplication of β-casein occurred in eutherians to produce α_{s2}-casein and further duplications have also occurred since in human and mouse lineages to produce α_{s2a}- and α_{s2b}-casein genes.

β-Lactoglobulin (β-Lg) is the major whey protein in ruminant milk and is present in milk from a variety of species; however, it is absent in rodents, such as the mouse, lagomorphs (such as the rabbit), and humans, although reports to the contrary exist. β-Lg in human milk probably arises either from spurious antibody cross-reactivity (Brignon *et al.*, 1985) or from the presence of ingested bovine β-Lg in the mother's milk (Fukushima *et al.*, 1997). Not all primate milks are devoid of β-Lg as both the macaque and the baboon have the protein within their milk (Hall *et al.*, 2001).

Because of the absence of β-Lg in human milk, the dairy industry has developed ways in which to convert the protein composition of bovine milk to something more similar to the protein composition of human milk. Genetic variants have been observed in essentially all of the species from which the protein has been observed. There are several genes encoding β-Lgs within the different species. Sequence analysis of species genomes suggests that some species such as the dog and the cat have three β-Lg genes, whereas others such as the horse and the donkey have two β-Lg genes, and that bovine, sheep and goat have one β-Lg gene and bovine has a pseudogene. Marsupials, monotremes and some primates have one β-Lg gene.

Comparison of β-Lg protein sequences shows high conservation of the protein during evolution. Bioactive peptides derived from β-Lg are currently being intensively studied. These peptides are active only once they are released from the precursor protein. Once released, these peptides are suggested to play important roles in human health, including having antimicrobial, antihypertensive and antioxidant activities (Hernandez-Ledesma *et al.*, 2007). The variation of gene number and protein content in milk represents a specific evolution of β-Lg for each lineage. It has not yet been explained how a major protein in cow's milk such as β-Lg can be completely abolished in human, rat or rabbit milk with little to no consequence to the offspring.

Whey acidic protein (WAP) is a major whey protein that is found in the milk of numerous species such as the mouse (Hennighausen and Sippel, 1982), rat (Campbell *et al.*, 1984), pig (Simpson *et al.*, 1998), rabbit (Devinoy *et al.*, 1988), camel (Beg *et al.*, 1986), wallaby (Simpson *et al.*, 2000) brushtail possum (Demmer *et al.*, 2001), echidna and platypus (Sharp *et al.*, 2007), but is absent in the milk of cows, sheep and humans (Hajjoubi *et al.*, 2006). WAP pseudogenes have been identified in the bovine and the human. A nucleotide deletion at the end of the first exon is reported to have caused a truncation of the bovine WAP protein (Hajjoubi *et al.*, 2006). This deletion is also found in ovine and caprine species. In comparison, it is the absence of an ATG initiation codon in human WAP that is suggested to be the cause of the pseudogene in this species. Interestingly, the polyadenylation signal AATAAA is still present in the ruminant sequence but not in the human sequence.

WAP proteins from different species show moderate sequence similarity; however, they differ in the number of four-disulfide core (4-DSC) motifs between the species. Each specific 4-DCS domain is recognizable by sequence similarity and it is possible to trace the origins of each domain during evolution (Figure 3.1). Although not expressed, both bovine and human pseudogenes have similar gene structure to the eutherian WAPs and it is suggested that these were once functional. As suggested in Chapter 1, the function of the WAP proteins appears to differ with the presence or absence of each of these domains, and it is interesting to speculate why this protein appears to be dispensable in some species.

The apparently dispensable role of proteins such as β-Lg and WAP may be related to the diet of the young, which is being adequately compensated in more domesticated animals or may be linked to improved care and hygiene in domesticated life.

Constraints and opportunities for evolution or manipulation of bovine milk proteins

Although the protein component of bovine milk is only ≈2–4% of the total weight, which typically is marginally less than both total fat and carbohydrate, the population of proteins present in milk is diverse and impacts on the physical properties

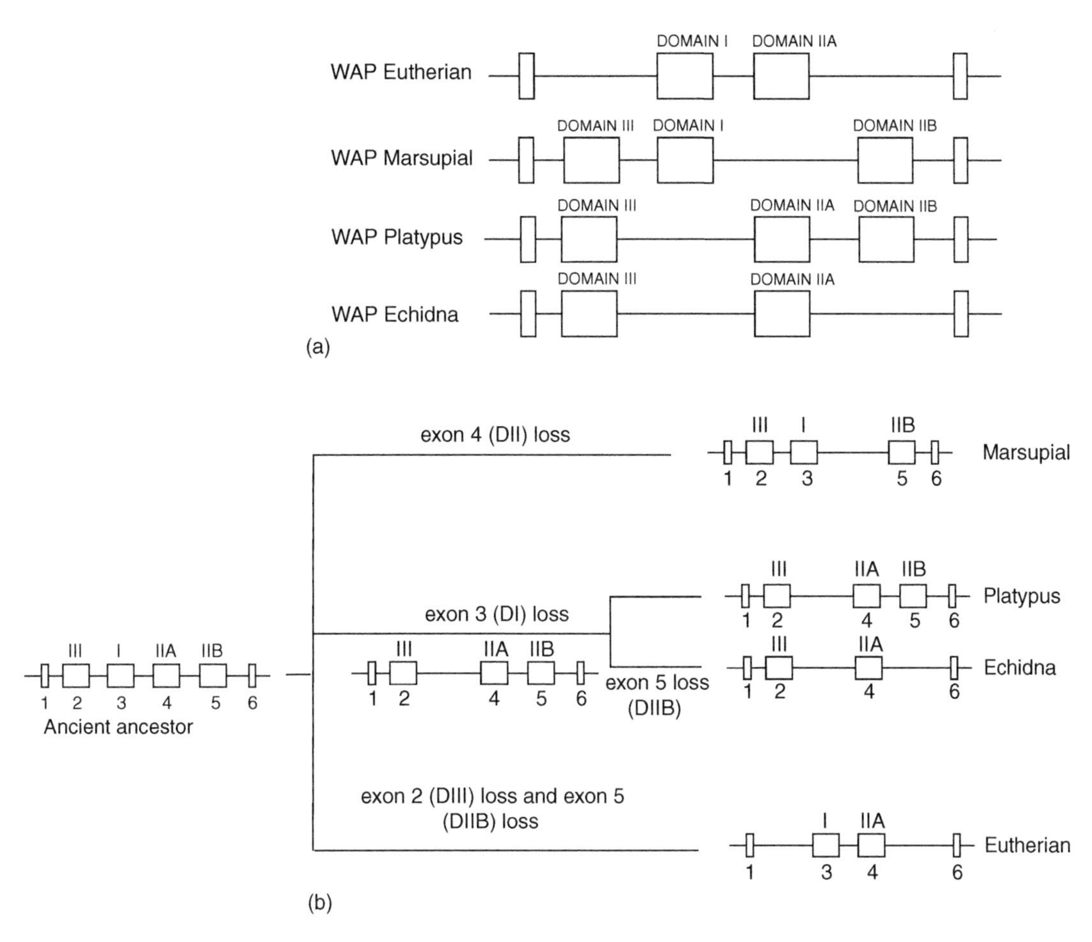

Figure 3.1 Schematic representation of the origins of WAP gene structure. (a). Boxes represent exons within the WAP genes. Each 4-DSC domain is encoded by different exons as indicated. In eutherian WAP, exons 2 and 3 encode the two 4-DCS domains (DI and DIIA). In marsupial WAP, there is an additional exon, and exons 2, 3 and 4 encode the three 4-DSC domains (DIII, DI and DIIB). In echidna WAP, exons 2 and 3 encode the two 4-DCS domains (DIII and DIIA). In platypus WAP, exons 2, 3 and 4 encode the three 4DSC domains (DIII, DIIA and DIIB). Comparison of sequence similarity suggests that DIIB of marsupial WAP and DIIB of platypus WAP are derived from similar exons and that DIIA of eutherian WAP and DIIA of monotreme WAP are derived from similar exons. (b). Schematic representation of stepwise evolutionary history of 4-DSC domains of WAP within the major lineages. The ancestral progenitor, represented by six exons (numbered) and four 4-DSC domains (DIII, DI, DIIA, DIIB), is shown, followed by events of exon loss leading to the evolution of the present day WAPs in platypus, echidna, marsupial and monotremes. Numbers represent the order of exons relative to the ancient WAP. Adapted from Sharp *et al.* (2007).

of milk, the nutrition of a neonate and the metabolic homeostasis and development of both the dam and the offspring. Approximately 80% of the protein present in milk consists of four structurally and functionally interrelated proteins called the caseins. These include α_{s1}- and α_{s2}-caseins as well as β- and κ-caseins. Some of these proteins show some conservation of gene and protein sequence and structure and there is similar evidence of relatedness between proteins observed in milks of different species. The remaining 20% of the milk protein population includes the major whey proteins β-Lg and α-lactalbumin (α-La) as well as other constituent proteins including immunoglobulins, serum proteins, milk fat globule proteins, transferrin, lactoferrin, β2-microglobulin, an array of enzymes and numerous proteolytic products.

The gene and protein structures and sequences of the six major bovine milk proteins (the four caseins, β-Lg and α-La) have been well characterized although the precise secondary and tertiary structures of the caseins have been difficult to demonstrate experimentally as they have proven to be challenging to investigate via X-ray crystallography. The casein phosphoproteins precipitate at pH 4.6, which is a significant characteristic that is utilized in cheese production but, in milk, these proteins aggregate to form protein aggregates or micelles of ≈50–300 nm in diameter. κ-Casein stabilizes the surface of the protein micelles with the other caseins forming the core, which is stabilized by the phosphorylation of seryl residues. The organic phosphate moieties in turn enable calcium binding, which also stabilizes the micelle structure and hence, as well as being a significant source of dietary amino acids, the micelle aggregates also form a valuable source of calcium and phosphorus for neonatal nutrition and development.

Apart from the neonatal nutritional and industrial (dairy products) significance of bovine milk proteins, a wide range of biological activities is observed in the neonate as well as the dam; these activities are attributed to both the intact proteins and more importantly their products of *in vivo* digestion (see the review by Meisel, 2005). When considered together, bovine milk proteins, which provide neonatal nutrition, regulate physiological processes, co-ordinate neonatal development and form the basis of an internationally significant agricultural commodity market, are globally important proteins. For this reason, their functional characteristics need to be considered when evaluating the potential for enhancement or modification of their structure or biosynthesis.

The detailed structure of milk proteins and their genes

An understanding of the detailed structure of milk proteins and their genes is critical in the consideration of the impact of modifying bovine milk proteins to enhance productivity, to enhance milk product manufacturing efficiency or to exploit inherent or proposed bioactivity. The physical and structural characteristics of bovine milk proteins have been extensively reviewed (Farrell *et al.*, 2004).

The bovine casein genes are present as single copy genes per chromosome and are clustered together on chromosome 6 in a 200 kb region at position q31–33 in the order α_{s1}-, β-, α_{s2}- and κ- (Threadgill and Womack, 1990). The bovine α_{s1}-casein gene is 17 508 bp in length and contains 19 exons. The protein is the most abundant of the bovine caseins, present at 12–15 g/L. There are up to eight reported genetic variants of this protein, with the B variant being the most common in *Bos taurus* species and comprising 199 amino acids with a molecular weight of 23.6 kDa. The α_{s1}-casein protein may contain up to eight serine monophosphate residues, which cluster in a hydrophilic domain between amino acids 43–80 and, through modelling studies, are thought to be connected to a hydrophobic domain (amino acids 100–199) by helical and sheet secondary structures. The less abundant bovine α_{s2}-casein gene is 18 483 bp in length and contains 18 exons with significant exon duplication (Mercier and Vilotte,1993). This gene codes for 207 amino acids with a molecular weight of 25.23 kDa. This protein exhibits a number of phosphorylation states as well as four variants that have been observed in common bovine populations.

In contrast, the bovine β-casein gene is significantly less complex, containing nine exons in 8498 bp. The protein is present at 9–11 g/L in modern dairy cattle, consists of 209 amino acids and has a molecular weight of approximately 24 kDa. Up to 12 genetic variants have been observed and the protein exhibits a flexible structure that can adopt multiple conformations, although a small amount of secondary structure in the form of α helices and β sheets has been predicted. There are five phosphoserine residues that reside at the N-terminal domain of the protein.

The bovine κ-casein gene exhibits fewer structural similarities than the other casein genes and contains five exons and is approximately 13 700 bp in length. It has diverged from the fibrinogen gene family but is more highly conserved between species than the other casein genes. Two common variants are observed in dairy cattle (although several others have been reported), with the B variant containing 169 amino acids with a molecular weight of 19 kDa. Up to six threonine residues may be glycosylated and, similar to the other caseins, κ-casein has a relatively high proline content that influences the predicted secondary structure of two sets of antiparallel β sheet structures exhibiting hydrophobic side chains that may be important for micelle formation. As with the other caseins, the detailed structure of bovine κ-casein is poorly characterized although the N-terminal domain of amino acids 1–44 has been experimentally shown to contain a helical structure (Bansal *et al.*, 2006).

The bovine α-La gene contains four exons and is approximately 3.5 kb in length. It is located on bovine chromosome 5 and exhibits significant interspecies homology in exon 4 (significant for its interaction with galactosyltransferase) and the 5′ untranslated region. Two variants have been characterized for the protein with the most common B variant exhibiting a molecular weight of 14.1 kDa and being present at 0.6–1.7 g/L in skim milk. The 123 amino acids are organized into four helical domains and a small sheet structure that is dependent on pH and metal ion concentrations (Sawyer and Holt, 1993). Different glycosylation states have been observed and structurally bovine α-La has some similarity to lysozyme proteins (Brew *et al.*, 1970). This protein, like the caseins, also has calcium-binding properties, as described by Hiraoka *et al.*, (1980).

Bovine β-Lg is located on chromosome 11, contains four exons in 4723 bp and is well conserved between species (Folch *et al.*, 1994). Structural similarities occur throughout the lipocalin protein family and β-Lg has been shown to bind to a number of hydrophobic ligands including palmitic acid (Kontopidis *et al.*, 2004; Konuma *et al.*, 2007). There are two common variants of this protein—A and B—with the protein itself containing 162 amino acids, with a molecular weight of 18.3 kDa. It exhibits a number of β strands, which form a β-barrel structure that contains the hydrophobic-ligand-binding site, and a single α helical domain is observed at the C-terminal end of the protein.

The secondary structures of the major bovine milk proteins are diverse, yet are clearly critical for the functions of the proteins including micelle formation, calcium and phosphate binding and other bioactivities. Although there may be opportunities for manipulation of structure and therefore function of these proteins, even modest changes to nucleotide and amino acid sequence may significantly alter protein and peptide functionality.

The function of bovine milk proteins

The structural complexity of these milk proteins suggests that they serve many functions in the neonate in addition to the mere provision of a source of amino acids for protein biosynthesis in the fast-developing neonate. Yet the range of functions is not likely to be as great as that in more altricial species such as the monotremes and marsupials, in which the milk constituents orchestrate almost the entire developmental process from the fetus through to attaining physiological independence (see Chapter 2). The contributions that individual proteins make to this developmental spectrum are poorly understood, although longitudinal studies on gene expression and related changes in milk composition will assist in unravelling this relationship.

The protracted emphasis on selection of cows for milk volume has probably also influenced the balance of biological activities residing within these proteins, because contemporary dairy herds are larger in stature than their predecessors. Thus, developmental activities encoded within the protein fraction may have been altered to emphasize growth promotion rather than other key functions such as the storage of body energy reserves to support reproductive activity and even to promote maturation of the immune system.

The complexity increases further once these proteins are digested enzymatically or are hydrolyzed in the developing gastrointestinal tract and then absorbed to be transported to their site of action. The precise balance between the peptide component that is absorbed as functional entities and the peptide component that remains for further processing into short peptides as a source of dietary amino acids is not well understood. However, the need to conserve a constant supply of developmental molecules is probably more important than the need for protein synthetic substrate in sustaining the development of the neonate.

The observation that over 100 hormones and growth factors have been identified in colostrum or milk (Koldovsky, 1996) would suggest that the milk proteins may play important roles, working in concert with these established endocrine/paracrine signals. Therefore, the sustained selection of animals for milk volume output will have altered the balance of bioactivities in colostrum and milk to account for this response. Thus, the balance of bioactive peptides may differ from those in species in which this selection pressure has been absent. From a commercial point of view, these milk constituents also play a role in determining the sensory properties of milk and its manufactured products. Of course, these same sensory factors probably also serve as attractants to the newborn as it seeks its nutrient supply from the udder.

The extraction of intact proteins of high commercial value from milk has been attractive to dairy farmers tied to tightly regulated commercial milk prices. The profitability of such ventures depends on the ease of extraction, the relative concentration of the protein in milk and the value placed on the bioactivity. A range of these proteins is presented in Table 3.1.

Both α-La and β-Lg have been isolated on a large scale using a range of chromatographic, ultra-high pressure and membrane separation techniques (Korhonen and Pihlanto, 2007). Both of these proteins convey important activities (Table 3.1) that provide the impetus for large-scale production. One of the most important clinical

Table 3.1 Major milk proteins with potential for extraction on a commercial scale to exploit bioactivities (adapted from Korhonen and Pihlanto, 2007)

Protein	Concentration in colostrum (g/L)	Concentration in milk (g/L)	Molecular weight	Exploitable bioactivity
Caseins (α_{s1}-, α_{s2}-, β- and κ-)	26	28	14000–22000	Ion carriers, bioactive peptide sources, immunomodulators
β-Lactoglobulin	8.0	3.3	18400	Antioxidant, vitamin-A-binding protein, bioactive peptide source
α-Lactalbumin	3.0	1.2	14200	Lactose synthesis and milk volume. Calcium binding, immunomodulator, bioactive peptide source
Immunoglobulins	20–150	0.5–1.0	150000–1000000	Specific immune protection, potential bioactive peptide source
Glycomacropeptide	2.5	1.2	8000	Antimicrobial, bifidogenic, gastric modulator, anticlotting factor
Lactoferrin	1.5	0.1	80000	Antimicrobial, antioxidant, anti-inflammatory, anticarcinogenic, bioactive peptide source, immunomodulator
Lactoperoxidase	0.02	0.03	78000	Antimicrobial, immunopotentiator
Lysozyme	0.0004	0.0004	14000	Antimicrobial, immunopotentiator
Serum albumin	1.3	0.3	66300	Bioactive peptide source

findings is the ability of these whey proteins to decrease the incidence of certain cancers. In most studies, crude whey protein mixes are used; these are thought to act through the provision of additional sulfur amino acids that act as substrates for the synthesis of glutathione in many tissues (Bounos and Gold, 1991). This tripeptide is a potent antioxidant, acting through selenium-dependent glutathione peroxidase that contributes to the removal of reactive oxygen species (Parodi, 2007).

Many of the highly expressed proteins clearly have important immunological functions. The passive immunity provided through the immunoglobulin complement of colostrum to the neonate is supplemented by a range of other bioactivities designed to countenance pathogen loads to which the newborn is exposed in its new environment. The activity of these proteins extends to the initiation of mucosal immunity in the gut and an optimal environment for beneficial microflora. Such proteins include immunoglobulins, many chemokines, mucin 1, lactoferrin, cathelicidin 1, lactoperoxidase, S100 calcium-binding proteins, complement proteins and polymeric immunoglobulin receptor.

It is beyond the scope of this review to detail all of the bioactivities identified in milk. However, some specific proteins are illustrative of principles that are found across the milk proteome.

Lactoferrin is one such protein; it has the dual roles of scavenging iron and acting as a bacteriostat. It is found in exocrine secretions such as milk, tears, tubotympanum and nasal exudate, saliva, bronchial mucus, gastrointestinal fluids, cervicovaginal mucus and seminal fluid, all of which require antibiotic activity. It is also found in neutrophils that are attracted to sites of infection where both iron sequestration and antibiotic activity assist in the healing process. However, its biological effects do not end there. Its commercial utility is increased through its antiviral, antioxidant, antitoxigenic, antithrombotic and, importantly, anti-carcinogenic actions. However, it is important to note that lactoferrin is just one component of a mix of protective molecules in biological fluids, as lysozyme, beta defensins and the surfactant proteins A and D (SP-A, SP-D) among others also contribute to this important function.

Typically, lactoferrin is found in concentrations of 100 mg/L in colostrum and is extracted by conventional fractionation and chromatography. However, the expression of milk proteins in cereal grains such as rice using recombinant DNA technology (Lonnerdal, 2006) seems to be a more effective way of providing these unique milk proteins in our diet. Similarly, recombinant human lactoferrin given orally has been suggested as a mechanism for the prevention of certain gastrointestinal infections in pre-term infants (Sherman and Petrak, 2005). There is little doubt that this will be a continuing trend, the potential for which was recognized in the last decade (Arakawa *et al.*, 1999); transgenic plants may well supersede the cow as a source of this and other unique milk proteins.

Bioactive peptides sequestered within milk proteins

Milk proteins contain latent biofunctional peptide sequences within their primary structures that exert physiological responses *in vivo*. They remain latent until released through the processes of enzymatic processing in the gastrointestinal tract. A large range of these peptides has been identified, including opioid, antimicrobial, immunomodulatory, mineral-transporting, growth-promoting, anticancer, proteinase and angiotensin-converting enzyme (ACE) inhibitory peptides (Shah, 2000; Ferranti *et al.*, 2004). In an evolutionary sense, the cow is targeting peptides to specific effector sites through this process. Presumably, if the availability of metabolic substrates to support maximal biological response is inadequate, then the animal has the ability to limit the quantities of peptides released by suppressing the release of rate-limiting enzymes.

The biofunctional peptides currently most studied in food proteins appear to be those that inhibit ACE. This enzyme plays a central role in the regulation of blood pressure through the production of the potent vasoconstrictor angiotensin (Ang) II and the degradation of the vasodilator bradykinin. ACE inhibitory peptides may therefore have the ability to lower blood pressure *in vivo* by limiting the vasoconstrictory effects of Ang II and by potentiating the vasodilatory effects of bradykinin (Murray and Fitzgerald, 2007). However, this peptide is just one of a number that contribute to the regulation of this important physiological parameter; in addition to the renin–angiotensin system, endothelins and their converting enzymes, the kinin nitric oxide pathway and the neutral endopeptidases all play a role.

The development of functional foods with antihypertensive properties provides an attractive and potentially commercially lucrative range of products. Although this enzyme is found widely through metabolic tissues, it is largely associated with the vascular epithelium (Ondetti and Cushman, 1982). These ACE inhibitory peptides can be enzymatically released from intact proteins *in vitro* and *in vivo* during food processing and gastrointestinal digestion respectively. ACE inhibitory peptides may be generated in or incorporated into functional foods in the development of "natural" beneficial health products (Murray and Fitzgerald, 2007). In view of the extensive range of peptides derived from various precursor proteins that are involved in the regulation of blood pressure, it may be too simplistic to expect a functional food containing just one or two of these peptides to decrease suppressor activity in the long term. However, several products are currently being evaluated as beneficial functional foods/food ingredients.

One major success story in deriving a commercial product from milk protein tryptic digests is the casein phosphopeptides used in the chewing gum "Recaldent" (Cross *et al.*, 2007). These peptides, containing the sequence −Pse–Pse–Pse–Glu–Glu–, where Pse is a phosphoseryl residue, stabilize calcium and phosphate ions in aqueous solution and make these essential nutrients bioavailable. Investigations of the chemistry of these intriguing peptides have shown that they can be altered to form casein phosphopeptide–amorphous calcium phosphate which in turn is capable of forming a calcium fluoride. The practical outcome from these outstanding discoveries is a product that is capable of remineralizing carious lesions in dental enamel and that can be incorporated into dental care products and foodstuffs (Cross *et al.*, 2007).

The potential for the discovery of new peptides is high and it would seem reasonable to suspect that careful screening of digesta from various segments of the gastrointestinal tract may prove to be a valuable approach for this discovery process.

Existing variation in bovine milk proteins and the impact on expression, function and milk quality

As described above, a number of genetic variants of the major bovine milk proteins have been observed and characterized. These variations range from minor amino acid substitutions to larger deletions. There are many studies that suggest that the presence or absence of a particular variant expressed within an individual may be associated with altered production or processing quality of milk (see the reviews of Jakob and Puhan 1992; Martin *et al.*, 2002). Many of these studies are based on statistical associations but others have been well characterized to elucidate cause−effect relationships. These naturally occurring variations in milk protein amino acid sequences serve as examples for the breadth of potential that modification of milk proteins may have on milk production and quality.

The relationship between genetic variants of bovine milk proteins and productive characteristics has been extensively reviewed. The characteristics observed to be associated with milk protein polymorphism include first lactation milk production, protein production and fat percentage in Californian Jersey cows (Ojala *et al.*, 1997); milk and protein yield and fat content in Finnish Ayrshire cows (Ikonen *et al.*, 2001); milk, protein and fat yield in Holstein and Ayrshire cows (Lin *et al.*, 1986); the effect of nutritional regimens on milk composition (Mackle *et al.*, 1999); somatic cell count (Ng-Kwai-Hang *et al.*, 1987); total solids and milk protein profile (McLean *et al.*, 1984); and lifetime performance (Lin *et al.*, 1989).

Yet, despite this body of evidence, milk protein genotype is not currently used in selection to enhance genetic gain for milk production characteristics, possibly because of the variable and, on occasion, conflicting observations or the potential cost relative to the sometimes modest production differential. Understandably, many of the genotypic variants associated with increases in milk yield are generally already present at a high frequency in modern dairy populations.

The B variant of bovine α_{s1}-casein is the most abundant variant in western dairy populations, possibly because of indirect selection as other variants may in some way be associated with decreased α_{s1}-casein synthesis, such as the A variant which may be subject to a potential decrease in translational efficiency because of observed polymorphism at a polyadenylation signal site (McKnight *et al.*, 1989) or because of promoter polymorphisms that may alter transcriptional efficiency (Prinzenberg *et al.*, 2003). Similarly, animals that are observed to have a lower bovine α_{s1}-casein protein concentration in milk (heterozygous for the G allele) exhibit a lower casein content and a slower clot formation time, yet display a faster curd-firming time and a higher curd firmness (Mariani *et al.*, 2001). In contrast to bovine α_{s1}-casein, although there are four known variants of the bovine α_{s2}-casein protein, few studies have shown any relationships between the genotype of this protein and milk production or quality traits.

The A1 and A2 variants of bovine β-casein are present in modern dairy populations at high frequencies and the A2 variant has been shown in defined populations of Danish dairy cattle breeds to be associated with higher milk, fat and protein yields when homozygous (Bech and Kristiansen, 1990).

Some controversy surrounds the A2 protein genotype, as it has been suggested that human populations who consume milk containing higher concentrations of the A2 variant exhibit a lower incidence of cardiovascular disease, a lower incidence of type 1 diabetes and decreased severity of symptoms of neurological diseases because the A2 variant does not liberate the opioid bioactive peptide casomorphin 7 upon proteolytic digestion (Laugesen and Elliott, 2003; Tailford *et al.*, 2003; Bell *et al.*, 2006; Kaminski *et al.*, 2007).

In a review of evidence, however, Truswell (2005) suggested that there was limited evidence to suggest that milk containing the A2 variant of β-casein had any significant human health advantage and, indeed, human trials suggested that there was no differential effect between dietary products containing either the A1 variant or the A2 variant on human plasma cholesterol concentrations (Venn *et al.*, 2006) or cardiovascular health (Chin-Dusting *et al.*, 2006).

With the significant structural role of bovine κ-casein in the casein micelle, it is not surprising that there are only two common variants—A and B—in modern dairy populations, although up to nine other variants have been characterized in bovine species (Farrell *et al.*, 2004). A number of studies have been conducted to investigate associations between κ-casein protein variants and milk production and quality characteristics. Of all the genetic polymorphisms of dairy cattle affecting milk composition, the κ-casein genotype is one of the more significant, with association differences identified between genotype and lifetime production (Lin *et al.*, 1989), concentrations of individual milk proteins (McLean *et al.*, 1984), protein yield and percentage (Tsiaras et al., 2005) as well as cheese production characteristics including rennet clotting time, curd formation and coagulation strength (Pagnacco and Caroli, 1987).

Of the 12 identified bovine β-Lg variants, only the A and B variants occur at high frequencies in dairy cow populations. The B variant has been associated with higher casein percentage (Lundén *et al.*, 1997), higher total solids (McLean *et al.*, 1984) and higher milk yield, fat yield, fat percentage and lactose yield (Tsiaras *et al.*, 2005), whereas the A variant is associated with higher whey protein content and lower casein content (Auldist *et al.*, 2000), possibly suggesting that it may be expressed at higher concentrations than the B variant. The B variant association with the casein composition of milk therefore has implications for milk quality for cheese production. Of the three bovine α-La variants, the B variant is the most common in modern dairy populations, yet the A and B variants differ by a single amino acid residue. The role of this protein in lactose biosynthesis is such that there seems to be little genetic divergence and little function or associated differences observed between the common variants.

Experimental modifications of bovine milk proteins

The experimental manipulation of patterns of expression or characteristics of bovine milk proteins has previously been reviewed (Bremel *et al.*, 1989; Yom and Bremel, 1993; Clark, 1996). Both over-expression and impairment or "knockout" of milk proteins have been conducted predominantly in rodent models, presumably because of the expense and technical challenges of conducting studies in bovine or other ruminant species. Although observations from experiments conducted in model species must be considered in context, valuable insight into the potential for future manipulation of bovine milk proteins in ruminant species has been gained.

Although a number of experiments have been conducted to either over-express or inhibit expression of milk proteins in mice and other rodents, only a few have further characterized the milk to assess the impact of manipulation of expression profiles on the quality of milk for dairy product manufacture. A transgenic mouse model over-expressing bovine κ-casein (Gutiérrez-Adán *et al.*, 1989) exhibited no changes in milk protein concentration, yet milk from these mice exhibited a smaller micelle size and a tendency toward stronger curd characteristics. Conversely, a mouse model in which the endogenous κ-casein was suppressed exhibited a loss in micelle stability, resulting in casein precipitation in alveolar lumens, which prevented lactation (Shekar *et al.*, 2006).

These investigations highlight the significance of the role of κ-casein in the tertiary structure of protein in milk regardless of species. In comparison, the manipulation of β-casein expression results in less dramatic consequences. The over-expression of bovine β-casein in a mouse model (Hitchin *et al.*, 1996) had little observed effect on lactation although the processing quality of milk was not assessed in this study. In another study, the inhibition of endogenous murine β-casein (Kumar *et al.*, 1994) changed micelle size and reduced pup growth.

Similarly, in cloned transgenic cattle in which both bovine β-casein and κ-casein were over-expressed (Brophy *et al.*, 2003), milk levels of both proteins were increased, demonstrating that milk protein composition can be altered in large ruminants using recombinant gene technologies. Although the processing characteristics of the milk from these animals were not established, the authors stated that the fat, lactose and mineral contents of the transgenic milk were within normal ranges.

The functionality of caseins can be altered through enzymatic modification of the proteins, e.g. through dephosphorylation. This results in increased pepsin hydrolysis, which in turn alters digestibility (Li-Chan and Nakai, 1989) and potentially the release of bioactive peptides. Importantly, none of the studies discussed identified defects or significant changes to mammary development or structure, suggesting that manipulation of milk proteins may not have undesirable effects on the mammary gland itself, compromising either the potential to lactate or animal health.

The whey proteins have also been investigated for manipulation to alter processing characteristics of milk. Because of its role in the synthesis of lactose and the implications for manipulation of osmotic potential of milk, α -La has been a target for manipulation. Mice transgenic for expression of bovine α-La exhibited increased lactose concentration in milk, and a slight increase in pup growth rate (Boston *et al.*, 2001).

Similarly, transgenic pigs over-expressing bovine α-La produced milk with lower total solids, higher milk yield in early lactation, higher lactose concentration and increased litter growth rates (Noble *et al.*, 2002). In contrast, a mouse model of α-La deficiency resulted in milk that was so viscous that pups were not able to suckle effectively (Stinnakre *et al.*, 1994). Bovine α-La itself has also been modified to alter its characteristics and various mutations have resulted in changes in affinity for galactosyltransferase, increased glucose binding (Grobler *et al.*, 1994) and changes in molten globule conformation (Uchiyama *et al.*, 1995).

The β-Lg protein has also been evaluated experimentally. Over-expression in transgenic mice resulted in normal mammary physiology and therefore normal milk secretion and pup growth (Hyttinen *et al.*, 1998; Gutiérrez-Adán *et al.*, 1999). However, a similar study in which ovine β-Lg was over-expressed reported an increase in total protein content (Simons *et al.*, 1987). Mutants of bovine β-Lg have also been studied, with variation in thermal stability (Cho *et al.*, 1994), rate of secretion (Katakura *et al.*, 1999) and rate of denaturation and digestion (Jayat *et al.*, 2004) being reported.

Adding value to milk through the use of milk protein genomics

The potential value of milk as a source of animal protein varying in characteristics that promote human health has been demonstrated by the range of bioactivities residing within the mammary protein phenome. The ease with which dairy products are distributed to consumers irrespective of their socio-economic status also makes this an ideal vehicle for the administration of therapeutics. The hyperimmune nature of colostrum, containing up to 40% by weight of immunoglobulin, suggests that immunization of cows against specific pathogens may provide a rich source of therapeutic antibodies. As human colostrum contains antibodies reactive to colonization factors 1 and 2 of enterotoxigenic *Escherichia coli* (Correa *et al.*, 2006), thereby preventing diarrhea in infants, the production of similar antibodies in cows seemed to be a logical commercial target. The Australian commercial biotechnology company Anadis has exploited this technology in producing an antibody-enriched colostrum tablet to inhibit travellers' diarrhea (www.anadis.com).

However, perhaps the greatest potential that milk proteins and peptides have for human health is through the addition of whey protein and casein peptides to foods to increase their functionality. Estimates of world whey protein production exceed 0.5 million tonnes, which provides a resource that should be used more effectively than by simply disposing of it as a liquid waste or as an animal feed. Considerable progress has been made in the development of methods for separation and then purification of specific proteins from the whey fraction. Native immunoglobulins, lactoferrin, lactoperoxidase, α-La and β-Lg have all been recovered in industrial quantities. Yet the major advances in recombinant technologies may ultimately provide a cost-competitive alternative supply of these proteins.

The development of transgenic dairy herds expressing altered protein composition and yield has been muted for the past decade. It was thought that the use of yeast and bacterial artificial chromosomes capable of conveying gene constructs of an Mb may be useful in supporting the expression of the 200 kb casein locus (with all four caseins) including the regulatory elements co-coordinating their expression (Zuelke, 1998). To date, the only report of substance is that of Brophy *et al.* (2003), cited above, in which both β- and κ-casein were over-expressed. Given the intricacy of the ultrastructure of the micelle, and its critical nature in providing calcium and phosphorus in the correct form for absorption by the calf, it is not surprising that any genetic manipulation of protein composition is going to upset this equilibrium, which has been refined through millions of years of evolution. However, our ability to manipulate milk composition through manufacturing technologies has effectively removed the need for such animals. Growing animal welfare concerns must also be considered when embarking on such ambitious projects.

Nevertheless, the development of somatic nuclear transfer techniques that facilitate targeted genetic modifications has driven transgenic research over the past 5 years. Small interfering RNA techniques and lentiviral vectors have also contributed to this cause, as have modified episomal vectors designed to promote high levels of expression of therapeutic genes (Manzini *et al.*, 2006). The use of transgenic animals as bioreactors to synthesize valuable proteins in large quantities is probably the major application for new transgenic animals. Goats and rabbits are the most popular target species, in addition to the cow, for producing a range of proteins including enzymes such as alpha-glucosidase, hormones such as human growth hormone and large proteins such as lactoferrin, albumin, collagen and vaccines (e.g. for malaria). The development of a recombinant human antithrombin III expressed in goat's milk is the closest of these products to commercial release (Niemannn and Kues, 2007).

Ideally, a milk enriched in peptides promoting immune function, controlling blood pressure, acting as a bacteriostat and minimizing oxidative stress and cancer risk, while at the same time relieving depression and preventing dental caries, would seem to have the makings of a highly valuable functional food. Combining this with an enrichment with n-3 fatty acids thought to increase insulin sensitivity and therefore prevent diabetes, together with certain milk carbohydrates capable of improving cognition, adds greatly to a product that already acts as a rich source of amino acids and energy to promote normal growth processes. Manipulation of these proteins in milk will inevitably occur in the factory and potentially in the cow. The challenge remains to turn this speculation into commercial reality for the benefit of societies in both the developed world and the developing world.

Conclusions

The bovine milk proteins form the basis for a global industry in dairy products, and as such have been intensively studied by both classical deconstructive observation and in dynamic whole animal systems utilizing post-genomic or functional genomics approaches. From these studies we have a greater understanding of the roles of these proteins in milk, their biological significance in the neonate and the biochemical characteristics of bovine milk proteins that affect efficiencies of manufacturing processes.

The emerging interest in elucidating the bioactivities inherent in bovine milk proteins is both challenging and potentially economically and socially rewarding. Initially the benefits of this research will be realized by dairy manufacturers promoting brand differentiation following from marketing trends in milk fortification yet the far greater potential of enhancing these bioactivities of native proteins into complementary health products or nutraceuticals will represent a major repositioning of dairy products in consumer consciousness and result in the re-evaluation of the value of dairy products.

The enthusiasm for exploitation of these bioactivities in milk and colostrum should also be tempered by our knowledge of the biological interrelatedness of the major milk proteins and the bioactivities of trace proteins and elements in milk. These products should be produced through a means that is acceptable to consumers to allay fears regarding food safety and concerns for animal welfare. Approaches ranging from exploitation of existing variation within dairy populations, manipulation during processing or even transgenic approaches may be desirable depending on the value of the end product to human health and nutrition.

References

Arakawa, T., Chong, D. K., Slattery, C. W. and Langridge, W. H. (1999). Improvements in human health through production of human milk proteins in transgenic food plants. *Advances in Experimental Medicine and Biology*, **464**, 149–59.

Auldist, M. J., Thomson, N. A., Mackle, T. R., Hill, J. P. and Prosser, C. G. (2000). Effects of pasture allowance on the yield and composition of milk from cows of different β-lactoglobulin phenotypes. *Journal of Dairy Science*, **83**, 2069–74.

Bansal, P. S., Grieve, P. A., Marschke, R. J., Daly, N. L., McGhie, E., Craik, D. J. and Alewood, P. F. (2006). Chemical synthesis and structure elucidation of bovine κ-casein (1–44). *Biochemical and Biophysical Research Communications*, **340**, 1098–103.

Bech, A. M. and Kristiansen, K. R. (1990). Milk protein polymorphism in Danish dairy cattle and the influence of genetic variants on milk yield. *Journal of Dairy Research*, **57**, 53–62.

Beg, O. U., von Bahr-Lindstrom, H., Zaidi, Z. H. and Jornvall, H. (1986). A camel milk whey protein rich in half-cystine, Primary structure, assessment of variations, internal repeat patterns, and relationships with neurophysin and other active polypeptides. *European Journal of Biochemistry*, **159**, 195–201.

Bell, S. J., Grochoski, G. T. and Clarke, A. J. (2006). Health implications of milk containing β-casein with the A2 genetic variant. *Critical Reviews in Food Science and Nutrition*, **46**, 93–100.

Birn, H. (2006). The kidney in vitamin B12 and folate homeostasis: characterization of receptors for tubular uptake of vitamins and carrier proteins. *American Journal of Physiology. Renal Physiology*, **291**(1), F22–36.

Boston, W. S., Bleck, G. T., Conroy, J. C., Wheeler, M. B. and Miller, D. J. (2001). Short communication: effects of increased expression of α-lactalbumin in transgenic mice on milk yield and pup growth. *Journal of Dairy Science*, **84**, 620–22.

Bounous, G. and Gold, P. (1991). The biological activity of undenatured dietary whey proteins: role of glutathione. *Clinical and Investigative Medicine*, **14**, 296–309.

Bremel, R. D., Yom, H. C. and Bleck, G. T. (1989). Alteration of milk composition using molecular genetics. *Journal of Dairy Science*, **72**, 2826–33.

Brew, K., Castellino, F. J., Vanaman, T. C. and Hill, R. L. (1970). The complete amino acid sequence of bovine α-lactalbumin. *Journal of Biological Chemistry*, **245**, 4570–82.

Brignon, G., Chtourou, A. and Ribadeau-Dumas, B. (1985). Does β-lactoglobulin occur in human milk? *Journal of Dairy Research*, **52**, 249–54.

Brophy, B., Smolenski, G., Wheeler, T., Wells, D., L'Huillier, P. and Laible, G. (2003). Cloned transgenic cattle produce milk with higher levels of β-casein and κ-casein. *Nature Biotechnology*, **21**, 157–62.

Campbell, S. M., Rosen, J. M., Hennighausen, L. G., Strech-Jurk, U. and Sippel, A. E. (1984). Comparison of the whey acidic protein genes of the rat and mouse. *Nucleic Acids Research*, **12**, 8685–97.

Cane, K. N., Arnould, J. P. and Nicholas, K. R. (2005). Characterisation of proteins in the milk of fur seals. *Comparative Biochemistry and Physiology—Part B: Biochemistry and Molecular Biology*, **141**, 111–20.

Chin-Dusting, J., Shennan, J., Jones, E., Williams, C., Kingwell, B. and Dart, A. (2006). Effect of dietary supplementation with β-casein A1 or A2 on markers of disease development in individuals at high risk of cardiovascular disease. *British Journal of Nutrition*, **95**, 136–44.

Cho, Y., Gu, W., Watkins, S., Lee, S. P., Kim, T. R., Brady, J. W. and Batt, C. A. (1994). Thermostable variants of bovine β-lactoglobulin. *Protein Engineering*, **7**, 263–70.

Clark, A. J. (1996). Genetic modification of milk proteins. *American Journal of Clinical Nutrition*, **63**, 633S–638S.

Correa, S., Palmeira, P., Carneiro-Sampaio, M. M. S., Nishimura, L. S. and Guth, B. E. C. (2006). Human colostrum contains antibodies reactive to colonization factors I and II of enterotoxigenic *Escherichia coli*. *FEMS Immunology and Medical Microbiology*, **47**, 199–216.

Cross, K. J., Huq, N. L. and Reynolds, E. C. (2007). Casein phosphopeptides in oral health chemistry and clinical applications. *Current Pharmaceutical Design*, **13**, 793–98.

Dalma-Weiszhausz, D. D., Warrington, J., Tanimoto, E. Y. and Miyada, C. G. (2006). The Affymetrix GeneChip platform: an overview. *Methods in Enzymology*, **410**, 3–28.

Demmer, J., Stasiuk, S. J., Grigor, M. R., Simpson, K. J. and Nicholas, K. R. (2001). Differential expression of the whey acidic protein gene during lactation in the brushtail possum (*Trichosurus vulpecula*). *Biochimica et Biophysica Acta*, **1522**, 187–94.

Devinoy, E., Hubert, C., Jolivet, G., Thepot, D., Clergue, N., Desaleux, M., Dion, M., Servely, J. L. and Houdebine, L. M. (1988). Recent data on the structure of rabbit milk protein genes and on the mechanism of the hormonal control of their expression. *Reproduction Nutrition Development*, **28**, 1145–64.

Farrell, H. M. Jr., Jimenez-Flores, R., Bleck, G. T., Brown, E. M., Butler, J. E., Creamer, L. K., Hicks, C. L., Hollar, C. M., Ng-Kwai-Hang, K. F. and Swaisgood, H. E. (2004). Nomenclature of the proteins of cows' milk sixth revision. *Journal of Dairy Science*, **87**, 1641–74.

Ferranti, P., Traisci, M. V., Picariello, G., Nasi, A., Boschi, V., Siervo, M., Falconi, C., Chianese, L. and Addeo, F. (2004). Casein proteolysis in human milk: tracing the pattern of casein breakdown and the formation of potential bioactive peptides. *Journal of Dairy Research*, **71**, 74–87.

Folch, J. M., Coll, A. and Sanchez, A. (1994). Complete sequence of the caprine β-lactoglobulin gene. *Journal of Dairy Science*, **77**, 3493–97.

Fukushima, Y., Kawata, Y., Onda, T. and Kitagawa, M. (1997). Consumption of cow milk and egg by lactating women and the presence of β-lactoglobulin and ovalbumin in breast milk. *American Journal of Clinical Nutrition*, **65**, 30–35.

Girard, C. L. and Matte, J. J. (1998). Dietary supplements of folic acid during lactation: effects on the performance of dairy cows. *Journal of Dairy Science*, **81**(5), 1412–19.

Girard, C. and Matte, J. (2005). Folic acid and vitamin B12 requirements of dairy cows: A concept to be revised. *Livestock Production Science*, **98**, 123–33.

Girard, C. L., Matte, J. J. and Temblay, G. F. (1989). Serum folates in gestating and lactating dairy cows. *Journal of Dairy Science*, **72**(12), 3240–46.

Girard, C. L., Lapierre, H., Matte, J. J. and Lobley, G. E. (2005). Effects of dietary supplements of folic acid and rumen-protected methionine on lactational performance and folate metabolism of dairy cows. *Journal of Dairy Science*, **88**(2), 660–70.

Graulet, B., Matte, J. J., Desrochers, A., Doepel, L., Palin, M. F. and Girard, C. L. (2007). Effects of dietary supplements of folic acid and vitamin B12 on metabolism of dairy cows in early lactation. *Journal of Dairy Science*, **90**(7), July, 3442–55.

Grobler, J. A., Wang, M., Pike, A. C. and Brew, K. (1994). Study by mutagenesis of the roles of two aromatic clusters of α-lactalbumin in aspects of its action in the lactose synthase system. *Journal of Biological Chemistry*, **269**, 5106–14.

Gutiérrez-Adán, A., Maga, E. A., Meade, H., Shoemaker, C. F., Medrano, J. F., Anderson, G. B. and Murray, J. D. (1989). Alterations of the physical characteristics of milk from transgenic mice producing bovine κ-casein. *Journal of Dairy Science*, **79**, 791–99.

Gutiérrez-Adán, A., Maga, E. A., Behboodi, E., Conrad-Brink, J. S., Mackinlay, A. G., Anderson, G. B. and Murray, J. D. (1999). Expression of bovine β-lactoglobulin in the milk of transgenic mice. *Journal of Dairy Research*, **66**, 289–94.

Hajjoubi, S., Rival-Gervier, S., Hayes, H., Floriot, S., Eggen, A., Piumi, F., Chardon, P., Houdebine, L. M. and Thepot, D. (2006). Ruminants genome no longer contains whey acidic protein gene but only a pseudogene. *Gene*, **370**, 104–12.

Hall, A. J., Masel, A., Bell, K., Halliday, J. A., Shaw, D. C. and VandeBerg, J. L. (2001). Characterization of baboon (*Papio hamadryas*) milk proteins. *Biochemical Genetics*, **39**, 59–71.

Hawken, R. J., Barris, W. C., McWilliam, S. M. and Dalrymple, B. P. (2004). An interactive bovine in silico SNP database (IBISS). *Mammalian Genome*, **15**, 819–27.

Hennighausen, L. G. and Sippel, A. E. (1982). Mouse whey acidic protein is a novel member of the family of 'four-disulfide core' proteins. *Nucleic Acids Research*, **10**, 2677–84.

Hernandez-Ledesma, B., Recio, I. and Amigo, L. (2007). β-Lactoglobulin as source of bioactive peptides. *Amino Acids* (online ahead of print).

Hiraoka, Y., Segawa, T., Kuwajima, K., Sugai, S. and Murai, N. (1980). α-Lactalbumin: a calcium metalloprotein. *Biochemical and Biophysical Research Communications*, **95**, 1098–104.

Hitchin, E., Stevenson, E. M., Clark, A. J., McClenaghan, M. and Leaver, J. (1996). Bovine beta-casein expressed in transgenic mouse milk is phosphorylated and incorporated into micelles. *Protein Expression and Purification*, **7**, 247–52.

Hyttinen, J. M., Korhonen, V. P., Hiltunen, M. O., Myöhänen, S. and Jänne, J. (1998). High-level expression of bovine β-lactoglobulin gene in transgenic mice. *Journal of Biotechnology*, **61**, 191–98.

Ikonen, T., Bovenhuis, H., Ojala, M., Ruottinen, O. and Georges, M. (2001). Associations between casein haplotypes and first lactation milk production traits in Finnish Ayrshire cows. *Journal of Dairy Science*, **84**, 507–14.

Jakob, E. and Puhan, Z. (1992). Technological properties of milk as influenced by genetic polymorphism of milk proteins—a review. *International Dairy Journal*, **2**, 157–78.

Jayat, D., Gaudin, J. C., Chobert, J. M., Burova, T. V., Holt, C., McNae, I., Sawyer, L. and Haertlé, T. (2004). A recombinant C121S mutant of bovine β-lactoglobulin is more susceptible to peptic digestion and to denaturation by reducing agents and heating. *Biochemistry*, **43**, 6312–21.

Kaminski, S., Cieslinska, A. and Kostyra, E. (2007). Polymorphism of bovine beta-casein and its potential effect on human health. *Journal of Applied Genetics*, **48**, 189–98.

Katakura, Y., Ametani, A., Totsuka, M., Nagafuchi, S. and Kaminogawa, S. (1999). Accelerated secretion of mutant β-lactoglobulin in *Saccharomyces cerevisiae* resulting from a single amino acid substitution. *Biochimica et Biophysica Acta*, **1432**, 302–12.

Khatkar, M. S., Thomson, P. C., Tammen, I. and Raadsma, H. W. (2004). Quantitative trait loci mapping in dairy cattle: review and meta-analysis. *Genetics Selection Evolution*, **36**, 163–90.

Khatkar, M. S., Zenger, K. R., Hobbs, M., Hawken, R. J., Cavanagh, J. A., Barris, W., McClintock, A. E., McClintock, S., Thomson, P. C., Tier, B., Nicholas, F. W. and Raadsma, H. W. (2007). A primary assembly of a bovine haplotype block map based on a 15,036-single-nucleotide polymorphism panel genotyped in Holstein-Friesian cattle. *Genetics*, **176**, 763–72.

Koldovsky, O. (1996). The potential physiological significance of milk-borne hormonally active substances for the neonate. *Journal of Mammary Gland Biology and Neoplasia*, **1**, 317–23.

Kontopidis, G., Holt, C. and Sawyer, L. (2004). Invited review: β-lactoglobulin: binding properties, structure, and function. *Journal of Dairy Science*, **87**, 785–96.

Konuma, T., Sakurai, K. and Goto, Y. (2007). Promiscuous binding of ligands by beta-lactoglobulin involves hydrophobic interactions and plasticity. *Journal of Molecular Biology*, **368**, 209–18.

Korhonen, H. and Pihlanto, A. (2007). Technological options for the production of health-promoting proteins and peptides derived from milk and colostrum. *Current Pharmaceutical Design*, **13**, 829–43.

Kumar, S., Clarke, A. R., Hooper, M. L., Horne, D. S., Law, A. J., Leaver, J., Springbett, A., Stevenson, E. and Simons, J. P. (1994). Milk composition and lactation of β-casein-deficient mice. *Proceedings of the National Academy of Sciences of the USA*, **91**, 6138–42.

Laugesen, M. and Elliott, R. (2003). Ischaemic heart disease, Type 1 diabetes, and cow milk A1 β-casein. *New Zealand Medical Journal*, **116**, U295.

Lefèvre, C. M., Digby, M. R., Mailer, S., Whitley, J. C., Strahm, S. J. Y. and Nicholas, K. R. (2007). Lactation transcriptomics in the Australian marsupial, *Macropus eugenii*: transcript sequencing and quantification. *BMC Genomics*, **8**, 417–30.

Li-Chan, E. and Nakai, S. (1989). Enzymic dephosphorylation of bovine casein to improve acid clotting properties and digestibility for infant formula. *Journal of Dairy Research*, **56**, 381–90.

Lin, C. Y., McAllister, A. J., Ng-Kwai-Hang, K. F. and Hayes, J. F. (1986). Effects of milk protein loci on first lactation production in dairy cattle. *Journal of Dairy Science*, **69**, 704–12.

Lin, C. Y., McAllister, A. J., Ng-Kwai-Hang, K. F., Hayes, J. F., Batra, T. R., Lee, A. J., Roy, G. L., Vesely, J. A., Wauthy, J. M. and Winter, K. A. (1989). Relationships of milk protein types to lifetime performance. *Journal of Dairy Science*, **72**, 3085–90.

Lonnerdal, B. (2006). Recombinant human milk proteins. *Nestle Nutrition Workshop Series. Paediatric Programme,*, **58**, 207–15.

Lundén, A., Nilsson, M. and Janson, L. (1997). Marked effect of β-lactoglobulin polymorphism on the ratio of casein to total protein in milk. *Journal of Dairy Science*, **80**, 2996–3005.

Mackle, T. R., Bryant, A. M., Petch, S. F., Hill, J. P. and Auldist, M. J. (1999). Nutritional influences on the composition of milk from cows of different protein phenotypes in New Zealand. *Journal of Dairy Science*, **82**, 172–80.

Manzini, S., Vargiolu, A., Stehle, I. M., Bacci, M. L., Cerrito, M. G., Giovannoni, R., Zannoni, A., Bianco, M. R., Forni, M., Donini, P., Papa, M., Lipps, H. J. and Lavitrano, M. (2006). Genetically modified pigs produced with a nonviral episomal vector. *Proceedings of the National Academy of Sciences of the USA*, **103**, 17672–77.

Mariani, B. P., Summer, A., Di Gregorio, P. D., Randoe, A., Fossa, E. and Pecorari, M. (2001). Effects of the CSN1A(G) allele on the clotting time of cow milk and on the rheological properties of rennet-curd. *Journal of Dairy Research*, **68**, 63–70.

Martin, P., Szymanowska, M., Zwierzchowski, L. and Leroux, C. (2002). The impact of genetic polymorphisms on the protein composition of ruminant milks. *Reproduction Nutrition Development*, **42**, 433–59.

McKnight, R. A., Jimenez-Flores, R., Kang, Y., Creamer, L. K. and Richardson, T. (1989). Cloning and sequencing of a complementary deoxyribonucleic acid coding for a bovine α_{s1}-casein A from mammary tissue of a homozygous B variant cow. *Journal of Dairy Science*, **72**, 2464–73.

McLean, D. M., Graham, E. R., Ponzoni, R. W. and McKenzie, H. A. (1984). Effects of milk protein genetic variants on milk yield and composition. *Journal of Dairy Research*, **51**, 531–46.

Meisel, H. (2005). Biochemical properties of peptides encrypted in bovine milk proteins. *Current Medicinal Chemistry*, **12**, 1905–19.

Mercier, J. C. and Vilotte, J. L. (1993). Structure and function of milk protein genes. *Journal of Dairy Science*, **76**, 3079–98.

Murray, B. A. and Fitzgerald, R. J. (2007). Angiotensin converting enzyme inhibitory peptides derived from food proteins: biochemistry, bioactivity and production. *Current Pharmaceutical Design*, **13**, 773–91.

Ng-Kwai-Hang, K. F., Hayes, J. F., Moxley, J. E. and Monardes, H. G. (1987). Variation in milk protein concentrations associated with genetic polymorphism and environmental factors. *Journal of Dairy Science*, **70**, 563–70.

Nicholas, K., Simpson, K., Wilson, M., Trott, J. and Shaw, D. (1997). The tammar wallaby: a model to study putative autocrine-induced changes in milk composition. *Journal of Mammary Gland Biology and Neoplasia*, **2**, 299–310.

Niemann, H. and Kues, W. A. (2007). Transgenic farm animals: an update. *Reproduction, Fertility and Development*, **19**, 762–70.

Noble, M. S., Rodriguez-Zas, S., Cook, J. B., Bleck, G. T., Hurley, W. L. and Wheeler, M. B. (2002). Lactational performance of first-parity transgenic gilts expressing bovine α-lactalbumin in their milk. *Journal of Animal Science*, **80**, 1090–96.

Ojala, M., Famula, T. R. and Medrano, J. F. (1997). Effects of milk protein genotypes on the variation for milk production traits of Holstein and Jersey cows in California. *Journal of Dairy Science*, **80**, 1776–85.

Ondetti, M. A. and Cushman, D. W. (1982). Enzymes of the renin-angiotensin system and their inhibitors. *Annual Review of Biochemistry*, **51**, 283–308.

Pagnacco, G. and Caroli, A. (1987). Effects of casein and β-lactoglobulin genotypes on renneting properties of milk. *Journal of Dairy Research*, **54**, 479–85.

Parodi, P. W. (2007). A role for milk proteins and their peptides in cancer prevention. *Current Pharmaceutical Design*, **13**, 813–28.

Prinzenberg, E. M., Weimann, C., Brandt, H., Bennewitz, J., Kalm, E., Schwerin, M. and Erhardt, G. (2003). Polymorphism of the bovine CSN1S1 promoter: linkage mapping, intragenic, haplotypes, and effects on milk production traits. *Journal of Dairy Science*, **86**, 2696–705.

Rijnkels, M. (2002). Multispecies comparison of the casein gene loci and evolution of casein gene family. *Journal of Mammary Gland Biology and Neoplasia*, **7**, 327–45.

Sawyer, L. and Holt, C. (1993). The secondary structure of milk proteins and their biological function. *Journal of Dairy Science*, **76**, 3062–78.

Shah, N. P. (2000). Effects of milk-derived bioactives: an overview. *British Journal of Nutrition*, **84**(Suppl 1), S3–S10.

Sharp, J. A., Lefevre, C. and Nicholas, K. R. (2007). Molecular evolution of monotreme and marsupial whey acidic protein genes. *Evolution and Development*, **9**, 378–92.

Shekar, P. C., Goel, S., Rani, S. D., Sarathi, D. P., Alex, J. L., Singh, S. and Kumar, S. (2006). κ-casein-deficient mice fail to lactate. *Proceedings of the National Academy of Sciences of the USA*, **103**, 8000–5.

Sherman, M. P. and Petrak, K. (2005). Lactoferrin-enhanced anoikis: a defense against neonatal necrotizing enterocolitis. *Med Hypotheses.*, **65**(3), 478–82.

Simons, J. P., McClenaghan, M. and Clark, A. J. (1987). Alteration of the quality of milk by expression of sheep β-lactoglobulin in transgenic mice. *Nature*, **328**, 530–32.

Simpson, K. J., Bird, P., Shaw, D. and Nicholas, K. (1998). Molecular characterisation and hormone-dependent expression of the porcine whey acidic protein gene. *Journal of Molecular Endocrinology*, **20**, 27–35.

Simpson, K. J., Ranganathan, S., Fisher, J. A., Janssens, P. A., Shaw, D. C. and Nicholas, K. R. (2000). The gene for a novel member of the whey acidic protein family encodes three four-disulfide core domains and is asynchronously expressed during lactation. *Journal of Biological Chemistry*, **275**, 23074–81.

Smolenski, G., Haines, S., Kwan, F. Y., Bond, J., Farr, V., Davis, S. R., Stelwagen, K. and Wheeler, T. T. (2007). Characterisation of host defence proteins in milk using a proteomic approach. *Journal of Proteome Research*, **6**, 207–15.

Stinnakre, M. G., Vilotte, J. L., Soulier, S. and Mercier, J. C. (1994). Creation and phenotypic analysis of α-lactalbumin-deficient mice. *Proceedings of the National Academy of Sciences of the USA*, **91**, 6544–48.

Suchyta, S. P., Sipkovsky, S., Halgren, R. G., Kruska, R., Elftman, M., Weber-Nielsen, M., Vandehaar, M. J., Xiao, L., Tempelman, R. J. and Coussens, P. M. (2003). Bovine mammary gene expression profiling using a cDNA microarray enhanced for mammary-specific transcripts. *Physiological Genomics*, **16**, 8–18.

Tailford, K. A., Berry, C. L., Thomas, A. C. and Campbell, J. H. (2003). A casein variant in cow's milk is atherogenic. *Atherosclerosis*, **170**, 13–19.

Tellam, R. (2007). Capturing benefits from the bovine genome sequence. *Australian Journal of Experimental Agriculture*, **47**, 1039–50.

Threadgill, D. W. and Womack, J. E. (1990). Genomic analysis of the major bovine milk protein genes. *Nucleic Acids Research*, **18**, 6935–42.

Truswell, A. S. (2005). The A2 milk case: a critical review. *European Journal of Clinical Nutrition*, **59**, 623–31.

Tsiaras, A. M., Bargouli, G. G., Banos, G. and Boscos, C. M. (2005). Effect of κ-casein and β-lactoglobulin loci on milk production traits and reproductive performance of Holstein cows. *Journal of Dairy Science*, **88**, 327–34.

Uchiyama, H., Perez-Prat, E. M., Watanabe, K., Kumagai, I. and Kuwajima, K. (1995). Effects of amino acid substitutions in the hydrophobic core of α-lactalbumin on the stability of the molten globule state. *Protein Engineering*, **8**, 1153–61.

Venn, B. J., Skeaff, C. M., Brown, R., Mann, J. I. and Green, T. J. (2006). A comparison of the effects of A1 and A2 β-casein protein variants on blood cholesterol concentrations in New Zealand adults. *Atherosclerosis*, **188**, 175–78.

Xue, G. P. and Snoswell, A. M. (1985). Regulation of methyl group metabolism in lactating ewes. *Biochemistry International*, **11**(3), 381–85.

Yom, H. C. and Bremel, R. D. (1993). Genetic engineering of milk composition: modification of milk components in lactating transgenic animals. *American Journal of Clinical Nutrition*, **58**, 299S–306S.

Zenger, K. R., Khatkar, M. S., Cavanagh, J. A., Hawken, R. J. and Raadsma, H. W. (2007). Genome-wide genetic diversity of Holstein Friesian cattle reveals new insights into Australian and global population variability, including impact of selection. *Animal Genetics*, **38**, 7–14.

Zuelke, K. A. (1998). Transgenic modification of cows milk for value-added processing. *Reproduction, Fertility and Development*, **10**, 671–76.

Post-translational modifications of caseins

4

John W. Holland

Abstract

The caseins exhibit a high degree of heterogeneity as a result of post-translational modifications (PTMs). Phosphorylation of the α- and β-caseins and glycosylation of κ-casein are the best-known modifications and are critical for the formation and stability of casein micelles. κ-Casein, in particular, has long been known to exhibit a high degree of variability in glycosylation. It is somewhat surprising to see so much variability in such important structural features. In recent years, the adoption of proteomic techniques has greatly enhanced our ability to unravel heterogeneity in proteins arising from complex and variable patterns of PTMs. In this chapter, a summary of our knowledge of the PTMs of caseins is attempted, with a particular emphasis on κ-casein and the implications that variations in PTMs have for dairy processors.

Introduction

The caseins are phosphoproteins and constitute about 80% of the protein in milk (Swaisgood, 2003; Farrell *et al.*, 2004). They are assembled in a colloidal complex with calcium phosphate and small amounts of other minerals. Although obviously important for the provision of amino acids, calcium and phosphorus for infant nutrition, the casein micelle structure is also critical in determining the physical properties of milk. The stability of the micelle, or its controlled destabilization in the case of cheese and yoghurt manufacture, is of primary concern to the dairy industry.

A number of reviews of micelle structure have been published in recent years (Rollema, 1992; Holt and Horne, 1996; Horne, 1998; Walstra, 1999; Horne, 2002;

Milk Proteins: From Expression to Food
ISBN: 978-0-12-374039-7

Chapter 5 in this volume). In simple terms, the micelle is a network of protein molecules held together by a combination of hydrophobic interactions between protein molecules and electrostatic interactions between phosphoserine-rich regions of the α_{s1}-, α_{s2}- and β-caseins and micellar calcium phosphate. Whereas the internal structure is still the subject of debate (Horne, 1998; Walstra, 1999), there is general acceptance of the "hairy micelle" concept, in which the hydrophilic C-terminal portion of κ-casein extends from the surface, providing steric and electrostatic repulsion, which prevents micelle aggregation.

A critical factor in micelle formation and stability is the presence of post-translational modifications (PTMs) such as phosphorylation on the α_{s1}-, α_{s2}- and β-caseins and glycosylation on κ-casein. PTMs on secreted proteins such as the caseins occur in the endoplasmic reticulum and/or golgi complex after synthesis of the polypeptide chain. As such, they are not encoded by the genes *per se* but may be dependent on protein (and hence gene) sequence motifs that are necessary, but of themselves not sufficient, for modification to take place. There are a number of other factors that determine whether or not a PTM occurs, including expression of the genes encoding the enzymes necessary for the modification, the availability of their substrates and the accessibility of the modification site on the protein, especially after folding. Therefore, although it may be possible to predict the theoretical sites of modification on proteins, determination of the actual sites and the degree to which they are modified requires considerable experimental characterization.

Advances in our understanding of complex systems such as the caseins micelle are frequently preceded by advances in technology. In recent years, the development of proteomic technologies has greatly enhanced our ability to analyze milk proteins, particularly with respect to PTMs. Two-dimensional electrophoresis (2-DE), in particular, provides a high-resolution methodology for displaying the heterogeneity of the major milk proteins. As can be seen in Figure 4.1, genetic variants, phospho-variants and glyco-variants of the caseins can be resolved on a single 2-D gel. Advances in mass spectrometry (MS) have enhanced our ability to analyze the proteins arrayed on 2-D gels. Therefore, not only is it possible to resolve many proteins and their isoforms but also it is possible to characterize them, particularly with respect to the many PTMs that affect their electrophoretic mobility.

As will be seen below, the four caseins are present in many diverse forms as a result of differential PTMs. The biological reasons behind the diversity are not clear. However, what is clear is that the PTMs of the caseins are critical for their function in micelle formation and stability. The first part of this chapter summarizes what is known about the PTMs of the α_{s1}-, α_{s2}- and β-caseins. The second part focuses on κ-casein, which has been the subject of recent proteomic studies, and includes an extended discussion on the functional significance of κ-casein heterogeneity.

Bovine casein

The α_{s1}-, α_{s2}- and β-caseins are phosphoproteins that are generally well characterized in terms of their PTMs and have been the subject of a number of reviews (Mercier, 1981;

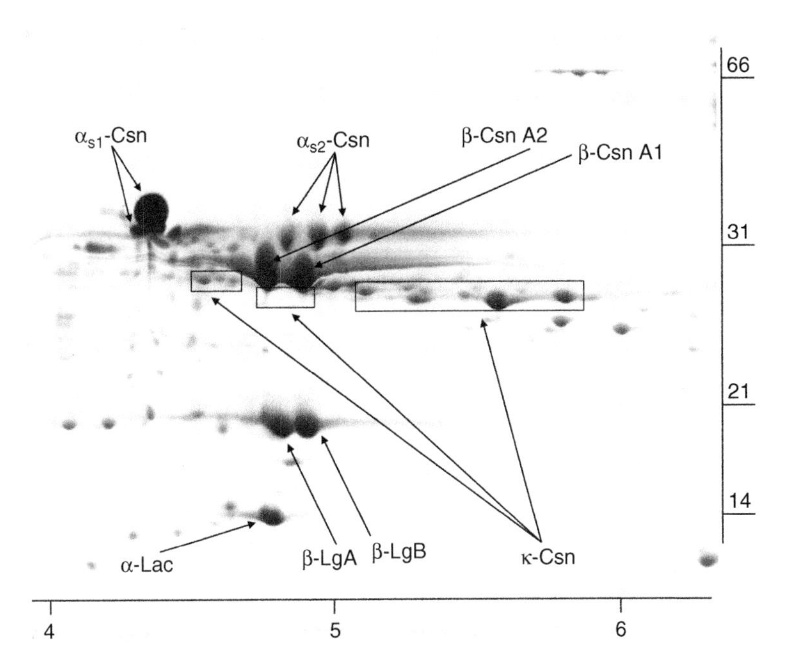

Figure 4.1 2-D gel of bovine milk. Section of a 2-D gel highlighting the major protein forms found in bovine milk. Adapted from Holland *et al.* (2004).

Ginger and Grigor, 1999; Farrell *et al.*, 2004). In fact, the α- and β-caseins have been used as model phosphoproteins in the development of proteomic techniques to examine global phosphorylation patterns (e.g. Cox *et al.*, 2005; Kapkova *et al.*, 2006; Sweet *et al.*, 2006; Zhou *et al.*, 2006; Wu *et al.*, 2007). The multitude of techniques developed for the analysis of phosphorylation is beyond the scope of this chapter but has been reviewed recently (Bodenmiller *et al.*, 2007; Collins *et al.*, 2007). The PTMs of the α_{s1}-, α_{s2}- and β-caseins are summarized in the sections below.

An important note concerns the numbering of the amino acids in the casein sequences described in subsequent sections. The numbering of residues is based on the Swiss-Prot database entry for the relevant protein (see Figures 4.2 and 4.3) and includes the signal peptide which is normally removed during processing to generate the mature protein. This differs from the numbering in most of the dairy literature, in which the N-terminal amino acid of the mature protein is numbered 1.

α_{s1}-Casein

The predominant form of α_{s1}-casein in bovine milk contains eight phosphate groups. The phosphates are attached to hydroxyamino acids occurring in the sequence motif Ser/Thr-Xxx-Glu/Asp/pSer (Mercier, 1981). However, the vast majority of casein phosphorylation sites, including the eight sites in the major form of α_{s1}-casein, occur in the more restricted Ser-Xxx-Glu/pSer motif. Phosphorylation of threonine or of serine in the Ser-Xxx-Asp motif is relatively uncommon. A minor form of α_{s1}-casein with nine phosphates, originally called α_{s0}-casein, also occurs in bovine milk. It contains one extra phosphate on Ser^{56}, which occurs in a Ser-Xxx-Asp motif (Manson *et al.*, 1977).

P02662|CASA1_BOVIN Alpha-s1-casein–Bos taurus (Bovine).

```
1   MKLLILTCLV AVALARPKHP IKHQGLPQEV LNENLLRFFV APFPEVFGKE  50
51  KVNELSKDIG SESTEDQAME DIKQMEAESI SSSEEIVPNS VEQKHIQKED 100
101 VPSERYLGYL EQLLRLKKYK VPQLEIVPNS AEERLHSMKE GIHAQQKEPM 150
151 IGVNQELAYF YPELFRQFYQ LDAYPSGAWY YVPLGTQYTD APSFSDIPNP 200
201 IGSENSEKTT MPLW                                       214
```

P02663|CASA2_BOVIN Alpha-s2-casein–Bos taurus (Bovine).

```
1   MKFFIFTCLL AVALAKNTME HVSSSEESII SQETYKQEKN MAINPSKENL  50
51  CSTFCKEVVR NANEEEYSIG SSSEESAEVA TEEVKITVDD KHYQKALNEI 100
101 NQFYQKFPQY LQYLYQGPIV LNPWDQVKRN AVPITPTLNR EQLSTSEENS 150
151 KKTVDMESTE VFTKKTKLTE EEKNRLNFLK KISQRYQKFA LPQYLKTVYQ 200
201 HQKAMKPWIQ PKTKVIPYVR YL                              222
```

P02666|CASB_BOVIN Beta-casein–Bos taurus (Bovine).

```
1   MKVLILACLV ALALARELEE LNVPGEIVES LSSSEESITR INKKIEKFQS  50
51  EEQQQTEDEL QDKIHPFAQT QSLVYPFPGP IPNSLPQNIP PLTQTPVVVP 100
101 PFLQPEVMGV SKVKEAMAPK HKEMPFPKYP VEPFTESQSL TLTDVENLHL 150
151 PLPLLQSWMH QPHQPLPPTV MFPPQSVLSL SQSKVLPVPQ KAVPYPQRDM 200
201 PIQAFLLYQE PVLGPVRGPF PIIV                            224
```

Figure 4.2 Amino acid sequences of the bovine α- and β-caseins. Swiss-Prot accession numbers, entry names, protein names, species and sequences for the α- and β-caseins. The database sequences are for the B variant of α_{s1}-casein, the A variant of α_{s2}-casein and the A2 variant of β-casein. Potential phosphorylation sites are shown in bold type and those that have been experimentally confirmed are underlined. The signal peptides are shown in italics.

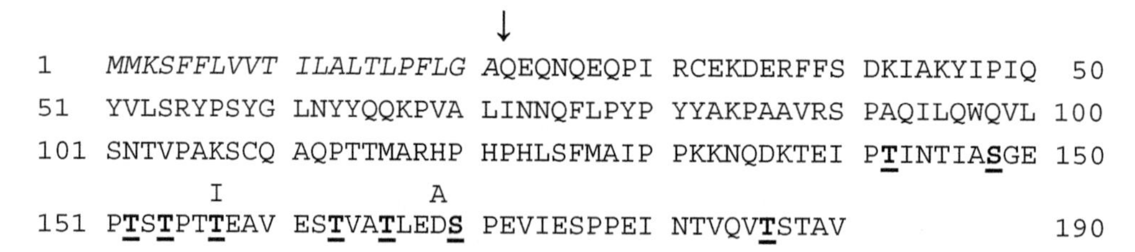

Figure 4.3 Amino acid sequence of κ-casein. Amino acid sequence of κ-casein A (Swiss-Prot accession number P02668). The N-terminal signal sequence (1–21) is shown in italics. The arrow indicates the amino-terminus of the mature protein. Recognized sites of potential phosphorylation and glycosylation are indicated in bold, underlined text. Amino acid substitutions distinguishing the B variant are shown above the main sequence.

The reference protein for α_{s1}-casein is α_{s1}-CN B-8P, where B-8P signifies the B genetic variant with eight phosphates (Farrell *et al.*, 2004).

A number of genetic variants of α_{s1}-casein have been described (Farrell *et al.*, 2004). Only the less common forms, D and F, are likely to have altered phosphorylation profiles. Variant D has Ala^{68} substituted with Thr, which generates a phosphorylation site of the form Thr-Xxx-Glu. Phosphorylation of this residue was detected when the variant was first identified (Grosclaude *et al.*, 1972a). Variant F has Ser^{81} substituted with Leu, which disrupts the serine cluster—Ser^{79}-Ile-Ser-Ser-Ser-Glu-Glu^{85}—and eliminates the secondary phosphorylation sites at Ser^{79} and Ser^{81}. The amino acid sequence and the modifications of α_{s1}-casein are shown in Figure 4.2.

α_{s2}-Casein

The α_{s2}-casein component of bovine milk is more varied than the α_{s1}-casein component. It generally presents as a mixture of four phosphoforms with 10–13 phosphates. The reference protein for α_{s2}-casein is α_{s2}-CN A-11P (Farrell *et al.*, 2004). The A variant has 12 serine residues in Ser-Xxx-Glu/pSer motifs and four threonine residues in Thr-Xxx-Glu motifs (Figure 4.2). Consequently, up to 16 phosphates are theoretically possible. Presumably, the 12 serine residues are the first to be phosphorylated. However, it is not known whether specific residues remain unphosphorylated in the different forms but, given the consistent appearance of the forms in milk, it is likely to be the case. Unfortunately, it is not possible to draw any conclusions with regard to phosphorylation site occupation until the individual phosphoforms are analyzed. This should be possible using gel-based proteomic techniques.

Only four genetic variants of α_{s2}-casein have been described (Farrell *et al.*, 2004) and, in each case, an altered phosphorylation profile would be expected. In variant B, Ser^{23} is changed to Phe as a result of a single nucleotide substitution (Ibeagha-Awemu *et al.*, 2007). This causes loss of a phosphorylation site in the first phosphoserine cluster. Variant C has three amino acid changes: Glu^{48} is changed to Gly, with loss of the phosphorylation site at Ser^{46}; Ala^{62} is changed to Thr, creating a potential site with the motif Thr^{62}-Xxx-Glu^{64}; Thr^{145} is changed to Ile, with loss of the potential site at Thr^{145}.

Variant D has a deletion of nine amino acids as a result of skipping exon VIII (Bouniol *et al.*, 1993). This results in loss of the first three serines from the second phosphoserine cluster. A second PTM on α_{s2}-casein is the formation of an intramolecular disulfide bond between the two cysteine residues in the protein (Rasmussen *et al.*, 1994). The functional role of disulfide bonding is not clear at this stage, but it may contribute to micelle stability and is discussed further in the section on κ-casein.

β-Casein

Bovine β-casein is usually present as a single form with five phosphates, indicating that all five Ser-Xxx-Glu/pSer sites in the sequence are constitutively phosphorylated. The reference protein is β-CN A^2-5P (Farrell *et al.*, 2004). Some 12 genetic variants of β-casein have been characterized, but only two variants appear to have altered phosphorylation profiles. Variant C has a Glu to Lys substitution at residue 52, which removes the phosphorylation site at Ser^{50}. Variant D has a Ser to Lys substitution at residue 33, which removes the primary phosphorylation site at Ser^{33}. Variation in β-casein arises primarily as a result of proteolysis rather than PTMs. The sequence and the phosphorylation sites of β-casein are summarized in Figure 4.2.

Although much of the focus has been on bovine milk, other species have not been entirely neglected (Ginger and Grigor, 1999). It is apparent that considerable variations in caseins and their PTMs occur between different species. For example, the β-casein of human milk exists as six different forms with 0 to 5 phosphates (Greenberg *et al.*, 1984). Equine β-casein also shows variation in phosphorylation, with typically 3 to 7 phosphates on full-length β-casein (Girardet *et al.*, 2006) and 1 to 7 phosphates on a low-molecular-weight form that arises from an internal deletion (Miclo *et al.*, 2007). Ovine β-casein has also been reported to be variably phosphorylated, with 0 to 7 phosphates (Ferranti *et al.*, 2001), although it is not clear where the seventh phosphorylation site is. Caprine β-casein appears to be more like bovine β-casein with the same five phosphorylation sites and an additional site on Thr^{27} (Neveu *et al.*, 2002).

κ-Casein

κ-Casein does not contain any phosphoserine clusters and probably plays little part in calcium binding. Its major feature is a variable degree of glycosylation. The keen interest in κ-casein arises largely from its key role as a stabilizer of the micelle structure. In mice, the absence of κ-casein causes a failure of lactation, as the lumina of the mammary gland become clogged with aggregated caseins (Shekar *et al.*, 2006). The PTMs of κ-casein have been the subject of more recent research and are covered here in much greater detail than those of the other caseins.

The full amino acid sequence of bovine κ-casein was first reported in 1973 (Mercier *et al.*, 1973). The mature protein consists of a single chain of 169 amino acids (Figure 4.3) and has a theoretical molecular weight of 18 974 Da and a theoretical pI of 5.93 (A variant). The amino terminal glutamine is cyclized to form a pyrrolidone glutamic acid residue. The bovine κ-casein gene sequence was published in 1988 (Alexander *et al.*, 1988). It consists of five exons spread over about 14 kilobases, with most of the protein-coding region located in exon 4. A cleavable amino terminal signal sequence of 21 amino acids directs secretion of the mature protein.

A number of polymorphisms in the κ-casein gene have been identified, resulting in one or more amino acid substitutions in the mature protein. The most common variants are the A and B variants, which differ by two amino acids (Asp^{169}/Thr^{157} in variant A and Ala^{169}/Ile^{157} in variant B). The genetic variants of κ-casein are summarized in Table 4.1. A number of polymorphisms in the non-coding region have also been identified (Schild *et al.*, 1994; Keating *et al.*, 2007). Although these do not affect the amino acid sequence, they have the potential to affect expression levels. The full amino acid sequences of κ-casein from 18 species are currently in the Swiss-Prot database, with another 24 entries covering sub-species and incomplete sequences (Table 4.2). The reference protein for κ-casein is κ-CN A-1P, Uniprot P02668 (Farrell *et al.*, 2004).

Phosphorylation

κ-Casein appears to be constitutively phosphorylated at Ser^{170} and only partially phosphorylated at Ser^{148} (Talbot and Waugh, 1970; Mercier *et al.*, 1973). A minor tri-phosphorylated form has also been detected (Vreeman *et al.*, 1986; Molle and Leonil, 1995). Other studies have managed to detect only mono-phosphorylated

Table 4.1 Genetic variants of bovine κ-casein

Variant	Amino acid changes (relative to A variant)	References
A		(Grosclaude *et al.*, 1972)
B	Thr^{157} to Ile Asp^{169} to Ala	(Mercier *et al.*, 1973)
B2	Thr^{157} to Ile Asp^{169} to Ala Ile^{174} to Thr	(Gorodetskii *et al.*, 1983)
C	Arg^{118} to His Thr^{157} to Ile Asp^{169} to Ala	(Miranda *et al.*, 1993)
E	Ser^{176} to Gly	(Schlieben *et al.*, 1991)
F1	Asp^{169} to Val	(Sulimova *et al.*, 1992)
F2	Arg^{31} to His	(Prinzenberg *et al.*, 1996)
G1	Arg^{118} to Cys	(Prinzenberg *et al.*, 1996)
G2	Asp^{169} to Ala	(Sulimova *et al.*, 1996)
H	Thr^{156} to Ile	(Prinzenberg *et al.*, 1999)
I	Ser^{125} to Ala	(Prinzenberg *et al.*, 1999)
J	Thr^{157} to Ile Asp^{169} to Ala Ser^{176} to Arg	(Mahe *et al.*, 1999)
A(1)	Silent (Pro150, CCA to CCG)	(Prinzenberg *et al.*, 1999)

Notes: Other variants have been described but not confirmed or have proven to be identical to those above.

forms (Rasmussen *et al.*, 1997; Riggs *et al.*, 2001) although, in one of these (Riggs *et al.*, 2001), the phosphorylation site appears to have been identified incorrectly.

Phosphorylation has also been examined by MS of intact κ-casein extracted from 2-D gels. Both mono- and di-phosphorylated forms were observed and phosphorylation at Ser^{170} was confirmed by MS/MS (Claverol *et al.*, 2003). However, the electrophoretic mobility of some phospho-forms was not consistent with the MS analysis and probably reflected artifactual modification (e.g. deamidation) during purification. Using 2-DE with isoelectric focusing as the first dimension, phosphorylation variants in whole milk can be easily resolved because of the pI shifts caused by the acidic phosphate groups (Holland *et al.*, 2004).

The two main phosphorylation sites have been confirmed by tandem MS sequencing of peptic peptides released from protein forms separated by 2-DE. Tri-phosphorylated forms of both the A and B variants have been observed and the third phosphorylation site has been identified recently as Thr^{166} (Holland *et al.*, 2006). This site is consistent with the general observation of casein phosphorylation on the Ser/Thr-Xxx-Glu/pSer motif, with only relative low levels of phosphorylation on threonine residues.

Glycosylation

Whereas about 40% of κ-casein has been estimated to be non-glycosylated, the remaining 60% has up to six glycans attached (Vreeman *et al.*, 1986). The presence

Table 4.2 κ-casein entries in the Swiss Protein Database (Release 54.1, 21/08/2007)

Name (accession number)	Full length or fragment	Species (common name)
CASK_BALPH (Q27952)	(Fragment)	*Balaenoptera physalus* (Finback whale) (Common rorqual)
CASK_BISBO (P42155)	(Fragment)	*Bison bonasus* (European bison)
CASK_BOVIN (P02668)	precursor	*Bos taurus* (Bovine)
CASK_BUBBU (P11840)	precursor	*Bubalus bubalis* (Domestic water buffalo)
CASK_CAMDR (P79139)	precursor	*Camelus dromedarius* (Dromedary) (Arabian camel)
CASK_CAPCA (Q95146)	(Fragment)	*Capreolus capreolus* (Roe deer)
CASK_CAPCR (P42156)	precursor	*Capricornis crispus* (Japanese serow)
CASK_CAPHI (P02670)	precursor	*Capra hircus* (Goat)
CASK_CAPSU (P50420)	precursor	*Capricornis sumatrensis*
CASK_CAPSW (P50421)	precursor	*Capricornis swinhoei*
CASK_CAVPO (P19442)	precursor	*Cavia porcellus* (Guinea pig)
CASK_CERDU (Q95147)	(Fragment)	*Cervus duvaucelii* (Swamp deer)
CASK_CEREL (Q95149)	(Fragment)	*Cervus elaphus* (Red deer)
CASK_CERNI (P42157)	precursor	*Cervus nippon* (Sika deer)
CASK_CERUN (Q95177)	(Fragment)	*Cervus unicolor* (Sambar)
CASK_ELADA (Q95184)	(Fragment)	*Elaphurus davidianus* (Pere David's deer)
CASK_EQUGR (Q28400)	(Fragment)	*Equus grevyi* (Grevy's zebra)
CASK_GIRCA (Q28417)	(Fragment)	*Giraffa camelopardalis* (Giraffe)
CASK_HIPAM (Q28441)	(Fragment)	*Hippopotamus amphibius* (Hippopotamus)
CASK_HORSE (P82187)	precursor	*Equus caballus* (Horse)
CASK_HUMAN (P07498)	precursor	*Homo sapiens* (Human)
CASK_LAMGU (Q28451)	(Fragment)	*Lama guanicoe* (Guanaco)
CASK_MAZAM (Q95191)	(Fragment)	*Mazama americana* (Red brocket)
CASK_MOUSE (P06796)	precursor	*Mus musculus* (Mouse)
CASK_MUNRE (Q95199)	(Fragment)	*Muntiacus reevesi* (Chinese muntjak)
CASK_NEMGO (P50422)	precursor	*Nemorhaedus goral* (Gray goral)
CASK_ODOHE (Q95225)	(Fragment)	*Odocoileus hemionus* (Mule deer) (Black-tailed deer)
CASK_ODOVI (Q95228)	(Fragment)	*Odocoileus virginianus virginianus* (Virginia white-tailed deer)
CASK_OREAM (P50423)	precursor	*Oreamnos americanus* (Mountain goat)
CASK_OVIDA (Q95224)	(Fragment)	*Ovis dalli* (Dall sheep)
CASK_OVIMO (Q95227)	(Fragment)	*Ovibos moschatus* (Muskox)
CASK_PIG (P11841)	precursor	*Sus scrofa* (Pig)
CASK_RABIT (P33618)	precursor	*Oryctolagus cuniculus* (Rabbit)
CASK_RANTA (Q95239)	(Fragment)	*Rangifer tarandus* (Reindeer) (Caribou)
CASK_RAT (P04468)	precursor	*Rattus norvegicus* (Rat)
CASK_RUPRU (P50424)	precursor	*Rupicapra rupicapra* (Chamois)
CASK_SAITA (P50425)	precursor	*Saiga tatarica*
CASK_SHEEP (P02669)	precursor	*Ovis aries* (Sheep)
CASK_TAPIN (Q29135)	(Fragment)	*Tapirus indicus* (Asiatic tapir) (Malayan tapir)
CASK_TAYTA (Q28794)	(Fragment)	*Tayassu tajacu* (Collared peccary) (Pecari tajacu)
CASK_TRAJA (Q29137)	(Fragment)	*Tragulus javanicus* (Lesser Malay chevrotain)
CASK_UNCUN (Q29150)	(Fragment)	*Uncia uncia* (Snow leopard) (Panthera uncia)

of sugars on κ-casein was recognized as long ago as 1961 (Alais and Jolles, 1961) and, during the 1970s and 1980s, a large number of studies were directed at elucidating the sugar composition, sequence and sites of attachment to the protein. The major glycan is a tetrasaccharide composed of galactose (Gal), N-acetylgalactosamine (GalNAc) and sialic or neuraminic acid (NeuAc) of the form

NeuAca(2-3)Galβ(1-3)[NeuAca(2-6)]GalNAc, but monosaccharide (GalNAc), disaccharide (Galβ(1-3)GalNAc) and trisaccharide (NeuAca(2-3)Galβ(1-3)GalNAc or Galβ(1-3)[NeuAca(2-6)]GalNAc) are also found (Fournet *et al.*, 1975, 1979; van Halbeek *et al.*, 1980; Fiat *et al.*, 1988; Saito and Itoh, 1992).

The relative amounts have been determined by high-performance liquid chromatography (HPLC) as 56.0% tetrasaccharide, 36.9% trisaccharide (18.4% linear and 18.5% branched), 6.3% disaccharide and 0.8% monosaccharide (Saito and Itoh, 1992). It is not known whether the minor forms arise from incomplete synthesis of the tetrasaccharide in mammary epithelial cells or are products of degradation after synthesis and/or secretion of κ-casein into the lumen of the mammary gland.

Establishment of the attachment site(s) of the glycans has been more controversial. On the basis of Edman sequencing of short glycopeptides obtained by enzymatic digestion, Jolles *et al.* (1973) proposed Thr^{152} or Thr^{154} as the glycan attachment site. Kanamori *et al.* (1980) proposed Thr^{152}, Thr^{154} and Thr^{156} (or Thr^{157}) after analyzing a glycopeptide that was derived from κ-casein and that contained three GalNAc residues. Work from the same laboratory on bovine κ-casein from colostrum also indicated glycosylation at Thr^{152}, Thr^{154} and Thr^{156} (Doi *et al.*, 1980). Subsequently, using a different peptide fraction prepared from κ-casein of normal milk, glycosylation at Thr^{154} and Ser^{162} was reported (Kanamori *et al.*, 1981). Meanwhile, further work from Jolles' laboratory identified Thr^{152} as the glycan attachment site on κ-casein from normal milk and Thr^{152} and Thr^{163} as the attachment sites on κ-casein from colostrum (Fiat *et al.*, 1981). Zevaco and Ribadeau-Dumas (1984) suggested that glycans could be attached to any of the previously identified sites (Thr^{152}, Thr^{154}, Thr^{156}, Thr^{157}, Ser^{162} or Thr^{163}) but their published study contained no conclusive evidence for any site.

All these studies were limited by the technology available at the time. In normal Edman sequencing, glycosylated serine or threonine residues are not detected and their presence is inferred from a blank in the sequencing cycle where serine or threonine is expected (for a more detailed discussion, see Pisano *et al.*, 1994). This is further complicated by the fact that serine and threonine are themselves low-yield amino acids. Thus, in assigning O-glycosylation sites, Edman sequencing data can easily be misinterpreted.

Conclusive identification of glycosylation sites in κ-casein was achieved using solid-phase Edman sequencing, which allows the direct detection of glycosylated serine and threonine residues (Pisano *et al.*, 1994). Variable levels of glycosylation at Thr^{142}, Thr^{152}, Thr^{154}, Thr^{157} (A variant only), Thr^{163} and Thr^{186} were detected and no evidence of glycosylation at any serine residue was obtained. However, even this study did not give the full picture of κ-casein glycosylation, as it could detect only average glycosylation site occupancy in a crude mixture of glycoforms.

We have shown that κ-casein glycoforms can be separated by 2-DE (Holland *et al.*, 2004) and the resolution obtained is shown in Figure 4.4. When the glycosylation site occupancy of individual glycoforms was investigated using tandem MS sequencing of chemically tagged peptides, an interesting pattern was observed. The mono-glycoform was glycosylated exclusively at Thr^{152}, the di-glycoform was glycosylated at Thr^{152} and Thr^{163} and the tri-glycoform was glycosylated at Thr^{152}, Thr^{154} and Thr^{163} (Holland *et al.*, 2005). Further studies using enriched fractions of κ-casein

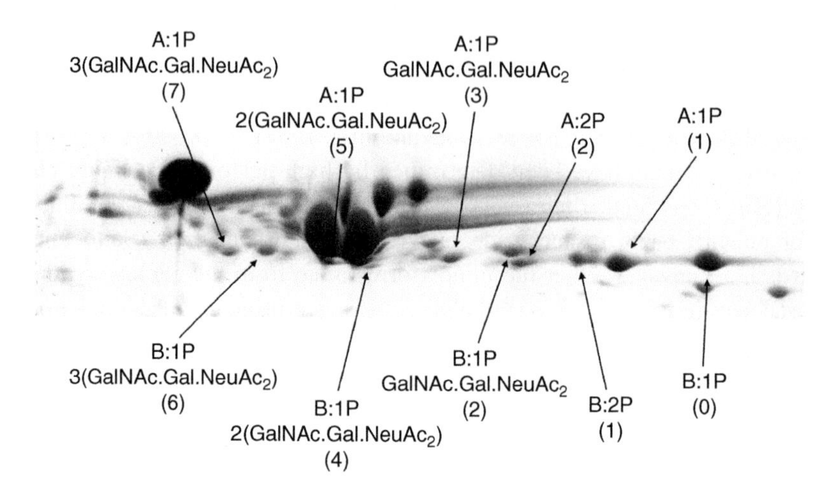

Figure 4.4 Heterogeneity of κ-casein in cow's milk. 2-D gel showing multiple forms of κ-casein. The genetic variant, number of phosphate residues and glycosylation state are indicated. Numbers in brackets indicate the extra negatively charged residues relative to κ-casein B-1P. Adapted from Holland *et al.* (2004).

separated on 2-D gels showed up to six glycans on κ-casein B where only five sites had been previously identified. Tandem MS analysis provided evidence for glycosylation at Thr^{166} on the tetra-glycoform (Holland *et al.*, 2006). The remaining two glycosylation sites were not confirmed but were presumably on Thr^{142} and Thr^{186}, as proposed earlier (Pisano *et al.*, 1994). This pattern is illustrated in Figure 4.5.

As Thr^{166} can be phosphorylated or glycosylated, there is potential for competition at this site. However, as both the tri-phosphate and tetra-glycoforms are very minor forms, it may be of little significance. Overall, the glycosylation of κ-casein in the mammary epithelial cells appears to take place in a highly controlled manner and this suggests that it is a rather important process with considerable functional significance.

Disulfide bonding

κ-Casein purified from bovine milk occurs as both monomeric forms and oligomeric forms with up to eight or more monomers linked by disulfide bonds (Swaisgood *et al.*, 1964; Talbot and Waugh, 1970). A more recent study has shown that reduced and carboxymethylated κ-casein can form large fibrillar structures, although these do not occur in milk (Farrell *et al.*, 2003). The nature of the disulfide-linked complexes has been examined by a number of authors (Groves *et al.*, 1992, 1998; Rasmussen *et al.*, 1992, 1994, 1999; Farrell *et al.*, 2003).

There are only two cysteine residues (Cys^{32} and Cys^{109}) in bovine κ-casein (Mercier *et al.*, 1973) and they appear to be randomly linked in disulfide bonds in oligomeric forms (Rasmussen *et al.*, 1992). In monomeric κ-casein, the two cysteines form an intramolecular disulfide bond (Rasmussen *et al.*, 1994). As can be seen

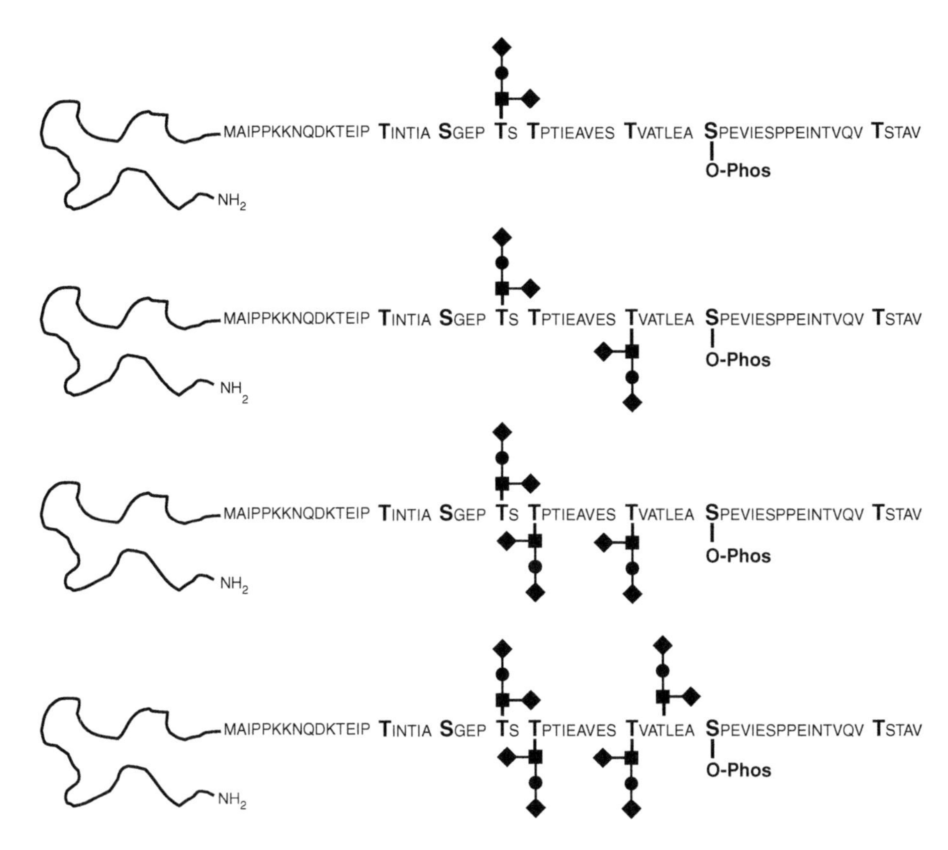

Figure 4.5 Attachment of tetrasaccharide units to the four major glycoforms of κ-casein. Schematic representation of κ-casein highlighting the glycomacropeptide portion and the attachment sites of phosphate and sugar residues in the major glycoforms. (■) GalNAc, (●) Gal, (◆) NeuAc.

in Figure 4.6, disulfide-bonded monomers, dimers and trimers can be resolved on 2-D gels of whole milk when reducing agents are omitted (Holland *et al.*, 2008). It is not clear whether these higher order complexes of κ-casein have any importance in micellar structure but it would be expected that they would be less likely to dissociate from the micelles.

The cysteine residues are not well conserved across species, with human, porcine and rodent κ-caseins containing only a single cysteine, precluding the formation of disulfide-linked oligomers larger than dimers (Rasmussen *et al.*, 1999; Bouguyon *et al.*, 2006). However, the ability of κ-casein to form disulfide-linked complexes with itself or with other proteins during heat treatment is relevant to dairy processing (see below). The combined PTMs of bovine κ-casein are summarized in Figure 4.7.

Sources and functional significance of κ-casein heterogeneity

Although the heterogeneity of κ-casein has been recognized for many decades and the structural elements are now fairly well defined, the source of the heterogeneity,

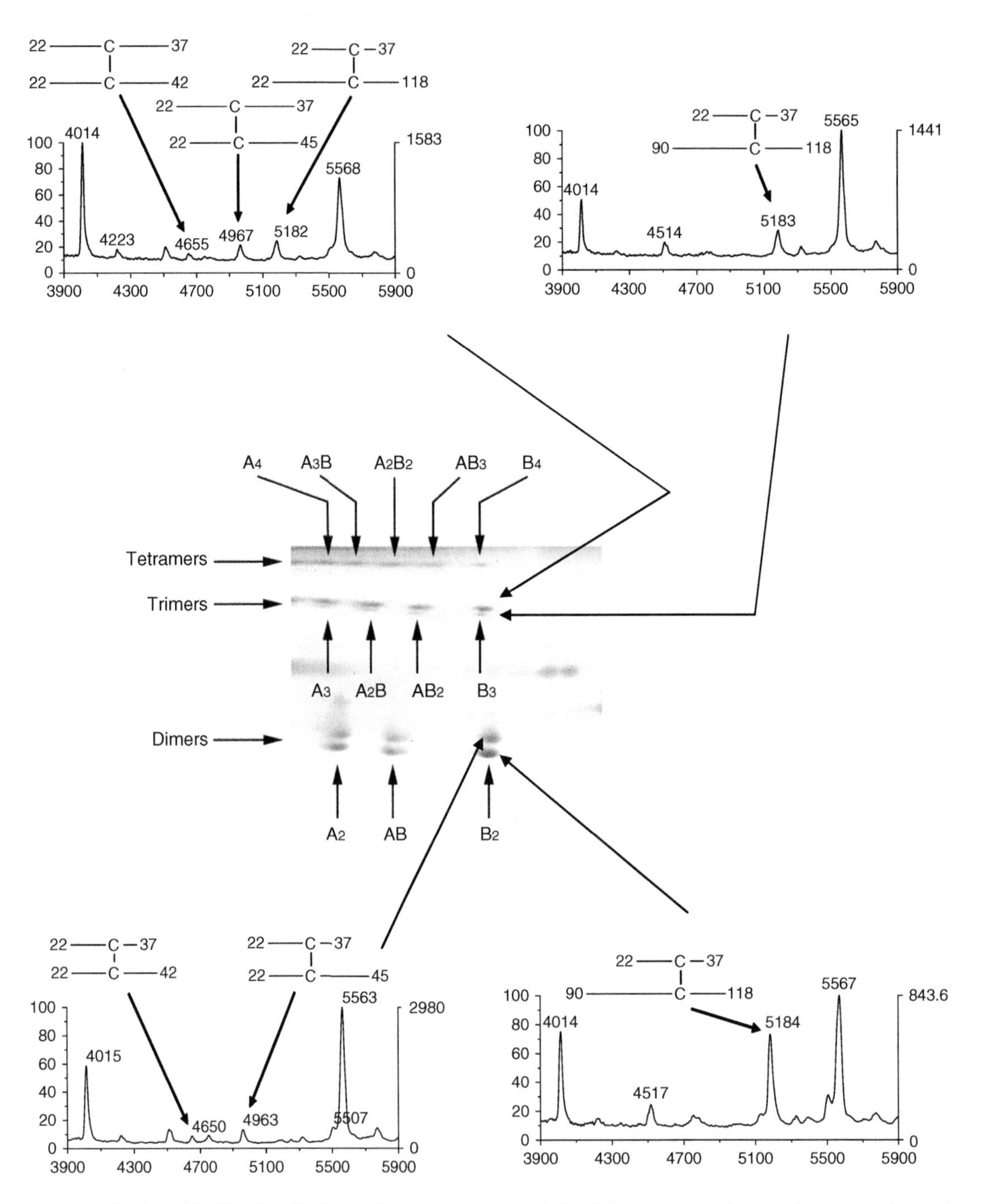

Figure 4.6 Distribution of disulfide-linked isoforms of κ-casein on a non-reducing 2-D gel. Dimers, trimers and tetramers of κ-casein are labelled to show the participating monomeric forms. The dimers and trimers run as doublets depending on the disulfide linkages. MALDI-TOF mass spectra show the disulfide-linked peptides obtained from tryptic digests of the homodimer and homotrimer of κ-casein B:1P.

particularly in glycosylation, and its functional role(s) are not known. Early studies on the influence of the glycosylation of κ-casein on its biological properties have been reviewed previously (Dziuba and Minikiewicz, 1996). They highlighted a number of studies addressing factors that could influence the degree of glycosylation

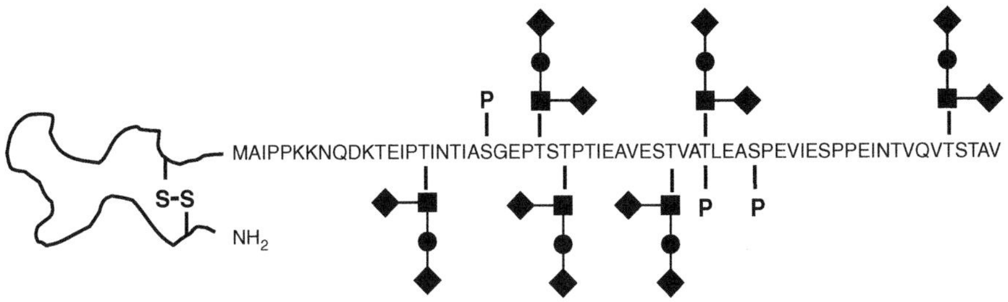

MAIPPKKNQDKTEIPTINTIASGEPTSTPTIEAVESTVATLEASPEVIESPPEINTVQVTSTAV

Figure 4.7 Potential modifications on κ-casein B. This schematic summarizes the potential PTMs on κ-casein B. (P) Phosphorylation sites, (■) GalNAc, (●) Gal, (◆) NeuAc. Note that trisaccharides and disaccharides lacking one or both NeuAc residues also occur. The monomer has an intramolecular disulfide bond between Cys^{32} and Cys^{109}.

and what influence glycosylation might have on micelle stability. The sections below cover some of those studies again and highlight more recent work related to the sources and functional significance of κ-casein heterogeneity.

Sources of heterogeneity

A large number of studies have examined the influence of milk protein polymorphism on milk composition and yield, and these have been extensively reviewed (e.g. Ng-Kwai-Hang, 1997; Martin *et al.*, 2002). However, in many cases, the results have been inconsistent, which is probably a reflection of the multi-factorial nature of milk production. It is difficult to isolate the effects of a single protein polymorphism from those of the other major milk proteins, especially as there appears to be a substantial degree of co-ordination of their expression. There are also a number of environmental or cow factors such as feed type and lactational stage that are frequently interrelated, as they all vary with the seasonal changes in dairy farming. Studies on specific effects of κ-casein variants have largely focused on the common A and B variants and there appears to be a consensus that milk from B variant cows contains more fat, protein, casein and κ-casein than milk from A variant cows (Bovenhuis *et al.*, 1992; Ng-Kwai-Hang, 1997; Bobe *et al.*, 1999).

Studies relating to the glycosylation status of κ-casein are more limited. Robitaille *et al.* (1991a) identified a number of factors that appeared to contribute to variation in the NeuAc content of bovine κ-casein. The NeuAc content was higher in cows with the κ-casein AB phenotype than in cows with the AA phenotype; it decreased with increasing parity and varied over the course of lactation, dropping to a minimum at 2–3 months after calving before increasing over the next 9–10 months. They also examined the association between κ-casein glycosylation and milk production/composition (Robitaille *et al.*, 1991b).

Although there appeared to be a statistically significant association between the NeuAc content of κ-casein and milk yield, the most striking result of these investigations was the variability of NeuAc/κ-casein measurements (mean, $64 \pm 21\,\mu g/mg$; range, $23–166\,\mu g/mg$), which suggests that other factors could have had a large impact on glycosylation or that the inherent variability in the assay masked any true associations. Limited 2-D gel analyses suggest that the pattern of glycosylation is far more consistent than these measurements indicate (Holland *et al.*, 2004, 2005).

It would be very informative to examine some of these supposed extremes in NeuAc and hence the glycosylation level of κ-casein on 2-D gels.

Significant differences in the content of non-glycosylated κ-casein in milk have been reported for cows of different κ-casein genotypes (Lodes *et al.*, 1996). Non-glycosylated κ-casein levels were higher (as a percentage of total protein) in milk from cows with the B variant than in milk from cows with the A variant. The rarer variants, C and E, were generally associated with lower levels. However, as no measurements of glycosylated κ-casein levels were reported, no effect of genetic variant on glycosylation can be inferred from this report.

Electrophoretic and chromatographic techniques have been used to profile the κ-casein macropeptide from cows of the AA and BB phenotypes (Coolbear *et al.*, 1996). They found that the B variant macropeptide was more highly glycosylated than the A variant, with an increased content of both hexosamine (i.e. GalNAc) and sialic acid (i.e. NeuAc). After anion-exchange HPLC on a MonoQ column, the elution profile of the B variant contained more peaks, suggesting that an increased number of oligosaccharide chains were attached. These results were consistent with earlier studies suggesting more extensive glycosylation of the B variant (Vreeman *et al.*, 1986) compared with the A variant (Molle and Leonil, 1995), despite the fact that the A variant contains an extra (potential) glycosylation site (Pisano *et al.*, 1994). From these and other results, Coolbear *et al.* (1996) suggested that there were generally consistent patterns of glycosylation for the genetic variants but that the overall extent of glycosylation could vary.

Variations in κ-casein glycosylation during pregnancy and lactation have been touched on above. Early studies indicated a higher degree of glycosylation of κ-casein in colostrum than in mature milk as well as the presence of an additional sugar moiety, N-acetylglucosamine (GlcNAc) (Guerin *et al.*, 1974; Fournet *et al.*, 1975). Subsequently, a number of studies addressed the structure of the oligosaccharides attached to colostral κ-casein and how they varied with time after parturition (Saito *et al.*, 1981a, 1981b, 1982; van Halbeek *et al.*, 1981; Fiat *et al.*, 1988).

As well as the structures already identified above in normal milk, the following structures have been reported: the acidic hexasaccharide, NeuAcα(2-3)Galβ(1-3)[NeuAcα(2-3)Galβ(1-4)GlcNAcβ(1-6)]GalNAc; the acidic pentasaccharide, NeuAcα(2-3)Galβ(1-3)[Galβ(1-4)GlcNAcβ(1-6)]GalNAc; the acidic tetrasaccharide, GlcNAcβ(1-3)Galβ(1-3)[NeuAcα(2-6)]GalNAc; the neutral pentasaccharide Galβ(1-3)[Galβ(1-4){Fucα(1-3)}GlcNAcβ(1-6)]GalNAc; the neutral tetrasaccharide, Galβ(1-3)[Galβ(1-4)GlcNAcβ(1-6)]GalNAc; and the neutral trisaccharide, Galβ(1-3)[GlcNAcβ(1-6)]GalNAc.

This extra complexity is already observable 15 min after parturition but decreases to normal over about 66 h (Fiat *et al.*, 1988). These results suggest changes in the expression profiles of the glycosyltransferases responsible for assembling the O-linked glycans on κ-casein. The initial step of attachment of GalNAc to a threonine residue is catalyzed by UDP-GalNAc:polypeptide *N*-acetylgalactosaminyltransferase (GalNAcT) and, although not much is known about the bovine enzymes, there are at least 12 mammalian GalNAcTs that have been cloned and functionally expressed (Ten Hagen *et al.*, 2003). In mice, the expression of several isoforms changes markedly during pregnancy and lactation (Young *et al.*, 2003). It is likely that expression of the other required glycosyltransferases also varies, presumably under control of lactogenic hormones.

Other seasonal factors related to climate, such as heat stress, drought and nutrition (e.g. pasture versus fodder), can have an impact on milk production and composition. However, we are not aware of any specific studies on their effect on κ-casein glycosylation.

Functional significance

κ-Casein plays a key role in micelle stability by acting as a hairy layer that provides both steric and electrostatic repulsion between micelles, preventing aggregation. Glycosylation of κ-casein increases both the size of the hydrophilic C-terminal "hairs" and their charge—because of the bulk of the hydrophilic sugar residues with their hydration shells and the negative charge of the neuraminic acid groups respectively. Theoretically, the higher the degree of glycosylation of κ-casein, the greater its stabilizing ability should be. As such, it might be expected that the degree of glycosylation of κ-casein would have a marked effect on both the size and the stability of the casein micelles.

However, whereas the size of the casein micelles has been shown to be inversely related to their κ-casein content (relative to total casein) by a number of authors (McGann *et al.*, 1980; Davies and Law, 1983; Donnelly *et al.*, 1984; Dalgleish *et al.*, 1989; O'Connell and Fox, 2000), there is no clear correlation between micelle size and degree of glycosylation. Slattery (1978) found an apparent inverse relationship between the proportion of glycosylated κ-casein and micelle size but it did not apply to all of the size fractions isolated.

In contrast, Dalgleish (1985, 1986) found that the proportions of glycosylated and non-glycosylated κ-casein did not vary with micelle size. More recently, O'Connell and Fox (2000) showed an apparent increase in κ-casein glycosylation with increasing micelle size. Some of these discrepancies are probably the result of the different experimental approaches adopted. Currently, there is no conclusive evidence for a distinct relationship between micelle size and κ-casein glycosylation.

As stated above, micellar stability, or controlled destabilization in the case of cheese and yoghurt manufacture, is of key importance in dairy manufacturing. A number of authors have looked for effects of κ-casein heterogeneity on micellar stability and the processing properties of milk. Takeuchi *et al.* (1985) used ion-exchange chromatography to prepare nine fractions of κ-casein A-1P that varied in the level of glycosylation. The ability of these sub-fractions to stabilize α_{s1}-casein was shown to increase with increasing carbohydrate content. In cheese manufacture, the initial step is the chymosin (rennet-)catalyzed cleavage of the Phe^{126}-Met^{127} bond in κ-casein, resulting in release of the hydrophilic glycomacropeptide from the micelle surface, which leads to micellar aggregation or clotting.

Doi *et al.* (1979) examined the susceptibility to chymosin action of κ-casein preparations with different degrees of glycosylation. They found that more highly glycosylated forms were less susceptible to hydrolysis not only by chymosin but also by other proteases. Others have also found an inverse relationship between glycosylation and chymosin susceptibility with purified κ-casein fractions (Addeo *et al.*, 1984; Vreeman *et al.*, 1986) and in model systems (Addeo *et al.*, 1984; Leaver and Horne, 1996).

In milk, the relationship is not so clear. Chaplin and Green (1980) claimed that all κ-casein molecules were hydrolyzed with equal efficiency, whereas van Hooydonk *et al.* (1984) found that the rate of chymosin-catalyzed hydrolysis decreased with increasing glycosylation. Again, differences in experimental approach were probably responsible for at least some of the apparent discrepancy. Rennet clotting time (RCT), rate of curd firming and curd firmness have been measured to assess the effect of κ-casein glycosylation on the coagulation properties of milk (Robitaille *et al.*, 1993). Whereas no effect on RCT was observed, the rate of curd firming decreased and the curd firmness increased at higher glycosylation levels.

Differences in rennet coagulation properties have also been observed for genetic variants of κ-casein. Shorter RCTs, higher curd-firming rates and higher curd firmness have been reported for milk from cows with the BB variant than for milk from cows with the AA variant (Walsh *et al.*, 1998; and references therein). These differences between A and B variant milks were maintained after heat treatments of up to 80°C for 2 min, despite an overall deterioration of the coagulation properties at elevated temperatures (Choi and Ng-Kwai-Hang, 2003). Milks containing the rarer κ-casein C variant form rennet gels even more slowly than the A or B variant milks, possibly because of the substitution of histidine for arginine at residue 118, which may affect chymosin binding (Smith *et al.*, 1997). Similar results have been observed for the κ-casein G variant, which has cysteine at residue 118 (Erhardt *et al.*, 1997), and a similar explanation has been proposed (Smith *et al.*, 1997).

Coagulation of milk can also be induced by acid, as is the case in yoghurt manufacture. There are fewer studies on the effect of the glycosylation of κ-casein on acid coagulability. Cases *et al.* (2003) found that partial deglycosylation with neuraminidase had little effect on micellar surface charge and solvation but caused a decrease in acid gelation time, a higher rate of gel firming and a higher final firmness.

Heat treatment of milk can also destabilize the casein micelle structure. The heat-induced coagulation of milk is a very complex process that is affected by many parameters (O'Connell and Fox, 2003). A number of studies have examined the influence of genetic variants of κ-casein on heat stability parameters and it is generally accepted that B variant milks are more stable than A variant milks (FitzGerald and Hill, 1997). The reason may be more related to the effects on κ-casein concentration and micelle size mentioned above than to the structural differences between the variants (Smith *et al.*, 2002). Again, there are fewer studies related to the influence of the glycosylation of κ-casein on heat stability. Using a model system composed of casein micelles in simulated milk ultrafiltrate, Minkiewicz *et al.* (1993) showed that enzymatic removal of neuraminic acid using neuraminidase caused a decrease in heat stability.

However, Robitaille and Ayers (1995), using whole milk, could not find a significant effect of neuraminidase treatment on heat stability. When milk is heated above 65°C, β-lactoglobulin denatures, exposing a previously buried sulfhydryl group that can participate in disulfide exchange reactions with other cysteine-containing proteins including κ-casein. This interaction has been recognized for many decades (Sawyer, 1969) and has been the subject of numerous investigations and reviews over the years, but a detailed analysis is beyond the scope of this review (for an extensive review, see O'Connell and Fox, 2003). Recent studies have addressed both the mechanism of formation (Guyomarc'h *et al.*, 2003) and the impact on product quality (Vasbinder *et al.*, 2003) of disulfide-linked complexes. Despite the vast amount of literature on this topic, there do not appear to be any studies that have addressed the impact of the variable glycosylation of κ-casein on its ability to form disulfide-linked complexes either with itself or with β-lactoglobulin.

Heat-induced changes in micelle structure are particularly relevant for UHT milk production and storage. The extremes of heat treatment (of the order of 140–145°C for 4–10 s) produce a number of changes in the milk, not least of which is the formation of κ-casein-β-lactoglobulin complexes. On storage, UHT-treated milks show a variable tendency to form gels and this phenomenon, known as age gelation, affects product shelf life (for a review, see Datta and Deeth, 2001). Again, despite extensive studies over many years on UHT processing and product performance, the influence of κ-casein heterogeneity, particularly heterogeneity with respect to glycosylation, has not been addressed.

From a theoretical perspective, higher initial levels of glycosylation may act to temper the deleterious effects of heat treatment through effects on micellar size, micellar stability and the formation of disulfide-linked complexes. The heat treatment itself may affect the glycosylation level at the surface of the micelle either indirectly, through loss of κ-casein in complex formation with β-lactoglobulin, or directly, through degradation of glycosidic residues (van Hooydonk *et al.*, 1987; as quoted in Dziuba and Minikiewicz, 1996). Subsequent changes in the glycosylation level during storage could be mediated by the action of heat-stable glycosidases originating from psychrotrophic bacteria present in the raw milk (Marin *et al.*, 1984).

Release of monosaccharides during the storage of UHT milk has been observed (Recio *et al.*, 1998; Belloque *et al.*, 2001). Thus, both the initial glycosylation level of the κ-casein and the residual amount after UHT treatment may affect the storage properties of UHT-treated milk. As the actions of heat-resistant proteases can contribute to the age gelation of UHT milk, the inhibitory effects of glycosylation on the activity of proteases such as plasmin (Doi *et al.*, 1979) may be important for prolonging shelf life. Unravelling specific effects will require the application of modern proteomic technologies for κ-casein analysis (Claverol *et al.*, 2003; Holland *et al.*, 2004, 2005, 2006; O'Donnell *et al.*, 2004). Using these technologies, it will be possible to elaborate the heterogeneous glycoforms of κ-casein in raw milk, after pre-treatment(s), after UHT processing and during storage leading up to gelation. This will allow a definitive assessment of the functional significance of κ-casein glycosylation.

One aspect of κ-casein heterogeneity that has not been considered above is its influence on the biological properties of milk. This area has been reviewed extensively (Dziuba and Minikiewicz, 1996). There are two main areas to consider. Firstly, there is the nutritional contribution of the carbohydrate residues in κ-casein, particularly NeuAc. The importance of NeuAc and its roles in numerous biological functions have been reviewed recently (Schauer, 2000). NeuAc is commonly found as the terminal sugar residue on mammalian glycoproteins. Although mammals can synthesize NeuAc, the high levels in milk and especially colostrum may be related to a high demand for neonatal growth and development. The normal glycans on κ-casein are part of a class known as the Thomsen-Friedenreich-related antigens (Dall'Olio and Chiricolo, 2001). The terminal NeuAc residues may play a key role in preventing colonization of the gut by pathogenic organisms by providing alternative binding sites that minimize binding to the normal gut epithelium.

The second aspect relates to the enormous interest in bioactive peptides derived from milk proteins (Clare and Swaisgood, 2000; Kilara and Panyam, 2003). Numerous *in vitro* activities have been ascribed to κ-casein, its glycomacropeptide or peptides derived from them (Dziuba and Minikiewicz, 1996; Brody, 2000). Some of these activities appear to be associated with particular forms of κ-casein (Malkoski *et al.*, 2001) and can be glycosylation dependent (Li and Mine, 2004).

Whether or not the same activities occur *in vivo* is not always clear because it requires both generation and absorption of the active component during digestion and this is not easy to detect. *In vivo* production of glycomacropeptide is known to occur after milk ingestion (Ledoux *et al.*, 1999; and references therein), and has been detected in the plasma of infants after the ingestion of milk (Chabance *et al.*, 1995). Any naturally occurring bioactivity of glycomacropeptide-derived peptides could be strongly influenced by the glycosylation status of κ-casein either directly, by modifying the activity of the peptide, or indirectly, by affecting proteolysis of κ-casein and hence release of the peptide.

Conclusions

PTMs such as phosphorylation, glycosylation and perhaps disulfide bond formation play a critical role in casein micelle formation and stability. It seems somewhat surprising then that so much variability occurs in these PTMs on the caseins. Whereas significant functional differences in milk properties have been consistently reported for milks with different genetic variants of the caseins, the effects reported for variable PTMs have been limited and not always consistent. Undoubtedly, a major contributing factor has been the lack of adequate methodology to definitively assess the heterogeneous casein variants present within a milk sample and the way these change as a result of treatment. High-resolution proteomic techniques may well hold the key to advancing our knowledge of milk protein PTMs and their influence on milk quality, processability and storage stability.

References

Addeo, F., Martin, P. and Ribadeau-Dumas, B. (1984). Susceptibility of buffalo and cow κ-caseins to chymosin action. *Milchwissenschaft*, **39**, 202–5.

Alais, C. and Jolles, P. (1961). Comparative study of the caseinoglycopeptides formed by the action of rennin on caseins from the cow, sheep and goat, II, Study of the non-peptide portion. *Biochimica et Biophysica Acta*, **51**, 315–22.

Alexander, L. J., Stewart, A. F., Mackinlay, A. G., Kapelinskaya, T. V., Tkach, T. M. and Gorodetsky, S. I. (1988). Isolation and characterization of the bovine κ-casein gene. *European Journal of Biochemistry*, **178**, 395–401.

Belloque, J., Villamiel, M., Lopez, F. R. and Olano, A. (2001). Release of galactose and N-acetylglucosamine during the storage of UHT milk. *Food Chemistry*, **72**, 407–12.

Bobe, G., Beitz, D. C., Freeman, A. E. and Lindberg, G. L. (1999). Effect of milk protein genotypes on milk protein composition and its genetic parameter estimates. *Journal of Dairy Science*, **82**, 2797–804.

Bodenmiller, B., Mueller, L. N., Mueller, M., Domon, B. and Aebersold, R. (2007). Reproducible isolation of distinct, overlapping segments of the phosphoproteome. *Nature Methods*, **4**, 231–37.

Bouguyon, E., Beauvallet, C., Huet, J-C. and Chanat, E. (2006). Disulphide bonds in casein micelle from milk. *Biochemical and Biophysical Research Communications*, **343**, 450–58.

Bouniol, C., Printz, C. and Mercier, J-C. (1993). Bovine α_{s2}-casein D is generated by exon VIII skipping. *Gene*, **128**, 289–93.

Bovenhuis, H., Van Arendonk, J. A. and Korver, S. (1992). Associations between milk protein polymorphisms and milk production traits. *Journal of Dairy Science*, **75**, 2549–59.

Brody, E. P. (2000). Biological activities of bovine glycomacropeptide. *British Journal of Nutrition*, **84**(Suppl 1), S39–S46.

Cases, E., Vidal, V. and Cuq, J. L. (2003). Effect of κ-casein deglycosylation on the acid coagulability of milk. *Journal of Food Science*, **68**, 2406–10.

Chabance, B., Jolles, P., Izquierdo, C., Mazoyer, E., Francoual, C., Drouet, L. and Fiat, A. M. (1995). Characterization of an antithrombotic peptide from κ-casein in newborn plasma after milk ingestion. *British Journal of Nutrition*, **73**, 583–90.

Chaplin, B. and Green, M. L. (1980). Determination of the proportion of κ-casein hydrolyzed by rennet on coagulation of skim-milk. *Journal of Dairy Research*, **47**, 351–58.

Choi, J. W. and Ng-Kwai-Hang, K. F. (2003). Effects of genetic variants of κ-casein and beta-lactoglobulin and heat treatment on coagulating properties of milk. *Asian-Australasian Journal of Animal Sciences*, **16**, 1212–17.

Clare, D. A. and Swaisgood, H. E. (2000). Bioactive milk peptides: a prospectus. *Journal of Dairy Science*, **83**, 1187–95.

Claverol, S., Burlet-Schiltz, O., Gairin, J. E. and Monsarrat, B. (2003). Characterization of protein variants and post-translational modifications: ESI-MSn analyses of intact proteins eluted from polyacrylamide gels. *Molecular and Cellular Proteomics*, **2**, 483–93.

Collins, M. O., Yu, L. and Choudhary, J. S. (2007). Analysis of protein phosphorylation on a proteome-scale. *Proteomics*, **7**, 2751–68.

Coolbear, K. P., Elgar, D. F. and Ayers, J. S. (1996). Profiling of genetic variants of bovine k-casein macropeptide by electrophoretic and chromatographic techniques. *International Dairy Journal*, **6**, 1055–68.

Cox, D. M., Zhong, F., Du, M., Duchoslav, E., Sakuma, T. and McDermott, J. C. (2005). Multiple reaction monitoring as a method for identifying protein post-translational modifications. *Journal of Biomolecular Techniques*, **16**, 83–90.

Dalgleish, D. G. (1985). Glycosylated κ-caseins and the sizes of bovine casein micelles. Analysis of the different forms of κ-casein. *Biochimica et Biophysica Acta*, **830**, 213–15.

Dalgleish, D. G. (1986). Analysis by fast protein liquid chromatography of variants of κ-casein and their relevance to micellar structure and renneting. *Journal of Dairy Research*, **53**, 43–51.

Dalgleish, D. G., Horne, D. S. and Law, A. J. R. (1989). Size-related differences in bovine casein micelles. *Biochimica et Biophysica Acta*, **991**, 383–87.

Dall'Olio, F. and Chiricolo, M. (2001). Sialyltransferases in cancer. *Glycoconjugate Journal*, **18**, 841–50.

Datta, N. and Deeth, H. C. (2001). Age gelation of UHT milk—a review. *Food and Bioproducts Processing*, **79**, 197–210.

Davies, D. T. and Law, A. J. R. (1983). Variation in protein composition of bovine casein micelles and serum casein in relation to casein micelle size and milk temperature. *Journal of Dairy Research*, **50**, 67–75.

Doi, H., Kawaguchi, N., Ibuki, F. and Kanamori, M. (1979). Susceptibility of κ-casein components to various proteases. *Journal of Nutritional Science and Vitaminology (Tokyo)*, **25**, 33–41.

Doi, H., Kobatake, H., Ibuki, F. and Kanamori, M. (1980). Attachment sites of carbohydrate portions to peptide chain of κ-casein from bovine colostrum. *Agricultural and Biological Chemistry*, **44**, 2605–11.

Donnelly, W. J., McNeill, G. P., Buchheim, W. and McGann, T. C. (1984). A comprehensive study of the relationship between size and protein composition in natural bovine casein micelles. *Biochimica et Biophysica Acta*, **789**, 136–43.

Dziuba, J. and Minikiewicz, P. (1996). Influence of glycosylation on micelle-stabilising ability and biological properties of C-terminal fragments of cow's κ-casein. *International Dairy Journal*, **6**, 1017–44.

Erhardt, G., Prinzenberg, E-M., Buchberger, J., Krick-Salek, H., Krause, I. and Miller, M. (1997). Bovine κ-casein G: detection, occurrence, molecular genetic characterisation, genotyping and coagulation properties, In *Milk Protein Polymorphism*, IDF Special Issue 9702, pp. 328–29. Brussels: International Dairy Federation.

Farrell, H. M. Jr., Cooke, P. H., Wickham, E. D., Piotrowski, E. G. and Hoagland, P. D. (2003). Environmental influences on bovine κ-casein: reduction and conversion to fibrillar (amyloid) structures. *Journal of Protein Chemistry*, **22**, 259–73.

Farrell, H. M. Jr., Jimenez-Flores, R., Bleck, G. T., Brown, E. M., Butler, J. E., Creamer, L. K., Hicks, C. L., Hollar, C. M., Ng-Kwai-Hang, K. F. and Swaisgood, H. E. (2004). Nomenclature of the proteins of cows' milk – sixth revision. *Journal of Dairy Science*, **87**, 1641–74.

Ferranti, P., Pizzano, R., Garro, G., Caira, S., Chianese, L. and Addeo, F. (2001). Mass spectrometry-based procedure for the identification of ovine casein heterogeneity. *Journal of Dairy Research*, **68**, 35–51.

Fiat, A. M., Jolles, J., Loucheux-Lefebvre, M. H., Alais, C. and Jolles, P. (1981). Localisation of the prosthetic sugar groups of bovine colostrum κ-casein. *Hoppe Seylers Zeitschrift fur Physiologische Chemie*, **362**, 1447–54.

Fiat, A. M., Chevan, J., Jolles, P., De Waard, P., Vliegenthart, J. F., Piller, F. and Cartron, J. P. (1988). Structural variability of the neutral carbohydrate moiety of cow colostrum κ-casein as a function of time after parturition. Identification of a tetrasaccharide with blood group I specificity. *European Journal of Biochemistry*, **173**, 253–59.

FitzGerald, R. J. Hill, J. P. (1997) The relationship between milk protein polymorphism and the manufacture and functionality of dairy products. In *Milk Protein Polymorphism,* IDF Special Issue 9702, pp. 355–71. Brussels: International Dairy Federation.

Fournet, B., Fiat, A. M., Montreuil, J. and Jolles, P. (1975). The sugar part of κ-caseins from cow milk and colostrum and its microheterogeneity. *Biochimie*, **57**, 161–65.

Fournet, B., Fiat, A. M., Alais, C. and Jolles, P. (1979). Cow κ-casein: structure of the carbohydrate portion. *Biochimica et Biophysica Acta*, **576**, 339–46.

Ginger, M. R. and Grigor, M. R. (1999). Comparative aspects of milk caseins. *Comparative Biochemistry and Physiology B*, **124**, 133–45.

Girardet, J-M., Miclo, L., Florent, S., Mollé, D. and Gaillard, J-L. (2006). Determination of the phosphorylation level and deamidation susceptibility of equine β-casein. *Proteomics*, **6**, 3707–17.

Gorodetskii, S. I., Kershulite, D. R. and Korobko, V. G. (1983). Primary structure of cDNA of Bos taurus κ-casein macropeptide. *Bioorganicheskaya Khimiya*, **9**, 1693–95.

Greenberg, R., Groves, M. L. and Dower, H. J. (1984). Human β-casein. Amino acid sequence and identification of phosphorylation sites. *Journal of Biological Chemistry*, **259**, 5132–38.

Grosclaude, F., Mahe, M. F., Mercier, J. C. and Ribadeau-Dumas, B. (1972a). Characterization of genetic variants of α-s1 and β bovine caseins. *European Journal of Biochemistry*, **26**, 328–37.

Grosclaude, F., Mahe, M-F., Mercier, J-C. and Ribadeau-Dumas, B. (1972b). Localization of amino-acid substitutions that differenciate bovine κ-casein variants A and B. *Annales de Genetique et de Selection Animale*, **4**, 515–21.

Groves, M. L., Dower, H. J. and Farrell, H. M. Jr (1992). Re-examination of the polymeric distributions of κ-casein isolated from bovine milk. *Journal of Protein Chemistry*, **11**, 21–28.

Groves, M. L., Wickham, E. D. and Farrell, H. M. Jr (1998). Environmental effects on disulfide bonding patterns of bovine κ-casein. *Journal of Protein Chemistry*, **17**, 73–84.

Guerin, J., Alais, C., Jolles, J. and Jolles, P. (1974). κ-casein from bovine colostrum. *Biochimica et Biophysica Acta*, **351**, 325–32.

Guyomarc'h, F., Law, A. J. and Dalgleish, D. G. (2003). Formation of soluble and micelle-bound protein aggregates in heated milk. *Journal of Agricultural and Food Chemistry*, **51**, 4652–60.

Holland, J. W., Deeth, H. C. and Alewood, P. F. (2004). Proteomic analysis of κ-casein micro-heterogeneity. *Proteomics*, **4**, 743–52.

Holland, J. W., Deeth, H. C. and Alewood, P. F. (2005). Analysis of O-glycosylation site occupancy in bovine κ-casein glycoforms separated by two-dimensional gel electrophoresis. *Proteomics*, **5**, 990–1002.

Holland, J. W., Deeth, H. C. and Alewood, P. F. (2006). Resolution and characterisation of multiple isoforms of bovine κ-casein by 2-DE following a reversible cysteine-tagging enrichment strategy. *Proteomics*, **6**, 3087–95.

Holland, J. W., Deeth, H. C. and Alewood, P. F. (2008). Analysis of disulfide linkages in bovine κ-casein oligomers using 2-dimensional electrophoresis. *Electrophoresis*, **29**, 2402–10.

Holt, C. and Horne, D. S. (1996). The hairy casein micelle: evolution of the concept and its implications for dairy technology. *Netherlands Milk and Dairy Journal*, **50**, 85–111.

Horne, D. S. (1998). Casein interactions: casting light on the black boxes, the structure in dairy products. *International Dairy Journal*, **8**, 171–77.

Horne, D. S. (2002). Casein structure, self-assembly and gelation. *Current Opinion in Colloid and Interface Science*, **7**, 456–61.

Ibeagha-Awemu, E. M., Prinzenberg, E. M., Jann, O. C., Luhken, G., Ibeagha, A. E., Zhao, X. and Erhardt, G. (2007). Molecular characterization of bovine CSN1S2*B and extensive distribution of zebu-specific milk protein alleles in European cattle. *Journal of Dairy Science*, **90**, 3522–29.

Jolles, J., Fiat, A. M., Alais, C. and Jolles, P. (1973). Comparative study of cow and sheep κ-caseinoglycopeptides: determination of the N-terminal sequences with a sequencer and location of the sugars. *FEBS Letters*, **30**, 173–76.

Kanamori, M., Kawaguchi, N., Ibuki, F. and Doi, H. (1980). Attachment sites of carbohydrate moieties to peptide chain of bovine κ-casein from normal milk. *Agricultural and Biological Chemistry*, **44**, 1855–61.

Kanamori, M., Doi, H., Ideno, S. and Ibuki, F. (1981). Presence of O-glycosidic linkage through serine residue in κ-casein component from bovine mature milk. *Journal of Nutritional Science and Vitaminology (Tokyo)*, **27**, 231–41.

Kapkova, P., Lattova, E. and Perreault, H. (2006). Nonretentive solid-phase extraction of phosphorylated peptides from complex peptide mixtures for detection by matrix-assisted laser desorption/ionization mass spectrometry. *Analytical Chemistry*, **78**, 7027–33.

Keating, A. F., Davoren, P., Smith, T. J., Ross, R. P. and Cairns, M. T. (2007). Bovine k-casein gene promoter haplotypes with potential implications for milk protein expression. *Journal of Dairy Science*, **90**, 4092–99.

Kilara, A. and Panyam, D. (2003). Peptides from milk proteins and their properties. *Critical Reviews in Food Science and Nutrition*, **43**, 607–33.

Leaver, J. and Horne, D. S. (1996). Chymosin-catalysed hydrolysis of glycosylated and nonglycosylated bovine κ-casein adsorbed on latex particles. *Journal of Colloid and Interface Science*, **181**, 220–24.

Ledoux, N., Mahé, S., Dubarry, M., Bourras, M., Benamouzig, R. and Tomé, D. (1999). Intraluminal immunoreactive caseinomacropeptide after milk protein ingestion in humans. *Nahrung*, **43**, 196–200.

Li, E. W. Y. and Mine, Y. (2004). Immunoenhancing effects of bovine glycomacropeptide and its derivatives on the proliferative response and phagocytic activities of human macrophagelike cells, U937. *Journal of Agricultural and Food Chemistry*, **52**, 2704–08.

Lodes, A., Krause, I., Buchberger, J., Aumann, J. and Klostermeyer, H. (1996). The influence of genetic variants of milk proteins on the composition and technological properties of milk. 1. Casein micelle size and the content of non-glycosylated κ-casein. *Milchwissenschaft*, **51**, 368–73.

Mahe, M. F., Miranda, G., Queval, R., Bado, A., Zafindrajaona, P. S. and Grosclaude, F. (1999). Genetic polymorphism of milk proteins in African Bos taurus and Bos indicus populations. Characterisation of variants α_{s1}-casein H and κ-casein. *Genetics Selection Evolution*, **31**, 239–53.

Malkoski, M., Dashper, S. G., O'Brien-Simpson, N. M., Talbo, G. H., Macris, M., Cross, K. J. and Reynolds, E. C. (2001). Kappacin, a novel antibacterial peptide from bovine milk. *Antimicrobial Agents and Chemotherapy*, **45**, 2309–15.

Manson, W., Carolan, T. and Annan, W. D. (1977). Bovine α_{s0} casein; a phosphorylated homologue of α_{s1} casein. *European Journal of Biochemistry*, **78**, 411–17.

Marin, A., Mawhinney, T. P. and Marshall, R. T. (1984). Glycosidic activities of *Pseudomonas fluorescens* on fat-extracted skim milk, buttermilk, and milk fat globule membranes. *Journal of Dairy Science*, **67**, 52–59.

Martin, P., Szymanowska, M., Zwierzchowski, L. and Leroux, C. (2002). The impact of genetic polymorphisms on the protein composition of ruminant milks. *Reproduction Nutrition Development*, **42**, 433–59.

McGann, T. C., Donnelly, W. J., Kearney, R. D. and Buchheim, W. (1980). Composition and size distribution of bovine casein micelles. *Biochimica et Biophysica Acta*, **630**, 261–70.

Mercier, J. C. (1981). Phosphorylation of caseins, present evidence for an amino acid triplet code post-translationally recognized by specific kinases. *Biochimie*, **63**, 1–17.

Mercier, J. C., Brignon, G. and Ribadeau-Dumas, B. (1973). Primary structure of bovine κ B casein. Complete sequence. *European Journal of Biochemistry*, **35**, 222–35.

Miclo, L., Girardet, J. M., Egito, A. S., Molle, D., Martin, P. and Gaillard, J. L. (2007). The primary structure of a low-M(r) multiphosphorylated variant of β-casein in equine milk. *Proteomics*, **7**, 1327–35.

Minkiewicz, P., Dziuba, J. and Muzinska, B. (1993). The contribution of N-acetylneuraminic acid in the stabilization of micellar casein. *Polish Journal of Food and Nutrition Sciences*, **2**, 39–48.

Miranda, G., Anglade, P., Mahe, M. F. and Erhardt, G. (1993). Biochemical characterization of the bovine genetic κ-casein C and E variants. *Animal Genetics*, **24**, 27–31.

Molle, D. and Leonil, J. (1995). Heterogeneity of the bovine κ-casein caseinomacropeptide, resolved by liquid chromatography on-line with electrospray ionization mass spectrometry. *Journal of Chromatography A*, **708**, 223–30.

Neveu, C., Molle, D., Moreno, J., Martin, P. and Leonil, J. (2002). Heterogeneity of caprine β-casein elucidated by RP-HPLC/MS: genetic variants and phosphorylations. *Journal of Protein Chemistry*, **21**, 557–67.

Ng-Kwai-Hang, K. F. (1997). A review of the relationship between milk protein polymorphism and milk composition/milk production, In *Milk Protein Polymorphism,* IDF Special Issue 9702, pp. 22–37. Brussels: International Dairy Federation.

O'Connell, J. E. and Fox, P. F. (2000). The two-stage coagulation of milk proteins in the minimum of the heat coagulation time-pH profile of milk: effect of casein micelle size. *Journal of Dairy Science*, **83**, 378–86.

O'Connell, J. E. and Fox, P. F. (2003). Heat-induced coagulation of milk. In *Advanced Dairy Chemistry,* Volume 1, *Protein*, 3rd edn (P. F. Fox and P. L. H. McSweeney, eds) pp. 879–945. New York: Kluwer Academic/Plenum Publishers.

O'Donnell, R., Holland, J. W., Deeth, H. C. and Alewood, P. (2004). Milk proteomics. *International Dairy Journal*, **14**, 1013–23.

Pisano, A., Packer, N. H., Redmond, J. W., Williams, K. L. and Gooley, A. A. (1994). Characterization of O-linked glycosylation motifs in the glycopeptide domain of bovine κ-casein. *Glycobiology*, **4**, 837–44.

Prinzenberg, E. M., Hiendleder, S., Ikonen, T. and Erhardt, G. (1996). Molecular genetic characterization of new bovine κ-casein alleles CSN3F and CSN3G and geno-typing by PCR-RFLP. *Animal Genetics*, **27**, 347–49.

Prinzenberg, E. M., Krause, I. and Erhardt, G. (1999). SSCP analysis at the bovine CSN3 locus discriminates six alleles corresponding to known protein variants (A, B, C, E, F, G) and three new DNA polymorphisms (H, I, A1). *Animal Biotechnology*, **10**, 49–62.

Rasmussen, L. K., Hojrup, P. and Petersen, T. E. (1992). The multimeric structure and disulfide-bonding pattern of bovine κ-casein. *European Journal of Biochemistry*, **207**, 215–22.

Rasmussen, L. K., Hojrup, P. and Petersen, T. E. (1994). Disulphide arrangement in bovine caseins: localization of intrachain disulphide bridges in monomers of κ- and α_{s2}-casein from bovine milk. *Journal of Dairy Research*, **61**, 485–93.

Rasmussen, L. K., Sorensen, E. S., Petersen, T. E., Nielsen, N. C. and Thomsen, J. K. (1997). Characterization of phosphate sites in native ovine, caprine, and bovine casein micelles and their caseinomacropeptides: a solid-state phosphorus-31 nuclear magnetic resonance and sequence and mass spectrometric study. *Journal of Dairy Science*, **80**, 607–14.

Rasmussen, L. K., Johnsen, L. B., Tsiora, A., Sorensen, E. S., Thomsen, J. K., Nielsen, N. C., Jakobsen, H. J. and Petersen, T. E. (1999). Disulphide-linked caseins and casein micelles. *International Dairy Journal*, **9**, 215–18.

Recio, M. I., Villamiel, M., Martínez-Castro, I. and Olano, A. (1998). Changes in free monosaccharides during storage of some UHT milks: a preliminary study. *Zeitschrift für Lebensmitteluntersuchung und -Forschung A*, **207**, 180–81.

Riggs, L., Sioma, C. and Regnier, F. E. (2001). Automated signature peptide approach for proteomics. *Journal of Chromatography A*, **924**, 359–68.

Robitaille, G. and Ayers, C. (1995). Effects of κ-casein glycosylation on heat stability of milk. *Food Research International*, **28**, 17–21.

Robitaille, G., Ng-Kwai-Hang, K. F. and Monardes, H. G. (1991a). Variation of the N-acetylneuraminic acid content of bovine κ-casein. *Journal of Dairy Research*, **58**, 107–14.

Robitaille, G., Ng-Kwai-Hang, K. F. and Monardes, H. G. (1991b). Association of κ-casein glycosylation with milk production and composition in holsteins. *Journal of Dairy Science*, **74**, 3314–17.

Robitaille, G., Ng-Kwai-Hang, K. F. and Monardes, H. G. (1993). Effect of κ-casein glycosylation on cheese yielding capacity and coagulating properties of milk. *Food Research International*, **26**, 365–69.

Rollema, H. S. (1992). Casein association and micelle formation. In *Advanced Dairy Chemistry,* (P. F. Fox, ed.) pp. 111–40. London: Elsevier Applied Science.

Saito, T. and Itoh, T. (1992). Variations and distributions of O-glycosidically linked sugar chains in bovine κ-casein. *Journal of Dairy Science*, **75**, 1768–74.

Saito, T., Itoh, T. and Adachi, S. (1981a). The chemical structure of a tetrasaccharide containing N-acetylglucosamine obtained from bovine colostrum κ-casein. *Biochimica et Biophysica Acta*, **673**, 487–94.

Saito, T., Itoh, T., Adachi, S., Suzuki, T. and Usui, T. (1981b). The chemical structure of neutral and acidic sugar chains obtained from bovine colostrum κ-casein. *Biochimica et Biophysica Acta*, **678**, 257–67.

Saito, T., Itoh, T., Adachi, S., Suzuki, T. and Usui, T. (1982). A new hexasaccharide chain isolated from bovine colostrum κ-casein taken at the time of parturition. *Biochimica et Biophysica Acta*, **719**, 309–17.

Sawyer, W. H. (1969). Complex between β-lactoglobulin and κ-casein. A review. *Journal of Dairy Science*, **52**, 1347–55.
Schauer, R. (2000). Achievements and challenges of sialic acid research. *Glycoconjugate Journal*, **17**, 485–99.
Schild, T. A., Wagner, V. and Geldermann, H. (1994). Variants within the 5′-flanking regions of bovine milk-protein-encoding genes: I. κ-casein-encoding gene. *Theoretical and Applied Genetics*, **89**, 116–20.
Schlieben, S., Erhardt, G. and Senft, B. (1991). Genotyping of bovine κ-casein (κ-CNA, κ-CNB, κ-CNC, κ-CNE) following DNA sequence amplification and direct sequencing of kappa-CNE PCR product. *Animal Genetics*, **22**, 333–42.
Shekar, P. C., Goel, S., Rani, S. D. S., Sarathi, D. P., Alex, J. L., Singh, S. and Kumar, S. (2006). κ-Casein-deficient mice fail to lactate. *Proceedings of the National Academy of Sciences of the USA*, **103**, 8000–5.
Slattery, C. W. (1978). Variation in the glycosylation pattern of bovine κ-casein with micelle size and its relationship to a micelle model. *Biochemistry*, **17**, 1100–4.
Smith, M. H., Hill, J. P., Creamer, L. K., and Plowman, J. E. (1997). Towards understanding the variant effect on the rate of cleavage by chymosin on κ-caseins A and C, In *Milk Protein Polymorphism,* IDF Special Issue 9702, pp. 185–88. Brussels: International Dairy Federation.
Smith, M. H., Edwards, P. J., Palmano, K. P. and Creamer, L. K. (2002). Structural features of bovine caseinomacropeptide A and B by 1H nuclear magnetic resonance spectroscopy. *Journal of Dairy Research*, **69**, 85–94.
Sulimova, G. E., Sokolova, S. S., Semikozova, O. P., Nguet, L. M. and Berberov, E. M. (1992). Analysis of DNA polymorphism of cluster genes in cattle: casein genes and major histocompatibility complex (MHC) genes. *Tsitologiia i Genetika*, **26**, 18–26.
Sulimova, G. E., Badagueva, Iu. N. and Udina, I. G. (1996). Polymorphism of the κ-casein gene in populations of the subfamily Bovinae. *Genetika*, **32**, 1576–82.
Swaisgood, H. E. (2003). Chemistry of the caseins. In *Advanced Dairy Chemistry, Volume 1, Proteins*, 3rd edn (P. F. Fox and P. L. H. McSweeney, eds) pp. 139–202. New York: Kluwer Academic/Plenum Publishers.
Swaisgood, H. E., Brunner, J. R. and Lillevik, H. A. (1964). Physical parameters of κ-casein from cow's milk. *Biochemistry*, **20**, 1616–23.
Sweet, S. M. M., Creese, A. J. and Cooper, H. J. (2006). Strategy for the identification of sites of phosphorylation in proteins: neutral loss triggered electron capture dissociation. *Analytical Chemistry*, **78**, 7563–69.
Takeuchi, M., Tsuda, E., Yoshikawa, M., Sasaki, R. and Chiba, H. (1985). Fractionation and characterization of 9 subcomponents of bovine κ-casein A. *Agricultural and Biological Chemistry*, **49**, 2269–76.
Talbot, B. and Waugh, D. F. (1970). Micelle-forming characteristics of monomeric and covalent polymeric κ-caseins. *Biochemistry*, **9**, 2807–13.
Ten Hagen, K. G., Fritz, T. A. and Tabak, L. A. (2003). All in the family: the UDP-GalNAc:polypeptide N-acetylgalactosaminyltransferases. *Glycobiology*, **13**, 1R–16R.
van Halbeek, H., Dorland, L., Vliegenthart, J. F., Fiat, A. M. and Jolles, P. (1980). A 360-MHz 1H-NMR study of three oligosaccharides isolated from cow κ-casein. *Biochimica et Biophysica Acta*, **623**, 295–300.

van Halbeek, H., Dorland, L., Vliegenthart, J. F., Fiat, A. M. and Jolles, P. (1981). Structural characterization of a novel acidic oligosaccharide unit derived from cow colostrum κ-casein. A 500 MHz 1H-NMR study. *FEBS Letters*, **133**, 45–50.

van Hooydonk, A. C. M., Olieman, C. and Hagedoorn, H. G. (1984). Kinetics of the chymosin-catalysed proteolysis of κ-casein in milk. *Netherlands Milk and Dairy Journal*, **38**, 207–22.

van Hooydonk, A. C. M., De Koster, P. G. and Boerrigter, I. J. (1987). The renneting properties of heated milk. *Netherlands Milk and Dairy Journal*, **41**, 3–18.

Vasbinder, A. J., Alting, A. C., Visschers, R. W. and De Kruif, C. G. (2003). Texture of acid milk gels: formation of disulfide cross-links during acidification. *International Dairy Journal*, **13**, 29–38.

Vreeman, H. J., Visser, S., Slangen, C. J. and Van Riel, J. A. (1986). Characterization of bovine κ-casein fractions and the kinetics of chymosin-induced macropeptide release from carbohydrate-free and carbohydrate-containing fractions determined by high-performance gel-permeation chromatography. *Biochemical Journal*, **240**, 87–97.

Walsh, C. D., Guinee, T. P., Reville, W. D., Harrington, D., Murphy, J. J., O'Kennedy, B. T. and FitzGerald, R. J. (1998). Influence of κ-casein genetic variant on rennet gel microstructure, Cheddar cheesemaking properties and casein micelle size. *International Dairy Journal*, **8**, 707–14.

Walstra, P. (1999). Casein sub-micelles: do they exist?. *International Dairy Journal*, **9**, 189–92.

Wu, H. Y., Tseng, V. S. and Liao, P. C. (2007). Mining phosphopeptide signals in liquid chromatography-mass spectrometry data for protein phosphorylation analysis. *Journal of Proteome Research*, **6**, 1812–21.

Young, W. W. Jr., Holcomb, D. R., Ten Hagen, K. G. and Tabak, L. A. (2003). Expression of UDP-GalNAc:polypeptide N-acetylgalactosaminyltransferase isoforms in murine tissues determined by real-time PCR: a new view of a large family. *Glycobiology*, **13**, 549–57.

Zevaco, C. and Ribadeau-Dumas, B. (1984). A study on the carbohydrate binding sites of bovine κ-casein using high performance liquid chromatography. *Milchwissenschaft*, **39**, 206–10.

Zhou, H., Xu, S., Ye, M., Feng, S., Pan, C., Jiang, X., Li, X., Han, G., Fu, Y. and Zou, H. (2006). Zirconium phosphonate-modified porous silicon for highly specific capture of phosphopeptides and MALDI-TOF MS analysis. *Journal of Proteome Research*, **5**, 2431–37.

Casein micelle structure and stability

5

David S. Horne

Abstract

The physico-chemical properties of the casein proteins are reviewed, highlighting the factors controlling the strength of those interactions most important to the assembly and structure of the casein micelle, namely electrostatic repulsion and hydrophobic attraction, with particular emphasis on their magnitude and range. The various strands are drawn together step-by-step to develop the dual-binding model of casein micelle assembly and structure as a polymerization of a block copolymer system. The model is then used to predict the behavioral properties of the micelle in fluid milk where, by considering the rheology of high concentration milks, the hard sphere colloidal approach is shown to be a special case limited to milk of normal concentration and pH. The necessity for a dual-binding approach is then forcefully demonstrated in its ability to provide full mechanistic explanations of observed behavior in the renneting, acid gelation and alcohol-induced destabilization of skim milk. Bond mobility is identified as a crucial factor but also important is bond location and whether bonding can be extended beyond the protective shell of the micelle—its hairy layer.

Introduction

The caseins are a family of phosphoproteins found in the milks of all mammals. They exist in these milks generally as complex aggregates or micelles of the proteins and mineral calcium phosphate. Because the caseins utilize the same calcium-sequestering mechanism to regulate the calcium phosphate concentration of their environment, they

Milk Proteins: From Expression to Food
ISBN: 978-0-12-374039-7

have recently been identified as members of a wider family of secretory calcium-binding phosphoproteins descended from a common ancestor gene (Kawasaki and Weiss, 2003, 2006). These secretory phosphoproteins include enamel matrix proteins, dentine, salivary proteins, bone extracellular matrix proteins and the caseins, amongst others. All are descended from early primordial genes by duplication and divergence to serve their specialized adaptive functions. Their genes retain common functional and sequence features even after this extensive divergence. It is thought that primordial calcium-sensitive casein genes diverged from enamel matrix protein genes before the appearance of monotremes in the Jurassic era (Kawasaki and Weiss, 2003).

Casein makes milk supersaturated with calcium phosphate. Essentially, it transports safely through the mammary gland the mineral calcium phosphate that is essential for the development of bones and teeth in the suckling infant, through the medium of the casein micelle. Locking up the calcium phosphate in this package is one aspect of the biological function of the micelle. Ensuring release of this same calcium phosphate in the gastric destination of the milk is a property of the micelle that is not generally given great consideration. More research effort has been put into trying to give mechanistic understanding to the technological behavior of the micelle, understanding necessary to achieve efficient conversion of milk into products such as cheese and yoghurt, or the behavior of milk components in emulsions or reconstituted dairy products.

Many of the physical and technological properties of the casein micelle (diffusion, viscosity, light scattering) can be described by treating the casein micelles as colloidal hard spheres (Alexander *et al.*, 2002). The initial stages, up to the onset of instability, in processes such as renneting, acid-induced gelation and flocculation in the presence of ethanol can apparently be well described by allowing these micellar hard spheres to become adhesive, essentially, as described later, treating their κ-casein outer layer as a salted polyelectrolyte brush (De Kruif and Zhulina, 1996; De Kruif and Holt, 2003). Beyond the critical point in all three processes, however, changes in micellar integrity and internal structure render the simple colloidal particle approach inadequate (Horne, 2003a, 2003b; Choi *et al.*, 2007a).

Neither is the adhesive sphere approach at all helpful in furthering our understanding of the processes of micellar assembly, the pathways to dissociation or the maintenance of micellar integrity. For that, we must turn to a structural model of the casein micelle, bearing in mind that, notwithstanding the inadequacy of the adhesive sphere approach, our micelle model has also to be adaptable enough to explain why that primitive approach has been so successful within its limitations.

Various models of casein micelle assembly and structure have appeared over the years and have been subjected to regular review and appraisal, most recently by Fox (2003). The three principal contenders are (i) the sub-micelle model of Slattery and Evard (Slattery and Evard, 1973; Slattery, 1977), subsequently elaborated by Schmidt (1980); (ii) the nanocluster model of Holt (Holt, 1992, 2004; De Kruif and Holt, 2003); (iii) the dual-binding model proposed by Horne (1998). There has been considerable debate in more recent years over the suitability and success of these models (Farrell *et al.*, 2006; Horne, 2006). It is not our intention to rehash those arguments here but rather to highlight the necessity for a dual-binding model by demonstrating its usefulness in providing mechanistic understanding of the production of renneted and acidified milk gels, of micellar assembly and dissociation, and of the physical properties of micellar suspensions including their rheological behavior at high concentration. We first summarize the physico-chemical properties and interactions of the caseins and show how these lead naturally to the dual-binding picture.

Casein primary structure and interactions

Just as casein micelles are aggregates of all of the casein proteins and micellar calcium phosphate, so does the dual-binding model involve the properties and interactions of all of the caseins. Central to this argument are those features of the proteins that are conserved across species and through millennia. The caseins were identified as members of the wider secretory calcium phosphate binding family by their possession of functional and sequence features common to that family (Kawasaki and Weiss, 2003, 2006). Among the conserved motifs is the SXE peptide (Ser-X_{aa}-Glu) where X_{aa} may be any amino acid. In the caseins, this peptide provides a recognition template for post-translational phosphorylation of the serine in the mammary gland by a casein kinase (Mercier, 1981).

Moreover, in the caseins, the serine residues are often found clustered in groups of two, three or four. Such clusters in the α_s- and β-caseins are highly conserved (Martin *et al.*, 2003) and their numbers attest to the significance of the calcium phosphate requirement for post-natal growth in mammals; even more so when it is noted by reference to their sequences (Swaisgood, 2003) that the α_s-caseins, e.g. the α_{s1}- and α_{s2}-caseins of bovine milk, themselves possess two or more such clusters. From now on, discussion will be confined largely to the behavior of bovine caseins because it is for these that the largest body of research data is available. However, the extensions to the behavior in the micelles of other milks will be obvious.

These clusters of phosphoserine residues and the necessary glutamic acid residues templating their existence give rise to massive downward spikes in the hydrophobicity profiles of the bovine α_{s1}- and β-caseins (Figure 5.1). Associated with these are significantly high densities of negative charge at normal milk pH. There is a charge density of $-9e$ within the span of residues 65–72 of α_{s1}-casein and a further $-6e$ along the sequence 48–53 of the same protein. A similarly high charge density of $-9e$ is found between residues 16–23 of β-casein, encompassing the phosphoserine cluster there. Similar high densities are found around the phosphoserine clusters of α_{s2}-casein. All of these estimates of charge density assume a contribution of $-1.5e$ from each phosphoserine residue at or near the natural pH of milk, 6.7.

Away from the phosphoserine clusters, the casein molecules are distinctly hydrophobic. This segregation of hydrophilic and hydrophobic residues confers on the caseins a definite amphipathic nature, which contributes to their ability to function successfully as stabilizers in oil-in-water emulsions. The topography of β-casein adsorbed at the oil–water interface was probed by testing the accessibility of the reactive sites to the proteolytic enzyme trypsin (Leaver and Dalgleish, 1990). In aqueous solution, the reactive sites of β-casein were attacked randomly at no preferential rate.

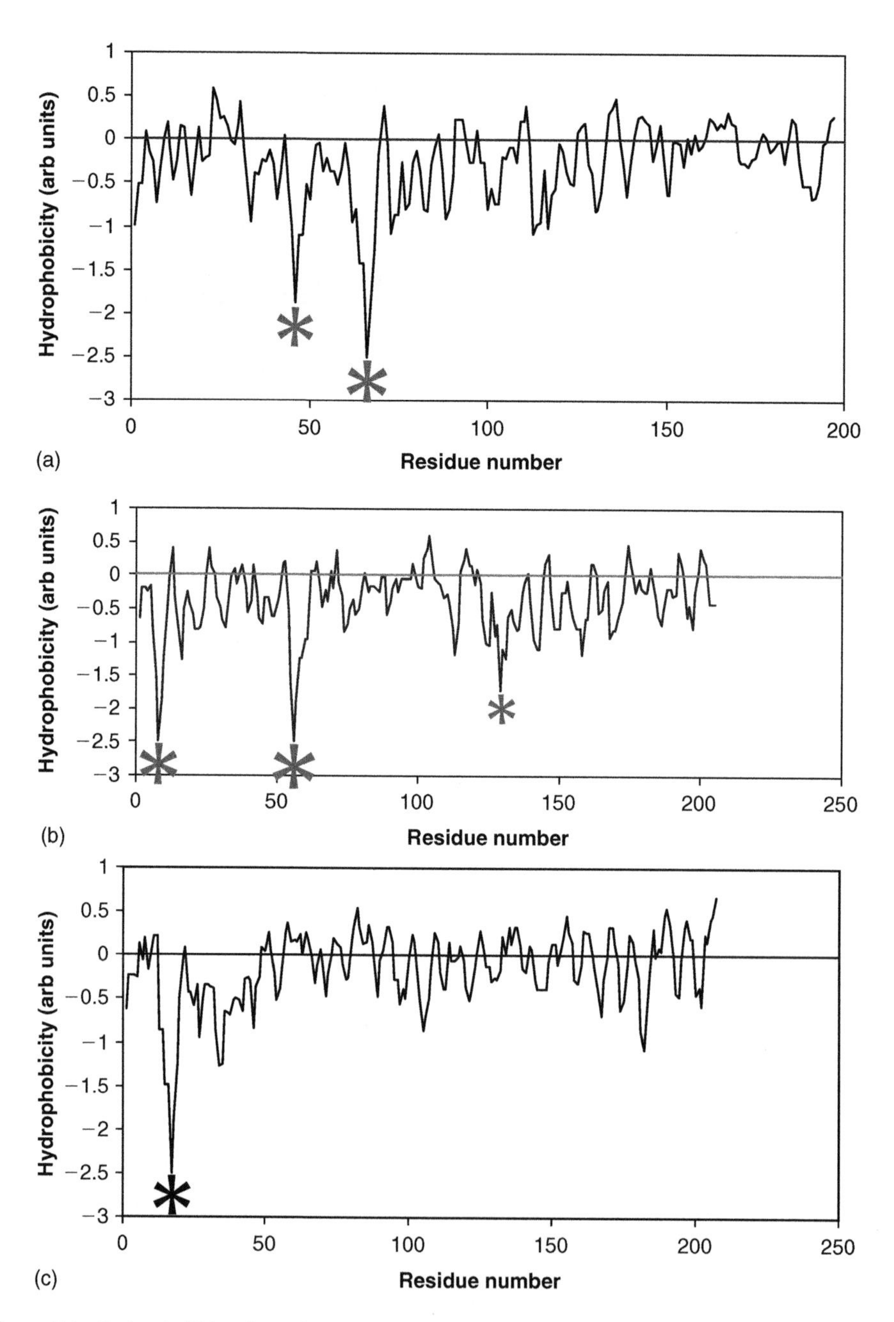

Figure 5.1 Hydrophobicity plots of (a) α_{s1}-casein, (b) α_{s2}-casein and (c) β-casein calculated as a moving average (window $n = 3$) of amino acid hydrophobicities taken from the consensus scale used by Horne (1988). Asterisks denote centers of electrostatic repulsion arising from phosphoserine cluster motifs, the size indicating the number of negative charges associated with each, as listed in the text.

With β-casein-stabilized emulsions, however, the peptides released showed the lysines at positions 25 and 28 of the sequence to be readily accessible to trypsin, whereas all other possible attack sites were less so (Leaver and Dalgleish, 1990). These residues lie in the center of the highly charged, hydrophilic N-terminal region containing the

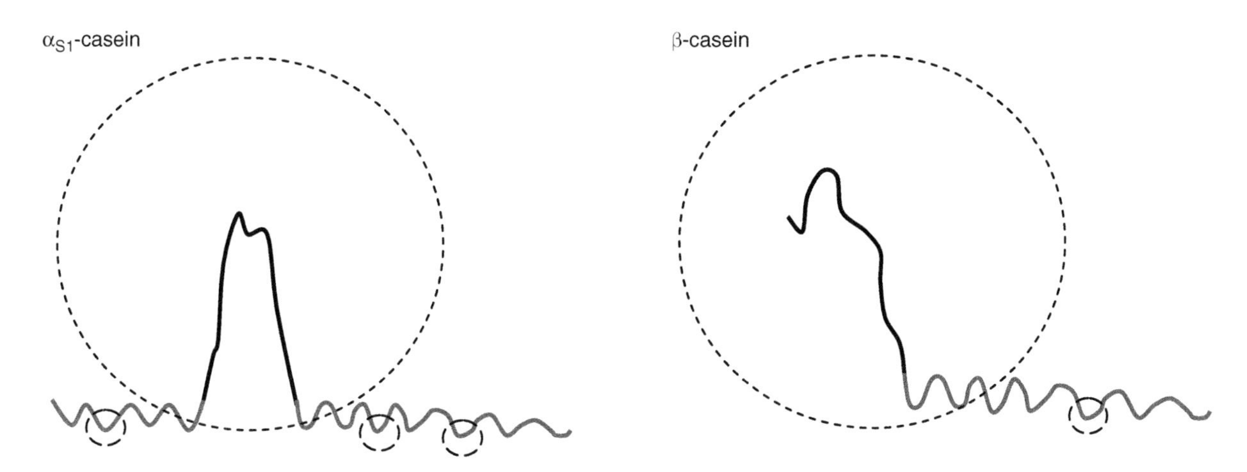

Figure 5.2 Schematic structures of α_{s1}- and β-casein, based on self-consistent-field calculations of the proteins adsorbed on to a hydrophobic interface, illustrating the train–loop–train structure of α_{s1}-casein and the loop–train structure of β-casein. The dashed circles give an idea of the range of the interaction potential components, the larger circle around the loops being the electrostatic repulsion arising from the negative charge centers thereon and the smaller circle being the regions of hydrophobic attraction in the train.

four-phosphoserine cluster in β-casein. Measured by dynamic light scattering, a decrease of approximately 13 nm in hydrodynamic radius of the emulsion droplets also accompanied the scission of these peptides, indicating the extent to which they stretched out into the aqueous phase from the emulsion droplet surface (Dalgleish and Leaver, 1991).

The remaining hydrophobic portion of the molecule was speculated to lie along the droplet surface, shielded from trypsin attack. Similar changes in hydrodynamic radius were observed when β-casein was adsorbed from aqueous buffers on to the surface of polystyrene latex particles, indicating a similar adsorption pattern (Dalgleish, 1990; Brooksbank *et al.*, 1993), a pattern that was replicated at the air–water interface, as observed by neutron reflectivity (Dickinson *et al.*, 1993).

The combined experimental evidence was therefore consistent with the view that much of the hydrophobic end of the adsorbed β-casein was directly associated with the hydrophobic interface, with the hydrophilic N-terminal tail extending significantly out into the aqueous phase. Self-consistent-field calculations of the conformation of β-casein adsorbed at a planar hydrophobic interface confirmed this picture of a tail–train structure, and also predicted a train–loop–train structure for adsorbed α_{s1}-casein with anchor points at both ends of the molecule (Leermakers *et al.*, 1996; Dickinson *et al.*, 1997a, 1997b). Schematic representations of these structures were drawn by Horne (1998) and were used in depicting assembly of the casein micelle via the dual-binding model. Perhaps rather than depicting the hydrophobic regions of the molecules as rectangular bars, it would have been more realistic to depict these regions as puckered, as in Figure 5.2, because not all amino acids therein are equally hydrophobic, as the profile plots of Figure 5.1 demonstrate.

Such representations were also used by Horne (1998) to picture the aggregates produced by self-association of β-casein or α_{s1}-casein (Figure 5.3). Thus β-casein was envisaged as a hedgehog-like micelle subject to a monomer/micelle equilibrium,

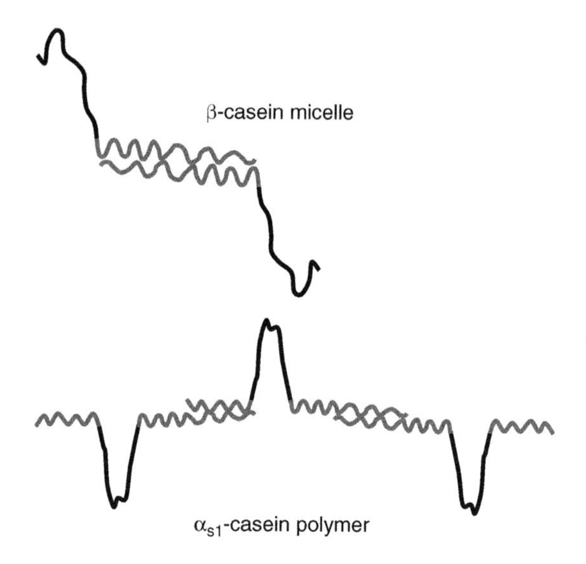

Figure 5.3 Diagramatic representations of the polymeric structures generated when the hydrophobic chains of the caseins interact: the worm-like chain of α_{s1}-casein; the micelle of β-casein, where only two molecules have been included to simplify the diagram.

a mechanism proposed by Payens and co-workers (Payens and van Markwijk, 1963; Payens *et al.*, 1969). In this picture, the hydrophobic trains are buried inside and the charged hydrophilic tails extend from the surface into solution.

More recently the self-association of β-casein has been revisited by De Kruif and collaborators (De Kruif and Grinberg, 2002; Mikheeva *et al.*, 2003; O'Connell *et al.*, 2003), who applied high sensitivity differential scanning calorimetry and static and dynamic light scattering techniques to the problem. These experiments again concluded that a micelle-like structure was adopted but, rather than being formed by the highly co-operative monomer/micelle equilibrium suggested previously, instead the micellization took place as a series of consecutive additions of monomer to a growing micelle, as suggested in the shell model of Kegeles (1979). In a like fashion, with its hydrophobic chains at opposite ends of the molecule and a central section containing the highly charged phosphoserine clusters, α_{s1}-casein self-associates to produce a worm-like chain polymer (Payens and Schmidt, 1966; Schmidt, 1970a, 1970b).

Though driven by hydrophobic interactions, electrostatic repulsive interactions are also very important in the self-association of these caseins. In particular, note how the equilibrium structures adopted by their polymers place the centers of charge as far apart as possible while still permitting the self-association to take place. Compared with hydrophobic interaction, electrostatic repulsion is a long-range force, an important factor now being recognized in studies of protein–protein interactions (Kegel and Van der Schoot, 2004; Piazza, 2004; Stradner *et al.*, 2004), and certainly manifesting itself in these reactions of the caseins. These electrostatic interactions define the degree of polymerization and limit further growth. Thus, increasing the pH, which increases the protein charge, decreases the polymer size in both α_{s1}-casein and β-casein solutions, whereas increasing the ionic strength, which decreases the range of the electrostatic repulsion component, allows the formation of larger polymers for both casein species (Payens *et al.*, 1969; Schmidt, 1970a, 1970b).

The importance of charge in controlling the extent of aggregation of the caseins cannot be stressed too highly. Precipitation of the caseins can be achieved by lowering the pH and titrating away sufficient of the charge of the phosphoseryl and carboxyl groups to reach the isoelectric points of the proteins. α_{s2}-Casein, α_{s1}-casein and β-casein are termed the calcium-sensitive caseins because they can be precipitated in the presence of ionic calcium, the order of sensitivity being as given (Swaisgood, 2003).

The most extensive data set is available for α_{s1}-casein. Here the aggregation shows a lag phase with little change in molecular weight with time until a critical time, beyond which rapid aggregation occurs. Horne and Dalgleish (1980) demonstrated that the logarithm of this critical coagulation time was a linear function of Q^2, where Q is the net negative charge of the protein. Thus Q is the algebraic sum of the negative and positive charges of the protein, reduced by twice the number of calcium ions bound to the protein, each calcium carrying two positive charges. Furthermore, this relationship held when changes in the net negative charge were produced by chemical modification, whether by conversion of positively charged lysine residues to neutral or negatively charged derivatives, or even by introduction of new negatively charged sites by iodination of tyrosine residues to the di-iodo form (Horne, 1983; Horne and Moir, 1984). Each of these modifications effectively increases the net negative charge of the protein, thereby reducing its propensity for calcium-induced precipitation and slowing down the rate of aggregation.

However, once the protein charge is corrected for the measured extent of modification, the logarithm of the rate of precipitation has been shown to remain linear in Q^2, all points lying on the same line as those obtained with the unmodified protein, all other reaction conditions being the same (Horne, 1983; Horne and Moir, 1984). The net charge of the protein therefore dominates its precipitation behavior. Farrell *et al.* (2006) have suggested that positively charged residues in the N-terminal hydrophobic chain of α_{s1}-casein could participate in binding to the phosphate groups of phosphoseryl residues, but such +/− bridging does nothing to reduce the *net* negative charge of the protein having already been accounted for in the algebraic summation leading to Q, which, as we have demonstrated, controls the level of aggregation in these proteins.

It is only when the local balance of electrostatic repulsion and hydrophobic attractive interaction is in favor of attraction that hydrophobic bonds are formed. Sequentially, the major centers of electrostatic repulsion—the phosphoserine clusters—are remote from the hydrophobic regions in the trains of Figure 5.2, though, depending on the adopted conformation, they may not be remote spatially. Figure 5.2 is a depiction of a possible conformation of each protein at a hydrophobic interface, the puckering displaying the tendency to form a multitude of weak, short-range bonds by the hydrophobic chain. Lowering the temperature weakens hydrophobic bonds; hence the tendency to find monomeric β-casein at low temperatures but a micellar aggregate at room temperature and above.

The aggregation of β-casein induced by calcium also shows a marked temperature dependence with no precipitation observed at 4°C (Parker and Dalgleish, 1981). At higher temperature, these hydrophobic bonds are in general stronger but, because they are relatively weak overall, statistically individual bonds are readily ruptured by the increased thermal energy available, leading to more mobile labile interactions between molecules.

Casein micelle properties

Almost all of the casein proteins present in bovine milk are incorporated into the casein micelles, together with a high proportion of the available calcium and inorganic phosphate. The calcium and phosphate within the micelle form low molecular mass species collectively known variously as colloidal calcium phosphate, micellar calcium phosphate and latterly calcium phosphate nanoclusters. The micelles are very open, highly hydrated structures, with typical hydration values of 2–3 g H_2O/g protein, depending on the method of measurement.

Electron microscopy shows that casein micelles are generally spherical in shape, with diameters ranging from 50 to 500 nm (average $\approx$ 150 nm) and a molecular weight ranging from 10^6 to $>10^9$ Da (average $\approx$ 10^8 Da) (Fox, 2003). For a casein content of 2.5 g/100 mL milk, there are some 10^{14}–10^{16} micelles/mL milk, which implies a relatively close packing with inter-surface separations less than one micelle diameter.

Milk is white largely because the colloidal dimensions of the casein micelles are such that they scatter significant amounts of light, an effect compounded by their high number density. Scattering of shorter wavelength radiation (neutrons and X-rays) reveals the internal structure to be heterogeneous, with a correlation length of variations in scattering length density within the particle of approximately 18 nm. This scattering behavior has been interpreted by Stothart and Cebula (1982) as due to a structure composed of closely packed spherical sub-units of this diameter, a picture that mirrors the raspberry-like appearance in early electron micrographs of the casein micelle (Schmidt, 1982).

More recent electron microscopy studies (McMahon and McManus, 1998) have suggested that these well-defined structures are likely to be artifacts of the fixation process, although micrographs recently obtained by field emission scanning electron microscopy (SEM) show a complex surface structure of cylindrical or tubular, but not spherical, protrusions between 10 and 20 nm in diameter, extending from the surface of the micelle (Dalgleish *et al.*, 2004). These samples were not metal coated, although they were of necessity subjected to a fixation and dehydration process, which might have introduced some collapse of more loosely bound protein on to a denser skeleton.

In a careful study using cryo-transmission electron microscopy (cryo-TEM) and small angle X-ray scattering (SAXS), Marchin *et al.* (2007) have extended previous structural studies to elucidate greater detail of micellar fine structure. Their cryo-TEM pictures show small regions of high electron density, approximately 2.5 nm in diameter, uniformly distributed in a homogeneous web of protein, giving the micelles a granular aspect that diminishes when the pH is reduced from 6.7 to 5.2. Paralleling this change in appearance, the SAXS scattering profile loses its characteristic shoulder at high wave vector when the pH is reduced. This clearly demonstrates that the shoulder is directly linked to the presence of micellar calcium phosphate, and its dissociation from the micelle on acidification to pH 5.2.

The results also demonstrate that this micellar calcium phosphate is uniformly distributed through the micelle in small particles, approximately 2.5 nm in diameter and that, after their removal, other forces/interactions must contrive to maintain micellar structural integrity. Their loss on acidification does not result in micellar disruption.

Any structural model of casein micelle assembly would have to lead to a structure that could reproduce both the cryo-TEM pictures and the field emission SEM pictures, allowing for possible changes in appearance brought about by techniques of sample preparation.

Casein micelle structure is not fixed, but dynamic. In various ways, it responds to changes in micellar environment, temperature and pressure. Cooling milk on release from the udder at 37°C to storage at refrigeration temperatures brings about significant solubilization of β-casein, some κ-casein and much lower amounts of α_{s1}- and α_{s2}-casein from the micelles (Dalgleish and Law, 1988). Raising the temperature back to 37°C reverses the process. None of this movement of β-casein does anything to disrupt the internal structure of the micelle, as observed by cryo-TEM and SAXS (Marchin *et al.*, 2007). Almost complete disruption of the micelles, manifested by a loss of their scattering power and removal of the white color of milk, can be achieved by:

- addition of a strong calcium sequestrant such as ethylene diamine tetraacetic acid (EDTA) (Griffin *et al.*, 1988);
- addition of urea (McGann and Fox, 1974);
- dialysis against a phosphate-free buffer (Holt *et al.*, 1986);
- increasing the pH, by exposure to high pressure (Huppertz *et al.*, 2006); or
- addition of ethanol at ≈70°C (O'Connell *et al.*, 2001).

Significantly, the colloidal calcium phosphate can also be solubilized by lowering the pH but, as confirmed by Marchin *et al.* (2007), without substantial disruption of the micelle structure.

Fractionation of the casein micelles according to size can be realized by a stepwise centrifugation protocol. The proportions of α_{s1}- and α_{s2}-casein remain constant with micelle size but κ-casein content increases inversely with that size (Donnelly *et al.*, 1984; Dalgleish *et al.*, 1989). For a solid sphere, the surface-to-volume ratio is inversely proportional to the radius of the sphere and these results imply that κ-casein resides on the micellar surface, where its content controls the micellar total surface area and hence the micelle size.

A surface location for the κ-casein component may also be inferred from the requirement that this protein be readily accessible for rapid and specific hydrolysis by chymosin and similar proteases, a reaction that destabilizes the micelles and leads to clot formation, which is exploited in cheese manufacture. A surface location is also required to enable the κ-casein to interact with β-lactoglobulin in milk to form a complex on heating, the formation of which modifies the rennet and acid coagulation properties of the micelles. It is evident that a principal requirement, which must be met by any micelle model, is that it should generate a surface location for κ-casein.

Models of casein micelle structure

Casein micelle structure and casein micelle models have been extensively reviewed (Schmidt, 1982; Walstra, 1990, 1998; Holt, 1992; Rollema, 1992; Horne, 1992,

1998, 2006; Fox, 2003; Farrell *et al.*, 2006). As mentioned previously, based on the biochemical and physical properties of the micelles and the casein proteins outlined above, three main models have been proposed: the sub-micelle model (Slattery and Evard, 1973; Schmidt, 1982; Walstra, 1998), the nanocluster model of Holt (Holt, 1992; De Kruif and Holt, 2003) and the dual-binding model (Horne, 1998, 2002).

In the first model, the casein micelles are composed of smaller proteinaceous sub-units—the sub-micelles—linked together via colloidal calcium phosphate. In the second model, the nanoclusters of colloidal calcium phosphate are randomly distributed, cross-linking a three-dimensional web of casein molecules. Both of these models have been severely criticized (Farrell *et al.*, 2006; Horne, 2006), and the dual-binding model arose first as an attempt to overcome their deficiencies. Rather than reiterate those arguments, we present first a summary of the dual-binding model as providing a rational mechanism for micelle assembly and structure, and demonstrate how this model may be exploited to explain various observations of micellar properties and behavior.

The dual-binding model for micelle assembly and structure[1]

In the dual-binding model, micellar assembly and growth take place by a polymerization process involving, as the name suggests, two distinct forms of bonding, namely, cross-linking through hydrophobic regions of the caseins and bridging across calcium phosphate nanoclusters. Central to the model is the concept that bond formation is facilitated, and hence micellar integrity and stability are maintained, by a local excess of hydrophobic attraction over electrostatic repulsion, bearing in mind the quite different ranges of these interaction components. The individual casein molecules behave and interact as they do in their self-association equilibria, as described previously.

Each casein molecule effectively functions as a block copolymer, as detailed in Figure 5.2, with the hydrophobic region(s) offering the opportunity for a multitude of individual, weak, hydrophobic interactions. The hydrophilic regions of the casein molecules contain the phosphoserine cluster (or clusters), with the exception of κ-casein, which has no such cluster, each offering multiple functionality for cross-linking. Thus, as we have seen, α_{s1}-casein can polymerize (self-associate) through the hydrophobic blocks, giving the worm-like chain of Figure 5.3. Further growth is limited by the strong electrostatic repulsion of the hydrophilic regions but, in the casein micelle situation, the negative charges of the phosphoserine clusters are neutralized by intercalating their phosphate groups into a facet of the calcium phosphate nanocluster.

This has two very important implications for the micelle. Firstly, by removal of a major electrostatic repulsion component, it increases the propensity for hydrophobic bonding upstream and downstream of the nanocluster link. It effectively permits and strengthens those bonds. Secondly, it allows for multiple protein binding to each nanocluster, allowing a different network to be built up. β-Casein, with only two blocks—a hydrophilic region containing its phosphoserine cluster and the hydrophobic C-terminal tail—can form polymer links into the network through both, allowing further chain extension through both. α_{s2}-Casein is envisaged in this model as having two of each

[1] The description here largely follows that found in Horne (2002) with minor refinements highlighted.

block, two (possibly three, see below) phosphoserine clusters and two hydrophobic regions. It is only a small fraction of the total bovine casein but, by being able to sustain growth through all its blocks, it is likely to be bound tightly into the network.

κ-Casein is the most important of the caseins in the dual-binding model of micellar assembly and structure. It can link into the growing chains through its hydrophobic N-terminal block but its C-terminal block is hydrophilic and cannot sustain growth by linking hydrophobically to another casein molecule. Neither does κ-casein possess a phosphoserine cluster and therefore it cannot extend the polymer cluster through a nanocluster link. Thus, chain and network growth are terminated wherever κ-casein joins the chain. This leaves the network with an outer layer of κ-casein, satisfying the prime requirement recognized earlier.

The nanocluster bridging pathway through the phosphoserine clusters is the only pathway allowed in the nanocluster micelle model of Holt (Holt, 1992; De Kruif and Holt, 2003), where around 50 casein molecules are considered to link into each calcium phosphate nanocluster. Horne *et al.* (Horne, 2006; Horne *et al.*, 2007a) have argued on the basis of mineral content and stoichiometry that the functionality of these nanoclusters will be much lower and is more likely to be four to six phosphoserine clusters, these not necessarily originating from different casein molecules.

Some consideration has also to be given to what constitutes a phosphoserine cluster capable of linking into the calcium phosphate nanocluster. Aoki *et al.* (1992) suggested a minimum of three phosphoserine residues but De Kruif and Holt (2003) argued that two might be sufficient. This would allow the phosphoserine pair at positions 46 and 48 of α_{s1}-casein, or those at positions 129 and 131 of α_{s2}-casein, to function as nanocluster linkage sites, particularly if the carboxyls of the neighboring glutamate residues acted as pseudo-phosphate groups. This would give α_{s1}-casein two linkage sites and α_{s2}-casein three linkage sites.

This level of functionality in these caseins is absolutely essential to the Holt model to build the required three-dimensional network as, without them, an α_{s2}-casein molecule with only two linkage sites and with such a low percentage of the total casein would probably prove to be insufficient. Although they are not essential to the dual-binding model, these mini-clusters of pairs of phosphoserines may provide for a weaker bridging link to the calcium phosphate nanocluster, allowing a range of nanocluster bond strengths to prevail.

Calcium phosphate equilibria in the dual-binding model

From the mineral viewpoint, casein micelle assembly is a frustrated crystallization of calcium phosphate. Milk is supersaturated in calcium and phosphate and, were it not for the presence and intervention of the highly phosphorylated caseins, a precipitation of calcium phosphate and potentially painful calcification of the mammary gland duct system would occur. The phosphoseryl clusters on the casein have the potential to act as a template for the crystal formation but can also act as a cap, cutting off growth at a particular facet. In this way, the overall total number of phosphoseryl clusters in a millilitre of milk controls the size and the number of nanoclusters present in the milk (Horne *et al.*, 2007a). This can happen only if all (or close to all) of the phosphoserine cluster motifs are involved in nanocluster stabilization.

In turn, this must mean that the binding of a serine phosphate group into the nanocluster must present a bonding advantage thermodynamically over that of a free phosphate group going into a growing calcium phosphate crystal in an equilibrium situation. It is a more favorable outcome thermodynamically.

But, just as the calcium phosphate crystal in equilibrium in solution with free calcium and phosphate is subject to environmental constraints shifting that equilibrium, so too is the micellar nanocluster species. The two equilibria may exist in parallel though when micelles are present in normal milk conditions, the nanocluster form is the favored option. However, conditions may change where "solution" crystal growth is favored, and we try to explore these speculations in some of the discussions following. The possibility of shifting calcium phosphate equilibria is an aspect of casein micelle structural behavior that has not been considered previously but, as the nanoclusters are involved, may offer alternative avenues to explore in addressing problems of micellar behavior associated with the application of high pressure, or of the addition of ethanol.

Application of the dual-binding model

Predictions of casein micelle properties

Size and appearance

A major failing of the earlier micelle models was their lack of a plausible mechanism for assembly, growth and, more importantly, termination of growth. All such elements are in place in the dual-binding model. Furthermore, the product of the dual-binding model satisfactorily represents the appearance and scattering behavior of the native casein micelle. Network growth is envisaged as a random process and its termination along any particular pathway depends on the serendipitous arrival of a κ-casein molecule. Micelle size will therefore depend on the proportion of κ-casein in the mix, but will also present a range of sizes dependent as it is on random events. The model also reproduces the heterogeneity in structure required by the X-ray and neutron scattering data. The dense calcium phosphate nanoclusters will be rather homogeneously distributed through the matrix and will give rise to the structures observed by cryo-TEM and inferred from SAXS (Marchin *et al.*, 2007).

The dual-binding model presents the casein micelle as a dynamic, "living" entity. The hydrophobic interactions are individually weak and capable of breaking and recombining on an almost continuous basis. To some extent, the molecular movements this allows will be restricted by the stronger nanocluster linkages and the low probability of rupturing simultaneously all hydrophobic bonds involving any particular molecule. Molecular movement has several consequences, however. The micelle may have an outer layer of individual κ-casein molecules when initially constructed but conditions within the Golgi vesicle are suitable for disulfide bond formation; otherwise β-lactoglobulin and the other whey proteins would not fold properly. Movements of that outer "hairy" layer may bring those κ-caseins into proximity and allow their polymerization through disulfide bridging, the size of the polymer depending on when the chain closes into a loop, but giving rise to the polymeric κ-casein entities observed on micellar dissociation.

Another consequence of the "living" nature of these hydrophobic bonds is that the dehydration of the micelle required in the preparation of a sample for electron microscopy would also tend to be accompanied by the collapse of the more mobile, weaker and less multitudinously bonded regions on to those more strongly cross-linked; hence, perhaps giving rise to the raspberry-like (Schmidt, 1982) or tubular (Dalgleish *et al.*, 2004) structures seen in some electron micrographs. Even the putative caps suggested for those tubules by Dalgleish *et al.* (2004) can be provided by the dual-binding model, as the disulfide bridging between the κ-casein molecules would enhance the Velcro effect of the weak hydrophobic bonding of an individual molecule to many such molecules in the chain.

Effects of urea, pH, sequestrants, temperature

The concept of a local excess of hydrophobic attraction over electrostatic repulsion, as well as permitting the visualization of micellar growth, successfully accommodates the response of the micelle to changes in pH, temperature, urea addition or removal of calcium phosphate by sequestrants, all in accordance with experimental observations.

Urea disrupts hydrophobic bonds and high concentrations will bring about micellar disintegration. In some regions of the micelle, this may be only partial because the nanocluster cross-links through the phosphoserines remain, unaffected by this reagent. Micellar fragments in a range of sizes may be produced, even some as large as some of the original micelles, though perhaps more open and swollen from their own starting state before urea treatment. Extensive disruption does occur, however, as is observed by the loss of the white appearance of skim milk (McGann and Fox, 1974). The dual-binding model fully accounts for these observations.

Removal of calcium from the calcium phosphate nanocluster by sequestrant addition, whether EDTA, citrate or oxalate, restores the negative charge of the hydrophilic region, if the pH is maintained at the native milk pH. This shifts the hydrophobic attraction/electrostatic repulsion balance in favor of repulsion and the micelle breaks up. Decreasing the milk pH solubilizes the colloidal calcium phosphate (Dalgleish and Law, 1989), but the negative charges associated with the cross-linking phosphoseryl groups are also titrated away. The strength of the hydrophobic bonds remains unaffected or may be enhanced if other carboxyl charges are also titrated away. The integrity of the micelles is maintained but their scattering behavior and their appearance in cryo-TEM micrographs reflect the loss of the nanoclusters (Marchin *et al.*, 2007).

Increasing the pH may be expected to be the reverse of the dissolution process and to favor the formation of calcium phosphate species. Fox (2003) noted that raising the pH to >9.0 does not dissolve colloidal calcium phosphate but rather increases its level. However, increasing the milk pH to these levels does lead to dissociation of the micelles and creation of a translucent solution. Whether this is due to conversion of the phosphoserine residues from singly to doubly charged units, which are no longer capable of binding to the calcium phosphate nanoclusters, and hence allowing greater amounts of colloidal calcium phosphate, or whether the increase in charge is sufficient to upset the balance of electrostatic repulsion and hydrophobic attraction in upstream and downstream hydrophobic bonds, it is not possible to say—both routes leading to micellar disruption.

Decreasing the temperature is known to decrease the strength of hydrophobic attraction and to shift the monomer/micelle equilibrium in β-casein solutions towards the monomer side at temperatures below 15°C (De Kruif and Grinberg, 2002). Lowering the temperature of milk to refrigeration levels also brings about dissociation of a large fraction of the β-casein from the casein micelle (Dalgleish and Law, 1988), possibly some of which is not bound into the micellar matrix through its phosphoserine cluster. Raising the temperature back to its initial value reverses the process and the β-casein is reincorporated into the micelle. It is to be anticipated that, even at room temperature, the ongoing equilibrium would allow exchange of micellar casein, albeit with low levels of β-casein in the serum phase.

There are also shifts in the calcium phosphate equilibria in milk associated with temperature change. Ultrafiltration permeate is a clear, straw-yellow liquid when prepared at 4°C but becomes turbid when heated to room temperature and above because of the precipitation of calcium phosphate. Even permeate collected at room temperature clouds on heating but reverts to clarity on cooling.

Hilgemann and Jenness (1951) noted that calcium phosphate also precipitates in milk. However, the calcium phosphate precipitate was only slowly resolubilized (Jenness and Patton, 1959). Weakening the calcium phosphate "solution" equilibrium would favor preservation of the nanoclusters but anything that pushes that "solution" equilibrium to the solid side could have an effect on the continuing existence of the nanoclusters. There are indications that heating milks in the temperature range 50–90°C brings about increasing mobility in the micelle (Rollema and Branches, 1989), which would be in line with partial disruption. The behavior of casein micelles in this temperature range merits further scrutiny, particularly as so many processes in the dairy industry are conducted just in this range.

The dual-binding model and micellar interactions

The ideas outlined earlier in this chapter allow us to schematically describe in Figure 5.4 how the casein micelle might appear as an interacting species at the various pH values indicated.

Internally, at pH 6.7 (Figure 5.4a), the micellar matrix is closely interlinked through a combination of nanocluster bridging bonds (the small black circles) and hydrophobic interactions, occurring randomly along any selected polymer chain. The hydrophobic interactions at this pH (indicated as crossover points in the tangled protein network in the diagrams in Figure 5.4) are many but relatively weak, being counterbalanced by the negative charges present on ionized carboxyl groups, dispersed along the chains and throughout the network. The micellar outer reaches are mainly κ-casein molecules, which have terminated polymer extension and limited micellar growth in the dual-binding model. The negative charges from the ionized carboxyls and sialic acid groups on the κ-casein macropeptides provide the electrostatic repulsion component in the inter-micellar interaction potential, which inhibits micellar aggregation. Its longer range, illustrated by the thickness of the shell around the micelle, prevents close approach of the hydrophobic regions buried beneath

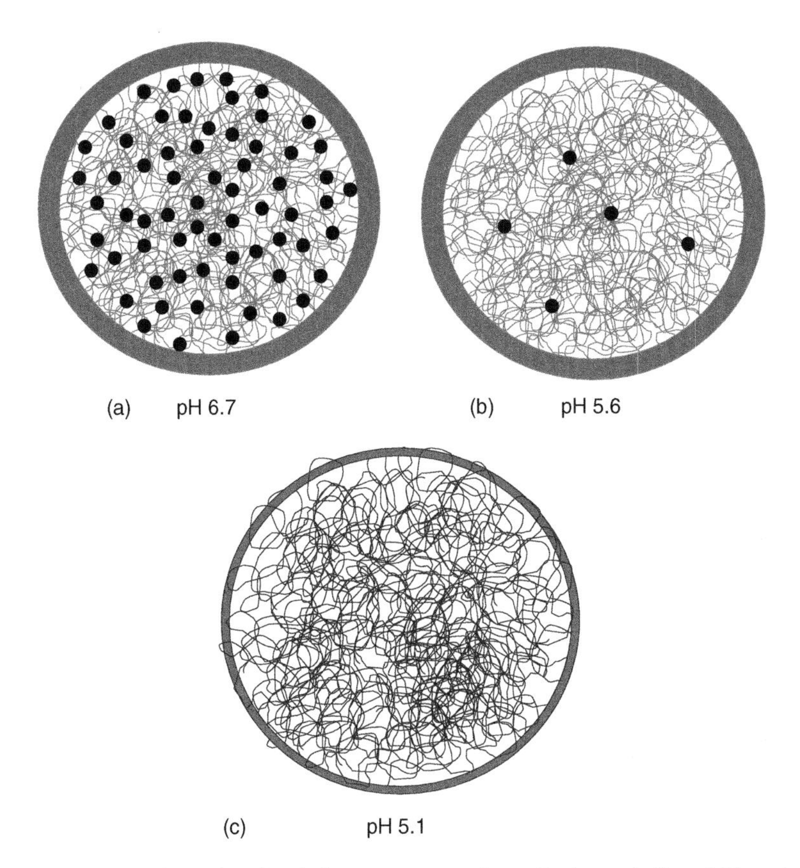

Figure 5.4 Representations of casein micelle structures at various pH values as indicated. The pale chains indicate protein molecules, where they cross being a hydrophobic interaction junction. The small black circles are the calcium phosphate nanoclusters that are solubilized when the pH is lowered. The outer circle is indicative of the range of steric repulsion generated between micelles and preventing interaction of the surface protein chains (see also Plate 5.4).

the shell, and amply fulfils the requirements of a hard sphere model colloid at this pH, 6.7.

At the lower intermediate pH of 5.6 (Figure 5.4b), the same shell continues to prevent close approach of the micelles. The pK values of the acidic groups giving rise to the negative charge are generally lower than 5.5 and have yet to be titrated away. Internally though, most of the micellar calcium phosphate nanoclusters have been solubilized and the bridges between phosphoserine cluster motifs have been lost, weakening the overall network structure of the micelle. The bond strengths of hydrophobic interactions remain relatively weak, still being counterbalanced by ionized carboxyl groups dispersed through the micelle.

By pH 5.1 (Figure 5.4c), the surface charges are being titrated away, the shell in Figures 5.4a and b is much thinner and aggregation begins. Internally, the hydrophobic interactions are effectively being strengthened because the counterbalancing electrostatic repulsions are also being removed from the equation, leading to reduced mobility within the micellar particles.

Rheology of micellar dispersions

In milk as produced from the cow at its natural pH of 6.7 and temperatures from ambient to blood heat, casein micelles closely follow the behavior of hard sphere colloids (De Kruif, 1998; Alexander *et al.*, 2002). Justification for this assertion comes from studies utilizing light and neutron scattering to measure micelle size and polydispersity (Hansen *et al.*, 1996), from sedimentation behavior (De Kruif, 1998) and from measurements of micellar voluminosity (De Kruif, 1998), diffusivity (De Kruif, 1992) and viscosity of micellar suspensions (Griffin *et al.*, 1989).

Paralleling colloidal hard sphere behavior holds only for a limited range of concentrations and, above a critical concentration, micellar suspensions show strong deviations from expected hard sphere behavior (Mezzenga *et al.*, 2005). The viscosity continues to increase but at a slower rate than that expected for hard spheres. This is accompanied by a transition from Newtonian viscosity behavior at natural milk concentration to non-Newtonian viscoelastic behavior in the high concentration regime. More enlightening demonstrations of the departure from hard sphere behavior come from studies of the rheology of high concentration micellar suspensions produced by ultrafiltration (Karlsson *et al.*, 2005), by evaporation to 45% total solids (Bienvenue *et al.*, 2003) and by centrifugal sedimentation and pelleting (Horne, 1998). In these instances, concentrated micellar suspensions are close packed and show a gel-like behavior, which can be interpreted with the assistance of the dual-binding model.

Karlsson *et al.* (2005) concentrated skim milk by ultrafiltration to produce a micellar suspension with 19.5% casein and studied the effects of pH and ionic strength on its viscoelastic properties. Their suspensions exhibited Newtonian viscosities at very low (Brownian) and very high (hydrodynamic) shear rates, with shear thinning at intermediate shear rates and stresses. The concentration of the micelles by ultrafiltration forced the micelles to interact, jamming them together at this high volume fraction and producing a honeycomb-like structure in freeze-fracture electron micrographs. The elastic modulus of these gels decreased as the pH was lowered from the value achieved in the ultrafiltration retentate. Addition of NaCl at levels of 0.33 and 0.66 mol/kg prior to ultrafiltration increased the elasticity of the gels but shifted their pHs to more acid values. Thereafter in the salt-added systems, lowering the pH produced a decrease in elasticity that paralleled the untreated suspension behavior, the higher salt level giving the greater elasticity throughout.

Karlsson *et al.* (2005) also measured the phase angle—the partitioning between viscous and elastic components in these gels as the pH was reduced. In the no-added-salt system, they found the phase angle to increase through a maximum close to 45° and thereafter decrease with decreasing pH. In the presence of added salt, the maximum in phase angle was again observed but shifted to much lower pH values: in the case of the higher salt level, to a pH value lower than for acid gel formation in milk of normal concentration and, in both cases, where elasticity had been observed to increase again in these salted concentrated suspensions.

The dual-binding model explains this behavior with reference to the schematic of the micellar interaction potential depicted in Figure 5.5. The increase in micellar concentration in the no-added-salt case forces the micelles together and into the secondary minimum generated by hydrophobic interactions. This is the source of the

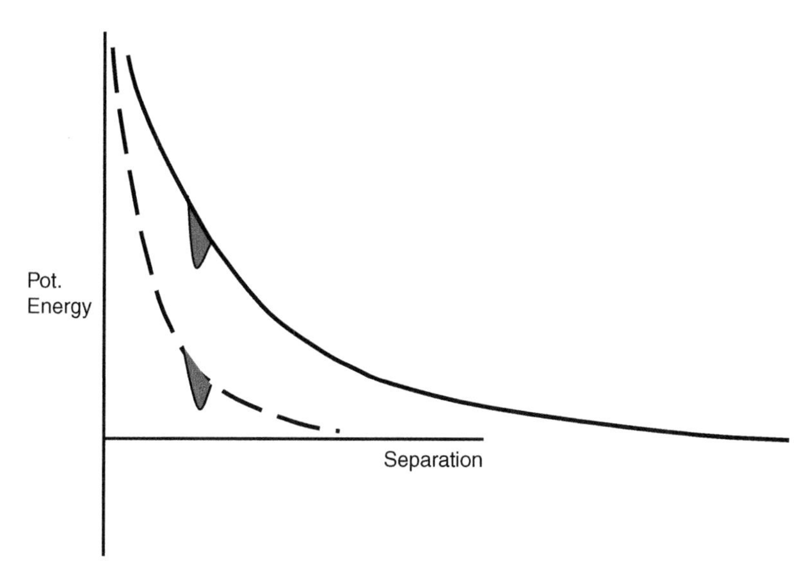

Figure 5.5 Repulsive inter-micellar interaction potential with inner hydrophobic interaction minimum. The dashed line shows the effect of salt addition on the range of the electrostatic repulsion component.

attractive interaction giving rise to the viscoelasticity observed. The micelles are also in a jammed structure and their internal bonding contributes to the measured elasticity. On lowering the pH, the loss of the calcium phosphate nanocluster bridges weakens this structure and the elasticity decreases, as observed. The bonding due to hydrophobic interactions is relatively weak and the loss of the nanocluster bridges further contributes to the mobility in the gel, as evidenced by the observed increase in phase angle. Dropping the pH further titrates away carboxyl groups; however, it reduces the counterbalancing electrostatic component and thereby strengthens hydrophobic bonds in the matrix, reducing mobility and producing the subsequent drop in phase angle.

The major effect of the addition of salt is to reduce the Debye-Huckel parameter and shorten the range of the electrostatic repulsion between micelles. This makes it easier to enter the secondary minimum in the interaction potential and increases the gel elasticity, as observed, with more salt producing the stronger gel. Again, though, the calcium phosphate nanocluster bridges contribute stress-carrying bonds and their removal by lowering the pH leads to the observed decrease in the elasticity of the gel. Throughout this titration, the bonds in the system are relatively stronger than in the no-salt case—the salt also contributes to decreasing the effectiveness of intramicellar electrostatic repulsion—and the phase angles are lower in comparison.

Karlsson *et al.* (2005) suggested that a significant effect of the salt addition is to exchange bound calcium within the micelle for monovalent ions, which would imply no nanoclusters in the system to be solubilized on decreasing the pH and thereby voiding the above explanation for the decrease in elasticity with pH. Huppertz and Fox (2006) did indeed find increased levels of calcium in serum when 600 mM NaCl was added to a two-times-concentrated milk but they found no increase in serum inorganic phosphate, suggesting that the increase in calcium came from displacement of casein-bound calcium rather than a salt-induced dissociation of the calcium phosphate nanoclusters, leaving these to be solubilized on acidification.

The evaporated milks produced by Bienvenue *et al.* (2003) had 45% total solids or were approximately concentrated from normal by a factor of four, rather than the eight times concentration of the micelles in the ultrafiltration retentates of Karlsson *et al.* (2005). The milks of Bienvenue *et al.* (2003) increased in viscosity on storage at 50°C, with salt addition accelerating the increase. Such behavior is in line with the predictions of the dual-binding model outlined above. The collision rate increased by concentration will be further increased by raising the temperature, bringing about a higher frequency of micelles attempting to enter the secondary minimum. A higher success rate and flocculation due to more thermal energy will give rise to the observed increase in viscosity. The weak flocs can be disrupted by higher shear stresses, giving the observed shear-thinning behavior. The effect of salt, as above, would be to render it easier to enter the secondary minimum and promote the flocculation reaction.

Finally, the casein micelle pellets produced by the centrifugation of skim milk at 19 000 *g* for 60 min had protein concentrations of approximately 20% (Horne, 1998). At high temperatures (40°C), this pellet flowed freely. Its viscosity was Newtonian, independent of shear rate or frequency. At low temperature (5°C), however, this micellar suspension exhibited all the properties of a classical viscoelastic gel, with elastic moduli independent of frequency and phase angles less than 45°. At intermediate temperatures, there was a crossover between viscous and elastic behavior. The behavior here is dominated by that of the hydrophobic interactions.

At low temperatures, the strength of these interactions is low. Both β-casein and κ-casein are known to depart from the micelle under such conditions (Dalgleish and Law, 1988) but, in the close-packed conditions prevailing in the pellet, they are liable to migrate or link to neighboring micelles or to become entangled with proteins loosened from those micelles, leading to the gel-like behavior. As the temperature is increased, the strength of the hydrophobic interaction increases but the ability to break bonds is also enhanced and more mobility is allowed. The strengthening of the bonding may also lead to a tightening up of the micelles and their becoming more compact may allow the suspension to flow more freely.

The dual-binding model and micellar destabilization

The concept of the casein micelle electrosterically stabilized by a “hairy layer” coat of κ-casein appears to enjoy universal acceptance (Holt, 1975; Walstra, 1979; Holt and Horne, 1996).

Because the dual-binding model of the casein micelle naturally provides a surface location for κ-casein in a growth-limiting role, it readily explains the destabilization of the casein micelle system on the proteolysis of κ-casein by chymosin and the loss of the steric-stabilizing hairs. Such proteolysis also leads to a significant drop in the micellar zeta potential (Dalgleish, 1984), and consequent reduction in the electrostatic repulsion between micelles. Further confirmation of the importance of electrostatic repulsion in inter-micellar interactions is evinced by the necessary presence of ionic calcium to bring about/promote the aggregation of the chymosin-treated micelles.

Notwithstanding the importance of electrostatics, hydrophobic interactions also play an important part, as evidenced by the fact that fully renneted micelles show no signs of aggregation at low temperatures (<10°C) (Dalgleish, 1983) or that rennet gels increase in elasticity as the incubation temperature is raised (Horne, 1998).

Similarly, the action of ethanol in collapsing the hairs and inducing micellar aggregation is a major coup for the "hairy micelle" model, translated into the adhesive sphere picture of De Kruif and Holt (2003). As the κ-casein hairs are also negatively charged, their neutralization on acidifying milk would also remove a component of the stabilizing barrier and induce aggregation (De Kruif and Holt, 2003). Attractive as these scenarios are for explaining these three routes to micellar destabilization, in none of them does the adhesive sphere/hairy micelle approach tell the whole story. To achieve this, the influence of reaction conditions on micellar integrity has to be considered and it is here that the full power of the dual-binding model comes into play.

Dual-binding model and rennet curd formation

The aggregation and gelation of casein micelles induced by the chymosin proteolysis of κ-casein is the reaction that comes closest to the colloidal aggregation model, especially in strictly controlled laboratory studies, many of which maintained the pH of the milk at its natural value of 6.7. It is under such conditions that the casein micelle exhibits most closely the properties of a colloidal hard sphere and, importantly for the aggregation observed, where the internal integrity of the micelle is undisturbed. It seems that the internal binding through calcium phosphate nanocluster bridges limits extensive rearrangements and constrains excursions of hydrophobic regions so that only those close to the surface behind the barrier of the charged macropeptide can take part in micellar aggregation once that barrier is removed. This explains the success of the reaction schemes for the initial stages of aggregate growth based on the particle model (reviewed by Hyslop, 2003, and previously by Dalgleish, 1992). It also explains the success of the gel strength model described by Horne (1995, 1996), with the particles remaining largely unchanged through the gel formation process.

However, cheeses are seldom manufactured at the natural pH of milk, because this is not the optimum pH for enzyme action (Dalgleish, 1992). Generally, some acidification of the milk is applied and this modifies the internal integrity of the micelles with consequent effects on the rennet coagulation and curd properties. Some of these properties, and their influence on the cheeses produced from such curds, have been studied by Choi *et al.* (2007, 2008), by varying the milk pH or by adding EDTA at a fixed milk pH of 6.0, the objective being to examine in detail the impact of removing micellar calcium phosphate.

In all samples, the elastic modulus of the rennet gel passed through a maximum as the gel was formed, declining during longer reaction times. Rennet gels produced at pH 6.4 had the highest maximum, probably due to a lower electrostatic repulsion because only low levels of micellar calcium phosphate were solubilized at this pH. The maximum elasticity in the gelation profile thereafter decreased with the decreasing pH of the preparation from 6.4 to 5.4. This is explained in the dual-binding model by the decline in the number of nanocluster bridges within the micelle that contribute to the overall strength of the gel matrix. There was also a decrease in the maximum gel elasticity with an increase in the added concentration of EDTA, which removed nanocluster links by sequestration of calcium. In both cases, pH adjustment and EDTA addition, the decrease in the maximum elasticity in the gel curds was accompanied by an increase in the rheological loss tangent. The removal of the nanocluster bridges was permitting greater mobility in these gels; a weaker, more flexible network was being produced.

The microstructure of these rennet-induced gels was also examined, near the point of their maximum elasticity and again some 2–10 h later using fluorescence microscopy (Choi *et al.*, 2007). When elasticity was at its maximum, the gels obtained at pH 6.4 manifested more branched, interconnected networks than those obtained at pH 5.4 where the strands/clusters were larger with more obvious open regions between. In all cases, there was a decrease in apparent interconnectivity between strands in the gel microstructure during aging, which agreed with the decrease in elasticity beyond the maximum.

It is apparent that gel strength is a function not only of the number and strength of potential bonds in a system but also of their spatial distribution. Here, the loss of the nanocluster bridges weakens the network but also introduces into it greater mobility. Rearrangements occur at a rate governed by that mobility and apparently towards a more compact clustering of the casein proteins, which weakens the matrix structure. Predicting the relationship between matrix morphology, bond strength and bond number is one of the challenges yet to be addressed in food materials science.

These changes in rennet gel matrix structure have a direct impact on the functional properties of the cheeses made from the gels. Choi *et al.* (2008) demonstrated that removal of micellar calcium phosphate contributed to a greater softening and ease of flow of these cheeses at higher temperature. However, they found that there was an optimum pH of preparation, below which the increasing influence of attractive hydrophobic interactions in the balance of forces reduced bond lability and inhibited curd stretching.

Dual-binding model and ethanol stability

Although studies of the response of dilute suspensions of casein micelles to the addition of ethanol constituted one of the greater successes of the hairy micelle model, they also provided pointers to the failings of the adhesive sphere concept developed from that model (Horne, 2003a).

Dynamic light scattering studies (Horne, 1984, 1986; Horne and Davidson, 1986) demonstrated the collapse of the hairy layer and the consequent loss of the steric-stabilizing component with sub-critical concentrations of ethanol, but attempts to measure layer thickness as a function of buffer pH or ionic calcium concentration were confounded by the observation that initial micelle size was a function of these parameters. Raising the pH or decreasing the ionic strength produced an increase in the hydrodynamic size of the micelle, presumably due to a loosening up of the micelle as either treatment increased the effective electrostatic repulsion between the hairs.

More importantly for the ethanol-induced collapse of the hairs, this loosening of the structure extended deeper into the micelle and greater apparent layer thickness was shown as the micelle structure was caused to expand. This behavior is accommodated within the dual-binding model as a manifestation of the control of internal binding by the balance of hydrophobic attraction and electrostatic repulsion, and also by the presence of the calcium phosphate nanocluster bridges, which can be lost on acidification. The model allows for changes in the rigidity of the micelle structure, particularly in the surface layers, which would then control the apparent size of the micelle when the steric-stabilizing hairs are collapsed by the non-solvent—ethanol.

The experiments described above confirm yet again the contribution of steric stabilization to micelle stability but relate to the behavior of the casein micelle in a highly dilute suspension in an artificial environment, devoid of inorganic phosphate. In milk, the results of the alcohol stability test, and particularly the behavior of the alcohol stability/pH profile as a function of mineral content, demand consideration of a different mechanism of destabilization (Horne, 1987), but a mechanism wholly consistent with the dual-binding model, although initially proposed well in advance of that model.

The mechanism, proposed by Horne (1987), suggests that there are two competing effects of ethanol on the micellar system: destabilization through loss of the hairy layer, and shifts in the calcium phosphate equilibria, first noted by Pierre (1985). The behavior of calcium phosphate and its colloidal form has been at the center of much discussion in this chapter. If ethanol promotes the precipitation of calcium phosphate external to the micelle, it would first of all reduce the concentration of free calcium, reduce the level of caseinate-bound calcium and disrupt the binding through calcium phosphate nanoclusters. Moderate losses would increase the negative charge of the caseins and increase the thickness of the steric-stabilizing layer. The higher the alcohol level, the faster and more extensive would be the precipitation of calcium phosphate. The ensuing adjustment in protein charge and conformation, although relatively rapid, still requires a finite response time.

Countering these changes are the effects of ethanol as a non-solvent for the proteins, promoting cross-linking and collapse of the hairy layer. When the coagulation reaction occurs faster than the adjustment of charge and conformation resulting from shifts in the calcium phosphate equilibria, or the extent of the latter is limited by insufficient ethanol, the aggregation reaction dominates and precipitation of micelles follows.

The origin of the sigmoidal ethanol stability/pH profile can also be explained through the effect of pH on calcium phosphate precipitation. Increasing the pH brings about increased calcium phosphate precipitation, possibly further enhanced by the ethanol, which means that more ethanol is required to precipitate the protein, i.e. to overcome the increased energy barrier being erected following the transfer of calcium phosphate from the nanocluster state. Conversely, decreasing the pH acts to diminish the influence of ethanol-induced precipitation of calcium phosphate by titrating away negative charge and reducing electrostatic repulsion between protein species. Other effects of milk serum composition, of forewarming the milk and of modifying the milk concentration and ionic strength, can all be explained in a similar fashion (Horne, 2002).

Finally, what is effectively a competition between mineral precipitation and protein aggregation explains the anomalous destabilization of milk by trifluoroethanol (TFE) (Horne and Davidson, 1987) and the behavior of ethanol in milks at high temperature (~70°C) (O'Connell *et al.*, 2001). With TFE, it was found that, after passing through a critical range of TFE concentrations, which caused protein precipitation, higher levels gave rise to micellar dissociation and produced translucent suspensions.

When ethanol/milk mixtures were heated by O'Connell *et al.* (2001), these too became translucent as the micelles dissociated. In both instances in the mechanism proposed here, this behavior would be seen as the result of the calcium phosphate precipitation proceeding so fast or to such an extent that the micelle would disintegrate before the micellar aggregation reaction could occur—any protein aggregates remaining small and giving rise to insignificant turbidity.

Dual-binding model and acid gel formation

It is in trying to describe milk acid gel formation in terms of molecular events that the dual-binding model is most useful. Superficially, the initial stages of acid-induced aggregation of casein micelles can be accommodated by the adhesive sphere model, where titration of micellar charge collapses the "hairy layer" and allows aggregation to proceed (De Kruif and Holt, 2003), but closer study, particularly of the kinetics of gel formation using glucono-d-lactone (gdl) as acidulant, reveals anomalies (Horne, 2003b).

Firstly, at any given temperature, increasing the quantity of gdl used, and hence the rate of acidification of the milk, leads to stiffer gels. There is no mechanism for this in the adhesive sphere model, which has pH as the only variable but, in the dual-binding model, we postulate that titration of the negative charges of the caseins will also affect internal bonds in the micelle, and that reduction of electrostatic repulsion will deepen the attraction between the molecules. If the acidification is proceeding slowly, then this may allow equilibration and rearrangement into localized denser structures with few linkages between, giving rise to weaker gels. More rapid drops in pH may lock the protein into a more dispersed structure with greater density of possibly stronger strands, as was observed microscopically when similar trends in elasticity were detected in rennet gels (Choi *et al.*, 2007a).

Secondly, if the skim milk is heat treated at 90°C for 10 min, a process practiced by the dairy industry and known as forewarming, prior to acid-induced gelation by gdl at 40°C, not only is the critical coagulation pH shifted to a higher pH value and a much stiffer gel produced, but also a distinct step is introduced on to the gelation profile. This is not an artifact introduced by slippage in the rheometer. It is fully reproducible and its presence is subject to the reaction conditions applied. It can be made to disappear with forewarmed milk, for example, by lowering the incubation temperature to 30°C and below (D S Horne, unpublished observations).

A similar step can be introduced into the profile of a non-heat-treated milk by raising the incubation temperature to 40°C or above and employing a high concentration of gdl. Such steps in the profiles were also seen following the acidification of milks in the presence of xanthan (Aichinger, 2005), where their magnitude and definition were dependent on the level of xanthan introduced. With heat-treated milks, the presence, the magnitude and the definition of the stepped gelation profile change in response to the temperature and duration of the preheating (Figure 5.6).

The critical pH for gelation and the maximum in complex modulus are monotonic functions of the level of denaturation of the whey proteins brought about by the prior heat treatment (Figure 5.7). Stepped gelation profiles like this are also seen during the formation of fermented milk gels, again with forewarmed milks, where

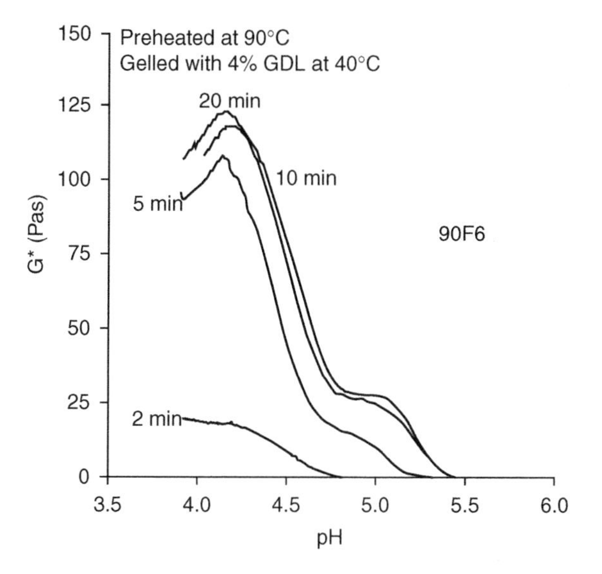

Figure 5.6 Exemplary gelation profiles showing the complex modulus of the developing gel as a function of the measured pH for skim milks preheated at 90°C for the durations indicated on each curve. All milks were gelled at 40°C using 4% gdl as acidulant.

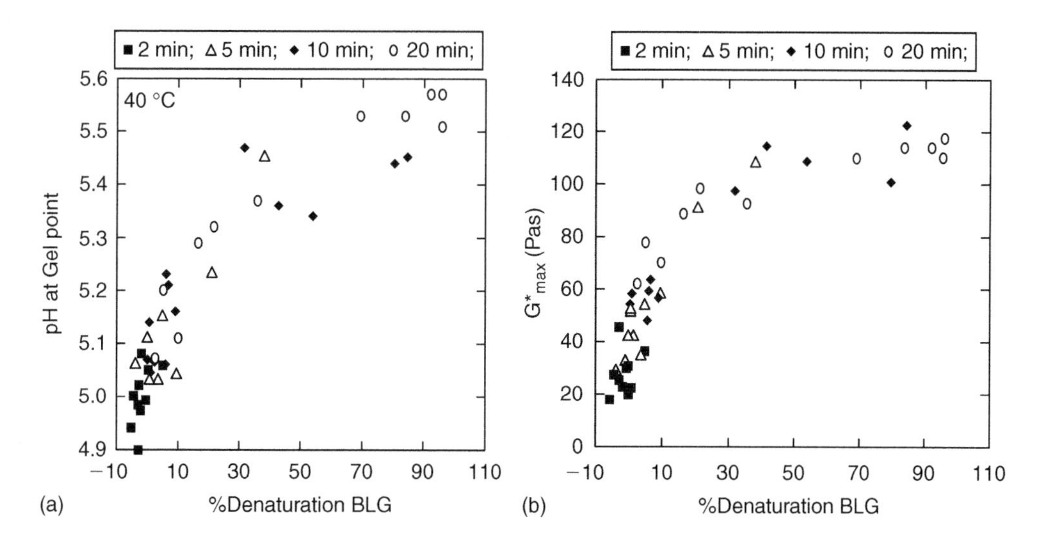

Figure 5.7 Gelation pH and maximum value of the complex modulus as a function of the level of denaturation of β-lactoglobulin in milks preheated at temperatures ranging from 60 to 90°C. Different symbols indicate different durations of heating. All milks were gelled at 40°C using 4% gdl as acidulant.

the step can be removed by suitable modest additions of trisodium citrate (TSC) (Oczan-Yilsay *et al.*, 2007). The critical pH and the loss tangent remained unaffected by additions of TSC up to 10 mM although the gel elasticity recorded at pH 4.6 was enhanced and the gelation profile lost its step and assumed a monotonic increase in G′ with decreasing pH.

All of the above observations can be reconciled within the following picture, which relies heavily on the precepts of the dual-binding model. Aggregation, and thereafter gelation, of the casein micelles on acidification begins only when the electrostatic repulsion component is reduced below a critical level, i.e. when a critical pH is reached. The rate of aggregation is a function of a number of factors: the nature of the surface, the energy (or temperature) involved and the collision rate. Modifying the surface by partially or wholly coating it with denatured whey proteins through formation of a β-lactoglobulin/κ-casein complex, raising the incubation temperature and increasing the micellar concentration by imposing a phase separation through addition of incompatible polysaccharide will all increase the rate of aggregation and shift the critical pH. These are all operational in the above examples, and in all cases we are forcing the micelles to interact, probing further into the inner reaches of the interaction potential with greater frequency and accessing the hydrophobic minimum there.

So far, this is no more than another adhesive hard sphere description but, during acidification, the internal integrity of the casein micelles is also being compromised through the solubilization of the micellar calcium phosphate nanoclusters. Pushing the critical gelation pH to higher values means that more nanoclusters are still present in the aggregating micelles, which are in consequence more rigid that at lower pH. Because the hydrophobic interactions are counterbalanced by any remaining electrostatic repulsion, these hydrophobic bonds are weaker the higher is the coagulation pH. Rearrangements and bond breakage are relatively rapid, as evidenced by the higher value of the phase angle at this point. As the pH continues to drop, the loss of the remaining nanoclusters allows more mobility in the gel, giving the increasing phase angle, but also permitting stronger strands to be formed in the rearrangements, and hence the greater elasticity in the gel.

However, the elasticity is not increasing so rapidly that it cannot be overtaken by its diminution on the loss of the nanocluster bridges; hence the maximum or inflexion point in G' with system pH. Eventually the titration of the carboxyl groups wins out and the increasing contribution from attractive hydrophobic interactions produces increasing elasticity in the gels. Compared with the gel produced in a reaction regime with a low critical gelation pH, this "hi-pH" gel is anticipated to have a more uniform distribution of contributing strands, as has been observed in yoghurt studies (Oczan-Tilsay *et al.*, 2007). When the critical pH is low, as for an unheated milk for example, the reactions described above relating to the behavior on the loss of the remaining calcium phosphate nanoclusters still occur but are now confined *within* the micellar particle, which remains unaggregated because the inter-micellar potential is still repulsive.

Mobility is introduced and bonds are strengthened but only within the micelle so that, when eventually aggregation and gelation do take place, fewer inter-cluster bonds are formed and a more open gel structure is obtained. Location and confinement are major factors in defining the final architecture and strength of the gels formed, and the dual-binding model helps us to get to grips with understanding what is going on.

Conclusions

This chapter has reviewed the physico-chemical properties and interactions of the caseins and blended them on to the construction of a mechanistic framework for a working model of the casein micelle—the dual-binding model. The various properties of the casein micelle have been rationalized in terms predicted by the model. Particular attention has been paid to those treatments leading to destabilization and gel formation, where the necessity for the dual-binding model to explain fully the behavior in the developing gels has been forcefully demonstrated.

Constraints of space have limited the number of cases we have been able to consider. A major omission has been explaining the dissociation and reassociation behavior of the casein micelle under high pressure, but this has recently been covered by Huppertz *et al.* (2006) in terms that can be aligned with little difficulty with the dual-binding model. Neither have we extensively mentioned the application of this model to explaining the behavior of non-bovine milks, although a recent study by Horne *et al.* (2007b) has demonstrated its usefulness in this context for marsupial milks. Our objective has not been to be exhaustive but to provide examples of how the model can be made to work in various situations. Every reader could come up with other examples where the model might prove to be useful. Testing it to failure can only result in superseding it with a better model.

References

Aichinger, P-A. (2005). Kinetic trapping of microstructures—control of gelation in dairy products. Thesis. University of Strathclyde, Glasgow, UK.

Alexander, M., Rojas-Ochoa, L. F., Leser, M. and Schurtenberger, P. (2002). Structure, dynamics and optical properties of concentrated milk suspensions: an analogy to hard-sphere liquids. *Journal of Colloid and Interface Science*, **253**, 34–46.

Aoki, T., Umeda, T. and Kako, Y. (1992). The least number of phosphate groups for cross-linking of casein by colloidal calcium phosphate. *Journal of Dairy Science*, **75**, 971–75.

Bienvenue, A., Jimenez-Flores, R. and Singh, H. (2003). Rheological properties of concentrated skim milk: importance of soluble minerals in the changes in viscosity during storage. *Journal of Dairy Science*, **88**, 3784–97.

Brooksbank, D. V., Davidson, C. M., Horne, D. S. and Leaver, J. (1993). Influence of electrostatic interactions on β-casein layers adsorbed on polystyrene lattices. *Journal of the Chemical Society, Faraday Transactions*, **89**, 3419–25.

Choi, J., Horne, D. S. and Lucey, J. A. (2007). Effect of insoluble calcium concentration on rennet coagulation properties of milk. *Journal of Dairy Science*, **90**, 2612–23.

Choi, J., Horne, D. S., Johnson, M. E. and Lucey, J. A. (2008). Effect of the concentration of insoluble calcium phosphate associated with casein micelles on cheese functionality. *Journal of Dairy Science*, **91**; 513–22.

Dalgleish, D. G. (1983). Coagulation of renneted casein micelles: dependence on temperature, calcium ion concentration and ionic strength. *Journal of Dairy Research*, **50**, 331–40.

Dalgleish, D. G. (1984). Measurement of electrophoretic mobilities and zeta potentials of particles from milk using laser Doppler electrophoresis. *Journal of Dairy Research*, **54**, 425–38.

Dalgleish, D. G. (1990). The conformations of proteins on solid/water interfaces: caseins and phosvitin on polystyrene lattices. *Colloids and Surfaces*, **46**, 141–45.

Dalgleish, D. G. (1992). The enzymatic coagulation of milk. In *Advanced Dairy Chemistry,* Volume 1, *Proteins*, 2nd edn (P. F. Fox, ed.) pp. 579–619. Barking, Essex: Elsevier Applied Science.

Dalgleish, D. G. and Law, A. J. R. (1988). pH-induced dissociation of bovine casein micelles. Analysis of liberated caseins. *Journal of Dairy Research*, **55**, 529–38.

Dalgleish, D. G. and Law, A. J. R. (1989). pH-induced dissociation of casein micelles. Mineral solubilization and its relation to casein release. *Journal of Dairy Research*, **56**, 727–35.

Dalgleish, D. G. and Leaver, J. (1991). The possible conformations of milk proteins adsorbed at oil/water interfaces. *Journal of Colloid and Interface Science*, **141**, 288–94.

Dalgleish, D. G., Horne, D. S. and Law, A. J. R. (1989). Size-related differences in bovine casein micelles. *Biochimica et Biophysica Acta*, **991**, 383–87.

Dalgleish, D. G., Spagnuolo, P. A. and Goff, H. D. (2004). A possible structure of the casein micelle based on high resolution field-emission scanning electron microscopy. *International Dairy Journal*, **14**, 1025–31.

De Kruif, C. G. (1992). Casein micelles: diffusivity as a function of renneting time. *Langmuir*, **8**, 2932–37.

De Kruif, C. G. (1998). Supra-aggregates of casein micelles as a prelude to coagulation. *Journal of Dairy Science*, **81**, 3019–28.

De Kruif, C. G. and Grinberg, V. Y. (2002). Micellization of β-casein. *Colloids and Surfaces A. Physicochemical and Engineering Aspects*, **210**, 183–90.

De Kruif, C. G. and Holt, C. (2003). Casein micelle structure, functions and interactions. In *Advanced Dairy Chemistry,* Volume 1, *Proteins*, 3rd edn (P. F. Fox and P. L. H. McSweeney, eds) pp. 213–76. New York: Kluwer Academic/Plenum Publishers.

De Kruif, C. G. and Zhulina, E. B. (1996). κ-Casein as a polyelectrolyte brush on the surface of casein micelles. *Colloids and Surfaces A. Physicochemical and Engineering Aspects*, **117**, 151–59.

Dickinson, E., Horne, D. S., Phipps, J. S. and Richardson, R. M. (1993). A neutron reflectivity study of the adsorption of β-casein at fluid interfaces. *Langmuir*, **9**, 242–48.

Dickinson, E., Horne, D. S., Pinfield, V. J. and Leermakers, F. A. M. (1997a). Self-consistent-field modelling of casein adsorption. Comparison of results for α_{s1}-casein and β-casein. *Journal of the Chemical Society, Faraday Transactions*, **93**, 425–32.

Dickinson, E., Pinfield, V. J., Horne, D. S. and Leermakers, F. A. M. (1997b). Self-consistent-field modelling of adsorbed casein. Interaction between two protein-coated surfaces. *Journal of the Chemical Society, Faraday Transactions*, **93**, 1785–90.

Donnelly, W. J., McNeill, G. P., Buchheim, W. and McGann, T. C. A. (1984). A comprehensive study of the relationship between size and protein composition in natural casein micelles. *Biochimica et Biophysica Acta*, **789**, 136–43.

Farrell, H. M. Jr, Malin, E. L., Brown, E. M. and Qi, P. X. (2006). Casein micelle structure: what can be learned from milk synthesis and structural biology? *Current Opinion in Colloid and Interface Science*, **11**, 135–47.

Fox, P. F. (2003). Milk proteins: general and historical review. In *Advanced Dairy Chemistry, Volume 1, Proteins*, 3rd end (P. F. Fox and P. L. H. McSweeney, eds) pp. 1–48. New York: Kluwer Academic/Plenum Publishers.

Griffin, M. C. A., Lyster, R. L. J. and Price, J. C. (1988). The disaggregation of calcium-depleted micelles. *European Journal of Biochemistry*, **174**, 339–43.

Griffin, M. C. A., Price, J. C. and Griffin, W. G. (1989). Variation of the viscosity of a concentrated, sterically stabilized colloid: effect of ethanol on casein micelles of bovine milk. *Journal of Colloid and Interface Science*, **128**, 223–29.

Hansen, S., Bauer, R., Lomholt, S. B., Qvist, K. B., Pedersen, J. S. and Mortensen, K. (1996). Structure of casein micelles studied by small-angle neutron scattering. *European Biophysics Journal*, **24**, 143–47.

Hilgemann, M. and Jenness, R. (1951). Observations on the effect of heat treatment upon the dissolved calcium and phosphorus in milk. *Journal of Dairy Science*, **34**, 483–84.

Holt, C. (1975). The stability of casein micelles, In *Proceedings of an International Conference on Colloid and Interface Science, Budapest,* Volume 1 (E. Wolfram, ed.) pp. 641–44.

Holt, C. (1992). Structure and stability of casein micelles. *Advances in Protein Chemistry*, **43**, 63–151.

Holt, C. (2004). An equilibrium thermodynamic model of the sequestration of calcium phosphate by casein micelles and its application to the calculation of the partition of milk salts in milk. *European Biophysics Journal*, **33**, 421–34.

Holt, C. and Horne, D. S. (1996). The hairy casein micelle: evolution of the concept and its implications for dairy technology. *Netherlands Milk and Dairy Journal*, **50**, 85–111.

Holt, C., Davies, D. T. and Law, A. J. R. (1986). Effects of colloidal calcium phosphate content and free calcium ion concentration in the milk serum on the dissociation of bovine casein micelles. *Journal of Dairy Research*, **53**, 557–72.

Horne, D. S. (1983). The calcium-induced precipitation of α_{s1}-casein: effect of modification of lysine residues. *International Journal of Biological Macromolecules*, **5**, 296–300.

Horne, D. S. (1984). Steric effects in the coagulation of casein micelles by ethanol. *Biopolymers*, **23**, 989–93.

Horne, D. S. (1986). Steric stabilization and casein micelle stability. *Journal of Colloid and Interface Science*, **111**, 250–60.

Horne, D. S. (1987). Ethanol stability of casein micelles – a hypothesis concerning the role of calcium phosphate. *Journal of Dairy Research*, **54**, 389–95.

Horne, D. S. (1988). Predictions of protein helix content from an autocorrelation analysis of sequence hydrophobicities. *Biopolymers*, **27**, 451–77.

Horne, D. S. (1992). Ethanol stability. In *Advanced Dairy Chemistry,* Volume 1, *Proteins*, 2nd edn (P. F. Fox, ed.) pp. 657–89. Barking, Essex: Elsevier Applied Science.

Horne, D. S. (1995). Scaling behavior of shear moduli during the formation of rennet milk gels. In *Food Macromolecules and Colloids* (E. Dickinson and D. Lorient, eds) pp. 456–61. Cambridge: Royal Society of Chemistry.

Horne, D. S. (1996). Aspects of scaling behavior in the kinetics of particle gel formation. *Journal de Chimie Physique et de Physico–Chimie Biologique*, **96**, 977–86.

Horne, D. S. (1998). Casein interactions: casting light on the black boxes, the structure in dairy products. *International Dairy Journal*, **8**, 171–77.

Horne, D. S. (2002). Caseins—molecular properties, casein micelle formation and structure. In *Encyclopedia of Dairy Sciences* (H. Roginski, J. W. Fuquay and P. F. Fox, eds) pp. 1902–9. New York: Elsevier.

Horne, D. S. (2003a). Ethanol stability. In *Advanced Dairy Chemistry,* Volume 1, *Proteins*, 3rd edn (P. F. Fox and P. L. H. McSweeney, eds) pp. 975–99. New York: Kluwer Academic/Plenum Publishers.

Horne, D. S. (2003b). Casein micelles as hard spheres: limitations of the model in acidified gel formation. *Colloids and Surfaces A. Physicochemical and Engineering Aspects*, **213**, 255–63.

Horne, D. S. (2006). Casein micelle structure: models and muddles. *Current Opinion in Colloid and Interface Science*, **11**, 148–53.

Horne, D. S. and Dalgleish, D. G. (1980). Electrostatic interactions and the kinetics of protein aggregation: α_{s1}-casein. *International Journal of Biological Macromolecules*, **2**, 154–60.

Horne, D. S. and Davidson, C. M. (1986). The effect of environmental conditions on the steric stabilization of casein micelles. *Colloid and Polymer Science*, **264**, 727–34.

Horne, D. S. and Davidson, C. M. (1987). Alcohol stability of bovine milk. Anomalous effects with trifluoroethanol. *Milchwissenschaft.*, **42**, 509–12.

Horne, D. S. and Moir, P. D. (1984). The iodination of α_{s1}-casein and its effect on the calcium-induced aggregation reaction of the modified protein. *International Journal of Biological Macromolecules*, **6**, 316–20.

Horne, D. S., Lucey, J. A. and Choi, J. (2007a). Casein interactions, does the chemistry really matter?. In *Food Colloids: Self-Assembly and Materials Science* (E. Dickinson and M. Leser, eds) pp. 155–66. Cambridge: Royal Society of Chemistry.

Horne, D. S., Anema, S., Zhu, X., Nicholas, K. R. and Singh, H. (2007b). A lactational study of the composition and integrity of casein micelles from the milk of the Tammar wallaby (*Macropus eugenii*). *Archives of Biochemistry and Biophysics*, **467**, 107–18.

Huppertz, T. and Fox, P. F. (2006). Effect of NaCl on some physicochemical properties of concentrated bovine milk. *International Dairy Journal*, **16**, 1142–48.

Huppertz, T., Kelly, A. L. and De Kruif, C. G. (2006). Disruption and reassociation of casein micelles under high pressure. *Journal of Dairy Research*, **73**, 294–98.

Hyslop, D. B. (2003). Enzymatic coagulation of milk. In *Advanced Dairy Chemistry,* Volume 1, *Proteins*, 3rd edn (P. F. Fox and P. L. H. McSweeney, eds) pp. 839–78. New York: Kluwer Academic/Plenum Publishers.

Jenness, R. and Patton, S. (1959). *Principles of Dairy Chemistry.* New York: Wiley.

Karlsson, A. O., Ipsen, R., Schrader, K. and Ardo, Y. (2005). Relationship between physical properties of casein micelles and rheology of skim milk concentrate. *Journal of Dairy Science*, **88**, 3784–97.

Kawasaki, K. and Weiss, K. M. (2003). Mineralized tissue and vertebrate evolution: the secretory calcium-binding phosphoprotein gene cluster. *Proceedings of the National Academy of Sciences of the USA*, **100**, 4060–65.

Kawasaki, K. and Weiss, K. M. (2006). Evolutionary genetics of vertebrate tissue mineralization: the origin and evolution of the secretory calcium-binding phosphoprotein family. *Journal of Experimental Zoology (Molecular Development and Evolution)*, **306B**, 295–316.

Kegel, W. K. and Van der Schoot, P. (2004). Competing hydrophobic and screened Coulomb interactions in hepatitis B virus capsid assembly. *Biophysical Journal*, **86**, 3905–13.

Kegeles, G. (1979). A shell model for size distribution in micelles. *Journal of Physical Chemistry*, **83**, 1728–32.

Leaver, J. and Dalgleish, D. G. (1990). The topography of bovine β-casein at an oil/water interface as determined from the kinetics of trypsin-catalyzed hydrolysis. *Biochimica et Biophysica Acta*, **1041**, 217–22.

Leermakers, F. A. M., Atkinson, P. J., Dickinson, E. and Horne, D. S. (1996). Self-consistent-field modelling of adsorbed β-casein: effect of pH and ionic strength. *Journal of Colloid and Interface Science*, **178**, 681–93.

Marchin, S., Putaux, J-L., Pignon, F. and Leonil, J. (2007). Effects of the environmental factors on the casein micelle structure studied by cryo-transmission electron microscopy and small-angle X-ray scattering/ultrasmall-angle X-ray scattering. *Journal of Chemical Physics*, **126**, 045101.

Martin, P., Ferranti, P., Leroux, C. and Addeo, F. (2003). Non-bovine caseins: quantitative variability and molecular diversity. In *Advanced Dairy Chemistry,* Volume 1, *Proteins*, 3rd edn (P. F. Fox and P. L. H. McSweeney, eds) pp. 277–317. New York: Kluwer Academic/Plenum Publishers.

McGann, T. C. A. and Fox, P. F. (1974). Physico-chemical properties of casein micelles reformed from urea-treated milk. *Journal of Dairy Research*, **41**, 45–53.

McMahon, D. J. and McManus, W. R. (1998). Rethinking casein micelle structure using electron microscopy. *Journal of Dairy Science*, **81**, 2985–93.

Mercier, J-C. (1981). Phosphorylation of casein. Present evidence for an amino acid triplet code post-translationally recognized by specific kinases. *Biochimie*, **68**, 1–17.

Mezzenga, R., Schurtenberger, P., Burbridge, A. and Michel, M. (2005). Understanding foods as soft materials. *Nature Materials*, **4**, 729–40.

Mikheeva, L. M., Grinberg, N. V., Grinberg, V. Y., Khokhlov, A. R. and De Kruif, C. G. (2003). Thermodynamics of micellization of bovine β-casein studied by high sensitivity differential scanning calorimetry. *Langmuir*, **19**, 2913–21.

O'Connell, J. E., Kelly, A. L., Fox, P. F. and De Kruif, C. G. (2001). Mechanism for the ethanol-dependent, heat-induced dissociation of casein micelles. *Journal of Agricultural and Food Chemistry*, **49**, 4424–28.

O'Connell, J. E., Grinberg, V. Y. and De Kruif, C. G. (2003). Association behavior of β-casein. *Journal of Colloid and Interface Science*, **258**, 33–39.

Oczan-Yilsay, T., Lee, W-J., Horne, D. and Lucey, J. A. (2007). Effect of tri-sodium citrate on the rheological and physical properties and microstructure of yogurt. *Journal of Dairy Science*, **90**, 1644–52.

Parker, T. G. and Dalgleish, D. G. (1981). Binding of calcium ions to bovine β-casein. *Journal of Dairy Research*, **48**, 71–76.

Payens, T. A. J. and Schmidt, D. G. (1966). Boundary spreading of rapidly polymerizing α_{s1}-casein B and C during sedimentation. Numerical solutions of the Lamm–Gilbert–Fujita equation. *Archives of Biochemistry and Biophysics*, **115**, 136–45.

Payens, T. A. J. and Van Markwijk, B. W. (1963). Some features of the self-association of β-casein. *Biochimica et Biophysica Acta*, **71**, 517–30.

Payens, T. A. J., Brinkhuis, J. A. and Van Markwijk, B. W. (1969). Self-association in non-ideal systems. Combined light scattering and sedimentation measurements in β-casein solutions. *Biochimica et Biophysica Acta*, **175**, 434–37.

Piazza, R. (2004). Protein interactions and association: an open challenge for colloid science. *Current Opinion in Colloid and Interface Science*, **8**, 515–22.

Pierre, A. (1985). Milk coagulation by alcohol. Studies on the solubility of the milk calcium and phosphate in alcoholic solutions. *Lait*, **65**, 201–12.

Rollema, H. S. (1992). Casein association and micelle formation. In *Advanced Dairy Chemistry,* Volume 1, *Proteins*, 2nd edn (P. F. Fox, ed.) pp. 111–40. Barking, Essex: Elsevier Applied Science.

Rollema, H. S. and Branches, J. A. (1989). A ^{1}H-NMR study of bovine casein micelles: influence of pH, temperature and calcium ions on micellar structure. *Journal of Dairy Research*, **46**, 417–25.

Schmidt, D. G. (1970a). The association of α_{s1}-casein at pH 6.6. *Biochimica et Biophysica Acta*, **207**, 130–38.

Schmidt, D. G. (1970b). Differences between the association of genetic variants B, C and D of α_{s1}-casein. *Biochimica et Biophysica Acta*, **221**, 140–42.

Schmidt, D. G. (1980). Colloidal aspects of the caseins. *Netherlands Milk and Dairy Journal*, **34**, 42–64.

Schmidt, D. G. (1982). Association of caseins and casein micelle structure, In *Developments in Dairy Chemistry*, Volume 1, (P. F. Fox, ed.) pp. 61–86. London: Applied Science Publishers.

Slattery, C. W. (1977). Model calculations of casein micelle size distribution. *Biophysical Chemistry*, **6**, 59–64.

Slattery, C. W. and Evard, R. (1973). A model for the formation and structure of casein micelles from subunits of variable composition. *Biochimica et Biophysica Acta*, **317**, 529–38.

Stradner, A., Sedgwick, H., Cardinaux, F., Poon, W. C. K., Egelhaaf, S. U. and Schurtenberger, P. (2004). Equilibrium cluster formation in concentrated protein solutions and colloids. *Nature*, **432**, 492–95.

Stothart, P. H. and Cebula, D. J. (1982). Small angle neutron scattering study of bovine casein micelles and sub-micelles. *Journal of Molecular Biology*, **160**, 391–95.

Swaisgood, H. E. (2003). Chemistry of the caseins. In *Advanced Dairy Chemistry,* Volume 1, *Proteins*, 3rd edn (P. F. Fox and P. L. H. McSweeney, eds) pp. 139–201. New York: Kluwer Academic/Plenum Publishers.

Walstra, P. (1979). The voluminosity of bovine casein micelles and some of its implications. *Journal of Dairy Research*, **46**, 317–23.

Walstra, P. (1990). On the stability of casein micelles. *Journal of Dairy Science*, **73**, 1965–79.

Walstra, P. (1998). Casein sub-micelles: do they exist? *International Dairy Journal*, **9**, 189–92.

6

Structure and stability of whey proteins

Patrick B. Edwards, Lawrence K. Creamer and Geoffrey B. Jameson

Abstract

The chemical and physical stability of the more common proteins of bovine and, where available, ovine, caprine and equine whey (β-lactoglobulin, α-lactalbumin, serum albumin, immunoglobulins and lactoferrin) is reviewed with regard to their molecular structures and dynamics. The behavior of the proteins separately and in combination, to temperature, pressure, pH, denaturants (such as guanidinium chloride and urea) and stabilizers (such as fatty acids, and metal ions) has been considered. Particular emphasis has been placed on studies that have utilized X-ray, NMR, fluorescence and circular dichroism techniques. Attention is directed to the role of cysteines and disulfide bridges with regard to chemical stability. Whereas there is considerable knowledge of structure–function relationships of individual proteins, there is a dearth of three-dimensional structural knowledge of combinations of proteins, despite clear importance of such knowledge to functionality, especially with regard to food processes.

Introduction

Information regarding whey protein structure and stability has great potential to facilitate knowledge-based product design. Recent reviews highlight the importance of knowledge of structure and stability, including the effects of pressure, temperature and

Milk Proteins: From Expression to Food
ISBN: 978-0-12-374039-7

Table 6.1 Typical protein composition of whey (based on Farrell *et al.*, 2004)

Protein	Proportion by mass (%)	No. amino acids	Molecular mass (Da)	Iso-ionic point	Disulfide bonds/thiols	Comments
β-Lactoglobulin (β-Lg)	60	162	18 363[a]	5.35	2/1	Two common variants, A and B
α-Lactalbumin (α-La)	20	123	14 178	4.80	4	About 10% of molecules are glycosylated
Bovine serum albumin (BSA)	3	583	66 399		17/1	Also present in blood serum
Immunoglobulin G (IgG)	10	>500	161 000 (G1)[b]	Many isoforms		Passive transfer of immunities
Lactoferrin (Lf)	<0.1	689	76 110	8.95	17	Bacteriostatic role; glycoprotein

[a] Molar mass for the A variant.
[b] G1 is the major immunoglobulin; two other classes, IgM and IgA, are present in much lower abundance.

chemical denaturants (Chatterton *et al.*, 2006; Lopez-Fandino, 2006a). In this chapter, we discuss the structures of the whey proteins shown in Table 6.1 under quiescent and destabilizing conditions (change in pH, temperature and pressure and addition of chaotropes) and in the presence or absence of small-molecule ligands. Particular emphasis is placed on information that has been obtained via high-resolution X-ray crystallographic and high-field nuclear magnetic resonance (NMR) studies. The results of these studies have been applied to many projects reported in the present volume (e.g. Chapters 7, 8, 9, 10 and 13).

Bovine β-lactoglobulin

β-Lg contains 162 amino acids and has a molecular weight of 18.3 kDa (Hambling *et al.*, 1992). It is a member of the lipocalin (a contraction of the Greek *lipos*, "fat, grease" and *calyx*, "cup") family of proteins (Banaszak *et al.*, 1994; Flower, 1996), so called because of their ability to bind small hydrophobic molecules into a hydrophobic cavity. This led to the proposal that β-Lg functions as a transport protein for retinoid species, such as vitamin A (Papiz *et al.*, 1986).

β-Lg is the most abundant whey protein in the milk of most mammals (≈10% of total protein or ≈50% of whey protein), but has not been detected in the milk of humans, rodents or lagomorphs. In the case of human milk, α-lactalbumin (see below) is the dominant whey protein. Bovine β-Lg is the most commonly studied milk protein.

There are ten known genetic variants of bovine β-Lg. The most abundant variants are labelled β-Lg A and β-Lg B (Farrell *et al.*, 2004) and differ by two amino acid substitutions, Asp64Gly and Val118Ala respectively. The quaternary structure of the protein varies among monomers, dimers or oligomers depending on the pH, temperature and ionic strength, with the dimer being the prevalent form under physiological conditions (Kumosinski and Timasheff, 1966; McKenzie and Sawyer, 1967; Gottschalk *et al.*, 2003). This variable state of association is likely to be the result of a delicate balance among hydrophobic, electrostatic and hydrogen-bond interactions (Sakurai *et al.*, 2001; Sakurai and Goto, 2002).

Molecular structure of bovine β-Lg

β-Lg was an early target of X-ray diffraction as newly applied at the Royal Institution to protein crystals. This was due to its high abundance and relatively easy purification from milk, and its propensity to form suitable crystals. In retrospect, this was a very ambitious project because β-Lg was not the easiest protein to analyze (Green *et al.*, 1979), partly because of the multiple crystal forms. Nevertheless, this study established that the protein monomer was near spherical with a block of electron density with a rod-like structure across one face.

The next attempt (Creamer *et al.*, 1983) to determine the structure was by calculation using sequence data and structural probabilities to estimate which portions of the amino acid sequence might form into the helices, strands and sheets. The secondary structure of β-Lg was predicted to comprise 15% α-helix, 50% β-sheet and 15–20% reverse turn (Creamer *et al.*, 1983). It is interesting to note that many of the residues that reside in the extended structures of the native protein have been shown to have a nascent propensity to form α-helical structures in the presence of trifluoroethanol or amphiphiles (Hamada *et al.*, 1995; Kuroda *et al.*, 1996; Chamani *et al.*, 2006).

In 1986, the first medium-resolution structure of β-Lg was published (Papiz *et al.*, 1986). Structural similarity to a seemingly different type of protein, plasma retinol-binding protein, has given rise to much speculation as to the role of β-Lg in bovine milk. Higher resolution structures subsequently revealed the now familiar eight-stranded β-barrel (calyx), flanked by a three-turn α-helix. A final ninth strand forms the greater part of the dimer interface at neutral pH (Papiz *et al.*, 1986; Bewley *et al.*, 1997; Brownlow *et al.*, 1997). The β-barrel is formed by two β-sheets, where strands A to D form one sheet and strands E to H form the other (with some participation from strand A, facilitated by a 90° bend at Ser21). Two disulfide bonds link Cys66 on loop CD (which, as its name suggests, connects strands C and D) with Cys160 near the C-terminus, and Cys106 on strand G with Cys119 on strand H, leaving Cys121 as a free, but unexposed, thiol. The loops connecting strands BC, DE and FG are relatively short whereas those at the open end of the barrel, strands AB, CD, EF and GH, are longer and more flexible. These features are illustrated in Figure 6.1.

The structures of the A and B variants are very similar. However, the Asp64Gly substitution results in the CD loop adopting different conformations (Qin *et al.*, 1999). The Val118Ala substitution causes no detectable change to the structures, but the void created by substituting the bulky isopropyl substituent with the smaller methyl group results in the hydrophobic core of the B variant being less well packed, and may account for the lower thermal stability of the B variant (Qin *et al.*, 1999).

Very careful titrimetric and thermodynamic measurements in the late 1950s (Tanford and Nozaki, 1959; Tanford *et al.*, 1959) established the presence of a carboxylate residue with an anomalously high pK_a value of 7.3. This was attributed to

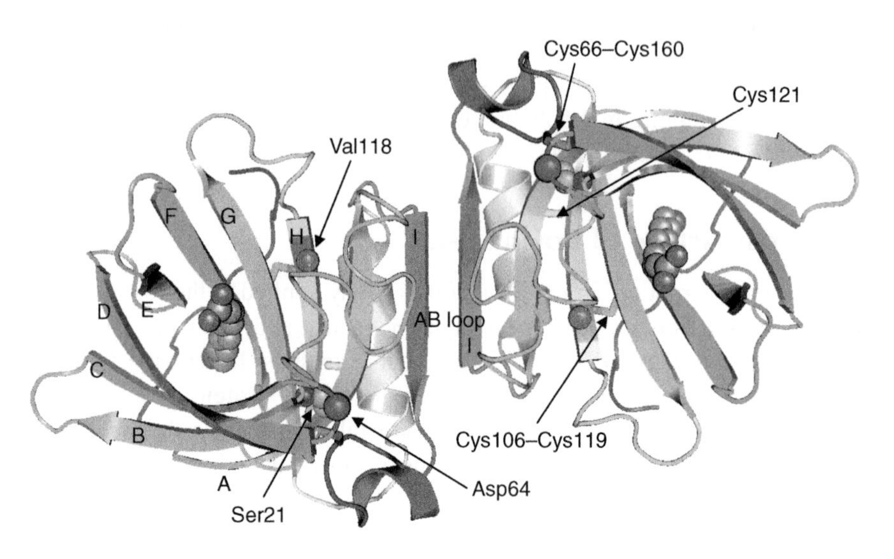

Figure 6.1 Diagram of the dimeric structure of bovine β-Lg A looking down the two-fold axis. The co-ordinates are taken from the structure of β-Lg A in the trigonal Z lattice with 12-bromododecanoic acid bound (PDB code: 1bso). The strands that form the β barrel are labeled A to H. The I strand, together with part of the AB loop, forms the dimer interface at neutral pH. The locations of the sites of difference between the A and B variants are also shown. The structure is rainbow colored, beginning with blue at the N-terminus and ending with red at the C-terminus. Ser21, which shows conformational flexibility, and the 12-bromododecanoate anion are shown as spheres. Figure drawn with PyMOL (Delano, 2002) (see also Plate 6.1).

a pH-dependent conformational change, a conclusion that rationalized earlier measurements of pH-dependent sedimentation coefficients (Pedersen, 1936) and specific optical rotation data (Groves *et al.*, 1951). Much later, X-ray structure analyzes (Qin *et al.*, 1998a) at pH values above and below this so-named Tanford transition established that, at pH 6.2, the EF loop is closed over the top of the barrel, burying Glu89 (the carboxylic acid with the anomalous pK_a) inside the calyx. At pH 8.1, this loop is articulated away from the barrel such that the formerly buried glutamic acid becomes exposed in the carboxylate form (Qin *et al.*, 1998a).

An early structure of bovine β-Lg crystallized in the presence of retinol appeared to show retinol bound externally to the protein (Monaco *et al.*, 1987), apparently later confirmed by a body of fluorescence data (Dufour *et al.*, 1994; Lange *et al.*, 1998; Narayan and Berliner, 1998). However, subsequent structural analyzes have shown that fatty acids, retinoid species (including vitamin A) and cholesterol (including vitamin D) all bind inside the calyx (Kontopidis *et al.*, 2004). Induced circular dichroism (CD) measurements and NMR measurements confirm the X-ray crystallographic observations. Ligand binding is discussed in more detail below, because it relates to the probable physiological function of β-Lg as well as to the stability of this molecule and current technological interest in the role of protein–ligand interactions in flavor perception (see Chapter 13). At this stage, there is no evidence for the binding of fatty acids or retinoid species outside the calyx. Except for very bulky ligands (see below), ligands bind inside the calyx of β-Lg at pH ≈7.

At about the same time as the Tanford transition and ligand-binding modes were elucidated by high-resolution X-ray crystallography (Qin *et al.*, 1998a, 1998b), NMR studies of β-Lg structure in low-pH solutions, where the protein is monomeric, were initiated (Ragona *et al.*, 1997; Fogolari *et al.*, 1998; Kuwata *et al.*, 1998; Uhrínová *et al.*, 1998). These NMR studies, described below, have provided proof of persistence of the tertiary structure down to pH 2 and have yielded a depth of insight into structural stability and protein dynamics that is not possible by standard X-ray crystallographic techniques.

Structure of bovine β-Lg in aqueous solution

NMR spectroscopy is used to obtain protein structures in solution (Cavanagh *et al.*, 1995). The technique is best suited to monomeric proteins with molecular weights $<\approx 25$ kDa and usually requires recombinant singly (^{15}N) or doubly ($^{15}N/^{13}C$) labeled material for molecules with molecular weights $>\approx 8$ kDa. Therefore, most NMR studies of bovine β-Lg have been at a pH of between 2 and 3 where the molecule is monomeric. Early studies using wild-type B-variant protein (Fogolari *et al.*, 1998) confirmed the presence of the eight-stranded β-barrel. However, the full structure (of the A variant) (Kuwata *et al.*, 1999; Uhrínová *et al.*, 2000) required the use of isotopically labeled recombinant material (Kim *et al.*, 1997; Denton *et al.*, 1998). As the full structure was determined by NMR techniques independently and near simultaneously by two groups from Tokyo and Edinburgh, this has provided objective and very useful comparisons (Jameson *et al.*, 2002a).

The solution structure was shown by both Kuwata *et al.* (1999) and Uhrínová *et al.* (2000) to have an overall similarity to that established earlier by X-ray crystallography at pH 6.2, despite the considerably lower pH (and concomitant increase in the protein's surface charge) necessary to obtain usable NMR spectra. The EF loop is firmly closed over the open end of the β-barrel at this pH and the side chain of the Glu89 "latch" is buried, as in the X-ray structures. The biggest difference, when compared with the Z lattice X-ray structure at pH 6.2 (Qin *et al.*, 1998a), is that the three-turn α-helix adopts a different position with respect to the β-barrel, possibly because of the pH-induced increase in positive charge on this part of the protein's surface (Uhrínová *et al.*, 2000). The lower pH was also found to move the conformation of the AB loop by up to 3.5 Å, which may be significant for the disruption of the dimer interface.

Further differences were found at the N- and C-termini, but these can be ascribed to limitations imposed by the use of recombinant protein with a non-native N-terminus for the NMR structure and possible crystal-packing effects at the C-terminus for the X-ray structure. In some crystal forms, much of the C-terminus from residues ≈152 to 162 is not observed or is very poorly defined in electron density maps.

Studies of bovine β-Lg by NMR at neutral pH

The large size of the bovine β-Lg dimer at pH 7 is expected to cause some broadening of the peaks in its ^{1}H NMR spectrum because of slower molecular reorientation. However, this problem is exacerbated by chemical exchange broadening of peaks in the vicinity of the dimer interface by the dynamic equilibrium of molecules in the associated or unassociated state. These factors render the resulting spectra unsuitable for structure determination. Several methods have been employed to allow NMR studies at neutral pH. The most straightforward of these has been to use a non-ruminant β-Lg that is intrinsically monomeric, yet with the same overall tertiary structure as the bovine protein, in this case equine β-Lg (Kobayashi *et al.*, 2000). Alternatively, the dimer interface may be disrupted by producing bovine β-Lg mutants with amino acid substitutions carefully chosen to disrupt the intermolecular interactions between either the I strands or the AB loops (Sakurai and Goto, 2002) (see Figure 6.1).

It is worth noting that an attempt to form dimeric equine β-Lg by producing a mutant with amino acid substitutions chosen to mimic those of the bovine protein at the interface was not successful (Kobayashi *et al.*, 2002), indicating that subtle features in β-Lg conformation remote from the interface have an impact on successful dimer formation. Indeed, reaction of the free thiol of Cys121 (located away from the interface in the H strand and covered by the main α-helix, Asp129–Lys141; see Figure 6.2) with 2-nitro-5-thiobenzoic acid produces a monomeric species with native structure at pH 2 and a monomeric but unfolded structure at pH 7 (Sakai *et al.*, 2000). The configuration of the α-helix is known to change with pH (Uhrínová *et al.*, 2000) and this may therefore also have an important influence on both the protein's stability and its quaternary state.

The third approach to overcome the problems of the rate constants for the dissociation/reassociation equilibrium of the bovine β-Lg dimer being in the intermediate exchange regime has been to covalently bond two monomers via an Ala34Cys mutant (Sakurai and Goto, 2006). This variant was used to study the dynamics of the EF loop across the Tanford transition (see the following section) and, more recently, to examine the nature of ligand binding to β-Lg (Konuma *et al.*, 2007). Although no full structure determination was reported, the amide chemical shifts of the mutant were within 0.1 ppm of those from monomeric β-Lg (except for seven residues that encompassed the substitution site). This fact, combined with the similarity of the mutant and wild-type β-Lg CD spectra, indicated that the tertiary structures of the mutant and wild-type proteins were similar.

Bovine β-Lg dynamics

Crystallographic atomic displacement parameters, often loosely referred to as temperature factors or just *B* factors, describe the spread of an atom's electron density in space and can therefore be used to infer residue-specific mobility. However, for surface residues, the *B* factors of both main-chain and side-chain atoms are highly sensitive to intermolecular crystal-packing contacts. Moreover, except where data to ultra-high resolution (better than 1.0 Å, which is not yet the case for any β-Lg) are available, similarity restraints are imposed on *B* values of adjacent atoms and residues along the polypeptide chain to ensure stable refinement.

Nonetheless, in the case of isomorphous structures at similar resolution (where structures share the same average *B* value, the same space group, very similar unit cell parameters and, hence, very similar intermolecular contacts), or in regions where non-isomorphous structures lack intermolecular contacts, some meaning can be

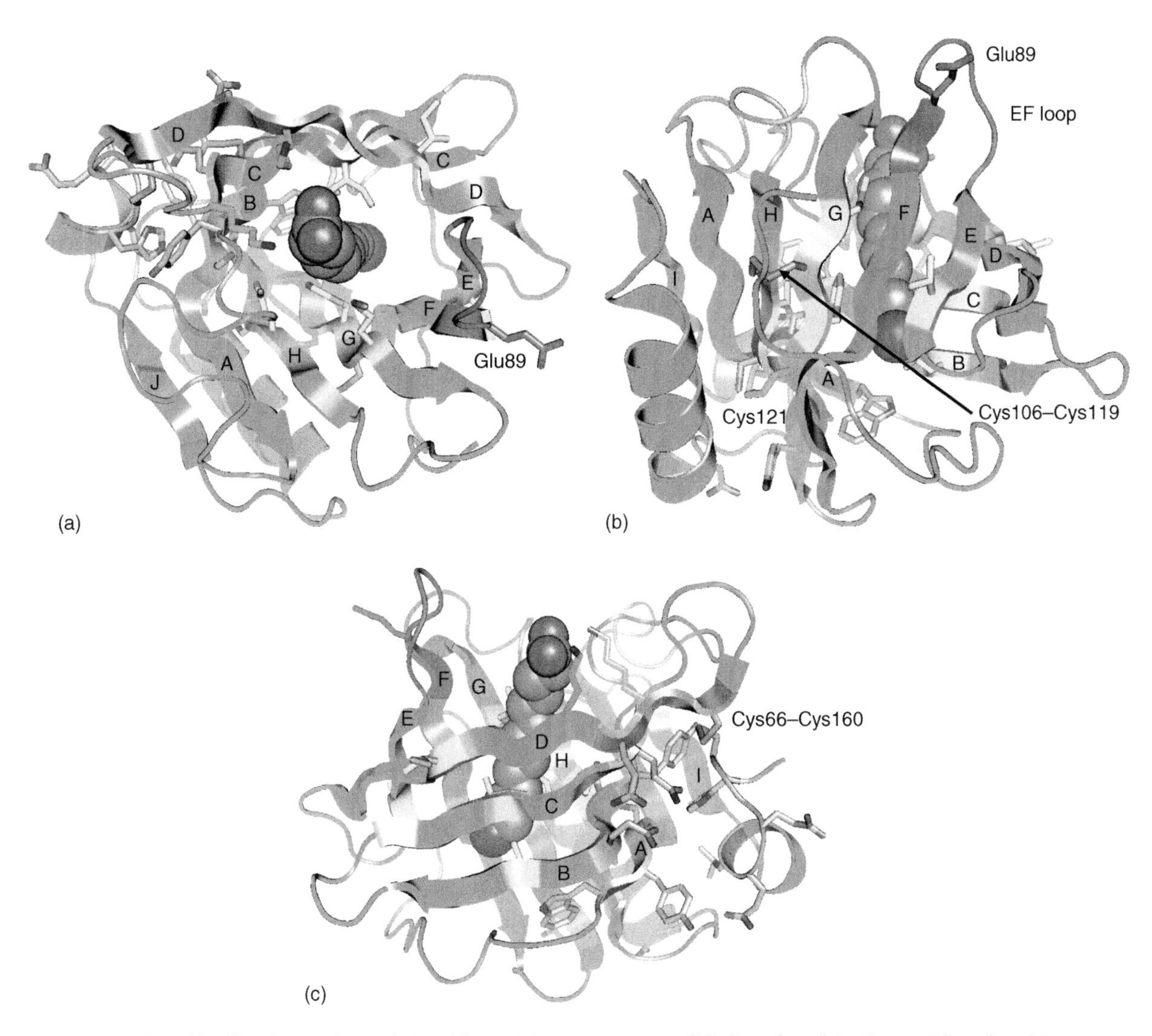

Figure 6.2 Ligand-binding sites on β-Lg as inferred from NMR measurements of binding of small (<12 atoms) ligands to β-Lg at acidic pH (Luebke *et al.*, 2002). The binding site of 12-bromododecanoic acid is shown for reference (Qin *et al.*, 1998b). (a) View into the calyx showing the primary binding site of fatty acids at pH ≈7 and of flavor components at pH ≈2, highlighted in yellow. (b) Secondary binding site for flavor components at pH 2 at N-terminal ends of strands A, B, C and D, and the C-terminal strand, highlighted in pink. (c) Secondary binding site for flavor components at pH 2 adjacent to the three-turn helix and strand G, highlighted in cyan. To show more clearly the attachment of side chains to the main chain, loops and strands have not been smoothed. The pH-sensitive EF loop is colored in magenta. Figure drawn with PyMOL (Delano, 2002) using co-ordinates with PDB code 1bso (see also Plate 6.2).

placed on differences observed in *B* factors. These differences can be both within a particular structure and between structures determined, for example, at different pH or in the presence/absence of added ligands. High *B* factors, indicating apparent high mobility, can also arise from a distribution of slightly different yet immobile conformations or from errors in model building. For these reasons, it is advantageous to study dynamics of the protein (particularly those of the backbone) by NMR techniques, using uniformly ^{15}N-labeled protein.

Flexibility on the nanosecond timescale can be inferred from low ^{15}N steady-state nuclear Overhauser effect values. As might be expected, mobile residues for β-Lg tend to have highly accessible surface areas (and such residues identified in NMR studies correlate in general with those that have relatively high *B* factors; Kuwata *et al.*, 1999). Slower conformational exchange processes can be indicated by large values for the ratio of the *T*1 (spin–lattice) and *T*2 (spin–spin) relaxation times of ^{15}N nuclei. Such residues include Ser21 at the midpoint (kink) of the A strand, possibly caused by fluctuations of the barrel, and residues 61 and 66 at either end of the CD loop, consistent with a slow segmental or hinging motion of this loop (Uhrínová *et al.*, 2000). All three sets of relaxation parameters can be analyzed in concert using the extended model-free formalism to give amplitudes and timescales for the internal motion of the backbone N–H bond vectors (Lipari and Szabo, 1982; Clore *et al.*, 1990).

When applied to variant A of β-Lg, this method confirms the above observations, but also identifies a number of residues in the EF loop undergoing substantial conformational change (Edwards *et al.*, 2003). Preliminary results also suggest that the B variant is more mobile relative to the A variant at the Asp64Gly (Gly in variant B) substitution site, whereas the dynamics at the Val118Ala (Ala in variant B) substitution site are very similar (Edwards *et al.*, 2003).

NMR measurements of the dynamics of the covalently bonded Ala34Cys mutant dimer have recently given complementary information regarding the structural changes associated with the Tanford transition established previously using X-ray crystallography (Qin *et al.*, 1998a). The ^{15}N dynamics of the EF loop measured either side of the transition indicate a three-step process. With increasing pH, the first event is a conformational change to the GH loop. This is followed by the breaking of hydrogen bonds at the hinges of the EF loop followed by the subsequent articulation of the EF loop away from the calyx (Sakurai and Goto, 2006). The dynamic flexibility of the EF loop at pH $\approx$ 2 is important, as it means that, at low pH, neither ingress into nor egress from the hydrophobic pocket is kinetically prevented.

Structures of β-Lgs from other species

Equine (horse) β-Lg, which shares 58% identity with bovine β-Lg, has been shown to be monomeric over a wide pH range, whereas porcine (pig) β-Lg, which shares 63% identity with bovine β-Lg, is dimeric below pH 5 and monomeric at pH 5 and above (in contrast to bovine β-Lg). At pH 7, both equine β-Lg and porcine β-Lg are monomeric and therefore amenable to NMR study. Equine β-Lg has been extensively studied by NMR with regard to denaturation processes, but a full structural characterization by either NMR or X-ray methods has yet to be published.

The X-ray crystal structure of porcine β-Lg at pH 3.2 clearly revealed a dimeric structure formed by domain swapping of N-terminal regions, a quaternary structure quite different from that observed for bovine β-Lg (Hoedemaeker *et al.*, 2002). The EF loop adopts the closed conformation over the calyx, as found also for bovine β-Lg at acidic pH, consistent with the notion that this loop acts as a lid to the calyx. However, the porcine protein is much less conformationally stable at acidic pH than its bovine counterpart (Burova *et al.*, 2002; Invernizzi *et al.*, 2006), which has led to questioning of the role of β-Lg as a transporter of hydrophobic molecules through the acidic environment of the gut (Burova *et al.*, 2002). Despite 63% identity between bovine β-Lg and porcine β-Lg, the RMS difference in C_α positions between these two structures is remarkably high at 2.8 Å, although inspection of the two structures shows that the core β-barrel structure superimposes closely and that these differences are concentrated in the flexible loop regions, which comprise nearly a third of the structure.

The 2.1 Å resolution structure of rangiferine (reindeer) β-Lg at pH ≈ 6.5 was published recently (Oksanen *et al.*, 2006). Both the monomeric tertiary structure and the dimeric quaternary structure are very similar to those of bovine β-Lg, which is not unexpected as polypeptide lengths are identical and sequence identity is greater than 94%. At pH ≈ 6.5, the EF loop is observed to be in the closed position. There are few structural data on ovine (sheep) and caprine (goat) β-Lgs, despite the commercial importance of their milk.

Although β-Lg is absent from human milk, two other secreted lipocalins share limited sequence identity (but close structural similarity) with bovine β-Lg. Tear lipocalin is the major protein in tears, and structural and functional studies indicate that it, like β-Lg, binds a broad range of hydrophobic molecules (Glasgow *et al.*, 1995). Glycodelin, a heavily glycosylated lipocalin, is found in the human endometrium in early pregnancy where its function remains unclear. It has been suggested (Kontopidis *et al.*, 2004) that β-Lg arose by gene duplication of glycodelin and exists in milk solely for nutritive value. However, there appears to have been considerable selection pressure to retain not only the Glu89 in all sequences of the β-Lgs found to date but also the pH-sensitive conformational switch for the EF loop of all β-Lgs (and indeed for tear lipocalin) that have been functionally characterized. Such preservation is not consistent with a solely nutritive role for β-Lg, although the origin of the β-Lg gene from glycodelin remains an intriguing possibility.

Ligand binding to β-Lg

β-Lg has the ability to bind a large number of small molecules (for a review, see Sawyer *et al.*, 1998), although the location of the bound ligand continues to generate controversy (Dufour *et al.*, 1994; Lange *et al.*, 1998; Narayan and Berliner, 1998). However, in recent years, NMR techniques have been used to map ligand-binding sites on the protein through changes in chemical shift and relaxation times of protein residues.

Another technique that provides reliable evidence of binding to the protein is induced CD, where an achiral chromophore "lights up" in the CD spectrum on being placed into a chiral environment on or in the protein. This technique was applied to bovine β-Lg by us (Creamer *et al.*, 2000) to show unequivocally that retinol and fatty-acid binding (e.g. palmitic and *cis*-parinaric acids) is competitive. The binding of these and other chirally active ligands, including *trans*-parinaric acid and retinoic acid, was explored over a range of pH and with non-chromophoric fatty-acid ligands to gain an understanding of the parameters surrounding the Tanford transition. More recently, induced CD has been used to study the binding of these (Zsila *et al.*, 2002) and other ligands (Zsila, 2003; Zsila *et al.*, 2005), including piperine, to β-Lg and to other lipocalins.

Ligand binding has also been extensively studied by fluorescence spectroscopy in which, typically, the fluorescence from tryptophan residues is monitored for changes, which may be positive or negative, that are interpreted as being due to ligand binding. For bovine β-Lg, one tryptophan (Trp19) is buried and is part of the highly conserved lipocalin Gly–X–Trp motif at the beginning of strand A (see Figure 6.2b), whereas the other (Trp61) is largely exposed and is part of the mobile CD loop. Fluorescence by both tryptophans is sensitive to small changes in the positions of nearby quenchers, respectively a nearby charged side chain of Arg124 and the Cys66–Cys160 disulfide bond. NMR spectroscopy, induced CD and X-ray results clearly indicate that interpretation of fluorescence measurements for evidence of ligand binding poses hazards.

Ragona *et al.* (2003) used a combination of electrostatic calculations, docking simulations and NMR measurements to suggest that the pH-dependent conformational change of the EF loop triggered by the protonation of Glu89 is common to all β-Lgs and that ligand binding (of palmitic acid) is determined by the opening of this loop. In earlier work using ^{13}C-labelled palmitic acid, this group had shown that the ligand also undergoes conformational change with increasing pH (Ragona *et al.*, 2000).

Recent NMR studies of the binding of palmitic acid to the "NMR friendly" Ala34Cys mutant dimer of bovine β-Lg have indicated that, although a rigid connection is made by the protein with the ligand at the bottom of the calyx, the interaction at the open end of the calyx is more dynamic (Konuma *et al.*, 2007). These observations complement those of Ragona *et al.* (2003) and also the X-ray studies on the binding of fatty acids (Qin *et al.*, 1998b; Wu *et al.*, 1999), which showed the carboxylate head group to be substantially less well ordered than the hydrophobic tail. The results of the study with the Ala34Cys mutant suggest that it is the plasticity of the D strand and the EF and GH loops that allows β-Lg to accommodate such a wide range of ligands (Konuma *et al.*, 2007). With the exception of changes in conformations of the side chains of Phe105 and Met107, NMR and X-ray studies show that the core lipocalin structure remains invariant upon ligand binding.

Crystallographic data clearly show that both fatty acids and retinol bind in the calyx (Qin *et al.*, 1998b; Sawyer *et al.*, 1998; Wu *et al.*, 1999). Although all crystallographic studies of ligand binding have been under conditions of high ionic strength, congruence of these data with NMR and induced CD data collected under conditions of low ionic strength indicate that the X-ray results are not an artifact of ionic strength. The preservation of the structure of β-Lg, in particular the hydrophobic cavity, at conditions of near-zero ionic strength at the pI of the protein (≈5.3) (Adams *et al.*, 2006) further demonstrates that the primary, and possibly only, ligand-binding site at ≈pH 7 is inside the calyx.

Although ligands such as palmitic acid appear to be released at acid pH (Ragona *et al.*, 2000), NMR evidence (based on perturbations of backbone chemical shifts) for the binding of the flavor compounds γ-decalactone and β-ionone at pH 2 has been reported (Luebke *et al.*, 2002; Tromelin and Guichard, 2006). Therefore, there is evidence for three binding sites to β-Lg: the canonical site inside the calyx;

a second site involving perturbation of residues Trp19, Tyr20, Tyr42, Glu44, Gln59, Gln68, Leu156, Glu157, Glu158 and His161; and a third site involving residues Tyr102, Leu104 and Asp129. These binding sites are illustrated in Figure 6.2.

Initially, it was thought that porcine β-Lg did not bind fatty acids (Frapin *et al.*, 1993). However, recent NMR studies have shown that the pH for 50% uptake of ligand has shifted by nearly 4 pH units from ≈5.8 for bovine β-Lg to 9.7 for porcine β-Lg, whereupon the EF loop undergoes a structural change analogous to that of its bovine counterpart (Ragona *et al.*, 2003).

Effect of temperature on bovine β-Lg

The thermal properties of β-Lg variants are of considerable commercial relevance because of their role in the fouling of processing equipment as well as the functional qualities that can be imparted to dairy products by thermally induced β-Lg aggregation. Consequently, this aspect of the protein's behavior has been the focus of extensive experimental work.

At neutral pH, the midpoint of the thermal unfolding transitions, as determined by differential scanning calorimetry (DSC), is ≈70°C (de Wit and Swinkels, 1980), whereupon the protein dimer dissociates and the constituent molecules begin to unfold. This reveals the free thiol of Cys121 (located at the C-terminal end of the H strand—see Figure 6.1) and a patch of hydrophobic residues, leading to the possibility of both covalent and hydrophobic intermolecular association (Qi *et al.*, 1995; Iametti *et al.*, 1996). The ensuing disulfide interchange reactions lead to the formation of a variety of mixed disulfide-bonded polymeric species (Creamer *et al.*, 2004).

Genetically engineered mutants with an extra cysteine positioned to allow a third disulfide bond to be formed to Cys121 have been shown both to retard thermal denaturation by 8–10°C and to resist heat-induced aggregation (Cho *et al.*, 1994). In mixtures of bovine β-Lg, α-lactalbumin (α-La) and bovine serum albumin (BSA), or of β-Lg and one or other of α-La and BSA at pH 6.8 subjected to high temperatures, homo- and heteropolymeric disulfide-bridged species were observed (Havea *et al.*, 2001). The formation of α-La–α-La disulfide links (α-La has no free cysteine; see below) is attributed to catalysis by BSA or β-Lg (Havea *et al.*, 2001). At low pH, where the protein is monomeric, denaturation is largely reversible at temperatures below 70°C (Pace and Tanford, 1968; Alexander and Pace, 1971; Mills, 1976; Edwards *et al.*, 2002). Heating above this temperature leads to the formation of large aggregates, but, in contrast to the behavior at neutral pH, the species are predominantly non-covalently bonded (Schokker *et al.*, 2000).

The precise denaturation process is complex and is influenced by factors such as pH, protein concentration, ionic environment, genetic variant and presence of ligands. Both lowering the pH (Kella and Kinsella, 1988; Relkin *et al.*, 1992) and adding calyx-bound ligands (Puyol *et al.*, 1994; Considine *et al.*, 2005a; Busti *et al.*, 2006) make the protein more resistant to thermal unfolding. The stability of the genetic variants (at pH 6.7) appears to decrease in the order C > A>B, with the A variant showing the least co-operative unfolding transition (Manderson *et al.*, 1997). The protein's susceptibility to thermal denaturation at pH 6.7–8 is strongly concentration dependent up to about 6 mM, being most susceptible to unfolding at a concentration of ≈1.4 mM (Qi *et al.*, 1995). It is possible that, at high protein concentration (≈6 mM), tertiary structure is lost directly from the native dimer state (Qi *et al.*, 1995).

There is some evidence that the thermal unfolding occurs in more than one step. Kaminogawa *et al.* (1989) used antibody binding affinities to propose that thermal unfolding of variant A of β-Lg occurs in at least two stages, starting with conformational changes near the N-terminus followed by changes in the region of the three-turn α-helix. Fourier-transform infrared (FT-IR) measurements by Casal *et al.* (1988) have also indicated a loss of helical content early in the denaturation process (using variant B of β-Lg in 50 mM phosphate buffer at pH 7). Qi *et al.* (1997) used FT-IR and CD measurements to propose that variant A of β-Lg forms a molten globule with reduced β structure when heated above 65°C in 30–60 mM NaCl at pH 6.5. NMR studies, observing hydrogen/deuterium (H/D) exchange of the backbone amide protons of β-Lg A at pH 2–3, have revealed a stable core comprising the FG and H strands, possibly stabilized by the Cys106–Cys119 disulfide bond between strands F and G (Belloque and Smith, 1998; Edwards *et al.*, 2002). Significant secondary structure even at a temperature as high as 90°C has been reported (Casal *et al.*, 1988; Qi *et al.*, 1997; Bhattacharjee *et al.*, 2005).

Effect of pressure on bovine β-Lg

High-pressure treatment of food is of increasing commercial importance because of increasing consumer demand for products that have been subjected to minimal processing damage. Pressure treatment as part of the processing regime has the potential to produce dairy products with improved functional and organoleptic properties compared with those produced by thermal treatment alone (Messens *et al.*, 2003).

Of the major whey proteins, β-Lg is the most susceptible to pressure-induced change (Stapelfeldt *et al.*, 1996; Patel *et al.*, 2005). Presumably this is due to its relatively inefficient packing, caused by the presence of the β-barrel with its large solvent-exposed hydrophobic pocket, and the lower number of disulfide bonds (two compared with four in, for example, the similar-sized α-La). A reduction in the molar volume of bovine β-Lg has been detected at pressures as low as 10 MPa, possibly because of a contraction of the calyx (Vant *et al.*, 2002). A number of studies have shown that β-Lg becomes more susceptible to enzymatic cleavage when exposed to pressure, possibly because of pressure-induced conformational change. The free cysteine has been shown to become exposed at between 50 and 100 MPa (Stapelfeldt *et al.*, 1996; Tanaka and Kunugi, 1996; Moller *et al.*, 1998).

Exposure of the protein to pressures in excess of ≈300 MPa causes irreversible changes to β-Lg's tertiary and quaternary structure. A combination of CD and fluorescence spectroscopy of β-Lg at neutral pH exposed to pressures as high as 900 MPa indicated that pressure induces monomer formation with subsequent aggregation, but with only small irreversible effects on β-Lg tertiary structure (Iametti *et al.*, 1997). However, more recent results from tryptic hydrolysis suggest that, whereas exposure to pressures below 150 MPa has no detectable permanent effect on β-Lg A's conformation, pressures above 300 MPa lead to the detachment of strands D and G from the β-barrel together with the formation of disulfide-bonded oligomers (Knudsen *et al.*, 2002).

In mixtures of bovine β-Lg with either α-La or BSA at pH 6.6 subjected to high pressures, intermolecular disulfide-bridged aggregates form only between β-Lg and itself. No β-Lg–α-La or β-Lg–BSA disulfide-bridged species are detected (Patel *et al.*, 2005), in contrast to heat-treated mixtures where such species are observed (Havea *et al.*, 2001).

In order to correlate the pressure-induced conformational changes with the protein's primary sequence, Belloque *et al.* (2000) made NMR amide H/D exchange observations of β-Lg A and B following exposure of solutions at neutral pH to pressures of up to 400 MPa. Little H/D exchange was reported at 100 MPa, which indicated that any conformational change that occurred did not increase the exposure of most amide protons to the solvent compared with their exposure in the native conformation at ambient pressure. A large increase in the extent of H/D substitution at 200 MPa and above indicated increased conformational flexibility, but the similarity of the spectra of control samples recorded in H_2O rather than D_2O before and after pressurization demonstrated that any pressure-induced conformational changes were largely reversible up to 400 MPa. The authors proposed that the structure of the A variant was more sensitive to changes in pressure than that of the B variant and that the F, G and H strands of the protein's β-barrel were the most resistant to conformational change, the latter conclusion paralleling the effects of temperature (Belloque and Smith, 1998; Edwards *et al.*, 2002).

FT-IR and small-angle X-ray scattering experiments suggest that, even at 1 GPa, the unfolded state contains significant secondary structure (Panick *et al.*, 1999). Combined application of pressure and heat has shown that changing the temperature over the range from 5 to 37°C has negligible effect on the susceptibility of β-Lg to pressures up to 200 MPa (Skibsted *et al.*, 2007). However, combined application of pressure and moderate temperature at 600 MPa/50°C (Yang *et al.*, 2001) and 294 MPa/62°C (Aouzelleg *et al.*, 2004) has indicated the formation of a molten globule with an α-helical structure on the basis of results obtained from CD spectroscopy.

It should be noted therefore that the potential for temperature increases induced by rapid pressurization of the sample needs to be considered when studying the effects of pressure on protein conformation and stability.

Enyzmatic proteolysis observations indicate that β-Lg is less susceptible to pressure-induced change at acidic pH (Dufour *et al.*, 1995) than at neutral or basic pH. Nevertheless, NMR measurements of monomeric β-Lg at pH 2 while under pressure at up to 200 MPa have shown that the two β-sheets unfold independently to form two intermediates in an unfolded state that still appears to contain significant secondary structure (Kuwata *et al.*, 2001).

A three-step mechanism has been proposed for β-Lg denaturation at neutral pH and ambient temperature, which broadly encompasses the above observations: a pressure of 50 MPa causes partial collapse of the calyx (with concomitant reduction in ligand-binding capacity) together with exposure of Cys121. Increasing the pressure to 200 MPa causes further (partially reversible) disruption to the hydrophobic structure together with a decrease in the molecular volume. Higher pressures cause irreversible aggregation reactions involving disulfide interchange reactions (Stapelfeldt and Skibsted, 1999; Considine *et al.*, 2005b).

Effect of chemical denaturants on bovine β-Lg

Chemical denaturants are often used to unfold proteins and to characterize mechanisms and transition states of protein-folding processes. Commonly used denaturants include alcohols, particularly 2,2,2-trifluoroethanol (TFE), urea and guanidinium chloride (GdmCl).

Theoretical calculations predict a significantly higher amount of α-helical secondary structure than is actually observed in native β-Lg (Creamer *et al.*, 1983; Nishikawa and Noguchi, 1991). That is, the native structure is the result of competition between α-helix-favoring local interactions and β-sheet-forming long-range interactions. However, addition of alcohols such as TFE can disturb this balance by weakening the hydrophobic interactions and strengthening the helical propensity of the peptide chain (Thomas and Dill, 1993). The ability to increase the α-helical content of bovine β-Lg by the addition of alcohols (ethanol, 1-propanol, 2-chloroethanol) was first demonstrated by Tanford *et al.* (1960) using optical rotary dispersion measurements.

Contemporary studies tend to favor the use of TFE, where the β-Lg β-sheet-to-α-helix transition has been shown to be highly co-operative, occurring over the range ≈15–20% v/v of cosolvent (Shiraki *et al.*, 1995; Hamada and Goto, 1997; Kuwata *et al.*, 1998). The higher proportion of α-helical structure in the so-called TFE-state is found in the N-terminal half of the molecule (Kuwata *et al.*, 1998). Magnetic relaxation dispersion measurements of the solvent nuclei have shown that this state is an open, solvent-permeated structure (unlike the collapsed state of a molten globule) and that its formation is accompanied by a progressive swelling of the protein with increasing TFE concentration (Kumar *et al.*, 2003). High protein and TFE concentration (8% v/w and 50% v/v respectively) can lead to fibrillar aggregation and gel formation of bovine β-Lg at both acid and neutral pH (Gosal *et al.*, 2002).

The ability of urea to induce protein unfolding is thought to be via a combination of hydrogen-bond formation with the protein backbone and a reduction in the magnitude of the hydrophobic effect (Bennion and Daggett, 2003). Therefore, in contrast to TFE, both the helical propensity and the hydrophobic effect are reduced. Urea-induced unfolding of bovine β-Lg at acidic pH was first reported as a two-state process (Pace and Tanford, 1968). Subsequent NMR H/D exchange measurements of bovine β-Lg B at pH 2.1 also allowed the urea-induced unfolding to be well approximated as a two-state transition between folded protein and the unfolded state via a co-operative unfolding of the β-barrel and the C-terminus of the major α-helix (Ragona *et al.*, 1999). However, Dar *et al.* (2007) have recently provided evidence that urea also causes unfolding via an intermediate, albeit with structural properties between those of the native and unfolded states. Addition of anionic amphiphiles, sodium dodecyl sulfate (SDS) or palmitate, causes β-Lg to resist urea-induced unfolding because of binding inside the calyx (Creamer, 1995).

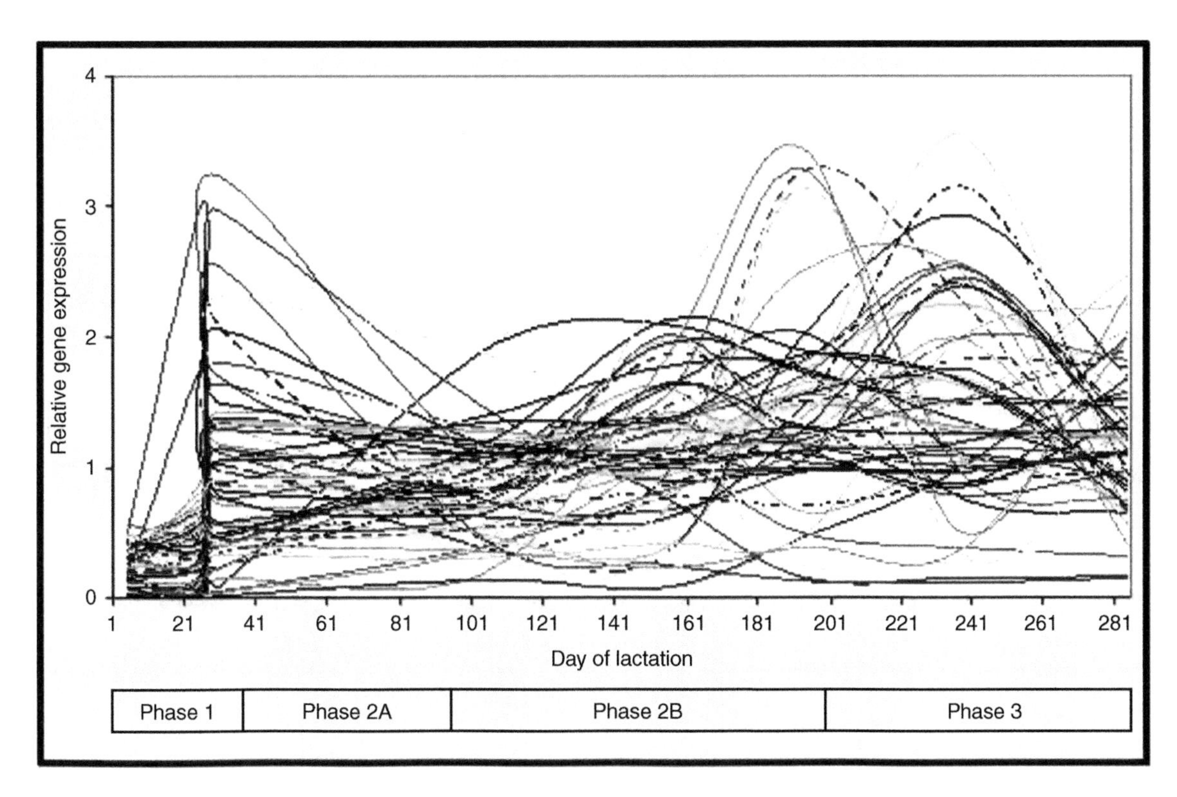

Plate 2.6 Microarray analysis of genes coding for secreted proteins in the tammar wallaby mammary gland during lactation. The phases of lactation are described in Figure 2.2. A total of 75 gene transcripts are shown (see also Figure 2.6).

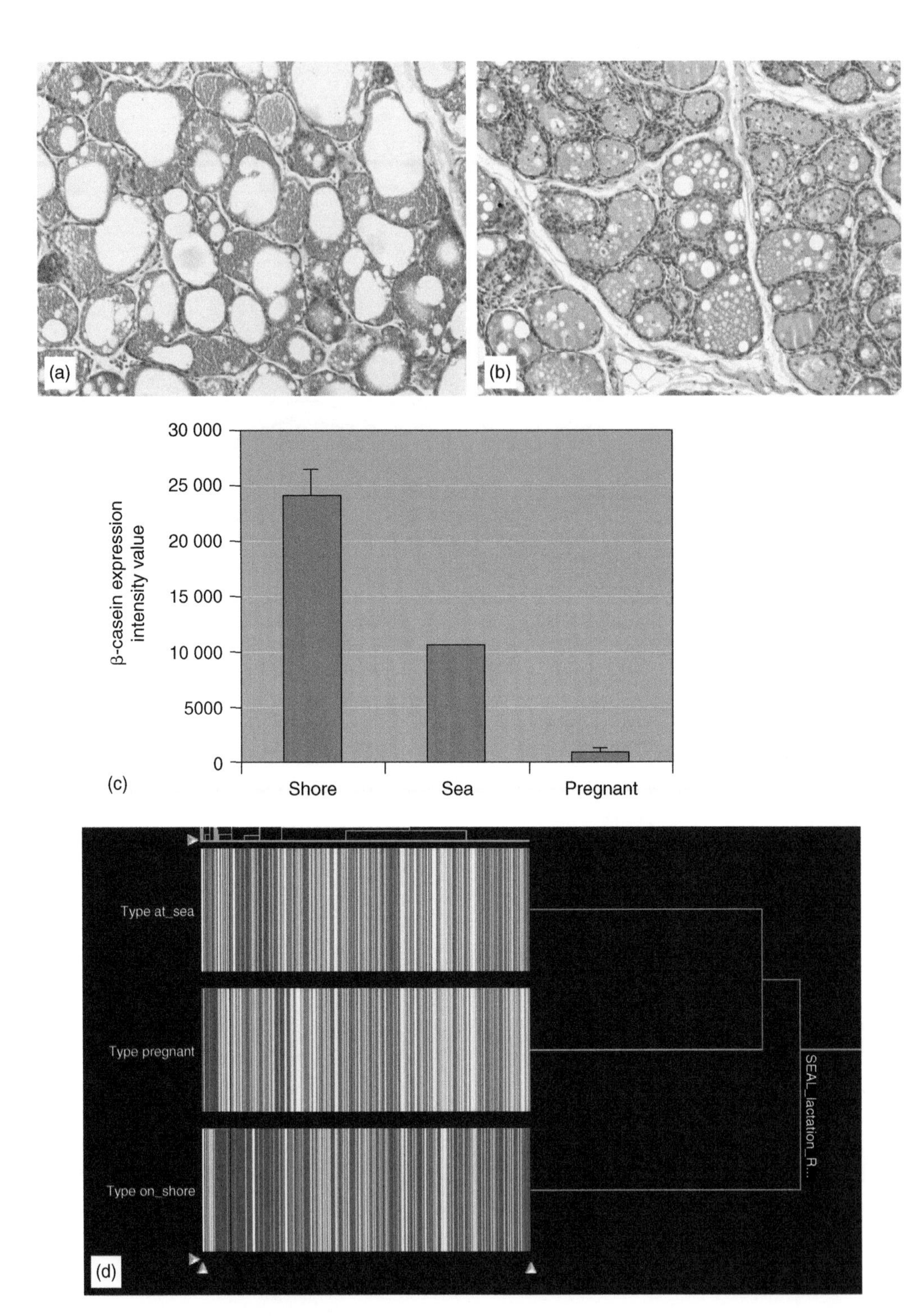

Plate 2.11 Histological sections of the mammary gland from Cape fur seals: (a) lactating while nursing on-shore and (b) lactating while foraging at sea. Sections are stained with hematoxylin and eosin. Magnification x 100. (c) Milk protein gene expression. β-Casein expression during Cape fur seal lactation cycle. Analysis of expression using canine Affymetrix chips hybridized to cDNA probes generated from RNA from pregnant (placental gestation and non-lactating, $n = 2$), lactating on-shore ($n = 2$) and lactating at sea ($n = 1$) (animals in embryonic diapause) Cape fur seals. (d) Cluster analysis of gene expression profiles from the Cape fur seal mammary gland during different stages of lactation. A total of 1020 Cape fur seal mammary messenger RNA (mRNA) transcripts were identified with expression levels above an intensity of 250 in any sample type. Hierarchical clustering was conducted using Euclidean distance. Pregnant and on-shore lactating data represent an average of two animals. Off-shore data represent a single sample. Reprinted from Current Topics in Developmental Biology, **72**, G. P. Schatten, snr ed., J. A. Sharp, K. N. Cane, C. Lefevre, J. P. Y. Arnold, and K. R. Nicholas, *Fur Seal Adaptations to Lactation: Insights into Mammary Gland Function*, pp. 276–308, New York: Academic Press. Copyright 2006, with permission from Elsevier (see also Figure 2.11).

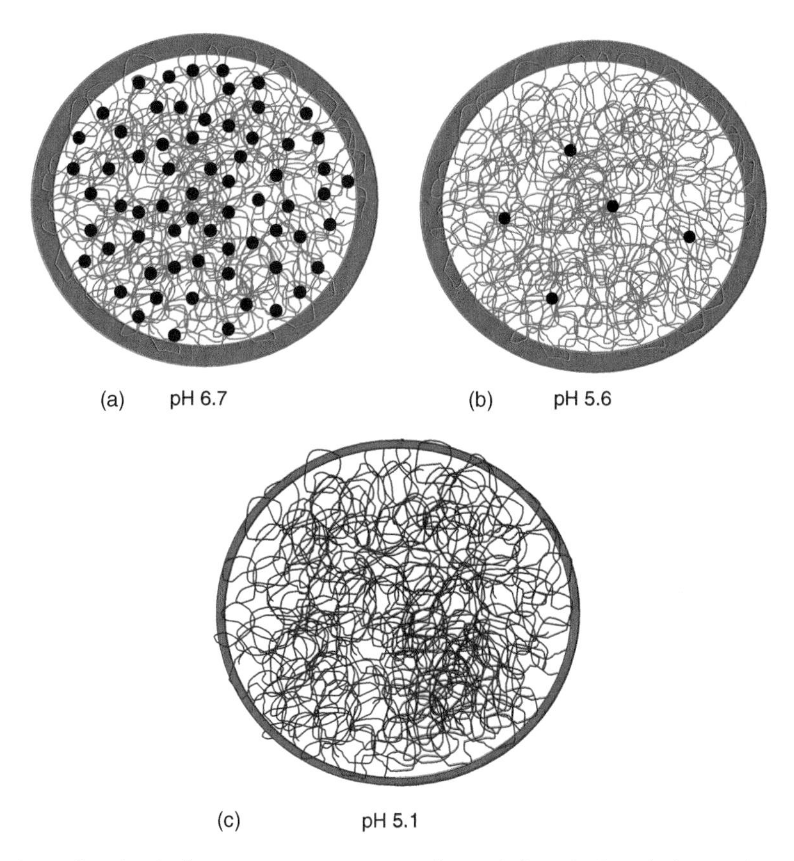

Plate 5.4 Representations of casein micelle structures at various pH values as indicated. The pale chains indicate protein molecules, where they cross being a hydrophobic interaction junction. The small black circles are the calcium phosphate nanoclusters that are solubilized when the pH is lowered. The outer circle is indicative of the range of steric repulsion generated between micelles and preventing interaction of the surface protein chains (see also Figure 5.4).

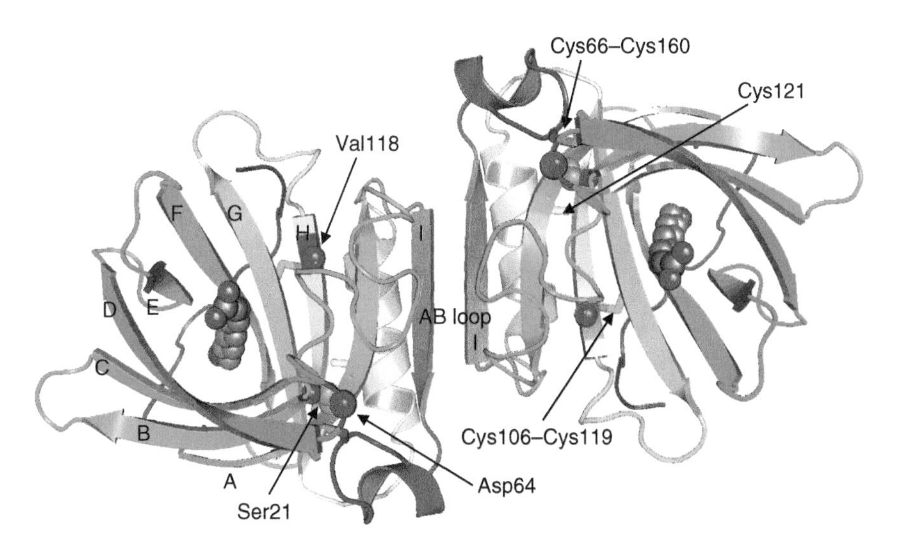

Plate 6.1 Diagram of the dimeric structure of bovine β-Lg A looking down the two-fold axis. The co-ordinates are taken from the structure of β-Lg A in the trigonal Z lattice with 12-bromododecanoic acid bound (PDB code: 1bso). The strands that form the β barrel are labeled A to H. The I strand, together with part of the AB loop, forms the dimer interface at neutral pH. The locations of the sites of difference between the A and B variants are also shown. The structure is rainbow colored, beginning with blue at the N-terminus and ending with red at the C-terminus. Ser21, which shows conformational flexibility, and the 12-bromododecanoate anion are shown as spheres. Figure drawn with PyMOL (Delano, 2002) (see also Figure 6.1).

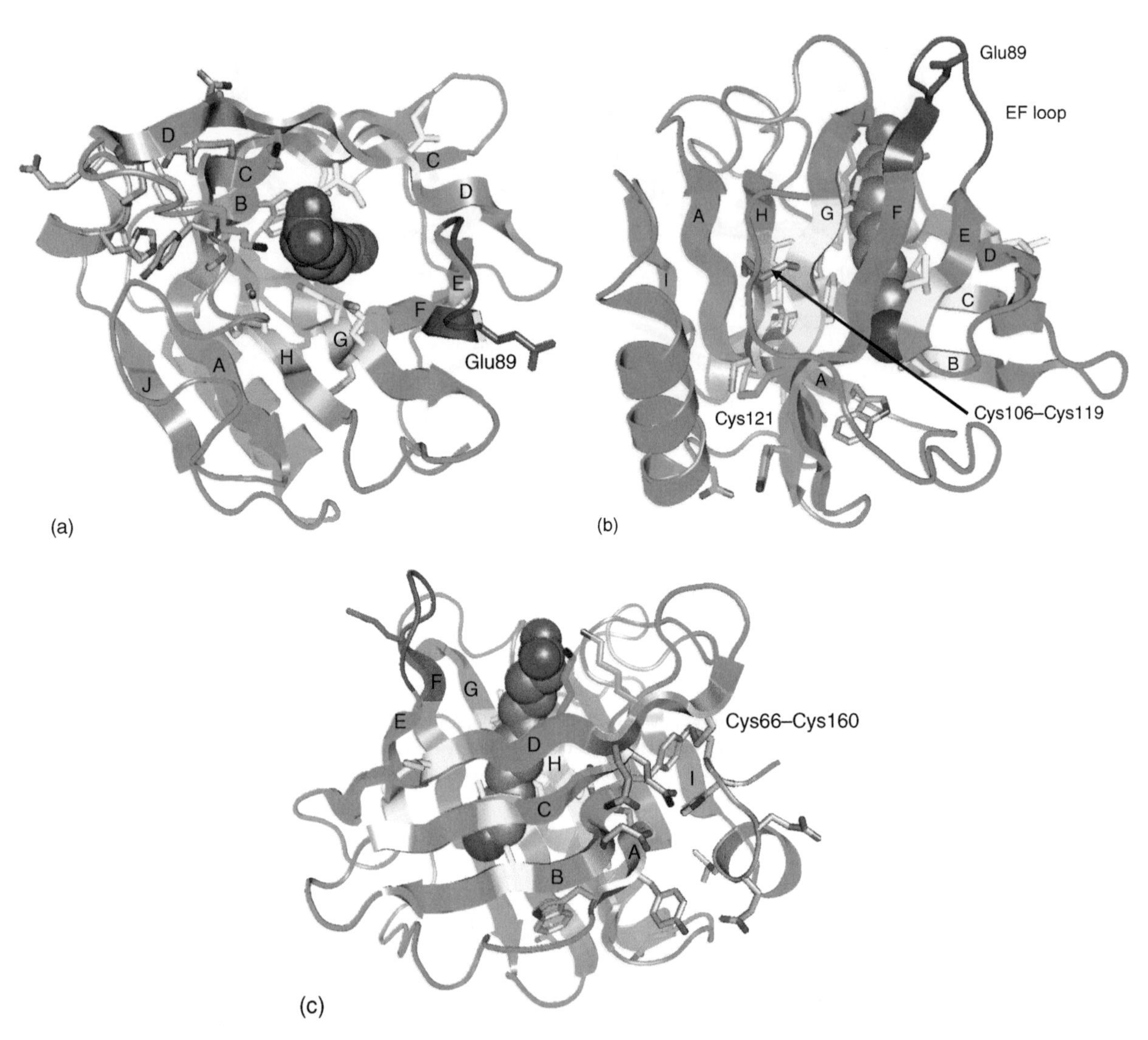

Plate 6.2 Ligand-binding sites on β-Lg as inferred from NMR measurements of binding of small (<12 atoms) ligands to β-Lg at acidic pH (Luebke *et al.*, 2002). The binding site of 12-bromododecanoic acid is shown for reference (Qin *et al.*, 1998b). (a) View into the calyx showing the primary binding site of fatty acids at pH ≈7 and of flavor components at pH ≈2, highlighted in yellow. (b) Secondary binding site for flavor components at pH 2 at N-terminal ends of strands A, B, C and D, and the C-terminal strand, highlighted in pink. (c) Secondary binding site for flavor components at pH 2 adjacent to the three-turn helix and strand G, highlighted in cyan. To show more clearly the attachment of side chains to the main chain, loops and strands have not been smoothed. The pH-sensitive EF loop is colored in magenta. Figure drawn with PyMOL (Delano, 2002) using co-ordinates with PDB code 1bso (see also Figure 6.2).

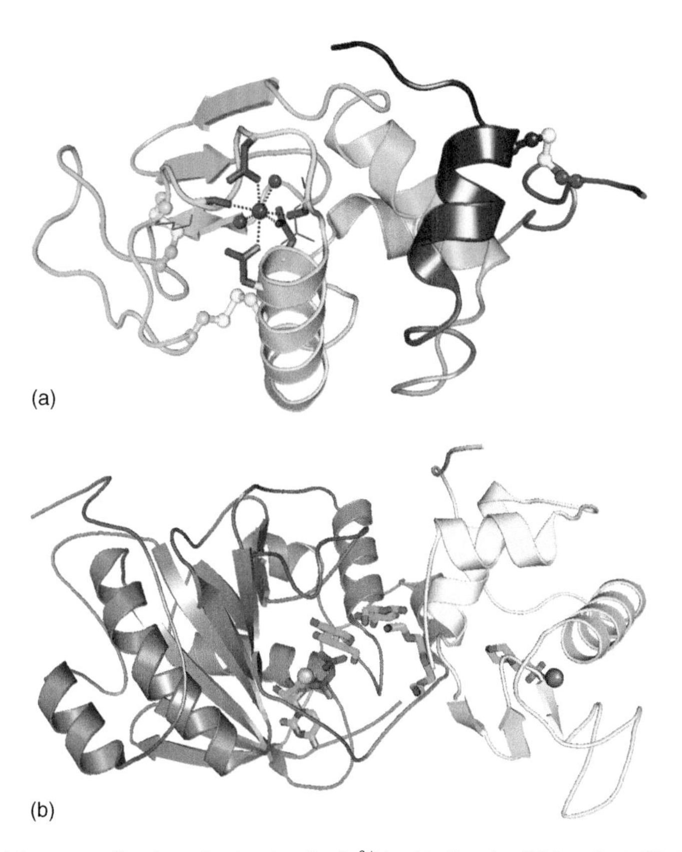

Plate 6.3 (a) Structure of bovine α-La showing the Ca^{2+} ion binding site (PDB code: 2yfd). The peptide chain is rainbow colored, beginning at the N-terminus in blue and progressing to the C-terminus in red, in order to show the assembly of the sub-domains. The Ca^{2+} ion is seven co-ordinate. Loop 79–84 provides three ligands, two from main-chain carbonyl oxygen atoms of Lys79 and Asp84 and one from the side chain of Asp82. Co-ordination about the Ca^{2+} ion is completed by carboxylate oxygen atoms from Asp87 and Asp88 at the N-terminal end of the main four-turn helix and by two water molecules. The four disulfide bonds are shown in ball-and-stick representation (one in the helical domain is obscured and the two linking the helical domain and the Ca^{2+} ion binding loop to the β domain are on the left half of the panel). (b) The lactose synthase complex formed from bovine α-La (yellow) with β-1,4-galactosyltransferase (gray) (PDB code: 1f6s). Several substrate molecules are observed, together with the cleaved nucleotide sugar moiety (cyan sticks). The Mn^{II} ion is shown as a pink sphere and the Ca^{2+} ion is shown as a gray sphere. For clarity, loop regions are given a smoothed representation. Figure drawn with PyMOL (Delano, 2002) (see also Figure 6.3).

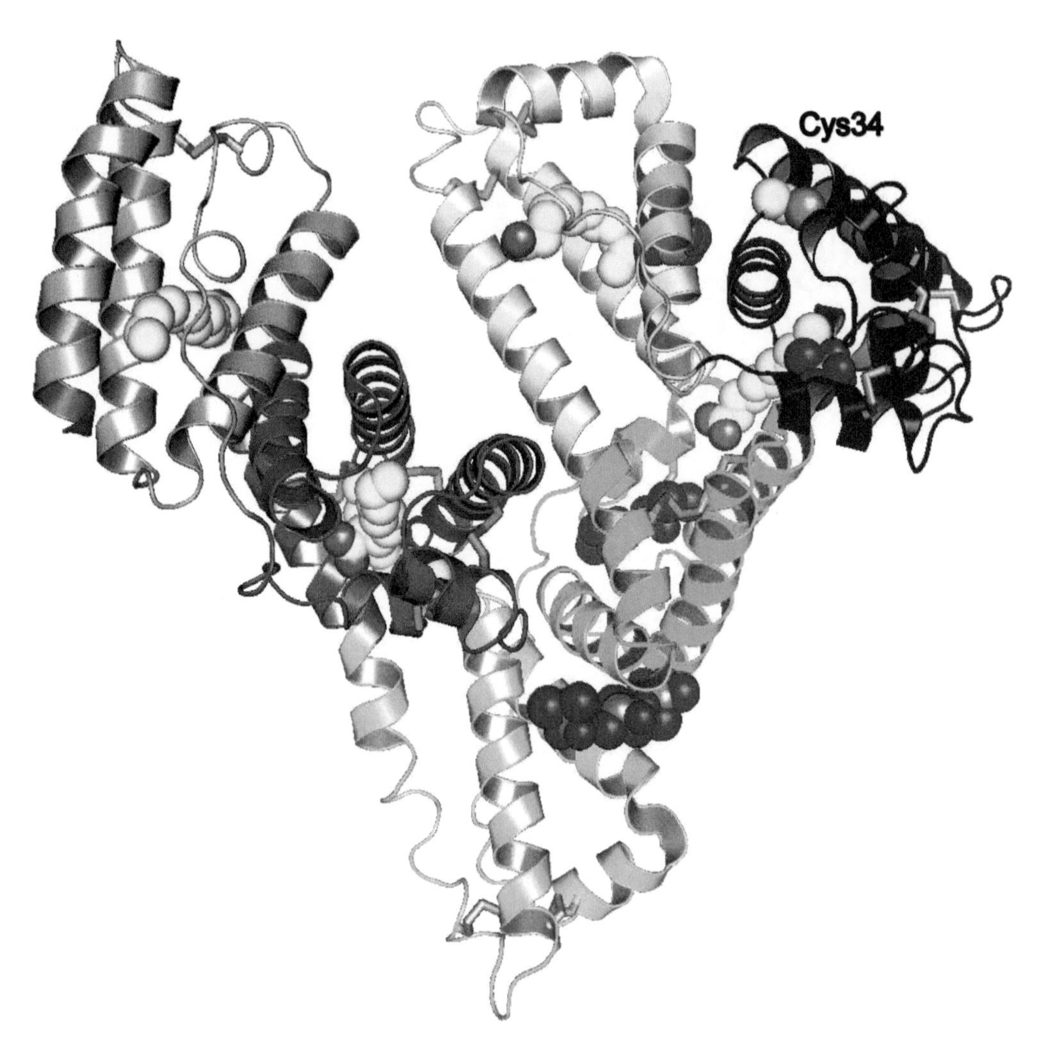

Plate 6.4 Structure of HSA complexed with halothane (slate/purple) partially occupying seven distinct sites and myristic acid (yellow/red) fully occupying five distinct sites (PDB code: 1e7c). Domain IA (residues 5–107) is shown in blue; domain IB (residues 108–196) is shown in light blue; domain IIA (residues 197–297) is shown in green; domain IIB (residues 297–383) is shown in light green; domain IIIA (residues 384–497) is shown in red; and domain IIIB (residues 498–582) is shown in light red. The single cysteine, Cys34, is labeled (Bhattacharya *et al.*, 2000). The 17 disulfide bonds, which tie together individual sub-domains, are represented in stick format. Figure drawn with PyMOL (Delano, 2002) (see also Figure 6.4).

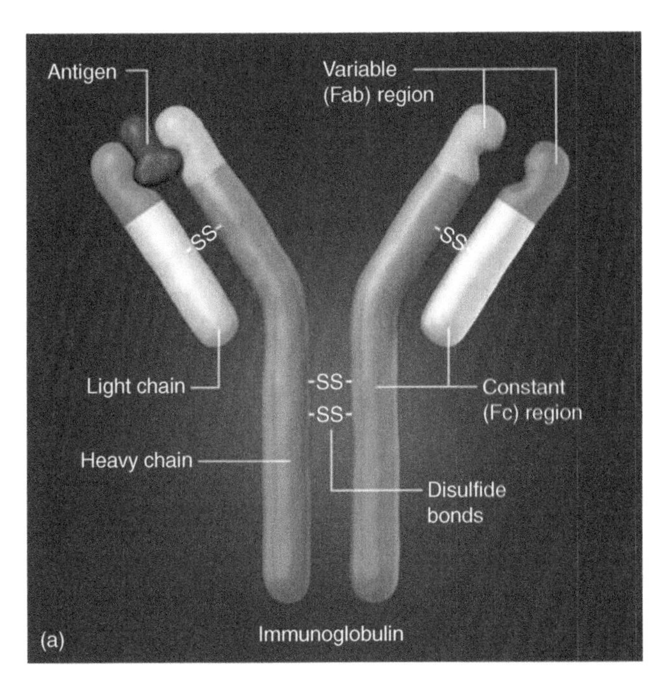

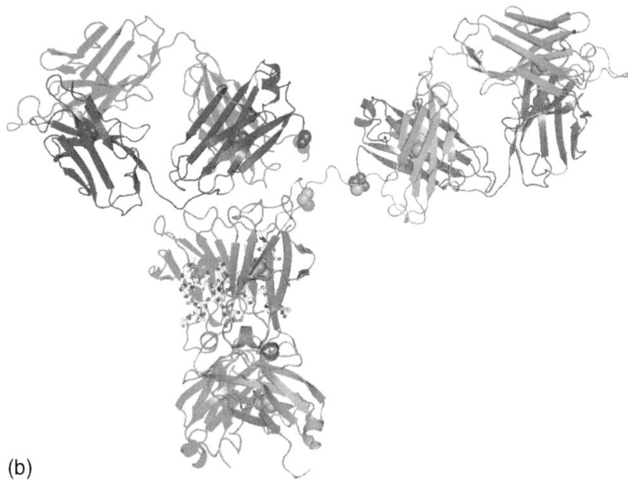

Plate 6.5 (a) Schematic of the general structure of Igs. Reproduced from Gapper *et al.* (2007) with permission. The different classes are distinguished by the constant or Fc regions of the heavy and light chains. (b) The X-ray structure of the human IgG1 molecule (PDB code: 1hzh). The heavy chains are in blue and green; the light chains are in magenta and pink. The asparagine N-linked glycan is shown in stick representation. The lack of two-fold symmetry indicates the extreme flexibility of the domains with respect to one another and the consequent sensitivity to crystal-packing effects. Disulfide bridges are shown as spheres. Each domain has a disulfide bridge joining the two sheets; additional disulfide bridges link the two heavy chains and the light chains to the heavy chains. The N-terminus of each chain is at the top left and top right of the diagram; the C-termini of the heavy chains are at the base of the molecule. Structures (e.g. PDB code: 1wej, 3hfm) where antigens are bound at the light chain–heavy chain interface indicate that binding of antigen occurs across the top of the molecule with little embedding of antigen between the domains, contrary to the mode implied by (a). Figure drawn with PyMOL (Delano, 2002) (see also Figure 6.5).

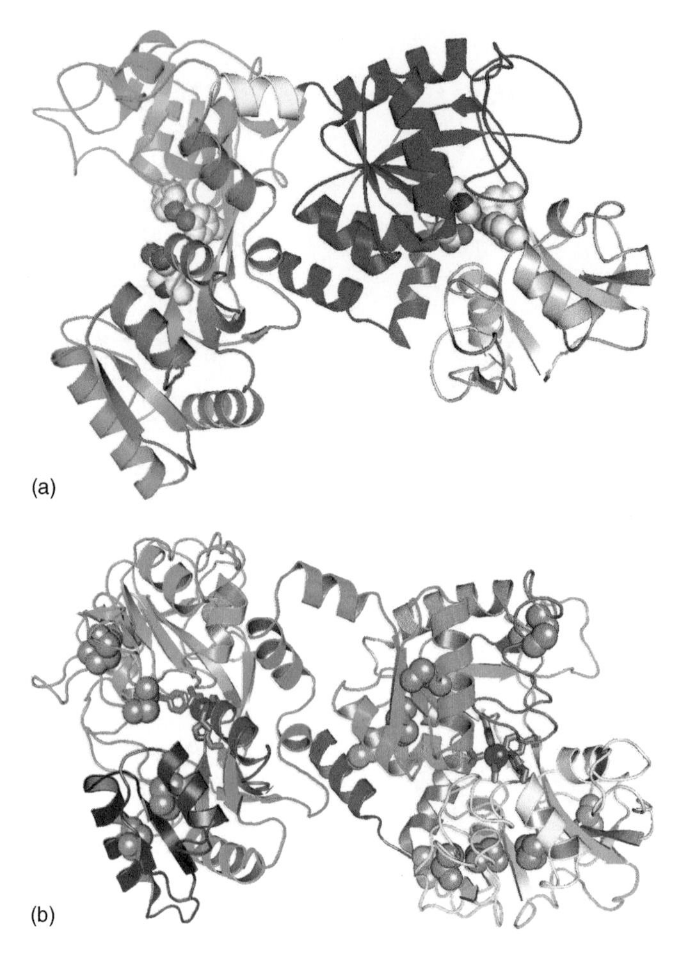

Plate 6.6 Structure of Lf. (a) Human apo-Lf (PDB code: 1cb6), showing the domain structure. The N lobe (blue for sub-domain N-I_N and cyan for sub-domain N-I_C) is in the open conformation; the C-lobe (magenta for sub-domain C-I_N and pink for sub-domain C-I_C), despite no metal ion present, is in the closed conformation. The helix connecting the two lobes is shown in yellow. Metal-binding ligands are shown as spheres. (b) Bovine holo-(Fe(III))-Lf (PDB code: 1blf; the human analog, 1b0l, is structurally very similar). The polypeptide is rainbow colored, blue at the N-terminus to red at the C-terminus, to highlight the manner in which the sub-domains N-I_N and C-I_N are formed from residues ≈1–90 and ≈250–320 (N-I_N) and ≈350–440 and ≈600–680 (C-I_N); for reference, these sub-domains are shown in approximately the same orientation in frames (a) and (b). The iron-binding ligands, including the synergistic bicarbonate ion in (b), are shown in stick form, the iron atom is shown as a red sphere and the cysteines are shown as orange spheres (16 Cys forming eight disulfides for human Lf and 34 Cys forming 17 disulfides for bovine Lf). Loops are not smoothed. Figure drawn with PyMOL (Delano, 2002) (see also Figure 6.6).

GdmCl is often used as an alternative to urea in studies of protein stability. At the neutral or acidic pH of most stability studies, GdmCl will be fully dissociated. At low GdmCl concentration ($<\approx 1$ M), chloride ions screen the electrostatic repulsion between positively charged groups of the protein (Hagihara *et al.*, 1993). The result is that the additional electrostatic interactions of GdmCl compared with the neutral urea molecule have the potential to both *stabilize* and destabilize protein structure depending on the concentration of GdmCl (Hagihara *et al.*, 1993).

D'Alfonso *et al.* (2002) have compared the denaturations of bovine β-Lg B with both GdmCl and urea between pH 2 and 8, as monitored by CD, UV differential absorption and fluorescence measurements. Discrepancies between unfolding free energies obtained using the two denaturants could be reconciled if GdmCl denaturation was assumed to occur via an intermediate state. The secondary structure of this state is similar to that of the native protein, but with greater rigidity in the vicinity of the tryptophan residues, consistent with the screening of electrostatic repulsion between charged residues (D'Alfonso *et al.*, 2002). The GdmCl-induced unfolding intermediate of bovine β-Lg A at pH 2 has been reported to have increased α-helical structure (Dar *et al.*, 2007).

Porcine β-Lg has also been shown to unfold via an intermediate state on addition of GdmCl. The stability of the porcine protein was lower than that of its bovine counterpart and the intermediate state was richer in α-helical structure. Most of the hydrophobic–hydrophobic interactions of the buried core of the native state are conserved between bovine β-Lg and porcine β-Lg. However, four pairwise interactions of the Phe105 side chain of bovine β-Lg are lost on the change to Leu in the porcine protein. This indicates that the presence of the aromatic residue may play an important role in the increased stability of the bovine protein (D'Alfonso *et al.*, 2004). It is interesting to note that this residue is particularly resistant to H/D exchange in heated β-Lg solutions (Edwards *et al.*, 2002).

α-Lactalbumin

α-La is a 123-amino-acid, 14.2-kDa globular protein that is found in the milk of all mammals. The bovine protein binds Ca^{2+}, with the holo form being the more abundant form in milk. Within the Golgi apparatus of the mammary epithelial cell, α-La is the regulatory component of the lactose synthase complex (in which it combines with N-acetyl lactosamine synthase, now named β-1,4-galactosyltransferase-I), the role of which is to transfer galactose from UDP-galactose to glucose (Brew, 2003). In the absence of α-La and in the presence of a transition metal ion such as manganese(II), the catalytic domain of bovine β–1,4-galactosyltransferase-1 (residues 130–402) transfers galactose (Gal) to N-acetylglucosamine (GlcNAc), which may be either free or linked to an oligosaccharide, generating a disaccharide unit, Gal-β-1,4-GlcNAc (N-acetyl lactosamine). The Ca^{2+} ion binding site is remote from the active site of the α-La–β-1,4-galactosyltransferase-I complex (Brew, 2003). α-La has been studied extensively, largely because of its formation of a molten globule state under mild denaturing conditions (Dolgikh *et al.*, 1981).

Molecular structure of bovine α-La

The tertiary structure of bovine α-La is typical of that of the protein from other mammalian species (Acharya *et al.*, 1991; Calderone *et al.*, 1996; Pike *et al.*, 1996) and is similar to that of lysozyme, with which it shares significant homology. As illustrated in Figure 6.3a, α-La is made up of two lobes: the α-lobe contains residues 1–34 and 86–123; and the smaller β-lobe spans residues 35–85. The α-lobe contains three α-helices (residues 5–11, 23–34 and 86–98) and two short 3_{10}-helices (residues 18–20 and 115–118). A small, three-stranded β-sheet (residues 41–44, 47–50 and

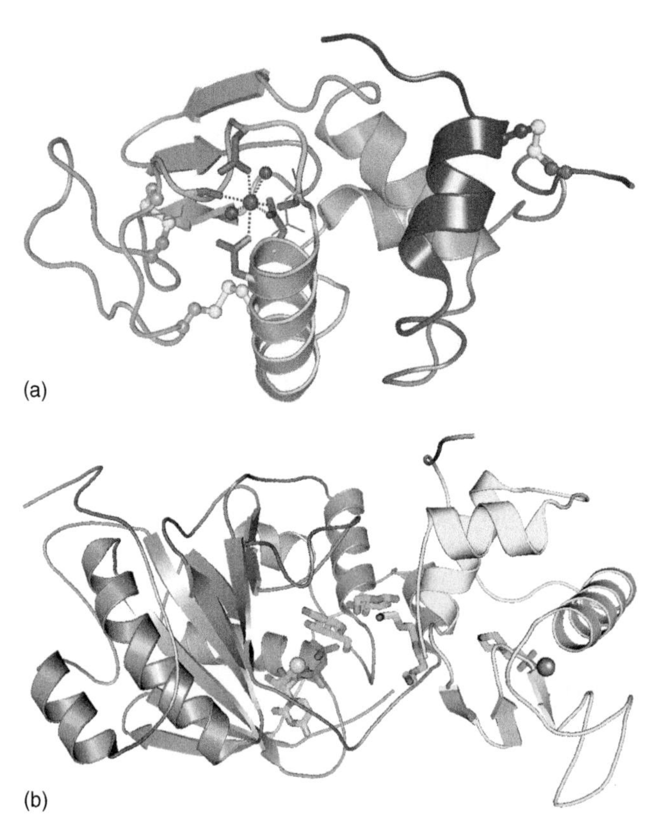

Figure 6.3 (a) Structure of bovine α-La showing the Ca^{2+} ion binding site (PDB code: 2yfd). The peptide chain is rainbow colored, beginning at the N-terminus in blue and progressing to the C-terminus in red, in order to show the assembly of the sub-domains. The Ca^{2+} ion is seven co-ordinate. Loop 79–84 provides three ligands, two from main-chain carbonyl oxygen atoms of Lys79 and Asp84 and one from the side chain of Asp82. Co-ordination about the Ca^{2+} ion is completed by carboxylate oxygen atoms from Asp87 and Asp88 at the N-terminal end of the main four-turn helix and by two water molecules. The four disulfide bonds are shown in ball-and-stick representation (one in the helical domain is obscured and the two linking the helical domain and the Ca^{2+} ion binding loop to the β domain are on the left half of the panel). (b) The lactose synthase complex formed from bovine α-La (yellow) with β-1,4-galactosyltransferase (gray) (PDB code: 1f6s). Several substrate molecules are observed, together with the cleaved nucleotide sugar moiety (cyan sticks). The Mn^{II} ion is shown as a pink sphere and the Ca^{2+} ion is shown as a gray sphere. For clarity, loop regions are given a smoothed representation. Figure drawn with PyMOL (Delano, 2002) (see also Plate 6.3).

55–56) and a short 3_{10}-helix (residues 77–80) make up the β-lobe (Pike *et al.*, 1996). The structure is stabilized by four disulfide bonds (Cys6–Cys120 and Cys28–Cys111 in the α-lobe, Cys60–Cys77 in the β-sheet and Cys73–Cys90 tethering the two lobes together) (Brew, 2003).

Unlike β-Lg, there is no free thiol. A Ca^{2+} ion binds with a sub-micromolar dissociation constant at the so-called binding elbow formed by residues 79–88 located in a cleft between the two lobes, with the metal ion co-ordinated in a distorted pentagonal bipyramidal configuration by the side-chain carboxylate groups of Asp82, Asp87 and Asp88, the carbonyl oxygens of Lys79 and Asp84 and the oxygen atoms of two water molecules (Pike *et al.*, 1996).

The structure of the apo form of bovine α-La is similar to that of the holo protein; the largest changes involve the movement of the Tyr103 side chain in the inter-lobe cleft with little change in the vicinity of the Ca^{2+} ion binding site (Chrysina *et al.*, 2000). The salient features of the structure of the holo protein are depicted in Figure 6.3, together with its complex with β-1,4-galactosyltransferase-I with bound substrates and, interestingly, a trapped intermediate species.

The structures of α-La from several other species, including baboon, human, guinea pig and buffalo, have also been characterized by X-ray diffraction methods. Consistent with high sequence identity, there are no significant differences among these structures, except for a flexible loop at residues 105–110 implicated in the formation of the lactose synthase complex (Acharya *et al.*, 1989; Calderone *et al.*, 1996; Pike *et al.*, 1996). It is worth noting that the recombinant goat protein, which has an added methionine at the N-terminus, is markedly less stable, by ≈14 kJ/mol, than the native protein, mostly the result of an increased rate of unfolding (but a preserved rate of refolding) (Chaudhuri *et al.*, 1999). Similar observations have been made on recombinant bovine α-La (Acharya *et al.*, 1989). Thus, native protein functionality and stability should not in general be inferred from measurements of recombinant proteins heterologously expressed in bacterial systems (which generally add an N-terminal methionine residue).

Effect of temperature on bovine α-La

In general, holo-α-La undergoes a thermal unfolding at a lower temperature than does β-Lg (Ruegg *et al.*, 1977). The role of bound Ca^{2+} ions appears to be to confer stability to the tertiary structure: with less than equimolar amounts of bound calcium, the thermal unfolding transition is lowered substantially, decreasing to about 35°C for the apo form (Relkin, 1996; Ishikawa *et al.*, 1998). The presence of Ca^{2+} ions also accelerates the rate of refolding of α-La by more than two orders of magnitude (Wehbi *et al.*, 2005). The presence of Ca^{2+} ions also aids in the refolding and the formation of the correct disulfide linkages of the denatured reduced protein (Belloque *et al.*, 2000). The structurally closely related, but functionally unrelated, enzyme lysozyme can be subdivided into two classes: a non-calcium-binding subclass, typified by egg-white lysozyme (Grobler *et al.*, 1994; Steinrauf, 1998), and a calcium-binding subclass, including equine and echidna lysozyme (Tsuge *et al.*, 1992; Guss *et al.*, 1997).

The thermal denaturation behavior of bovine α-La from three different sources has been studied; significant differences have been reported (McGuffey *et al.*, 2005), which has provided some resolution of the apparently discordant denaturation data from different groups. In the presence of β-Lg or BSA, each of which has an unpaired cysteine, β-Lg–α-La, α-La–BSA and even α-La–α-La oligomers form at high temperature. Because α-La (which lacks a free thiol) by itself fails under similar conditions to form disulfide-linked oligomers, intermolecular disulfide–sulfydryl interchange reactions appear to play a role in forming α-La–α-La oligomers (Havea *et al.*, 2001; Hong and Creamer, 2002).

Effect of pressure on bovine α-La

Again, the absence of free thiol groups renders α-La intrinsically less susceptible to irreversible structural and functional change induced by high pressure. Reversible unfolding to a molten globule state begins at 200 MPa and loss of native structure becomes irreversible beyond 400 MPa (McGuffey *et al.*, 2005) (corresponding figures for β-Lg are 50 and 150 MPa [Stapelfeldt and Skibsted, 1999; Considine *et al.*, 2005b]). In the presence of Ca^{2+} ions, the denaturation pressure increases by 200 MPa for α-La (Dzwolak *et al.*, 1999). Only in the presence of thiol reducers does oligomerization of α-La occur at high pressures (Jegouic *et al.*, 1996).

Effect of chemical denaturants on bovine α-La

At neutral pH, the calcium-depleted, or apo, form of α-La reversibly denatures to a variety of partially folded or molten globule states upon moderate heating (45°C), or, at room temperature, by dissolving the protein in aqueous TFE (15% TFE) or by adding oleic acid (7.5 equivalents) (Svensson *et al.*, 2000; de Laureto *et al.*, 2002). Under these various conditions, the UV-CD spectra of apo-α-La are essentially identical to those of the most studied molten globule form of α-La—the A-state found at pH 2.0 (Kuwajima, 1996). At 4°C and pH 8.3, proteolysis of apo-α-La by proteinase-K occurs slowly and non-specifically, leading to small peptides only. In contrast, at 37°C, preferential cleavage by proteinase-K is observed at peptide bonds located in loop regions of the β-sheet sub-domain of the β-domain of the protein (residues 35–85), creating peptides in which disulfide bridges link N-terminal residues 1–34 to C-terminal fragments, residues 54–123 or 57–123.

Preferential cleavage at similar sites and similar disulfide-bridged fragments have also been observed for proteolysis of the molten globule states induced by TFE and oleic acid. Polypeptides formed from the molten globule A-state of α-La comprise, therefore, a well-structured native-like conformation of the α-domain and a disordered conformation of the β-sub-domain, residues 34–57 (de Laureto *et al.*, 2002).

Oleic acid treatment leads to a kinetically trapped folding variant of the protein, which can also bind Ca^{2+} ions, called HAMLET (**h**uman **α**-lactalbu**m**in made **le**thal to **t**umor cells, and its bovine analog BAMLET), which has been shown to induce apoptosis in tumor cells (Svensson *et al.*, 2000). Under conditions where thermal denaturation of α-La is reversible, thermal denaturation of HAMLET is irreversible, with respect to loss of its apoptotic effect on tumor cells (Fast *et al.*, 2005).

Human α-La and bovine α-La also weakly bind a second Ca^{2+} ion, which has been structurally characterized for human α-La (Chandra *et al.*, 1998). In addition, Zn^{2+} ion binding to possibly structurally inequivalent sites has been characterized for human α-La and bovine α-La by fluorescence spectroscopic techniques (Permyakov and Berliner, 2000). The binding of Zn^{2+} ions to calcium-loaded α-La has been shown to destabilize the native structure to heat denaturation (Permyakov and Berliner, 2000). However, the weak binding of zinc (sub-millimolarity dissociation constant) means that it is probably not physiologically relevant.

Serum albumin

Serum albumin (SA) is an approximately 580-residue protein that is found in both the blood serum and the milk of all mammals and appears to function as a promiscuous transporter of hydrophobic molecules. However, as with many proteins, this transport role appears not to be the only physiological function for SAs. The structure of human serum albumin (HSA), described in more detail in the next sub-section, is notable also for the number of disulfide bridges, 17 in total. There is one unpaired cysteine, Cys34 in HSA (highlighted in Figure 6.4), and also Cys34 in BSA. This cysteine is part of a highly conserved QQCP(F/Y) motif. It is susceptible to various oxidations, including a two-electron oxidation to sulfenic acid (−SOH) (Carballal *et al.*, 2007).

Recently, evidence that, at least in blood serum, this cysteine is involved in HSA's role in the control of redox properties has been found (Kawakami *et al.*, 2006). A similar role can be postulated for BSA in blood serum but, in both human milk and bovine milk, this redox role has not been established (or even investigated). In terms of milk flavor and the flavor of milk products, control of the redox states of milk components is obviously of importance. It appears also that, in blood serum, where HSA is the major protein component, present at a concentration of 0.6 mM, HSA is the first line of defence against radicals, including reactive oxygen species and nitric oxide.

In vitro studies showing the reactivity of Cys34 in HSA have been complemented by *in vivo* studies that show that, in primary nephrotic syndrome, Cys34 is oxidized to sulfonate, $-SO_3^-$ (Musante *et al.*, 2006). Again, in milk, defences against reactive oxygen species are essential to preserve milk quality, but HSA and BSA are at much lower concentrations in milk than in blood. HSA has also been shown to have esterase activity (Sakurai *et al.*, 2004). Whether this is physiologically important in blood (or in milk) has not been established for either BSA or HSA. Finally, an active role for HSA in the transport of fatty acids across membranes has been characterized (Cupp *et al.*, 2004). It is in this process that SAs are introduced into mammalian milks.

Structure of SAs

Although the three-dimensional structure of BSA has not been determined, the structure of HSA, with which BSA shares 75% sequence identity, has been well characterized for the apo protein as well as for a variety of complexes with a variety of long-chain fatty acids and other more compact hydrophobic molecules. The structure of HSA complexed with the anaesthetic halothane ($C_2F_3Cl_2Br$) and myristic acid ($CH_3(CH_2)_{12}COOH$) is shown in Figure 6.4.

The structure of HSA comprises three structurally homologous domains, each of just under 200 residues, denoted I–III (Curry *et al.*, 1998) and involving residues 5−196, 197−383 and 384−582 respectively. Each domain has two sub-domains, each of ≈100 residues, denoted A and B. The structure lacks β-strands and is predominantly (68%) α-helical, with several lengthy loops connecting the A and B sub-domains. On ligand binding, there is substantial movement of the domains with respect to each other, but the tertiary structure of each domain undergoes only small changes (Curry *et al.*, 1998). Medium- and long-chain fatty acids occupy five distinct

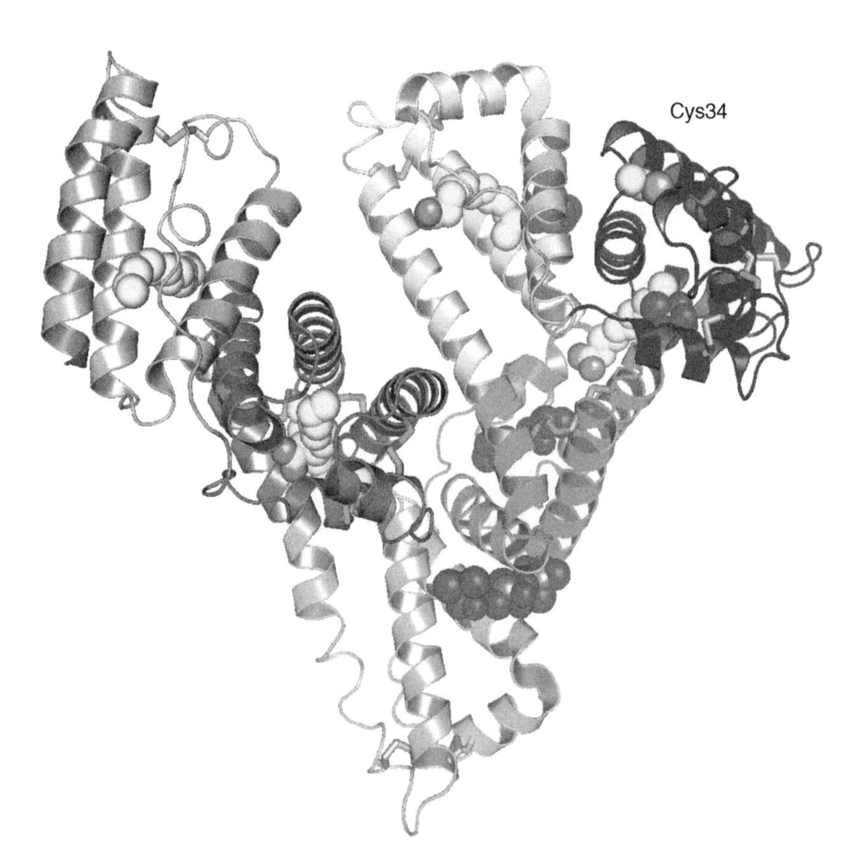

Figure 6.4 Structure of HSA complexed with halothane (slate/purple) partially occupying seven distinct sites and myristic acid (yellow/red) fully occupying five distinct sites (PDB code: 1e7c). Domain IA (residues 5-107) is shown in blue; domain IB (residues 108-196) is shown in light blue; domain IIA (residues 197-297) is shown in green; domain IIB (residues 297-383) is shown in light green; domain IIIA (residues 384-497) is shown in red; and domain IIIB (residues 498-582) is shown in light red. The single cysteine, Cys34, is labeled (Bhattacharya *et al.*, 2000). The 17 disulfide bonds, which tie together individual sub-domains, are represented in stick format. Figure drawn with PyMOL (Delano, 2002) (see also Plate 6.4).

sites (dissociation constants 0.05–1 μM) (Spector, 1975), one in domain I, a second between domains I and II and the remaining three in domain III, as characterized by X-ray techniques for HSA. In the case of halothane binding, two sites are located in domain I and five are located in domain II.

A comprehensive study of the binding of 17 distinct drugs to HSA, in the presence and absence of myristate, has been published recently (Ghuman *et al.*, 2005). Whereas the binding of steroids to BSA is influenced by the binding of fatty acids, for HSA, there is much less influence (Watanabe and Sato, 1996). NMR titrations have shown that BSA, like HSA, binds five myristates; four of the five sites appear to be structurally homologous to those identified crystallographically for HSA (Hamilton *et al.*, 1984, 1991; Cistola *et al.*, 1987; Simard *et al.*, 2005). The X-ray structure of equine serum albumin (ESA) is very similar to that of HSA, consistent with ≈75% sequence identity between these two proteins (Ho *et al.*, 1993).

Effect of temperature on SAs

Careful DSC measurements of defatted HSA and its binding to short- to medium-chain fatty acids have been made. In the absence of fatty acids, a single sharp endotherm is observed, yielding a midpoint temperature for denaturation, T_m, of 64.7°C, consistent with a concerted unfolding. In the presence of fatty acids, the endotherm broadens and there is a steady increase in T_m as the chain length of the fatty acid increases from *n*-butanoate (T_m = 77.6°C) to *n*-octanoate (T_m = 87.2°C); T_ms for *n*-nonanoate and *n*-decanoate are very similar to that for *n*-octanoate.

The short-chain fatty acids formate, acetate and *n*-propionate show evidence for inducing increased stability in HSA through binding of the fatty acids at secondary sites that are inaccessible to the longer-chain fatty acids. For a 30 mg/mL (≈0.5 mM) solution of HSA at pH 7.0, the concentration of fatty acid at which maximum stability of HSA is achieved decreases from more than 2900 mM for formate to less than 15 mM for *n*-dodecanoate (Shrake *et al.*, 2006).

For the native protein, DSC data could be fitted to a two-state model with about seven more or less equivalent binding sites for *n*-decanoate. This number may be contrasted with the value of five observed by X-ray and NMR methods for the binding of myristate (tetradecanoate acid) to HSA and also to BSA (see above) (Simard *et al.*, 2005).

Effect of pressure on SAs

BSA, despite the unpaired cysteine, is relatively stable to high pressures (800 MPa) (Lopez-Fandino, 2006b). BSA undergoes substantial secondary structure changes but, unlike β-Lg, these changes are reversible. It appears that the large number of disulfide bonds protects the hydrophobic core of the protein, including the largely buried Cys34 in sub-domain IA, which is held together by three disulfide bonds (Lopez-Fandino, 2006b). The effects of binding partners, such as fatty acids or other whey components, such as β-Lg, on the structure and stability of BSA at high pressure remain uncharacterized.

Combined pressure/temperature infrared studies of the amide vibrational modes of ESA have shown very recently that high pressure (400 MPa) can convert an intermolecular β-sheet aggregate formed by heating ESA to 60°C at 0.1 MPa (i.e. ambient pressure) to a disordered structure, which reverts to the native structure upon the release of pressure. The activation volume of ≈+92 mL/mol and the partial molar volume difference between the native and heat-denatured states ($\Delta V_{N\rightarrow HA} = +32$ mL/mol) are consistent with decreasing stability of the heat-denatured intermolecular β-sheet with increasing pressure (Okuno *et al.*, 2007).

Effect of chemical denaturants on SAs

The denaturation of BSA and HSA by urea and GdmCl has been extensively studied (e.g. Lapanje and Skerjanc, 1974; Khan *et al.*, 1987; Guo and Qu, 2006). In the presence of fatty acids and other molecules, especially molecules that bind to domain III (e.g. diazepam), denaturation of BSA by urea changes from a three-step process to a two-step process, indicating that the initial denaturation involves changes in the tertiary and secondary structure of domain III (Ahmad and Qasim, 1995; Tayyab *et al.*, 2000). For HSA, denaturation appears to be an intrinsically two-step process with fatty acids converting denaturation to a one-step process (Muzammil *et al.*, 2000; Shrake *et al.*, 2006). In both cases, binding of ligands to BSA or HSA stabilizes the protein against urea-induced denaturation. There is evidence that the urea-induced denatured state and that induced by high pressure are similar, at least for HSA (Tanaka *et al.*, 1997).

In the presence of cations, such as guanidinium and cetylpyridinium salts, domain III of SA is again the most susceptible to denaturation (Ahmad *et al.*, 2005; Sun *et al.*, 2005). Consistent with the high helical content of SA, perfluorinated alcohols and alcohols with bulky hydrophobic heads stabilize HSA (and presumably BSA) against denaturation by both urea and GdmCl (Kumar *et al.*, 2005).

The stability of HSA in the presence of polyethylene glycols (PEGs) has also been examined (Farruggia *et al.*, 1999); low-molecular-weight PEG affects ionization of surface tyrosines and high-molecular-weight PEGs lower the thermal transition temperatures.

Immunoglobulins

The immunoglobulin (Ig) proteins form a diverse family whose members, when in milk, protect the gut mucosa against pathogenic micro-organisms. In bovine milk, the predominant species of Ig proteins are members of the IgG subfamily, in particular IgG1. Colostrum contains 40–300 times the concentration of IgG proteins than milk does; their role is to confer passive immunity to the neonate while its own immune system is developing (Gapper *et al.*, 2007).

IgG proteins have multiple functions, including complement activation, bacterial opsonization (rendering bacterial cells susceptible to immune response) and agglutination. They inactivate bacteria by binding to specific sites on the bacterial surface.

Given the significance attributed to bovine milk and milk products in human nutrition and health, it is important to note that there are significant differences in the levels of the various subfamilies of Igs in milks from different species. Human colostrum and milk contain relatively low levels of the IgG subfamily compared with bovine milk; the reverse occurs for the IgA subfamily. The properties and the accurate quantitation of bovine Ig proteins have been reviewed recently and in detail (Gapper *et al.*, 2007).

Structure of IgG

The structure of IgG is illustrated schematically in Figure 6.5a and as revealed by X-ray techniques in Figure 6.5b. Both the heavy chain and the light chain are predominantly β-sheet structures. Disulfide bridges link pairs of molecules, as well as the heavy chain to the light chain. The protein is generally glycosylated at a number of sites. However, the actual structure is much less tidy than the schematic Y-shaped figure; in particular, the disulfide bonds are at the base of the light chain–heavy chain associations.

Effects of temperature, pressure and chemical denaturants on Ig structure and stability

The response of bovine IgG (isoform not specified) to temperature and chemical denaturants, urea and GdmCl, has been reported (Ye *et al.*, 2005). Thermal denaturation and thermal denaturation in the presence of denaturants was irreversible, producing, via a series of steps, an incompletely unfolded aggregate. Isothermal chemical denaturation produced, also by a series of steps, a completely unfolded random coil state (Ye *et al.*, 2005). The response of IgG to high pressure (200–700 MPa) in the presence of the kosmotrope sucrose has also been reported (Zhang *et al.*, 1998).

A comparative study of the thermal denaturation of bovine IgG, IgA and IgM has been published, with stability in the order just given (Mainer *et al.*, 1997). As retention of immunological properties under standard milk processing conditions was of interest, the activity of the heat-treated protein was determined by an immunological assay using antibodies raised against these Ig proteins. Relatively recently, the response of IgG to pulsed electric fields (and to heat) was reported (Li *et al.*, 2005). Little change in secondary structure or in immunoactivity was reported for samples subjected to a pulsed electric field of ≈41 kV/cm.

Lactoferrin

Lactoferrin (Lf) is a monomeric, globular, Fe^{3+}-binding glycoprotein comprising ≈680 amino acids, giving a molecular mass of ≈80 kDa. It is a member of the transferrin family but, unlike the eponymous protein, to date there is no strong evidence

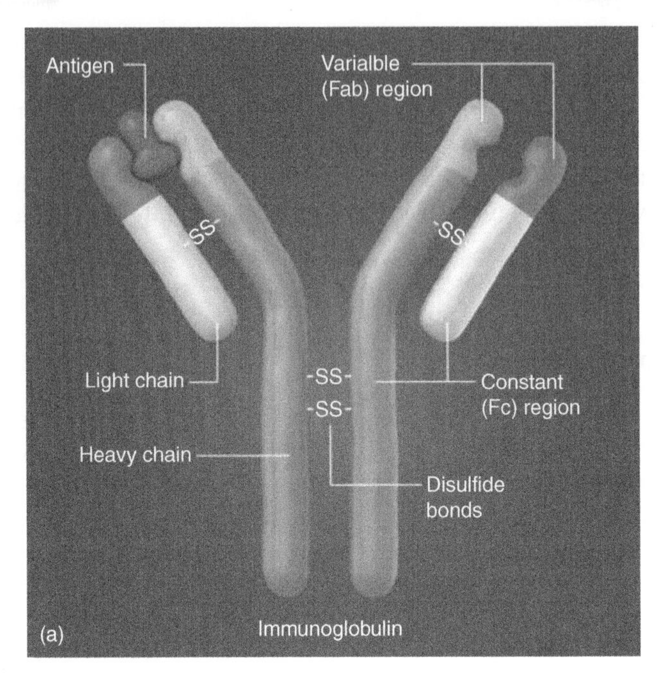

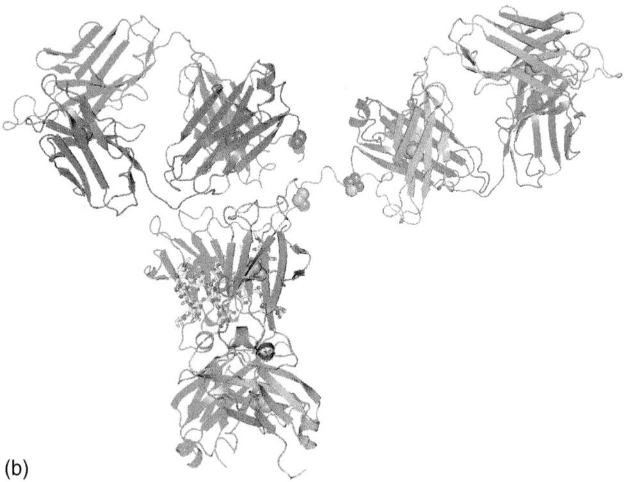

Figure 6.5 (a) Schematic of the general structure of Igs. Reproduced from Gapper *et al.* (2007) with permission. The different classes are distinguished by the constant or Fc regions of the heavy and light chains. (b) The X-ray structure of the human IgG1 molecule (PDB code: 1hzh). The heavy chains are in blue and green; the light chains are in magenta and pink. The asparagine N-linked glycan is shown in stick representation. The lack of two-fold symmetry indicates the extreme flexibility of the domains with respect to one another and the consequent sensitivity to crystal-packing effects. Disulfide bridges are shown as spheres. Each domain has a disulfide bridge joining the two sheets; additional disulfide bridges link the two heavy chains and the light chains to the heavy chains. The N-terminus of each chain is at the top left and top right of the diagram; the C-termini of the heavy chains are at the base of the molecule. Structures (e.g. PDB code: 1wej, 3hfm) where antigens are bound at the light chain–heavy chain interface indicate that binding of antigen occurs across the top of the molecule with little embedding of antigen between the domains, contrary to the mode implied by (a). Figure drawn with PyMOL (Delano, 2002) (see also Plate 6.5).

that Lf is involved in iron transport or metabolism under normal circumstances. Indeed, the protein is only lightly loaded with iron(III) in milk, allowing it to perform a major bacteriostatic role by sequestering iron(III) despite bacteria producing iron(III)-sequestering agents (siderophores) that have affinities for iron(III) many orders of magnitude higher than that of Lf. However, Lf is a multifunctional protein, with evidence to suggest that it, and especially its N-terminal arginine-rich fragments, called lactoferricin(s), obtained by pepsin hydrolysis of the entire protein, have active antimicrobial activity as well as activity as antiviral and antiparasitic agents (Strom *et al.*, 2000, 2002).

Accordingly, commercial applications utilizing bovine Lf and its partially digested peptides are appearing as nutraceuticals in infant formulas, health supplements, oral care products and animal feeds. In addition, reputed antioxidant properties are being utilized in cosmetics (van Hooijdonk and Steijns, 2002). For a review on the remarkable properties of Lf and their commercial applications, see Brock (2002). The protein is synthesized in the mammary gland, but is also found in other exocrine fluids besides milk. Bovine Lf is sometimes used as a supplement in bovine-milk-based infant formulas (to offset the lower abundance relative to that found in human milk) and is potentially useful as a constituent in functional foods, with marked effects on bone-cell activity (Cornish *et al.*, 2004).

Structure of bovine Lf

Bovine Lf has a tertiary structure (Moore *et al.*, 1997) that is very similar to that of the Lfs of other species determined so far: human (Anderson *et al.*, 1987, 1989; Haridas *et al.*, 1995), buffalo (Karthikeyan *et al.*, 1999) and horse (Sharma *et al.*, 1999). All Lfs and transferrins contain two lobes, which share internal homology ($\approx$40% sequence identity between the N- and C-terminal lobes) and a common fold (see Figure 6.6). The homology between lobes for a given species is less than that between corresponding lobes from different species, indicating that the gene duplication event is of ancient origin. Each lobe contains two α/β domains divided by a cleft that incorporates an iron-binding site. Huge structural changes accompany iron binding (each lobe closes over the iron(III), encapsulating a synergistic bicarbonate anion).

Notwithstanding these structural changes, for the apo protein, at least for human Lf, there is little difference in free energy between the open and closed forms (indeed the structure of human apo-Lf has one lobe open and the other lobe closed). This delicate balance is achieved by both open and closed conformations of the protein having, remarkably, the same surface area exposed to solvent. The same applies, but by a different structural mechanism, to the open and closed forms of the N-terminal recombinantly produced half molecule of human Lf (Jameson *et al.*, 1998, 1999, 2002b). Like β-Lg, holo-Lf undergoes a pH-dependent conformational change (in this case at as low as pH $\approx$ 3) that releases the bound ferric ion; the structurally related, but genetically and functionally distinct, serum transferrin releases its cargo at significantly higher pH (pH $\approx$ 5.5) (Baker and Baker, 2004).

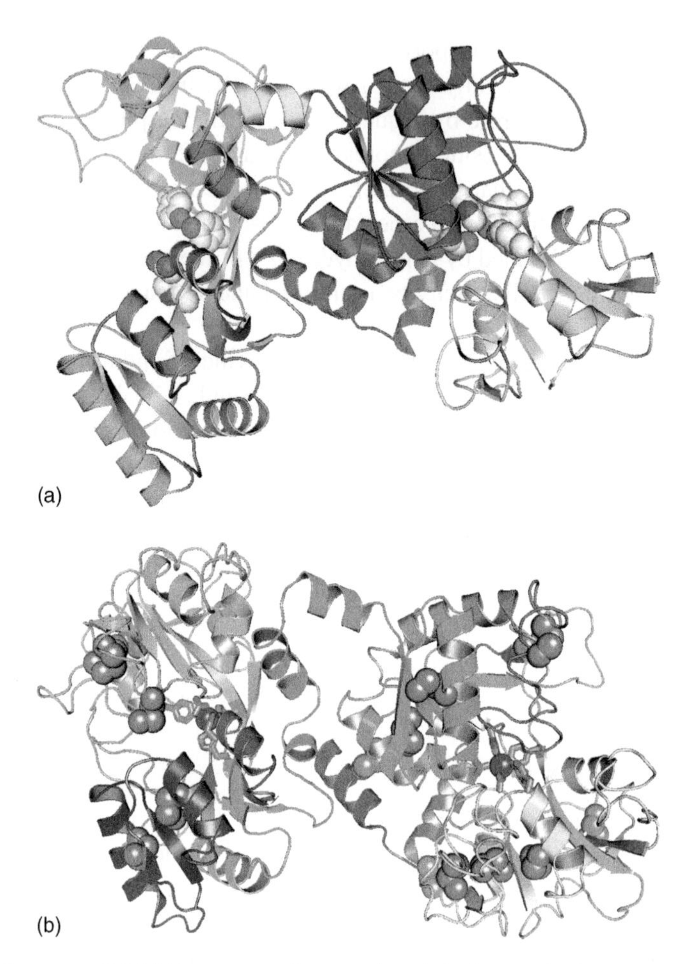

Figure 6.6 Structure of Lf. (a) Human apo-Lf (PDB code: 1cb6), showing the domain structure. The N lobe (blue for sub-domain N-I_N and cyan for sub-domain N-I_C) is in the open conformation; the C-lobe (magenta for sub-domain C-I_N and pink for sub-domain C-I_C), despite no metal ion present, is in the closed conformation. The helix connecting the two lobes is shown in yellow. Metal-binding ligands are shown as spheres. (b) Bovine holo-(Fe(III))-Lf (PDB code: 1blf; the human analog, 1b0l, is structurally very similar). The polypeptide is rainbow colored, blue at the N-terminus to red at the C-terminus, to highlight the manner in which the sub-domains N-I_N and C-I_N are formed from residues ≈1–90 and ≈250–320 (N-I_N) and ≈350–440 and ≈600–680 (C-I_N); for reference, these sub-domains are shown in approximately the same orientation in frames (a) and (b). The iron-binding ligands, including the synergistic bicarbonate ion in (b), are shown in stick form, the iron atom is shown as a red sphere and the cysteines are shown as orange spheres (16 Cys forming eight disulfides for human Lf and 34 Cys forming 17 disulfides for bovine Lf). Loops are not smoothed. Figure drawn with PyMOL (Delano, 2002) (see also Plate 6.6).

Effects of temperature, pressure and chemical denaturants on Lf structure and stability

As a minor component of milk proteins, Lf seems largely to have escaped detailed study of its intrinsic response to temperature, pressure and chemical denaturants or of its influence, if any, on the stability of either itself, or other whey proteins, in the presence of other whey proteins. Calcium ions have been reported to bind to bovine

Lf, probably to the sialic acid groups of the asparagine N-linked glycan, with micromolar dissociation constants; both the apo and holo forms of calcium-bound bovine Lf are more stable to heat and chemical denaturants (Rossi *et al.*, 2002). The calcium-bound forms appear to reduce the Lf-induced release of lipopolysaccharide moieties from bacterial membranes (Rossi *et al.*, 2002).

The heat stability of bovine Lf in isolation (Sánchez *et al.*, 1992) and in association with the major bovine whey proteins, β-Lg (weak association), α-La (no detectable association) and BSA (weak 1:1 association) (Lampreave *et al.*, 1990) has also been studied. The stability of bovine Lf is such that standard pasteurization conditions (but not UHT treatment) are likely to have little effect on its structure and properties, especially its immunological properties (Oria *et al.*, 1993).

Conclusions

Three-dimensional structure determines physiological function and also functionality of whey proteins in their varied applications, both in whole milk and in whey itself. In this review, we have tried to look beyond individual proteins to see what structural features may be important to protein–protein interactions under stresses of temperature, pressure and chemical denaturants. In particular, we have attempted to uncover more recent results that challenge existing paradigms or offer new perspectives. In general, changes that are brought about as a consequence of pH, heat, pressure, chaotropes, etc., shift equilibrium points (e.g. between monomer and dimer, native and unfolded states, etc.) and it is only when a new covalent bond is formed (or removed) that there is irreversible change.

One key interaction between like and unlike proteins arises from disulfide bond interchange, generally facilitated by a single cysteine residue. For this reason, we have focused, especially in the figures, on the observed locations of cysteines. However, detailed structural information, at a level comparable with that of individual partners, is generally lacking on intermolecular assemblies, even when only pairwise associations are formed and isolated.

Whereas considerable attention has been directed to understanding interactions among components of milk and of whey, relatively little attention has been directed to date on the redox state and redox changes during the storage and processing of milk proteins and their effects on milk processing and the flavors of milk-derived products. This remark originates from recent studies on blood sera, where SAs play key roles in redox states and protection against reactive oxygen species. It is important, then, that these structures and functional states are characterized and understood, especially if milk processing, particularly relatively new methods such as pressure treatment, results in refolded proteins that may have highly undesirable properties, such as amyloidogenesis.

Acknowledgments

The authors would like to thank Mike Boland and Lindsay Sawyer for helpful discussions.

References

Acharya, K. R., Stuart, D. I., Walker, N. P. C., Lewis, M. and Phillips, D. C. (1989). Refined structure of baboon α-lactalbumin at 1.7 Å resolution. Comparison with C type lysozyme. *Journal of Molecular Biology*, **208**, 99–127.

Acharya, K. R., Ren, J. S., Stuart, D. I., Phillips, D. C. and Fenna, R. E. (1991). Crystal structure of human α-lactalbumin at 1.7 Å resolution. *Journal of Molecular Biology*, **221**, 571–81.

Adams, J. J., Anderson, B. F., Norris, G. E., Creamer, L. K. and Jameson, G. B. (2006). Structure of bovine β-lactoglobulin (variant A) at very low ionic strength. *Journal of Structural Biology*, **154**, 246–54.

Ahmad, N. and Qasim, M. A. (1995). Fatty acid binding to bovine serum albumin prevents formation of intermediate during denaturation. *European Journal of Biochemistry*, **227**, 563–65.

Ahmad, B., Ahmed Md, Z., Haq, S. K. and Khan, R. H. (2005). Guanidine hydrochloride denaturation of human serum albumin originates by local unfolding of some stable loops in domain III. *Biochimica et Biophysica Acta*, **1750**, 93–102.

Alexander, S. S. and Pace, C. N. (1971). A comparison of the denaturation of bovine β-lactoglobulins A and B and goat β-lactoglobulin. *Biochemistry*, **10**, 2738–43.

Anderson, B. F., Baker, H. M., Dodson, E. J., Norris, G. E., Rumball, S. V., Waters, J. M. and Baker, E. N. (1987). Structure of human lactoferrin at 3.2 Å resolution. *Proceedings of the National Academy of Sciences of the United States of America*, **84**, 1769–73.

Anderson, B. F., Baker, H. M., Norris, G. E., Rice, D. W. and Baker, E. N. (1989). Structure of human lactoferrin – crystallographic structure-analysis and refinement at 2.8 Å resolution. *Journal of Molecular Biology*, **209**, 711–34.

Aouzelleg, A., Bull, L-A., Price, N. C. and Kelly, S. M. (2004). Molecular studies of pressure/temperature-induced structural changes in bovine β-lactoglobulin. *Journal of the Science of Food and Agriculture*, **84**, 398–404.

Baker, H. M. and Baker, E. N. (2004). Lactoferrin and iron: structural and dynamic aspects of binding and release. *BioMetals*, **17**, 209–16.

Banaszak, L., Winter, N., Xu, Z., Bernlohr, D. A., Cowan, S. and Jones, T. A. (1994). Lipid-binding proteins: a family of fatty acid and retinoid transport proteins. *Advances in Protein Chemistry*, **45**, 89–151.

Belloque, J. and Smith, G. M. (1998). Thermal denaturation of β-lactoglobulin. A ^{1}H NMR study. *Journal of Agricultural and Food Chemistry*, **46**, 1805–13.

Belloque, J., Lopez-Fandino, R. and Smith, G. M. (2000). A ^{1}H-NMR study on the effect of high pressures on β-lactoglobulin. *Journal of Agricultural and Food Chemistry*, **48**, 3906–12.

Bennion, B. J. and Daggett, V. (2003). The molecular basis for the chemical denaturation of proteins by urea. *Proceedings of the National Academy of Sciences of the United States of America*, **100**, 5142–47.

Bewley, M. C., Qin, B. Y., Jameson, G. B., Sawyer, L., and Baker, E. N. (1997). Bovine β-lactoglobulin and its variants: a three-dimensional structural perspective. In: *Milk Protein Polymorphism*, IDF Special Issue 9702, pp. 100–9. Brussels: International Dairy Federation.

Bhattacharjee, C., Saha, S., Biswas, A., Kundu, M., Ghosh, L. and Das, K. P. (2005). Structural changes of β-lactoglobulin during thermal unfolding and refolding—an FT-IR and circular dichroism study. *Protein Journal*, **24**, 27–35.

Bhattacharya, A. A., Curry, S. and Franks, N. P. (2000). Binding of the general anesthetics propofol and halothane to human serum albumin – high resolution crystal structures. *Journal of Biological Chemistry*, **275**, 38731–38.

Brew, K. (2003). α-Lactalbumin. In *Advanced Dairy Chemistry*, Volume 1, *Proteins*, 3rd edn. (P. F. Fox and P. L. H. McSweeney, eds) pp. 387–419. New York: Kluwer Academic/Plenum Publishers.

Brock, J. H. (2002). The physiology of lactoferrin. *Biochemistry and Cell Biology*, **80**, 1–6.

Brownlow, S., Cabral, J. H. M., Cooper, R., Flower, D. R., Yewdall, S. J., Polikarpov, I., North, A. C. and Sawyer, L. (1997). Bovine β-lactoglobulin at 1.8 Å resolution – still an enigmatic lipocalin. *Structure (London)*, **5**, 481–95.

Burova, T. V., Grinberg, N. V., Visschers, R. W., Grinberg, V. Y. and de Kruif, C. G. (2002). Thermodynamic stability of porcine β-lactoglobulin—a structural relevance. *European Journal of Biochemistry*, **269**, 3958–68.

Busti, P., Gatti, C. A. and Delorenzi, N. J. (2006). Binding of alkylsulfonate ligands to bovine β-lactoglobulin: effects on protein thermal unfolding. *Food Research International*, **39**, 503–9.

Calderone, V., Giuffrida, M. G., Viterbo, D., Napolitano, L., Fortunato, D., Conti, A. and Acharya, K. R. (1996). Amino acid sequence and crystal structure of buffalo α-lactalbumin. *FEBS Letters*, **394**, 91–95.

Carballal, S., Alvarez, B., Turell, L., Botti, H., Freeman, B. A. and Radi, R. (2007). Sulfenic acid in human serum albumin. *Amino Acids*, **32**, 543–51.

Casal, H. L., Kohler, U. and Mantsch, H. H. (1988). Structural and conformational changes of β-lactoglobulin B: an infrared spectroscopic study of the effect of pH and temperature. *Biochimica et Biophysica Acta*, **957**, 11–20.

Cavanagh, J., Fairbrother, W., Palmer, A. G. III and Skelton, N. (1995). *Protein NMR Spectroscopy: Principles and Practice.* San Diego, California: Academic Press.

Chamani, J., Moosavi-Movahedi, A. A., Rajabi, O., Gharanfoli, M., Momen-Heravi, M., Hakimelahi, G. H., Neamati-Baghsiah, A. and Varasteh, A. R. (2006). Cooperative α-helix formation of β-lactoglobulin induced by sodium n-alkyl sulfates. *Journal of Colloid and Interface Science*, **293**, 52–60.

Chandra, N., Brew, K. and Acharya, K. R. (1998). Structural evidence for the presence of a secondary calcium binding site in human α-lactalbumin. *Biochemistry*, **37**, 4767–72.

Chatterton, D. E. W., Smithers, G., Roupas, P. and Brodkorb, A. (2006). Bioactivity of β-lactoglobulin and α-lactalbumin – technological implications for processing. *International Dairy Journal*, **16**, 1229–40.

Chaudhuri, T. K., Horii, K., Yoda, T., Arai, M., Nagata, S., Terada, T. P., Uchiyama, H., Ikura, T., Tsumoto, K., Kataoka, H., Matsushima, M., Kuwajima, K. and Kumagai, I. (1999). Effect of the extra N-terminal methionine residue on the stability and folding of recombinant α-lactalbumin expressed in Escherichia coli. *Journal of Molecular Biology*, **285**, 1179–94.

Cho, Y., Gu, W., Watkins, S., Lee, S. P., Kim, T. R., Brady, J. W. and Batt, C. A. (1994). Thermostable variants of bovine β-lactoglobulin. *Protein Engineering*, **7**, 263–70.

Chrysina, E. D., Brew, K. and Acharya, K. R. (2000). Crystal structures of apo- and holo-bovine α-lactalbumin at 2.2 Å resolution reveal an effect of calcium on inter-lobe interactions. *Journal of Biological Chemistry*, **275**, 37021–29.

Cistola, D. P., Small, D. M. and Hamilton, J. A. (1987). Carbon 13 NMR studies of saturated fatty acids bound to bovine serum albumin. 1. The filling of individual fatty-acid binding sites. *Journal of Biological Chemistry*, **262**, 10971–79.

Clore, G. M., Szabo, A., Bax, A., Kay, L. E., Driscoll, P. C. and Gronenborn, A. M. (1990). Deviations from the simple two-parameter model-free approach to the interpretation of ^{15}N nuclear magnetic relaxation of proteins. *Journal of the American Chemical Society*, **112**, 4989–91.

Considine, T., Patel, H. A., Singh, H. and Creamer, L. K. (2005a). Influence of binding of sodium dodecyl sulfate, all-trans-retinol, palmitate, and 8-anilino-1-naphthalene-sulfonate on the heat-induced unfolding and aggregation of β-lactoglobulin B. *Journal of Agricultural and Food Chemistry*, **53**, 3197–205.

Considine, T., Singh, H., Patel, H. A. and Creamer, L. K. (2005b). Influence of binding of sodium dodecyl sulfate, all-trans-retinol, and 8-anilino-1-naphthalenesulfonate on the high-pressure-induced unfolding and aggregation of β-lactoglobulin B. *Journal of Agricultural and Food Chemistry*, **53**, 8010–18.

Cornish, J., Callon, K. E., Naot, D., Palmano, K. P., Banovic, T., Bava, U., Watson, M., Lin, J-M., Tong, P. C., Chen, Q., Chan, V. A., Reid, H. E., Fazzalari, N., Baker, H. M., Baker, E. N., Haggarty, N. W., Grey, A. B. and Reid, I. R. (2004). Lactoferrin is a potent regulator of bone cell activity and increases bone formation *in vivo*. *Endocrinology*, **145**, 4366–74.

Creamer, L. K. (1995). Effect of sodium dodecyl sulfate and palmitic acid on the equilibrium unfolding of bovine β-lactoglobulin. *Biochemistry*, **34**, 7170–76.

Creamer, L. K., Richardson, T. and Parry, D. A. D. (1981). Secondary structure of bovine α_{S1}- and β-casein in solution. *Archives of Biochemistry and Biophysics*, **211**, 689–96.

Creamer, L. K., Parry, D. A. D. and Malcolm, G. N. (1983). Secondary structure of bovine β-lactoglobulin B. *Archives of Biochemistry and Biophysics*, **227**, 98–105.

Creamer, L. K., Blair, M., Korte, R. and Jameson, G. B. (2000). Binding of small amphipathic molecules to β-lactoglobulin. Abstract 424. *Journal of Dairy Science*, **83**(Suppl.1), 99.

Creamer, L. K., Bienvenue, A., Nilsson, H., Paulsson, M., Van Wanroij, M., Lowe, E. K., Anema, S. G., Boland, M. J. and Jimenez-Flores, R. (2004). Heat-induced redistribution of disulfide bonds in milk proteins. 1. Bovine β-lactoglobulin. *Journal of Agricultural and Food Chemistry*, **52**, 7660–68.

Cupp, D., Kampf, J. P. and Kleinfeld, A. M. (2004). Fatty acid-albumin complexes and the determination of the transport of long chain free fatty acids across membranes. *Biochemistry*, **43**, 4473–81.

Curry, S., Mandelkow, H., Brick, P. and Franks, N. (1998). Crystal structure of human serum albumin complexed with fatty acid reveals an asymmetric distribution of binding sites. *Nature Structural Biology*, **5**, 827–35.

D'Alfonso, L., Collini, M. and Baldini, G. (2002). Does β-lactoglobulin denaturation occur via an intermediate state?. *Biochemistry*, **41**, 326–33.

D'Alfonso, L., Collini, M., Ragona, L., Ugolini, R., Baldini, G. and Molinari, H. (2004). Porcine β-lactoglobulin chemical unfolding: identification of a non-native α-helical intermediate. *Proteins: Structure Function and Bioinformatics*, **58**, 70–79.

Dar, T. A., Singh, L. R., Islam, A., Anjum, F., Moosavi-Movahedi, A. A. and Ahmad, F. (2007). Guanidinium chloride and urea denaturations of β-lactoglobulin A at pH 2.0 and 25°C: the equilibrium intermediate contains non-native structures (helix, tryptophan and hydrophobic patches). *Biophysical Chemistry*, **127**, 140–48.

de Laureto, P. P., Frare, E., Gottardo, R. and Fontana, A. (2002). Molten globule of bovine α-lactalbumin at neutral pH induced by heat, trifluoroethanol, and oleic acid: a comparative analysis by circular dichroism spectroscopy and limited proteolysis. *Proteins: Structure Function and Genetics*, **49**, 385–97.

de Wit, J. N. and Swinkels, G. A. M. (1980). A differential scanning calorimetric study of the thermal denaturation of bovine β-lactoglobulin. Thermal behavior at temperatures up to 100 °C. *Biochimica et Biophysica Acta, Protein Structure*, **624**, 40–50.

Delano, W. L. (2002). *PyMOL*. Palo Alto, California: Delano Scientific.

Denton, H., Smith, M., Husi, H., Uhrín, D., Barlow, P. N., Batt, C. A. and Sawyer, L. (1998). Isotopically labeled bovine β-lactoglobulin for NMR studies expressed in Pichia pastoris. *Protein Expression and Purification*, **14**, 97–103.

Dolgikh, D. A., Gilmanshin, R. I., Brazhnikov, E. V., Bychkova, V. E., Semisotnov, G. V., Venyaminov, S. Y. and Ptitsyn, O. B. (1981). α-Lactalbumin: compact state with fluctuating tertiary structure?. *FEBS Letters*, **136**, 311–15.

Dufour, E., Genot, C. and Haertlé, T. (1994). β-Lactoglobulin binding properties during its folding changes studied by fluorescence spectroscopy. *Biochimica et Biophysica Acta*, **1205**, 105–12.

Dufour, E., Hervé, G. and Haertlé, T. (1995). Hydrolysis of β-lactoglobulin by thermolysin and pepsin under high hydrostatic pressure. *Biopolymers*, **35**, 475–83.

Dzwolak, W., Kato, M., Shimizu, A. and Taniguchi, Y. (1999). Fourier-transform infrared spectroscopy study of the pressure-induced changes in the structure of the bovine α-lactalbumin: the stabilizing role of the calcium ion. *Biochimica et Biophysica Acta Protein Structure and Molecular Enzymology*, **1433**, 45–55.

Edwards, P. J. B., Jameson, G. B., Palmano, K. P. and Creamer, L. K. (2002). Heat-resistant structural features of bovine β-lactoglobulin A revealed by NMR H/D exchange observations. *International Dairy Journal*, **12**, 331–44.

Edwards, P. J. B., Uhrín, D., Jameson, G. B., Loo, T., Norris, G. E. and Barlow, P. N. (2003). Backbone dynamics of bovine β-lactoglobulin from ^{15}N NMR measurements. Unpublished results.

Farrell, H. M. Jr., Jimenez-Flores, R., Bleck, G. T., Brown, E. M., Butler, J. E., Creamer, L. K., Hicks, C. L., Hollar, C. M., Ng-Kwai-Hang, K. F. and Swaisgood, H. E. (2004). Nomenclature of the proteins of cows' milk – sixth revision. *Journal of Dairy Science*, **87**, 1641–74.

Farruggia, B., Nerli, B., Di Nuci, H., Rigatusso, R. and Pico, G. (1999). Thermal features of the bovine serum albumin unfolding by polyethylene glycols. *International Journal of Biological Macromolecules*, **26**, 23–33.

Fast, J., Mossberg, A-K., Svanborg, C. and Linse, S. (2005). Stability of HAMLET—a kinetically trapped α-lactalbumin oleic acid complex. *Protein Science*, **14**, 329–40.

Flower, D. R. (1996). The lipocalin protein family: structure and function. *Biochemical Journal*, **318**, 1–14.

Fogolari, F., Ragona, L., Zetta, L., Romagnoli, S., de Kruif, K. G. and Molinari, H. (1998). Monomeric bovine β-lactoglobulin adopts a β-barrel fold at pH 2. *FEBS Letters*, **436**, 149–54.

Frapin, D., Dufour, E. and Haertlé, T. (1993). Probing the fatty acid binding site of β-lactoglobulins. *Journal of Protein Chemistry*, **12**, 443–49.

Gapper, L., Copestake, D., Otter, D. and Indyk, H. (2007). Analysis of bovine immunoglobulin G in milk, colostrum and dietary supplements: a review. *Analytical and Bioanalytical Chemistry*, **389**, 93–109.

Ghuman, J., Zunszain, P. A., Petitpas, I., Bhattacharya, A. A., Otagiri, M. and Curry, S. (2005). Structural basis of the drug-binding specificity of human serum albumin. *Journal of Molecular Biology*, **353**, 38–52.

Glasgow, B. J., Abduragimov, A. R., Farahbakhsh, Z. T., Faull, K. F. and Hubbell, W. L. (1995). Tear lipocalins bind a broad array of lipid ligands. *Current Eye Research*, **14**, 363–72.

Gosal, W. S., Clark, A. H., Pudney, P. D. A. and Ross-Murphy, S. B. (2002). Novel amyloid fibrillar networks derived from a globular protein: β-lactoglobulin. *Langmuir*, **18**, 7174–81.

Gottschalk, M., Nilsson, H., Roos, H. and Halle, B. (2003). Protein self-association in solution: the bovine β-lactoglobulin dimer and octamer. *Protein Science*, **12**, 2404–11.

Green, D. W., Aschaffenburg, R., Camerman, A., Coppola, J. C., Dunnill, P., Simmons, R. M., Komorowski, E. S., Sawyer, L., Turner, E. M. C. and Woods, K. F. (1979). Structure of bovine β-lactoglobulin at 6 Å resolution. *Journal of Molecular Biology*, **131**, 375–97.

Grobler, J. A., Rao, K. R., Pervaiz, S. and Brew, K. (1994). Sequences of two highly divergent canine type c lysozymes: implications for the evolutionary origins of the lysozyme/α-lactalbumin superfamily. *Archives of Biochemistry and Biophysics*, **313**, 360–66.

Groves, M. L., Hipp, N. J. and McMeekin, T. L. (1951). Effect of pH on the denaturation of β-lactoglobulin and its dodecyl sulfate derivative. *Journal of the American Chemical Society*, **73**, 2790–93.

Guo, L-H. and Qu, N. (2006). Chemical-induced unfolding of cofactor-free protein monitored by electrochemistry. *Analytical Chemistry*, **78**, 6275–78.

Guss, J. M., Messer, M., Costello, M., Hardy, K. and Kumar, V. (1997). Structure of the calcium-binding echidna milk lysozyme at 1.9 Å resolution. *Acta Crystallographica, Section D: Biological Crystallography*, **D53**, 355–63.

Ha, E. and Zemel, M. B. (2003). Functional properties of whey, whey components, and essential amino acids: mechanisms underlying health benefits for active people. *Journal of Nutritional Biochemistry*, **14**, 251–58.

Hagihara, Y., Aimoto, S., Fink, A. L. and Goto, Y. (1993). Guanidine hydrochloride-induced folding of proteins. *Journal of Molecular Biology*, **231**, 180–84.

Hamada, D. and Goto, Y. (1997). The equilibrium intermediate of β-lactoglobulin with non-native α-helical structure. *Journal of Molecular Biology*, **269**, 479–87.

Hamada, D., Kuroda, Y., Tanaka, T. and Goto, Y. (1995). High helical propensity of the peptide fragments derived from β-lactoglobulin, a predominantly β-sheet protein. *Journal of Molecular Biology*, **254**, 737–46.

Hambling, S. G., McAlpine, A. S. and Sawyer, L. (1992). β-Lactoglobulin. In *Advanced Dairy Chemistry,* Volume 1, *Proteins*, 2nd edn (P. F. Fox, ed.) pp. 141–90. Barking, Essex: Elsevier Applied Science.

Hamilton, J. A., Cistola, D. P., Morrisett, J. D., Sparrow, J. T. and Small, D. M. (1984). Interactions of myristic acid with bovine serum albumin: a ^{13}C NMR study. *Proceedings of the National Academy of Sciences of the United States of America—Biological Sciences*, **81**, 3718–22.

Hamilton, J. A., Era, S., Bhamidipati, S. P. and Reed, R. G. (1991). Locations of the 3 primary binding sites for long-chain fatty acids on bovine serum albumin. *Proceedings of the National Academy of Sciences of the United States of America*, **88**, 2051–54.

Haridas, M., Anderson, B. F. and Baker, E. N. (1995). Structure of human diferric lactoferrin refined at 2.2 Å resolution. *Acta Crystallographica Section D: Biological Crystallography*, **51**, 629–46.

Havea, P., Singh, H. and Creamer, L. K. (2001). Characterization of heat-induced aggregates of β-lactoglobulin, α-lactalbumin and bovine serum albumin in a whey protein concentrate environment. *Journal of Dairy Research*, **68**, 483–97.

Ho, J. X., Holowachuk, E. W., Norton, E. J., Twigg, P. D. and Carter, D. C. (1993). X-ray and primary structure of horse serum-albumin (Equus-caballus) at 0.27 nm resolution. *European Journal of Biochemistry*, **215**, 205–12.

Hoedemaeker, F. J., Visschers, R. W., Alting, A. C., de Kruif, K. G., Kuil, M. E. and Abrahams, J. P. (2002). A novel pH-dependent dimerization motif in β-lactoglobulin from pig (Sus scrofa). *Acta Crystallographica Section D: Biological Crystallography*, **D58**, 480–86.

Hong, Y-H. and Creamer, L. K. (2002). Changed protein structures of bovine β-lactoglobulin B and α-lactalbumin as a consequence of heat treatment. *International Dairy Journal*, **12**, 345–59.

Iametti, S., De Gregori, B., Vecchio, G. and Bonomi, F. (1996). Modifications occur at different structural levels during the heat denaturation of β-lactoglobulin. *European Journal of Biochemistry*, **237**, 106–12.

Iametti, S., Transidico, P., Bonomi, F., Vecchio, G., Pittia, P., Rovere, P. and Dall'Aglio, G. (1997). Molecular modifications of β-lactoglobulin upon exposure to high pressure. *Journal of Agricultural and Food Chemistry*, **45**, 23–29.

Invernizzi, G., Samalikova, M., Brocca, S., Lotti, M., Molinari, H. and Grandori, R. (2006). Comparison of bovine and porcine β-lactoglobulin: a mass spectrometric analysis. *Journal of Mass Spectrometry*, **41**, 717–27.

Ishikawa, N., Chiba, T., Chen, L. T., Shimizu, A., Ikeguchi, M. and Sugai, S. (1998). Remarkable destabilization of recombinant α-lactalbumin by an extraneous N-terminal methionyl residue. *Protein Engineering*, **11**, 333–35.

Jameson, G. B., Anderson, B. F., Norris, G. E., Thomas, D. H. and Baker, E. N. (1998). Structure of human apolactoferrin at 2.0 Å resolution. Refinement and analysis of ligand-induced conformational change. *Acta Crystallographica, Section D: Biological Crystallography*, **54**, 1319–35.

Jameson, G. B., Anderson, B. F., Norris, G. E., Thomas, D. H. and Baker, E. N. (1999). Structure of human apolactoferrin at 2.0 Å resolution. Refinement and analysis of ligand-induced conformational change (corrigendum 54:1319). *Acta Crystallographica, Section D: Biological Crystallography*, **55**, 1108.

Jameson, G. B., Adams, J. J. and Creamer, L. K. (2002a). Flexibility, functionality and hydrophobicity of bovine β-lactoglobulin. *International Dairy Journal*, **12**, 319–29.

Jameson, G. B., Anderson, B. F., Breyer, W. A., Day, C. L., Tweedie, J. W. and Baker, E. N. (2002b). Structure of a domain-opened mutant (R121D) of the human lactoferrin N-lobe refined from a merohedrally twinned crystal form. *Acta Crystallographica, Section D: Biological Crystallography*, **58**, 955–62.

Jegouic, M., Grinberg, V. Y., Guingant, A. and Haertlé, T. (1996). Thiol-induced oligomerization of α-lactalbumin at high pressure. *Journal of Protein Chemistry*, **15**, 501–9.

Kaminogawa, S., Shimizu, M., Ametani, A., Hattori, M., Ando, O., Hachimura, S., Nakamura, Y., Totsuka, M. and Yamauchi, K. (1989). Monoclonal antibodies as probes for monitoring the denaturation process of bovine β-lactoglobulin. *Biochimica et Biophysica Acta*, **998**, 50–56.

Karthikeyan, S., Paramasivam, M., Yadav, S., Srinivasan, A. and Singh, T. P. (1999). Structure of buffalo lactoferrin at 2.5 Å resolution using crystals grown at 303 K shows different orientations of the N and C lobes. *Acta Crystallographica Section D: Biological Crystallography*, **55**, 1805–13.

Kawakami, A., Kubota, K., Yamada, N., Tagami, U., Takehana, K., Sonaka, I., Suzuki, E. and Hirayama, K. (2006). Identification and characterization of oxidized human serum albumin. A slight structural change impairs its ligand-binding and antioxidant functions. *FEBS Journal*, **273**, 3346–57.

Kella, N. K. and Kinsella, J. E. (1988). Enhanced thermodynamic stability of β-lactoglobulin at low pH. A possible mechanism. *Biochemical Journal*, **255**, 113–18.

Khan, M. Y., Agarwal, S. K. and Hangloo, S. (1987). Urea-induced structural transformations in bovine serum albumin. *Journal of Biochemistry (Tokyo, Japan)*, **102**, 313–17.

Kim, T-R., Goto, Y., Hirota, N., Kuwata, K., Denton, H., Wu, S-Y., Sawyer, L. and Batt, C. A. (1997). High-level expression of bovine β-lactoglobulin in Pichia pastoris and characterization of its physical properties. *Protein Engineering*, **10**, 1339–45.

Knudsen, J. C., Otte, J., Olsen, K. and Skibsted, L. H. (2002). Effect of high hydrostatic pressure on the conformation of β-lactoglobulin A as assessed by proteolytic peptide profiling. *International Dairy Journal*, **12**, 791–803.

Kobayashi, T., Ikeguchi, M. and Sugai, S. (2000). Molten globule structure of equine β-lactoglobulin probed by hydrogen exchange. *Journal of Molecular Biology*, **299**, 757–70.

Kobayashi, T., Ikeguchi, M. and Sugai, S. (2002). Construction and characterization of β-lactoglobulin chimeras. *Proteins: Structure Function and Genetics*, **49**, 297–301.

Kontopidis, G., Holt, C. and Sawyer, L. (2004). β-Lactoglobulin: binding properties, structure, and function. *Journal of Dairy Science*, **87**, 785–96.

Konuma, T., Sakurai, K. and Goto, Y. (2007). Promiscuous binding of ligands by β-lactoglobulin involves hydrophobic interactions and plasticity. *Journal of Molecular Biology*, **368**, 209–18.

Kumar, S., Modig, K. and Halle, B. (2003). Trifluoroethanol-induced β → α transition in β-lactoglobulin: hydration and cosolvent binding studied by ^{2}H, ^{17}O, and ^{19}F magnetic relaxation dispersion. *Biochemistry*, **42**, 13708–16.

Kumar, Y., Muzammil, S. and Tayyab, S. (2005). Influence of fluoro, chloro and alkyl alcohols on the folding pathway of human serum albumin. *Journal of Biochemistry (Tokyo, Japan)*, **138**, 335–41.

Kumosinski, T. F. and Timasheff, S. N. (1966). Molecular interactions in β-lactoglobulin. X. The stoichiometry of the β-lactoglobulin mixed tetramerization. *Journal of the American Chemical Society*, **88**, 5635–42.

Kuroda, Y., Hamada, D., Tanaka, T. and Goto, Y. (1996). High helicity of peptide fragments corresponding to β-strand regions of β-lactoglobulin observed by 2D-NMR spectroscopy. *Folding and Design*, **1**, 255–63.

Kuwajima, K. (1996). The molten globule state of α-lactalbumin. *FASEB Journal*, **10**, 102–9.

Kuwata, K., Hoshino, M., Era, S., Batt, C. A. and Goto, Y. (1998). α → β transition of β-lactoglobulin as evidenced by heteronuclear NMR. *Journal of Molecular Biology*, **283**, 731–39.

Kuwata, K., Hoshino, M., Forge, V., Era, S., Batt, C. A. and Goto, Y. (1999). Solution structure and dynamics of bovine β-lactoglobulin A. *Protein Science*, **8**, 2541–45.

Kuwata, K., Li, H., Yamada, H., Batt, C. A., Goto, Y. and Akasaka, K. (2001). High pressure NMR reveals a variety of fluctuating conformers in β-lactoglobulin. *Journal of Molecular Biology*, **305**, 1073–83.

Lampreave, F., Piñeiro, A., Brock, J. H., Castillo, H., Sánchez, L. and Calvo, M. (1990). Interaction of bovine lactoferrin with other proteins of milk whey. *International Journal of Biological Macromolecules*, **12**, 2–5.

Lange, D. C., Kothari, R., Patel, R. C. and Patel, S. C. (1998). Retinol and retinoic acid bind to a surface cleft in bovine β-lactoglobulin: a method of binding site determination using fluorescence resonance energy transfer. *Biophysical Chemistry*, **74**, 45–51.

Lapanje, S. and Skerjanc, J. (1974). Dilatometric study of the denaturation of bovine serum albumin by guanidine hydrochloride. *Biochemical and Biophysical Research Communications*, **56**, 338–42.

Li, S. Q., Bomser, J. A. and Zhang, Q. H. (2005). Effects of pulsed electric fields and heat treatment on stability and secondary structure of bovine immunoglobulin G. *Journal of Agricultural and Food Chemistry*, **53**, 663–70.

Lipari, G. and Szabo, A. (1982). Model-free approach to the interpretation of nuclear magnetic resonance relaxation in macromolecules. 1. Theory and range of validity. *Journal of the American Chemical Society*, **104**, 4546–59.

Lopez-Fandino, R. (2006a). Functional improvement of milk whey proteins induced by high hydrostatic pressure. *Critical Reviews in Food Science and Nutrition*, **46**, 351–63.

Lopez-Fandino, R. (2006b). High pressure-induced changes in milk proteins and possible applications in dairy technology. *International Dairy Journal*, **16**, 1119–31.

Luebke, M., Guichard, E., Tromelin, A. and Le Quere, J. L. (2002). Nuclear magnetic resonance spectroscopic study of β-lactoglobulin interactions with two flavor compounds, γ-decalactone and β-ionone. *Journal of Agricultural and Food Chemistry*, **50**, 7094–99.

Mainer, G., Sanchez, L., Ena, J. M. and Calvo, M. (1997). Kinetic and thermodynamic parameters for heat denaturation of bovine milk IgG, IgA and IgM. *Journal of Food Science*, **62**, 1034–38.

Manderson, G. A., Hardman, M. J., and Creamer, L. K. (1997). Spectroscopic examination of the heat-induced changes in β-lactoglobulin A, B and C. In *Milk Protein Polymorphism*, IDF Special Issue 9702, pp. 204–11. Brussels: International Dairy Federation.

McGuffey, M. K., Epting, K. L., Kelly, R. M. and Foegeding, E. A. (2005). Denaturation and aggregation of three α-lactalbumin preparations at neutral pH. *Journal of Agricultural and Food Chemistry*, **53**, 3182–90.

McKenzie, H. A. and Sawyer, W. H. (1967). Effect of pH on β-lactoglobulins. *Nature (London)*, **214**, 1101–4.

Messens, W., Van Camp, J., and Dewettinck, K. (2003). High-pressure processing to improve dairy product quality. In *Dairy Processing: Improving Quality*, pp. 310–32. Boca Raton, Florida: CRC Press LLC.

Mills, O. E. (1976). Effect of temperature on tryptophan fluorescence of β-lactoglobulin B. *Biochimica et Biophysica Acta*, **434**, 324–32.

Moller, R. E., Stapelfeldt, H. and Skibsted, L. H. (1998). Thiol reactivity in pressure-unfolded β-lactoglobulin. Antioxidative properties and thermal refolding. *Journal of Agricultural and Food Chemistry*, **46**, 425–30.

Monaco, H. L., Zanotti, G., Spadon, P., Bolognesi, M., Sawyer, L. and Eliopoulos, E. E. (1987). Crystal structure of the trigonal form of bovine β-lactoglobulin and of its complex with retinol at 2.5 Å resolution. *Journal of Molecular Biology*, **197**, 695–706.

Moore, S. A., Anderson, B. F., Groom, C. R., Haridas, M. and Baker, E. N. (1997). Three-dimensional structure of diferric bovine lactoferrin at 2.8 Å resolution. *Journal of Molecular Biology*, **274**, 222–36.

Musante, L., Bruschi, M., Candiano, G., Petretto, A., Dimasi, N., Del Boccio, P., Urbani, A., Rialdi, G. and Ghiggeri, G. M. (2006). Characterization of oxidation end product of plasma albumin '*in vivo*'. *Biochemical and Biophysical Research Communications*, **349**, 668–73.

Muzammil, S., Kumar, Y. and Tayyab, S. (2000). Anion-induced stabilization of human serum albumin prevents the formation of intermediate during urea denaturation. *Proteins: Structure Function and Genetics*, **40**, 29–38.

Narayan, M. and Berliner, L. J. (1998). Mapping fatty acid binding to β-lactoglobulin: ligand binding is restricted by modification of Cys 121. *Protein Science*, **7**, 150–57.

Nishikawa, K. and Noguchi, T. (1991). Predicting protein secondary structure based on amino acid sequence. *Methods in Enzymology*, **202**, 31–44.

Oksanen, E., Jaakola, V. P., Tolonen, T., Valkonen, K., Aakerstroem, B., Kalkkinen, N., Virtanen, V. and Goldman, A. (2006). Reindeer β-lactoglobulin crystal structure with pseudo-body-centered noncrystallographic symmetry. *Acta Crystallographica Section D: Biological Crystallography*, **D62**, 1369–74.

Okuno, A., Kato, M. and Taniguchi, Y. (2007). Pressure effects on the heat-induced aggregation of equine serum albumin by FT-IR spectroscopic study: secondary structure, kinetic and thermodynamic properties. *Biochimica et Biophysica Acta Proteins and Proteomics*, **1774**, 652–60.

Oria, R., Ismail, M., Sánchez, L., Calvo, M. and Brock, J. H. (1993). Effect of heat treatment and other milk proteins on the interaction of lactoferrin with monocytes. *Journal of Dairy Research*, **60**, 363–69.

Pace, C. N. and Tanford, C. (1968). Thermodynamics of the unfolding of β-lactoglobulin A in aqueous urea solutions between 5 and 55°. *Biochemistry*, **7**, 198–208.

Panick, G., Malessa, R. and Winter, R. (1999). Differences between the pressure and temperature induced denaturation and aggregation of β-lactoglobulin A, B, and AB monitored by FT-IR spectroscopy and small-angle x-ray scattering. *Biochemistry*, **38**, 6512–19.

Papiz, M. Z., Sawyer, L., Eliopoulos, E. E., North, A. C. T., Findlay, J. B. C., Sivaprasadarao, R., Jones, T. A., Newcomer, M. E. and Kraulis, P. J. (1986). The structure of β-lactoglobulin and its similarity to plasma retinol-binding protein. *Nature (London)*, **324**, 383–85.

Patel, H. A., Singh, H., Havea, P., Considine, T. and Creamer, L. K. (2005). Pressure-induced unfolding and aggregation of the proteins in whey protein concentrate solutions. *Journal of Agricultural and Food Chemistry*, **53**, 9590–601.

Pedersen, K. O. (1936). Ultracentrifugal and electrophoretic studies on the milk proteins: the lactoglobulin of Palmer. *Biochemical Journal*, **30**, 961–70.

Permyakov, E. A. and Berliner, L. J. (2000). α-Lactalbumin: structure and function. *FEBS Letters*, **473**, 269–74.

Pike, A. C. W., Brew, K. and Acharya, K. R. (1996). Crystal structures of guinea pig, goat and bovine α-lactalbumin highlight the enhanced conformational flexibility of regions that are significant for its action in lactose synthase. *Structure (London)*, **4**, 691–703.

Puyol, P., Perez, M. D., Peiro, J. M. and Calvo, M. (1994). Effect of binding of retinol and palmitic acid to bovine β-lactoglobulin on its resistance to thermal denaturation. *Journal of Dairy Science*, **77**, 1402–94.

Qi, X. L., Brownlow, S., Holt, C. and Sellers, P. (1995). Thermal denaturation of β-lactoglobulin: effect of protein concentration at pH 6.75 and 8.05. *Biochimica et Biophysica Acta*, **1248**, 43–49.

Qi, X. L., Holt, C., McNulty, D., Clarke, D. T., Brownlow, S. and Jones, G. R. (1997). Effect of temperature on the secondary structure of β-lactoglobulin at pH 6.7, as determined by CD and IR spectroscopy: a test of the molten globule hypothesis. *Biochemical Journal*, **324**, 341–46.

Qin, B. Y., Bewley, M. C., Creamer, L. K., Baker, H. M., Baker, E. N. and Jameson, G. B. (1998a). Structural basis of the Tanford transition of bovine β-lactoglobulin. *Biochemistry*, **37**, 14014–23.

Qin, B. Y., Creamer, L. K., Baker, E. N. and Jameson, G. B. (1998b). 12-Bromododecanoic acid binds inside the calyx of bovine β-lactoglobulin. *FEBS Letters*, **438**, 272–78.

Qin, B. Y., Bewley, M. C., Creamer, L. K., Baker, E. N. and Jameson, G. B. (1999). Functional implications of structural differences between variants A and B of bovine β-lactoglobulin. *Protein Science*, **8**, 75–83.

Ragona, L., Pusterla, F., Zetta, L., Monaco, H. L. and Molinari, H. (1997). Identification of a conserved hydrophobic cluster in partially folded bovine β-lactoglobulin at pH 2. *Folding and Design*, **2**, 281–90.

Ragona, L., Fogolari, F., Romagnoli, S., Zetta, L., Maubois, J. L. and Molinari, H. (1999). Unfolding and refolding of bovine β-lactoglobulin monitored by hydrogen exchange measurements. *Journal of Molecular Biology*, **293**, 953–69.

Ragona, L., Fogolari, F., Zetta, L., Perez, D. M., Puyol, P., de Kruif, K. G., Lohr, F., Ruterjans, H. and Molinari, H. (2000). Bovine β-lactoglobulin: interaction studies with palmitic acid. *Protein Science*, **9**, 1347–56.

Ragona, L., Fogolari, F., Catalano, M., Ugolini, R., Zetta, L. and Molinari, H. (2003). EF loop conformational change triggers ligand binding in β-lactoglobulins. *Journal of Biological Chemistry*, **278**, 38840–46.

Relkin, P. (1996). Thermal unfolding of β-lactoglobulin, α-lactalbumin, and bovine serum albumin. A thermodynamic approach. *Critical Reviews in Food Science and Nutrition*, **36**, 565–601.

Relkin, P., Eynard, L. and Launay, B. (1992). Thermodynamic parameters of β-lactoglobulin and α-lactalbumin. A DSC study of denaturation by heating. *Thermochimica Acta*, **204**, 111–21.

Rossi, P., Giansanti, F., Boffi, A., Ajello, M., Valenti, P., Chiancone, E. and Antonini, G. (2002). Ca^{2+} binding to bovine lactoferrin enhances protein stability and influences the release of bacterial lipopolysaccharide. *Biochemistry and Cell Biology*, **80**, 41–48.

Ruegg, M., Moor, U. and Blanc, B. (1977). Calorimetric study of thermal-denaturation of whey proteins in simulated milk ultrafiltrate. *Journal of Dairy Research*, **44**, 509–20.

Sakai, K., Sakurai, K., Sakai, M., Hoshino, M. and Goto, Y. (2000). Conformation and stability of thiol-modified bovine β-lactoglobulin. *Protein Science*, **9**, 1719–29.

Sakurai, K. and Goto, Y. (2002). Manipulating monomer–dimer equilibrium of bovine β-lactoglobulin by amino acid substitution. *Journal of Biological Chemistry*, **277**, 25735–40.

Sakurai, K. and Goto, Y. (2006). Dynamics and mechanism of the Tanford transition of bovine β-lactoglobulin studied using heteronuclear NMR spectroscopy. *Journal of Molecular Biology*, **356**, 483–96.

Sakurai, K., Oobatake, M. and Goto, Y. (2001). Salt-dependent monomer–dimer equilibrium of bovine β-lactoglobulin at pH 3. *Protein Science*, **10**, 2325–35.

Sakurai, Y., Ma, S. F., Watanabe, H., Yamaotsu, N., Hirono, S., Kurono, Y., Kragh-Hansen, U. and Otagiri, M. (2004). Esterase-like activity of serum albumin: characterization of its structural chemistry using p-nitrophenyl esters as substrates. *Pharmaceutical Research*, **21**, 285–92.

Sánchez, L., Peiró, J. M., Castillo, H., Pérez, M. D., Ena, J. M. and Calvo, M. (1992). Kinetic parameters for denaturation of bovine milk lactoferrin. *Journal of Food Science*, **57**, 873–79.

Sawyer, L., Brownlow, S., Polikarpov, I. and Wu, S-Y. (1998). β-Lactoglobulin: structural studies, biological clues. *International Dairy Journal*, **8**, 65–72.

Schokker, E. P., Singh, H., Pinder, D. N. and Creamer, L. K. (2000). Heat-induced aggregation of β-lactoglobulin AB at pH 2.5 as influenced by ionic strength and protein concentration. *Journal of Agricultural and Food Chemistry*, **10**, 233–40.

Sharma, A. K., Paramasivam, M., Srinivasan, A., Yadav, M. P. and Singh, T. P. (1999). Three-dimensional structure of mare diferric lactoferrin at 2.6 Å resolution. *Journal of Molecular Biology*, **289**, 303–17.

Shiraki, K., Nishikawa, K. and Goto, Y. (1995). Trifluoroethanol-induced stabilization of the α-helical structure of β-lactoglobulin: implication for non-hierarchical protein folding. *Journal of Molecular Biology*, **245**, 180–94.

Shrake, A., Frazier, D. and Schwarz, F. P. (2006). Thermal stabilization of human albumin by medium- and short-chain n-alkyl fatty acid anions. *Biopolymers*, **81**, 235–48.

Simard, J. R., Zunszain, P. A., Ha, C. E., Yang, J. S., Bhagavan, N. V., Petitpas, I., Curry, S. and Hamilton, J. A. (2005). Locating high-affinity fatty acid-binding sites on albumin by X-ray crystallography and NMR spectroscopy. *Proceedings of the National Academy of Sciences of the United States of America*, **102**, 17958–63.

Skibsted, L. H., Orlien, V. and Stapelfeldt, H. (2007). Temperature effects on pressure denaturation of β-lactoglobulin. *Milchwissenschaft*, **62**, 13–15.

Spector, A. A. (1975). Fatty-acid binding to plasma albumin. *Journal of Lipid Research*, **16**, 165–79.

Stapelfeldt, H. and Skibsted, L. H. (1999). Pressure denaturation and aggregation of β-lactoglobulin studied by intrinsic fluorescence depolarization, Rayleigh scattering, radiationless energy transfer and hydrophobic fluoroprobing. *Journal of Dairy Research*, **66**, 545–58.

Stapelfeldt, H., Petersen, P. H., Kristiansen, K. R., Qvist, K. B. and Skibsted, L. H. (1996). Effect of high hydrostatic pressure on the enzymic hydrolysis of β-lactoglobulin B by trypsin, thermolysin and pepsin. *Journal of Dairy Research*, **63**, 111–18.

Steinrauf, L. K. (1998). Structures of monoclinic lysozyme iodide at 1.6 Å and of triclinic lysozyme nitrate at 1.1 Å. *Acta Crystallographica Section D: Biological Crystallography*, **D54**, 767–79.

Strom, M. B., Rekdal, O. and Svendsen, J. S. (2000). Antibacterial activity of 15-residue lactoferricin derivatives. *Journal of Peptide Research*, **56**, 265–74.

Strom, M. B., Haug, B. E., Rekdal, O., Skar, M. L., Stensen, W. and Svendsen, J. S. (2002). Important structural features of 15-residue lactoferricin derivatives and methods for improvement of antimicrobial activity. *Biochemistry and Cell Biology*, **80**, 65–74.

Sun, C., Yang, J., Wu, X., Huang, X., Wang, F. and Liu, S. (2005). Unfolding and refolding of bovine serum albumin induced by cetylpyridinium bromide. *Biophysical Journal*, **88**, 3518–24.

Svensson, M., Hakansson, A., Mossberg, A. K., Linse, S. and Svanborg, C. (2000). Conversion of alpha-lactalbumin to a protein inducing apoptosis. *Proceedings of the National Academy of Sciences of the United States of America*, **97**, 4221–26.

Tanaka, N. and Kunugi, S. (1996). Effect of pressure on the deuterium exchange reaction of α-lactalbumin and β-lactoglobulin. *International Journal of Biological Macromolecules*, **18**, 33–39.

Tanaka, N., Nishizawa, H. and Kunugi, S. (1997). Structure of pressure-induced denatured state of human serum albumin: a comparison with the intermediate in urea-induced denaturation. *Biochimica et Biophysica, Acta Protein Structure and Molecular Enzymology*, **1338**, 13–20.

Tanford, C. and Nozaki, Y. (1959). Physico-chemical comparison of β-lactoglobulins A and B. *Journal of Biological Chemistry*, **234**, 2874–77.

Tanford, C., Bunville, L. G. and Nozaki, Y. (1959). Reversible transformation of β-lactoglobulin at pH 7.5. *Journal of the American Chemical Society*, **81**, 4032–36.

Tanford, C., De, P. K. and Taggart, V. G. (1960). The role of the α-helix in the structure of proteins. Optical rotatory dispersion of β-lactoglobulin. *Journal of the American Chemical Society*, **82**, 6028–34.

Tayyab, S., Sharma, N. and Mushahid Khan, M. (2000). Use of domain specific ligands to study urea-induced unfolding of bovine serum albumin. *Biochemical and Biophysical Research Communications*, **277**, 83–88.

Thomas, P. D. and Dill, K. A. (1993). Local and nonlocal interactions in globular proteins and mechanisms of alcohol denaturation. *Protein Science*, **2**, 2050–65.

Tromelin, A. and Guichard, E. (2006). Interaction between flavor compounds and β-lactoglobulin: approach by NMR and 2D/3D-QSAR studies of ligands. *Flavour and Fragrance Journal*, **21**, 13–24.

Tsuge, H., Ago, H., Noma, M., Nitta, K., Sugai, S. and Miyano, M. (1992). Crystallographic studies of a calcium binding lysozyme from equine milk at 2.5 Å resolution. *Journal of Biochemistry*, **111**, 141–43.

Uhrínová, S., Uhrín, D., Denton, H., Smith, M., Sawyer, L. and Barlow, P. N. (1998). Complete assignment of ^{1}H, ^{13}C and ^{15}N chemical shifts for bovine β-lactoglobulin: secondary structure and topology of the native state is retained in a partially unfolded form. *Journal of Biomolecular NMR*, **12**, 89–107.

Uhrínová, S., Smith, M. H., Jameson, G. B., Uhrín, D., Sawyer, L. and Barlow, P. N. (2000). Structural changes accompanying pH-induced dissociation of the β-lactoglobulin dimer. *Biochemistry*, **39**, 3565–74.

van Hooijdonk, T. and Steijns, J. (2002). Incorporation of bioactive substances into products: technological challenges. *Bulletin of the International Dairy Federation*, **375**, 65–69.

Vant, S. C., Glen, N. F., Kontopidis, G., Sawyer, L. and Schaschke, C. J. (2002). Volumetric changes to the molecular structure of β-lactoglobulin processed at high pressure. *High Temperatures–High Pressures*, **34**, 705–12.

Watanabe, S. and Sato, T. (1996). Effects of free fatty acids on the binding of bovine and human serum albumin with steroid hormones. *Biochimica et Biophysica Acta—General Subjects*, **1289**, 385–96.

Wehbi, Z., Perez, M-D., Sanchez, L., Pocovi, C., Barbana, C. and Calvo, M. (2005). Effect of heat treatment on denaturation of bovine α-lactalbumin: determination of kinetic and thermodynamic parameters. *Journal of Agricultural and Food Chemistry*, **53**, 9730–36.

Wu, S. Y., Perez, M. D., Puyol, P. and Sawyer, L. (1999). β-Lactoglobulin binds palmitate within its central cavity. *Journal of Biological Chemistry*, **274**, 170–74.

Yalcin, A. S. (2006). Emerging therapeutic potential of whey proteins and peptides. *Current Pharmaceutical Design*, **12**, 1637–43.

Yang, J., Dunker, A. K., Powers, J. R., Clark, S. and Swanson, B. G. (2001). β-Lactoglobulin molten globule induced by high pressure. *Journal of Agricultural and Food Chemistry*, **49**, 3236–43.

Ye, M-Q., Yi, T-Y., Li, H-P., Guo, L-L. and Zou, G-L. (2005). Study on thermal and thermal chemical denaturation of bovine immunoglobulin G. *Huaxue Xuebao*, **63**, 2047–54.

Zhang, H., Deligeersang, Guo, J., Mu, Z., Zhang, Y. and Zhu, H. (1998). Denaturation of bovine milk IgG at high pressure and its stabilization. *Shipin Kexue (Beijing)*, **19**, 10–12.

Zsila, F. (2003). A new ligand for an old lipocalin: induced circular dichroism spectra reveal binding of bilirubin to bovine β-lactoglobulin. *FEBS Letters*, **539**, 85–90.

Zsila, F., Bikadi, Z. and Simonyi, M. (2002). Retinoic acid binding properties of the lipocalin member β-lactoglobulin studied by circular dichroism, electronic absorption spectroscopy and molecular modeling methods. *Biochemical Pharmacology*, **64**, 1651–60.

Zsila, F., Hazai, E. and Sawyer, L. (2005). Binding of the pepper alkaloid piperine to bovine β-lactoglobulin: circular dichroism spectroscopy and molecular modeling study. *Journal of Agricultural and Food Chemistry*, **53**, 10179–85.

7

High-pressure-induced interactions involving whey proteins

Hasmukh A. Patel and Lawrence K. Creamer

Abstract

High-pressure processing (HPP) has been reported to bring about particular changes in the molecular structure of proteins and thus gives rise to new properties that are inaccessible via conventional methods of protein modification. The aim of this chapter is to provide relevant current knowledge, mainly focusing on how HPP can affect the structural conformational, unfolding and aggregation of whey proteins and their interactions with other proteins. The denaturation and aggregation pathways and mechanisms underlying pressure-induced denaturation and aggregation of β-lactoglobulin, α-lactalbumin and bovine serum albumin and their mixtures in different systems are explained. Heat and HPP have been reported to have some similar effects and some different effects on proteins. In some instances, pressure-induced changes in various proteins are compared and contrasted with the heat-induced changes in milk proteins.

Introduction

Processing treatments such as heat and high pressure are normally applied in the food industry for the purpose of microbial destruction or shelf-life extension, or to achieve

Milk Proteins: From Expression to Food
ISBN: 978-0-12-374039-7

desired functionality in the final product (Jelen and Rattray, 1995; Singh, 1995). It is well known that the heat-induced interactions of milk proteins have a marked impact on the functionality of the final products and such interactions are of considerable commercial importance in the dairy and food industries. Therefore, considerable research efforts have been directed to studying the detailed pathways/mechanisms of heat-induced functionality and the effects of heat treatments on milk proteins (denaturation, aggregation and gelation of whey proteins) and protein–protein interactions, including interactions of caseins and whey proteins, have been studied in great detail over 5–6 decades (e.g. Haque *et al.*, 1987; Haque and Kinsella, 1988; Hill, 1989; Noh *et al.*, 1989; Matsudomi *et al.*, 1992, 1993, 1994; Hines and Foegeding, 1993; McSwiney *et al.*, 1994a, 1994b; Gezimati *et al.*, 1996, 1997; Havea *et al.*, 1998, 2000, 2001, 2004; Manderson *et al.*, 1998; Schokker *et al.*, 1999, 2000; Hong and Creamer, 2002; Anema and Li, 2003a, 2003b; Cho *et al.*, 2003; Livney *et al.*, 2003; Livney and Dalgleish, 2004; Patel *et al.*, 2004, 2006, 2008) and the subject has often been reviewed (e.g. Mulvihill and Donovan, 1987; de Wit, 1990; Singh and Creamer, 1992; Jelen and Rattray, 1995; Singh, 1995; De la Fuente *et al.*, 2002; O'Connell and Fox, 2003; Singh and Havea, 2003; Singh, 2004; Considine *et al.*, 2007b).

Although thermal processing is effective, economical and readily available, in many cases, it has undesirable effects on the sensory and nutritional qualities of food (Balny and Masson, 1993; Trujillo *et al.*, 2002). More than a century ago, pioneering work by Hite (1899) showed the potential of high-pressure processing (HPP) as a non-thermal (alternative) preservation process and Bridgman (1914) demonstrated the effects of HPP on the denaturation and functional properties of egg proteins. However, it was not until 1990 that equipment advances and growing consumer demand for minimally processed, high-quality, nutritious and safe foods led to considerable research interest in HPP technology (Hayashi, 1988; Farr, 1990; Hoover, 1993; Datta and Deeth, 1999, 2003; Needs *et al.*, 2000a, 2000b; Velazquez de la Cruz *et al.*, 2002).

The recent interest in the HPP of food materials as an alternative to or in addition to temperature treatment led to many fundamental studies on the pressure behavior of proteins. It is also evident from the literature that, in many cases, heat treatment and high-pressure treatment have different effects on different milk proteins, suggesting that HPP has the potential for both preservation and modification of the structure of proteins, alteration to their functional properties and the creation of value-added products (see Ohmiya *et al.*, 1989; López-Fandiño *et al.*, 1996; García-Risco *et al.*, 1998, 2000; Datta and Deeth, 1999, 2003; Patel *et al.*, 2004, 2005, 2006; López-Fandiño, 2006a, 2006b).

Today, HPP is considered to be a possible alternative to heat treatment in many cases and has reached the consumer in a variety of products, such as high-pressure-treated fresh fruit jams, jellies, juices, salad dressings, rice, cakes and guacamole (Farr, 1990; Cheftel, 1992; Earnshaw, 1992; Hoover, 1993; López-Fandiño, 2006a, 2006b), but no high-pressure-treated commercial dairy products are available as yet. However, as there is increased interest in the high-pressure treatment of dairy products, it is currently a major focus of investigation and this subject has been extensively reviewed (see Heremans, 1982; Weber and Drickamer, 1983; Balny *et al.*, 1989, 1992; Masson, 1992; Balny and Masson, 1993; Mozhaev *et al.*, 1994; Johnston, 1995; Heremans *et al.*, 1997;

Messens *et al.*, 1997; Balci and Wilbey, 1999; Datta and Deeth, 1999, 2003; Farkas and Hoover, 2000; Boonyaratanakornkit *et al.*, 2002; Huppertz *et al.*, 2002; Lullien-Pellerin and Balny, 2002; Royer, 2002; Trujillo *et al.*, 2002; Claeys *et al.*, 2003; Huppertz *et al.*, 2006a, 2006b; López-Fandiño, 2006a, 2006b; Considine *et al.*, 2007b; Rastogi *et al.*, 2007; Patel *et al.*, 2008).

As whey proteins are widely used as nutritional and functional ingredients in the food industry, there is obvious interest in studying the effects of HPP on whey proteins and their interactions with other proteins. The specific detailed structural aspects and some fundamental aspects related to the structural stability of β-lactoglobulin (β-LG), α-lactalbumin (α-LA), bovine serum albumin (BSA), immunoglobulins (Igs) and lactoferrin (LF) have been dealt with in Chapter 6. The present chapter builds on these fundamental aspects and extends the discussion to pathways and mechanisms of pressure-induced denaturation, aggregation and interactions of individual whey proteins and their mixtures in various systems.

Characterization of heat- and pressure-induced changes to proteins

Several different methods have been used in the study and characterization of protein denaturation and aggregation, including X-ray crystallography, far-UV circular dichroism (CD), Fourier transform infrared (FTIR) spectroscopy, Fourier transform of Raman spectra and hydrogen/deuterium (H/D) exchange using a nuclear magnetic resonance (NMR) method for determining changes in the secondary structure (Tanaka and Kunugi, 1996; Subirade *et al.*, 1998; Panick *et al.*, 1999; Belloque *et al.*, 2000; Edwards *et al.*, 2002; Ngarize *et al.*, 2004). ^{1}H NMR is a technique that is very sensitive to structural changes and can give structural and dynamic information at an atomic level.

Methods such as intrinsic and extrinsic fluorescence and induced CD have been widely used for tertiary structure determination (Pearce, 1975; Masson, 1992; Dufour *et al.*, 1994; Heremans *et al.*, 1997; Narayan and Berliner, 1997; Kelly and Price, 2000; Collino *et al.*, 2003; Kontopidis *et al.*, 2004). Scattering methods such as small-angle X-ray scattering (SAXS) and light scattering (LS) have also been used recently (Pessen *et al.*, 1989; Holt *et al.*, 2003). In particular, SEC–MALLS (size exclusion chromatography coupled to multi-angle laser light scattering) can provide excellent information on aggregate size (Schokker *et al.*, 1999). SEC–MALLS combined with electrophoretic techniques is a powerful tool for characterizing intermediates and aggregates formed during heating (Schokker *et al.*, 1999, 2000). The colorimetric methods used in the determination of protein sulfydryl groups in milk (Owusu-Apenten, 2005) and other methods used in the study of protein aggregation have been reviewed (Wang, 1999, 2005; De la Fuente *et al.*, 2002; Considine *et al.* 2007b).

It is evident that less commonly used polyacrylamide gel electrophoresis (PAGE) separation in different environments (e.g. sodium dodecyl sulfate [SDS] solution, in the presence or absence of a disulfide bond reducing agent) allows the separation and differentiation of the various polypeptide chains. SDS-PAGE dissociates the processing-induced non-covalent bonds and leaves the covalent bonds intact, whereas native- or alkaline-PAGE generally separates the whey proteins with hydrophobic and disulfide bonds intact; therefore, it is possible to show whether the processing-induced protein aggregates are hydrophobically linked aggregates, reducible disulfide-cross-linked aggregates etc. in heated solutions of pure β-LG (Manderson *et al.*, 1998) or commercial whey protein concentrate (WPC) solutions (Havea *et al.*, 1998, 2000, 2001). The same techniques have been used successfully to examine the pressure-induced aggregation of various dairy proteins, revealing the subtle and not-so-subtle differences (Patel *et al.*, 2004, 2005, 2006), which are discussed in detail in the later part of this chapter.

When heat-treated or pressure-treated samples were analyzed using various PAGE techniques, many changes in the PAGE patterns of these samples were noted (McSwiney *et al.*, 1994b; Gezimati *et al.*, 1996, 1997; Havea, 1998; Havea *et al.*, 1998, 2000, 2001, 2002, 2004; Manderson *et al.*, 1998; Cho *et al.*, 2003; Patel *et al.*, 2004, 2005, 2006; Considine *et al.*, 2005a, 2005b, 2007a). It became necessary to identify each of these changes and aggregates observed in different PAGE environments. For simplification, some specific nomenclature had to be formulated, e.g. "native monomer," "native dimer," "SDS monomer," "SDS dimer," etc. (see Table 7.1).

Two-dimensional (2D) PAGE (2D native- and then SDS-PAGE, and 2D SDS- and then reduced SDS-PAGE) can be applied to further characterize various intermediate species of protein aggregates and the high-molecular-weight aggregates formed as a consequence of heat or pressure treatment. A combination of two PAGE techniques can be an even more powerful technique; it can determine the composition of the protein aggregates and/or the types of bonds by which the protein aggregates are held together in the heat-treated (Havea *et al.*, 1998, 2000, 2001; Manderson *et al.*, 1998) or pressure-treated (Patel *et al.*, 2004, 2005, 2006) samples.

In the 2D native and then SDS-PAGE procedure, a sample containing protein complexes is analyzed using one-dimensional (1D) PAGE in a Tris HCl buffer at pH 8.7,

Table 7.1 Nomenclature of different forms of proteins and protein aggregates analyzed using various electrophoretic techniques

Nomenclature	Applicable to	Description
Native	All	Has all the characteristics of native proteins
Native dimer	β-LG	Normal state of native β-LG between pH 4 and pH 7 (at 30°C and at 0.05 M NaCl)
Native-like	All	Behaves like native protein on alkaline (native)-PAGE
Non-native monomer	All	Behaves like native protein on SDS-PAGE, but not on alkaline-PAGE
SDS monomer	All	Has mobility on SDS-PAGE close to that expected from molecular monomer
SDS dimer, trimer, etc.	All	Has mobility on SDS-PAGE close to that expected from molecular dimer, trimer, etc.
Apo	α-LA	Deficient in calcium (or other divalent cation)
Holo	α-LA	Same as native α-LA

called alkaline- or native-PAGE. The proteins or protein complexes separated on native-PAGE consist of all complexes including covalent (e.g. disulfide bonds) and non-covalent (e.g. hydrophobic)-bonded aggregates. The 1D native-PAGE gel strip with its separated protein bands is transferred to SDS-PAGE in the second dimension (i.e. transferred into a dissociating environment). Once the proteins in the strip are partially equilibrated with the SDS to form SDS-protein complexes, they are electrophoresed into a new (SDS) environment in a second dimension SDS-PAGE. SDS-PAGE in the second dimension dissociates non-covalent bonds (mainly hydrophobic aggregates) from the non-reduced native-PAGE gel strip, whereas the covalent bonds (disulfide bonds) remain unaffected.

The other type of 2D PAGE is SDS- and then reduced SDS-PAGE. This type of 2D PAGE technique separates the initial mixture of proteins using SDS-PAGE with all the native and process-induced disulfide bonds intact. These separated proteins and protein aggregates are treated with a disulfide bond reducing agent, 2-mercaptothanol (2-ME), while still in the gel strip. This gel strip is then used as the sample source for the second dimension PAGE analysis. SDS-PAGE in the second dimension separates the proteins as the reduced SDS-protein species. Thus the components of each of the various disulfide-bonded aggregates can be identified (for example, see Figures 7.3 and 7.4; refer to Patel *et al.*, 2004, 2005, 2006 and Considine *et al.*, 2007b for detailed descriptions of these PAGE methods).

Effects of high pressure on milk proteins

The native three-dimensional structure of a protein is maintained by a variety of non-covalent interactions (such as hydrogen bonding, electrostatic, van der Waals' and hydrophobic interactions) between amino acid residues within the polypeptide chain and between residues and solvent molecules (Singh, 1995). Three-dimensional structure has a very important role to play in the stability and functional properties of a protein. Excellent reviews describing the molecular basis of whey protein functionalities (Holt, 2000), the structure–function relationship (Qi *et al.*, 2001) and the possible usefulness of knowledge of the three-dimensional structure of milk proteins (Sawyer *et al.*, 2002) are available. The number and the position of free thiols and disulfide bonds present in the structure of different proteins are greatly significant to their stability, denaturation, interactions and functional properties (Holt, 2000; Sawyer *et al.*, 2002).

Applications of high pressure cause proteins to lose their native three-dimensional structure and lead to denaturation and aggregation of whey proteins and/or their interactions with each other (whey protein–whey protein interactions) or with the caseins (casein–whey protein interactions). According to Le Chatelier-Braun's principle, under pressure, reactions with a negative volume change are enhanced and reactions with a positive volume change are suppressed (Buchheim and Prokopek, 1992; Johnston, 1995; Balci and Wilbey, 1999). Thus, the basic thermodynamic approach to pressure-induced changes to proteins is based on the compressibility of molecules and on changes in volume (ΔV) during pressure treatment as, unlike temperature, pressure affects only the volume of a system (Balny and Masson, 1993; Gross and Jaenicke, 1994; Heremans and Smeller, 1998; Lullien-Pellerin and Balny, 2002). Changes in the solvation volume are caused mainly by pressure-induced ionization, changes in solvent exposure of amino acid side chains and peptide bonds, and diffusion of water into cavities located in the hydrophobic core of the protein (Heremans, 1982, 1992; Balny and Masson, 1993; Messens *et al.*, 1997; Hendrickx *et al.*, 1998; Claeys *et al.*, 2003).

In contrast to heat treatments, where covalent bonds and non-covalent bonds are affected, it has been reported that HPP at room temperature (≈20°C) disrupts only relatively weak bonding such as intramolecular hydrophobic and electrostatic interactions (Balny and Masson, 1993; Silva and Weber, 1993), whereas hydrogen bonds are relatively insensitive to pressure, suggesting that high pressure affects the tertiary (three-dimensional configuration held together mainly by hydrophobic and ionic interactions) and quaternary (the spatial arrangement by non-covalent interactions into a multimeric protein) structures of globular proteins and has little effect on their secondary structure. There are views that covalent bonds are largely insensitive to pressure treatment at relatively low temperature (Hayakawa *et al.*, 1996), which means that the primary protein structure (the amino acid sequence) will remain intact. This partly explains why HPP has been reported to have slightly different effects on protein structure compared with heat treatments.

The effects of high pressure on denaturation, aggregation and interactions of whey proteins have been studied extensively under various conditions and in different systems such as milk (see Felipe *et al.*, 1997; Law *et al.*, 1998; Arias *et al.*, 2000; Hinrichs, 2000; Scollard *et al.*, 2000; Huppertz *et al.*, 2002, 2004a, 2004b; Anema *et al.*, 2005a), WPC (see Van Camp *et al.*, 1997a, 1997b) and whey protein isolate (WPI) (Hinrichs *et al.*, 1996a, 1996b; Hinrichs, 2000; Michel *et al.*, 2001), and using pure whey proteins (Dumay *et al.*, 1994, 1998; Funtenberger *et al.*, 1995; Galazka *et al.*, 1996a; Jegouic *et al.*, 1996; Olsen *et al.*, 1999).

It has been reported that pressure-induced reactions of whey proteins lead to unfolding of monomeric proteins, aggregation and gelation (Zipp and Kautzmann, 1973; Heremans, 1982; Weber and Drickamer, 1983; Silva and Weber, 1993; Van Camp and Huyghebaert, 1995a, 1995b; Van Camp *et al.*, 1997a, 1997b; Balci and Wilbey, 1999; Tedford *et al.*, 1999a, 1999b; Huppertz *et al.*, 2002; Fertsch *et al.*, 2003), by reformation of intra- and intermolecular bonds within or between the molecules, linked by hydrophobic interactions and disulfide bridges (see Cheftel, 1992; Masson, 1992; Hoover, 1993; Galazka *et al.*, 1996a; Messens *et al.*, 1997; Trujillo *et al.*, 2002), depending on the type of protein, protein concentration, pH, ionic strength, applied pressure, pressurizing temperature and duration of the pressure treatment, etc. (Masson, 1992; Messens *et al.*, 1997; Fertsch *et al.*, 2003; Huppertz *et al.*, 2004a).

Also, different pressurizing temperatures may have different effects on the denaturation and aggregation of proteins (see Huppertz *et al.*, 2004a; Patel, 2007), because of combined effects of pressure and temperature, which can have different effects on the interactions that maintain protein structures. Many studies related to the combined effects of pressure and temperature on milk proteins have been published (e.g. Tedford *et al.*, 1999b; Huppertz *et al.*, 2004a; Patel, 2007). However, the present review focuses mainly on the HPP-induced changes in milk proteins that occur at ambient temperature, unless specified otherwise.

Selected reports on the effects of high pressure on β-LG, α-LA, BSA or their combinations in various systems are summarized in the following sections.

Denaturation and aggregation of pure whey proteins in model systems

β-Lactoglobulin (β-LG)

Many reports have suggested that β-LG is the most sensitive of the major whey proteins to high pressure and that it dominates the pressure-induced denaturation, aggregation and gelation of the whey protein system (Van Camp and Huyghebaert, 1995a, 1995b; Stapelfeldt *et al.*, 1996; Van Camp *et al.*, 1996, 1997a, 1997b; Kanno *et al.*, 1998; Belloque *et al.*, 2000; Patel *et al.*, 2004, 2005; López-Fandiño, 2006a, 2006b; Considine *et al.*, 2007b). Therefore, the majority of studies have concentrated on the effects of high pressure on β-LG in order to gain insight into the mechanisms of unfolding and aggregation that occur during pressurization or after pressure release.

Different effects of high pressure on proteins have been observed when the samples are analyzed under high pressure ("in situ" analysis) and when analyzed after pressure release. With increasing pressure, protein molecules undergo a sequence of conformational changes because of alterations in stabilizing interactions (Johnston *et al.*, 1992). During the pressure release phase and after pressure treatment, new intermolecular interactions are formed and the proteins may be newly structured (Fertsch *et al.*, 2003). The majority of the reports included in the present review deal with the interactions of proteins after pressure release.

At relatively low pressures (50 MPa), analysis of thiol reactivity (Møller *et al.*, 1998; Stapelfeldt *et al.*, 1999) and NMR studies (Tanaka and Kunugi, 1996) suggested the existence of a "pre-denatured" state of β-LG. This "pre-denatured" form corresponded to a not-completely-unfolded structure, which preceded reversible denaturation. Belloque *et al.* (2000), using ^{1}H NMR, showed that the degree of deuterium exchange was very small at 100 MPa and that there were no variations in the resonances belonging to the strongly bonded "core" of β-LG.

These observations might suggest that the regions of β-LG affected by the "pre-denatured" state are likely to be different from the "core," as the core was still tight and remained unaltered at 100 MPa.Whereas pressures ranging between 0 and 140 MPa did not affect β-sheets (Subirade *et al.*, 1998), the reactivity of the free sulfydryl group of β-LG increased with pressure up to 150 MPa (Tanaka *et al.*, 1996a, 1996b). These results suggested that, in spite of having a similar overall conformation, the architectures of β-LG before and after dynamic high pressure were stabilized by slightly different interactions (Subirade *et al.*, 1998).

Pressure-induced unfolding and refolding as studied by deuterium exchange (Belloque *et al.*, 2000) further demonstrated that the conformational flexibility of β-LG increased at 200 MPa. It was reported that, even though the core of β-LG was highly flexible at 400 MPa, its structure was found to be identical to the native structure after equilibration back to atmospheric pressure. It was also suggested that pressure-induced aggregates are formed by β-LG molecules maintaining most of their structure, and that the intermolecular disulfide bonds, formed by sulfydryl–disulfide exchange reactions, are likely to involve Cys66–Cys160 rather than Cys106–Cys119 (Belloque *et al.*, 2000), which is different from the possibility of thiol–disulfide interchange reactions as discussed by Considine *et al.* (2007b).

In addition, it was reported that the β-LG variants A and B could be distinguished in a ^{1}H NMR spectrum of a solution made with the AB mixed variant by the differences in chemical shifts of Met107 and Cys106 and that, under pressure, the core of β-LG A seemed to unfold faster than that of β-LG B and the structural recovery of the core was full for both variants (Belloque *et al.*, 2000). These results are somewhat in agreement with those of Iametti *et al.* (1997), who found that only 10% of the structure was lost at 600 or 900 MPa.

The pressure denaturation of β-LG was believed to be a simple two-step mechanism until Jonas and Jonas (1994) reported pressure-induced pre-denaturation transitions and thus demonstrated that the pressure denaturation of β-LG could be a stepwise process. A few years later, Stapelfeldt and Skibsted (1999) proposed a three-step denaturation process and recently Considine *et al.* (2005b) published a three-stage model of the pressure denaturation of β-LG (Figure 7.1; for a detailed description, refer to Considine *et al.*, 2007b). It has also been reported that addition of hydrophobic ligands such as all-*trans*-retinol, palmitic acid, SDS and 8-anilino-1-naphthalenesulfonate (ANS) to β-LG solution before pressure treatment (see Figure 7.1) affects the pathways of denaturation (Considine *et al.*, 2005b, 2007b).

The overall major changes that occur to the structure of β-LG include monomerization of the dimeric state (Iametti *et al.*, 1997), a decrease in α-helix and β-sheet content (Hayakawa *et al.*, 1996; Panick *et al.*, 1999) and irreversible changes involving the formation of intermolecular disulfide bonds (Funtenberger *et al.*, 1997; Iametti *et al.*, 1997; López-Fandiño *et al.*, 1997; Møller *et al.*, 1998). In the pressure-induced mechanism proposed (Iametti *et al.*, 1997), release of monomers represents one of the earliest events, whereas association of transiently modified monomers stabilizes the denatured forms of the protein. In addition, it has been reported that inter- and intramolecular reactions of sulfydryl groups can occur (Tanaka *et al.*, 1996a, 1996b),

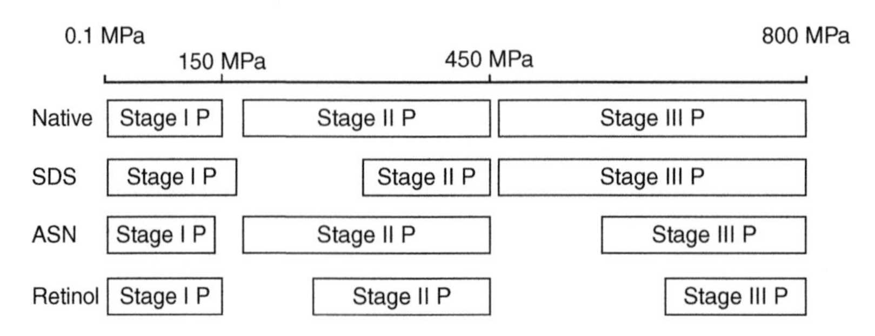

Figure 7.1 Proposed three-stage model of the pressure denaturation of β-LG A and β--LG B with added ANS, retinol or SDS. (Reproduced with the permission of Considine *et al.* (2005b), copyright 2005 *American Chemical Society.*)

leading to the formation of new disulfide bonds through sulfydryl-disulfide interchange reactions (Funtenberger *et al.*, 1997) when samples are pressure treated at 450 MPa. At 800 MPa, most of the β-LG present in the system becomes involved in hydrophobic and disulfide-linked aggregates (Patel, 2007); this behavior is quite similar to the effects of heat treatment on β-LG (Havea *et al.*, 1998, 2000, 2001).

Moreover, it has been reported that factors such as protein concentration (Dumay *et al.*, 1994), pH, ionic strength, type and molarity of the buffer used for preparation of the protein solutions (Funtenberger *et al.*, 1995; Cheftel and Dumay, 1996), pressure intensity, pressurizing time and pressurizing temperature (Yang *et al.*, 2001; Patel, 2007), and binding of hydrophobic ligands (Considine *et al.*, 2005b, 2007a) and small molecules such as sucrose (Dumay *et al.*, 1994) can affect the pressure-induced denaturation and aggregation of β-LG. It has been found that high-pressure-induced denaturation is partially reversible at lower (2.5%) protein concentration but that the denaturation is irreversible and that aggregation occurs at higher (5.0%) concentration (Dumay *et al.*, 1994).

The progressive formation of intermolecular disulfide-bonded dimers to hexamers or higher polymers of β-LG (pH 7.0) has been reported to be a function of the pressure level and of the buffer type and molarity (Funtenberger *et al.*, 1995). It was suggested that high pressure induced the formation of intermolecular disulfide bonds, especially at neutral pH. When the combined effects of pressure, temperature and time were evaluated, the pressure intensity was found to have major effects on the structure of β-LG (Aouzelleg *et al.*, 2004), which is somewhat different from the finding that the combined effects of pressure intensity and temperature have the most effect on the denaturation of β-LG (Patel, 2007).

Different aspects of the pressure-induced unfolding and aggregation of β-LG have also been reviewed in detail (see López-Fandiño, 2006b; Considine *et al.*, 2007b), including the effects of high pressure on the functional properties of β-LG (López-Fandiño, 2006b).

α-Lactalbumin (α-LA)

Several studies reported that, compared with β-LG, α-LA is resistant to pressure denaturation (López-Fandiño *et al.*, 1996; Tanaka and Kunugi, 1996; Scollard *et al.*, 2000; Huppertz *et al.*, 2004a; Patel *et al.*, 2004, 2006). A comparison of pressure-induced changes of two major whey proteins, α-LA and β-LG, at neutral pH showed that the reversible unfolding to a molten globule state of α-LA begins at 200 MPa and loss of native structure becomes irreversible only beyond 400 MPa, as compared with 50 and 150 MPa respectively for β-LG (Tanaka and Kunugi, 1996; Tanaka *et al.*, 1996c; Stapelfeldt and Skibsted, 1999; McGuffey *et al.*, 2005).

These different behaviors of the proteins were variously explained, including the differences in their secondary structures (which lead to a higher effective hydrophobicity in β-LG) and/or in the number of disulfide bonds (four in α-LA and two in β-LG) and also the Ca^{2+} binding sites (Tanaka and Kunugi, 1996). In fact, the binding of calcium is reported to remarkably stabilize α-LA to pressure, by a 200 MPa increase in the pressure value at which denaturation occurs (Dzwolak *et al.*, 1999; Hosseini-nia *et al.*, 2002). This observation is partly supported by the finding that there was a difference in ΔV values between apo- and holo-α-LA (Kobashigawa *et al.*, 1999).

Fluorescent measurement of dansylated (prepared at atmospheric pressure) proteins, especially the energy transfer from the intrinsic tryptophan residue to the dansyl group, showed that the protein structure was deformed by pressure and that the energy transfer mechanisms of the two proteins were differently affected by high pressure, probably reflecting the degree of compactness of their pressure-perturbed structures (Tanaka *et al.*, 1996c). It has been reported that α-LA is present in a molten globule state beyond 200 MPa and up to 400 MPa (Jonas, 2002) and that α-LA changes its conformation from the molten globule state to the unfolded state without volume changes (Kobashigawa *et al.*, 1999). The volume of α-LA changes only at the transition from the native state to the molten globule state (Kobashigawa *et al.*, 1999). Lasselle *et al.* (2003) reported that heat and high pressure had similar effects, supporting the view that the molten globule state is stabilized by hydrophobic interactions.

In samples of severely heated solutions of α-LA, dimers and larger aggregates of α-LA were formed (Lyster, 1970; Havea *et al.*, 2001). However, no effects on monomeric α-LA were noticeable when pure α-LA was pressure treated at 800 MPa (Patel, 2007), except that some changes in the structure of α-LA were found when α-LA samples were pressure treated at 1000 MPa (Jegouic *et al.*, 1996).

Bovine serum albumin (BSA)

BSA has been found to be quite resistant to pressure treatment up to 400 MPa (Hayakawa *et al.*, 1992; López-Fandiño *et al.*, 1996; Patel *et al.*, 2004, 2005, 2006; López-Fandiño, 2006b). Several reports explaining the pressure stability of BSA are available. There are views that pressure-induced changes in the secondary structure of BSA are mainly reversible (Hosseini-nia *et al.*, 2002) and that the greater stability of BSA is probably related to the fact that this molecule, through its 17 intramolecular disulfide bonds and the presence of several separate domains, has an extremely rigid structure (Hayakawa *et al.*, 1992; López-Fandiño *et al.*, 1996; López-Fandiño, 2006b). It is possible that the relatively high number of disulfide linkages in BSA may impede pressure-induced aggregation by protecting the hydrophobic core/groups present inside the molecule from exposure to the solvent (Hosseini-nia *et al.*, 2002).

Ceolín (2000) studied the hydrodynamic behavior of BSA, using a perturbed angular correlation technique, as a function of high pressure up to 410 MPa. It was reported that, at moderate pressure ($\approx$150 MPa), the BSA molecule suffers structural modifications that produce an increase in the molecular volume and the rotational correlation time of the molecule. However, it may be possible that, unlike β-LG, the changes in the secondary structure of BSA are largely reversible (López-Fandiño, 2006b). However, processing at 800 MPa was reported to have a substantial effect on the secondary structure of BSA, and BSA was polymerized through disulfide bonding involving the free sulfydryl residue (Galazka *et al.*, 1996b, 1997; Patel, 2007).

Immunoglobulins (Igs) and lactoferrin (LF)

Both IgG and LF are more stable under pressure than under heat (Patel *et al.*, 2005, 2006; Carroll *et al.*, 2006; Palmano *et al.*, 2006; Indyk *et al.*, 2008). This finding has great commercial significance for using HPP in the manufacture of nutritional products containing IgG and LF. However, little attention has been paid to this subject. It was reported that the pressure stability of IgG was better in colostrum solutions than in pure IgG solutions (Indyk *et al.*, 2008), suggesting that some other colostrum milk components had protective effects on the denaturation of IgG.

A study on the response of IgG to high pressure (200−700 MPa) in the presence of the kosmotrope sucrose has been reported (Zhang *et al.*, 1998). Recently, Brisson *et al.* (2007) studied the effects of iron saturation on the thermal aggregation of LF at neutral pH and found that iron saturation markedly increased the thermal stability of LF and decreased aggregation. A similar observation was made by Palmano *et al.* (2006) for pressure-treated iron-saturated LF solutions.

Mixtures of α-LA and β-LG

α-LA does not form aggregates when pressure treated alone at 800 MPa (Patel, 2007), but forms high-molecular-weight disulfide-bonded oligomers at high pressure only in the presence of thiol reducers (Jegouic *et al.*, 1996) and the oligomers of α-LA are stabilized mainly by non-native interchain disulfide bridges. As reported for heat-treated mixtures of α-LA and β-LG (see Havea *et al.*, 2001; Hong and Creamer, 2002), mixed aggregates of denatured α-LA and β-LG were readily formed in pressure-treated whey protein solutions (Jegouic *et al.*, 1997). This observation supports the view that the presence of reactive thiol groups is a prerequisite for pressure-induced denaturation and aggregation of α-LA (Jegouic *et al.*, 1996, 1997). Therefore, in mixtures of α-LA and β-LG, because of its free sulfydryl group, β-LG can induce the oligomerization of α-LA, resulting in the formation of a large heterogeneous population of oligomers (Jegouic *et al.*, 1997; Grinberg and Haertlé, 2000).

Yet another possibility is that the interactions of α-LA and β-LG occur in a hydrophobic environment. However, it has been reported that, under pressure, the volume change of α-LA is much less (Lassalle *et al.*, 2003) than that of β-LG (Royer, 2002) and therefore there is little possibility for such interactions to take place in a hydrophobic environment. This partly explains why α-LA retains most of its structure under high pressure. At very high pressure (e.g. 800 MPa), the irreversible denaturation of α-LA was much less than that of β-LG, which was assigned to the difference in the number of bonds stabilizing the structure of each protein (Hinrichs *et al.*, 1996b; Messens *et al.*, 1997).

Mixtures of β-LG, α-LA and BSA

Little has been reported on the effects of high pressure on mixtures of β-LG, α-LA and BSA in pure protein systems. Recently, Patel (2007) reported that, when mixtures of β-LG, α-LA and BSA were pressure treated, there was a somewhat similar aggregation trend to that reported for heat-treated mixtures of β-LG, α-LA and BSA (Gezimati *et al.*, 1996, 1997; Havea *et al.*, 2001). However, in the pressure-treated samples, it appeared that β-LG, being the most pressure-sensitive whey protein, formed early aggregates, prior to the unfolding of either α-LA or BSA.

Pressure treatment of a ternary mixture of BSA, β-LG and α-LA generated aggregates comprising a mixture of hydrophobically linked and disulfide-linked aggregates (Patel, 2007), whereas when pure BSA solutions or combinations/mixtures of BSA and other whey proteins (e.g. a binary mixture of BSA and α-LA) were pressure treated, almost all the aggregates were disulfide linked and only a small proportion of the aggregates were hydrophobically linked (Patel, 2007). This may be due to structural differences in each of these proteins, and/or the effects of high-pressure treatment on the structure of each of these proteins may be responsible for such differences.

Commercial whey protein solutions

In addition to the above studies using pure protein systems, several studies conducted on heat- and pressure-induced denaturation, aggregation and gelation of whey proteins using commercial whey protein ingredients such as WPC or WPI are available.

Characterization of pressure-treated WPC solutions using 2D PAGE (Patel *et al.*, 2004, 2005) suggested that HPP generated both hydrophobically bonded and disulfide-bonded aggregates consisting of all whey proteins including β-LG, IgG, LF, BSA and α-LA (Figure 7.2), similar to those reported by Havea *et al.* (1998) for heat-treated WPC solutions. Heated WPC solutions contained 1:1 disulfide-bonded adducts of α-LA and β-LG, which were more obvious at low concentrations. Almost all of the β-LG was incorporated into the aggregates via disulfide bonds and to a lesser extent via hydrophobic interactions (Havea *et al.*, 1998).

However, when similar samples were pressure treated, the β-LG dimer was predominant (Patel *et al.*, 2004, 2005). The detailed characterization and identification of the disulfide-linked aggregates formed in pressure-treated WPC solutions is shown in Figure 7.3, which clearly shows that severe pressure treatment of WPC solutions generated disulfide-bonded dimer, trimer, tetramer, etc. as well as 1:1 complexes of β-LG:α-LA and that higher molecular weight disulfide-linked aggregates consisting BSA, LF, Ig, β-LG and α-LA were formed.

It was found that the sensitivities of each of the whey proteins to heat treatment (Ig > LF > BSA > β-LG B > β-LG A > α-LA) and pressure treatment (β-LG B > β-LG A > IgG > LF > BSA > α-LA) were considerably different (Patel *et al.*, 2004, 2005). Also, high-pressure treatment generated a comparatively greater proportion of smaller aggregates than did heat treatment (Patel *et al.*, 2004). These results confirm and support the view that there are some similarities and some differences between the heat- and high-pressure-induced aggregations of whey proteins. Similar differences were found when whey protein gel formation was induced by either heat treatment or pressure treatment (Van Camp and Huyghebaert, 1995a, 1995b; Van Camp *et al.*, 1996; Dumay *et al.*, 1998).

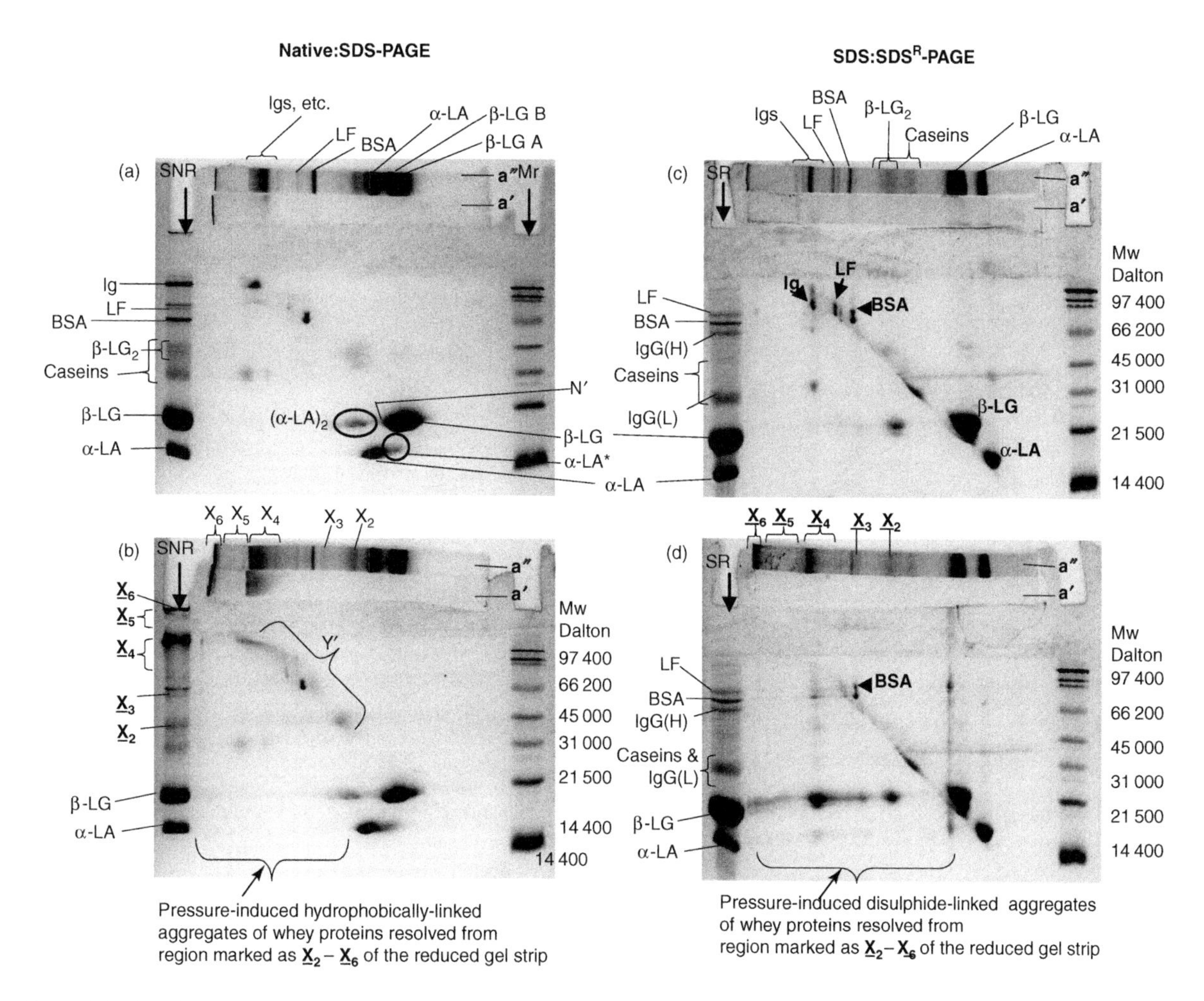

Figure 7.2 2D PAGE patterns of control and pressure-treated WPC solutions (12% w/v). Native- and then non-reduced SDS-PAGE patterns of (a) the control and (b) a sample pressure treated for 20 min at 800 MPa. Similarly, SDS- and then reduced SDS-PAGE patterns of (c) the control and (d) a sample pressure treated for 20 min at 800 MPa. Gel strips marked as a' and a" represent the sample strip and the stained strip respectively. X2 and X3 are dimer and trimer of β-LG respectively and X4, X5 and X6 are high-molecular-weight aggregates, which were caught up at the beginning of the resolving gel, caught up within the stacking gel and could not enter the gel respectively. For a detailed description, refer Patel *et al.* 2005. (Reproduced with permission from Patel *et al.* (2005), copyright 2005 *American Chemical Society.*)

At high protein concentration (10%), intermolecular interactions and irreversible aggregation are favored (Wong and Heremans, 1988; Dumay *et al.*, 1994). High-pressure treatment of concentrated (80 – 160 g/kg) β-LG isolate solutions (pH 7.0) prepared in water or various buffers induces β-LG gelation at low temperature (Zasypkin *et al.*, 1996; Dumay *et al.*, 1998). The decreasing solubility (in various dissociating media) of the protein constituents of pressure-induced gels as a function of storage time after pressure release suggests that the aggregation and gelation result from hydrophobic interactions, and also disulfide bonds, and that a progressive build-up of these interactions takes place after pressure release (Dumay *et al.*, 1998).

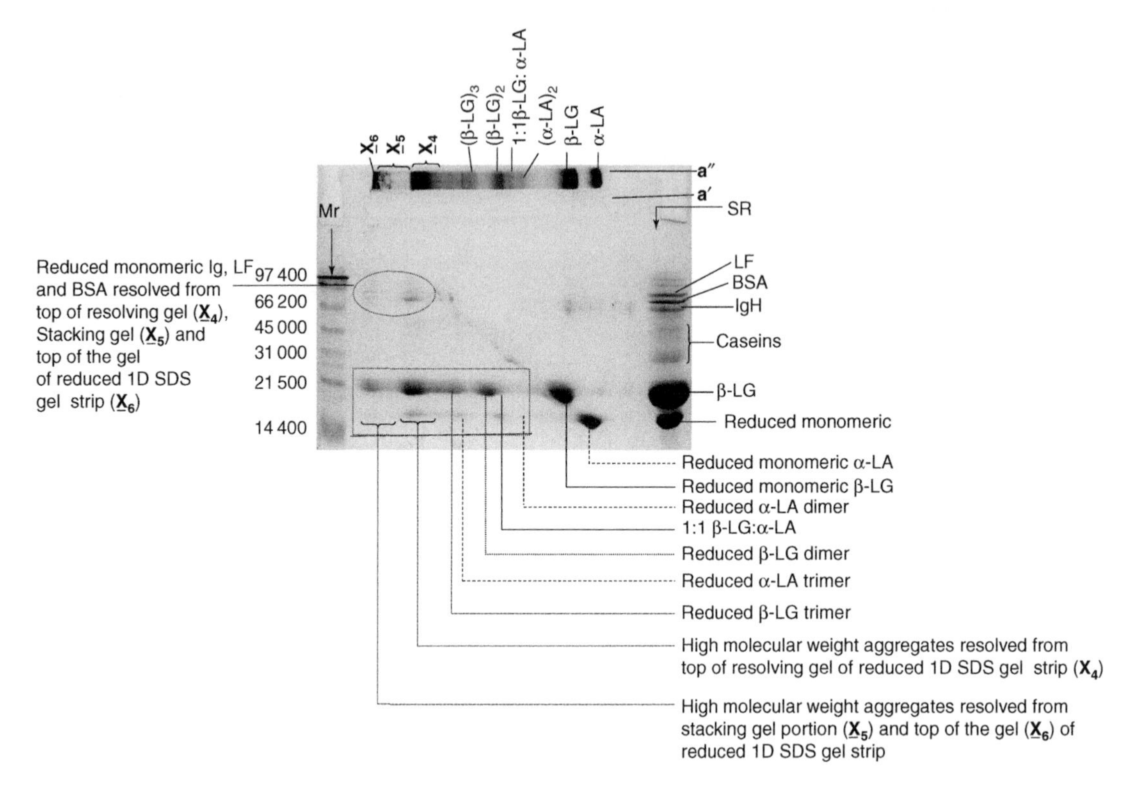

Figure 7.3 Detailed identification of the disulfide-linked aggregates and protein interactions on a 2D SDS- and then reduced SDS-PAGE pattern of a pressure-treated (800 MPa for 30 min) WPC solution.

Hinrichs *et al.* (1996b) determined orders of reaction of $n = 2.0$ for α-LA and $n = 2.5$ for β-LG in a WPI solution. These reaction rate constants were found to vary slightly at higher protein concentrations (Keim and Hinrichs, 2004). β-LG, α-LA and BSA participate in pressure-induced aggregation and gelation through disulfide bonding. Moreover, it has been reported that the number of stabilizing disulfide bonds directly influences the texture properties of pressure-induced whey protein gels (Keim and Hinrichs, 2004).

The effects of protein concentration, intensity of pressure treatment, holding time and pressurizing temperature on whey protein aggregation in WPC solutions have been investigated (Patel, 2007). The rate of aggregation of the whey proteins increased with an increase in the concentration of protein in the WPC solution and the pressurizing temperature. The combination of low protein concentration, mild pressure treatment (200 MPa) and low pressurizing temperature (20°C) led to minimal loss of native-like and SDS-monomeric β-LG, whereas the combination of high protein concentration, severe pressure treatment (600 MPa) and higher pressurizing temperature (40°C and higher) led to significant loss of both native-like and SDS-monomeric β-LG. The sensitivity of the pressure-resistant whey proteins, such as α-LA and BSA, to aggregation was significantly increased at pressurizing temperatures of 40°C and higher. Self-supporting gels were formed when 8 or 12% (w/v) WPC solutions were pressure treated at 600–800 MPa and 20°C.

Pressure-induced gelation of whey proteins

At protein concentrations sufficiently high for gel formation, WPC was found to produce pressure-induced gels in the pressure range 200–400 MPa (Van Camp and Huyghebaert, 1995a, 1995b; Van Camp *et al.*, 1996). In addition to protein concentration, applied pressure, holding time and pressurizing temperature were found to be major factors in the formation of pressure-induced gels and the properties of the gels (Van Camp and Huyghebaert, 1995a).

As discussed earlier, significant differences in protein denaturation and aggregation induced by heat compared with high pressure have been demonstrated (Heremans *et al.*, 1997; Patel *et al.*, 2004, 2005), which might suggest that the gels produced from whey proteins by high-pressure treatment may have different properties from those made by heat treatment. For example, it was found that heat-set gels, for equal protein concentrations, were firmer than pressure-induced gels (Van Camp and Huyghebaert, 1995a; Zasypkin *et al.*, 1996). High-pressure treatment generated gels with a more porous structure and lower firmness (Van Camp and Huyghebaert, 1995b; Zasypkin *et al.*, 1996; Dumay *et al.*, 1998) that were weaker, less elastic and more exudative than heat-induced gels (Cheftel and Dumay, 1996; Dumay *et al.*, 1998).

In contrast to heat-induced gels, pressure-induced gels of β-LG underwent mechanical and protein solubility changes when stored at 4°C following pressure release, clearly indicating a time-dependent strengthening of protein–protein interactions, probably because the primary aggregates of β-LG further aggregated during storage through hydrophobic interactions and disulfide bonds (Dumay *et al.*, 1998). Also, the WPC gels produced by high pressure (400 MPa for 30 min) at protein concentrations ranging from 110 g/L up to 183 g/L differed significantly from heat-induced protein gels (80°C for 30 min) with respect to gel strength and appearance (Van Camp and Huyghebaert, 1995a, 1995b).

The gel strength (Van Camp and Huyghebaert, 1995a; Van Camp *et al.*, 1996; Kanno *et al.*, 1998; Fertsch *et al.*, 2003), the storage modulus (G') and the loss modulus (G?) (Van Camp and Huyghebaert, 1995b) and the breaking stress (Kanno *et al.*, 1998) increased with increasing protein concentration and pressure intensity because of increased probability of interactions between denatured proteins (Van Camp *et al.*, 1996; Dumay *et al.*, 1998). The ratio G/G' decreased with increasing protein concentration and the gels became more elastic (Van Camp and Huyghebaert, 1995a). In addition, the gel strength increased with increasing pressure and prolonged holding time (Van Camp and Huyghebaert, 1995a; Van Camp *et al.*, 1996; Kanno *et al.*, 1998). Longer pressure holding times improve the strength of the gel network, stimulating the formation of more intensive intermolecular interactions (Van Camp and Huyghebaert, 1995a).

Many such studies on the microstructure and rheological analysis of pressure-induced gels have been reported. β-LG gels made by pressure treatment also exhibited different rheological properties from those made by heating (Dumay *et al.*, 1998). Electron microscopy suggested a higher level of cross-links in the heat-induced gels; high pressure generated a more porous network with a lower level of intermolecular cross-links (Van Camp and Huyghebaert, 1995a, 1995b; Van Camp *et al.*, 1996).

Pressurization of β-LG isolate solutions (70 g/kg protein, pH 7.0) at 450 MPa and 25°C for 30 min formed gels with a sponge-like texture and a porous microstructure that was prone to exudation (Zasypkin *et al.*, 1996). The rigidity and the elasticity of pressure-induced gels increased with increasing β-LG concentration, but remained lower than those of heat-induced gels (87°C for 40–45 min) at the same protein concentration (Zasypkin *et al.*, 1996). These heat-induced gels also displayed a finely stranded network and high water retention. Some of the previous studies also suggested that, at neutral pH, β-LG forms transparent, fine-stranded heat-induced gels (Paulsson, 1990; Stading and Hermansson, 1991; Langton and Hermansson, 1992; Stading *et al.*, 1993; Foegeding *et al.*, 1995). These results suggested that comparatively weaker intermolecular or interparticulate forces are formed by pressure treatment.

A recent study characterized the interactions of whey proteins during pressure-induced gel formation using combinations of techniques such as transmission electron microscopy (TEM), SEC and 1D and 2D PAGE (Patel *et al.*, 2006). Using SEC and 1D PAGE, the pressure-treated samples showed a time-dependent loss of native whey proteins, and a corresponding increase in non-native proteins and protein aggregates of different sizes. These aggregates altered the viscosity and opacity of the samples and were shown, using 1D PAGE (native, SDS and SDS reduced PAGE) and 2D PAGE (native then SDS and then reduced SDS PAGE), to be cross-linked by intermolecular disulfide bonds and by non-covalent interactions. It was concluded that the large internal hydrophobic cavity of β-LG may have been partially responsible for its sensitivity to high-pressure treatment. Conversely, α-LA responds to pressure by modifying its structure to be more molten globule-like and does not fully unfold at very high pressures.

Various possible hypotheses in support of pressure-induced gel formation have been discussed (Patel *et al.*, 2006). It seems likely that, at 800 MPa, the formation of a β-LG disulfide-bonded network precedes the formation of disulfide bonds between α-LA or BSA and β-LG to form multi-protein aggregates, possibly because the disulfide bonds of α-LA and BSA are less exposed than those of β-LG either during or after pressure treatment. It may be possible that intermolecular disulfide bond formation occurs at high pressure and that hydrophobic association becomes important after the high-pressure treatment, i.e. a novel pathway of whey protein gel formation using high pressure.

It was postulated that β-LG plays a major role in the aggregation and gel formation of WPC under pressure (Van Camp *et al.*, 1997a, 1997b; Patel *et al.*, 2006), which suggested that the major whey protein component in WPC primarily determines its functional behavior under high pressure. However, it was suggested that some additional studies will be needed to confirm this hypothesis, as well as to deduce the role of other whey proteins (i.e. α-LA, BSA and Ig) in gel formation (Van Camp and Huyghebaert, 1995b; Van Camp *et al.*, 1996).

Similar to heat-induced gelation of whey proteins, it has been reported that factors such as combinations of pressure and different temperatures (Walkenström and Hermansson, 1997), pHs (Van Camp and Huyghebaert, 1995b; Arias *et al.*, 2000) and calcium contents (Van Camp *et al.*, 1997b) affect the aggregation behavior, pressure-induced functionality such as gel formation, and physical, rheological and microstructural properties of whey proteins. Protein–protein interactions are favored near the isoelectric point of the whey proteins, and neutral and alkaline pHs stimulate the formation of intermolecular disulfide bonds (Van Camp and Huyghebaert, 1995b).

Pressure-induced β-LG denaturation increases considerably at alkaline pH and decreases at acidic pH (Arias *et al.*, 2000). Further, it has been reported that the role of calcium in the aggregation and gelation of whey proteins under pressure may be explained in a similar manner to the heat-induced effects on whey proteins (Mulvihill and Kinsella, 1988; Kinsella and Whitehead, 1989; Van Camp *et al.*, 1997b). A combination of pressure and higher pressurizing temperature (up to 70°C) has been recommended for inactivating microbial spores and therefore it is important to determine its effects on the proteins in food systems. However, little information on the effects of pressurizing temperature on whey protein aggregation and gelation is available.

HPP-induced changes in milk

Studies on the effects of high pressure on milk can be broadly grouped into several topics, including the effects of high pressure on casein micelle size and its dissociation, changes in the appearance of pressure-treated milks, denaturation of whey proteins and their interaction with the casein micelles in milk, and effects of high pressure on milk from various species. Some of these topics have been reviewed recently (Huppertz *et al.*, 2002, 2006a, 2006b; López-Fandiño, 2006a, 2006b; Considine *et al.*, 2007b), including the effects of HPP on technological properties including rennet coagulation and cheesemaking properties, acid coagulation properties, etc. This section covers the main aspects of the pressure-induced denaturation and aggregation of whey proteins and their interactions with casein in the milk system.

Denaturation of whey proteins in the milk system

Compared with heat treatment, fewer studies on whey protein denaturation and aggregation in pressure-treated milk samples have been reported.

Like heat-induced denaturation, the pressure intensity and the holding time have been reported to affect the level of denaturation of whey proteins in milk (López-Fandiño *et al.*, 1996; López-Fandiño and Olano, 1998a, 1998b; Huppertz *et al.*, 2004a; Anema *et al.*, 2005b; Hinrichs and Rademacher, 2005). However, considerable differences in the sensitivities of the different proteins to heat (LF > Ig > BSA > β-LG > α-LA) and pressure (β-LG > LF > Ig > BSA > α-LA) have been reported (Patel *et al.*, 2006), showing that β-LG is the most pressure sensitive among all the whey proteins. About 70–80% denaturation of β-LG occurs at 400 MPa (López-Fandiño *et al.*, 1996; López-Fandiño and Olano, 1998a; Arias *et al.*, 2000; García-Risco *et al.*, 2000; Scollard *et al.*, 2000). Relatively little further denaturation of β-LG occurs at 400–800 MPa (Scollard *et al.*, 2000). Compared with β-LG, α-LA is stable to pressures up to about 400–500 MPa in the milk environment at ambient temperature (Hinrichs *et al.*, 1996a, 1996b; López-Fandiño *et al.*, 1996; Felipe *et al.*, 1997; Gaucheron *et al.*, 1997; López-Fandiño and Olano, 1998a; Arias *et al.*, 2000; García-Risco *et al.*, 2000; Needs *et al.*, 2000a; Scollard *et al.*, 2000; Huppertz *et al.*, 2002, 2004b).

Various studies have reported different extents of denaturation of β-LG following high-pressure treatment at 600 MPa of pasteurized milk (Needs *et al.*, 2000a) or reconstituted skim milk powder (Gaucheron *et al.*, 1997); this may be attributed to the level of denaturation caused by treatments before pressurization, which may influence the amount of denaturation measured afterwards. The reaction order of pressure-induced denaturation of β-LG is 2.5 (Hinrichs *et al.*, 1996b), indicating that the denaturation process is concentration dependent and that a lower initial concentration of native β-LG should reduce the extent of denaturation of β-LG under pressure. Also, β-LG and α-LA are reported to be comparatively more pressure resistant in whey than in milk, which may be attributed to the absence of casein micelles and colloidal calcium phosphate in whey (Huppertz *et al.*, 2004b).

Differences in the pressure stabilities of α-LA and β-LG may be linked to the more rigid molecular structure of the former (López-Fandiño *et al.*, 1996; Gaucheron *et al.*, 1997), caused probably by differences in the secondary structure and in the number of disulfide bonds and Ca^{2+} binding sites. The pressure resistance of α-LA is partially caused by the different numbers of intramolecular disulfide bonds in the two proteins (Hinrichs *et al.*, 1996a, 1996b; Gaucheron *et al.*, 1997) or by the lack of a free sulfydryl group in α-LA (López-Fandiño *et al.*, 1996; Funtenberger *et al.*, 1997).

It has also been reported that the molecular structure of α-LA is more stable than that of β-LG, and that oligomerization takes place only if, during unfolding, free sulfydryl groups from other molecules are available (Hinrichs *et al.*, 1996b; López-Fandiño *et al.*, 1996; Gaucheron *et al.*, 1997; Jegouic *et al.*, 1997). This difference in pressure sensitivity can also be explained by the types of bonds stabilizing the conformational structures of β-LG and α-LA (Hinrichs *et al.*, 1996b; Messens *et al.*, 1997).

BSA has also been found to be resistant to pressures up to 400 MPa in raw milk (Hinrichs *et al.*, 1996b; López-Fandiño *et al.*, 1996) or 600 MPa (Hayakawa *et al.*, 1992). The high stability of BSA can be explained by the fact that BSA carries one sulfydryl group and 17 disulfide bonds. The energy received under pressure treatment is too small to disrupt all the disulfide bonds and to change the molecular structure of BSA. IgG in caprine milk (Felipe *et al.*, 1997) and bovine milk (Carroll *et al.*, 2006; Patel *et al.*, 2006) has been reported to be more resistant to pressure denaturation than to heat denaturation.

Interactions of whey proteins with casein micelles

One of the major reactions of interest is the interaction between the denatured whey proteins and the casein micelles, particularly the interactions of denatured β-LG with κ-casein (κ-CN) at the micelle surface. There has been considerable research on the specific interactions that occur, the composition of the interaction products and the sequence of events involved in this reaction in heated milks (see Chapter 8), but little information on such interactions on effects of HPP treatments is available.

On high-pressure treatment of milk at 300–600 MPa, β-LG may form small aggregates (Felipe *et al.*, 1997) or may interact with the casein micelles (Needs *et al.*, 2000a; Scollard *et al.*, 2000). It was reported that, when mixtures of κ-CN and β-LG were pressure treated at 400 MPa, the presence of β-LG reduced the susceptibility of κ-CN to subsequent hydrolysis by chymosin, indicating interactions between the proteins (López-Fandiño *et al.*, 1997). SDS-PAGE studies of pressure-treated and untreated milks or solutions containing κ-CN or β-LG or both in the presence or absence of denaturing agents showed evidence of the formation of aggregates linked by intermolecular disulfide bonds (López-Fandiño *et al.*, 1997). Although α_{s2}-casein (α_{s2}-CN) occurs at the same concentration as κ-CN and has one disulfide bond, it has not been reported to interact with β-LG in either heat- or pressure-treated milk systems.

Recently, using modified 2D SDS- and then reduced SDS-PAGE, Patel *et al.* (2006) showed that the effects of heat treatment and high-pressure treatment on the interactions of the caseins and whey proteins in milk were significantly different. The 2D SDS- and then reduced SDS-PAGE patterns (Figure 7.4) showed that pressure treatment of milk at 200 MPa (Figure 7.4D) caused β-LG to form disulfide-bonded dimers and incorporated β-LG into aggregates, probably disulfide bonded to κ-CN, suggesting that preferential reaction occurred at this pressure (Patel *et al.*, 2006). The other whey proteins appeared to be less affected at 200 MPa.

In contrast, pressure treatment at 800 MPa incorporated β-LG and most of the minor whey proteins (including Ig and LF), as well as κ-CN and much of the α_{s2}-CN, into large aggregates (Figure 7.4E). However, only a proportion of the α-LA was denatured or incorporated into the large aggregates. The relatively lower degree of α-LA reactivity at high pressures is probably related to the relative stability of this protein compared with β-LG, as discussed earlier, and is based on the unusual pressure-dependent behavior of α-LA (Kuwajima *et al.*, 1990; Kobashigawa *et al.*, 1999; Lasalle *et al.*, 2003). These and other results show that the differences between the stabilities of the proteins and the accessibilities of the disulfide bonds of the proteins at high temperature or pressure affect the formation pathways that result in differences among the compositions of resultant aggregation or interaction products (including their sizes) that ultimately may affect product functionalities.

The formation of disulfide-linked complexes involving α_{s2}-CN, κ-CN and whey proteins in heat- and pressure-treated milk has been demonstrated (see Figure 7.4) and has been explained by possible proposed reactions of the caseins and whey proteins in heat- and pressure-treated milk (Figure 7.5; see also Patel *et al.*, 2006).

The virtual absence of α_{s2}-CN from the heat-induced aggregates formed at 85–90°C in milk, as reported in previous studies, might be because α_{s2}-CN is not a surface component of the micelle and therefore its disulfide bond(s) are inaccessible to the denaturing or denatured β-LG. On the other hand, κ-CN is on the surface of the micelles, and its disulfide bond(s) could be readily accessible to a thiol group of β-LG. Based on this fact, a diagrammatic representation of the consequences of the various inter-protein reactions that might take place during the heat treatment of milk at about 90°C has been reported (Patel *et al.*, 2006). However, it has been reported

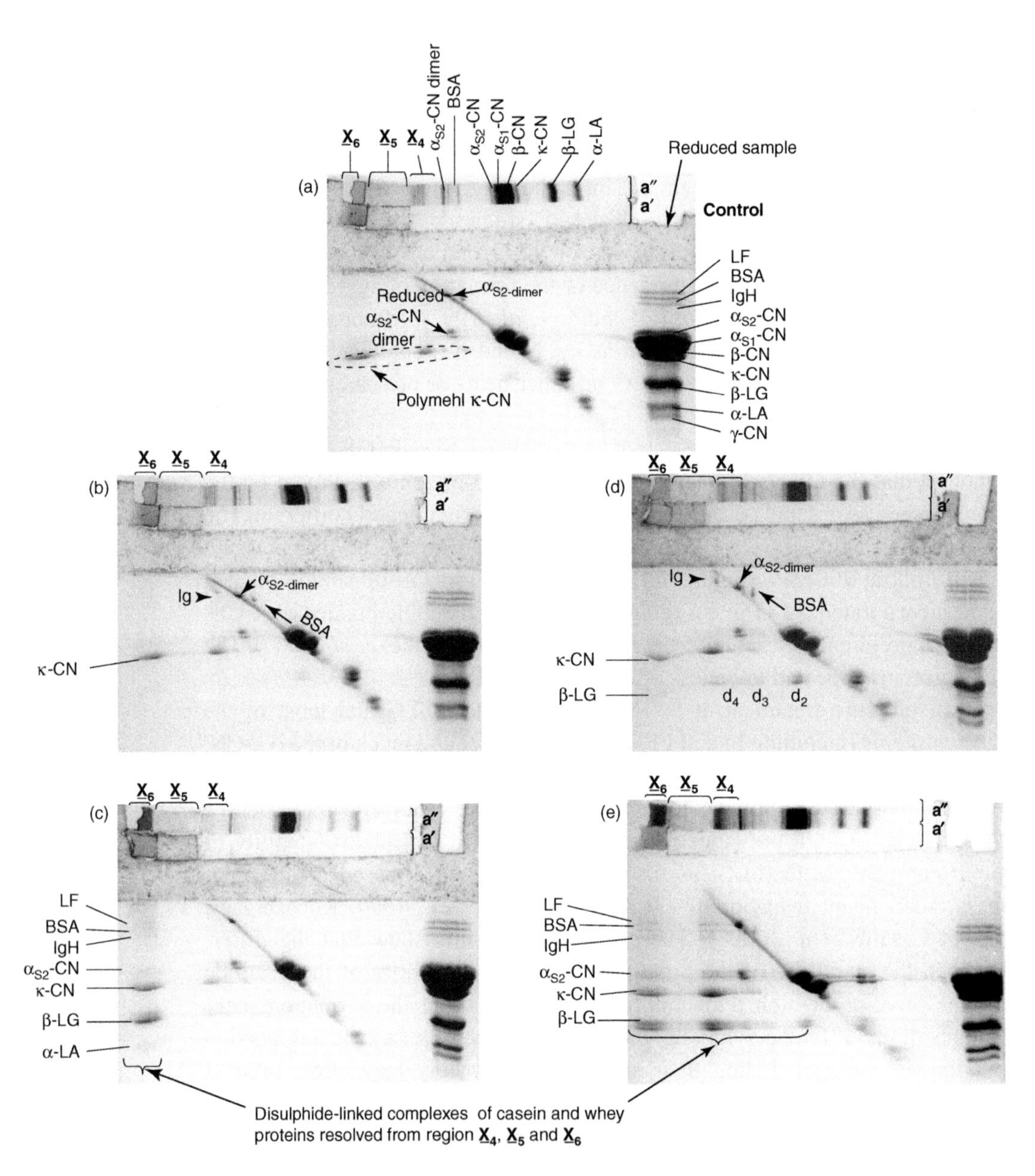

Figure 7.4 2D SDS- and then reduced SDS-PAGE patterns of the control sample (a) and samples heat treated at 72(C for 15 s (b) and 140(C for 5 s (c). Similarly, 2D PAGE patterns of samples pressure treated at 200 MPa for 30 min (d) and 800 MPa for 30 min (e). Gel strips marked as a' and a" represent the sample strip and the stained strip respectively and X4, X5 and X6 are high-molecular-weight aggregates, which were caught up at the beginning of the resolving gel, caught up within the stacking gel and could not enter the gel respectively. For a detailed description, refer Patel *et al.* 2006. (Reproduced in part with the permission of Patel *et al.*, 2006, copyright 2006 *American Chemical Society.*)

(Patel *et al.*, 2006) that α_{s2}-CN is apparent in the large aggregates in milk samples heated at temperatures above 100°C, supporting an earlier report by Snoeren and Van der Spek (1977), who reported that processing at severe temperatures can affect the proteins in a qualitatively different way.

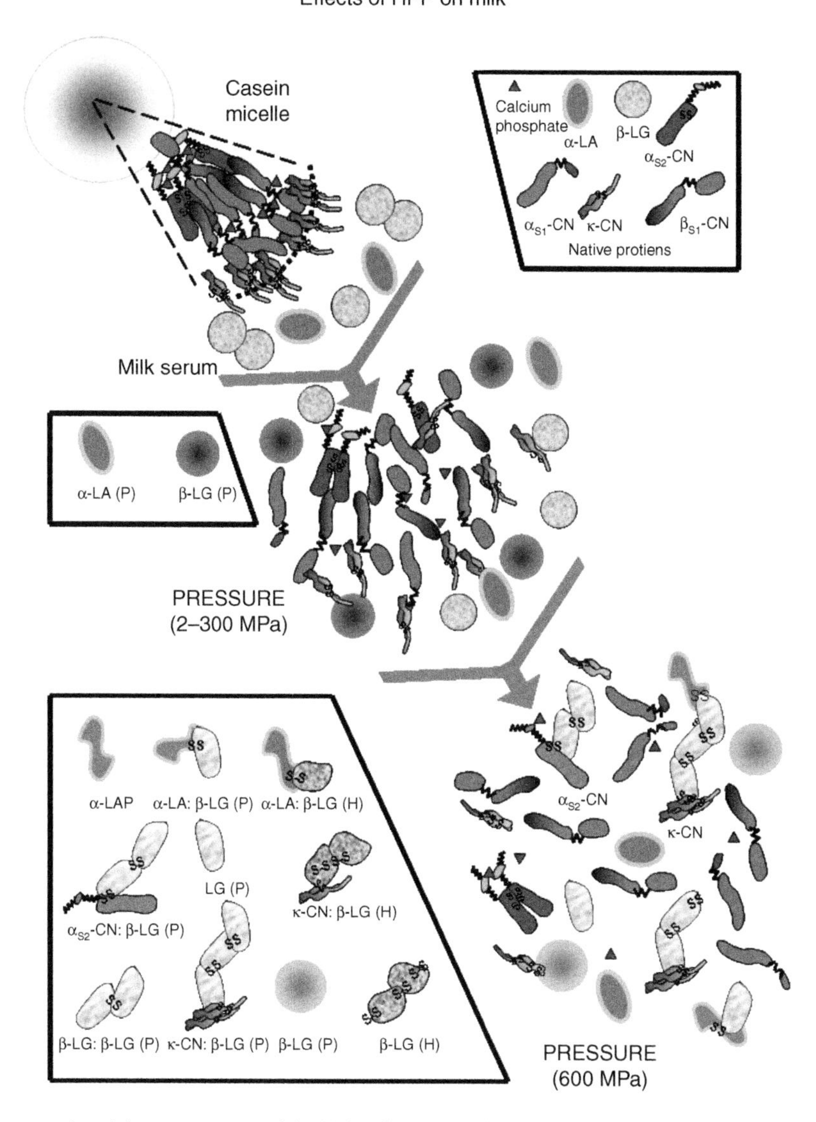

Figure 7.5 Pictorial representation of the likely effects of medium (≈250 MPa) and high (>600 MPa) pressure treatment at ≈22°C. The casein micelle swells at ≈250 MPa and the β-LG unfolds and aggregates via disulfide bonds. β-LG forms disulfide-bonded dimers at lower pressure and probably aggregates with κ-CN, but does not form larger β-LG aggregates. The proportion of α-LA that is included in the aggregates is less than that of β-LG because it does not readily unfold. At pressures >600 MPa, α_{s2}-CN becomes available for thiol interchange reactions, assisted by the permeation of water into the micelle and the dissolution of the calcium phosphate. Also the β-LG molecules can polymerize into larger aggregates than dimers. (P) = Pressure and (H) = Heat. (Redrawn with the permission of Patel *et al.*, 2006, copyright 2006 *American Chemical Society*; and Considine *et al.* (2007b), copyright Elsevier.)

In contrast, the effects of the pressure treatment of milk (Figure 7.5) are different from the effects of the heat treatment of milk. It can be postulated that both the casein micelles and the whey proteins are compressed at pressures up to about 150 MPa (Anema *et al.*, 2005a, 2005b). Thereafter, the micelles swell as the pressure is increased up to 400 MPa, and hydrogen bonds and hydrophobic associations are diminished and colloidal calcium phosphate dissolves. As a consequence, the micelles absorb more water, swell and subsequently disperse (Anema *et al.*, 2005a, 2005b). At 200 MPa, β-LG does not appear to become involved in interactions but is constrained to forming a stable, inactive, disulfide-bonded dimer (Figure 7.4) at <400 MPa, supporting the results of Considine *et al.* (2007b).

At higher pressures (>400 MPa), the polymerization of β-LG becomes the norm and pressure-induced β-LG aggregation becomes similar to heat-induced β-LG aggregation (Figures 7.4 and 7.5). The β-LG in WPC or in milk is not significantly modified by the other components, i.e. β-LG dominates the denaturation and aggregation pathway during pressure (>400 MPa) treatment, as it has been shown to dominate the reaction at high-temperature heat treatments.

Moreover, it has been reported that large quantities of very large aggregates that cannot enter the gel are present to a greater extent in heat-treated milk than in pressure-treated milk (Figure 7.4; Patel *et al.*, 2006), indicating that the sizes of the aggregates are comparatively smaller in pressure-treated milks than in heat-treated milks. Such differences can be attributed to different effects of heat treatment and pressure treatment on the structure of the proteins, which may ultimately lead to different textures of the final products.

Unlike studies on the effects of heat treatment on casein–whey protein interactions and the distribution of casein–whey protein complexes between the colloidal and serum phases as a function of pH, only a limited number of studies have examined the effects of high-pressure treatment on casein–whey protein interactions as a function of pH (Arias *et al.*, 2000; Huppertz *et al.*, 2004a).

Conclusions

HPP is a rapidly growing non-thermal preservation technology that can potentially be used to create novel protein structures by bringing about particular changes in the molecular structure of proteins and thus may give rise to innovative, new generation value-added food products and new properties that are inaccessible via conventional methods of protein modification. In order to realize the full potential of HPP technology, it will be useful to have comparable levels of understanding of the science underlying the differences in the effects of heat treatment and pressure treatment on denaturation, aggregation and interaction of the different milk proteins.

In this chapter, we attempted to review the mechanisms and pathways of pressure-induced denaturation and aggregation, and of interactions of whey proteins in various systems including simple model systems using purified milk proteins and complex systems such as WPC and milk. The selected examples of differences in the heat- and pressure-induced interactions of whey proteins were discussed based on recent research findings. Based on the current literature, it appears that more focus on identifying the specific pressure-induced interactions of milk proteins in actual systems is needed. The effects of a range of processing conditions and the detailed mechanisms of possible synergistic interactions also need to be addressed appropriately.

Acknowledgments

The authors are grateful to Therese Considine, Skelte Anema and Harjinder Singh for critical reading of this manuscript and Claire Woodhall for excellent editorial assistance.

References

Anema, S. G. and Li, Y. (2003a). Association of denatured whey proteins with casein micelles in heated reconstituted skim milk and its effect on casein micelle size. *Journal of Dairy Research*, **70**, 73–83.

Anema, S. G. and Li, Y. (2003b). Effect of pH on the association of denatured whey proteins with casein micelles in heated reconstituted skim milk. *Journal of Agricultural and Food Chemistry*, **51**, 1640–46.

Anema, S. G., Stockmann, R. and Lowe, E. K. (2005a). Denaturation of β-lactoglobulin in pressure-treated skim milk. *Journal of Agricultural and Food Chemistry*, **53**, 7783–91.

Anema, S. G., Lowe, E. K. and Stockmann, R. (2005b). Particle size changes and casein solubilisation in high-pressure-treated skim milk. *Food Hydrocolloids*, **19**, 257–67.

Aouzelleg, A., Bull, L. A., Price, N. C. and Kelly, S. M. (2004). Molecular studies of pressure/temperature-induced structural changes in bovine β-lactoglobulin. *Journal of the Science of Food and Agriculture*, **84**, 398–404.

Arias, M., López-Fandiño, R. and Olano, A. (2000). Influence of pH on the effects of high pressure on milk proteins. *Milchwissenschaft*, **55**, 191–94.

Balci, A. T. and Wilbey, R. A. (1999). High pressure processing of milk – the first 100 years in the development of new technology. *International Journal of Dairy Technology*, **52**, 149–55.

Balny, C. and Masson, P. (1993). Effects of high pressure on proteins. *Food Reviews International*, **9**, 611–28.

Balny, C., Masson, P. and Travers, F. (1989). Some recent aspects of the use of high-pressure for protein investigations in solution. *High Pressure Research*, **2**, 1–28.

Balny, C., Hayashi, R., Heremans, K. and Masson, P. (eds). (1992). *High Pressure and Biotechnology*. Colloque INSERM, Volume. 224. Montrouge, France: John Libbey Eurotext

Belloque, J., López-Fandiño, R. and Smith, G. M. (2000). A ^{1}H-NMR study on the effect of high pressures on β-lactoglobulin. *Journal of Agricultural and Food Chemistry*, **48**, 3906–12.

Boonyaratanakornkit, B. B., Park, C. B. and Clark, D. S. (2002). Pressure effects on intra- and intermolecular interactions within proteins. *Biochimica et Biophysica Acta*, **1595**, 235–49.

Bridgman, P. W. (1914). The coagulation of albumen by pressure. *Journal of Biological Chemistry*, **19**, 511–12.

Brisson, G., Britten, M. and Pouliot, Y. (2007). Heat-induced aggregation of bovine lactoferrin at neutral pH: effect of iron saturation. *International Dairy Journal*, **17**, 617–24.

Buchheim, W. and Prokopek, D. (1992). Die Hochdruckbehandlung. *Deutsche Milchwirtschaft*, **43**, 1374–78.

Carroll, T. J., Patel, H. A., Gonzalez, M. A., Dekker, J. W., Collett, M. A. and Lubbers, M. W. (2006). High pressure processing of metal ion lactoferrin. International Patent Publication Number WO 2006/096073 A1.

Ceolín, M. (2000). Perturbed angular correlation experiments on the pressure-induced structural modification of bovine serum albumin. *Journal of Biochemical and Biophysical Methods*, **45**, 117–25.

Cheftel, J. C. (1992). Effects of high hydrostatic pressure on food constituents: an overview, In *High Pressure and Biotechnology* (C. Balny, R. Hayashi, K. Heremans and P. Masson, eds.) pp. 195–209. Colloque INSERM, Volume 224. Montrouge, France: John Libbey Eurotext.

Cheftel, J. and Dumay, E. (1996). Effects of high pressure on dairy proteins: a review. In *High Pressure Bioscience and Biotechnology* (R. Hayashi and C. Balny, eds) pp. 299–308. Amsterdam: Elsevier Science.

Cho, Y., Singh, H. and Creamer, L. K. (2003). Heat-induced interactions of β-lactoglobulin A and κ-casein B in a model system. *Journal of Dairy Research*, **70**, 61–71.

Claeys, W. L., Indrawati, O., Van Loey, A. M. and Hendrickx, M. E. (2003). Review: are intrinsic TTIs for thermally processed milk applicable for high-pressure processing assessment? *Innovative Food Science and Emerging Technologies*, **4**, 1–14.

Collini, M., D'Alfonso, L., Molinari, H., Ragona, L., Catalano, M. and Baldini, G. (2003). Competitive binding of fatty acids and the fluorescent probe 1–8-anilinonaphthalene sulfonate to bovine β-lactoglobulin. *Protein Science*, **12**, 1596–603.

Considine, T., Patel, H. A., Singh, H. and Creamer, L. K. (2005a). Influence of binding of sodium dodecyl sulfate, all-*trans*-retinol, palmitate, and 8-anilino-1-naphthalenesulfonate on the heat-induced unfolding and aggregation of β-lactoglobulin B. *Journal of Agricultural and Food Chemistry*, **53**, 3197–205.

Considine, T., Singh, H., Patel, H. A. and Creamer, L. K. (2005b). Influence of binding of sodium dodecyl sulfate, all-*trans*-retinol, and 8-anilino-1-naphthalenesulfonate on high-pressure-induced unfolding and aggregation of β-lactoglobulin B. *Journal of Agricultural and Food Chemistry*, **53**, 8010–18.

Considine, T., Patel, H. A., Singh, H. and Creamer, L. K. (2007a). Influence of binding of conjugated linoleic acid and myristic acid on the heat- and high-pressure-induced unfolding and aggregation of β-lactoglobulin B. *Food Chemistry*, **102**, 1270–80.

Considine, T., Patel, H. A., Anema, S. G., Singh, H. and Creamer, L. K. (2007b). Interactions of milk proteins during heat and high hydrostatic pressure treatments a review. *Innovative Food Science and Emerging Technologies*, **8**, 1–23.

Datta, N. and Deeth, H. C. (1999). High pressure processing of milk and dairy products. *Australian Journal of Dairy Technology*, **54**, 41–48.

Datta, N. and Deeth, H. C. (2003). High pressure processing. In *Encyclopedia of Dairy Sciences* (H. Roginski, J. W. Fuquay and P. F. Fox, eds) pp. 1327–33. London: Academic Press.

De la Fuente, M. A., Singh, H. and Hemar, Y. (2002). Recent advances in the characterization of heat-induced aggregates and intermediates of whey proteins. *Trends in Food Science and Technology*, **13**, 262–74.

de Wit, J. N. (1990). Thermal stability and functionality of whey proteins. *Journal of Dairy Science*, **73**, 3602–12.

Dufour, E., Hoa, G. H. and Haertlé, T. (1994). High-pressure effects on β-lactoglobulin interactions with ligands studied by fluorescence. *Biochimica et Biophysica Acta*, **1206**, 166–72.

Dumay, E. M., Kalichevsky, M. T. and Cheftel, J. C. (1994). High-pressure unfolding and aggregation of β-lactoglobulin and the baroprotective effects of sucrose. *Journal of Agricultural and Food Chemistry*, **42**, 1861–68.

Dumay, E. M., Kalichevsky, M. T. and Cheftel, J. C. (1998). Characteristics of pressure-induced gels of β-lactoglobulin at various times after pressure release. *Lebensmittel-Wissenschaft und -Technologie*, **31**, 10–19.

Dzwolak, W., Kato, M., Shimizu, A. and Taniguchi, Y. (1999). Fourier-transform infrared spectroscopy study of the pressure-induced changes in the structure of the bovine α-lactalbumin: the stabilizing role of the calcium ion. *Biochimica et Biophysica Acta*, **1433**, 45–55.

Earnshaw, R. G. (1992). High pressure as a cell sensitiser: new opportunities to increase the efficacy of preservation processes, In *High Pressure and Biotechnology*. (C. Balny, R. Hayashi, K. Heremans and P. Masson, eds) pp. 261–67. Colloque INSERM, Volume 224. Montrouge, France: John Libbey Eurotext.

Edwards, P. J. B., Jameson, G. B., Palmano, K. P. and Creamer, L. K. (2002). Heat-resistant structural features of bovine β-lactoglobulin A revealed by NMR H/D exchange observations. *International Dairy Journal*, **12**, 331–44.

Farkas, D. F. and Hoover, D. G. (2000). High pressure processing. Kinetics of microbial inactivation for alternative foord processing technologies. *Journal of Food Science*, **65**(Suppl), 47–64.

Farr, D. (1990). High pressure technology in the food industry. *Trends in Food Science and Technology*, **1**, 14–16.

Felipe, X., Capellas, M. and Law, A. J. R. (1997). Comparison of the effects of high-pressure treatments and heat pasteurization on the whey proteins in goat's milk. *Journal of Agricultural and Food Chemistry*, **45**, 627–31.

Fertsch, B., Müller, M. and Hinrichs, J. (2003). Firmness of pressure-induced casein and whey protein gels modulated by holding time and rate of pressure release. *Innovative Food Science and Emerging Technologies*, **4**, 143–50.

Foegeding, E. A., Bowland, E. L. and Hardin, C. C. (1995). Factors that determine the fracture properties and microstructure of globular protein gels. *Food Hydrocolloids*, **9**, 237–49.

Funtenberger, S., Dumay, E. and Cheftel, J. C. (1995). Pressure-induced aggregation of β-lactoglobulin in pH 7.0 buffers. *Lebensmittel-Wissenschaft und -Technologie*, **28**, 410–18.

Funtenberger, S., Dumay, E. and Cheftel, J. C. (1997). High pressure promotes β-lactoglobulin aggregation through SH/S S interchange reactions. *Journal of Agricultural and Food Chemistry*, **45**, 912–21.

Galazka, V. B., Dickinson, E. and Ledward, D. A. (1996a). Effect of high pressure on the emulsifying behavior of β-lactoglobulin. *Food Hydrocolloids*, **10**, 213–19.

Galazka, V. B., Sumner, I. G. and Ledward, D. A. (1996b). Changes in protein–protein and protein–polysaccharide interactions induced by high pressure. *Food Chemistry*, **57**, 393–98.

Galazka, V. B., Ledward, D. A., Sumner, I. G. and Dickinson, E. (1997). Influence of high pressure on bovine serum albumin and its complex with dextran sulfate. *Journal of Agricultural and Food Chemistry*, **45**, 3465–71.

García-Risco, M. R., Cortés, E., Carrascosa, A. V. and López-Fandiño, R. (1998). Microbiological and chemical changes in high-pressure-treated milk during refrigerated storage. *Journal of Food Protection*, **61**, 735–37.

García-Risco, M. R., Olano, A., Ramos, M. and López-Fandiño, R. (2000). Micellar changes induced by high pressure. Influence in the proteolytic activity and organoleptic properties of milk. *Journal of Dairy Science*, **83**, 2184–89.

Gaucheron, F., Famelart, M. H., Mariette, F., Raulot, K., Michel, F. and Le Graet, Y. (1997). Combined effects of temperature and high pressure treatments on physicochemical characteristics of skim milk. *Food Chemistry*, **59**, 439–47.

Gezimati, J., Singh, H. and Creamer, L. K. (1996). Heat-induced interactions and gelation of mixtures of bovine β-lactoglobulin and serum albumin. *Journal of Agricultural and Food Chemistry*, **44**, 804–10.

Gezimati, J., Creamer, L. K. and Singh, H. (1997). Heat-induced interactions and gelation of mixtures of β-lactoglobulin and α-lactalbumin. *Journal of Agricultural and Food Chemistry*, **45**, 1130–36.

Grinberg, V. Y. and Haertlé, T. (2000). Reducer driven baric denaturation and oligomerisation of whey proteins. *Journal of Biotechnology*, **79**, 205–9.

Gross, M. and Jaenicke, R. (1994). Proteins under pressure. The influence of high hydrostatic pressure on structure, function and assembly of proteins and protein complexes. *European Journal of Biochemistry*, **221**, 617–30.

Haque, Z. and Kinsella, J. E. (1988). Interaction between heated κ-casein and β-lactoglobulin: predominance of hydrophobic interactions in the initial stages of complex formation. *Journal of Dairy Research*, **55**, 67–80.

Haque, Z., Kristjansson, M. M. and Kinsella, J. E. (1987). Interaction between κ-casein and β-lactoglobulin: possible mechanism. *Journal of Agricultural and Food Chemistry*, **35**, 644–49.

Havea, P. (1998). *Studies on heat-induced interactions and gelation of whey protein*. PhD Thesis. Massey University, Palmerston North, New Zealand.

Havea, P., Singh, H., Creamer, L. K. and Campanella, O. H. (1998). Electrophoretic characterization of the protein products formed during heat treatment of whey protein concentrate solutions. *Journal of Dairy Research*, **65**, 79–91.

Havea, P., Singh, H. and Creamer, L. K. (2000). Formation of new protein structures in heated mixtures of BSA and α-lactalbumin. *Journal of Agricultural and Food Chemistry*, **48**, 1548–56.

Havea, P., Singh, H. and Creamer, L. K. (2001). Characterization of heat-induced aggregates of β-lactoglobulin, α-lactalbumin and bovine serum albumin in a whey protein concentrate environment. *Journal of Dairy Research*, **68**, 483–97.

Havea, P., Singh, H. and Creamer, L. K. (2002). Heat-induced aggregation of whey proteins: comparison of cheese WPC with acid WPC and relevance of mineral composition. *Journal of Agricultural and Food Chemistry*, **50**, 4674–81.

Havea, P., Carr, A. J. and Creamer, L. K. (2004). The roles of disulphide and non-covalent bonding in the functional properties of heat-induced whey protein gels. *Journal of Dairy Research*, **71**, 330–39.

Hayakawa, I., Kajihara, J., Morikawa, K., Oda, M. and Fujio, Y. (1992). Denaturation of bovine serum albumin (BSA) and ovalbumin by high pressure, heat and chemicals. *Journal of Food Science*, **57**, 288–92.

Hayakawa, I., Linko, Y-Y. and Linko, P. (1996). Mechanism of high pressure denaturation of proteins. *Lebensmittel-Wissenschaft und -Technologie*, **29**, 756–62.

Hayashi, R. (1988). Utilisation of the effects of high pressures on organisms, organic compounds and biopolymers (as applied to food processing, e.g. preservation, maturation, enzyme regulation, pest control and microbial inhibition). *Bioscience and Industry*, **46**, 3931–33.

Hendrickx, M., Ludikhuyze, L., Van den Broek, I. and Weemaes, C. (1998). Effects of high pressure on enzymes related to food quality. *Trends in Food Science and Technology*, **9**, 197–203.

Heremans, K. (1982). High pressure effects on proteins and other biomolecules. *Annual Review of Biophysics and Bioengineering*, **11**, 1–21.

Heremans, K. (1992). From living system to biomolecules, In *High Pressure and Biotechnology*. (C. Balny, R. Hayashi, K. Heremans and P. Masson, eds) pp. 37–44. Colloque INSERM, Volume 224. Montrouge, France: John Libbey Eurotext.

Heremans, K. and Smeller, L. (1998). Protein structure and dynamics at high pressure. *Biochimica et Biophysica Acta*, **1386**, 353–70.

Heremans, K., Van Camp, J. and Huyghebaert, A. (1997). High-pressure effects on proteins. In *Food Proteins and their Applications* (S. Damodaran and A. Paraf, eds) pp. 473–502. New York: Marcel Dekker, Inc.

Hill, A. R. (1989). The β-lactoglobulin-κ-casein complex. *Canadian Institute of Food Science and Technology Journal*, **22**, 120–23.

Hines, M. E. and Foegeding, E. A. (1993). Interactions of α-lactalbumin and bovine serum albumin with β-lactoglobulin in thermally induced gelation. *Journal of Agricultural and Food Chemistry*, **41**, 341–46.

Hinrichs, J. (2000). Ultrahochdruckbehandlung von Lebensmitteln mit Schwerpunkt Milch und Milchprodukte Phänomene, Kinetik und Methodik. Fortschritt-Berichte VDI, Reihe 3, Nr. 656. VDI-verlag, Düsseldorf.

Hinrichs, J. and Rademacher, B. (2005). Kinetics of combined thermal and pressure-induced whey protein denaturation in bovine skim milk. *International Dairy Journal*, **15**, 315–23.

Hinrichs, J., Rademacher, B. and Kessler, H. G. (1996a). Food processing of milk products with ultra high pressure. In *Heat Treatments and Alternative Methods,* pp. 185–201. IDF Special Issue No 9602. Brussels: International Dairy Federation.

Hinrichs, J., Rademacher, B. and Kessler, H. G. (1996b). Reaction kinetics of pressure-induced denaturation of whey proteins. *Milchwissenschaft*, **51**, 504–9.

Hite, B. H. (1899). The effect of pressure in the preservation of milk. *Bulletin of the West Virginia University Agricultural Experimental Station*, **58**, 15–35.

Holt, C. (2000). Molecular basis of whey protein food functionalities. *Australian Journal of Dairy Technology*, **55**, 53–55.

Holt, C., de Kruif, C. G., Tuinier, R. and Timmins, P. A. (2003). Substructure of bovine casein micelles by small-angle X-ray and neutron scattering. *Colloids and Surfaces A*, **213**, 275–84.

Hong, Y-H. and Creamer, L. K. (2002). Changed protein structures of bovine β-lactoglobulin B and α-lactalbumin as a consequence of heat treatment. *International Dairy Journal*, **12**, 345–59.

Hoover, D. G. (1993). Pressure effects on biological systems. *Food Technology*, **47**(6), 150–55.

Hosseini-nia, T., Ismail, A. A. and Kubow, S. (2002). Effect of high hydrostatic pressure on the secondary structures of BSA and apo- and holo-α-lactalbumin employing Fourier transform infrared spectroscopy. *Journal of Food Science*, **67**, 1341–47.

Huppertz, T., Kelly, A. L. and Fox, P. F. (2002). Effects of high pressure on constituents and properties of milk. *International Dairy Journal*, **12**, 561–72.

Huppertz, T., Fox, P. F. and Kelly, A. L. (2004a). High pressure treatment of bovine milk: effects of casein micelles and whey proteins. *Journal of Dairy Research*, **71**, 97–106.

Huppertz, T., Fox, P. F. and Kelly, A. L. (2004b). High pressure-induced denaturation of α-lactalbumin and β-lactoglobulin in bovine milk and whey: a possible mechanism. *Journal of Dairy Research*, **71**, 489–95.

Huppertz, T., Fox, P. F., de Kruif, K. G. and Kelly, A. L. (2006a). High pressure-induced changes in bovine milk proteins: a review. *Biochimica et Biophysica Acta*, **1764**, 593–98.

Huppertz, T., Smiddy, M. A., Upadhyay, V. K. and Kelly, A. L. (2006b). High pressure-induced changes in bovine milk: a review. *International Journal of Dairy Technology*, **59**, 58–66.

Iametti, S., Transidico, P., Bonomi, F., Vecchio, G., Pittia, P., Rovere, P. and Dall'Aglio, G. (1997). Molecular modifications of β-lactoglobulin upon exposure to high pressure. *Journal of Agricultural and Food Chemistry*, **45**, 23–29.

Indyk, H. E., Williams, J. W. and Patel, H. A. (2008). Analysis of denaturation of bovine IgG by heat and high pressure using an optical biosensor. *International Dairy Journal*, in press

Jegouic, M., Grinberg, V. Y., Guingant, A. and Haertlé, T. (1996). Thiol-induced oligomerization of α-lactalbumin at high pressure. *Journal of Protein Chemistry*, **15**, 501–9.

Jegouic, M., Grinberg, V. Y., Guingant, A. and Haertlé, T. (1997). Baric oligomerization in α-lactalbumin/β-lactoglobulin mixtures. *Journal of Agricultural and Food Chemistry*, **45**, 19–22.

Jelen, P. and Rattray, W. (1995). Thermal denaturation of whey proteins. In *Heat-induced Changes in Milk*, 2nd edn (P. F. Fox, ed.) pp. 66–85. IDF Special Issue No 9501. Brussels: International Dairy Federation.

Johnston, D. E. (1995). High pressure effects on milk and meat. In *High Pressure Processing of Foods* (D. A. Ledward, D. E. Johnston, R. G. Earnshaw and A. P. M. Hasting, eds) pp. 99–121. Nottingham: Nottingham University Press.

Johnston, D.E., Austin, B.A., Murphy, R.J. (1992). The effects of high pressure treatment of skim milk In *High Pressure and Biotechnology* (C. Balny, R. Hayashi,

K. Heremans and P. Masson, eds) pp. 243–247. Colloque INSERM, Volume. 224. Montrouge, France: John Libbey Eurotext.

Jonas, J. (2002). High-resolution nuclear magnetic resonance studies of proteins. *Biochimica et Biophysica Acta*, **1595**, 145–59.

Jonas, J. and Jonas, A. (1994). High-pressure NMR spectroscopy of proteins and membranes. *Annual Review of Biophysics and Biomolecular Structure*, **23**, 287–318.

Kanno, C., Mu, T-H., Hagiwara, T., Ametani, M. and Azuma, N. (1998). Gel formation from industrial milk whey proteins under hydrostatic pressure: effect of hydrostatic pressure and protein concentration. *Journal of Agricultural and Food Chemistry*, **46**, 417–24.

Keim, S. and Hinrichs, J. (2004). Influence of stabilizing bonds on the texture properties of high-pressure-induced whey protein gels. *International Dairy Journal*, **14**, 355–63.

Kelly, S. M. and Price, N. C. (2000). The use of circular dichroism in the investigation of protein structure and function. *Current Protein and Peptide Science*, **1**, 349–84.

Kinsella, J. E. and Whitehead, D. M. (1989). Proteins in whey: chemical, physical, and functional properties. *Advances in Food and Nutrition Research*, **33**, 343–438.

Kobashigawa, Y., Sakurai, M. and Nitta, K. (1999). Effect of hydrostatic pressure on unfolding of α-lactalbumin: volumetric equivalence of the molten globule and unfolded state. *Protein Science*, **8**, 2765–72.

Kontopidis, G., Holt, C. and Sawyer, L. (2004). Invited review: β-Lactoglobulin: binding properties, structure, and function. *Journal of Dairy Science*, **87**, 785–96.

Kuwajima, K., Ikeguchi, M., Sugawara, T., Hiraoka, Y. and Sugai, S. (1990). Kinetics of disulfide bond reduction in α-lactalbumin by dithiothreitol and molecular basis of superreactivity of the Cys6–Cys120 disulfide bond. *Biochemistry*, **29**, 8240–49.

Langton, M. and Hermansson, A-M. (1992). Fine-stranded and particulate gels of β-lactoglobulin and whey protein at varying pH. *Food Hydrocolloids*, **5**, 523–39.

Lassalle, M. W., Li, H., Yamada, H., Akasaka, K. and Redfield, C. (2003). Pressure-induced unfolding of the molten globule of all-Ala α-lactalbumin. *Protein Science*, **12**, 66–72.

Law, A. J. R., Leaver, J., Felipe, X., Ferragut, V., Pla, R. and Guamis, B. (1998). Comparison of the effects of high pressure and thermal treatments on the casein micelles in goat's milk. *Journal of Agricultural and Food Chemistry*, **46**, 2523–30.

Livney, Y. D. and Dalgleish, D. G. (2004). Specificity of disulfide bond formation during thermal aggregation in solutions of β-lactoglobulin B and κ-casein A. *Journal of Agricultural and Food Chemistry*, **52**, 5527–32.

Livney, Y. D., Verespej, E. and Dalgleish, D. G. (2003). Steric effects governing disulfide bond interchange during thermal aggregation in solutions of β-lactoglobulin B and α-lactalbumin. *Journal of Agricultural and Food Chemistry*, **51**, 8098–106.

López-Fandiño, R. (2006a). High pressure-induced changes in milk proteins and possible applications in dairy technology. *International Dairy Journal*, **16**, 1119–31.

López-Fandiño, R. (2006b). Functional improvement of milk whey proteins induced by high hydrostatic pressure. *Critical Reviews in Food Science and Nutrition*, **48**, 351–63.

López-Fandiño, R. and Olano, A. (1998a). Effects of high pressures combined with moderate temperatures on the rennet coagulation properties of milk. *International Dairy Journal*, **8**, 623–27.

López-Fandiño, R. and Olano, A. (1998b). Cheese-making properties of bovine and caprine milks submitted to high pressures. *Lait*, **78**, 341–50.

López-Fandiño, R., Carrascosa, A. V. and Olano, A. (1996). The effects of high pressure on whey protein denaturation and cheese-making properties of raw milk. *Journal of Dairy Science*, **79**, 929–36.

López-Fandiño, R., Ramos, M. and Olano, A. (1997). Rennet coagulation of milk subjected to high pressures. *Journal of Agricultural and Food Chemistry*, **45**, 3233–37.

Lullien-Pellerin, V. and Balny, C. (2002). High-pressure as a tool to study some proteins' properties: conformational modification, activity and oligomeric dissociation. *Innovative Food Science and Emerging Technologies*, **3**, 209–21.

Lyster, R. L. J. (1970). The denaturation of α-lactalbumin and β-lactoglobulin in heated milk. *Journal of Dairy Research*, **37**, 233–43.

Manderson, G. A., Hardman, M. J. and Creamer, L. K. (1998). Effect of heat treatment on the conformation and aggregation of β-lactoglobulin A, B, and C. *Journal of Agricultural and Food Chemistry*, **46**, 5052–61.

Masson, P. (1992). Pressure denaturation of proteins. In *High Pressure and Biotechnology* (C. Balny, R. Hayashi, K. Heremans and P. Masson, eds) pp. 89–99. Colloque INSERM, Volume 224. Montrouge, France: John Libbey Eurotext.

Matsudomi, N., Oshita, T., Sasaki, E. and Kobayashi, K. (1992). Enhanced heat-induced gelation of β-lactoglobulin by α-lactalbumin. *Bioscience, Biotechnology, Biochemistry*, **56**, 1600–97.

Matsudomi, N., Oshita, T., Kobayashi, K. and Kinsella, J. E. (1993). α-Lactalbumin enhances the gelation properties of bovine serum albumin. *Journal of Agricultural and Food Chemistry*, **41**, 1053–57.

Matsudomi, N., Oshita, T. and Kobayashi, K. (1994). Synergistic interaction between β-lactoglobulin and bovine serum albumin in heat-induced gelation. *Journal of Dairy Science*, **77**, 1487–93.

McGuffey, M. K., Epting, K. L., Kelly, R. M. and Foegeding, E. A. (2005). Denaturation and aggregation of three α-lactalbumin preparations at neutral pH. *Journal of Agricultural and Food Chemistry*, **53**, 3182–90.

McSwiney, M., Singh, H. and Campanella, O. H. (1994a). Thermal aggregation and gelation of bovine β-lactoglobulin. Food Hydrocolloids, **8**, 441–53.

McSwiney, M., Singh, H., Campanella, O. H. and Creamer, L. K. (1994b). Thermal gelation and denaturation of bovine β-lactoglobulins A and B. *Journal of Dairy Research*, **61**, 221–32.

Messens, W., Van Camp, J. and Huyghebaert, A. (1997). The use of high pressure to modify the functionality of food proteins. *Trends in Food Science and Technology*, **8**, 107–12.

Michel, M., Leser, M. E., Syrbe, A., Clerc, M-F., Bauwens, I., Bovetto, L., von Schack, M-L. and Watzke, H. J. (2001). Pressure effects on whey protein-pectin mixtures. *Lebensmittel-Wissenschaft und -Technologie*, **34**, 41–52.

Møller, R. E., Stapelfeldt, H. and Skibsted, L. H. (1998). Thiol reactivity in pressure-unfolded β-lactoglobulin. Antioxidative properties and thermal refolding. *Journal of Agricultural and Food Chemistry*, **46**, 425–30.

Mozhaev, V. V., Heremans, K., Frank, J., Masson, P. and Balny, C. (1994). Exploiting the effects of high hydrostatic pressure in biotechnological applications. *Trends in Biotechnology*, **12**, 93–501.

Mulvihill, D. M. and Donovan, M. (1987). Whey proteins and their thermal denaturation – a review. *Irish Journal of Food Science and Technology*, **11**, 43–47.

Mulvihill, D. M. and Kinsella, J. E. (1988). Gelation of β-lactoglobulin: effects of sodium chloride and calcium chloride on rheological and structural properties of gels. *Journal of Food Science*, **53**, 231–36.

Narayan, M. and Berliner, L. J. (1997). Fatty acids and retinoids bind independently and simultaneously to β-lactoglobulin. *Biochemistry*, **36**, 1906–11.

Needs, E. C., Capellas, M., Bland, A. P., Manoj, P., MacDougal, D. and Paul, G. (2000a). Comparison of heat and pressure treatments of skim milk, fortified with whey protein concentrate, for set yogurt preparation: effects of milk proteins and gel structure. *Journal of Dairy Research*, **67**, 329–48.

Needs, E. C., Stenning, R. A., Gill, A. L., Ferragut, V. and Rich, G. T. (2000b). High-pressure treatment of milk: effects on casein micelle structure and on enzymic coagulation. *Journal of Dairy Research*, **67**, 31–42.

Ngarize, S., Herman, H., Adams, A. and Howell, N. (2004). Comparison of changes in the secondary structure of unheated, heated, and high-pressure-treated β-lactoglobulin and ovalbumin proteins using Fourier transform Raman spectroscopy and self-deconvolution. *Journal of Agricultural and Food Chemistry*, **52**, 6470–77.

Noh, B., Richardson, T. and Creamer, L. K. (1989). Radio labelling study of heat-induced interactions between α-lactalbumin, β-lactoglobulin and κ-casein in milk and in buffer solutions. *Journal of Food Science*, **54**, 889–93.

O'Connell, J. E. and Fox, P. F. (2003). Heat-induced coagulation of milk. In *Advanced Dairy Chemistry*, Volume 1, *Proteins*, 3rd edn (P. F. Fox and P. L. H. McSweeney, eds) pp. 879–930. New York: Kluwer Publishers Academic/Plenum.

Ohmiya, K., Kajino, T., Shimizu, S. and Gekko, K. (1989). Dissociation and reassociation of enzyme-treated caseins under high pressure. *Journal of Dairy Research*, **56**, 435–42.

Olsen, K., Ipsen, R., Otte, J. and Skibsted, L. H. (1999). Effect of high pressure on aggregation and thermal gelation of β-lactoglobulin. *Milchwissenschaft*, **54**, 543–45.

Owusu-Apenten, R. (2005). Colorimetric analysis of protein sulfydryl groups in milk: applications and processing effects. *Critical Reviews in Food Science and Nutrition*, **45**, 1–23.

Palmano, K. P., Patel, H. A., Carroll, T. J., Elgar, D. F. and Gonzalez-Martin, M. A. (2006). High pressure processing of bioactive compositions. International Patent Publication Number WO 2006/096074 A1.

Panick, G., Malessa, R. and Winter, R. (1999). Differences between the pressure- and temperature-induced denaturation and aggregation of β-lactoglobulin A, B, and AB monitored by FT-IR spectroscopy and small-angle X-ray scattering. *Biochemistry*, **38**, 6512–19.

Patel, H. A. (2007). *Studies on heat- and pressure-induced interactions of milk proteins*. PhD Thesis, Massey University, Palmerston North, New Zealand.

Patel, H. A., Singh, H., Anema, S. G. and Creamer, L. K. (2004). Effects of heat and high-hydrostatic pressure treatments on the aggregation of whey proteins in whey protein concentrate solutions. *Food New Zealand*, **4**(3), 29–35.

Patel, H. A., Singh, H., Havea, P., Considine, T. and Creamer, L. K. (2005). Pressure-induced unfolding and aggregation of the proteins in whey protein concentrate solutions. *Journal of Agricultural and Food Chemistry*, **53**, 9590–601.

Patel, H. A., Singh, H., Anema, S. G. and Creamer, L. K. (2006). Effects of heat and high hydrostatic pressure treatments on disulfide bonding interchanges among the proteins in skim milk. *Journal of Agricultural and Food Chemistry*, **54**, 3409–20.

Patel, H. A., Carroll, T. and Kelly, A. L. (2008). Nonthermal preservation technologies for dairy applications. In *Dairy Processing and Quality Assurance* (R. C. Chandan, A. Kilara and N. P. Shah, eds), Chapter 21. Ames, Iowa: Blackwell Publishing.

Paulsson, M. (1990). *Thermal denaturation and gelation of whey proteins and their adsorption at air/water interfaces,* PhD Dissertation. Department of Food Technology, University of Lund. Lund, Sweden.

Pearce, K. N. (1975). A fluorescence study of the temperature-dependent polymerization of bovine β-casein A1. *European Journal of Biochemistry*, **58**, 23–29.

Pessen, H., Kumosinski, T. F. and Farrell, H. M. Jr (1989). Small-angle X-ray scattering investigation of the micellar and submicellar forms of bovine casein. *Journal of Dairy Research*, **56**, 443–51.

Qi, P. X., Brown, E. M. and Farrell, H. M. Jr (2001). 'New-views' on structure–function relationships in milk proteins. *Trends in Food Science and Technology*, **12**, 339–46.

Rastogi, N. K., Raghavarao, K. S. M. S., Balasubramaniam, V. M., Niranjan, K. and Knorr, D. (2007). Opportunities and challenges in high pressure processing of foods. *Critical Reviews in Food Science and Nutrition*, **47**, 9–112.

Royer, C. A. (2002). Revisiting volume changes in pressure-induced protein unfolding. *Biochimica et Biophysica Acta*, **1595**, 201–9.

Sawyer, L., Barlow, P. N., Boland, M. J., Creamer, L. K., Denton, H., Edwards, P. J. B., Holt, C., Jameson, G. B., Kontopidis, G., Norris, G. E., Uhrínová, S. and Wu, S-Y. (2002). Milk protein structure – what can it tell the dairy industry? *International Dairy Journal*, **12**, 299–310.

Schokker, E. P., Singh, H., Pinder, D. N., Norris, G. E. and Creamer, L. K. (1999). Characterization of intermediates formed during heat-induced aggregation of β-lactoglobulin AB at neutral pH. *International Dairy Journal*, **9**, 791–800.

Schokker, E. P., Singh, H. and Creamer, L. K. (2000). Heat-induced aggregation of β-lactoglobulin A and B with α-lactalbumin. *International Dairy Journal*, **10**, 843–53.

Scollard, P. G., Beresford, T. P., Needs, E. C., Murphy, P. M. and Kelly, A. L. (2000). Plasmin activity, β-lactoglobulin denaturation and proteolysis in high pressure treated milk. *International Dairy Journal*, **10**, 835–41.

Silva, J. L. and Weber, G. (1993). Pressure stability of proteins. *Annual Review of Physical Chemistry*, **44**, 89–113.

Singh, H. (1995). Heat-induced changes in casein, including interactions with whey proteins. In *Heat-induced Changes in Milk*, 2nd edn (P. F. Fox, ed.) pp. 86–104. Brussels: International Dairy Federation.

Singh, H. (2004). Heat stability of milk. *International Journal of Dairy Technology*, **57**, 111–19.

Singh, H. and Creamer, L. K. (1992). Heat stability of milk. In *Advanced Dairy Chemistry – 1: Proteins* (P. F. Fox, ed.) pp. 621–56. London: Elsevier Applied Science Publishers.

Singh, H. and Havea, P. (2003). Thermal denaturation, aggregation and gelation of whey proteins. In *Advanced Dairy Chemistry*, Volume 1, *Proteins*, 3rd edn (P. F. Fox and P. L. H. McSweeney, eds) pp. 1257–83. New York: Kluwer Academic/Plenum Publishers.

Snoeren, T. H. M. and Van der Spek, C. A. (1977). The isolation of a heat-induced complex from UHTST milk. *Netherlands Milk and Dairy Journal*, **31**, 352–55.

Stading, M. and Hermansson, A-M. (1991). Large deformation properties of β-lactoglobulin gel structures. *Food Hydrocolloids*, **5**, 339–52.

Stading, M., Langton, M. and Hermansson, A-M. (1993). Microstructure and rheological behavior of particulated β-lactoglobulin gels. *Food Hydrocolloids*, **7**, 195–212.

Stapelfeldt, H. and Skibsted, L. H. (1999). Pressure denaturation and aggregation of β-lactoglobulin studied by intrinsic fluorescence depolarization, Rayleigh scattering, radiationless energy transfer and hydrophobic fluoroprobing. *Journal of Dairy Research*, **66**, 545–58.

Stapelfeldt, H., Petersen, P. H., Kristiansen, K. R., Qvist, K. B. and Skibsted, L. H. (1996). Effect of high hydrostatic pressure on the enzymic hydrolysis of β-lactoglobulin B by trypsin, thermolysin and pepsin. *Journal of Dairy Research*, **63**, 111–18.

Stapelfeldt, H., Olsen, C. E. and Skibsted, L. H. (1999). Spectrofluorometric characterization of β-lactoglobulin B covalently labeled with 2-(4′-maleimidylanilino)naphthalene-6-sulfonate. *Journal of Agricultural and Food Chemistry*, **47**, 3986–90.

Subirade, M., Loupil, F., Allain, A-F. and Paquin, P. (1998). Effect of dynamic high pressure on the secondary structure of β-lactoglobulin and on its conformational properties as determined by Fourier transform infrared spectroscopy. *International Dairy Journal*, **8**, 135–40.

Tanaka, N. and Kunugi, S. (1996). Effect of pressure on the deuterium exchange reaction of α-lactalbumin and β-lactoglobulin. *International Journal of Biological Macromolecules*, **18**, 33–39.

Tanaka, N., Koyasu, A., Kobayashi, I. and Kunugi, S. (1996a). Pressure-induced change in proteins studied through chemical modifications. *International Journal of Biological Macromolecules*, **18**, 275–80.

Tanaka, N., Nakajima, K. and Kunugi, S. (1996b). The pressure-induced structural change of bovine α-lactalbumin as studied by a fluorescence hydrophobic probe. *International Journal of Peptide and Protein Research*, **48**, 259–64.

Tanaka, N., Tsurui, Y., Kobayashi, I. and Kunugi, S. (1996c). Modification of the single unpaired sulfhydryl group of β-lactoglobulin under high pressure and the role of intermolecular S-S exchange in the pressure denaturation (single SH of β-lactoglobulin and pressure denaturation). *International Journal of Biological Macromolecules*, **19**, 63–68.

Tedford, L-A., Smith, D. and Schaschke, C. J. (1999a). High pressure processing effects on the molecular structure of ovalbumin, lysozyme and β-lactoglobulin. *Food Research International*, **32**, 101–6.

Tedford, L-A., Kelly, S. M., Price, N. C. and Schaschke, C. J. (1999b). Interactive effects of pressure, temperature and time on molecular structure of β-lactoglobulin. *Journal of Food Science*, **64**, 396–99.

Trujillo, A. J., Capellas, M., Saldo, J., Gervilla, R. and Guamis, B. (2002). Applications of high-hydrostatic pressure on milk and dairy products: a review. *Innovative Food Science and Emerging Technologies*, **3**, 295–307.

Van Camp, J. and Huyghebaert, A. (1995a). A comparative rheological study of heat and high pressure induced whey protein gels. *Food Chemistry*, **54**, 357–64.

Van Camp, J. and Huyghebaert, A. (1995b). High pressure-induced gel formation of a whey protein and haemoglobin protein concentrate. *Lebensmittel-Wissenschaft und -Technologie*, **28**, 111–17.

Van Camp, J., Feys, G. and Huyghebaert, A. (1996). High pressure induced gel formation of haemoglobin and whey proteins at elevated temperatures. *Lebensmittel-Wissenschaft und -Technologie*, **29**, 49–57.

Van Camp, J., Messens, W., Clément, J. and Huyghebaert, A. (1997a). Influence of pH and sodium chloride on the high pressure-induced gel formation of a whey protein concentrate. *Food Chemistry*, **60**, 417–24.

Van Camp, J., Messens, W., Clément, J. and Huyghebaert, A. (1997b). Influence of pH and calcium chloride on the high-pressure-induced aggregation of a whey protein concentrate. *Journal of Agricultural and Food Chemistry*, **45**, 1600–7.

Velazquez de la Cruz, G., Gandhi, K. and Torres, J. A. (2002). Hydrostatic pressure processing: a review. *Biotam*, **12**(2), 71–78.

Walkenström, P. and Hermansson, A-M. (1997). High-pressure treated mixed gels of gelatin and whey proteins. *Food Hydrocolloids*, **11**, 195–208.

Wang, W. (1999). Instability, stabilization, and formulation of liquid protein pharmaceuticals. *International Journal of Pharmaceutics*, **185**, 129–88.

Wang, W. (2005). Protein aggregation and its inhibition in biopharmaceutics. *International Journal of Pharmaceutics*, **289**, 1–30.

Weber, G. and Drickamer, H. G. (1983). The effect of high pressure upon proteins and other biomolecules. *Quarterly Reviews of Biophysics*, **16**, 89–112.

Wong, P. T. T. and Heremans, K. (1988). Pressure effects on protein secondary structure and hydrogen deuterium exchange in chymotrypsinogen: a Fourier transform infrared spectroscopic study. *Biochimica et Biophysica Acta*, **956**, 1–9.

Yang, J., Dunker, A. K., Powers, J. R., Clark, S. and Swanson, B. G. (2001). β-Lactoglobulin molten globule induced by high pressure. *Journal of Agricultural and Food Chemistry*, **49**, 3236–43.

Zasypkin, D. V., Dumay, E. and Cheftel, J. C. (1996). Pressure- and heat-induced gelation of mixed β-lactoglobulin/xanthan solutions. *Food Hydrocolloids*, **10**, 203–11.

Zhang, H., Deligeersang, Y., Guo, J., Mu, Z., Zhang, Y. and Zhu, H. (1998). Denaturation of bovine milk IgG at high pressure and its stabilization. *Shipin Kexue (Beijing)*, **19**, 10–12.

Zipp, A. and Kautzmann, W. (1973). Pressure denaturation of metamyoglobin. *Biochemistry*, **12**, 4217–28.

8

The whey proteins in milk: thermal denaturation, physical interactions and effects on the functional properties of milk

Skelte G. Anema

Abstract

This chapter reviews the literature on the denaturation of the whey proteins in milk, their interaction reactions with other milk protein components, and provides some examples on the relationships between denaturation/interaction reactions of the whey proteins and the functional behavior of the milk in selected applications. Early studies on whey protein denaturation in milk were aimed at developing methods to assess denaturation levels and determining the relationships between the denaturation of the whey proteins and the functional behavior of milk products in bakery and other applications. Subsequent studies were directed towards modeling the whey protein denaturation processes through kinetic and thermodynamic evaluations and determining the role of various milk components on these denaturation processes.

Milk Proteins: From Expression to Food
ISBN: 978-0-12-374039-7

Although the denaturation of whey proteins is critical in modifying the functional behavior of dairy products, it has become increasing apparent that a measure of the denaturation level of the whey proteins was not in itself a good predictor of functional performance. Therefore further studies have investigated the interactions of the denatured whey proteins with other milk protein components, in particular the interactions between the denatured whey proteins and the casein micelles (including identification of the specific disulfide bonds involved in complex formation between the denatured β-lactoglobulin and κ-casein). A limited number of recent studies have indicated that manipulation of the interactions of the denatured whey proteins with the other milk protein components may provide a significant tool for modifying or controlling the functional performance of milk protein products in some applications.

Introduction

Milk is produced in the mammary gland of female mammals and is intended for the feeding of the neonate from birth to weaning. Milk is a highly nutritious readily digested food, rich in protein, minerals and energy in an aqueous solution. It also provides the neonate with many other essential compounds such as protective agents, hormones and growth factors. Milk is a highly perishable fluid and was intended by nature to be consumed soon after production. However, humans have used milk and dairy-derived foods to supplement their diet for centuries and dairy products are still a major food source. Because of the commercial and nutritional significance of dairy products, manufacturing processes to preserve the food value of milk long after its initial production have been developed.

Over the last century, modern dairy milk processing has been transformed from an art into a science. Traditional products such as cheeses and yoghurts combine centuries-old knowledge with modern science, technology and processing techniques; in contrast, more recently developed products (such as spray-dried milk products, milk protein concentrates and whey protein concentrates) have been based on modern technologies of the time. Internationally, the majority of the milk processed is of bovine (cow) origin; however, significant quantities of buffalo, goat and sheep milk are also manufactured into dairy products (Fox, 2003).

The behavior of the protein components to various treatments largely dictates how milk will behave during processing. In milk, the lactose, some of the mineral components and the native whey proteins are in true molecular solution. However, the casein and most of the calcium and phosphate are found in large macromolecular assemblies called casein micelles. The colloidal suspension of casein micelles in milk serum is a remarkably stable food protein system, which allows the milk to be processed under the rigorous conditions used in modern dairy factories.

Milk can be subjected to high temperatures and pressures, high shear and variations in concentration without appreciable damage to the casein micelle system. Even the extreme action of drying milk to a powder does not significantly alter the milk system as milk powders can be reconstituted to produce liquid milks which have many properties that are similar to those of the milk from which they were derived (Singh and Newstead, 1992; Kelly *et al.*, 2003; Nieuwenhuijse and van Boekel, 2003; O'Connell and Fox, 2003).

The casein micelle

In order to understand and rationalize any changes to the properties and stability of milk, it is necessary to have some knowledge of the casein micelle structure and assembly. Despite extensive research efforts, the detailed structure and assembly of the casein micelle has not been unequivocally established, although several models have been proposed (Figure 8.1; Schmidt, 1982; Walstra and Jenness, 1984; Walstra, 1990; Holt and Horne, 1996; Horne, 1998; Walstra, 1999).

Evidence from electron microscopy and light scattering suggested that the casein micelle was assembled from smaller sub-units and, as a consequence, sub-micelle models of the casein micelle structure were developed. In the later iterations of these sub-micelle models, the casein proteins were hydrophobically aggregated to form the sub-micelle units, and these sub-micelle units were linked by colloidal calcium phosphate (CCP) to form the casein micelle. The distribution of κ-casein (κ-CN) between sub-micelles was heterogeneous, and the sub-micelles with high levels of κ-CN were located at the micelle surface whereas those with low levels of κ-CN were in the interior, thus giving a surface location to κ-CN that was consistent with experiment (Figure 8.1a; Schmidt, 1982; Walstra and Jenness, 1984; Walstra, 1990).

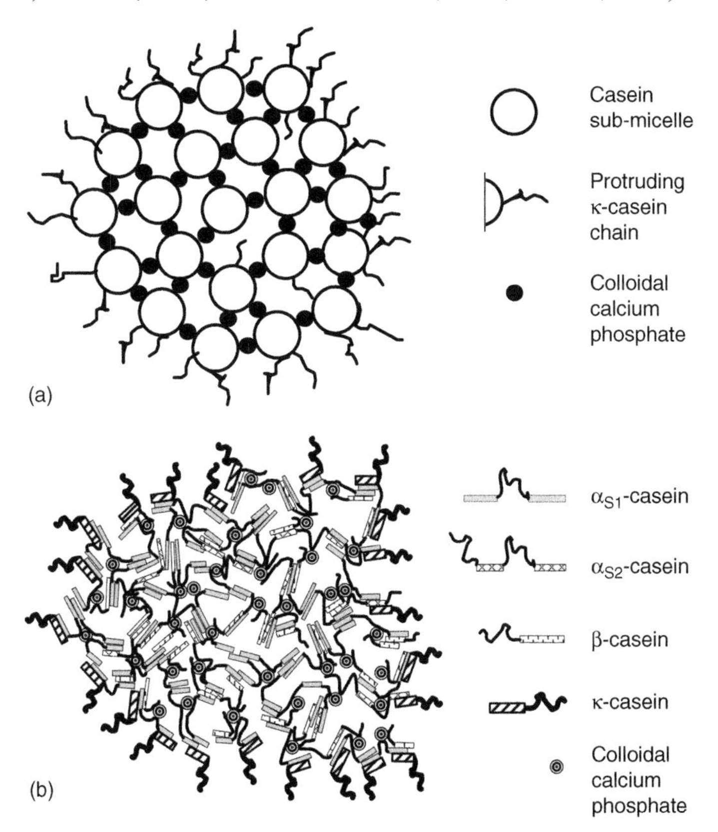

Figure 8.1 Recent models of the casein micelle. (a) The sub-micelle model of the casein micelle. Adapted with permission from Walstra and Jenness (1984). Copyright (1984) John Wiley and Sons. (b) The dual-binding model of the casein micelle. Adapted with permission from Horne (1998). Copyright (1998) Elsevier.

In recent years, there has been growing speculation over the existence of sub-micelles and the validity of the sub-micelle models of the casein micelles (Holt and Horne, 1996; Horne, 1998; Walstra, 1999; Horne, 2002, 2006). In particular, there was evidence to show that the CCP was uniformly distributed through the casein micelle, which precluded sub-micelles being linked by CCP to form the micelle. In addition, it was considered to be unlikely that there would be heterogeneous populations of casein sub-micelles with different levels of κ-CN, or that assembly into casein micelles via calcium phosphate occurs only after the casein sub-micelles have been formed.

Recent electron micrograph images of casein micelles did not display the internal sub-structure expected for casein sub-micelles and it was considered that the appearance of sub-micelles in earlier electron micrographs may have been artifacts of the early preparation techniques for electron microscopy (McMahon and McManus, 1998; Horne, 2006). The sub-micelle model of the casein micelle was refined to change the role of CCP from that of linking the sub-micelles to a charge-neutralizing agent to allow for a uniform distribution of CCP, and the sub-micelles were now linked together via hydrophobic interactions (Walstra, 1999). However, new models for the casein micelle that do not rely on the formation of sub-micelles have been proposed (Holt and Horne, 1996; Horne, 1998), the most recent of which is commmonly known as the dual-binding model, as shown in Figure 8.1b and described by David Horne in Chapter 5 of this volume.

Although various models of the casein micelles have been proposed, these have largely been derived from the same pool of research data and are therefore different depictions or interpretations of similar information. As a consequence, many of the salient features of the structure, assembly and stability of the different models are similar (Schmidt, 1982; Walstra, 1990; Holt and Horne, 1996; Horne, 1998; Walstra, 1999). Hydrophobic interactions and CCP are important in maintaining micelle integrity. Therefore, micelle integrity can be modified or destroyed by disruption to hydrophobic interactions or by the dissolution of the CCP. In all recent models, κ-CN has a preferential surface location with the C-terminal region protruding from the surface layer as a flexible hair.

These models of the casein micelles, with the surface layers of κ-CN and an internal structure maintained by hydrophobic interactions and CCP, have been used to explain micelle stability and the destabilization by the enzymes in rennet, by acidification or by the addition of alcohol (Walstra, 1990; Holt and Horne, 1996; Horne, 1998, 2003). However, less studied and less well understood are the mechanisms responsible for the changes occurring to the casein micelles during the heating of milk, in particular the interactions with the denatured whey proteins, the heat-induced, pH-dependent dissociation of the casein from the micelles and the eventual heat-induced coagulation of the casein micelles.

The heat treatment of milk

The effect of heat on the milk system is an important consideration in dairy chemistry as a heat treatment is involved in the manufacture of almost all milk products.

The heat treatment may range from thermization (about 65°C for 15 s) to sterilization (about 120°C for 10–20 min) or ultra-high temperature (UHT) treatment (typically 138–142°C for several seconds). As the thermal history of milk influences the behavior of the milk in subsequent applications, the effects of heat on milk have been the subject of intensive, if somewhat intermittent, research, and many reviews and books on the subject are available (Singh and Creamer, 1992; International Dairy Federation, 1995, 1996; O'Connell and Fox, 2003; Singh, 2004).

When milk is heated, a number of competitive and often interdependent reactions occur and the importance of each reaction is determined by the heating conditions as well as by factors such as milk composition or concentration. When considering the protein components of milk, reactions of particular importance are whey protein denaturation, the interactions of denatured whey proteins with other proteins (including those of the casein micelles) and casein micelle dissociation reactions. These three reaction processes can markedly modify the physico-chemical properties of milk and may play a major role in determining the stability of milk and the functional performance of heated milk products.

Whey protein denaturation

The whey proteins as found in milk are typical globular proteins with well-defined secondary and tertiary structures. In contrast to the highly stable caseins, the globular whey proteins (especially α-lactalbumin [α-LA] and β-lactoglobulin [β-LG]) retain their native conformations only within relatively limited temperature ranges. Exposing the whey proteins to extremes of temperature results in the denaturation and aggregation of the proteins; this process can be expressed using the simple reaction scheme as shown in Equation 8.1.

For protein species where the native protein is in the form of non-covalently linked oligomers (such as dimeric β-LG), the first step in the denaturation process is the reversible dissociation of the oligomer into monomeric species (Equation 8.1a). The monomeric protein can then undergo partial unfolding of the native structure (Equation 8.1b). In principle, this unfolding step is reversible; however, in reality, and in complex mixtures such as milk, the unfolding process is accompanied by the exposure of reactive amino side-chain groups, which allows irreversible aggregation reactions to occur. The unfolded whey protein can undergo aggregation reactions with other (unfolded) whey proteins, with aggregates or with the casein micelles (represented by A in Equation 8.1c).

At a fundamental level, protein denaturation is often defined as any non-covalent change to secondary or tertiary structure of the protein molecule (Equation 8.1b). From this denatured state, the protein can revert to its native state (refold) or interact with other components in the system (aggregate). Under this definition, α-LA is generally regarded as one of the most heat-labile whey proteins whereas β-LG is one of the most heat-stable whey proteins (Ruegg *et al.*, 1977).

However, for the dairy industry, it is the irreversible aggregation processes that follow the unfolding of the native whey proteins that largely determine the functional properties of dairy products. Hence, it is common practice to define whey protein denaturation as the formation of irreversibly denatured and aggregated whey proteins (Sanderson, 1970a; Singh and Newstead, 1992; Kelly *et al.*, 2003), and therefore this encompasses only the irreversible process shown in Equation 8.1c. Unless otherwise stated, the irreversible denaturation process is the definition used in this chapter. Using this definition for the denaturation of whey proteins in milk, the immunoglobulins are the most heat labile and α-LA is the most heat stable of the whey proteins, with β-LG and bovine serum albumin being intermediate (Larson and Rolleri, 1955). In general, significant denaturation of the major whey proteins—α-LA and β-LG—occurs only on heating milk at temperatures above ≈70°C.

$$(P_N)_n \leftrightarrows nP_N \tag{8.1a}$$

$$P_N \leftrightarrows P_U \tag{8.1b}$$

$$P_U + A \rightarrow (P - A) \tag{8.1c}$$

Assessment of the denaturation of whey proteins in milk

A considerable amount of research has been directed towards determining and understanding the denaturation processes of the major whey proteins when milk is heated. Early studies precipitated the casein and denatured whey protein by adjustment of the pH to the isoelectric point of the casein (≈pH 4.6), and analyzed the supernatant and the original milk for protein nitrogen, which gave estimates of initial native whey protein levels and levels after heat treatment (Rowland, 1933). However, a rapid method for determining the native whey protein levels was required for assessing milk powders for suitability in applications in the bakery industry (Harland and Ashworth, 1947), and also for categorizing milk powders based on the heat treatments received during the powder manufacturing process (Sanderson, 1970b, 1970c). From these requirements, the whey protein nitrogen index (WPNI) method was developed.

In the WPNI method, the casein and the denatured whey protein were precipitated and separated from the native whey proteins by saturation of the milk with salt, and the supernatant containing the native whey proteins was analyzed for protein content. This was originally achieved by dilution and pH adjustment of the supernatant to produce a turbid solution, with the turbidity proportional to the level of native whey protein originally present (Harland and Ashworth, 1947; Kuramoto *et al.*, 1959; Leighton, 1962). However, the WPNI method displayed considerable variability in the degree of turbidity developed for samples with similar levels of whey protein denaturation. To overcome this, Sanderson (1970a) combined a dye-binding method for determining the total protein content of milk with the original WPNI method, thus giving a more accurate and reliable method for determining the WPNI.

The original WPNI method or one of its variants is still the industry standard for determining the native whey protein levels of milk powder products, and is still widely used to classify milk powders according to the heat treatments received (Singh and Newstead, 1992; Kelly *et al.*, 2003). However, a recent report indicates that Fourier transform near-infrared spectroscopy may have potential as a rapid method for the determination of the WPNI of milk powders in the dry state, eliminating the necessity of reconstitution, precipitation and filtration (Patel *et al.*, 2007).

Although the WPNI method can give an estimate of the level of whey protein denaturation, research into the denaturation and interactions of the individual whey proteins requires more accurate separation and analysis procedures. There are numerous quantitative methods for separating and determining the level of the individual whey proteins in milk and these can be used directly or adapted to determine the level of denaturation after defined heat treatments. Methods that have been used include polyacrylamide gel electrophoresis (PAGE; e.g. Hillier and Lyster, 1979; Dannenberg and Kessler, 1988c; Kessler and Beyer, 1991; Anema and McKenna, 1996), capillary electrophoresis (e.g. Fairise and Cayot, 1998; Butikofer *et al.*, 2006), differential scanning calorimetry (e.g. Ruegg *et al.*, 1977; Manji and Kakuda, 1987), high-performance liquid chromatography (HPLC; e.g. Kessler and Beyer, 1991) and various immuno-based assays (e.g. Lyster, 1970).

In general, good correlations have been observed when the various methods for determining whey protein denaturation have been compared (Manji and Kakuda, 1987; Kessler and Beyer, 1991; Anema and Lloyd, 1999; Patel *et al.*, 2007). These methods are more time-consuming than the traditional WPNI methods and therefore cannot be used for routine analysis and classification of milk products. However, they have higher accuracy and reproducibility and can be used to determine the denaturation behavior of the individual whey proteins. In addition, variations of the techniques or coupling to additional detection devices can provide further information on the interactions of the denatured whey proteins with other components in the milk (e.g. Lowe *et al.*, 2004; Patel *et al.*, 2006, 2007).

Kinetic evaluation and modelling of whey protein denaturation

Early studies showed that the denaturation of whey proteins was a kinetic phenomenon, and therefore dependent on both the temperature and the time of the heat treatment (Rowland, 1933; Harland and Ashworth, 1945). Even though these early studies considered the whey protein components as a single entity, it was noted that the denaturation of the whey proteins did not follow a simple exponential law and was not a first-order (unimolecular) process; therefore, it was concluded that the effect of temperature on the rate constants was difficult to determine. In addition, there was a change in temperature dependence at temperatures above about 80°C, which was probably the first indication of the complex nature of the irreversible denaturation of the whey proteins in milk (Rowland, 1933).

Although early studies on the effect of temperature and heating time on the denaturation of the individual whey proteins had been performed (e.g. Harland and Ashworth, 1945; Gough and Jenness, 1962), it was the kinetic study of Lyster (1970) over a wide temperature range (68–155°C) that conclusively demonstrated the complexity of the denaturation process of the individual whey proteins. Lyster (1970) found that the denaturation of α-LA appeared to follow first-order kinetics and that the denaturation of β-LG was second order. Arrhenius plots for the denaturation of both α-LA and β-LG indicated that the irreversible denaturation reaction was not a simple process, as a change in temperature dependence was observed at about 80–90°C for both α-LA and β-LG (Figure 8.2). The rate constants increased more rapidly with an increase in temperature in the low temperature ranges than at higher temperatures.

Further studies on the kinetic evaluation of the denaturation of whey proteins in milk confirmed this complexity of the denaturation process and provided relationships between compositional aspects and the rate of denaturation (Lyster, 1970; Hillier and Lyster, 1979; Manji and Kakuda, 1986). However, it was the kinetic and thermodynamic studies of Dannenberg and Kessler (1988c) that provided insights into the possible mechanisms responsible for the complex temperature dependences of the denaturation of α-LA and β-LG. Dannenberg and Kessler (1988c) found that, in milk, the denaturation of β-LG had an order of ≈1.5, which is now generally accepted, and that the denaturation of α-LA was pseudo first order. On the basis of the thermodynamic evaluations of the denaturation reactions of β-LG and α-LA in the two temperature ranges (i.e. at temperatures above and below the marked change in temperature

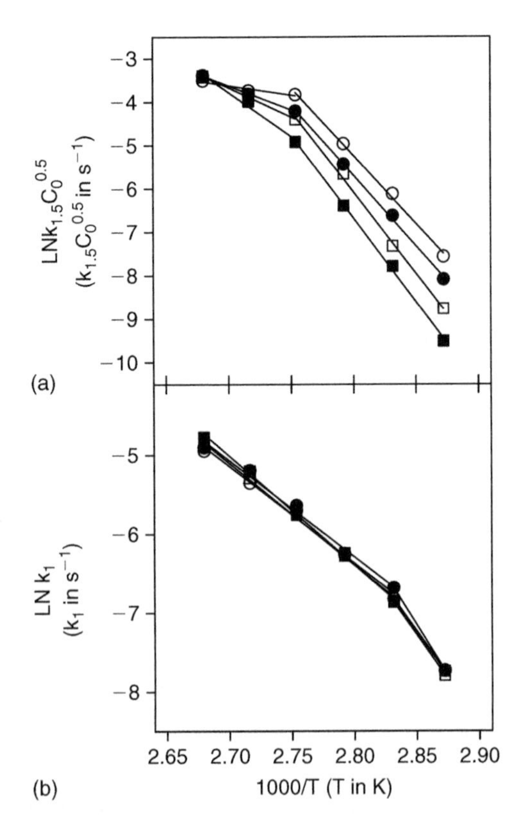

Figure 8.2 Effect of milk concentration on the Arrhenius plot for the thermal denaturation of β-LG (a) and α-LA (b) over a 75–100°C temperature range. ○: 9.6% total solids milk; ●: 19.2% total solids milk; □: 28.8% total solids milk; ■: 38.4% total solids milk. Part A was adapted with permission from Anema (2000). Copyright (2000) American Chemical Society. Part B was adapted with permission from Anema (2001). Copyright (2001) Blackwell Publishing.

dependence for the denaturation reactions; Figure 8.2), information on the possible rate-determining steps in the denaturation reactions was obtained.

At temperatures below about 90°C for β-LG and 80°C for α-LA, the high values for the activation energies and enthalpies indicated that a large number of bonds were disrupted and the positive activation entropies indicated a lower state of order of the reaction products. These kinetic and thermodynamic parameters were interpreted as indicating that the unfolding (reversible denaturation) of the whey proteins was the rate-determining step in the lower temperature ranges.

At higher temperatures, above 80°C for α-LA and above 90°C for β-LG, considerably lower activation energies and enthalpies were typical of chemical reactions and the negative activation entropies indicated a higher state of order. These parameters suggested that chemical (aggregation) reactions were the rate-determining step in the higher temperature ranges. Subsequent studies on the kinetic and thermodynamic evaluation of the denaturation reactions have supported these interpretations in skim and whole milk under industrial processing conditions (Anema and McKenna, 1996; Oldfield *et al.*, 1998a).

The denaturation reactions of both β-LG and α-LA are enhanced when the pH of the milk is increased from the natural pH and are retarded when the pH is decreased (Law and Leaver, 2000). The denaturation of β-LG was retarded when all components in the milk were concentrated, although the effect was less pronounced as the temperature was increased (Figure 8.2a; Anema, 2000). In contrast, the denaturation of α-LA was hardly affected by milk concentration, with similar rates of denaturation at all milk concentrations regardless of the heating temperature (Figure 8.2b; Anema, 2001).

The seemingly contrasting effects of milk concentration on the denaturations of α-LA and β-LG have been explained by detailed studies on the effect of the concentrations of the individual components of milk on the denaturation reactions. Increasing the protein concentration of milk, while maintaining essentially constant non-protein soluble component concentrations, increased the rate of denaturation of both α-LA and β-LG (Figure 8.3; Law and Leaver, 1997; Anema *et al.*, 2006), with a similar effect at all temperatures (Anema *et al.*, 2006).

Increasing the non-protein soluble components while maintaining constant protein concentrations retarded the denaturation of both β-LG and α-LA; however, the effects on these two proteins were somewhat different (Figure 8.3; Anema *et al.*, 2006). For β-LG, increasing the non-protein soluble components caused a substantial retardation of denaturation in the lower temperature range and this effect became less pronounced at higher temperatures.

In contrast, the effect of increasing the non-protein soluble components on α-LA denaturation was less pronounced than for β-LG and was similar at all temperatures investigated (Figure 8.3). The increasing lactose concentration, the major component of the non-protein soluble components, could explain much of the effect of increasing non-protein soluble components; however, clearly other composition factors such as pH and ionic components also have an effect (Figure 8.3; Anema *et al.*, 2006).

From these results, it was possible to explain the effects of milk concentration on the denaturation of β-LG and α-LA. For α-LA, on increasing the total solids concentration of the milk (both protein and non-protein soluble components), the retardation

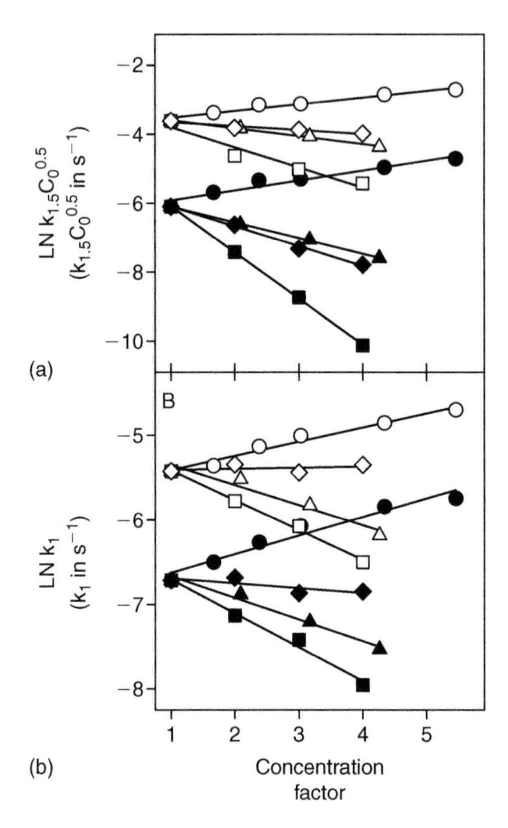

Figure 8.3 Comparison of the effects of the concentrations of protein (●, ○), non-protein soluble components (■, □), lactose (▲, △) and total solids (◆, ◇) on the rate constants for the denaturation of β-LG (a) and α-LA (b) at 80°C (filled symbols) and 95°C (open symbols). Reproduced with permission from Anema *et al.* (2006). Copyright (2006) American Chemical Society.

of the reaction rate by increasing the non-protein soluble components concentration was almost exactly offset by the increase in the denaturation rate for α-LA on increasing the protein concentration. As this effect was similar at all temperatures, increasing total solids appeared to have no effect on the rate of denaturation of α-LA (Figures 8.2b and 8.3b; Anema, 2001; Anema *et al.*, 2006).

However, for β-LG, the retardation in the rate of denaturation on increasing the concentration of the non-protein soluble components was not completely offset by the increasing rate of denaturation on increasing the protein concentration; therefore β-LG denaturation was retarded by increasing the total solids concentration of the milk. However, the non-protein soluble components were less effective in retarding the denaturation of β-LG at higher temperatures and, as a consequence, the increase in total solids concentration appeared to have a smaller effect on the denaturation of β-LG at the higher temperatures and particularly above about 90°C (Figures 8.2a and 8.3a; Anema, 2000; Anema *et al.*, 2006). The effects of the non-protein soluble components concentration or the lactose concentration on the denaturation of β-LG and α-LA have been discussed in terms of the preferential hydration theory (Anema, 2000; Anema *et al.*, 2006).

Interactions between denatured whey proteins and κ-CN/casein micelles

An understanding of the denaturation reactions of the whey proteins provides information on the initial steps of a complex series of aggregation reactions that can occur when milk is heated. This aggregation process can involve other milk protein components and may involve numerous reaction pathways or interaction processes. Although the reactions of the denatured whey proteins with other milk protein components are important, these types of reactions are considerably more difficult to measure than the irreversible denaturation processes, particularly in a complex mixture of components such as is found in (skim) milk.

Interactions between denatured whey proteins and κ-CN in model systems

One of the major reactions of interest is the interaction between the denatured whey proteins and the casein micelles, particularly interactions of denatured β-LG with κ-CN at the micelle surface. Early studies on model systems indicated that there was an interaction between β-LG and κ-CN when these components were heated together (Zittle *et al.*, 1962; Long *et al.*, 1963; Sawyer *et al.*, 1963). These conclusions were drawn from electrophoretic studies, which showed that the discrete bands assigned to κ-CN and β-LG observed in unheated solutions produced species of intermediate mobility when the solutions were heated together. Sedimentation velocity experiments also confirmed complex formation, as the β-LG–κ-CN complex formed on heating had markedly higher sedimentation coefficients than did the individual proteins when heated separately (Zittle *et al.*, 1962).

Once interaction between κ-CN and denatured β-LG had been confirmed, subsequent investigations in heated model systems were aimed at determining the types of bonds involved in complex formation, the stoichiometry of the complexes formed and the involvement of other whey proteins (particularly α-LA). It was shown that reducing agents dissociated the heat-induced complexes and that thiol-blocking agents prevented the formation of the complexes (Sawyer *et al.*, 1963). These results supported earlier suggestions that the free thiol group of β-LG was involved in the interactions (Trautman and Swanson, 1958; Zittle *et al.*, 1962) and it was suggested that intermolecular disulfide bonds were formed between κ-CN and denatured β-LG (Sawyer *et al.*, 1963). This has been corroborated by numerous subsequent studies (e.g. Grindrod and Nickerson, 1967; Purkayastha *et al.*, 1967; Sawyer, 1969; Tessier *et al.*, 1969).

Some studies indicated that the heat-induced self-aggregation of β-LG was limited when κ-CN was present, which suggested that κ-CN formed complexes with intermediate species of aggregated β-LG (Sawyer, 1969; McKenzie *et al.*, 1971). In contrast, other studies indicated that the aggregation of β-LG was not a prerequisite for interaction with κ-CN (Euber and Brunner, 1982). The reason for these apparently conflicting observations may have been resolved through the detailed study of Cho *et al.* (2003), in which many of the possible pathways involved in the aggregation of β-LG with κ-CN in heated protein model systems were elucidated. Cho *et al.* (2003) proposed that, when mixtures of β-LG and κ-CN were heated, the free thiol of β-LG was exposed and this initiated a series of thiol–disulfide exchange reactions of β-LG with other denatured β-LG molecules or with κ-CN. The products formed ranged from 1:1 β-LG–κ-CN complexes to large heterogeneous aggregates and the product mix was dependent on the ratio of κ-CN to β-LG. The aggregate species were held together by either or both disulfide bonds and hydrophobic interactions.

Although there have been some indications that there may be an interaction between α-LA and κ-CN on heating (Shalabi and Wheelock, 1976; Doi *et al.*, 1983), other studies have reported that interaction between these proteins does not occur (Baer *et al.*, 1976; Elfagm and Wheelock, 1978). It is now generally believed that interactions between α-LA and κ-CN will occur only if β-LG (or another whey protein with a free thiol) is present during heating, and this may require the initial formation of a β-LG–α-LA complex, which subsequently interacts with κ-CN (Baer *et al.*, 1976; Elfagm and Wheelock, 1978).

There is considerable evidence to show that disulfide bonds are involved in the aggregated species formed between the denatured whey proteins and κ-CN; however, there are reports that suggest that non-covalent bonding may be important in these interactions, particularly in the early stages of heating and at lower heating temperatures (Sawyer, 1969; Haque *et al.*, 1987; Haque and Kinsella, 1988; Hill, 1989). In addition, other studies have shown that, although a substantial part of the denatured whey proteins in heated milk are involved in disulfide-bonded aggregates, there is a significant proportion that can be recovered as monomeric protein under dissociating but non-reducing conditions, indicating that non-covalent interactions are also involved (Oldfield *et al.*, 1998b; Anema, 2000). As Cho *et al.* (2003) have suggested, it is likely that both hydrophobic and disulfide interactions are important in the early stages of aggregate formation, with the interaction mechanism dependent on the composition of the system and the conditions of heating.

Interactions between denatured whey proteins and κ-CN/casein micelles in milk systems

Most of the early studies examining the heat-induced interactions between denatured whey proteins and κ-CN were in model systems using purified proteins in buffer systems. Milk is considerably more complex, with numerous protein species that could potentially interact on heating. A number of the milk proteins have free thiol groups and/or disulfide bonds. Although β-LG is the major whey protein component, denatured α-LA and bovine serum albumin can also be involved in thiol–disulfide exchange reactions and therefore can be incorporated in the aggregated products. For the caseins, both κ-CN and α_{s2}-CN have disulfide bonds and therefore both could participate in thiol–disulfide exchange reactions with denatured β-LG or other denatured thiol-bearing whey proteins. As a consequence of this complexity, there are numerous potential thiol–disulfide interaction pathways, as well as non-covalent interactions, and therefore the separation and analysis of the reaction products can be a difficult process.

However, the studies on the interactions between the proteins in heated milk suggest that, despite the complexity of the system, the reactions between β-LG and κ-CN may be similar to those occurring in the model system. In early electrophoretic studies on heated milk, it was noted that the bands corresponding to β-LG disappeared, along with a reduction in the intensity of the bands corresponding to casein. This was accompanied by the formation of bands corresponding to new (heterogeneous) components (Slatter and van Winkle, 1952; Tobias *et al.*, 1952). When sulfydryl-blocking agents were added, the band pattern was comparable with that of the original skim milk, indicating that sulfydryl–disulfide exchange reactions were involved in the interaction mechanisms (Trautman and Swanson, 1958). Subsequent studies confirmed that an interaction between denatured β-LG and κ-CN on the casein micelles occurred on heating milk although, as expected, the other denatured whey proteins were also involved in the interactions (Snoeren and van der Spek, 1977; Elfagm and Wheelock, 1978; Smits and van Brouwershaven, 1980; Noh *et al.*, 1989a, 1989b; Corredig and Dalgleish, 1996a, 1996b; Oldfield *et al.*, 1998b; Corredig and Dalgleish, 1999).

Unlike κ-CN, α_{s2}-CN does not readily interact with denatured whey proteins when milk is heated, although some interactions in UHT milks have been reported (Snoeren and van der Spek, 1977; Patel *et al.*, 2006). This low reactivity may be due to the location of α_{s2}-CN in the interior of the casein micelles, which makes it less accessible for interaction, whereas κ-CN is located at the casein micelle surface and therefore may be more accessible for interaction (Walstra, 1990; Horne, 1998). Interestingly, in pressure-treated skim milk, disulfide-bonded aggregates between α_{s2}-CN and the denatured whey proteins are observed, suggesting that the disulfide bonds of α_{s2}-CN may become accessible to thiol groups of the denatured whey proteins when the casein micelle structure is disrupted under pressure (Patel *et al.*, 2006).

The degree of interaction of the denatured whey proteins with the casein micelles is dependent on many variables including the time, temperature and rate of heating, the milk and individual protein concentrations, the milk pH and the concentration of the milk salts (Smits and van Brouwershaven, 1980; Corredig and Dalgleish, 1996a, 1996b; Oldfield *et al.*, 2000; Anema and Li, 2003b; Oldfield *et al.*, 2005). For example, when the temperature of milk is gradually increased above 70°C, as in indirect heating systems, most of the denatured β-LG and α-LA associates with the casein micelles, presumably as disulfide-bonded complexes with κ-CN at the micelle surface (Smits and van Brouwershaven, 1980; Corredig and Dalgleish, 1996b).

In contrast, when milk is heated rapidly, as in direct heating systems, only about half of the denatured β-LG and α-LA associates with the casein micelles, with the rest remaining in the milk serum (Singh and Creamer, 1991a; Corredig and Dalgleish, 1996a; Oldfield *et al.*, 1998b). Corredig and Dalgleish (1999) suggested that, on heating milk, α-LA and β-LG initially aggregate in the serum phase at a ratio dependent on the initial individual whey protein concentrations. These complexes subsequently associate with κ-CN at the casein micelle surface on prolonged heating. However, Oldfield *et al.* (1998b) proposed that, under rapid heating rates, β-LG forms aggregates in the serum before interacting with the casein micelles and this limits the level of association with the casein micelles, whereas, at slower heating rates, monomers or smaller aggregates of β-LG may interact with the micelles and this may allow higher association with the casein micelles.

The pH of the milk at heating is important in determining the level of interaction between the denatured whey proteins and the casein micelles. Many of the early studies on the effect of pH were attempts to explain the unusual heat stability characteristics of the milk at very high temperatures. When milk is heated at high temperatures (≈140°C), the heat coagulation time/pH profiles (HCT/pH profiles) of most milks show increasing heat stability with increasing pH to a maximum at about pH 6.7, followed by decreasing stability to a minimum at about pH 6.9, and increasing stability again as the pH is increased further (Rose, 1961). Considerable research has been undertaken over decades in an attempt to explain this unusual pH-dependent heat stability of milk and numerous factors are known to influence the heat stability behavior. This has been covered in many review papers on the heat stability of milk (Singh and Creamer, 1992; International Dairy Federation, 1995; O'Connell and Fox, 2003; Singh, 2004).

The results from these studies on the heat stability of milk have influenced the direction of the future research on the effects of heat on milk and in particular the interactions between denatured whey proteins and κ-CN/casein micelles. Therefore, it is appropriate to briefly review aspects of the pH-dependence of heat stability that are relevant to understanding the interaction between denatured whey proteins and κ-CN/casein micelles. Electron microscopic studies showed that, when milk was heated at high temperatures (90–140°C) for long times (30 min) at pH below 6.7, the denatured whey proteins complexed on to the micelle surfaces as filamentous appendages. However, when the milk was heated at higher pH, the denatured whey proteins were found in the serum phase as aggregated complexes (Creamer *et al.*, 1978; Creamer and Matheson, 1980). These were the first indications that the pH at heating may influence the interactions between the denatured whey proteins and the casein micelles when milk is heated at high temperatures.

Kudo (1980) showed that the amount of non-sedimentable protein in milk heated at pH 6.5 was lower than that in unheated milk; however, the level of non-sedimentable protein increased with the pH at heating so that, above pH 6.7, the level was markedly higher than in the unheated milk and increased with increasing pH. Kudo (1980) concluded that the denatured whey proteins co-sedimented with the casein micelles at low pH (≈pH 6.5), whereas most of the denatured whey proteins along with some casein (particularly κ-CN) was released from the casein micelles at pH above 6.8. It was also proposed that the transition from whey-protein-coated casein micelles to protein-depleted forms with changing pH at heating could explain the pH-dependence of the heat stability of milk at high temperatures.

Singh and Fox (1985a, 1985b, 1986, 1987a, 1987b, 1987c), in a series of extensive studies, showed that the dissociation of κ-CN-rich protein on heating was dependent on the pH at heating. At pH below about 6.8, little dissociation of micellar κ-CN occurred whereas, at higher pH, particularly above pH 6.9, high levels of κ-CN dissociated from the micelles, with the level increasing proportionally with increased pH. The whey proteins, particularly β-LG, played an important role in the heat-induced pH-dependent dissociation of κ-CN (Singh and Fox, 1987b), as did mineral components such as calcium and phosphate (Singh and Fox, 1987c). The results from these extensive studies have been used to develop detailed mechanisms for the pH-dependent heat stability of milk and concentrated milk systems (O'Connell and Fox, 2003; Singh, 2004).

Initially, it was reported that the dissociation of κ-CN from micelles occurred only when milk at high pH (above ≈pH 6.8) was heated at high temperatures, particularly 90°C or above (Singh and Fox, 1985b). However, subsequent studies demonstrated that, at these pH values, the dissociation of κ-CN occurred as soon as the temperature was raised above ambient, with the level of dissociated κ-CN increasing proportionally with temperature up to 90°C. In these studies, the dissociation of α_s-CN (α_{s1}-CN and α_{s2}-CN combined) and β-CN showed an unusual temperature dependence. Increasing levels of these caseins dissociated as the temperature was increased up to about 70°C, with the levels then decreasing again at higher temperatures (Figure 8.4; Anema and Klostermeyer, 1997; Anema, 1998).

A subsequent study showed that the unusual temperature dependence of α_s-CN and β-CN was a consequence of the whey proteins, particularly β-LG. When whey-protein-depleted milk was heated, the levels of α_s-CN and β-CN dissociating from the casein micelles increased with increasing temperature up to 90°C. When compared with heating standard milk, this indicated that higher levels of α_s-CN and β-CN dissociated from the micelles in the whey-protein-depleted milks at temperatures above about 70°C (Anema and Li, 2000).

It was postulated that all the caseins dissociated from the micelles on heating. On subsequent cooling, the dissociated κ-CN stabilized the dissociated α_s-CN and β-CN as small serum phase aggregates if the heating temperature was below about 70°C. However, above about 70°C, κ-CN was associated with denatured whey proteins. It was already known that the complex formed between κ-CN and denatured β-LG was less effective at stabilizing α_s-CN and β-CN in the presence of calcium ions than uncomplexed κ-CN (Zittle *et al.*, 1962); therefore, this interaction may have prevented κ-CN from stabilizing the other caseins and they either reassociated with the casein micelles or formed larger aggregates on subsequent cooling (Anema and Li, 2000).

Early studies on the effect of the pH at heating on the interaction of denatured whey proteins with the casein micelles tended to use relatively large pH steps. In a model milk system containing casein micelles and β-LG, about 80% of the denatured β-LG associated with the casein micelles when the milk was heated at pH 5.8 or pH 6.3, whereas only about 20% associated with the casein micelles at pH 6.8 or pH 7.1 (Smits and van Brouwershaven, 1980).

The studies on the heat-induced, pH-dependent dissociation of κ-CN from the casein micelles showed that this dissociation of κ-CN was accompanied by increases in the levels of denatured whey proteins remaining in the serum (Singh and Creamer, 1991b), and this was confirmed by Anema and Klostermeyer (1997) and Oldfield *et al.* (2000), who reported that 80–90% of the denatured whey proteins associated with the casein micelles when milk was heated at pH below 6.7, whereas only about 20% of the denatured whey proteins associated with the casein micelles at pH above 6.8. Corredig and Dalgleish (1996b) measured the ratio of β-LG or α-LA to κ-CN in the colloidal phase obtained from heated milk adjusted to pH 5.8, 6.2 or 6.8. Although the denatured whey proteins interacted with the casein micelles at a faster rate at lower pH and at higher temperatures, the ratios of denatured whey proteins to

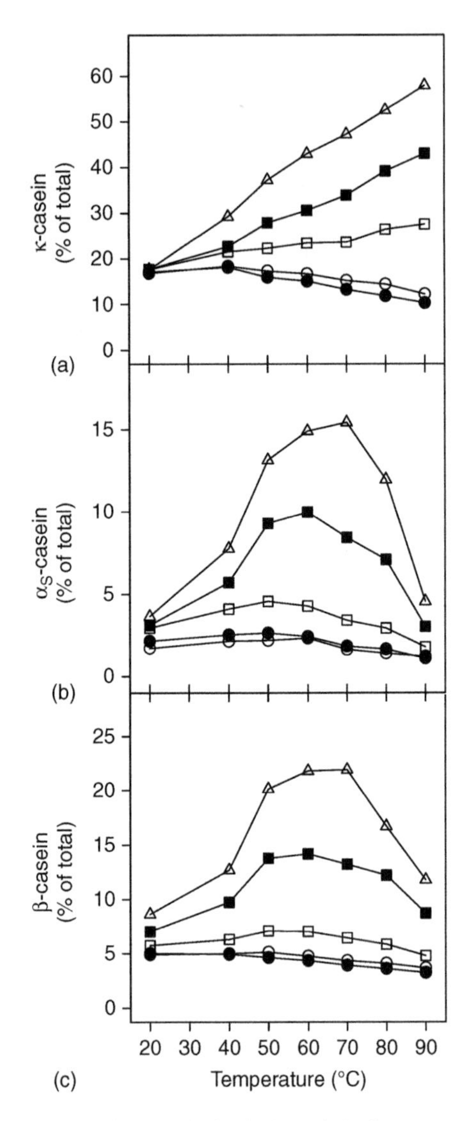

Figure 8.4 Effect of temperature and pH on the level of protein in the supernatants obtained from 10% total solids reconstituted skim milk samples heated for 30 min: κ-casein (a); α_s-casein (b); β-casein (c), ○: pH 6.3; ●: pH 6.5; □: pH 6.7; ■: pH 6.9; △: pH 7.1. Adapted with permission from Anema and Klostermeyer (1997). Copyright (1997). American Chemical Society.

κ-CN on the casein micelles were not markedly different under the different heating conditions.

Recent studies demonstrated the extreme importance of pH on the association of denatured whey proteins (α-LA and β-LG) with the casein micelles when milk was heated above 70°C, particularly at pH 6.7 or below, where differences in association behavior could be measured at pH differences as small as 0.05 pH units (Anema and Li, 2003a, 2003b; Vasbinder and de Kruif, 2003). From these studies, it was shown that about 80% of the denatured whey protein was associated with the casein

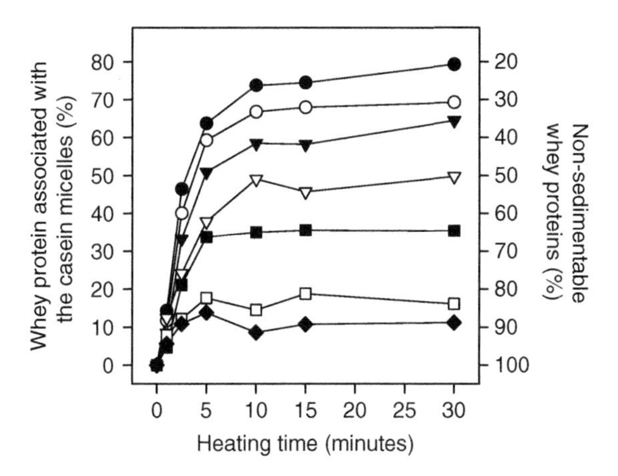

Figure 8.5 Level of whey proteins associated with the casein micelles/non-sedimentable whey proteins in skim milk samples that were heated at 90°C for various times. The pH values of the milk samples prior to heating were: ●: pH 6.5; ○: pH 6.55; ▼: pH 6.6; ▽: pH 6.65; ■: pH 6.7; □: pH 6.9; ◆: pH 7.1. Adapted with permission from Anema *et al.* (2004a). Copyright (2004) American Chemical Society.

micelles at pH 6.5 and that this level of association decreased linearly as the pH at heating was increased, so that only about 30% was associated at pH 6.7. At higher pH (above pH 6.7), very low levels of denatured whey proteins associated with the casein micelles (Figure 8.5).

Although the heat-induced pH-dependent dissociation of κ-CN from the casein micelles could explain the low levels of denatured whey proteins interacting with the casein micelles at pH above 6.8, it had been reported that very little κ-CN dissociated from the casein micelles at pH below 6.8 (Singh and Fox, 1985b; Nieuwenhuijse *et al.*, 1991; Singh, 2004). Therefore, it was initially unknown why small shifts in pH between pH 6.5 and pH 6.7 affected the association of denatured whey proteins with the casein micelles when milk was heated. The level of κ-CN in the serum phase was low; therefore, it was initially believed that κ-CN was not involved in this partition of the whey proteins between the serum and colloidal phases (Oldfield *et al.*, 1998b; Anema and Li, 2003b; Vasbinder and de Kruif, 2003).

However, more recent studies showed that the heat-induced dissociation of κ-CN was pH dependent from pH 6.5 to pH 7.1, with a linear increase in serum phase κ-CN as the pH was increased throughout the pH range from 6.5 to 7.1 (Figure 8.6a), and that the level of serum phase κ-CN was correlated with the level of serum phase denatured whey protein (Figure 8.6b; Anema, 2007). The differences in the level of dissociated κ-CN between the earlier studies and the later studies may be related to the centrifuging conditions, which may have masked the effects at the lower pH, especially under conditions where the particles are less hydrated and more readily deposited.

Although the level of κ-CN in the serum phase at pH below 6.7 was relatively low (less than ≈30% of the total κ-CN), the ratio of denatured whey protein to κ-CN was high and relatively constant (at about 2.5 whey proteins to each monomeric κ-CN) for the serum phase proteins at all pH. In contrast, the ratio of denatured whey protein to

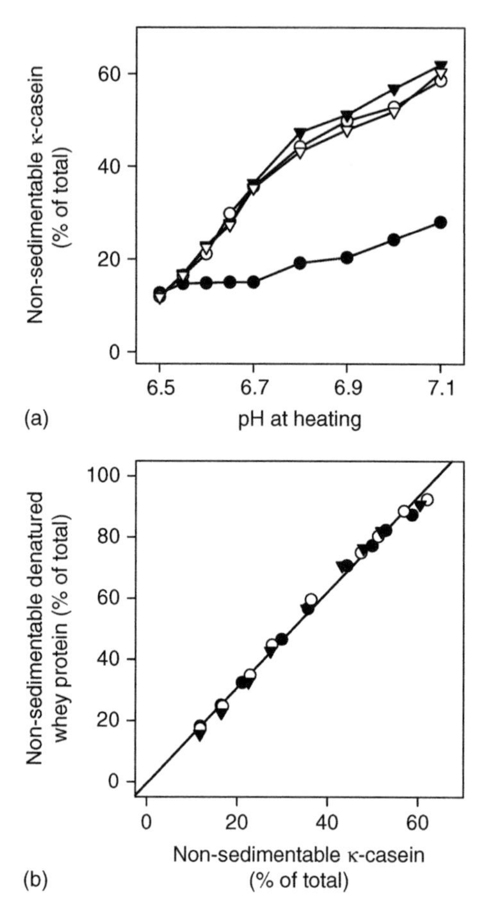

Figure 8.6 (a) Effect of the pH at heating on the level of non-sedimentable κ-casein in milk: ●: serum phase κ-casein in unheated milk; serum phase κ-casein in milk heated at 90°C for 20 min; ▼: serum phase κ-casein in milk heated at 90°C for 25 min; ▽: serum phase κ-casein in milk heated at 90°C for 30 min. (b) Relationship between the serum phase denatured whey protein and the level of serum phase κ-casein for the heated milk samples: milk heated at 90°C for 20 min; ○: milk heated at 90°C for 25 min; ◆: milk heated at 90°C for 30 min. Adapted with permission from Anema (2007). Copyright (2007) American Chemical Society.

κ-CN was only about 1:1 for the whey protein associated with the casein micelles at pH 6.5 and this decreased to about 0.5:1 at pH 7.1 (Anema, 2007). Intensive studies on the soluble whey-protein–κ-CN complexes formed when milk was heated at the natural pH also showed that κ-CN was intimately involved in the serum phase aggregates and that a high ratio of denatured whey proteins to κ-CN was observed (Guyomarch *et al.*, 2003). Electron micrographs of the serum phase whey protein–κ-CN aggregates indicated that these particles were roughly spherical with a relatively uniform size of about 20–50 nm (Parker *et al.*, 2005).

There is still some debate over the sequence of events in the interaction reactions between the denatured whey proteins and κ-CN. Some reports suggest that κ-CN dissociates from the micelles early in the heating process and that the denatured whey proteins subsequently interact with the κ-CN either in the serum phase or on the micelles, with a preferential serum phase reaction (Anema and Li, 2000; Anema, 2007). This proposal was supported by the observations that the dissociation of κ-CN is a rapid process and that significant dissociation of κ-CN can occur at temperatures below those where the denaturation of whey proteins occurs (Anema and Klostermeyer, 1997). In addition, significant dissociation of κ-CN occurs in systems that have been depleted of whey proteins (Anema and Li, 2000). The higher ratio of denatured whey protein to κ-CN for the serum phase, regardless of the pH at heating or the level of dissociated κ-CN, may also suggest a preferential serum phase reaction between the denatured whey proteins and κ-CN (Anema, 2007).

However, other reports suggest that, on heating milk, the denatured whey proteins first interact with the casein micelles and that the whey protein–κ-CN complex subsequently dissociates from the casein micelles (Parker *et al.*, 2005; Donato and Dalgleish, 2006). This proposal was supported by the observation that the addition of sodium caseinate to milk did not increase the level of serum phase complexes between the denatured whey proteins and κ-CN, which was interpreted as indicating that the complexes between the denatured whey proteins and κ-CN were formed on the casein micelle surface regardless of the pH at heating (Parker *et al.*, 2005).

In addition, it was suggested that two mechanisms occur depending on the pH at heating. This was based on observations that the protein composition of the serum phase appeared to vary markedly depending on whether the milk was heated at pH above or below the natural pH of the milk (Donato and Dalgleish, 2006), although other studies did not display a marked difference in composition (Anema, 2007). At this stage, further detailed investigations are required to clarify whether dissociation of κ-CN occurs before or after interaction with the denatured whey proteins.

The pH-dependent changes in the association of the denatured whey proteins with the casein micelles, and the dissociation of κ-CN from the micelles, can have a marked effect on some of the physical properties of the milk. A marked increase in casein micelle size was observed when high levels of denatured whey protein were associated with the colloidal phase, as is observed on heating milk at pH 6.5. This change in size was less pronounced as the pH at heating the milk was increased to pH 6.7 and a decrease in casein micelle size was observed when significant levels of κ-CN were dissociated from the colloidal phase, as is observed on heating milk at pH above 6.7 (Figure 8.7a; Anema and Li, 2003b). Similar changes in viscosity (Figure 8.7b) and turbidity (Figure 8.7c) with the pH at heating were also observed (Anema *et al.*, 2004c).

The difficulty in interpreting these changes in size, viscosity and turbidity is determining whether the association of the denatured whey proteins with the casein micelles is directly responsible for the change in size/volume of the casein micelles by increasing the diameters of the individual particles as the proteins interact, or whether there is some associated phenomenon, such as aggregation of the casein micelles, that is related to the level of whey protein or κ-CN that is in the serum phase or associated with the casein micelles. The strong relationship between the level of whey protein associating with the colloidal phase and the size/volume of the casein micelles, the observation that the size change plateaus on prolonged heating, and the relationships between the protein composition of the micelles and size, viscosity and turbidity seem

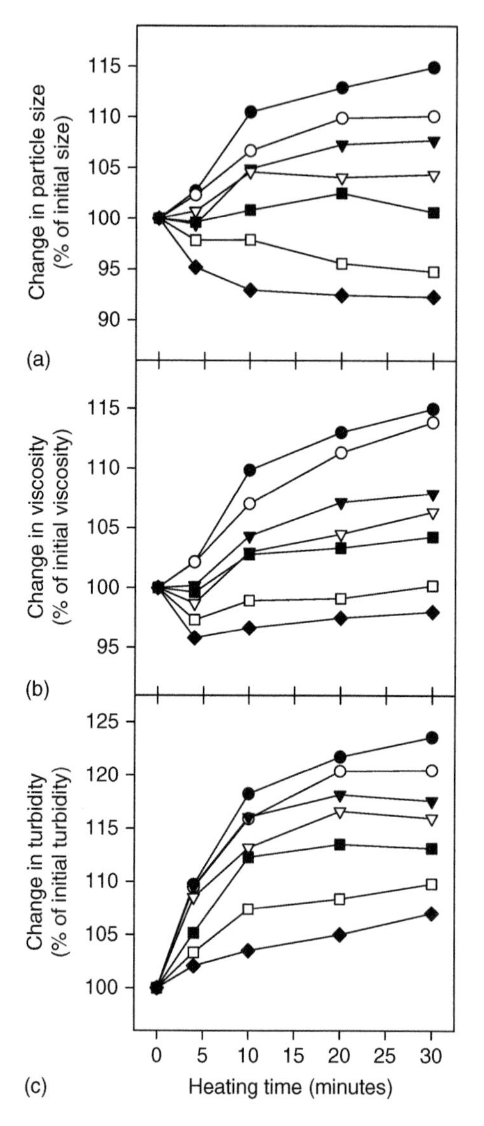

Figure 8.7 Effects of the pH at heating on the changes in the size of casein micelles (a), the viscosity of the milk (b) and the turbidity of the milk (c). The milk samples were heated at 90°C for various times and the pH values of the milk samples prior to heating were: ●: pH 6.5; ○: pH 6.55; ▼: pH 6.6; ▽: pH 6.65; ■: pH 6.7; □: pH 6.9; ◆: pH 7.1. Some of the particle size and viscosity results were adapted with permission from Anema *et al.* (2004c). Copyright (2004) Elsevier.

to suggest that the size changes are a direct consequence of the distribution of protein between the colloidal and serum phases, rather than an associated aggregation reaction (Anema and Li, 2003b; Anema *et al.*, 2004c).

Determination of the specific disulfide bonds formed between κ-CN and β-LG

Although many types of bonding may be involved in the early stages of interactions between the denatured whey proteins and κ-CN, there is clear evidence that disulfide bonds are involved in complex formation when model systems and milk are heated. Recent studies have focused on determining the specific thiol groups of κ-CN and, in particular, β-LG that are involved in the disulfide bonding between these two protein species when they are heated in model systems or milk. Understanding the specific disulfide bonds involved in the interaction process may provide useful insights into the mechanisms for the denaturation and subsequent aggregation reactions of the whey proteins in milk.

Native β-LG has two disulfide bonds and one free thiol group at Cys^{121} (Qin *et al.*, 1999), whereas κ-CN is found as a heterogeneous polymeric protein cross-linked by disulfide bonding via the two Cys groups in the monomer protein (Rasmussen *et al.*, 1999). It was believed that the formation of disulfide bonds between the denatured β-LG and other milk proteins, including κ-CN, during heating first involved the dissociation of the β-LG dimer to monomer species, followed by the unfolding of the native structure, exposing the buried side groups including the reactive free sulfydryl at Cys^{121}. The exposure of this free Cys^{121} then initiated a series of intermolecular thiol–disulfide exchange reactions with other denatured whey proteins and with κ-CN on the casein micelle surface (Snoeren and van der Spek, 1977; Iametti *et al.*, 1996; Hoffmann and van Mil, 1997; Verheul *et al.*, 1998; Vasbinder and de Kruif, 2003; Creamer *et al.*, 2004).

However, studies on pure β-LG indicated that, in the early stages of the denaturation process, non-native monomeric β-LG species are formed, which may be intermediates in the intermolecular aggregation processes. These non-native monomeric species are stable on subsequent cooling and can be separated by alkaline PAGE techniques (Manderson *et al.*, 1998; Hong and Creamer, 2002) or by gel permeation chromatography (Iametti *et al.*, 1996; Croguennec *et al.*, 2003, 2004). It was hypothesized that the non-native monomers were formed from intramolecular thiol–disulfide exchange reactions between the free Cys^{121} of β-LG and the Cys^{106}–Cys^{119} and/or Cys^{66}–Cys^{160} disulfide bonds within the same β-LG monomer (Iametti *et al.*, 1996; Manderson *et al.*, 1998; Hong and Creamer, 2002; Croguennec *et al.*, 2003, 2004).

With the advent of sensitive mass spectrometric techniques, the identification of the Cys residues involved in disulfide-bonded protein species was possible. The strategy for the identification of specific disulfide bonds involves a number of steps (GilleceCastro and Stults, 1996; Gorman *et al.*, 2002; Lowe *et al.*, 2004). For the sample under analysis, the protein species are first hydrolyzed under conditions where no further thiol–disulfide bond exchange reactions are likely to occur. The peptides formed from this hydrolysis are separated, usually by reverse phase HPLC, and the mass of individual peptides is determined by mass spectrometry (MS). The identification of individual peptides can be achieved by comparing the measured masses with those of expected peptides for the hydrolysis of the protein under study. For the disulfide-bonded peptides, usually other criteria also need to be satisfied, such as the peptides being present in non-reduced hydrolysates but absent in the reduced system.

Further confirmation can be gained by the use of tandem MS, where single molecular ions are isolated and analyzed in the first mass analyzer and then passed into a collision cell where fragmentation of the peptide is induced by collision with an inert gas (collision-induced dissociation [CID]) and the fragments are characterized in the second mass analyzer. From the mass of the fragments, the sequence of the amino acids in the peptides can be achieved, providing conclusive characterization of the peptides (Gorman *et al.*, 2002; Lowe *et al.*, 2004).

Using these types of mass spectrometric techniques, a stable non-native monomeric β-LG with a free sulfydryl group at position Cys^{119} rather than the natural position of Cys^{121} was found in heated β-LG solutions (Croguennec *et al.*, 2003), confirming that intramolecular thiol–disulfide exchange within monomeric β-LG could occur. It was suggested that this β-LG with the free thiol at Cys^{119} may be the activated monomer that was proposed as the starting point for intermolecular aggregation reactions leading to large polymers, although it was equally possible that unfolded protein with a free thiol at the natural position of Cys^{121} was what activated monomer (Croguennec *et al.*, 2003, 2004; Creamer *et al.*, 2004).

A more recent investigation on the disulfide bonding patterns in heated β-LG found that a significant proportion of Cys^{160} was in the reduced form after heating β-LG in solution, indicating that the Cys^{66}–Cys^{160} disulfide bond was broken during the early stages of heating, and that this may occur concurrently with the interchange of the free thiol from Cys^{121} to Cys^{119} (Creamer *et al.*, 2004). It was suggested that a monomeric β-LG species with a free thiol at Cys^{160} may be (one of) the reactive species involved in the intermolecular thiol–disulfide bonding responsible for cross-linking in heat-induced whey protein aggregates because of its position near the C-terminal end of the protein.

Attempts have been made to identify the specific Cys residues involved in disulfide bonds formed between κ-CN and β-LG when these proteins are heated together. Livney and Dalgleish (2004) compared masses of peptides from tryptic digests of heated κ-CN/β-LG mixtures with theoretical values and concluded that $Cys^{106/119/121}$ of β-LG were involved in disulfide bonds with both Cys^{11} and Cys^{88} of κ-CN (note that the hydrolysis pattern does not allow the separation of the three $Cys^{106/119/121}$ residues of β-LG unless CID is used for sequencing). Although some peptides involving Cys^{66} and Cys^{160} of β-LG and the two Cys residues of κ-CN were also identified based on mass comparisons, the high abundance of disulfide-bonded peptides containing $Cys^{106/119/121}$ led these authors to conclude that β-LG with a free thiol at $Cys^{119/121}$ was the predominant species that was involved in intermolecular disulfide bonding. The potential disulfide-bonded species were characterized based on mass analysis alone; no confirmatory experiments, such as comparing reduced with non-reduced systems to ensure that the proposed intermolecular peptides were disulfide bonded, or confirming sequences by CID–MS to preclude mis-identification of similarly massed peptides, were performed.

In a novel study, Lowe *et al.* (2004) used an activated monomeric κ-CN where the reduced thiol groups were blocked with thionitrobenzoate (TNB). β-LG was added to the mixture and the system was heated under very mild conditions (60°C). The TNB groups on the thiols of κ-CN are good leaving groups and, when a reactive thiol from β-LG is exposed, it is capable of interacting with the activated TNB groups on κ-CN in a specific 1:1 oxidative reaction forming a disulfide-bonded complex and releasing the TNB as a brightly colored compound. This approach allowed the formation of specific disulfide bonds between κ-CN and β-LG under mild heating conditions.

Because of the chemical nature of the reaction, it limited further thiol–disulfide exchange reactions, which allowed specific interactions between β-LG and κ-CN to be monitored during the early stages of the denaturation of β-LG.

The interacted β-LG–κ-CN complexes were hydrolyzed with trypsin and separated by reverse phase HPLC followed by MS. In addition, disulfide bonding was confirmed by comparing HPLC traces of non-reduced systems with those of reduced systems, and the identities of some peptides were confirmed by sequencing using CID–MS. Although it was possible to identify disulfide bonds between $Cys^{106/119/121}$ of β-LG and Cys^{88} of κ-CN, Cys^{160} of β-LG was found to have formed disulfide bonds with both Cys^{11} and Cys^{88} of κ-CN as major products (Figure 8.8). This supported the earlier findings on pure β-LG, that intramolecular thiol–disulfide exchange may precede the intermolecular reactions and that the non-native monomeric form of β-LG species with a free thiol at Cys^{160} is likely to be (one of) the reactive monomer species that initiates intermolecular thiol–disulfide exchange reactions (Creamer *et al.*, 2004).

Lowe *et al.* (2004), using the techniques developed for the model system, expanded the study to examine the specific disulfide bonds involved in aggregation between β-LG and κ-CN in heated milk systems. Interestingly, no disulfide bonds between $Cys^{106/119/121}$ of β-LG and the two Cys residues of κ-CN could be found, even though the disulfide bond between Cys^{88} of κ-CN and $Cys^{106/119/121}$ of β-LG was readily identified in the model system. In the heated milk system, it was found that Cys^{160} of β-LG formed disulfide bonds with both Cys^{88} and Cys^{11} of κ-CN, as was found in the model system (Figure 8.8). In independent studies, a similar disulfide bond between Cys^{160} of β-LG and Cys^{88} of κ-CN was identified in a heated model goat milk system consisting of isolated casein micelles and β-LG in milk ultrafiltrate, although it appears that no attempts were made to isolate and characterize other intermolecular disulfide bonds between these protein species (Henry *et al.*, 2002).

From these observations, it was concluded that, in the model system of β-LG and activated κ-CN, non-native monomeric β-LG species with a free thiol and either Cys^{119} or Cys^{121} (but probably not Cys^{106}) could be the reactive monomer that is involved in intermolecular thiol–disulfide exchange reactions, as $Cys^{119/121}$ was involved in disulfide bonds with κ-CN. However, further intramolecular thiol–disulfide exchange reactions in heated β-LG must precede or occur concurrently with the intermolecular reactions, as disulfide bonds between Cys^{160} of β-LG and the two Cys residues of κ-CN were also observed as major products in the model system (Figure 8.8).

In the heated milk system, no peptides involving $Cys^{106/119/121}$ and the two Cys residues of κ-CN were isolated and only peptides involving Cys^{160} and Cys^{66} with both Cys residues of κ-CN were found (Figure 8.8). As Cys^{160} and Cys^{66} are involved in a disulfide bond in native β-LG, this indicates that intramolecular thiol–disulfide exchange reactions in β-LG precede the intermolecular thiol–disulfide exchange reactions and that a β-LG (monomeric) species with a free thiol group at Cys^{160} may play a significant role in the inter-protein disulfide bonding that occurs in heat-induced milk or whey protein systems.

The differences in reaction products between the model systems and milk may be a consequence of factors such as the heating conditions, the nature of the reactions

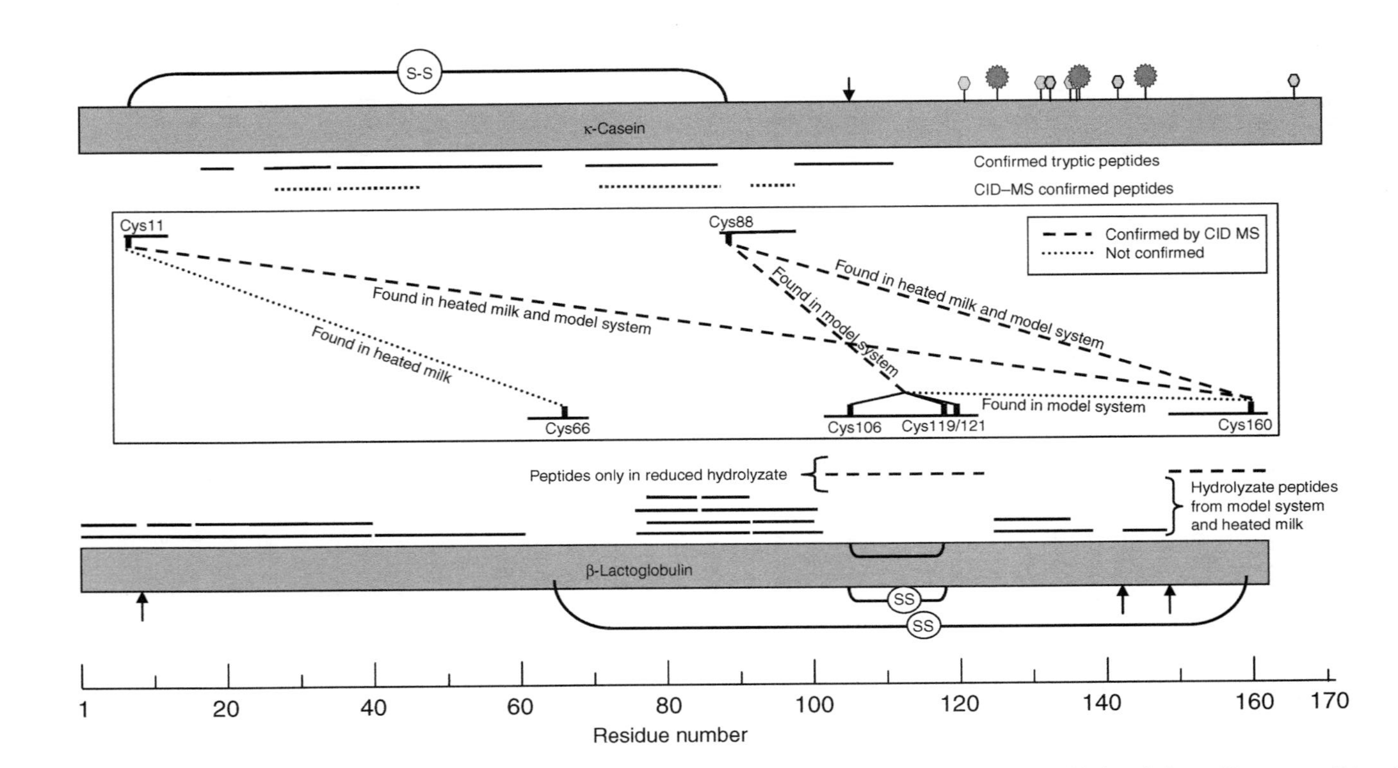

Figure 8.8 Diagram indicating the identified peptides on the linear sequences of κ-CN and β-LG and the intermolecular disulfide bonds formed between κ-CN and β-LG on heating model systems and milk. The horizontal box lines represent the protein sequence, the lines over the boxes represent the peptides and S–S indicates the presence of a disulfide bond. Arrows indicate the major proteolytic sites, the chymosin site for κ-CN and the rapid tryptic sites for β-LG. Potential glycosylation and phosphorylation sites are indicated for κ-CN. Reproduced with permission from Lowe *et al.* (2004). Copyright (2004) American Chemical Society.

(oxidative interaction compared with thiol–disulfide interchange reactions) and the fact that the κ-CN in milk is found within the casein micelles whereas in the model system it is not (Lowe *et al.*, 2004). Because of the C-terminal location of Cys^{160} in β-LG, when this Cys is in the free thiol form and not linked to Cys^{66}, it may be able to productively react with the disulfide bonds of κ-CN to give stable κ-CN–whey protein aggregates (Lowe *et al.*, 2004).

Relationships between denaturation/interactions of the whey proteins in heated milk and the functional properties of milk products

When milk is heated, there are numerous changes to the milk system, including changes to the proteins, the milk salts (including mineral equilibria between the colloidal and serum phases) and lactose; many of the changes can involve more than one of the milk constituents (International Dairy Federation, 1995). The changes can be irreversible or reversible to various extents depending on the changes being monitored and the conditions of the heat treatment. Although the changes to the protein system are an important determinant of the functional properties of milk products, all other changes to the milk system should also be considered to obtain a full understanding of the relationship between heat treatments, interactions and functional performance. However, there are limited examples of changes to components other than the proteins and the functional behavior of milk products, and therefore this review is restricted to some examples of the relationships between the changes in the milk protein system and the functional performance of the milk.

Examples of the relationships between whey protein denaturation and the functional properties of milk

In the early days of milk powder manufacture, it was recognized that the level of whey protein denaturation could be used as an index for the extent of heat treatment the milk had received during the manufacture of the milk powders, and that the functional properties of the milk products were related to some extent to the heat treatment that the milk had received during processing and therefore the level of whey protein denaturation (Harland and Ashworth, 1947; Larson *et al.*, 1951; Harland *et al.*, 1952).

Even as early as 1952, the concept of "tailor-made" milk powders was discussed, where powders were processed to provide specific requirements, such as low-heat powders for beverage applications and cottage cheese manufacture, and high-heat powders for bakery applications (Harland *et al.*, 1952). Although there were no standards of quality for processing at this time, it was recognized that the proper control of processing conditions, particularly preheating of the milk, was necessary to produce satisfactory products and that measurement of the level of whey protein denaturation could be used as an objective method for determining the suitability of milk (powder) products for particular commercial and functional applications.

The heat treatment of milk, whether in liquid milk applications or prior to drying for milk powder manufacture, remains one of the major processes for manipulating the functional properties of milk products, and products such as milk powders are still generally classified according to the heat treatments applied using one of the derivatives of the WPNI test (Singh and Newstead, 1992; Kelly *et al.*, 2003). With the extensive research on the denaturation of the whey proteins, and the ability to predict the denaturation levels after defined heat treatments, it would be envisaged that the level of denaturation of the whey proteins could be used as an indicator of the functional properties of milk products. In a broad sense, this is true. For example, certain heat classifications of milk powders will give improved functionality for particular applications over other classes of milk powders. Some of the general applications of different heat-classified milks and their functional uses, in particular for milk powder products, have been summarized in numerous publications (Singh and Newstead, 1992; International Dairy Federation, 1995, 1996; Kelly *et al.*, 2003).

However, a huge range of temperature and heating time combinations are available to denature the whey proteins when milk is heated. As a consequence, specific correlations between the level of whey protein denaturation and the functional properties of milk across all possible heating conditions and milk sources do not exist. For example, the WPNI method was developed for assessing the suitability of milk powders for use in bakery applications; however, it was noted that a powder with a low WPNI did not always correspond to good baking qualities (Harland and Ashworth, 1947). Some of these variations are due to factors such as natural variations in the initial whey protein levels in the milk (Harland *et al.*, 1955; Sanderson, 1970b); however, others are due to the methods of heat treatment during milk processing. As such, the WPNI or level of whey protein denaturation is at best a guide for the suitability of powders for specific applications, or an in-factory guide on processing conditions. Many manufacturers impose additional specifications to the milk powders to ensure suitability in their specific applications (Sanderson, 1970c; Singh and Creamer, 1991a).

Some of the most detailed studies on the relationship between the functional performance of milk and the heat treatment conditions or whey protein denaturation levels have been reported for acid gel or yoghurt systems. Parnell-Clunies *et al.* (1986) showed correlations between the level of whey protein denaturation and the firmness and apparent viscosity of yoghurt, regardless of the method used to heat the milk (batch [85°C], high-temperature short-time [98°C] and UHT [140°C] heating systems for different holding times). However, other properties, such as water-holding capacity/syneresis, were more dependent on the heating system used and it was concluded that high levels of whey protein denaturation in milk were not necessarily associated with an improved water-holding capacity in yoghurt.

In extensive studies, Dannenberg and Kessler (1988a, 1988b) examined the relationship between the denaturation level of whey proteins in milk and the functional performance (firmness, flow properties and syneresis) of the milk in set yoghurt applications. There was a clear relationship between the level of whey protein denaturation in the milk and the firmness, with a higher firmness at higher levels of whey protein denaturation. Similar results were obtained for the flow properties of the yoghurt. However, very high levels of whey protein denaturation appeared to be detrimental, with a decrease in the firmness and flow properties at denaturation levels above about 95% (Dannenberg and Kessler, 1988b).

For syneresis of the yoghurt, a negative relationship between the level of whey protein denaturation in the milk and the level of serum expelled from the yoghurt was observed (Dannenberg and Kessler, 1988a). Despite the apparent correlations between denaturation and firmness, flow properties and syneresis, there were significant variations at each denaturation level, indicating that the temperature of heating used to denature the whey proteins, rather than just the whey protein denaturation level, may be an important factor in determining the functional performance in acid gels.

In a study on reconstituted whole milk, McKenna and Anema (1993) also observed a positive correlation between the denaturation of the whey proteins in the milk and the firmness of the yoghurt made from the milk regardless of whether the heat treatment was performed before or after powder manufacture (Figure 8.9a). However, when individual heating conditions were examined, it was also noted that excessive heat treatment/denaturation of the milk could be detrimental to the firmness of the set yoghurt (McKenna and Anema, 1993). A less clear relationship between syneresis and the level

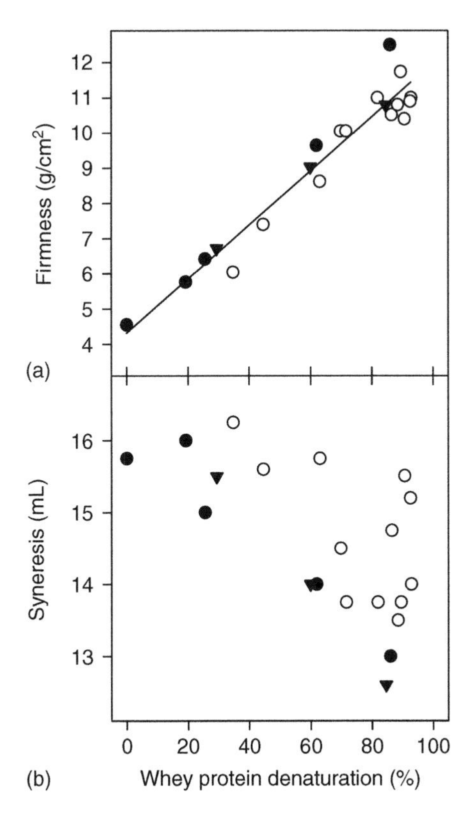

Figure 8.9 Relationship between the level of whey protein denaturation in reconstituted whole milk and the firmness (a) and syneresis (b) of acid gels prepared from the heated milks. The milks were heated only before powder manufacture (●), heated only after reconstitution (▼) or heated both before powder manufacture and after reconstitution (○). Adapted with permission from the results of McKenna and Anema (1993). Copyright (1993) International Dairy Federation (http://www.fil-idf.org).

of whey protein denaturation was observed, with the level of syneresis appearing to have a greater dependence on the heating conditions (temperature, time and before/after reconstitution) than on the level of denaturation itself (Figure 8.9b; McKenna and Anema, 1993), which supports the findings of Parnell-Clunies *et al.* (1986).

Examples of the relationships between the level of interactions of whey proteins with κ-CN/casein micelles and the functional properties of milk

A major limitation in using whey protein denaturation as an index of the functional properties of milk is that it does not consider the subsequent interaction reactions of the denatured whey proteins. These interactions will be dependent on the conditions of denaturation such as temperature and time as well as on the properties of the milk such as pH, concentration and composition. It is considerably more complex to investigate the subsequent aggregation reactions, as there are potentially numerous pathways and there is great difficulty in isolating and characterizing the specific reaction products. However, in recent years, some effort has been made in identifying the interaction reactions of the denatured whey proteins with other components in milk and in some cases their effects on the functional properties of the milk.

Anema *et al.* (2004a) showed that small changes in pH from the natural pH of milk at the time of heating markedly affected the properties of acid gels prepared from these heated milks. During acidification of the heated pH-adjusted milks, the acid gelation curves were progressively shifted to higher firmness as the pH at heating was increased (Figure 8.10a) so that the final firmness of the acid gels (at pH 4.2) was almost doubled as the pH at heating was increased from pH 6.5 to pH 7.1 (Figure 8.10). The effect was particularly pronounced in the milks that were heated for times sufficient to fully denature the whey proteins (Figure 8.10b). This effect of small changes in the pH of the milk at the time of heating on the firmness of acid gels prepared from the heated milk has been independently confirmed (Lakemond and van Vliet, 2005; Rodriguez del Angel and Dalgleish, 2006).

In addition to influencing the final firmness of the acid gels, the pH at the heat treatment of the milk also influenced the pH at which the milk started gelling/aggregating during acidification (Vasbinder and de Kruif, 2003; Anema *et al.*, 2004a, 2004b; Lakemond and van Vliet, 2005; Rodriguez del Angel and Dalgleish, 2006). On subsequent acidification of milk samples that were heated over a pH range from pH 6.5 to 7.1, those samples heated at higher pH (pH 7.1) started gelling at significantly higher pH on acidification than those samples heated at a lower pH. These effects were very dependent on the temperature at which the milks were acidified (Anema *et al.*, 2004b).

The changes in acid gel firmness on changing the pH at heating of the milk could not be related solely to the level of whey protein denaturation (Figure 8.11a). Small changes in the pH of the milk before heating markedly affect the distribution of the denatured whey proteins and κ-CN between the colloidal and serum phases (Figure 8.5; Anema and Li, 2003a, 2003b; Vasbinder and de Kruif, 2003; Rodriguez del Angel and Dalgleish, 2006). Although heating milk prior to acidification markedly

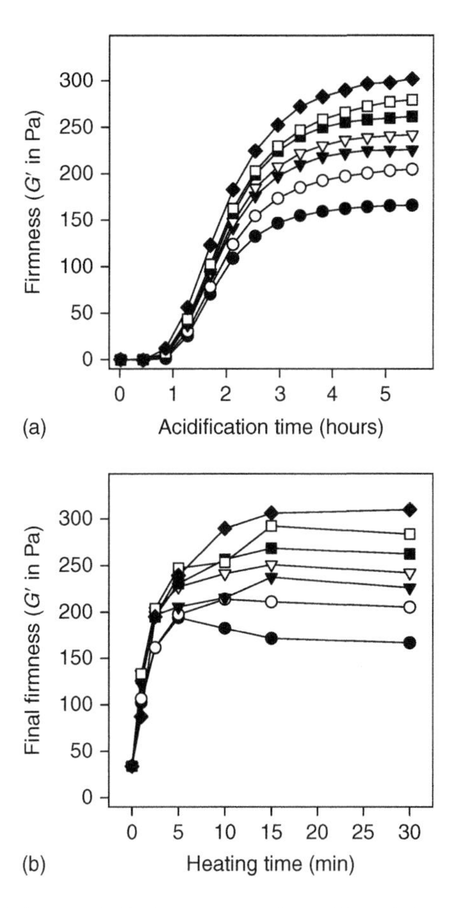

Figure 8.10 (a) Changes in firmness (G') with time after the addition of glucono-δ-lactone (GDL) for heated (90°C/30 min) skim milk samples. (b) Changes in the final firmness (final G') for acid gels prepared from milk samples heated for various times at 90°C. The pH values of the milk samples prior to heating were: ●: pH 6.5; ○: pH 6.55; ▼: pH 6.6; ▽: pH 6.65; ■: pH 6.7; □: pH 6.9; ◆: pH 7.1. In all samples, the pH was readjusted back to pH 6.7 before addition of GDL and this pH reduced to ≈pH 4.2 after 5.5 h. Reproduced with permission from Anema *et al.* (2004a). Copyright (2004) American Chemical Society.

increased the firmness of the acid gels (i.e. acid gels prepared from heated milks always had a considerably higher firmness than acid gels prepared from unheated milks [Dannenberg and Kessler, 1988b; Lucey *et al.*, 1997; Lucey and Singh, 1998]), the distribution of the denatured whey proteins and κ-CN between the colloidal and serum phases also appeared to influence the firmness of the acid gels. When the final firmness of the acid gels was plotted against the level of non-sedimentable denatured whey proteins in the milk, the results for all pH values were close to a single line (Figure 8.11b).

Anema *et al.* (2004a) concluded that, although the denatured whey proteins associated with the micelles have a significant effect on the final firmness of the acid gels, those denatured whey proteins that remain in the serum appear to have a more dominant influence over the final firmness than those associated with the casein micelles. For samples where virtually all the whey proteins were denatured, the final gel

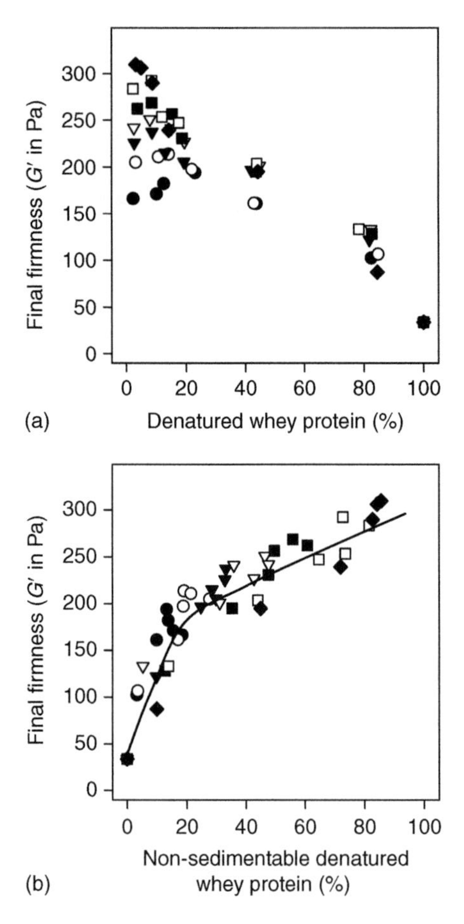

Figure 8.11 (a) Relationship between the level of total denatured whey protein in heated skim milk and the final firmness (final G') for acid gels prepared from the heated milks. (b) Relationship between the level of non-sedimentable denatured whey proteins in the heated skim milk samples and the final firmness (final G') for acid gels prepared from the heated milks. The pH values of the milk samples prior to heating were: ●: pH 6.5; ○: pH 6.55; ▼: pH 6.6; ▽: pH 6.65; ■: pH 6.7; □: pH 6.9; ◆: pH 7.1. Reproduced with permission from Anema *et al.* (2004a). Copyright (2004) American Chemical Society.

strength for acid gels prepared from milks in which all the denatured whey proteins were in the serum phase was found to be essentially a factor of two higher than that for acid gels prepared from milks in which all the whey proteins were associated with the casein micelles (Figure 8.12; Anema *et al.*, 2004a).

Rodriguez del Angel and Dalgleish (2006) separated the non-sedimentable whey protein–κ-CN aggregates from milks heated at different pH using size exclusion chromatography and related the peak area of these aggregates to the firmness of the acid gels. They also concluded that the gel firmness appeared to be strongly dependent on the formation of soluble complexes in the milks and that there appeared to be a linear relationship between the level of soluble aggregates in the heated milk and the final strength of the acid gels.

Based on these results, a hypothesis on the roles of the non-sedimentable and micelle-bound denatured whey protein–κ-CN has been developed (Anema *et al.*,

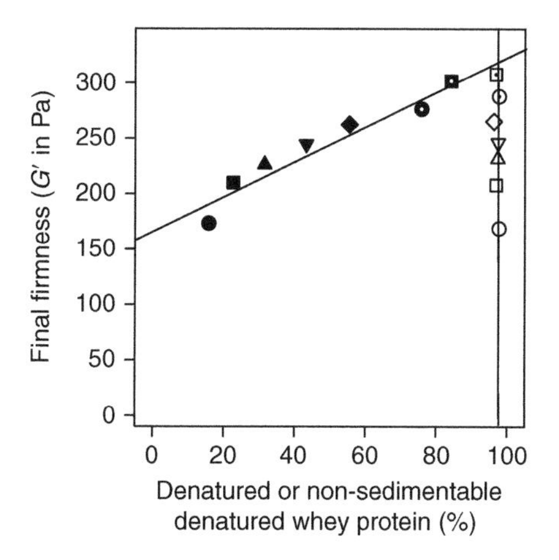

Figure 8.12 Comparison between the final firmness (final G') and the level of denatured whey protein (open symbols) and the level of soluble denatured whey protein (filled symbols) for acid gels prepared from heated (90°C/30 min) skim milk samples. The pH values at heating of the milks were: ●, ○: pH 6.5; ■, □: pH 6.55; ▲, △: pH 6.6; ▼, ▽: pH 6.65; ◆, ◇: pH 6.7; dotted, ●, ○: pH 6.9; dotted, ■, □: pH 7.1. Adapted with permission from Anema *et al.* (2004a). Copyright (2004) American Chemical Society.

2004a, 2004b; Rodriguez del Angel and Dalgleish, 2006; Donato *et al.*, 2007). The increased pH of gelation and the increased acid gel strength of heated milk when compared with unheated milk has been attributed to the incorporation of the whey proteins as well as casein (micelles) in the acid gel structure during the acidification of milk (Lucey *et al.*, 1997; Lucey, 2002; Graveland-Bikker and Anema, 2003).

In milk, the casein is insoluble at its isoelectric point (≈pH 4.6), whereas the native whey proteins remain soluble at all pH. Therefore, unheated milk starts aggregating when the milk pH approaches the isoelectric point of casein and visible gelation is observed at about pH 4.9. However, for heated milk, the denatured whey proteins are insoluble at their isoelectric points (≈pH 5.3 for β-LG, the major whey protein). Therefore, on acidification of heated milk, the proteins will start aggregating at a much higher pH, closer to the isoelectric points of the whey proteins. As a consequence, the contribution of the denatured whey proteins to the acid gel structure and the firmness of the acid gels is markedly higher than that observed for unheated milk (Lucey *et al.*, 1997; Graveland-Bikker and Anema, 2003).

The pH at heating the milk will produce casein micelle particles with markedly different compositions (Figures 8.5 and 8.6; Anema and Li, 2003b; Vasbinder and de Kruif, 2003). The isoelectric point of the casein micelles is ≈pH 4.6, whereas the isoelectric point of the whey proteins is about pH 5.3. Therefore, on the acidification of milks heated at different pH, different aggregation and gelation behavior is observed. For the milks heated at high pH, the serum phase denatured whey proteins/κ-CN can aggregate separately and at a higher pH than the casein micelles. As the isoelectric point of these serum phase protein components will be higher than that of the casein micelles, the pH at which aggregation occurs will be progressively shifted to higher pH as the heating pH and the concentration of the serum phase denatured whey proteins/κ-CN are increased (Anema *et al.*, 2004a; Rodriguez del Angel and Dalgleish, 2006).

In addition, the dissociation of κ-CN from the casein micelles may also contribute to the higher aggregation pH as the pH at heating is increased, particularly above pH 6.7. Lower levels of κ-CN on the micelles will reduce the density of the surface hairy layer. This may cause the surface hairy layer to collapse at a higher pH, or this layer may have a reduced efficiency in stabilizing the casein micelles. Either effect will allow the κ-CN-depleted micelles to aggregate at a pH that is markedly higher than that observed for the native casein micelles or for casein micelles in milk heated at a lower pH.

The firmness of acid gels can be related to the number and properties of the contact points between the protein components in the acid gel (van Vliet and Keetals, 1995; Lucey *et al.*, 1997). As the pH at heating of the milk is increased, the level of serum phase denatured whey protein–κ-CN complexes increases and therefore there are a greater number of particles to aggregate during the subsequent acidification to form the acid gels. There is also the potential for the formation of a more complex acid gel structure when the milk is heated at high pH, where there are high levels of serum phase denatured whey protein–κ-CN complexes, than when the milk is heated at low pH, where most of the denatured whey protein and κ-CN are associated with the casein micelles. In the latter case, the acid gel process will probably involve only entire whey protein–casein micelle complexes. Therefore, there may be fewer contact points in the acid gels formed from milk with the denatured whey proteins associated with the micelles than in those formed from milk with soluble denatured whey proteins and hence a gel with a lower firmness is observed (Anema *et al.*, 2004a, 2004b).

The large strain deformation properties also gave some indication of the types of bonds involved in the acid gel network. In these experiments, the strain was increased at a constant rate and the stress was monitored until the gel structure yielded and the stress decreased. The maximum in the strain versus stress curves was considered to be the point at which the gel structure broke (Figure 8.13). As the pH at heating was increased, the breaking stress of the acid gels prepared from the heated milks was found to increase markedly; however, the breaking strain was virtually unchanged.

For a gel to break on increasing the strain, the strands within the gel network are first straightened and then stretched until rupture of the strands or the bonds within (van Vliet and Keetals, 1995; Mellema *et al.*, 2002). Therefore, the breaking strain is dependent on factors such as the degree of curvature of the strands, with a higher breaking strain with higher strand curvature. As the breaking strain of the acid gels did not change with the pH at heating of the milk, despite the marked change in final firmness (Figures 8.10 and 8.12), this indicates that the relative curvature of the individual strands within the gel network was the same for all acid gel samples.

The types of bonds involved in the acid gel network will have an influence on the breaking stress (Mellema *et al.*, 2002). The breaking of strands containing covalent bonds would require a greater force than the breaking of strands held together by non-covalent bonds, as covalent bonds have higher bond energies. Therefore, a change in the number or distribution of covalent bonds within the gel network may explain the differences in breaking stress as the pH of the milk at heating was changed (Figure 8.13). It seems unlikely that the difference in breaking stress can be due to a greater degree of disulfide bonding within the gelled sample; although there may be continuing thiol–disulfide exchange reactions occurring during acidification (Vasbinder

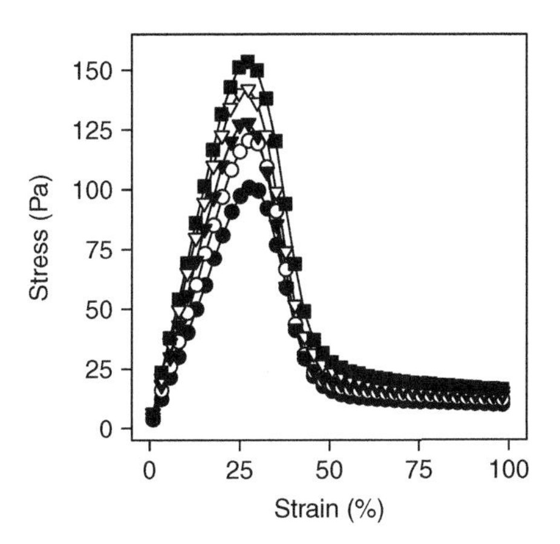

Figure 8.13 Stress versus strain curves for acid gels prepared from heated (90°C/15 min) skim milk samples. The pH values of the milk samples prior to heating were: ●: pH 6.5; ○: pH 6.6; ▼: pH 6.7; ▽: pH 6.9; ■: pH 6.7; □: pH 6.9; ◆: pH 7.1. The maximum in the stress represents the breaking point of the gel and the stress and strain at this point are considered to be the breaking stress and breaking strain respectively.

et al., 2003), the physical number of disulfide bonds is unlikely to be markedly different between the samples.

The denatured whey proteins, along with some of the κ-CN, are progressively transferred to the serum phase when the pH of the milk is increased before heating (Figure 8.5). As these interactions involve disulfide bonding, this indicates that the interaction between the denatured whey proteins and κ-CN is transferred from the colloidal phase (casein micelle) to the serum phase as the pH of the milk at heating is increased. On subsequent acidification, both non-sedimentable and colloidal phase denatured whey proteins are incorporated in the acid gel structure. The non-sedimentable denatured whey protein–κ-CN complexes can form strands that may be involved in interconnecting the colloidal particles. As the non-sedimentable aggregates are disulfide bonded, those samples heated at high pH and with high levels of non-sedimentable whey protein–κ-CN aggregates will have a greater number of these strands interconnecting the residual casein micelles.

In contrast, the samples heated at lower pH will have the denatured whey proteins predominantly associated with the casein micelles and therefore fewer of the whey protein–κ-CN aggregates interconnecting the colloidal particles. Therefore, the samples heated at higher pH may have a greater number of disulfide bonds interconnecting the colloidal particles and therefore a higher breaking stress whereas, for the samples heated at lower pH, most of the disulfide bonds are on the colloidal particles and fewer disulfide bonds interconnect the colloidal particles, which may explain the lower breaking stress (Figure 8.13).

Examples of the effect of denaturing whey proteins separately from casein micelles on the functional properties of milk

Interestingly, different effects are observed when the whey proteins are denatured and aggregated separately from the casein micelles than when the milk is heated.

For example, Lucey *et al.* (1998) showed that acid gels prepared from milk samples where the whey proteins were denatured in the presence of casein micelles had a markedly higher firmness than acid gels prepared from milk samples where the whey proteins were pre-denatured and added back to the casein micelles. In fact, in many cases, the samples with denatured whey proteins added back to the casein micelles produced acid gels with firmness similar to or only slightly higher than those prepared from unheated milks.

In a separate study, Schorsch *et al.* (2001) prepared model milk systems in which the whey proteins were either heated in the presence of casein micelles or heated separately and added back to the casein micelles. Acid gels were prepared from these model milk systems. It was shown that the acid-induced gelation occurred at a higher pH and in a shorter time when the whey proteins were denatured separately from the casein micelles than when the whey proteins were heated in the presence of casein micelles. However, the gels formed were weaker and more heterogeneous because of the particulate nature of the denatured whey proteins.

It was suggested that the large denatured whey protein aggregates, as formed when the whey proteins were heated separately from the casein micelles, hinder the formation of a casein gel network when the milk is subsequently acidified and that a weak acid gel with a heterogeneous structure results. When the whey proteins are heated in the presence of the casein micelles, the denatured whey proteins interact with the κ-CN at the casein micelle surface and, on subsequent acidification, the denatured whey protein–casein micelle complexes aggregate to form a firmer acid gel with a more homogeneous structure (Schorsch *et al.*, 2001). This proposal is supported by early studies, which showed that the aggregation of the denatured whey proteins, and in particular β-LG, formed large aggregate species when heated in the absence of κ-CN, whereas aggregation was limited when the whey proteins were heated in the presence of κ-CN (McKenzie *et al.*, 1971).

Conclusions

A considerable amount of work has gone into understanding the irreversible denaturation reactions of the whey proteins in heated milk systems. These detailed studies have produced models that allow reasonably accurate prediction of the level of whey protein denaturation in milks under a wide range of heating conditions, even in milk samples with markedly modified concentrations and compositions. However, with a few exceptions, monitoring of the whey protein denaturation levels provides only a crude indication of the functionality of the milk system. As a consequence, more recent research efforts have focused on trying to understand the specific interaction reactions of the denatured whey proteins with other proteins in the milk system. Early indications suggest that these types of studies on the interactions of denatured whey proteins may provide greater insights into the functional properties of heated milk products than can be obtained by monitoring just whey protein denaturation levels.

These initial studies on protein interactions have been conducted under relatively narrowly defined conditions (temperatures, heating times, pH, milk concentrations and milk compositions). It is likely that changes to these variables will markedly influence the interaction behavior and will explain the changes in functional behavior when the heating conditions are changed (even though the whey protein denaturation levels may be similar). Although studies on understanding the specific interactions between milk proteins, particularly in complex systems such as milk, are extremely difficult, these types of studies should continue to give useful insights into the behavior of milk proteins during heating and the functional behavior of the heated milk products.

References

Anema, S. G. (1998). Effect of milk concentration on heat-induced, pH-dependent dissociation of casein from micelles in reconstituted skim milk at temperatures between 20 and 120°C. *Journal of Agricultural and Food Chemistry*, **46**, 2299–305.

Anema, S. G. (2000). Effect of milk concentration on the irreversible thermal denaturation and disulfide aggregation of β-lactoglobulin. *Journal of Agricultural and Food Chemistry*, **48**, 4168–75.

Anema, S. G. (2001). Kinetics of the irreversible thermal denaturation and disulfide aggregation of α-lactalbumin in milk samples of various concentrations. *Journal of Food Science*, **66**, 2–9.

Anema, S. G. (2007). Role of κ-casein in the association of denatured whey proteins with casein micelles in heated reconstituted skim milk. *Journal of Agricultural and Food Chemistry*, **55**, 3635–42.

Anema, S. G. and Klostermeyer, H. (1997). Heat-induced, pH-dependent dissociation of casein micelles on heating reconstituted skim milk at temperatures below 100°C. *Journal of Agricultural and Food Chemistry*, **45**, 1108–15.

Anema, S. G. and Li, Y. (2000). Further studies on the heat-induced, pH-dependent dissociation of casein from the micelles in reconstituted skim milk. *Lebensmittel-Wissenschaft und -Technologie*, **33**, 335–43.

Anema, S. G. and Li, Y. (2003a). Association of denatured whey proteins with casein micelles in heated reconstituted skim milk and its effect on casein micelle size. *Journal of Dairy Research*, **70**, 73–83.

Anema, S. G. and Li, Y. (2003b). Effect of pH on the association of denatured whey proteins with the casein micelles in heated reconstituted skim milk. *Journal of Agricultural and Food Chemistry*, **51**, 1640–46.

Anema, S. G. and Lloyd, R. J. (1999). Analysis of whey protein denaturation: a comparative study of alternative methods. *Milchwissenschaft*, **54**, 206–10.

Anema, S. G. and McKenna, A. B. (1996). Reaction kinetics of thermal denaturation of whey proteins in heated reconstituted whole milk. *Journal of Agricultural and Food Chemistry*, **44**, 422–28.

Anema, S. G., Lee, S. K., Lowe, E. K. and Klostermeyer, H. (2004a). Rheological properties of acid gels prepared from heated pH-adjusted skim milk. *Journal of Agricultural and Food Chemistry*, **52**, 337–43.

Anema, S. G., Lowe, E. K. and Lee, S. K. (2004b). Effect of pH at heating on the acid-induced aggregation of casein micelles in reconstituted skim milk. *Lebensmittel-Wissenschaft und -Technologie*, **37**, 779–87.

Anema, S. G., Lowe, E. K. and Li, Y. (2004c). Effect of pH on the viscosity of heated reconstituted skim milk. *International Dairy Journal*, **14**, 541–48.

Anema, S. G., Lee, S. K. and Klostermeyer, H. (2006). Effect of protein, nonprotein-soluble components, and lactose concentrations on the irreversible thermal denaturation of β-lactoglobulin and α-lactalbumin in skim milk. *Journal of Agricultural and Food Chemistry*, **54**, 7339–48.

Baer, A., Oroz, M. and Blanc, B. (1976). Serological studies on heat-induced interactions of α-lactalbumin and milk proteins. *Journal of Dairy Research*, **43**, 419–32.

Butikofer, U., Meyer, J. and Rehberger, B. (2006). Determination of the percentage of α-lactalbumin and β-lactoglobulin of total milk protein in raw and heat treated skim milk. *Milchwissenschaft*, **61**, 263–66.

Cho, Y., Singh, H. and Creamer, L. K. (2003). Heat-induced interactions of β-lactoglobulin A and κ-casein B in a model system. *Journal of Dairy Research*, **70**, 61–71.

Corredig, M. and Dalgleish, D. G. (1996a). The binding of α-lactalbumin and β-lactoglobulin to casein micelles in milk treated by different heating systems. *Milchwissenschaft*, **51**, 123–27.

Corredig, M. and Dalgleish, D. G. (1996b). Effect of temperature and pH on the interactions of whey proteins with casein micelles in skim milk. *Food Research International*, **29**, 49–55.

Corredig, M. and Dalgleish, D. G. (1999). The mechanisms of the heat-induced interaction of whey proteins with casein micelles in milk. *International Dairy Journal*, **9**, 233–36.

Creamer, L. K. and Matheson, A. R. (1980). Effect of heat-treatment on the proteins of pasteurized skim milk. *New Zealand Journal of Dairy Science and Technology*, **15**, 37–49.

Creamer, L. K., Berry, G. P. and Matheson, A. R. (1978). Effect of pH on protein aggregation in heated skim milk. *New Zealand Journal of Dairy Science and Technology*, **13**, 9–15.

Creamer, L. K., Bienvenue, A., Nilsson, H., Paulsson, M., van Wanroij, M., Lowe, E. K., Anema, S. G., Boland, M. J. and Jimenez-Flores, R. (2004). Heat-induced redistribution of disulfide bonds in milk proteins. 1. Bovine β-lactoglobulin. *Journal of Agricultural and Food Chemistry*, **52**, 7660–68.

Croguennec, T., Bouhallab, S., Molle, D., O'Kennedy, B. T. and Mehra, R. (2003). Stable monomeric intermediate with exposed Cys–119 is formed during heat denaturation of β-lactoglobulin. *Biochemical and Biophysical Research Communications*, **301**, 465–71.

Croguennec, T., Molle, D., Mehra, R. and Bouhallab, S. (2004). Spectroscopic characterization of heat-induced nonnative β-lactoglobulin monomers. *Protein Science*, **13**, 1340–46.

Dannenberg, F. and Kessler, H. G. (1988a). Effect of denaturation of β-lactoglobulin on texture properties of set-style nonfat yoghurt. 1. Syneresis. *Milchwissenschaft*, **43**, 632–35.

Dannenberg, F. and Kessler, H. G. (1988b). Effect of denaturation of β-lactoglobulin on texture properties of set-style nonfat yoghurt. 2. Firmness and flow properties. *Milchwissenschaft*, **43**, 700–4.

Dannenberg, F. and Kessler, H. G. (1988c). Reaction kinetics of the denaturation of whey proteins in milk. *Journal of Food Science*, **53**, 258–63.

Doi, H., Tokuyama, T., Kuo, F. H., Ibuki, F. and Kanamori, M. (1983). Heat-induced complex-formation between κ-casein and α-lactalbumin. *Agricultural and Biological Chemistry*, **47**, 2817–24.

Donato, L. and Dalgleish, D. G. (2006). Effect of the pH of heating on the qualitative and quantitative compositions of the sera of reconstituted skim milks and on the mechanisms of formation of soluble aggregates. *Journal of Agricultural and Food Chemistry*, **54**, 7804–11.

Donato, L., Alexander, M. and Dalgleish, D. G. (2007). Acid gelation in heated and unheated milks: interactions between serum protein complexes and the surfaces of casein micelles. *Journal of Agricultural and Food Chemistry*, **55**, 4160–68.

Elfagm, A. A. and Wheelock, J. V. (1978). Heat interaction between α-lactalbumin, β-lactoglobulin and casein in bovine milk. *Journal of Dairy Science*, **61**, 159–63.

Euber, J. R. and Brunner, J. R. (1982). Interaction of κ-casein with immobilized β-lactoglobulin. *Journal of Dairy Science*, **65**, 2384–87.

Fairise, J. F. and Cayot, P. (1998). New ultrarapid method for the separation of milk proteins by capillary electrophoresis. *Journal of Agricultural and Food Chemistry*, **46**, 2628–33.

Fox, P. F. (2003). Milk proteins: general and historical aspects. In *Advanced Dairy Chemistry,* Volume 1, *Proteins*, 3rd edn (P. F. Fox and P. L. H. McSweeney, eds) pp. 1–48. New York: Kluwer Academic/Plenum Publishers.

GilleceCastro, B. L. and Stults, J. T. (1996). Peptide characterization by mass spectrometry. *High Resolution Separation and Analysis of Biological Macromolecules, Pt B*, **271**, 427–48.

Gorman, J. J., Wallis, T. P. and Pitt, J. J. (2002). Protein disulfide bond determination by mass spectrometry. *Mass Spectrometry Reviews*, **21**, 183–216.

Gough, P. and Jenness, R. J. (1962). Heat denaturation of β-lactoglobulins A and B. *Journal of Dairy Science*, **45**, 1033–39.

Graveland-Bikker, J. F. and Anema, S. G. (2003). Effect of individual whey proteins on the rheological properties of acid gels prepared from heated skim milk. *International Dairy Journal*, **13**, 401–8.

Grindrod, J. and Nickerson, T. A. (1967). Changes in milk proteins treated with hydrogen peroxide. *Journal of Dairy Science*, **50**, 142–46.

Guyomarch, F., Law, A. J. R. and Dalgleish, D. G. (2003). Formation of soluble and micelle-bound protein aggregates in heated milk. *Journal of Agricultural and Food Chemistry*, **51**, 4652–60.

Haque, Z. and Kinsella, J. E. (1988). Interaction between heated κ-casein and β-lactoglobulin – predominance of hydrophobic interactions in the initial stages of complex formation. *Journal of Dairy Research*, **55**, 67–80.

Haque, Z., Kristjansson, M. M. and Kinsella, J. E. (1987). Interaction between κ-casein and β-lactoglobulin – possible mechanism. *Journal of Agricultural and Food Chemistry*, **35**, 644–49.

Harland, H. A. and Ashworth, U. S. (1945). The preparation and effect of heat treatment on the whey proteins of milk. *Journal of Dairy Science*, **28**, 879–86.

Harland, H. A. and Ashworth, U. S. (1947). A rapid method for estimation of whey proteins as an indication of baking quality of nonfat dry-milk solids. *Food Research*, **12**, 247–51.

Harland, H. A., Coulter, S. T. and Jenness, R. (1952). The effect of the various steps in the manufacture on the extent of serum protein denaturation in nonfat dry milk solids. *Journal of Dairy Science*, **35**, 363–68.

Harland, H. A., Coulter, S. T. and Jenness, R. (1955). Natural variation of milk serum proteins as a limitation of their use in evaluating the heat treatment of milk. *Journal of Dairy Science*, **38**, 858–69.

Henry, G., Molle, D., Morgan, F., Fauquant, J. and Bouhallab, S. (2002). Heat-induced covalent complex between casein micelles and β-lactoglobulin from goat's milk: identification of an involved disulfide bond. *Journal of Agricultural and Food Chemistry*, **50**, 185–91.

Hill, A. R. (1989). The β-lactoglobulin-κ-casein complex. *Canadian Institute of Food Science and Technology Journal*, **22**, 120–23.

Hillier, R. M. and Lyster, R. L. J. (1979). Whey protein denaturation in heated milk and cheese whey. *Journal of Dairy Research*, **46**, 95–102.

Hoffmann, M. A. M. and van Mil, P. J. J. M. (1997). Heat-induced aggregation of β-lactoglobulin: role of the free thiol group and disulfide bonds. *Journal of Agricultural and Food Chemistry*, **45**, 2942–48.

Holt, C. and Horne, D. S. (1996). The hairy casein micelle: evolution of the concept and its implications for dairy technology. *Netherlands Milk and Dairy Journal*, **50**, 85–111.

Hong, Y. H. and Creamer, L. K. (2002). Changed protein structures of bovine β-lactoglobulin B and α-lactalbumin as a consequence of heat treatment. *International Dairy Journal*, **12**, 345–59.

Horne, D. S. (1998). Casein interactions: casting light on the black boxes, the structure in dairy products. *International Dairy Journal*, **8**, 171–77.

Horne, D. S. (2002). Casein structure, self-assembly and gelation. *Current Opinion in Colloid and Interface Science*, **7**, 456–61.

Horne, D. S. (2003). Casein micelles as hard spheres: limitations of the model in acidified gel formation. *Colloids and Surfaces A: Physicochemical and Engineering Aspects*, **213**, 255–63.

Horne, D. S. (2006). Casein micelle structure: models and muddles. *Current Opinion in Colloid and Interface Science*, **11**, 148–53.

Iametti, S., Gregori, B., Vecchio, G. and Bonomi, F. (1996). Modifications occur at different structural levels during the heat denaturation of β-lactoglobulin. *European Journal of Biochemistry*, **237**, 106–12.

International Dairy Federation (1995). *Heat-induced Changes in Milk*, 2nd edn, IDF Special Issue 9501 (P. F. Fox, ed.). Brussels: International Dairy Federation.

International Dairy Federation (1996). *Heat Treatments and Alternative Methods*, 2nd edn. IDF Special Issue 9602, Brussels: International Dairy Federation.

Kelly, A. L., O'Connell, J. E. and Fox, P. F. (2003). Manufacture and properties of milk powders. In *Advanced Dairy Chemistry,* Volume 1, *Proteins*, 3rd edn (P. F. Fox and P. L. H. McSweeney, eds) pp. 1027–61. New York: Kluwer Academic/Plenum Publishers.

Kessler, H. G. and Beyer, H. J. (1991). Thermal denaturation of whey proteins and its effect in dairy technology. *International Journal of Biological Macromolecules*, **13**, 165–73.

Kudo, S. (1980). The heat stability of milk: formation of soluble proteins and protein-depleted micelles at elevated temperatures. *New Zealand Journal of Dairy Science and Technology*, **15**, 255–63.

Kuramoto, S., Jenness, R., Coulter, S. T. and Choi, R. P. (1959). Standardization of the Harland–Ashworth test for whey protein nitrogen. *Journal of Dairy Science*, **42**, 28–38.

Lakemond, C. M. M. and van Vliet, T. (2005). Rheology of acid skim milk gels. In *Food Colloids: Interactions, Microstructure and Processing, Special Publication 298* (E. Dickinson, ed.) pp. 26–36. Cambridge: Royal Society of Chemistry.

Larson, B. L. and Rolleri, G. D. (1955). Heat denaturation of the specific serum proteins in milk. *Journal of Dairy Science*, **38**, 351–60.

Larson, B. L., Jenness, R., Geddes, W. F. and Coulter, S. T. (1951). An evaluation of the methods used for determining the baking quality of nonfat dry milk solids. *Cereal Chemistry*, **28**, 51–70.

Law, A. J. R. and Leaver, J. (1997). Effect of protein concentration on rates of thermal denaturation of whey proteins in milk. *Journal of Agricultural and Food Chemistry*, **45**, 4255–61.

Law, A. J. R. and Leaver, J. (2000). Effect of pH on the thermal denaturation of whey proteins in milk. *Journal of Agricultural and Food Chemistry*, **48**, 672–79.

Leighton, F. R. (1962). Determination of the whey protein index of skim milk powder. *Australian Journal of Dairy Technology*, **17**, 166–69.

Livney, Y. D. and Dalgleish, D. G. (2004). Specificity of disulfide bond formation during thermal aggregation in solutions of β-lactoglobulin B and κ-casein A. *Journal of Agricultural and Food Chemistry*, **52**, 5527–32.

Long, J. E., Gould, I. A. and van Winkle, Q. (1963). Heat-induced interaction between crude κ-casein and β-lactoglobulin. *Journal of Dairy Science*, **46**, 1329–34.

Lowe, E. K., Anema, S. G., Bienvenue, A., Boland, M. J., Creamer, L. K. and Jimenez-Flores, R. (2004). Heat-induced redistribution of disulfide bonds in milk proteins. 2. Disulfide bonding patterns between bovine β-lactoglobulin and κ-casein. *Journal of Agricultural and Food Chemistry*, **52**, 7669–80.

Lucey, J. A. (2002). Formation and physical properties of milk protein gels. *Journal of Dairy Science*, **85**, 281–94.

Lucey, J. A. and Singh, H. (1998). Formation and physical properties of acid milk gels: a review. *Food Research International*, **30**, 529–42.

Lucey, J. A., Teo, C. T., Munro, P. A. and Singh, H. (1997). Rheological properties at small (dynamic) and large (yield) deformations of acid gels made from heated milk. *Journal of Dairy Research*, **64**, 591–600.

Lucey, J. A., Tamehana, M., Singh, H. and Munro, P. A. (1998). Effect of interactions between denatured whey proteins and casein micelles on the formation and rheological properties of acid skim milk gels. *Journal of Dairy Research*, **65**, 555–67.

Lyster, L. J. (1970). The denaturation of α-lactalbumin and β-lactoglobulin in heated milk. *Journal of Dairy Research*, **37**, 233–43.

Manderson, G. A., Hardman, M. J. and Creamer, L. K. (1998). Effect of heat treatment on the conformation and aggregation of β-lactoglobulin A, B, and C. *Journal of Agricultural and Food Chemistry*, **46**, 5052–61.

Manji, B. and Kakuda, Y. (1986). Thermal denaturation of whey proteins in skim milk. *Canadian Institute of Food Science and Technology Journal*, **19**, 161–66.

Manji, B. and Kakuda, Y. (1987). Determination of whey protein denaturation in heat-processed milks: comparison of three methods. *Journal of Dairy Science*, **70**, 1355–61.

McKenna, A. B. and Anema, S. G. (1993). The effect of thermal processing during whole milk powder manufacture and after its reconstitution on set-yoghurt properties. In *Protein and Fat Globule Modifications by Heat Treatment, Homogenization and Other Technological Means for High Quality Dairy Products*, IDF Special Issue 9303, pp. 307–16. Brussels: International Dairy Federation.

McKenzie, G. H., Norton, R. S. and Sawyer, W. H. (1971). Heat-induced interaction of β-lactoglobulin and κ-casein. *Journal of Dairy Research*, **38**, 343–51.

McMahon, D. J. and McManus, W. R. (1998). Rethinking casein micelle structure using electron microscopy. *Journal of Dairy Science*, **81**, 2985–93.

Mellema, M., van Opheusden, J. H. J. and van Vliet, T. (2002). Categorization of rheological scaling models for particle gels applied to casein gels. *Journal of Rheology*, **46**, 11–29.

Nieuwenhuijse, J. A. and van Boekel, M. A. J. S. (2003). Protein stability in sterilised milk and milk products. In *Advanced Dairy Chemistry,* Volume 1, *Proteins*, 3rd edn (P. F. Fox and P. L. H. McSweeney, eds) pp. 947–74. New York: Kluwer Academic/Plenum Publishers.

Nieuwenhuijse, J. A., van Boekel, M. A. J. S. and Walstra, P. (1991). On the heat-induced association and dissociation of proteins in concentrated skim milk. *Netherlands Milk and Dairy Journal*, **45**, 3–22.

Noh, B., Richardson, T. and Creamer, L. K. (1989a). Radiolabelling study of the heat-induced interactions between α-lactalbumin, β-lactoglubulin and κ-casein in milk and in buffer solutions. *Journal of Food Science*, **54**, 889–93.

Noh, B. S., Creamer, L. K. and Richardson, T. (1989b). Thermally induced complex-formation in an artificial milk system. *Journal of Agricultural and Food Chemistry*, **37**, 1395–400.

O'Connell, J. E. and Fox, P. F. (2003). Heat-induced coagulation of milk. In *Advanced Dairy Chemistry,* Volume 1, *Proteins*, 3rd edn (P. F. Fox and P. L. H. McSweeney, eds) pp. 879–945. New York: Kluwer Academic/Plenum Publishers.

Oldfield, D. J., Singh, H., Taylor, M. and Pearce, K. (1998a). Kinetics of denaturation and aggregation of whey proteins in skim milk heated in an ultra-high temperature (UHT) pilot plant. *International Dairy Journal*, **8**, 311–18.

Oldfield, D. J., Singh, H. and Taylor, M. W. (1998b). Association of β-lactoglobulin and α-lactalbumin with the casein micelles in skim milk heated in an ultra-high temperature plant. *International Dairy Journal*, **8**, 765–70.

Oldfield, D. J., Singh, H., Taylor, M. and Pearce, K. (2000). Heat-induced interactions of β-lactoglobulin and α-lactalbumin with the casein micelle in pH-adjusted skim milk. *International Dairy Journal*, **10**, 509–18.

Oldfield, D. J., Singh, H. and Taylor, M. W. (2005). Kinetics of heat-induced whey protein denaturation and aggregation in skim milks with adjusted whey protein concentration. *Journal of Dairy Research*, **72**, 369–78.

Parker, E. A., Donato, L. and Dalgleish, D. G. (2005). Effects of added sodium caseinate on the formation of particles in heated milk. *Journal of Agricultural and Food Chemistry*, **53**, 8265–72.

Parnell-Clunies, E. M., Kakuda, Y., Mullen, K., Arnott, D. R. and Deman, J. M. (1986). Physical properties of yogurt – a comparison of vat versus continuous heating-systems of milk. *Journal of Dairy Science*, **69**, 2593–603.

Patel, H. A., Singh, H., Anema, S. G. and Creamer, L. K. (2006). Effects of heat and high hydrostatic pressure treatments on disulfide bonding interchanges among the proteins in skim milk. *Journal of Agricultural and Food Chemistry*, **54**, 3409–20.

Patel, H. A., Anema, S. G., Holroyd, S. E., Singh, H. and Creamer, L. K. (2007). Methods to determine denaturation and aggregation of proteins in low-, medium- and high-heat skim milk powders. *Lait*, **87**, 251–68.

Purkayastha, R., Tessier, H. and Rose, D. (1967). Thiol-disulfide interchange in formation of β-lactoglobulin-κ-casein complex. *Journal of Dairy Science*, **50**, 764–66.

Qin, B. Y., Bewley, M. C., Creamer, L. K., Baker, E. N. and Jameson, G. B. (1999). Functional implications of structural differences between variants A and B of bovine β-lactoglobulin. *Protein Science*, **8**, 75–83.

Rasmussen, L. K., Johnsen, L. B., Tsiora, A., Sorensen, E. S., Thomsen, J. K., Nielsen, N. C., Jakobsen, H. J. and Petersen, T. E. (1999). Disulphide-linked caseins and casein micelles. *International Dairy Journal*, **9**, 215–18.

Rodriguez del Angel, C. and Dalgleish, D. G. (2006). Structure and some properties of soluble protein complexes formed by the heating of reconstituted skim milk powder. *Food Research International*, **39**, 472–79.

Rose, D. (1961). Variations in the heat stability and composition of milk from individual cows during lactation. *Journal of Dairy Science*, **44**, 430–41.

Rowland, S. J. (1933). The heat denaturation of albumin and globulin in milk. *Journal of Dairy Research*, **5**, 46–53.

Ruegg, M., Moor, U. and Blanc, B. (1977). Calorimetric study of thermal-denaturation of whey proteins in simulated milk ultrafiltrate. *Journal of Dairy Research*, **44**, 509–20.

Sanderson, W. B. (1970a). Determination of undenatured whey protein nitrogen in skim milk powder by dye binding. *New Zealand Journal of Dairy Science and Technology*, **5**, 46–48.

Sanderson, W. B. (1970b). Seasonal variations affecting the determination of the whey protein nitrogen index of skim milk powder. *New Zealand Journal of Dairy Science and Technology*, **5**, 48–52.

Sanderson, W. B. (1970c). Reconstituted and recombined dairy products. *New Zealand Journal of Dairy Science and Technology*, **5**, 139–43.

Sawyer, W. H. (1969). Complex between β-lactoglobulin and κ-casein. A review. *Journal of Dairy Science*, **52**, 1347–55.

Sawyer, W. H., Coulter, S. T. and Jenness, R. (1963). Role of sulfhydryl groups in the interaction of κ-casein and β-lactoglobulin. *Journal of Dairy Science*, **46**, 564–65.

Schmidt, D. G. (1982). Association of caseins and casein micelle structure. In *Developments in Dairy Chemistry,* Volume 1, *Proteins* (P. F. Fox, ed.) pp. 61–86. London: Elsevier Applied Science.

Schorsch, C., Wilkins, D. K., Jones, M. G. and Norton, I. T. (2001). Gelation of casein–whey mixtures: effects of heating whey proteins alone or in the presence of casein micelles. *Journal of Dairy Research*, **68**, 471–81.

Shalabi, S. I. and Wheelock, J. V. (1976). Role of α-lactalbumin in the primary phase of chymosin action on heated casein micelles. *Journal of Dairy Research*, **43**, 331–35.

Singh, H. (2004). Heat stability of milk. *International Journal of Dairy Technology*, **57**, 111–19.

Singh, H. and Creamer, L. K. (1991a). Denaturation, aggregation and heat-stability of milk protein during the manufacture of skim milk powder. *Journal of Dairy Research*, **58**, 269–83.

Singh, H. and Creamer, L. K. (1991b). Influence of concentration of milk solids on the dissociation of micellar κ-casein on heating reconstituted milk at 120°C. *Journal of Dairy Research*, **58**, 99–105.

Singh, H. and Creamer, L. K. (1992). Heat stability of milk. In *Advanced Dairy Chemistry,* Volume 1, *Proteins* (P. F. Fox, ed.) pp. 621–56. London: Elsevier Applied Science.

Singh, H. and Fox, P. F. (1985a). Heat stability of milk – the mechanism of stabilization by formaldehyde. *Journal of Dairy Research*, **52**, 65–76.

Singh, H. and Fox, P. F. (1985b). Heat stability of milk: pH-dependent dissociation of micellar κ-casein on heating milk at ultra high temperatures. *Journal of Dairy Research*, **52**, 529–38.

Singh, H. and Fox, P. F. (1986). Heat stability of milk – further studies on the pH-dependent dissociation of micellar κ-casein. *Journal of Dairy Research*, **53**, 237–48.

Singh, H. and Fox, P. F. (1987a). Heat stability of milk – influence of modifying sulfhydryl-disulfide interactions on the heat coagulation time pH profile. *Journal of Dairy Research*, **54**, 347–59.

Singh, H. and Fox, P. F. (1987b). Heat stability of milk – role of β-lactoglobulin in the pH-dependent dissociation of micellar κ-casein. *Journal of Dairy Research*, **54**, 509–21.

Singh, H. and Fox, P. F. (1987c). Heat stability of milk – influence of colloidal and soluble salts and protein modification on the pH-dependent dissociation of micellar κ-casein. *Journal of Dairy Research*, **54**, 523–34.

Singh, H. and Newstead, D. F. (1992). Aspects of proteins in milk powder manufacture. In *Advanced Dairy Chemistry,* Volume 1, *Proteins* (P. F. Fox, ed.) pp. 735–65. London: Elsevier Applied Science.

Slatter, W. L. and van Winkle, Q. (1952). An electrophoretic study of the protein in skim milk. *Journal of Dairy Science*, **35**, 1083–93.

Smits, P. and van Brouwershaven, J. H. (1980). Heat-induced association of β-lactoglobulin and casein micelles. *Journal of Dairy Research*, **47**, 313–25.

Snoeren, T. H. M. and van der Spek, C. A. (1977). Isolation of a heat-induced complex from UHTST milk. *Netherlands Milk and Dairy Journal*, **31**, 352–55.

Tessier, H., Yaguchi, M. and Rose, D. (1969). Zonal ultracentrifugation of β-lactoglobulin and κ-casein complexes induced by heat. *Journal of Dairy Science*, **52**, 139–45.

Tobias, J., Whitney, R. M. and Tracy, P. H. (1952). Electrophoretic properties of milk proteins. II. Effect of heating to 300°F by means of the Mallory small-tube heat exchanger on skim milk proteins. *Journal of Dairy Science*, **35**, 1036–45.

Trautman, J. C. and Swanson, A. M. (1958). Additional evidence of a stable complex between β-lactoglobulin and α-casein. *Journal of Dairy Science*, **41**, 715.

van Vliet, T. and Keetals, C. J. A. M. (1995). Effect of preheating of milk on the structure of acidified milk gels. *Netherlands Milk and Dairy Journal*, **49**, 27–35.

Vasbinder, A. J. and de Kruif, C. G. (2003). Casein-whey protein interactions in heated milk: the influence of pH. *International Dairy Journal*, **13**, 669–77.

Vasbinder, A. J., Alting, A. C., Visschers, R. W. and de Kruif, C. G. (2003). Texture of acid milk gels: formation of disulfide cross-links during acidification. *International Dairy Journal*, **13**, 29–38.

Verheul, M., Roefs, S. and de Kruif, K. G. (1998). Kinetics of heat-induced aggregation of β-lactoglobulin. *Journal of Agricultural and Food Chemistry*, **46**, 896–903.

Walstra, P. (1990). On the stability of casein micelles. *Journal of Dairy Science*, **73**, 1965–79.

Walstra, P. (1999). Casein sub-micelles: do they exist?. *International Dairy Journal*, **9**, 189–92.

Walstra, P. and Jenness, R. (1984). *Dairy Chemistry and Physics*. New York: John Wiley and Sons.

Zittle, C. A., Custer, J. H., Cerbulis, J. and Thompson, M. P. (1962). κ-Casein-β-lactoglobulin interaction in solution when heated. *Journal of Dairy Science*, **45**, 807–10.

9

Effects of drying on milk proteins

Pierre Schuck

Abstract

Dehydration by spray drying is a valuable technique for water evaporation, using hot air to stabilize the majority of dairy ingredients. In view of the increased development of filtration processes, the dairy industry requires greater understanding of the effects of spray drying on the quality of protein powders. Several publications have reported that the proteins have an important role in the mechanisms of water transfer during drying and rehydration. The residence time of the droplet and then the powder is so short that it is very difficult to study the mechanism of the structural change in the protein without fundamental research into relationships with the process/product interactions. Following an introduction to spray drying, this chapter on the effects of drying on milk proteins covers five areas, i.e. the world dairy powder situation, properties of spray-dried milk products, principles of spray drying, drying of proteins and rehydration of dried protein powders.

Introduction

The purpose of the dehydration of milk and whey is to stabilize these products for their storage and later use. Milk and whey powders are used mostly in animal feeding. With changes in agricultural policies (such as the implementation of the quota system and the dissolution of the price support system in the European Union), the dairy industry was forced to look for better uses for the dairy surplus and for the

Milk Proteins: From Expression to Food
ISBN: 978-0-12-374039-7

by-products of cheese (whey) produced from milk and buttermilk produced from cream. Studies on the reuse of protein fractions with nutritional qualities and functionality led us to believe that they could have multiple applications.

In the past 25 years, the dairy industry has developed new technological processes for extracting and purifying proteins (casein, caseinates, whey proteins, etc.) (Kjaergaard *et al.*, 1987; Maubois, 1991) such as:

- dairy proteins and whey concentrates (Le Graët and Maubois, 1979; Goudédranche *et al.*, 1980; Madsen and Bjerre, 1981; Maubois *et al.*, 1987; Caron *et al.*, 1997);
- micellar casein concentrates (Fauquant *et al.*, 1988; Schuck *et al.*, 1994a);
- micellar casein (MC) (Pierre *et al.*, 1992; Schuck *et al.*, 1994b);
- whey concentrates;
- selectively demineralized concentrates (Jeantet *et al.*, 1996); and
- super-clean skim milk concentrates (Piot *et al.*, 1987; Vincens and Tabard, 1988; Trouvé *et al.*, 1991; Schuck *et al*, 1994a).

mainly because of the emergence of filtration technology (microfiltration, ultrafiltration, nanofiltration and reverse osmosis).

Most of these proteins, used as either nutritional or functional ingredients, are marketed in dehydrated form (Figure 9.1). The application of different processing steps allows the production of a wide range of different dried and stable intermediate dairy products. Many new uses for these constituents emerged with the manufacture of formula products, substitutes and adapted raw materials.

The most frequently used technique for the dehydration of dairy products is spray drying. It became popular in the dairy industry in the 1970s but, at that time, there were few scientific or technical studies on spray drying and in particular none on the effects of spray drying parameters or on the effects of the physico-chemical composition and microbiology of the concentrates on the powder quality. Manufacturers acquired expertise in milk drying and eventually in whey drying processes through trial and error. Because of the variety and complexity of the mixes to be dried, a more rigorous method based on physico-chemical and thermodynamic properties has become necessary. A better understanding of the biochemical properties of milk products before drying, water transfer during spray drying, the properties of powders and influencing factors is now essential in the production of milk powder. The lack of technical and economic information and of scientific methods prevents the manufacturer from optimizing dairy plants in terms of energy costs and powder quality.

The aim of this chapter is to give a brief summary of the process of the spray drying of dairy products and to review present knowledge on the properties of spray-dried milk products, modeling and simulation of water transfer processes (drying and rehydration), dairy powders and spray-drying equipment and energy consumption.

World dairy powder situation

There has been a change in the nature of dairy powders over the last 15–20 years (CNIEL, 1991, 2005). A decrease in production has occurred mainly for skim and

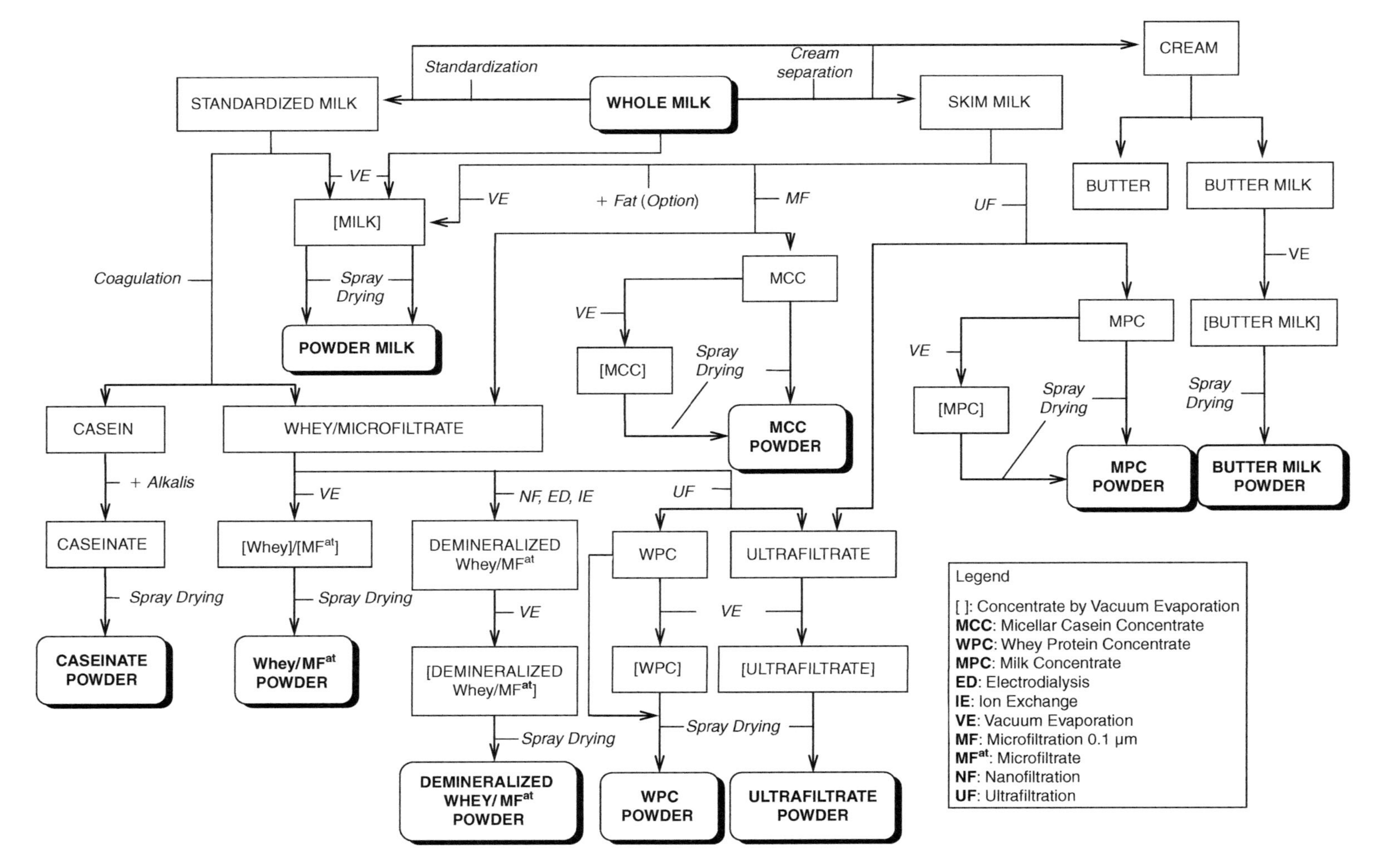

Figure 9.1 Fractionation of milk.

fat-filled milk powders but the production of whole milk powder and whey powder increased between 1986 and 2004. This increase was reflected in the types of whey and derived powders (protein concentrates) produced. Cheese production from cow's milk increased between 1986 and 2004, with a corresponding increase in whey production. Having fallen in 2004, the production of dry milk products did not recover significantly in the first months of 2005, the decline being due mainly to slower growth of milk supplies in many parts of the world.

Whole milk powder (WMP)

According to the International Dairy Federation (2005), the production of WMP was different in different parts of the world: WMP production increased, mainly in China, Latin America and New Zealand, but was almost unchanged in Europe and Australia. The most important producer is now China, with 0.9 million tonnes, accounting for more than one-quarter of recorded production and ahead of the EU25 (with 0.85 million tonnes). The next highest producers are New Zealand, Brazil, Argentina and Australia: all these countries together cover over 90% of world production. In the long term, the EU and Australia are likely to lose their share of world production.

In the medium and long term, the production of WMP is driven mainly by demand. WMP provides milk where raw milk is in very short supply compared with demand. In countries where reconstitution of dried milk is common, WMP is ideal for bridging regional and seasonal deficits and this is also the reason why WMP is the best product to meet this demand. WMP does not play a significant role in the international dairy trade, but the function of the WMP produced is just to balance the market over long distances and across seasons within a country.

Skim milk powder (SMP)

With SMP, the situation is somewhat different from WMP. As well as the function of balancing seasonal and regional markets, there is also the need to clear temporary surpluses when there is no other use for skim milk. This explains the decline in SMP production in 2004, because the overall availability of milk as a raw material fell short of the demand. As a result, world production of SMP fell by 0.3 million tonnes in 2004. Most of this reduction took place in the EU, with production falling by 250 000 tonnes in one year, to the extent that EU25 production in 2004 was less than EU15 production in 2003. The USA and New Zealand also reduced their output. A degree of recovery occurred in 2005, but it was only modest. This recovery is unlikely to replace the volumes made available in 2003 and 2004 by the clearing of stocks, in particular in Oceania, the USA and the EU.

Whey products, casein and other dairy ingredients

More liquid whey is now generated from cheese production, following the general growth in this sector. Although the increased application of new technologies, e.g. ultrafiltration, microfiltration (Maubois, 1991) and an increasing number of new processing techniques to exploit milk components, has changed the composition of the liquid that remains, the major part is still ordinary sweet whey. Thus, the main product is still sweet whey powder, but other derivatives such as whey protein concentrate, lactose and demineralized and delactosed products are also being produced and are evolving even faster than ordinary whey powder. However, the statistics are difficult to obtain. For the EU15, whey powder production was estimated at 1.3 million tonnes.

Other important producers are the USA and Canada, with a combined production of 500 000 tonnes. In 2004, production stagnated or even declined, attributable in part to the growing use of liquid whey for other derivatives and a reduction in casein manufacture, as well as low prices.

Casein production increased in the EU15 in 2004, but fell again in 2005. However, in Eastern Europe and in Australia, recovery can be expected.

Properties of spray-dried milk products

A dairy powder is characterized not only by its composition (proteins, carbohydrates, fats, minerals and water) but also by its microbiological and physical properties (bulk and particle density, instant characteristics, flowability, floodability, hygroscopicity, degree of caking, whey protein nitrogen index, thermostability, insolubility index, dispersibility index, wettability index, sinkability index, free fat, occluded air, interstitial air and particle size), which form the basic elements of quality specifications; there are well-defined test methods for their determination according to international standards (Pisecky, 1986, 1990, 1997; American Dairy Products Institute, 1990; Masters, 1991). These characteristics depend on drying parameters (type of tower spray drier, nozzles/wheels, pressure, agglomeration and thermodynamic conditions of the air, such as temperature, relative humidity and velocity) and the characteristics of the concentrate before spraying (composition/physico-chemical characteristics, viscosity, thermo-sensitivity and availability of water). Several scientific papers on the effects of technological parameters on these properties have been published (Hall and Hedrick, 1966; De Vilder *et al.*, 1979; Baldwin *et al.*, 1980; Pisecky, 1980, 1981, 1986; Kessler, 1981; Bloore and Boag, 1982; De Vilder, 1986; Tuohy, 1989; Ilari and Loisel, 1991; Masters, 1991; Mahaut *et al.*, 2000). Water content, water dynamics and water availability are among the most important properties (Figure 9.2).

The nutritional quality of dairy powders depends on the intensity of the thermal processing during the technological process. The thermal processing induces physico-chemical changes that tend to decrease the availability of the nutrients (loss of vitamins, reduction of available lysine content and whey protein denaturation) or to produce nutritional compounds such as lactulose (Straatsma *et al.*, 1999a, 1999b).

Principles of spray drying

According to Pisecky (1997), spray drying is an industrial process for the dehydration of a liquid containing dissolved and/or dispersed solids (e.g. dairy products), by

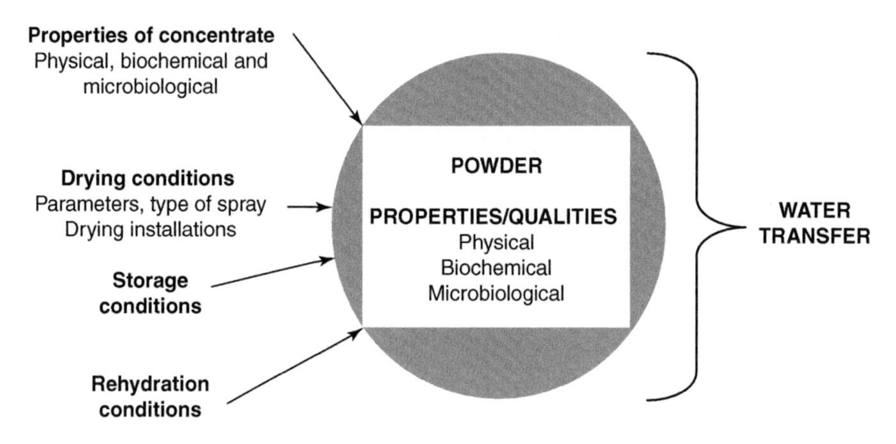

Figure 9.2 Properties and qualities of powders.

transforming the liquid into a spray of small droplets and exposing these droplets to a flow of hot air. The very large surface area of the spray droplets causes evaporation of the water to take place very quickly, converting the droplets into dry powder particles.

Indeed, when a wet droplet is exposed to hot dry gas, variations in the temperature and the partial pressure of water vapor are established spontaneously between this droplet and the air:

- heat transfer from the air to the droplet occurs under the influence of the temperature variation;
- water transfer occurs in the opposite direction, explained by variation in the partial pressure of water vapor between the air and the droplet surface.

Air is thus used both for fluid heating and as a carrier gas for the removal of water. The air enters the spray drier hot and dry and leaves wet and cool. Spray drying is a phenomenon of surface water evaporation maintained by the movement of capillary water from the interior to the surface of the droplet. As long as the average moisture is sufficient to feed the surface regularly, the evaporation rate is constant. If not, it decreases.

The drying kinetics are related to three factors:

1. Evaporation surface created by the diameter of the particles. Spraying increases the exchange surface: 1 L of liquid sprayed in particles of 100 μm diameter develops a surface area of 60 m^2, whereas the surface area is only approximately 5 dm^2 for one sphere of the same volume.
2. Difference in the partial pressure of water vapor between the particle and the drying air: a decrease in the absolute humidity of the air and/or an increase in the air temperature tend to increase the difference in the partial pressure of water vapor between the particle and the drying air.
3. Rate of water migration from the center of the particle towards its surface: this parameter is essential for the quality of dairy powders. Indeed, it is important

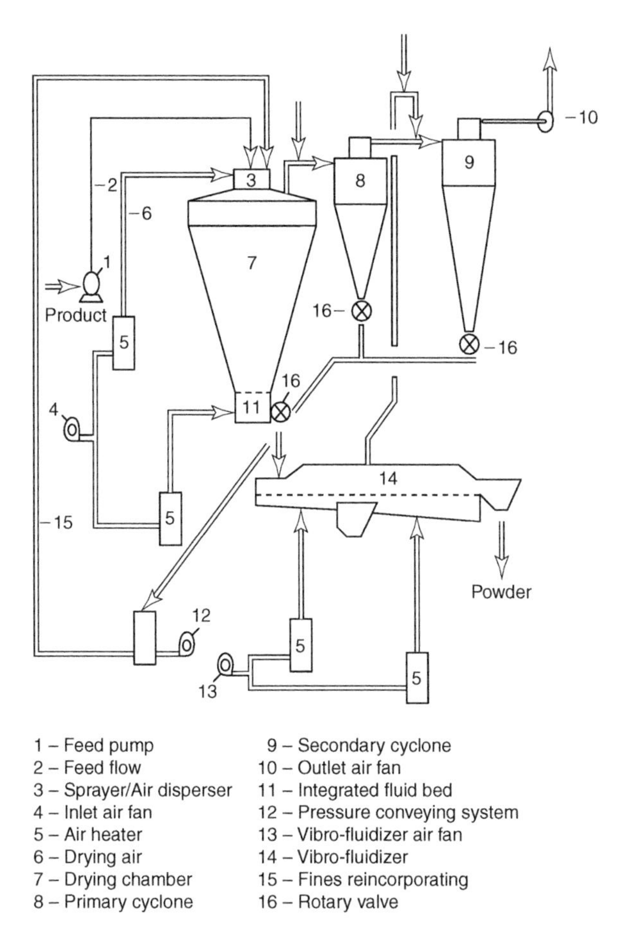

Figure 9.3 Multiple effect spray drier.

that there is always water on the surface of the product so that the powder surface remains at the wet bulb temperature for as long as possible. The rate of water migration depends on the water diffusion coefficient, which varies according to the biochemical composition, water content and droplet temperature. This is why calculation of this coefficient is complex and the mathematical models suggested are not easily exploitable by the dairy industry.

To define the components of a spray-drying installation, according to Masters (1991) and Pisecky (1997), the main components of the spray drier shown in Figure 9.3 are as follows:

- A drying chamber (Figure 9.3, 7). The chamber can be horizontal (box drier) although, in the dairy industry, the chamber design is generally vertical with a conical or flat base.
- An air disperser with a hot air supply system such as a main air filter, supply fan, air heater and air disperser (Figure 9.3, 3). The air aspiration is performed through filters, the type depending on the local conditions and the nature of

the product to be treated. The air can be heated in two different ways: by direct heating (gas) and/or by indirect heating (vapor, gas, oil or electricity). The air flow chamber can be in co-current, counter-current or mixed mode.

- An atomizing device with a feed supply system such as feed tank, feed pump, water tank, concentrate heater and atomizing device. There are three types of atomizing device: rotary atomizer (wheel or disk), nozzle atomizer (pressure, pneumatic or sonic) and combined (rotary and pneumatic) (Figure 9.3, 3).
- A powder recovery system. Separation of the dried product can be achieved by a primary discharge from the drying chamber followed by a secondary discharge from a particulate collector (using a cyclone, bag filter or electrostatic precipitation), followed by total discharge from the particulate collector and finishing with final exhaust air cleaning in a wet scrubber and dry filter (Figure 9.3, 8 and 9).

According to Sougnez (1983), Masters (1991) and Pisecky (1997), the simplest types of installation are single-stage systems with a very short residence time (20–60 s). Thus, there is no real balance between the relative humidity of the air and the moisture content of the powder. The outlet temperature of the air must therefore be higher and as a result the thermal efficiency of the single-stage spray drier is then reduced. This type of drying chamber was the standard equipment for drying milk in the 1960s. Space requirements were small and building costs were low. Generally, installations without any post-treatment system are suitable only for non-agglomerated powders not requiring cooling. If necessary, a pneumatic conveying system could be added to cool the powder while transporting the chamber fraction and the cyclone fraction to a single discharge point.

The two-stage drying system consists of limiting the spray-drying process to a process with a longer residence time (several minutes) to provide a better thermodynamic balance. This involves a considerable reduction in the outlet air temperature and also an increase in the inlet air temperature. A second final drying stage is necessary to optimize the moisture content by using an integrated fluid bed (static) or an external fluid bed (vibrating), the air temperatures of which are 15–25°C lower than with a single-stage system to improve and/or preserve the quality of the dairy powder (Figure 9.3, 11 and 14). Consequently, the surrounding air temperature at the critical drying stage and the particle temperature are also correspondingly lower, thus contributing to further economy improvement. The integrated fluid bed can be either circular (e.g. Multi Stage Drier [MSD™] chamber) or annular (e.g. Compact Drier [CD] chamber).

Two-stage drying has its limitations but it can be applied to products such as skim milk, whole milk, pre-crystallized whey, caseinates, whey proteins and derivatives. The moisture content of the powder leaving the first stage is limited by the thermoplasticity of the wet powder, i.e. by its stickiness in relation to the water activity and the glass transition temperature (Roos, 2002). The moisture content must be close to 7–8, 9–10 and 2–3% for skim/whole milk, caseinate/whey protein and pre-crystallized whey powders respectively. The two-stage drying techniques can be applied to the production of both non-agglomerated and agglomerated powders, but this technique is very suitable for the production of agglomerated powders, by separating the non-agglomerated particles from the agglomerates (i.e. collecting the cyclone fractions and reintroducing these fine fractions [called fines] into the wet zone around the atomizer of the chamber).

The three-stage drying systems, with an internal fluid bed as a second stage in combination with an external vibrating fluid bed as a third-stage drier, first appeared at the beginning of the 1980s and were called Compact Drier Instantization (CDI) or MSD™. Today, they dominate the dairy powder industry (Figure 9.3). Three-stage systems combine all the advantages of extended two-stage drying, using spray drying as the primary stage, fluid bed drying of a static fluid as the second drying stage and drying on an external vibrating fluid bed as the third drying stage. The final drying stage terminates with cooling to under the glass transition temperature. Evaporation performed at each stage can be optimized to achieve both gentle drying conditions and good thermal economy.

The CD is suitable for producing both non-agglomerated and agglomerated powders of practically any kind of dried dairy product. It can also cope successfully with whey powders, fat-filled milk and whey products as well as caseinates, both non-agglomerated and agglomerated. It has a fat content limit of about 50% fat in total solids. Powder quality and appearance are comparable with those of products from two-stage drying systems but they have considerably better flowability and the process is more economical. In comparison with the CD, the MSD™ can process an even wider range of products and can handle an even higher fat content. The main characteristic of MSD™ powder is related to very good agglomeration and mechanical stability, low particle size fractions (below 125 μm) and very good flowability.

Optimization of the process allows considerable improvement in the drying efficiency and the quality of the product obtained is generally better. The various advantages are:

- improved thermal efficiency: significant reduction in the outlet air temperature, permitting an increase in the inlet air temperature;
- reduction in material obstruction: the capacity in one volume is two or three times higher than for a traditional unit;
- considerable reduction in powder emission to the atmosphere: a reduction in the drying air flow and an increase in powder moisture content decrease the loss of fine particles in the outlet air;
- improved powder quality in relation to the agglomeration level, solubility, dispersibility, wettability, particle size, density, etc.

There are other examples of drying equipment such as the "tall form drier", the "Filtermat® drier," the "Paraflash® drier" and the "Tixotherm® drier." All these towers have characteristics related to the specific properties of the product being dried (e.g. high fat content, starch, maltodextrin, egg and hygroscopic products).

Drying of proteins

The native properties of milk components are substantially unaffected by moderate drying conditions. Depending on the preheating conditions, drier design and temperature operation, the properties of spray-dried powder may vary significantly. An evaporating milk droplet in a spray drier in co-current air flow initially does not appreciably exceed the wet bulb temperature and can be held effectively at temperatures below 60°C. As the falling temperature period is approached in the course of further evaporation, the temperature rises to a final value determined by the final temperature of the drying gas and the residence time in the drier. Under properly controlled spray drying conditions, the changes in milk protein structure and solubility are minor. Spray drying does not denature the whey protein significantly and the levels of denatured whey protein in dairy powders are more or less equal to those of condensed milk and heated milk, which is substantially more denatured than during spray drying. The best example is in relation to the whey protein nitrogen index (WPNI).

According to Pisecky (1997), the WPNI expresses the amount of undenatured whey protein (milligrams of whey protein nitrogen per gram of powder). It is a measure of the sum of heat treatments to which the milk has been subjected prior to evaporation and spray drying. The heat treatment of a concentrate and, after, of a powder has only a negligible effect on the WPNI. The main operation to adjust the required value is the preheating process, i.e. time/temperature combination. However, there are many other factors that influence the WPNI, including the total amount of whey protein and the overall composition of the processed milk as influenced by animal breed and seasonal variations. The individual design of the processing equipment, i.e. the pasteurizer and holding tubes, also has great significance. Therefore, it is difficult to predict the conditions of achieving the required WPNI on a general basis. Obviously, the primary purpose of preheat treatment is to ensure the microbiological quality of dairy products.

In milk powder production, the influence of the heat treatment on the denaturation of whey proteins for achieving the desired properties of the final products is just as important. SMP for cheese manufacture should have as much undenatured protein as possible, i.e. it should be *low heat* (WPNI > 6), whereas for bakeries *high heat* powder with high denaturation is required (WPNI < 1.5). For ice cream, chocolate and confectionery, *medium heat* powder is required. According to Schuck *et al.* (1994a), the use of microfiltration (pore diameter, 1.4 μm), coupled with a low heat treatment during vacuum evaporation, allows the production of a "*low low heat*" SMP with a WPNI close to 9 mg of whey protein nitrogen/g of powder, a bacterial count $<$1000 CFU/g powder, a solubility index $>$99.5%, a dispersibility index $>$98.5% and a wettability index $<$15 s. Such a powder after water rehydration has the same renneting time as the original raw milk and can be used as a reference powder for either industrial or scientific purposes.

The stability of protein powders during storage is critically affected by the moisture content and the storage temperature. More precisely, such stability is governed by the water activity (a_w) and the glass transition (T_g) temperature. The a_w should be close to 0.2 at 25°C for optimal preservation, with an ideal moisture content determined by using the sorption isotherm of some dairy powders. For example, the corresponding moisture contents for skim milk, whey and protein powders must be close to 4, 2–3 and 6% respectively. The optimal storage temperature must be below the T_g temperature, which is close to 40–50°C at 0.2 a_w.

Example of dairy protein concentrates and powders

In this study, micellar casein concentrate (MCC) was prepared by microfiltration and diafiltration (pore diameter, 0.1 μm) on an MFS 19 (Tetra Laval, Åarhus, Denmark; 4.6 m^2) at 50°C, according to Fauquant *et al.* (1988) and Pierre *et al.* (1992) at 200 g/kg total solids. The whey protein concentrate (WPC) was obtained by ultrafiltration and diafiltration of the microfiltrate (0.1 μm) obtained on a DDS module (GEA, Soeborg, Denmark) with the plane membrane (10 000 g/mol molecular weight cut-off, 9 m^2, 50°C) at 200 g/kg total solids. The sodium, calcium and potassium caseinate concentrates were reconstituted from sodium, calcium and potassium caseinate powder at 190 g/kg total solids. The microfiltration retentate (R4 MF) was obtained by microfiltration (0.1 μm) on an MFS 19 (Tetra Laval, Åarhus, Denmark; 4.6 m^2) at 50°C. The volume reduction ratio was 4.

The ultrafiltration retentate (R4 UF) was manufactured by ultrafiltration on a 2-S37 module with M_1 membranes (Tech Sep, Rhône Poulenc, St Maurice de Beynost, France; 100 000 g/mol nominal molecular weight cutoff, 6.8 m^2) at 50°C. The volume reduction ratio was 4. NaCl solution, $CaCl_2$ solution, sodium phosphate solution at pH 7.1 (for MCC) or pH 6.6 (for WPC) and sodium citrate solution at pH 7.1 (for MCC) or pH 6.6 (for WPC) in 205 5 g/kg total solids were added to the MCC or the WPC to obtain a concentrate with 12% (w/w) (NaCl, $CaCl_2$ and sodium phosphate solution) and 30% (w/w) (sodium citrate) of mineral salts/total solids. After addition of salt, the pH was adjusted to 7.1 (for MCC) or 6.6 (for WPC) with 1 N KOH (NaCl, $CaCl_2$ and sodium phosphate solution) or with 1 N HCl (sodium citrate) at 20°C.

The spray drying of the concentrates was performed at Bionov (Rennes, France) in a three-stage pilot plant spray drier (GEA, Niro Atomizer, St Quentin en Yvelines, France) according to Schuck *et al.* (1998a) and Bimbenet *et al.* (2002) to obtain a micellar casein powder (MCP) or a whey protein powder (WPP). The temperature of the concentrate before drying was 40°C for MCC and 20°C for WPC. The atomizer was equipped with a pressure nozzle (0.73 mm diameter orifice) and a four-slot core (0.51 mm nominal width), providing a 60 ° spray angle. The evaporation capacity was 70–120 kg/h (depending on the inlet and outlet air temperatures and the air flow). The pressure in the nozzle was 16 MPa.The inlet temperature was 208°C for WPC and 215°C for MCC, the integrated fluid bed air temperature was 70°C for MCC and WPC and the outlet temperature was 80°C for WPC and 70°C for MCC. The inlet air humidity was controlled and adjusted by a dehumidifier (Munters, Sollentuna, Sweden). For each MCP or WPP, two granulations were obtained (i.e. non-granulated [NG] and granulated [G] powders) by reintroduction of the fine particles after the cyclones at the top of the spray drier.

Research approach using drying by desorption

Principles

The concentrates were dried in a water activity meter (Novasina RTD 200/0, Pfäffikon, Switzerland) at 20°C (constant temperature). The concentrate (100 mg) was placed in a plastic support (area, 95 mm^2) with a zeolite WE 291 drier below (7 g) (Bayer, Puteaux, France). This method was used to simulate the conditions of spray drying by establishing a difference in vapor pressure equilibrium between the dairy concentrate and the drying air, and to determine the water transfer from inside the dairy concentrate to the surface. The relative humidity (RH) was measured versus time following water transfer from the dairy concentrate to the zeolite. The final slope of the absolute value of the decrease in RH (β) represented the ability to remove bound water from the solute at the end of the drying phase (Schuck *et al.*, 1998b, 1999); the lower the β slope, the greater the difficulty to remove water at the end of drying and the higher the bound water content.

Desorption results

The drying slopes (β) of the various dairy products tested are shown in Table 9.1. The dairy products, ranging from the highest to the lowest absolute value of the slope, were skim milk (β = 0.90%/min), R4 UF (β= 0.75%/min), R4 MF (β= 0.70%/min) and MCC (β= 0.34%/min). These results could be explained by a decrease in water diffusion through the dried product, i.e. the final residue obtained at the end of drying, when the micellar casein concentration increased. Water transport was probably affected by the high micellar casein content of the sprayed droplet in the atomization tower, and, similarly, when the powder granule was dissolved in water. These results are in accordance with the results of Schuck *et al.* (1994a, 1994b).

Drying of the caseinates showed that sodium caseinate and potassium caseinate dried more easily (β= 0.64 and 0.65%/min respectively) than calcium caseinate (β= 0.51%/min) (Table 9.1). The limitation of water diffusion through the calcium caseinate may be explained by the structure of this colloidal dispersion. Whereas, in calcium caseinate, the casein sub-units are more aggregated because of calcium binding, in sodium and potassium caseinates, the caseins are more soluble. These results

Table 9.1 Drying by desorption of dairy concentrates

Dairy concentrate	β Slope (%/min)
Skim milk	0.90
R4 UF	0.75
R4 MF	0.70
MCC	0.34
WPC	0.68
Sodium caseinate	0.64
Potassium caseinate	0.65
Calcium caseinate	0.51
WPC + NaCl	0.24
WPC + $CaCl_2$	0.46
WPC + Phosphate	0.41
WPC + Citrate	0.48
MCC + NaCl	0.19
MCC + $CaCl_2$	0.36
MCC + Phosphate	0.49
MCC + Citrate	0.45

R, Retentate; UF, Ultrafiltration; MF, Microfiltration; MCC, Micellar Casein Concentrate; WPC, Whey Protein Concentrate showed that the water bound in a micellar structure was more strongly bound than that bound to the soluble caseins in sodium caseinate. The situation was intermediate for calcium caseinate. We assumed that these differences in water transfer during drying could be explained by the casein structure. The decrease in water inside the dairy concentrate led to a decrease in the water concentration on the surface of the concentrate in the water activity meter or on the surface of the droplet during spray drying and decreased the drying kinetics.

These results were confirmed by the desorption drying of two different classes of proteins, i.e. MCC (micellar structure), with a β value of 0.34%/min, and WPC (globular structure), with a β value of 0.68%/min (Table 9.1). These two different types of protein had the same protein content (89% of total solids) and the same water content before desorption drying, but not the same drying time or β value. All these results show that the drying rate is dependent on the nature and the structure of the casein. Water may be less available during the drying of a protein with a micellar structure than during the drying of a protein with a globular structure.

Mineral additions to WPC decreased the β value, the smallest decrease being with citrate (29% reduction) and the greatest decrease being with NaCl (65% reduction) (Table 9.1). A decrease in the β value means a lower rate of water transfer at the end of drying (Schuck *et al.*, 1998b). The decrease in water transfer during the drying of modified WPC could be explained by the high hygroscopicity of the added mineral salts. This result suggested that the water is more closely bound to the mineral salts than to the whey proteins at the end of drying. The mineral addition to WPC under the test conditions had little effect on the whey protein structure but probably had some effect on the increase in bound water in the modified WPC.

Addition of NaCl to MCC decreased the β value (0.19%/min) (Table 9.1). Water in a NaCl-containing casein system is more rotationally mobile than water in a casein model system without NaCl, with the same water activity (Curme *et al.*, 1990). For high electrolyte concentrations, the amount of bound water decreases because of the suppression of the electrical double layer surrounding the protein molecule; this is directly related to the hydration of ions (Na^+, Cl^-) and hence to the ability to separate water molecules from the protein molecules (Robin *et al.*, 1993). Water is less closely bound to micellar casein in the presence of NaCl (Cayot and Lorient, 1998). The decrease in water transfer at the end of drying can be explained by the hygroscopicity of NaCl. The water is more closely bound to NaCl than to the micellar caseins.

Addition of $CaCl_2$ to MCC increased the β value (0.36%/min) (Table 9.1). Moreover, addition of $CaCl_2$ to milk decreases micellar solvation (Tarodo de la Fuente, 1985; Van Hooydonk *et al.*, 1986; Jeurnink and de Kruif, 1995; Le Ray *et al.*, 1998). Firstly, the water inside the micellar structure in the case of MCC without salt addition might be less available than the water inside the micellar structure in the case of MCC with $CaCl_2$. Secondly, the lower water content inside micellar structures with the addition of calcium might lead to an increase in water transfer during drying.

Addition of phosphate ions to MCC increased the β value (0.49%/min) (Table 9.1). The increase in water transfer was explained by the partial solubilization of caseins (Le Ray *et al.*, 1998), although an increase in casein micelle solvation and an increase in viscosity were observed. These results may be discussed in terms of the strength of water binding to caseins, either in a soluble form as in sodium caseinate (Schuck *et al.*, 1998b) or in the micellar structure (MCC) as in the current experiments reported here. The water bound to micellar caseins was probably less easy to remove than the water bound to soluble caseins. Partial solubilization of caseins improved the water transfer during dehydration of phosphate solution + MCC.

Addition of citrate solution to MCC at 30% total solids increased the β value (0.45%/min) (Table 9.1). Addition of citrate induces the release of large amounts of soluble casein from the micellar phase because of solubilization of colloidal calcium phosphate (Le Ray *et al.*, 1998). Similar results occurred with the addition of phosphate ion. Solubilization of the micellar casein improved the water transfer during dehydration of solution with added citrate.

Industrial implications

Introduction

Several studies (Masters, 1991; Pisecky, 1997) have reported that the moisture content can vary according to the product for the same outlet air temperature in the spray drier. For example, Pisecky (1997) has reported that the moisture content is close to 4% and 5% for WMP and SMP respectively for an outlet air temperature close to 85°C. On the other hand, to produce WMP and SMP at the same moisture content (4%, for example), the outlet air temperatures must be different (80 and 90°C respectively). All these differences can be explained by the effects of the chemical composition of the ingredients on the availability of the water that must be transferred from the droplet to the drying air.

The aims of this chapter were to evaluate water transfer during the spray drying of different dairy concentrates using thermodynamic and chemical approaches. Whey protein concentrates and isolates (WPC35, WPC50, WPC70, WPI90) with or without heat denaturation, MC, sodium caseinate (NaCas) and milk with and without whey protein enrichment were dried in a three-stage pilot plant spray drier. When the concentrate temperature, air flow rate, concentrate flow rate, total solids content of the concentrate, inlet air temperature absolute humidity, inlet air temperature before and after heating, and outlet air temperature after drying are known, it is possible to determine the specific energy consumption (SEC), which is the ratio of the energy consumed to the evaporation of 1 kg of water (measured in kJ/kg water) (Bimbenet *et al.*, 2002).

Thus, if free water is spray dried the energy spent in terms of the SEC would be close to 2500 kJ/kg water. If the concentrate has greater and greater amounts of bound water to free water, the SEC increases up to 10 000 kJ/kg water. The significance of a very high SEC is related to the water, which is less and less available, limiting water transfer, and thus increasing the surface temperature of the droplet and hence increasing the risk of protein denaturation of the powder.

Whey proteins

The results presented in Table 9.2 show that water transfer during spray drying decreased when the whey proteins were native proteins. For the same moisture, the

Table 9.2 Specific Energy Consumption at 4% moisture content for the drying of dairy proteins

	SM	SM + WPI	WPC						WPI		MC	NaCas
Protein content (%)	34	50	35		50		70		90		90	90
Heat treatment (72°C/4 min)	N	N	N	Y	N	Y	N	Y	N	Y	N	N
SEC (±3%) (kJ/kg water)	5900	6400	5950	7700	6800	6550	7050	6600	7200	6500	6900	5900

Y, heat treatment; N, no heat treatment; SM, skim milk; WPI, whey protein isolate; WPC, whey protein concentrate; MC, micellar casein; NaCas, sodium caseinate; SEC, specific energy consumption at 4% moisture content

SEC for drying was higher when (a) the native whey protein content increased in WPC and in milk and (b) the whey proteins were heat denatured in WPC35, but the SEC was lower when (c) the whey proteins were heat denatured in WPC50, WPC70 and WPI90. These results may be explained by the availability of the water (bound and unbound) in the concentrate in relation to the nature and the content of the whey proteins.

Caseins

The results presented in Table 9.2 show that water transfer during spray drying decreased when the micellar casein content increased. For the same moisture, the SEC for drying was higher when (a) the micellar casein content increased in MC compared with skim milk and (b) casein remained in a micellar state (as in MC) compared with a soluble state (e.g. in NaCas). These results may be explained by the availability of the water in the concentrate in relation to the content and the structure of the caseins. Water is more available when the caseins are soluble than when they are in a micellar state.

All these results also show that water transfer depends on the relationship between the water and the protein components and that these components should be taken into account when optimizing spray-drying parameters for proteins.

Rehydration of protein powders

Most food additives are prepared in powder form and need to be dissolved before use. Water interactions in dehydrated products and dissolution are thus important factors in food development and formulation (Hardy *et al.*, 2002). Dissolution is an essential quality attribute of a dairy powder as a food ingredient (King, 1966). Many sensors and analytical methods such as the insolubility index (International Dairy Federation, 1988; American Dairy Products Institute, 1990), nuclear magnetic resonance (NMR) spectroscopy (Davenel *et al.*, 1997), turbidity, viscosity and particle size distribution (Gaiani *et al.*, 2006) can now be used to study water transfer in dairy protein concentrates during rehydration. Using combinations of these methods, it is very easy to determine the different stages of the rehydration process, i.e. wettability, swellability, sinkability, dispersibility and solubility.

The insolubility index (ISI, in %), described by the IDF standard (International Dairy Federation, 1988), for skim milk, is the volume of sediment (for 50 mL) after rehydration (10 g of powder in 100 mL of distilled water, at 25°C), mixing (for 90 s, at 4000 rev/min) and centrifugation (for 300 s, at 160 *g*). With this method, the quantity of insoluble material (true and false not differentiated) can be determined.

NMR spectroscopy is a technique for determining the rate of dissolution, the time required for complete reconstitution of powders and the transverse relaxation rate of reconstituted solutions. A 40 mm diameter glass tube filled with 20 mL of water at 40°C was put into the gap of the magnet of a Minispec Bruker PC 10 NMR spectrometer operating at a resonance frequency of 10 MHz. A suitably designed funnel and an electric stirrer (glass spatula) were inserted into the tube. The method was first described by Davenel *et al.* (1997). They showed that the solubilization rate was independent of the quantity of powder poured (up to 20 g powder/100 mL water) and increased with the stirring rate. In subsequent experiments, the rotation rate of the stirrer was adjusted after starting to 1150 rev/min for spray-dried powders and 1 g of powder was poured into the water.

The NMR measurements were generally continued until the solution was completely reconstituted, except if insoluble material was formed. Each decay curve was obtained by sampling a maximum of 845 spin echoes of a Carr-Purcell-Meiboom-Gill (CPMG) sequence every 20 s during the reconstitution period. Interpulse spacing between 180° pulses was fixed at 2 ms to limit the effect of diffusion caused by stirring. The NMR kinetic method was used in triplicate. The CPMG curves were well approximated by the sum of two exponential curves to determine the protons attributed to water protons in fast exchange with exchangeable protons of non-dissolved powder particles, and the protons attributed to water protons and exchangeable protons in the reconstituted phase (Davenel *et al.*, 1997). With this method, it is possible to differentiate between the truly insoluble material and the falsely insoluble material. The falsely insoluble material can be explained by the low water transfer during rehydration and not by denatured protein, which is truly insoluble (Schuck *et al.*, 1994b).

For viscosity measurement, a rheometer can be used to obtain viscosity profiles. In our study, the blades were placed at right angles to each other to provide good homogenization. Industrial dissolution processes usually include stirring at a constant speed and the experiments were therefore designed to provide a constant shear rate (100 s^{-1}). MCP was added to the rheometer cup manually. The aqueous phase used was distilled water at a volume of 18 mL.The powder was dispersed in the rheometer cup 50 s after starting the rheometer. Dissolution is highly dependent on temperature and concentration. The total nitrogen concentration employed to study these effects was about 5% (w/v) and the temperature was about 24°C (Gaiani *et al.*, 2005, 2006).

The experiments to provide the turbidity profiles were carried out in a 2 L vessel equipped with a four-blade 45° impeller rotating at 400 rev/min. A double-walled jacket vessel maintained the temperature at 24°C. The turbidity sensor was placed 3 cm below the surface of the water and was positioned through the vessel wall to avoid disturbance during stirring. Turbidity changes accompanying powder rehydration were followed using a turbidity meter. The apparatus used light in the near-infrared region (860 nm), the incident beam being reflected back at 180° by any particle in suspension in the fluid to a sensitive electronic receptor (Gaiani *et al.*, 2005).

A laser light diffraction apparatus with a 5 mW He–Ne laser operating at a wavelength of 632.8 nm can be used to record particle size distributions. The particle size distribution of dried particles was determined using a dry powder feeder attachment and the standard optical model presentation for particles dispersed in air was used. To measure the particle size distribution of micellar casein in concentrates, 0.5 mL of suspension was taken from the rheometer cup and introduced into 100 mL of pre-filtered distilled water (membrane diameter, 0.22 μm) to reach the correct obscuration. The results obtained corresponded to average diameters calculated according to the Mie theory. The criterion selected was d(50), meaning that 50% of the particles had diameters lower than this criterion (midpoint of cumulative volume distribution) (Gaiani *et al.*, 2005, 2006).

Using this combination of three methods, it was possible to follow the water transfer during rehydration and to obtain the wetting time, determined using the first peak of increased viscosity and turbidity, and the swelling time, determined using the second peak of viscosity in relation to the increase in particle size. The rehydration time was then determined according to stabilization of the viscosity, turbidity and particle size values.

The results in Table 9.3 show that rehydration of MCP occurs in different stages: first there is wetting and swelling of the particles, followed by slow dispersion to reach a homogeneous fluid, in agreement with Gaiani *et al.* (2005, 2006). Using an NMR method, Davenel *et al.* (1997) also demonstrated two stages during MC rehydration, attributed to water absorption by powder and solubilization of particles (i.e. swelling and dispersion stages). They estimated the water uptake by the powder at around 5 g water/g powder during the first 20 min of rehydration but could not identify a wetting stage with this method.

MCPs with a high ISI (14.5 mL) are generally considered to be poorly soluble powders in which rehydration of the micelle remains incomplete (Jost, 1993).

Table 9.3 Reconstitution period, insolubility index and rehydration time of dairy protein powders

Powders	RP using NMR (min)	ISI using IDF Standard (mL)	WT (min)	ST (min)	DT + SolT (min)	RT (min)
MCP G	22	14.5	1	2	804	807
MCP NG	8	3.5	3	17	551	571
MCP + Carbohydrate G	18	5.0	1	nm	nm	116
MCP + Carbohydrate NG	nm	nm	2	0	95	97
MCP + NaCl G	9.5	0.9	nm	nm	nm	nm
MCP + $CaCl_2$ G	∞	14.5	nm	nm	nm	nm
MCP + SCS/SPS G	6/5	<0.5	nm	nm	nm	nm
WPP G	5	<0.5	4	0	0	4
WPP NG	15	<0.5	17	0	0	17

MCP, micellar casein powder; G, granulated; NG, non granulated; SCS, sodium citrate solution; SPS, sodium phosphate solution; WPP, whey protein powder; RP, reconstitution period; ISI, insolubility index; WT, wetting time; ST, swelling time; DT, dispersibility time; SolT, solubility time; RT, rehydration time = WT + ST + DT + SolT; ∞, infinite delay; nm, not measured

Addition of NaCl to the MC concentrate before spray drying considerably reduced the ISI and reconstitution period (RP) values (ISI 0.9 mL; RP 9.5 min) (Table 9.3). It has been hypothesized that the significant decrease in the RP value is more probably related to the hygroscopic strength of NaCl.

Addition of sodium citrate solution (SCS) or sodium phosphate solution (SPS) resulted in fast solubilization, as shown by the very low RP value and by the ISI value lower than 0.5 mL (Table 9.3). The resulting solution consisted of casein micelles in the form of sodium caseinate, associated with the occurrence of a single proton population, characterized by NMR relaxation rates that were much lower than the relaxation rate measured with reconstituted MC. This could be attributed to a decrease in the amount of hydration water induced by the change in micelle structure. The transparency of the solution indicated the formation of soluble caseins related to greater quantities of calcium complexes.

Reconstitution of MCP in the presence of $CaCl_2$ led to considerable changes in protein structure, associated with instability of the casein micelles, which began to precipitate just after mixing, as shown by the high ISI and the non-measurable RP value due to experimental delay. This precipitate probably resulted from aggregation of casein micelles or sub-micelles through a decreasingly negative charge on the protein by additional calcium binding, leading to a reduction in electrostatic repulsion (Dalgleish, 1982). In this case, the high ISI of these solutions was related to the presence of insoluble substances whereas, in the case of rehydration of MCP, the high ISI represented only the low water transfer rate in the casein (Table 9.3).

On the other hand, the rehydration of whey powders was totally different (Table 9.3). As the wettability of whey powders is poor, the instability at the beginning of the profile could be due to lump formation going past the sensor, as reported by Freudig *et al.* (1999). For non-granulated (NG) WPI powder, the very long signal instability could be explained by a tendency for the lumps to be stuck together in a thick layer of wet particles, due to the small size of the particles (Kinsella, 1984). Powder swelling was not reported for WPI powders, probably because globular protein powders bind less water than intact casein micelle powders (Kinsella, 1984; Robin *et al.*, 1993). De Moor and Huyghebaert (1983) also reported that whey powders have a lower water holding capacity than casein powder.

As expected, granulation had a positive effect on wetting. The wetting time was systematically better for granulated particles. This phenomenon is well known, as fast wetting is enhanced, with large particles forming large pores, high porosity and small contact angles between the powder surface and the penetrating water (Pisecky, 1986; Freudig *et al.*, 1999; Gaiani *et al.*, 2005). A surprising influence of granulation on the rehydration time was observed. Depending on the nature of the protein, the granulation influence involved opposite effects. WPI rehydration was enhanced for granulated particles whereas the rehydration time was shorter for non-granulated particles of MCP. The controlling stage for whey proteins is wetting (Baldwin and Sanderson, 1973; Schubert, 1993). As granulation improves the wetting stage, the rehydration of whey powders is enhanced for granulated particles. In contrast, the controlling stage for casein proteins is dispersion. Indeed, even with a shorter wetting time, a granulated powder is slower to rehydrate than a non-granulated powder (Gaiani *et al.*, 2005).

These results are not compatible with those of other studies, in which it was generally accepted that a single particle size around 200 μm (Neff and Morris, 1968) or 400 μm (Freudig *et al.*, 1999) represented optimal dispersibility and sinkability. In fact, this optimal particle size depends on the composition of the dairy powder. As shown in Table 9.3, if the industry wishes to optimize powder rehydration, it seems to be better to rehydrate granulated powders if the protein is whey and to rehydrate non-granulated powders if the protein is casein.

Conclusions

The aim of this chapter was to explain the process of dehydration, i.e. spray drying, in order to understand the effects of spray drying on the quality of protein powders (micellar caseins [MC] and globular proteins [WPC]) during drying and rehydration. We then demonstrated that the quality of these powders depends on the chemical environment.

It is very important for the dairy industry to understand that enrichment of milk in micellar casein (by ultrafiltration or microfiltration) decreases water transfer during the drying and rehydration processes. Insolubility (International Dairy Federation, 1988) is related to the decrease in water transfer required for rehydration and not to thermal denaturation, and decrease in water transfer is related to the micellar structure. The destabilization of the micellar structure induced by the addition of phosphate or citrate solution to MC increases water transfer during drying and during rehydration. Water transfer in WPC or MC with added carbohydrates is improved during rehydration. Addition of NaCl to MC decreases water transfer during drying but increases water transfer during rehydration, and thus is related to the hygroscopicity of the carbohydrate and the NaCl.

The industrial requirement for protein powders with specific properties is expanding. As powder is the easiest way to carry and store milk derivatives, an understanding of the rehydration behavior of a dairy powder will become more and more important in the future.

Moreover, it is essential for both dairy powder producers and dairy powder users to have a method to evaluate the rehydration behavior of dairy powders. As demonstrated in several studies, the industry should take into account certain technological factors such as granulation and the incorporation mode, and also the nature of the protein being rehydrated, to optimize the rehydration of a dairy powder. In contrast to other studies, we found that improving the wetting stage by using granulated powders did not systematically improve the total rehydration. Depending on the nature of the protein, it seems to be better to work with granulated (for whey) or non-granulated (for micellar casein) powders to obtain more rapid rehydration (Gaiani *et al.*, 2007).

In conclusion, water transfer in dairy protein concentrates during dehydration and during rehydration depends on the aqueous environment, the nature of the mineral salts and the structure of the dairy proteins (MC or WPC). The water–protein interaction requires further study, to understand the effects of preheat treatment and spray drying on the functional properties of protein powders.

References

American Dairy Products Institute (1990). *Standards for Grades of Dry Milk Including Methods of Analysis.* Chicago, Illinois: American Dairy Products Institute.

Baldwin, A. J. and Sanderson, W. B. (1973). Factors affecting the reconstitution properties of whole milk powder. *New Zealand Journal of Dairy Science and Technology*, **8**, 92–100.

Baldwin, A. J., Baucke, A. G. and Sanderson, W. B. (1980). The effect of concentrate viscosity on the properties of spray dried skim milk powder. *New Zealand Journal of Dairy Science and Technology*, **15**, 289–97.

Bimbenet, J. J., Schuck, P., Roignant, M., Brulé, G. and Méjean, S. (2002). Heat balance of a multistage spray-dryer: principles and example of application. *Lait*, **82**, 541–51.

Bloore, C. G. and Boag, I. F. (1982). The effect of processing variables on spray dried milk powder. *New Zealand Journal of Dairy Science and Technology*, **17**, 103–20.

Caron, A., St-Gelais, D. and Pouliot, Y. (1997). Coagulation of milk enriched with ultrafiltered or diafiltered microfiltered milk retentate powders. *International Dairy Journal*, **7**, 445–51.

Cayot, P. and Lorient, D. (1998). Modifications chimiques (biochimiques) de l'environnement des protéines du lait. In *Structures et Technofonctions des Protéines du Lait* pp. 159–78. Paris: Technique et Documentation Lavoisier.

CNIEL (1991). Centre National Interprofessionnel de l'Economie Laitière, L'économie laitière en chiffres, Paris.

CNIEL (2005). Centre National Interprofessionnel de l'Economie Laitière, L'économie laitière en chiffres, Paris.

Curme, A. G., Schmidt, S. J. and Steinberg, M. P. (1990). Mobility and activity of water in casein model systems as determined by ^{2}H NMR and sorption isotherms. *Journal of Food Science*, **55**, 430–33.

Dalgleish, D. G. (1982). Milk proteins: chemistry and physics. In *Food Proteins* (P. F. Fox and J. J. Condon, eds) pp. 155–78. London: Applied Science Publishers.

Davenel, A., Schuck, P. and Marchal, P. (1997). A NMR relaxometry method for determining the reconstitutability and the water-holding capacity of protein-rich milk powders. *Milchwissenschaft*, **52**, 35–39.

De Moor, H. and Huyghebaert, A. (1983). Functional properties of dehydrated protein-rich milk products. In *Physico-chemical Aspects of Dehydrated Protein-rich Milk Products*, Proceedings of IDF Symposium, Helsingor, Denmark, pp. 276–301. Statens Forsogsmejeri, Hillerod.

De Vilder, J. (1986). La fabrication de poudre de lait écrémé instantanée. I. Les caractéristiques physiques et chimiques. *Revue Agricole*, **39**, 865–77.

De Vilder, J., Martens, R. and Naudts, M. (1979). The influence of the dry matter content, the homogenization and the heating of concentrate on physical characteristics of whole milk powder. *Milchwissenschaft*, **34**, 78–84.

Fauquant, J., Maubois, J. L. and Pierre, A. (1988). Microfiltration du lait sur membrane minérale. *Techniques Laitières*, **1028**, 21–23.

Freudig, B., Hogekamp, S. and Schubert, H. (1999). Dispersion of powders in liquid in a stirred vessel. *Chemical Engineering and Processing*, **38**, 525–32.

Gaiani, C., Banon, S., Scher, J., Schuck, P. and Hardy, J. (2005). Use of a turbidity sensor to characterize casein powders rehydration: influence of some technological effects. *Journal of Dairy Science*, **88**, 2700–06.

Gaiani, C., Scher, J., Schuck, P., Hardy, J., Desobry, S. and Banon, S. (2006). The dissolution behaviour of native phosphocaseinate as a function of concentration and temperature using a rheological approach. *International Dairy Journal*, **16**, 1427–34.

Gaiani, C., Schuck, P., Scher, J., Hardy, J., Desobry, S. and Banon, S. (2007). Dairy powder rehydration: influence of proteins and some technological effects. *Journal of Dairy Science*, **90**, 570–81.

Goudédranche, H., Maubois, J. L., Ducruet, P. and Mahaut, M. (1980). Utilization of the new mineral UF membrane for making semi-hard cheeses. *Desalination*, **35**, 243–58.

Hall, C. W. and Hedrick, T. I. (1966). Quality control and sanitation. In *Drying Milk and Milk Products* (C. W. Hall and T. I. Hedrick, eds) pp. 197–231. Westport, Connecticut: AVI Publishing Co.

Hardy, J., Scher, J. and Banon, S. (2002). Water activity and hydration of dairy powders. *Lait*, **82**, 441–52.

Ilari, J. L. and Loisel, C. (1991). La maîtrise de la fonctionnalité des poudres. *Process*, **1063**, 39–43.

International Dairy Federation (1998). Dried milk and milk products — determination of insolubility index. IDF Standard 129A. Brussels: International Dairy Federation.

International Dairy Federation (2005). *World Dairy Situation*. CD-Rom. Brussels: International Dairy Federation.

Jeantet, R., Schuck, P., Famelart, M. H. and Maubois, J. L. (1996). Intérêt de la nanofiltration dans la production de poudres de lactosérum déminéralisées. *Lait*, **76**, 283–301.

Jeurnink, T. J. M. and de Kruif, K. G. (1995). Calcium concentration in milk in relation to heat stability and fouling. *Netherlands Milk and Dairy Journal*, **49**, 151–165.

Jost, R. (1993). Functional characteristics of dairy proteins. *Trends in Food Science and Technology*, **4**, 283–88.

Kessler, H-G. (1981). Drying — instantizing In *Food Engineering and Dairy Technology*, pp. 269–328. Freising: Verlag A Kessler.

King, N. (1966). Dispersibility and reconstitutability of dried milk. *Dairy Science Abstracts*, **28**(3), 105–18.

Kinsella, J. E. (1984). Milk proteins: physicochemical and functional properties. *Critical Reviews in Food Science and Nutrition*, **21**, 197–262.

Kjaergaard, J. G., Ipsen, R. H. and Ilsoe, C. (1987). Functionality and application of dairy ingredients in dairy products. *Food Technology*, **41**, 66–71.

Le Graët, Y. and Maubois, J. L. (1979). Fabrication de fromages à pâte fraîche à partir de poudres de rétentat et de préfromage. *Revue Laitière Française*, **373**, 23–26.

Le Ray, C., Maubois, J. L., Gaucheron, F., Brulé, G., Pronnier, P. and Garnier, F. (1998). Heat stability of reconstituted casein micelle dispersions: changes induced by salt addition. *Lait*, **78**, 375–90.

Madsen, R. F. and Bjerre, P. (1981). Production of cheese-base. *Nordeuropaeisk Mejeri Tidsskrift*, **5**, 135–39.

Mahaut, M., Jeantet, R., Brulé, G. and Schuck, P. (2000). *Les Produits Industriels Laitiers*. Paris: Technique et Documentation Lavoisier.

Masters, K. (1991). *Spray Drying.* Essex: Longman Scientific and Technical.

Maubois, J. L. (1991). New applications of membrane technology in the dairy industry. *Australian Journal of Dairy Technology*, **46**, 91–95.

Maubois, J. L., Pierre, A., Fauquant, J. and Piot, M. (1987). Industrial fractionation of main whey proteins. *Bulletin of the International Dairy Federation*, **212**, 154–59.

Neff, E. and Morris, H. A. (1968). Agglomeration of milk powder and its influence on reconstitution properties. *Journal of Dairy Science*, **51**, 330–38.

Pierre, A., Fauquant, J., Le Graët, Y., Piot, M. and Maubois, J. L. (1992). Préparation de phosphocaséinate natif par microfiltration sur membrane. *Lait*, **72**, 461–74.

Piot, M., Vachot, J. C., Veaux, M., Maubois, J. L. and Brinkman, G. E. (1987). Ecrémage et épuration bactérienne du lait entier cru par microfiltration sur membrane en flux tangentiel. *Technique Laitière and Marketing*, **1016**, 42–46.

Pisecky, J. (1980). Bulk density of milk powders. *Australian Journal of Dairy Technology*, **35**, 106–11.

Pisecky, J. (1981). Technology of skimmed milk drying. *Journal of the Society of Dairy Technology*, **34**, 57–62.

Pisecky, J. (1986). Standards, specifications and test methods for dry milk products. In *Concentration and Drying of Food* (D. MacCarthy, ed.) pp. 203–20. London: Elsevier.

Pisecky, J. (1990). 20 years of instant whole milk powder. *Scandinavian Dairy Information*, **4**, 74.

Pisecky, J. (1997). *Handbook of Milk Powder Manufacture.* Copenhagen: Niro A/S.

Robin, O., Turgeon, S. and Paquin, P. (1993). Functional proteins of milk proteins. In *Dairy Science and Technology Handbook. 1. Principles and Properties*, pp. 277–353. New York: VCH Publishers.

Roos, Y. H. (2002). Importance of glass transition and water activity to spray drying and stability of dairy powders. *Lait*, **82**, 478–84.

Schubert, H. (1993). Instantization of powdered food products. *International Chemical Engineering*, **33**(1), 28–45.

Schuck, P., Piot, M., Méjean, S., Fauquant, J., Brulé, G. and Maubois, J. L. (1994a). Déshydratation des laits enrichis en caséine micellaire par microfiltration; comparaison des propriétés des poudres obtenues avec celles d'une poudre de lait ultra-propre. *Lait*, **74**, 47–63.

Schuck, P., Piot, M., Méjean, S., Le Graët, Y., Fauquant, J., Brulé, G. and Maubois, J. L. (1994b). Déshydratation par atomisation de phosphocaséinate natif obtenu par microfiltration sur membrane. *Lait*, **74**, 375–88.

Schuck, P., Roignant, M., Brulé, G., Méjean, S. and Bimbenet, J. J. (1998a). Caractérisation énergétique d'une tour de séchage par atomisation multiple effet. *Industries Alimentaires et Agricoles*, **115**, 9–14.

Schuck, P., Roignant, M., Brulé, G., Davenel, A., Famelart, M. H. and Maubois, J. L. (1998b). Simulation of water transfer in spray drying. *Drying Technology*, **16**, 1371–93.

Schuck, P., Briard, V., Méjean, S., Piot, M., Famelart, M. H. and Maubois, J. L. (1999). Dehydration by desorption and by spray drying of dairy proteins: influence of the mineral environment. *Drying Technology*, **17**, 1347–57.

Sougnez, M. (1983). L'évolution du séchage par atomisation. *Chimie Magazine*, **1**, 1–4.

Straatsma, J., Vanhouwelingen, G., Steenbergen, A. E. and Dejong, P. (1999a). Spray drying of food products. 1. Simulation model. *Journal of Food Engineering*, **42**, 67–72.

Straatsma, J., Vanhouwelingen, G., Steenbergen, A. E. and Dejong, P. (1999b). Spray drying of food products. 2. Prediction of insolubility index. *Journal of Food Engineering*, **42**, 73–77.

Tarodo de la Fuente, B. (1985). Solvation of casein in bovine milk. *Journal of Dairy Science*, **58**, 293–300.

Trouvé, E., Maubois, J. L., Piot, M., Madec, M. N., Fauquant, J., Rouault, A., Tabard, J. and Brinkman, G. (1991). Rétention de différentes espèces microbiennes lors de l'épuration du lait par microfiltration en flux tangentiel. *Lait*, **71**, 1–13.

Tuohy, J. J. (1989). Some physical properties of milk powders. *Irish Journal of Food Science and Technology*, **13**, 141–52.

Van Hooydonk, A. C. M., Hagedoorn, H. G. and Boerrigter, I. J. (1986). The effect of various cations on the renneting of milk. *Netherlands Milk and Dairy Journal*, **40**, 369–90.

Vincens, D. and Tabard, J. (1988). L'élimination des germes bactériens sur membranes de microfiltration. *Techniques Laitières*, **1033**, 62–64.

10

Changes in milk proteins during storage of dry powders

Kerianne Higgs and Mike Boland

Abstract

Milk proteins undergo chemical changes, even in dried powders. This chapter reviews the changes undergone by caseins and whey proteins in milk powders and in purified protein products. Maillard compounds are of particular importance in milk powders, milk protein concentrates and whey protein concentrates, i.e. products where lactose is present. Caseins undergo isopeptide bond formation, as the result of dephosphorylation of phosphoserine and subsequent reactions of the dehydroalanine produced.

We discuss the nutritional significance of these changes. Both the Maillard reaction and the formation of isopeptides lead to loss of bioavailable lysine. This is not a problem in dairy proteins, which contain an excess of lysine, but it can affect the nutritional value of protein blends which may be limiting in lysine.

Introduction

Milk is an unstable foodstuff, prone in particular to microbiological degradation, but also to long-term chemical change. A large part of the world's dairy production occurs in areas remote from the markets in which it is consumed and production is often seasonal, requiring storage to smooth out supply. The production of milk powders and other dried milk-protein-containing products has been the method of choice for over

Milk Proteins: From Expression to Food
ISBN: 978-0-12-374039-7

100 years for the storage and shipping of milk over long distances and/or times, as it confers stability and massively reduces weight and bulk.

Milk powders were known to the Chinese and were described by Marco Polo. The production of milk powders was described by Nicolas Appert in the early nineteenth century and commercial processes for the spray drying of milk were patented in the USA in 1872 and 1905. This opened the way for large-scale industrial production of milk powders throughout the twentieth century.

Table 10.1 summarizes the biggest exporters and importers of milk powders in 2006. In the same period, more than 7 million tonnes of milk powder was produced globally.

In addition to milk powders, dried dairy protein products that are traded on the world market, largely as food ingredients, include casein and caseinate, whey powders, whey protein concentrates, whey protein isolates, milk protein concentrates, milk protein isolates and specialist nutritional powders and blends, which may also contain hydrolyzed dairy proteins.

Milk powders are used primarily for making reconstituted and recombined milks, usually sold to consumers in UHT format, although substantial amounts are used to make other dairy products such as yoghurt and ice cream as well as being minor ingredients in a wide range of non-dairy foods.

The main uses for dried milk protein products are nutritional and products include infant formulas, medical foods, specialist foods for weight management and foods for muscle building (where high protein and high levels of branched-chain amino acids are desirable). They are also used in non-nutritional applications including desserts, confectionery, toppings, imitation cheeses, sauces and dressings and coffee whiteners, where their functional properties are important.

These products, especially those from New Zealand and Australia, often have long storage times because of geographic distance to market and seasonal production. Research from our laboratories and that of others has shown that a range of reactions can occur in the dry powders and that they can affect powder functionality and nutritional value. The quality of dried milk protein products deteriorates on storage at ambient temperatures because of two main reactions: the Maillard reaction and isopeptide bond formation. There are also some minor reactions, which are covered later.

Table 10.1 Main exporters and importers of milk powders in 2006 (000 tonnes)

Export	New Zealand	European Union	Australia	USA	Argentina	Ukraine
SMP	245	185	204	250	–	67
WMP	630	470	170	–	190	–
Total	875	655	374	250	190	67
Import	**Philippines**	**Mexico**	**Algeria**	**Indonesia**	**Russia**	**China**
SMP	135	172	110	135	75	–
WMP	52	–	60	–	30	60
Total	187	172	170	135	105	60

Source: *Barry Wilson's Dairy Industry News*, 18:8, August 2006

Standard abbreviations for many of the dried milk products are used throughout the world and are used in this chapter: SMP = skim milk powder; WMP = whole milk powder; WPC = whey protein concentrate; WPI = whey protein isolate (usually >85% protein); MPC = milk protein concentrate; MPI = milk protein isolate. Milk powders sold internationally must conform to standard protein levels and fat levels for WMP. A number following the abbreviation for the other products is the percentage of protein by weight in the dry powder. Caseins and caseinates are not usually abbreviated and are typically >90% protein by dry weight.

A range of chemical reactions that modify proteins can occur in dried milk products, particularly at elevated temperature. The most important of these both involve lysyl side chains and are the formation of Maillard and pre-Maillard compounds, in the presence of sugars, and the formation of isopeptide bonds, particularly in products containing phosphoseryl residues, such as casein.

The formation of Maillard and pre-Maillard compounds

The Maillard reaction occurs when lysine-containing proteins interact with reducing sugars. The first stable compound formed during the Maillard reaction is the Amadori product (so-called because it is the result of a class of reaction called the Amadori rearrangement), shown in Figure 10.1. These compounds block the ε-amino groups of lysine residues, reducing the bioavailability of that essential amino acid. This reaction is dependent on a reducing sugar, usually lactose, being present as the co-reactant (Erbersdobler, 1986).

Highly pure protein products such as casein and caseinate do not suffer significantly from this reaction, as not enough lactose is present. The reaction is particularly important in WMPs, SMPs, WPCs and MPCs and can occur to a limited extent in some MPIs and WPIs. The rate of reaction is critically dependent on the level of moisture (water activity) and the temperature as well as the lactose content.

Advanced Maillard reaction products are partly responsible for the development of aromas and color during food processing and preparation.

Measuring lactulosyl lysine levels

Lactulosyl lysine formation is conveniently monitored by measuring furosine concentrations and can be indirectly measured by monitoring the amount of available amine. The results from both these methods have been shown to correlate well with the lactulosyl lysine measured in WPCs directly by mass spectrometry (Figure 10.2).

Rates of formation of lactulosyl lysine

We studied the rate of lactosylation for WPC products as dry powders. The rates were found to be dependent on the lactose concentration (a consequence of processing to reach desired protein levels), water activity (a_W) and temperature (T). The trials lasted

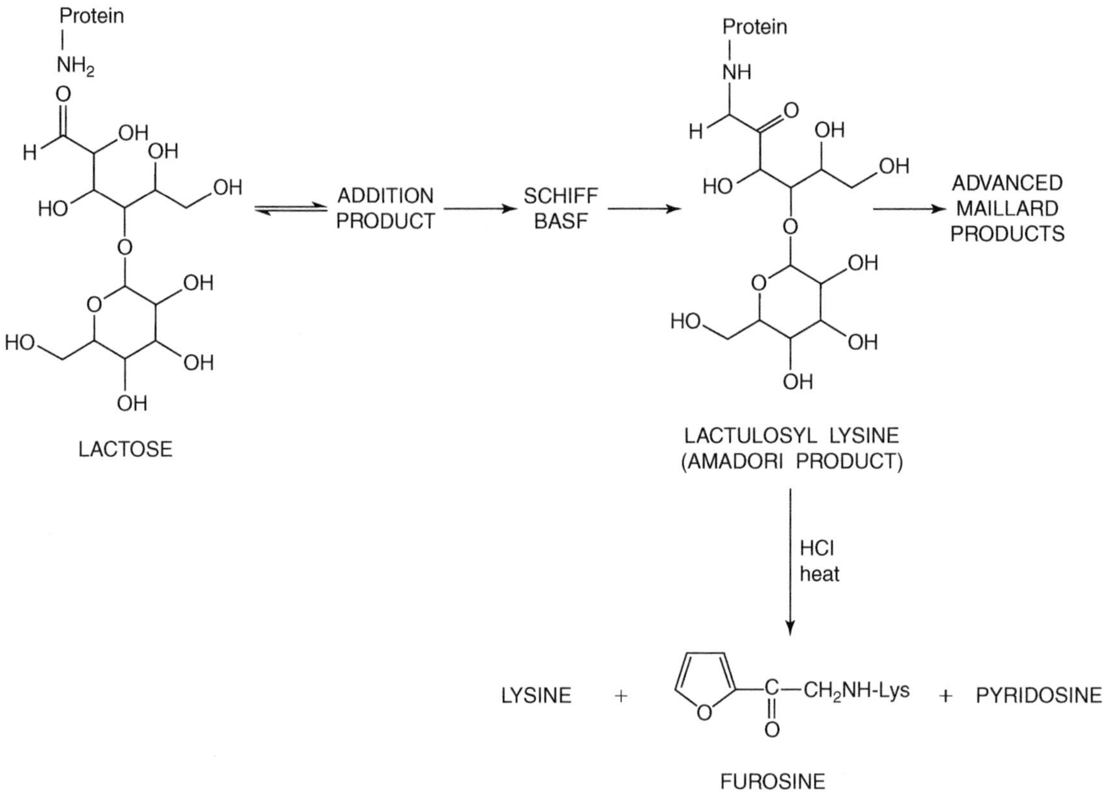

Figure. 10.1 Formation of lactulosyl lysine.

only a few months and extrapolation beyond that timeframe cannot be done with confidence.

Detailed kinetics developed using WPC80 for a range of a_W and T values allowed the rates of lactosylation to be predicted for periods up to 4 months.

For the kinetic evaluation, Figure 10.1 can be simplified to

$$\text{reactants} \xrightarrow{k_1} \text{lactulosyl lysine} \xrightarrow{k_2} \text{advanced Maillard products}$$

where k_1 and k_2 are the rate constants for the formation and degradation of lactulosyl lysine.

The rate equation for the formation of lactulosyl lysine is

$$\frac{dL}{dt} = k_1[\text{reactants}]^{n_1} - k_2[L]^{n_2} \tag{10.1}$$

where L represents lactulosyl lysine and n_1 and n_2 are the rate orders for the formation and degradation of lactulosyl lysine respectively.

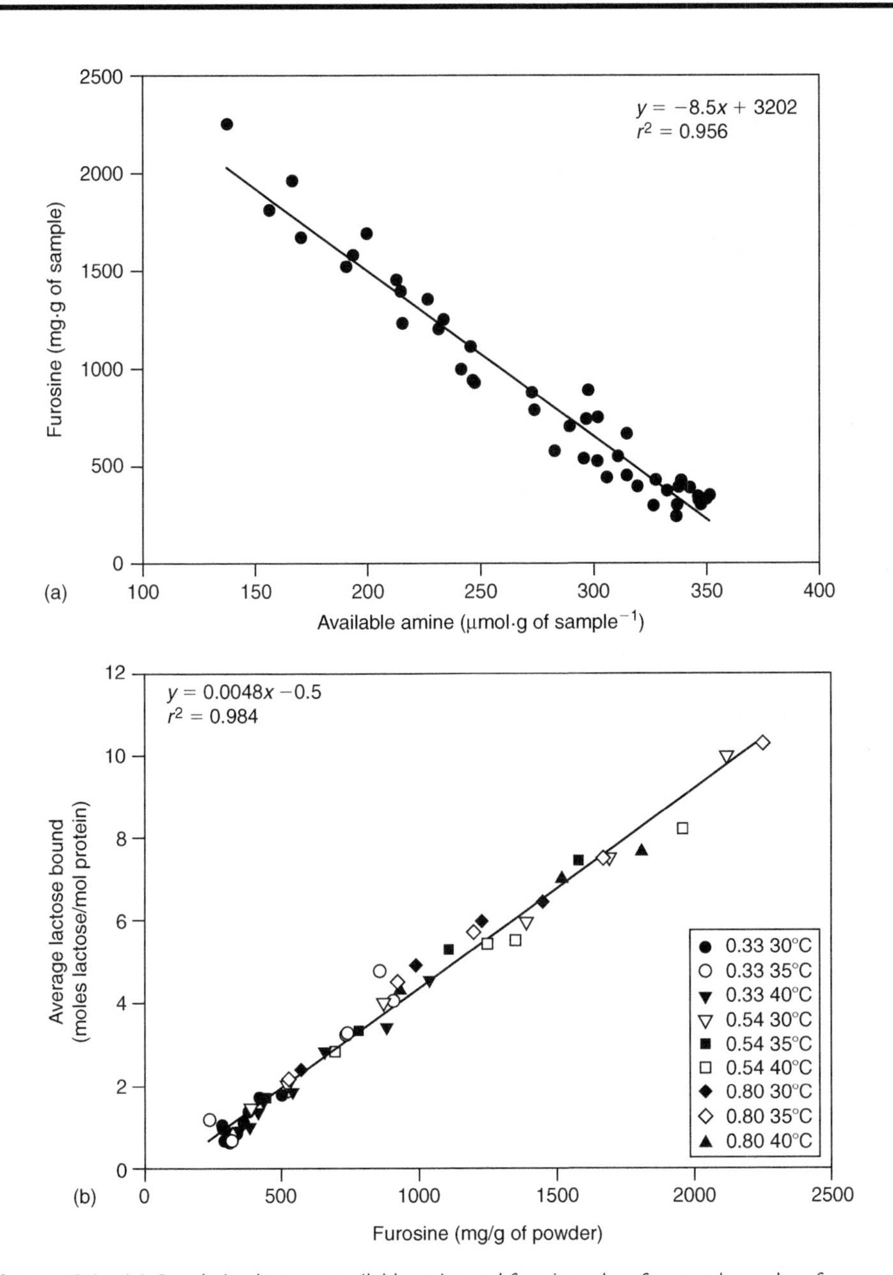

Figure. 10.2 (a) Correlation between available amine and furosine values for stored samples of a WPC56. (b) Average number of lactose molecules bound, determined by mass spectrometry, against furosine for a WPC56 with different a_w values (K. Higgs, unpublished data).

Because the formation of lactulosyl lysine in dairy powders uses only a fraction of the available reactants, the first reaction can be considered to be zero order and the degradation of lactulosyl lysine is a first-order process. This simplifies the equation to

$$\frac{dL}{dt} = k_1 - k_2[L] \tag{10.2}$$

Rearrangement of this equation and solving for $[L]_t$ gives

$$[L]_t = \left[\frac{k_1}{k_2} - \left(\frac{k_1}{k_2} - [L]_0\right)\exp(-k_2 t)\right] \tag{10.3}$$

Furosine is a hydrolysis product of lactulosyl lysine and gives a direct measure of its concentration (Figure 10.2). Therefore, furosine can be substituted for lactulosyl lysine in the rate equations.

The samples used were individual samples that were removed from controlled storage at seven time points (2, 6, 12, 24, 40, 78 and 116 days). They were analyzed for furosine and the values were plotted against time. The rates were determined using non-linear regression with Sigma Plot 8.0. At 40°C, the samples at time points after 40 days for water activities of 0.54 and 0.80 showed advanced Maillard browning and were not included in the regression analysis. R^2 values for the regressions were between 0.95 and 0.99.

The lactosylation rate constants were greater at higher temperatures and at higher water activities. The rates were low at low water activity ($a_W = 0.33$), with small increases with temperature; however, at higher water activities, the rates increased substantially with increasing temperature (Figure 10.3).

It was observed for a number of WPC80 specifications that lactosylation appeared to stop when only 2.5 (average) or 3 lactose molecules had been bound per β-lactoglobulin molecule. This corresponded to about 20% of the total lysine being blocked. This condition was specific to the 80% protein products, which contained

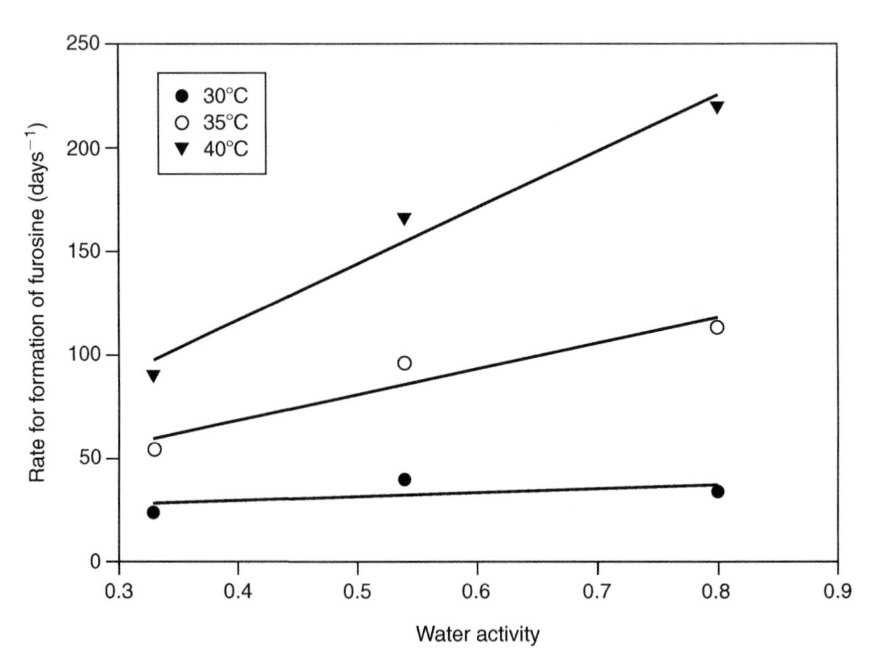

Figure. 10.3 Rate constants for the formation of furosine in a WPC80.

about 12% lactose. WPC56 products showed a much greater degree of lactosylation. However, it should be noted that much higher levels of lactosylation have reportedly been seen in overseas laboratories in WPC80 samples stored for long periods (W. J. Harper, Ohio State University, 2003, personal communication).

Formation of isopeptide bonds

Isopeptide bonds are formed largely by the breakdown of the phosphoseryl side chains that are present in products containing casein, to form dehydroalanyl side chains (Friedman, 1999). The latter are reactive and will form cross-links, mainly with adjacent lysyl (but also with histidinyl or cysteinyl) side chains to form lysinoalanyl, histidinoalanyl or lanthionyl isopeptides respectively. This reaction is not known to be significant in whey products, which do not contain significant amounts of phosphoseryl residue.

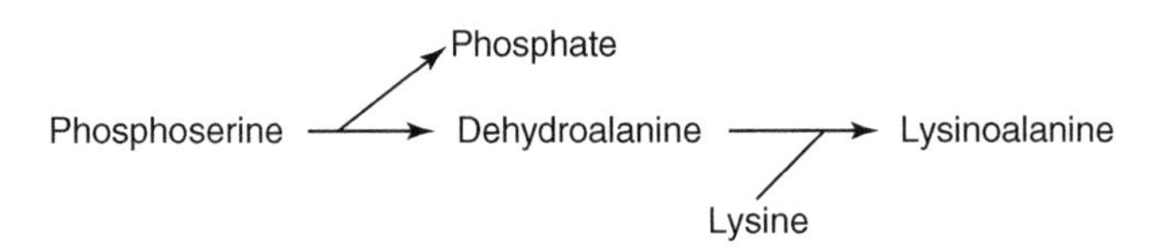

Figure. 10.4 Formation of lysinoalanine.

The main isopeptide product on digestion gives lysinoalanine, which renders lysine non-bioavailable. Additional minor reactions form histidinoalanine and lanthionine on digestion. The latter compound, although only a minor component, is also important because it renders cysteine partially non-bioavailable and the sulfur amino acids are often nutritionally limiting in milk proteins. (Note—lanthionine formation blocks the bioavailability of cysteine, which is not normally considered to be an essential amino acid, because it can be synthesized from methionine; however, methionine is itself a nutritionally limiting amino acid in casein.) Studies have indicated that, although lanthionine and histidinoalanine linkages are formed under the alkaline conditions encountered during processing, it is only lysinoalanine that is formed in the neutral conditions encountered in powders.

Lysinoalanine is usually measured directly in protein hydrolysates as part of an extended amino acid analysis. In casein products, measurement of available amine is a good alternative.

Rates of formation of lysinoalanine

The rates of formation of lysinoalanine in caseinates have been investigated by W. Thresher (1996, 1997, personal communications), for a range of temperatures and water activities. The rate constants for lysinoalanine formation as a function of temperature are shown in Figure 10.5.

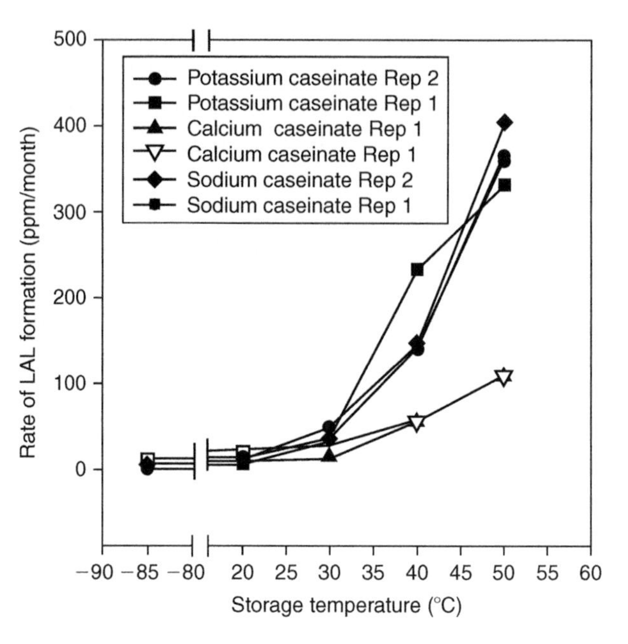

Figure. 10.5 Rate constants for lysinoalanine formation in caseinates stored at various temperatures (W. Thresher, 1996, personal communication).

Amino acids other than lysine

Cysteine, methionine and tryptophan are other essential amino acids that could be rendered non-bioavailable by reacting during processing and/or storage. Cysteine undergoes β-elimination to give dehydroalanine when treated with alkali. This can then react with a lysine residue to give lysinoalanine. Cysteine can also react with dehydroalanine to give lanthionine. Chemically determined values for cysteine and lysine availability have been found to correlate well with rat protein efficiency ratios for heat- and alkali-treated caseinates (W. Thresher, 1996, personal communication).

Tryptophan residues are relatively stable during processing and storage. They are not easily oxidized and have been found to be relatively resistant to oxidizing lipids, alkali, quinones and reducing sugars (Nielsen *et al.*, 1985). Any losses are small and not significant when compared with losses of other amino acids such as methionine and lysine. However, we have seen small amounts of oxidized tryptophan residues in digests of trim milk purchased from the supermarket (Figure 10.6).

Methionine is relatively easily oxidized to the sulfoxide, but methionine in the sulfoxide form is still bioavailable (Nielsen *et al.*, 1985).

Table 10.2 indicates levels of key essential amino acids following either alkali treatment of casein or extensive lactosylation of WPC. Note that the losses of amino acids other than lysine are considerably lower than the losses of lysine. The conditions used for the casein are well beyond any normal exposure during processing or storage.

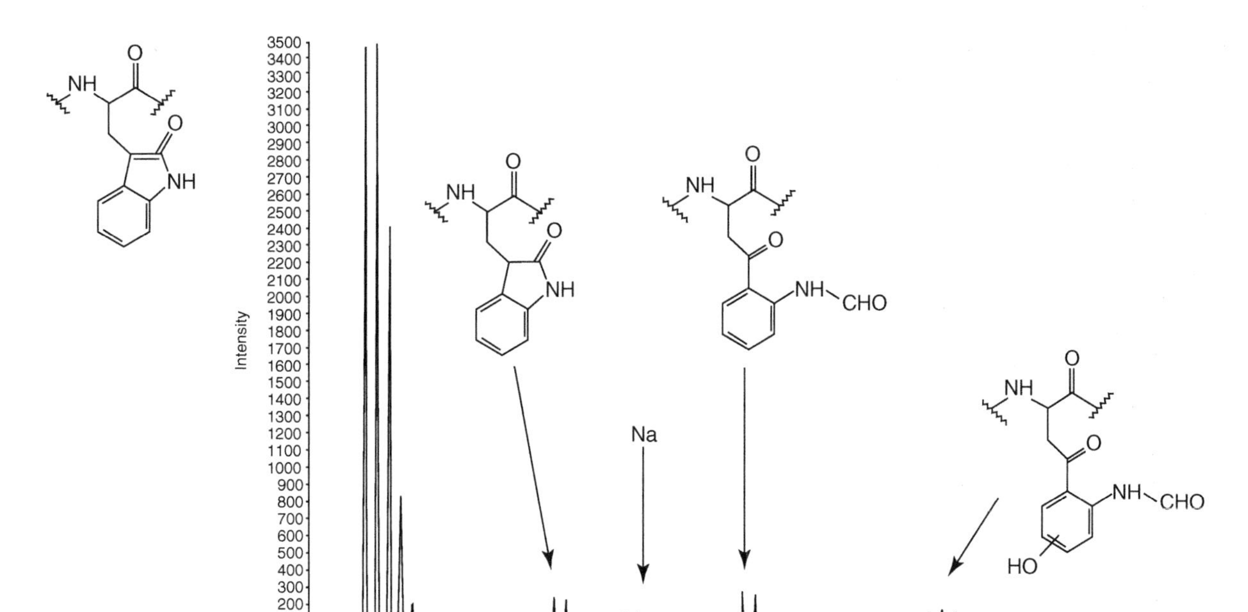

Figure. 10.6 Time-of-flight mass spectrometry of the M^{2+} ion from the tryptophan-containing κ-casein peptide SPAQILQWQVLSNTVPAK. The chemical structures in the figure show the various levels of oxidation found.

Table 10.2 Amino acid concentrations of a control and an alkali-treated casein, and a control and a lactosylated WPC56 (Sources: Nielsen *et al.*, 1985; K Higgs, unpublished results)

Amino acid	Concentration (mg/g crude protein)		Concentration(mg/g sample)	
	Control casein	Alkali-treated casein[a]	Control WPC	Lactosylated WPC[b]
Tryptophan	13.4	12.1 (90%)	8.4	8.1 (96%)
Lysine	91.0	60.8 (67%)	50.3	44.5 (88%)
Methionine	31.4	25.3 (80%)	12.3	12.5 (102%)
Cysteine	4.4	3.3 (75%)	14.9	14.8 (99%)

[a] Casein was heated for 4 h at 80°C in 0.15 M NaOH (Nielsen *et al.*, 1985)
[b] WPC was heated at 40°C, a_W 0.75 for 100 h. The median number of lactosyl groups on β-lactoglobulin was 5 (K Higgs, unpublished results)

Implications for nutritional value of milk proteins

Milk protein is rich in essential amino acids, with many, including lysine, well exceeding recommended requirements (Table 10.3). Use of these proteins as the predominant nutritional source thus poses few problems if some of the lysine is non-bioavailable.

Table 10.3 Essential amino acid content of milk and other proteins

Essential amino acid (AA)	Recommended requirement[a] (mg/g protein)	Caseinate	MPI	WPC	Soy	Wheat
Isoleucine	28	46	44	54	47	33
Leucine	66	91	103	119	85	68
Lysine	58	77	81	94	63	27
Sulfur AA[b]	25	33	39	52	24	39
Aromatic AA	63	106	102	68	97	78
Threonine	34	43	45	66	8	29
Tryptophan	11	12	14	20	11	11
Valine	35	57	57	51	49	43
Histidine	19	29	27	21	29	–

[a]For 2–5 year olds
[b]Includes cysteine, cystine and methionine

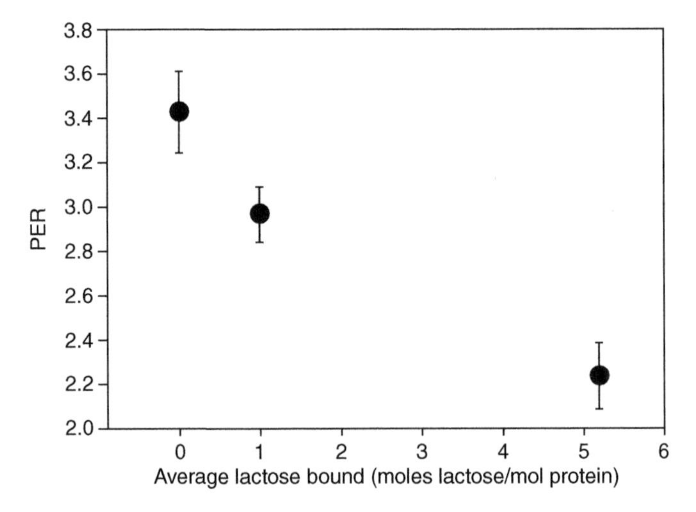

Figure. 10.7 Plot of lactosylation level against PER. The PER value for ANRC casein was used for zero lactosylation.

Lysine

The main concern during the storage of most nutritional proteins is the loss of lysine as a result of Maillard reactions or isopeptide bond formation.

Lactulosyl lysine renders lysine non-bioavailable. The protein efficiency ratio (PER) was found to be decreased in a WPC56, with an average of five lactulosyl lysine residues per protein molecule (Figure 10.7). A more detailed study using skim milk diets with pigs confirmed that lactulosyl lysine was non-bioavailable (Rerat *et al.*, 2002). That study also found a decrease in the digestibility of lysine, phenylalanine, valine, cystine, aspartic acid, glycine and methionine residues. This decrease suggests that lactulosyl lysine residues hinder the release and therefore utilization of adjacent amino acids.

Lysinoalanine is not a bioavailable source of lysine (Robbins *et al.*, 1980; Friedman, 1999). Alkali treatment of casein with 0.2 N NaOH at 80°C for 1 h reduced the PER of casein from 3.09 to 0.02 for a diet containing 10% casein (Possompes *et al.*, 1989; cited in Friedman, 1999).

Milk proteins are unusually rich in lysine (Table 10.3) and can stand to lose a significant proportion of the lysine before it becomes limiting (around 50% in the case of whey proteins and 25% in the case of casein). The real concern arises when milk proteins, and particularly whey proteins, are being added to a mixture to provide a source of lysine supplementation. When this is the case, it is particularly important to ensure that the lysine content has not been compromised after long storage. Steps to ensure this include keeping the product at temperatures less than 25°C for most of the time and ensuring that the water activity in the product is kept low, preferably at or below 0.3. Experiments carried out in our laboratories to determine bioavailable lysine showed that loss of lysine on storage was negligible at 30°C, whereas losses were severe at 40°C (Figure 10.8).

Sulfur amino acids

Modification of sulfur amino acids is possible through loss of cysteine via lanthionine formation and through oxidation of methionine to methionine sulfoxide. Whey proteins are relatively rich in sulfur amino acids and casein has more than adequate quantities. It should be noted that these scores are relative to World Health Organization (WHO) requirements for children aged 2–5 years. The requirement for rat diets is higher and this is thought to be because of the requirement of the rat for hair production. (Hair is rich in sulfur amino acids.)

Lanthionine is a problem only for caseinates and lanthionine formation is known to occur during caseinating. The rates of lanthionine formation in dry powder have not been studied. This should be done, if only to briefly investigate lanthionine levels in old stocks or library samples.

Oxidation of methionine to the sulfoxide does not alter its bioavailability *per se*; however, it has been claimed that the presence of the sulfoxide side chain in intact proteins may hamper digestion of the protein, thus affecting its overall bioavailability (Anon, 1973).

Other amino acids

One essential amino acid known to be destroyed during some processes is tryptophan. Extensive investigations by Nielsen *et al.* (1985) on whey proteins, casein and WMP showed that tryptophan remained relatively intact even when substantial lysine modification had occurred. Tryptophan analysis gave a result of 100% (within experimental error) and a bioavailability of >90% in milk powder that had been stored at 60°C for 5 weeks, which resulted in a loss of 80% of available amine. This powder was considered to show "advanced" Maillard browning.

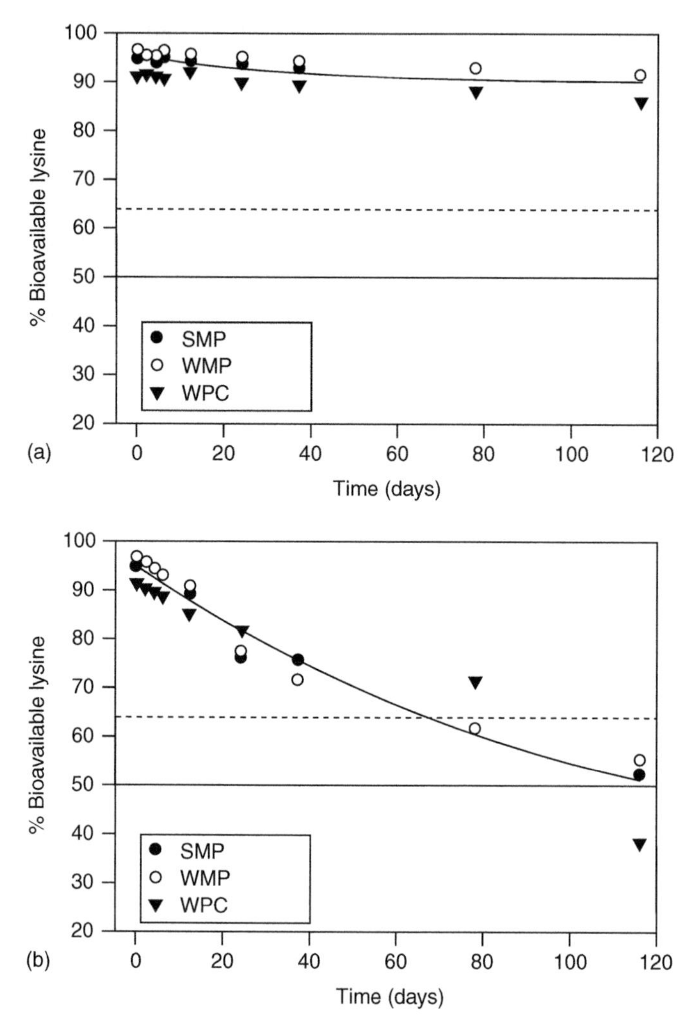

Figure. 10.8 Loss of bioavailable lysine with time for SMP, WMP and WPC with a water activity (a_w) of 0.3 during storage at (a) 30°C and (b) 40°C. The solid and dashed lines on the graphs show the levels at which lysine becomes limiting in WPC, and in SMP and WMP respectively. (K. Higgs, unpublished results.)

Product-specific storage trials

Samples of WPC80 powders that had earlier been shipped from New Zealand to the USA or Europe were obtained and analyzed for lactosylation levels. The powders were at the time less than 2 years old and thus were considered to be current stock. The powders had average bound lactose levels of between 0.7 and 1.2. This correlates to between 87–95% of the lysine in the products being bioavailable (estimated by extrapolation from the remaining lysine in β-lactoglobulin). This compares well with freshly produced WPC80 powders, which had an average number of lactose bound of 0.6, or 96% of available lysine remaining (Figure 10.9).

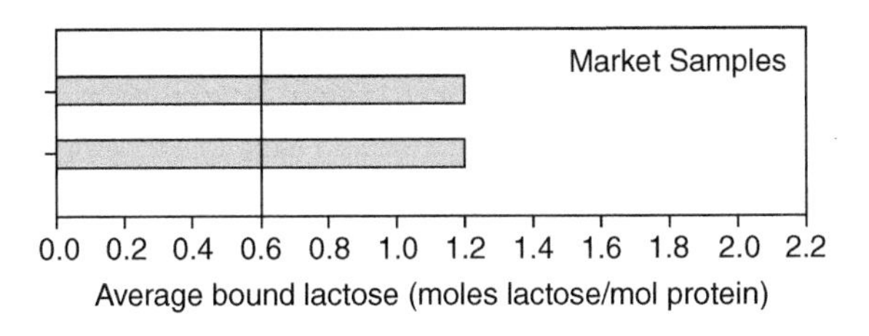

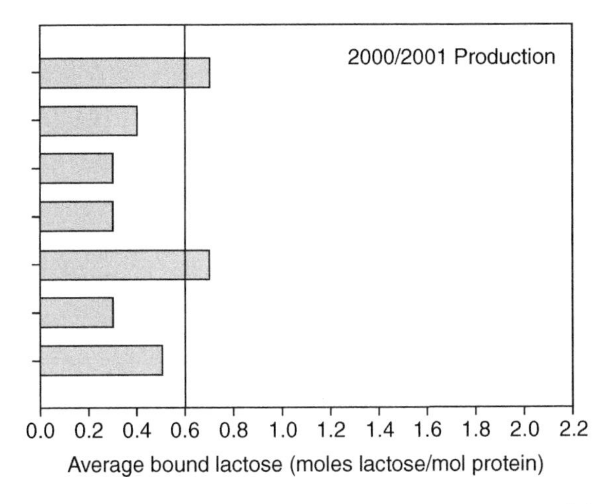

Figure. 10.9 Levels of lactosylation in market samples of a WPC80 compared with levels in freshly produced WPC80 in a subsequent season.

We did a 2-year storage study on >80% protein powders to determine changes, if any, in nutritional properties, but these powders were kept at constant temperature and were not exposed to any of the temporal variations possible during shipping and storage in overseas warehouses.

A decrease in available amine of 5% was seen in MPC85 when stored at 20°C for 2 years; this increased to 10% when the storage temperature was increased to 30°C. A single MPI stored for 2 years gave consistent available amine results over the storage period.

The caseinates and caseins in the 2-year study showed no definitive trend in available amine values. Most values remained consistent over the 2 years. This was expected as a previous 3-month study showed that storage temperatures in excess of 30°C were required for significant levels of lysinoalanine formation.

Conclusions

Milk proteins do undergo change on the storage of dry powders. Powders containing appreciable amounts of lactose (milk powders and WPCs) form pre-Maillard reaction products, rendering lysine non-bioavailable, whereas those containing casein undergo formation of isopeptide bonds, reducing the availability of lysine and sulfur amino acids. Because milk proteins are rich in lysine, loss of some lysine will not be a significant problem; however, when milk proteins are used as "balancers" in formulations with other proteins that are poor in lysine, attention should be given to the storage history of the protein.

Both the Maillard reaction and the formation of isopeptide bonds are undesirable and are best avoided by ensuring that shipping and storage temperatures do not exceed 30°C for significant periods. This should not be a problem in temperate climates; however, in climates where high temperatures routinely occur, consideration should be given to storage in a cool store.

References

Anon., (1973). Nutritional implications of sulfur amino acid oxidation. *Nutrition Reviews*, **31**, 220–21.

Erbersdobler, H. (1986). Twenty years of furosine – better knowledge about the biological significance of the Maillard reaction in food and nutrition. In *Amino-Carbonyl Reactions in Food and Biological Systems. Proceedings of the 3rd International Symposium on the Maillard Reaction, Susono, Shizuoka, Japan, 1–5 July 1985* (M. Fujimaki, M. Namiki and H. Kato, eds) pp. 481–91. Amsterdam: Elsevier.

Friedman, M. (1999). Chemistry, biochemistry, nutrition and microbiology of lysinoalanine, lanthionine, and histidinoalanine in food and other proteins. *Journal of Agricultural and Food Chemistry*, **47**, 1295–319.

Nielsen, H. K., de Weck, D., Finot, P. A., Liardon, R. and Hurrell, R. F. (1985). Stability of tryptophan during food processing and storage. 1. Comparative losses of tryptophan, lysine and methionine in different model systems. *British Journal of Nutrition*, **53**, 281–92.

Possompes, B., Berger, J., Diaz, B., Cuq, J. L. (1989). Factors affecting the evaluation of the nutritional value of severely alkali-treated casein. *Food Chemistry*, **32**, 189–99.

Rerat, A., Calmes, R., Vaissade, P. and Finot, P-A. (2002). Nutritional and metabolic consequences of the early Maillard reaction of heat treated milk in the pig. Significance for man. *European Journal of Nutrition*, **41**, 1–11.

Robbins, K. R., Baker, D. H. and Finley, J. W. (1980). Studies on the utilization of lysinoalanine and lanthionine. *Journal of Nutrition*, **110**, 907–15.

11

Interactions and functionality of milk proteins in food emulsions

Harjinder Singh and Aiqian Ye

Abstract

Because of their high nutritional quality and versatile functional properties, milk proteins are widely used as ingredients in many manufactured food colloids, e.g. dairy desserts, nutritional beverages, ice cream, yoghurt, spreads, confectionery and baked goods. Milk proteins perform a wide range of key functions in prepared foods, including emulsification, thickening, gelling and foaming. The functionality of milk proteins is a consequence of their molecular structures and interactions. An important functionality of milk proteins in food colloids is their ability to facilitate the formation and stabilization of oil droplets in emulsions. The ability of milk protein products to adsorb at the oil–water interface and to stabilize emulsions is influenced by the structures, flexibility and aggregation state of the constituent proteins.

This chapter deals mainly with the properties and functionalities of food emulsions formed with a range of milk protein products, and how they are influenced by different environmental and processing conditions. Of particular importance are the effects of pH, calcium ions and protein content and the influences of thermal and high-pressure processing. The chapter

Milk Proteins: From Expression to Food
ISBN: 978-0-12-374039-7

focuses on the structure and composition of adsorbed protein layers, competition between proteins and the creaming and flocculation behaviors of emulsion droplets.

Introduction

Milk proteins possess functional properties that provide desirable textural and other attributes to the final product and, for this reason, have found numerous applications in traditional dairy products and other foods. The functional properties of milk proteins, such as emulsification, thickening, gelling, flavor binding and foaming, contribute to the sensory characteristics and the stability of the manufactured foods. Several types of milk protein products, e.g. caseins and caseinates, whey protein concentrates (WPCs) and whey protein isolates (WPIs), milk protein concentrate (MPC) powders and hydrolyzed proteins, are manufactured from milk by the dairy industry.

Caseinates are produced from skim milk by adding acid (hydrochloric acid or lactic acid) or microbial cultures to precipitate the casein from the whey at pH 4.6. The acid-precipitated casein can then be resolubilized with alkali or an alkaline salt (using calcium, sodium, potassium or magnesium hydroxide) to about pH 6.7 and spray dried to form caseinate. Caseinates have exceptional water-binding capacity, fat emulsification properties and whipping ability, and a bland flavor. Emulsion-type products, e.g. coffee whiteners, whipped toppings, cream liqueurs and low-fat spreads, are an important application of caseinates in the food industry. In recent years, the use of casein and caseinate in dietary preparations, nutritional products and medical applications has increased; many of these preparations are also oil-in-water emulsions containing relatively small amounts of fat.

WPC and WPI are concentrated forms of whey protein components. Ultrafiltration, diafiltration and ion-exchange technology are used to concentrate and separate the protein from other components. The whey protein is then dried to obtain WPC or WPI, both of which are highly soluble, with protein levels ranging from 80–95%. Both WPC and WPI have a wide range of food applications and, because of their high protein content, can function as water-binding, gelling, emulsifying and foaming agents. Processing treatments used in the manufacture of WPC and WPI may sometimes cause some protein denaturation, which tends to affect their functionality.

MPCs are processed directly from skim milk by ultrafiltration/diafiltration and can have a range of protein contents from 56–82%. MPCs are used as functional ingredients in a wide variety of foods, including beverages, processed cheese and confectionery.

The functionality of milk proteins in processed foods is determined by their molecular structures and interactions with other food components, such as fats, sugars, polysaccharides, salts, flavors and aroma compounds. The type and the strength of various interactions determine the structure, texture, rheology, sensory properties and shelf life of manufactured food products. Much knowledge on the structure and properties of individual milk protein components has been gained, but less is known about interactions between different components that occur in a food system as a result of processing and formulation. Controlling these interactions is of key

Table 11.1 Functional properties of milk proteins in food systems

Functional property	Food system
Solubility	Beverages
Emulsification	Coffee whiteners, cream liqueurs, salad dressings, desserts
Foaming	Whipped toppings, shakes, mousses, cakes, meringues
Water binding	Bread, meats, bars, custard, soups, sauces, cultured foods
Heat stability	UHT- and retort-processed beverages, soups and sauces, custard
Gelation	Meats, curds, cheese, surimi, yoghurt
Acid stability	Acid beverages, fermented drinks

significance for the development of novel products and processes as well as for the improvement of conventional products and processes (Table 11.1).

This chapter focuses on the emulsifying properties of milk proteins, as this functional property is very important in all the food applications of milk protein products. The adsorption behavior of different milk protein products at oil-in-water interfaces and the stability of the resulting emulsions are considered, focusing on the work carried out in our laboratory at Massey University.

Adsorption of milk proteins during the formation of emulsions

Emulsions are composed of oil droplets (average range 0.5–5 μm diameter) enveloped by a continuous film of surfactant material that stabilizes the droplets. In the food industry, homogenization is widely used for finely dispersing oils in food products and proteins are most commonly used as emulsifying agents. The state of the droplet size distribution after homogenization reflects the emulsifying capacity of the proteins, the energy input during formation as well as the effects of various factors, such as pH, temperature, ionic strength and ratio of the two phases, on the surface activity of the proteins (Walstra, 1993; Dickinson, 1998a). In addition, the droplet size distribution influences markedly the properties of food emulsions, such as stability, viscosity, texture and mouthfeel.

During homogenization, the milk protein, in the form of individual molecules or protein aggregates, becomes rapidly adsorbed at the surface of the newly formed oil droplets. The amount of protein present at the interface per unit surface of dispersed phase is defined as the protein load, which is usually expressed as milligrams of protein per unit area of the dispersed phase (mg/m^2). The protein load determines the amount of protein required to make an emulsion with a desired oil volume and droplet size and is dependent on the concentration and the type of protein as well as on the conditions used for emulsion formation. The factors that affect the protein load include protein concentration, volume of oil, energy input, state of protein aggregation, pH, ionic strength, temperature and calcium ions (Dickinson and Stainsby, 1988; Walstra, 1993).

The properties of the adsorbed layers depend on the amounts and structures of the proteins present during homogenization. Proteins are amphipathic molecules containing both polar and non-polar parts and orientate at the interface in such a way that a substantial proportion of the non-polar amino acids remains in contact with the oil phase and the polar groups are in contact with the aqueous phase (Dickinson, 1992, 1998a). The main thermodynamic driving force for the adsorption of proteins is the removal of hydrophobic residues from the unfavorable environment of the bulk aqueous phase by displacement of structured water molecules from the close vicinity of the interface. An additional important driving force is the unfolding and reorganization of the native protein structure, which is due to interaction with the interface. By adsorbing at the interface, the protein reduces the free energy of the system and hence the interfacial tension.

The effectiveness of any particular protein in lowering the surface tension depends on the number and type of contacts it makes with the interface (Dickinson *et al.*, 1988a; Dickinson, 1999). A protein molecule that spreads out a lot, and thus has a substantial proportion of its non-polar residues in contact with the surface, is one that is also very effective in reducing the interfacial tension. Flexible proteins (caseins) with a higher proportion of non-polar groups are more effective at reducing the interfacial tension than rigid proteins with fewer non-polar groups (Dickinson and McClements, 1995). The order of surface activity that has been reported for the individual milk proteins is: β-casein > monodispersed casein micelles > serum albumin > α-lactalbumin > α_s-caseins = κ-casein > β-lactoglobulin (Mulvihill and Fox, 1989).

Once a protein is adsorbed at an interface, it undergoes unfolding and rearrangement to form a stabilizing adsorbed layer (Dickinson, 1992; Dalgleish, 1996) and the extent of unfolding depends on the flexibility of the protein molecule, i.e. on the strength of the forces maintaining the secondary and tertiary structures. Because the caseins have rather flexible structures, they unfold rapidly at the interface and may form extended layers up to about 10 nm thick (Dalgleish, 1990). Dalgleish (1999) suggested that casein molecules are stretched to their maximum extent when their overall surface coverage is less than about 1 mg/m^2. Conversely, the presence of excess casein increases the monolayer coverage to a maximum value of 3 mg/m^2, the parts of the molecules in contact with the interface adopt a more compact conformation and the hydrophilic moieties protrude further from the interface.

Whey proteins (such as β-lactoglobulin), which give adsorbed layers that are only about 2 nm thick, change conformation and unfold their structure to some extent at the surface (Dalgleish and Leaver, 1993; Mackie *et al.*, 1993; Dalgleish, 1995; Dickinson and McClements, 1995; Dalgleish, 1996; Fang and Dalgleish, 1998). The adsorbed whey protein structure lies somewhere intermediate between the native structure and the fully denatured state, which may have a native-like secondary structure and an unfolded tertiary structure (Dickinson, 1998a). Additionally, the partial unfolding of the globular whey protein structure following adsorption causes exposure of the reactive sulfydryl group, leading to slow polymerization of the adsorbed protein in the aged layer via sulfydryl−disulfide interchange (Dickinson and Matsumura, 1991; McClements *et al.*, 1993).

The amount of protein adsorbed on the interface of an emulsion droplet suggests the state of the protein adsorbed at the interface. If the protein load is $<1\,mg/m^2$, it suggests that the protein molecules are fully unfolded. If the protein load is $1–3\,mg/m^2$, a monolayer of globular proteins may be present or unfolded molecules may be adsorbed in the conformation of trains, loops and tails. Protein load values above $5\,mg/m^2$ suggest the adsorption of aggregates of proteins or multilayers of proteins. Some proteins of higher molecular weight may also give higher protein loads (Phillips, 1981; Hunt and Dalgleish, 1994; Dam *et al.*, 1995; Srinivasan *et al.*, 1996).

Extensive studies on purified milk protein systems show that a disordered casein monomer may be regarded as a complex linear copolymer that adsorbs to give an entangled monolayer of flexible chains, having some sequences of segments in direct contact with the surface (trains) and other sequences of segments protruding into the aqueous phase (loops and tails) (Dickinson, 1998a). Based on various experimental studies and molecular modeling, the β-casein molecule has been shown to adsorb with an extensive hydrophobic region anchored directly at the surface and a hydrophilic region (40−50 residues at the N-terminus) protruding extensively into the aqueous phase. This is probably also the portion of the molecule that forms the hydrodynamically thick layer.

In contrast to the dangling tail predicted for β-casein, a loop-like conformation has been predicted for α_{s1}-casein and it does not have such a pronounced inequality in the distribution of hydrophobic and hydrophilic residues in its primary structure. It has been suggested that α_{s1}-casein adsorbs to the oil–water interface via peptides towards the middle of its sequence, rather than the end as in β-casein, and it may be this that causes the protein to form a thinner adsorbed layer than does β-casein (Dalgleish, 1996). The simple train–loop–tail model is not adequate to describe the molecular configuration of adsorbed β-lactoglobulin. A closely packed, dense and rather thin (2–3 nm at neutral pH) layer of β-lactoglobulin is formed, which can be modeled as a dense two-dimensional assembly of highly interacting deformable particles.

Milk proteins used by the food industry contain complex mixtures of proteins in various states of aggregation. The structures of the adsorbed layers formed with these complex mixtures are not understood in molecular detail. The most commonly used milk proteins, i.e. sodium caseinate and whey proteins, show excellent emulsifying ability and it is possible to make stable emulsions at a relatively low protein-to-oil ratio (about 1:60). In emulsions formed with sodium caseinate or whey proteins, the protein load increases with an increase in protein concentration until it reaches a plateau value of about $2.0–3.0\,mg/m^2$ (Singh, 2005) (Figure 11.1). The emulsifying ability of "aggregated" milk protein products, such as MPC and calcium caseinate, is much lower than that of whey protein or sodium caseinate, i.e. much higher concentrations of protein are required to make stable emulsions and larger droplets are formed under similar homogenization conditions.

The surface protein concentration of emulsions formed with MPC is in the range $5–20\,mg/m^2$ depending on the protein concentration used in making the emulsions (Euston and Hirst, 1999). At low protein-to-oil ratios, protein aggregates are shared by adjacent droplets, resulting in bridging flocculation and consequently a marked increase in droplet size. In addition, the spreading of protein at the interface is

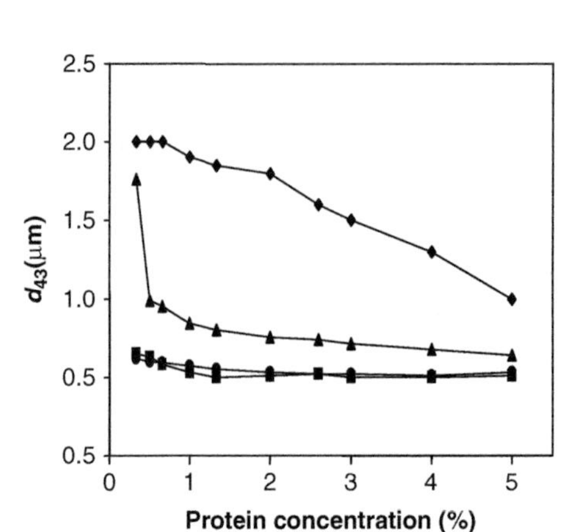

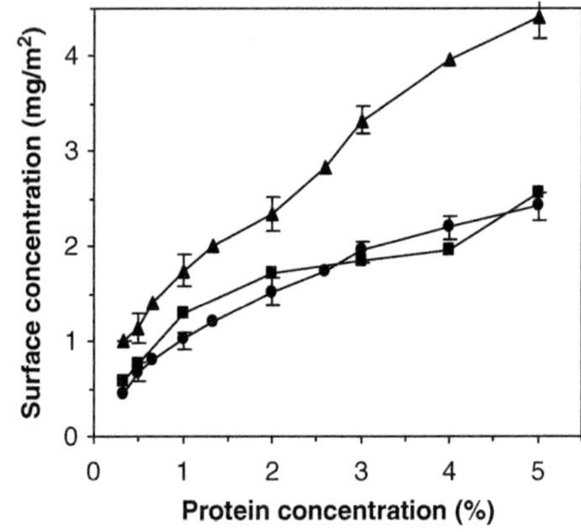

Figure 11.1 Influence of protein concentration on average droplet size d_{43} (left) and surface protein coverage (right) in emulsions (30% soya oil) made with sodium caseinate (●), calcium caseinate (▲), WPC (■) or MPC (◆) (from Singh, 2005, reproduced with the permission of The Royal Society of Chemistry).

limited in these emulsions, because the aggregates are held together by calcium bonds and/or colloidal calcium phosphate and these bonds are unlikely to be affected during the emulsification process. The higher conformational stability of these aggregates will also contribute to their reduced emulsifying ability (Euston and Hirst, 1999; Srinivasan *et al.*, 1999).

The composition of the interfacial layer is determined by the quantities and structures of the proteins present at the moment the emulsion is formed. If proteins are the only emulsifiers present, they will adsorb to the oil–water interface, generally in proportion to their concentration in the aqueous phase (Dalgleish, 1997). However, certain mixtures of caseins show competition during adsorption at oil–water interfaces and rapid exchanges between adsorbed and unadsorbed caseins after emulsion formation. Studies have demonstrated that β-casein, because of its greater surface activity, adsorbs in preference to α_{s1}-casein in emulsions stabilized by mixtures of these proteins and that β-casein displaces α_{s1}-casein rapidly from the droplet surface (Dickinson *et al.*, 1988b). In binary mixtures containing β-lactoglobulin and α-lactalbumin, some limited competitive adsorption does occur, but little exchange between the adsorbed and unadsorbed protein occurs. The protein that arrives at the interface first during homogenization is the protein that predominates there afterwards.

In contrast to model systems, no competitive adsorption has been observed in emulsions stabilized by more complex casein mixtures, such as sodium caseinate (Hunt and Dalgleish, 1994). Interestingly, this behavior appears to be related to the ratio of protein to oil in the emulsions (Euston *et al.*, 1995; Srinivasan *et al.*, 1999). Srinivasan *et al.* (1999) has shown that, in sodium caseinate emulsions, when the ratio of protein to oil is very low (about 1:60), β-casein is preferentially adsorbed at the droplet surface but, when the total amount of protein is greatly in excess of the amount needed for full surface coverage, α_{s1}-casein is adsorbed in preference to the other caseins. At all concentrations, κ-casein from sodium caseinate appears to be less readily adsorbed (Figure 11.2). The concentration dependence of the competitive

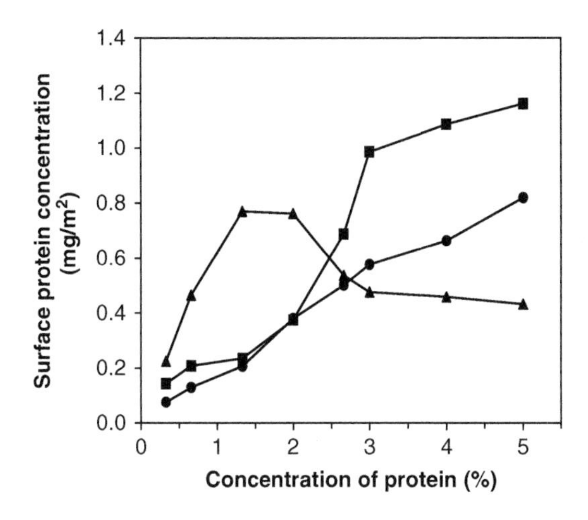

Figure 11.2 Surface concentrations of α_{s1}-casein (■), β-casein (▲) and κ-casein (●) in sodium caseinate emulsions (30% oil) (from Srinivasan *et al.*, 1999, reproduced with the permission of Elsevier Inc).

adsorption of α_{s1}-casein and β-casein in sodium caseinate emulsions may be a consequence of the different complexes that can be formed by caseins in solution (Rollema, 1992).

The preferential adsorption of β-casein, because of its high surface activity, appears to exist only at low concentrations where caseins may exist as monomers. With increasing protein concentration, caseins aggregate to form various complexes (Lucey *et al.*, 2000) and it is likely that β-casein loses its competitive ability because of its self-aggregation to form micelles or through the formation of complexes with other caseins. Therefore, the surface composition of emulsions formed using a relatively high sodium caseinate concentration is likely to be determined by the surface activities and flexibilities of the casein aggregates and complexes. Although extensive information on the surface activity and hydrophobicity of individual caseins is available, little is known about how these characteristics are modified when casein molecules undergo self-association under different environmental conditions.

When the casein is in the highly aggregated form of casein particles, as in calcium caseinate or MPC, there is very little competitive adsorption and protein exchange (Euston and Hirst, 1999; Srinivasan *et al.*, 1999). In these systems, the average surface composition is probably determined by the adsorption of protein aggregates of fixed composition. For instance, calcium caseinate solution consists of large α_{s1}-casein-rich aggregates, which appear to dominate the droplet surface after emulsification (Srinivasan *et al.*, 1999). When WPCs or WPIs are used to make emulsions, there is no preferential adsorption between β-lactoglobulin and α-lactalbumin regardless of the protein-to-oil ratio in the emulsion (Euston *et al.*, 1996; Ye and Singh, 2000, 2006a).

The aggregation state and the flexibility of protein molecules can be altered by changes in pH, addition of divalent cations and various processing treatments prior to emulsification. These changes will inevitably influence the adsorption behavior of milk proteins at the oil–water interface. For example, addition of $CaCl_2$ at above a certain critical concentration to a sodium caseinate or whey protein solution before homogenization increases the droplet size, increases the surface protein coverage and, in sodium caseinate emulsions, also affects the competition between different proteins (Ye and Singh, 2000, 2001). The proportions of β-lactoglobulin and α-lactalbumin at the droplet surface remain unaffected by the addition of $CaCl_2$ to a whey protein solution prior to emulsification. In contrast, addition of $CaCl_2$ to a sodium caseinate solution markedly enhances the adsorption of α_{s1}-casein at the droplet surface, with a much lesser effect on β-casein adsorption.

The effects of calcium on surface coverage and composition can be explained by the binding of the ions to the negatively charged amino acid residues on the protein. This reduces electrostatic repulsions between the protein molecules and increases the potential for intermolecular associations. Because of the presence of clusters of phosphoserine residues, the caseins have stronger affinity than the whey proteins to bind calcium. Consequently, the caseins (except κ-casein) are precipitated by calcium, with α_{s1}-casein being the most sensitive to aggregation and precipitation by calcium. In sodium caseinate emulsions, the increased surface coverage upon addition of calcium prior to emulsification is probably due to adsorption of casein aggregates on to the droplet surface (Ye and Singh, 2001). Greater α_{s1}-casein adsorption reflects its stronger tendency to be aggregated by calcium ions in solution or at the interface.

The native whey proteins do not bind much calcium and are not precipitated in the presence of calcium (Baumy and Brule, 1988), although heat-denatured whey proteins are able to bind considerable amounts of calcium and undergo aggregation (Pappas and Rothwell, 1991). The increase in surface protein coverage suggests the formation of aggregates of whey proteins in the presence of calcium, which subsequently become adsorbed during emulsification (Ye and Singh, 2000). This has been attributed to a decrease in the denaturation temperature of the whey proteins in the presence of calcium.

All these results confirm that, under a given set of homogenization conditions, the surface composition is largely dependent on the protein-to-oil ratio and the aggregation state of the proteins in solution. It appears that the structure of the interfacial layer in emulsions can be manipulated by controlling the protein concentration, the protein type and the ionic environment. Because of their different interfacial structures, these droplets would be expected to exhibit different reactivities, which could be exploited to develop new food textures. Further studies are required for an understanding of the relationship between droplet surface structures and the sensitivity of the droplets to different environments and processing conditions.

Stability of milk-protein-based emulsions

The term “emulsion stability” refers to the ability of an emulsion to resist any alteration in its properties over the timescale of observation (McClements, 1999; Dickinson, 2003; McClements, 2005). An emulsion is thermodynamically unstable as the free energy of mixing is positive because of the large interfacial area between the oil and the aqueous phase. Therefore, the kinetic stability, i.e. the time period for which the emulsion is stable, is important (Damodaran, 1997; McClements, 1999; Dickinson, 2003; McClements, 2005). For instance, an emulsion can be considered to be "stable" if the inevitable process of separation has been slowed to an extent that it is not of practical importance during the shelf life of the product. An emulsion may become unstable because of a number of different types of physical and chemical processes. Physical instability refers to the change in spatial arrangement or size distribution of emulsion droplets, such as creaming, flocculation or coalescence, whereas chemical instability includes change in the composition of the emulsion droplet itself, such as oxidation, hydrolysis, etc. (McClements and Decker, 2000; McClements, 2005).

Creaming is the movement of oil droplets, under gravity or in a centrifuge, to form a concentrated layer at the top of an oil-in-water emulsion sample, with no accompanying change in the droplet size distribution. Creaming is reversible and the original uniform distribution of droplets can usually be obtained by gentle mixing. The creaming process can be explained by Stokes' Law (Hunter, 1986; McClements, 2005):

$$\nu_{stokes} = \frac{2r^2(\rho_1 - \rho_2)}{9\eta} \tag{11.1}$$

where ν_{stokes} = velocity of creaming, r = emulsion droplet radius, ρ_1 and ρ_2 = density of the continuous phase and the dispersed phase respectively and η = shear viscosity of the continuous phase. The creaming rate can be reduced by lowering the radius, increasing the continuous phase viscosity or decreasing the difference in density between the two phases. However, this law often fails to define the rate of creaming due to flocculation or coalescence.

Coalescence, i.e. an increase in droplet size by accretion, gradually results in separation of the oil and the aqueous phase and is always irreversible. Coalescence requires rupture of the stabilizing film at the oil–water interface, but this occurs only when the layer of continuous phase between the droplets has thinned to a certain critical thickness (Dickinson and Stainsby, 1988; Britten and Giroux, 1991; Das and Kinsella, 1993; Walstra, 1993).

Flocculation has been defined as the reversible aggregation mechanism that arises when droplets associate as a result of unbalanced attractive and repulsive forces (Dalgleish, 1997). Generally, two types of flocculation are distinguished, i.e. depletion flocculation and bridging flocculation (Dickinson, 2003). The type of mechanism prevailing depends upon the interaction between the interfacial layer and the emulsion droplets.

Bridging flocculation normally occurs when a high-molecular-weight biopolymer at a significantly low concentration adsorbs to two or more emulsion droplets, forming bridges (Dickinson and Pawlowsky, 1998; Dickinson, 1998b; McClements, 1999; Dickinson, 2003; McClements, 2005; Fellows and Doherty, 2006). Depletion flocculation occurs as a result of the presence of unadsorbing biopolymer in the continuous phase, which can promote association of oil droplets by inducing an osmotic pressure gradient within the continuous phase surrounding the droplets (de Hek and Vrij, 1981; Dickinson, 1999; Tuinier and de Kruif, 1999; McClements, 2005).

Essentially, if the added biopolymer is either unadsorbed or poorly adsorbed, the biopolymer is squeezed out of the area between two approaching emulsion droplets. The concentration of biopolymer between the emulsion droplets becomes less than its overall solution concentration, resulting in osmotic imbalance. The net effect is that the particles are attracted towards each other, resulting in flocculation. The attraction energy is determined by the concentration of the polymer and the range of interaction depends on the radius of gyration of the polymer molecule. The bonds formed through the depletion flocculation mechanism are generally weak, flexible and reversible.

The ability of proteins to stabilize emulsions is the most important criterion besides the emulsion formation in most food applications. The forces involved in stabilizing and destabilizing emulsions include van der Waals' attractive forces, electrostatic interactions and steric factors. At pH values away from their isoelectric point, as proteins are electrically charged, there is an electrostatic repulsion, which prevents dispersed droplets from closely approaching one another. With the possible exception of highly charged proteins, a predominant contribution to emulsion stabilization by protein comes from the steric stabilization mechanism. Interactions between droplets stabilized by proteins may be influenced by the presence of certain ions, particularly calcium, as proteins are capable of binding ions.

As long as sufficient protein is present during homogenization to cover the oil droplets, emulsions stabilized by milk proteins are generally very stable to coalescence over prolonged storage. However, these emulsions are susceptible to different types of flocculation, which in turn leads to enhanced creaming or serum separation. At low protein-to-oil ratios, there is insufficient protein to fully cover the oil–water interface during homogenization and this results in bridging flocculation. Another consequence of insufficient protein is coalescence of droplets during or immediately after emulsion formation. Bridging flocculation is commonly observed in emulsions formed with aggregated milk protein products, such as calcium caseinate or MPC, in which the droplets are bridged by casein aggregates or micelles. Optimum stability can generally be attained at protein concentrations high enough to allow full saturation coverage at the oil–water interface. However, at very high protein-to-oil ratios, the presence of excess, unadsorbed protein may lead to depletion flocculation in some emulsions. Both depletion flocculation and bridging flocculation cause an emulsion to cream more rapidly.

Depletion flocculation has been observed in sodium-caseinate-based emulsions but not in emulsions formed with calcium caseinate, MPC or whey proteins (Dickinson and Golding, 1997; Euston and Hirst, 1999; Srinivasan *et al.*, 2001; Singh, 2005) (Figure 11.3). In sodium-caseinate-based emulsions, it was shown that, at a protein content of nearly 2.0 wt%, the emulsion droplets were protected from flocculation by a thick steric-stabilizing layer of sodium caseinate. The emulsion was stable against flocculation, coalescence and creaming for several weeks. However, when the protein content was increased to above 3.0 wt%, unadsorbed protein gave rise to depletion flocculation. Because of this depletion flocculation, the effective diameter of the droplets increased, resulting in a marked decrease in creaming stability with an increase in the caseinate concentration from 3 to 5 wt%. Further increasing the protein content to 6.0 wt% and above resulted in very high depletion flocculation, leading to a strong emulsion droplet network that was stable to creaming.

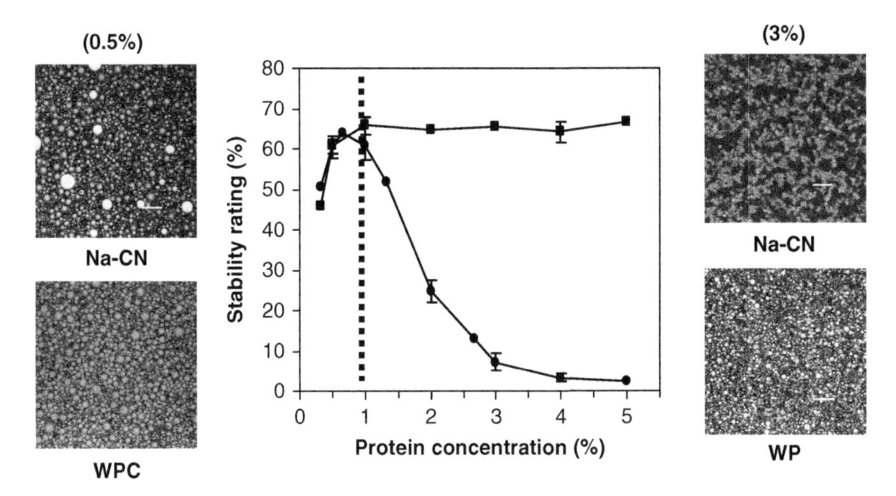

Figure 11.3 Creaming stability and microstructure of emulsions made with sodium caseinate (●) or WPC (■) (30% oil). Scale bar represents 10 μm (from Singh, 2005, reproduced with the permission of The Royal Society of Chemistry).

The differences in the creaming stabilities of emulsions made with different kinds of milk protein products are largely related to depletion flocculation effects (Singh, 2005). The depletion interaction free energy (ΔG_{DEP}), of the order of a few kT, can be estimated using Equation (11.2) (Walstra, 1993):

$$\Delta G_{\mathrm{DEP}} = -2\pi\gamma^2{}_m\Pi(\gamma_d - 2\gamma_m/3) \tag{11.2}$$

where Π is the osmotic pressure of the polymer solution, represented as a fluid of hard spheres of radius γ_m, and γ_d is the mean droplet radius. The osmotic pressure under ideal conditions is given by the following equation:

$$\Pi = cRT/M \tag{11.3}$$

where R is the molar gas constant, T is the temperature, M is the molecular mass of the polymer and c is the number concentration of the polymer.

For depletion flocculation to occur, the polymer has to have a fairly high M, so that the γ_m is relatively large. However, at a given c, M is inversely proportional to Π. Therefore, an increase in the polymer molecular mass will reduce the osmotic pressure driving the depletion interaction. Hence, at a given concentration, the depletion interaction free energy is low for a polymer of low molecular mass, increases with an increase in molecular mass until it reaches a maximum and then decreases with a further increase in molecular mass. Similarly, a reduction in the polymer number concentration will reduce the osmotic pressure.

Although the exact state of the casein molecules in concentrated sodium caseinate solutions is unknown, a sodium caseinate solution has been reported to have a radius of gyration of about 20–30 nm, as determined by static light scattering (Lucey *et al.*, 2000). It is likely that depletion flocculation in sodium caseinate emulsions is caused by the presence of these casein aggregates formed from self-assembly of sodium caseinate in the aqueous phase of the emulsion at concentrations above 2 wt% (Dickinson and Golding, 1998).

Emulsions formed with whey proteins, MPC and calcium caseinate do not show depletion flocculation, probably because there are no suitably sized protein particles at the required concentrations in the aqueous phase. The molecular size of whey proteins is less than the optimum, whereas the casein micelles in MPC are too large to induce depletion flocculation. Calcium caseinate consists of mixtures of casein aggregates of different sizes, but the concentration of aggregates capable of inducing depletion flocculation is probably too low. The extent of creaming in these emulsions is largely determined by the particle size of the droplets. Generally, in these emulsion systems, the creaming stability increases with increasing protein concentration up to a certain concentration and then remains almost constant (Euston and Hirst, 1999; Srinivasan *et al.*, 2001). However, the creaming stability of emulsions formed with calcium caseinate or MPC at relatively high protein concentration tends to be higher than that of whey-protein-stabilized emulsions. This can be attributed to an increase in the droplet density as a result of the presence of a much thicker and denser adsorbed protein layer at the droplet surface.

The addition of moderate amounts of $CaCl_2$ to emulsions containing excess sodium caseinate has been shown to eliminate depletion flocculation and to improve the creaming stability (Ye and Singh, 2001). This effect appears to be due to an increase in the average size of the casein aggregates in the aqueous phase, resulting in a large increase in the molecular mass of the caseins (Dickinson *et al.*, 2001). In addition, there is a reduction in the concentration of unadsorbed caseinate. Both these effects are expected to cause a substantial reduction in the concentration of small particles, which are assumed to be the main depleting species responsible for inducing reversible flocculation in the calcium-free systems.

Presumably, the substantial reduction in osmotic pressure makes the magnitude of ΔG_{DEP} predicted from Equation (11.2) too small to cause depletion flocculation. Similarly, addition of NaCl at above a certain concentration reduces the extent of depletion flocculation of sodium caseinate emulsions and improves the creaming stability (Srinivasan *et al.*, 2000). This effect is due to increased adsorption of protein at the droplet surface and hence a lower concentration of unadsorbed protein remaining in the solution. Decreasing the pH of emulsions formed with excess sodium caseinate also gradually eliminates depletion flocculation, through aggregation of adsorbed protein and a transfer of more protein to the droplet surface (Singh, 2005). Therefore, it seems to be possible to switch depletion flocculation off and on by controlling the concentration and the aggregation state of the casein molecules in the aqueous phase.

Heat-induced changes in milk-protein-based emulsions

Food emulsions are often heat treated at relatively high temperatures to provide a long shelf life to the product via microbial sterility. These heat treatments can cause denaturation and aggregation of adsorbed and unadsorbed proteins, resulting in aggregation or coalescence of droplets and gel formation. Emulsions formed with whey proteins at neutral pH are stable against heating when the ionic strength and/or the concentration of protein in the emulsions are low. Addition of KCl at 100 mM or above has been shown to cause destabilization of whey protein emulsions, leading to gel formation (Hunt and Dalgleish, 1995).

Both the unadsorbed protein and the adsorbed protein are necessary for the heat-induced aggregation of whey-protein-stabilized emulsions. Aggregation of emulsion droplets is more extensive and proceeds more rapidly as the concentration of protein in the emulsion is increased, whereas removal of unadsorbed protein from the emulsion decreases the rate of droplet aggregation (Euston *et al.*, 2000). During heat treatment, the protein-covered droplet appears to interact more readily with the unadsorbed protein than with another emulsion droplet. This has been explained by assuming that the relative surface hydrophobicities of the emulsion droplet and the unadsorbed denatured whey proteins are different. Interaction of two emulsion droplets through their respective adsorbed protein layers will have a relatively low probability because the surface hydrophobicity is likely to be relatively low. When an emulsion droplet and an unadsorbed protein molecule aggregate, at least one of them (the denatured protein molecule) has a relatively high surface hydrophobicity and this will increase the probability of interaction and aggregation (Euston *et al.*, 2000).

In emulsions made with 3.0% WPI and 25% soya oil, the amount of adsorbed protein was shown to increase from 2.9 to 3.7 mg/m^2 within the first 10 min of heating at 75°C, but further heating had no effect (Sliwinski *et al.*, 2003). At 90°C, the plateau value of about 4 mg/m^2 was reached within 5 min of heating. Studies on the effects of heating temperature in the range 50–90°C on WPI emulsions (pH 7.0) (Monohan *et al.*, 1996; Demetriades and McClements, 1998) show that droplet aggregation occurs on heating in the range 75–80°C, which causes an increase in viscosity and a loss of creaming stability, but the degree of aggregation and the susceptibility to creaming decrease on heating at temperatures above 80°C.

It has been suggested that, in the temperature range 75–80°C, the whey protein molecules at the droplet surface are only partly unfolded and that not all of the hydrophobic amino acid residues are directed towards the oil phase. Consequently, the surface of the droplet is more hydrophobic, making it susceptible to droplet aggregation. At higher temperatures, the proteins become fully unfolded with all of the hydrophobic residues being directed into the oil phase, which makes the droplets less prone to aggregation. The role of sulfydryl–disulfide interchange reactions in droplet aggregation is not clear. It has been suggested that disulfide-mediated interactions during heat treatment are not critical during the initial stages of aggregation but they tend to strengthen the aggregates (Demetriades and McClements, 1998).

Recently, Dickinson and Parkinson (2004) and Parkinson and Dickinson (2004) reported that addition of a very small proportion of caseinate (0.03−0.15% of the total emulsion) can stabilize a whey-protein-based emulsion against heat treatment. The magnitude of the effect is dependent on the type of casein, with the order of effectiveness being β-casein > sodium caseinate > α_{s1}-casein. The stabilizing effect of the casein in these mixed milk protein systems is strongly synergistic. The casein polymer appears to be acting in a colloidal stabilizing capacity at a surface concentration very much lower than that at which it could be used as an emulsifying or stabilizing agent simply on its own. It has been suggested that adsorbed casein molecules keep the emulsion droplet surfaces sufficiently far apart to prevent the "normal" cross-linking processes that occur between whey-protein-coated droplets during heat-induced aggregation and gelation, because of the steric hindrance from the loops and tails of the disordered casein polymers (Parkinson and Dickinson, 2004).

In contrast to whey proteins, emulsions formed with sodium caseinate (2 wt% protein, 20% soya oil) are stable to heating at 90°C for 30 min or 121°C for 15 min, as determined by droplet size analysis (Hunt and Dalgleish, 1995; Srinivasan *et al.*, 2002). However, the protein coverage and the adsorbed casein composition change upon heat treatment, indicating that interactions between unadsorbed caseinate molecules and caseinate at the droplet surface may occur during heating (Srinivasan *et al.*, 2002).

Analysis of adsorbed caseins isolated from emulsions heated at 121°C for 15 min has shown that a substantial proportion of the adsorbed caseinate is polymerized to form high-molecular-weight aggregates (Srinivasan *et al.*, 2002), held together through covalent bonds other than disulfide bonds. These covalent bonds appear to form mainly between caseinate molecules at the surface of the same droplet because of the higher local concentrations of casein molecules at the droplet surface.

Interestingly, the adsorbed caseins also appear to undergo thermal degradation, resulting in the formation of low-molecular-weight species. Relatively high proportions of casein degradation products present at the droplet surface indicate that the adsorbed caseinate molecules are more susceptible to fragmentation during heating than those in solution and that these peptides remain adsorbed. This is probably due to different structures and conformations of the caseins at the droplet surface than of those in the solution.

The creaming stability of sodium caseinate emulsions has been found to improve upon heating, with the onset of depletion flocculation occurring at higher protein concentration than in unheated emulsions (Srinivasan *et al.*, 2002). This can be attributed to a reduction in the number of unadsorbed caseinate molecules/aggregates in the aqueous phase as a result of increased surface coverage and heat-induced polymerization and degradation of the casein molecules. The improvement in the creaming stability in heated emulsions at low protein concentrations may be attributed to an increase in droplet density because of the presence of greater amounts of polymerized protein at the droplet surface.

The surface protein composition of emulsion droplets may also change during heat treatment in emulsions formed with whey proteins. For WPI-stabilized emulsions, the amount of β-lactoglobulin at the droplet surface was found to increase during heat treatment, whereas the amount of adsorbed α-lactalbumin decreased markedly (Ye and Singh, 2006a). It seems that β-lactoglobulin displaces α-lactalbumin from the interface on heating at temperatures up to 90°C. The reason for this is not clear.

Similar phenomena were observed in studies of exchanges of caseins and whey proteins at the interfaces of oil-in-water emulsion droplets (Dalgleish *et al.*, 2002). It was found that, at temperatures above 40°C, addition of WPI to the aqueous phase of caseinate-stabilized emulsions caused a displacement of adsorbed caseins. As the

β-lactoglobulin and α-lactalbumin adsorbed, α_{s1}- and β-caseins were desorbed, principally the former, whereas the α_{s2}- and κ-caseins were not displaced. The rate of the displacement or exchange reaction was temperature dependent, being almost undetectable at room temperature, but complete within 2 min at 80°C. The displacement reaction was not affected by ionic strength; neither were any of the reactions apparently dependent on sulfydryl exchange reactions (Dalgleish *et al.*, 2002). However, no exchange of proteins occurred when an emulsion prepared with WPI was treated with caseinate and heat treated at 80°C for 2 min (Brun and Dalgleish, 1999). This was surprising in view of the known interactions of whey proteins with α_{s2}- and κ-caseins, involving sulfydryl–disulfide interchange reactions.

Pressure-induced changes in milk-protein-based emulsions

The effect of ultra-high pressure (100 – 1000 MPa) on the structures of milk proteins in aqueous solution has received considerable attention over the last few years (see Chapter 7). High pressure can disrupt the quaternary and tertiary structures of globular proteins with relatively little influence on their secondary structure. In addition, proteinaceous colloidal aggregates (e.g. casein micelles), which are held together by ionic and hydrophobic interactions, can be dissociated by high-pressure treatment (Gaucheron *et al.*, 1997; Huppertz *et al.*, 2004).

Whey proteins are sensitive to high-pressure treatments (López-Fandinó *et al.*, 1996; Anema *et al.*, 2005). Solution studies (Patel *et al.*, 2005) of native β-lactoglobulin and whey protein products have shown that high-pressure treatment has a marked effect on the protein's conformation and consequently its aggregation behavior; the aggregation is more extensive at high protein concentrations (Patel *et al.*, 2005). The formation of aggregates is most probably due to the generation of intermolecular disulfide bridges through sulfydryl – disulfide interchange reactions (Patel *et al.*, 2006).

In model oil-in-water emulsions, high-pressure treatment has been shown to have no effect on the droplet size distribution or the emulsion viscosity of sodium-caseinate-based emulsions at pH 7 (Dumay *et al.*, 1996). However, high-pressure treatment significantly induced flocculation of emulsion droplets and increased the emulsion viscosity of oil-in-water emulsions stabilized by β-lactoglobulin or WPC at neutral pH (Dumay *et al.*, 1996; Dickinson and James, 1998). The unfolded unadsorbed whey proteins in the emulsion treated by high pressure appear to be the major contributor to the cross-linking or flocculation of emulsion droplets because greater emulsion flocculation was observed in emulsions with higher proportions of unadsorbed protein in the aqueous phase.

As in the case of emulsions treated by heat processing, whey-protein-stabilized emulsions are more sensitive to pressure and temperature at pH values closer to the isoelectric point and at high ionic strength. In terms of the change in emulsion rheology, severe high-pressure treatment (800 MPa for 60 min) is equivalent to relatively mild thermal treatment (65°C for 5 min) (Dickinson and James, 1998). In a concentrated emulsion formed with β-lactoglobulin (1% protein and 40% vol% *n*-tetradecane), an emulsion gel was produced following high-pressure treatment. When β-lactoglobulin or WPC solution was treated by high pressure before emulsion formation, the emulsions had larger droplet sizes than emulsions made with the native protein (Galazka *et al.*, 1995).

The results indicated a modification of protein structure, leading to the loss of emulsifying efficiency as a result of protein aggregation, despite an increase in surface hydrophobicity. After adsorption on the surface, the protein probably became partially unfolded at the interface and subsequent pressure treatment caused no further conformational change. No studies on the behavior of emulsions formed with aggregated milk proteins, such as micellar casein, upon high-pressure treatment have been reported.

Milk protein hydrolysates and oil-in-water emulsions

Milk protein hydrolysates have been used extensively in infant and specialized adult nutritional formulations. Extensively hydrolyzed proteins are more easily digested and have substantially reduced immunological reactivities. These formulations are essentially multicomponent emulsion systems and therefore the emulsifying properties of protein hydrolysates are important.

The flexibility and thus the availability of hydrophobic and hydrophilic segments within the protein chain can be improved by moderate enzymatic hydrolysis of globular proteins (e.g. whey proteins), thus improving the emulsifying properties of the protein. However, extensive hydrolysis (above 20% degree of hydrolysis), because of the production of many short peptides, has been found to be detrimental to the emulsifying and stabilizing properties of whey proteins (Singh and Dalgleish, 1998).

The main form of instability in emulsions formed with highly hydrolyzed whey proteins is the coalescence that arises because of the inability of the predominantly short peptides to adequately stabilize the large oil surface generated during homogenization (Singh and Dalgleish, 1998; Agboola *et al.*, 1998a, 1998b). Nevertheless, it seems to be possible to make a fairly stable emulsion using highly hydrolyzed whey proteins at high peptide concentrations (peptide-to-oil ratio about 1:1 w/w), and at low homogenization pressures, as the sole emulsifier (Agboola *et al.*, 1998a, 1998b). Under these conditions, there is a sufficient amount of high-molecular-weight peptides (>5000 Da) in the emulsion to cover and stabilize the emulsion droplets.

Addition of calcium at above 20 mM has been shown to reduce the emulsion stability of emulsions formed with whey protein hydrolysates (Ramkumar *et al.*, 2000). This instability arises mainly from the binding of calcium to the adsorbed peptides, leading to a reduction in the charge density at the droplet surface, which would reduce the inter-droplet repulsion and enhance the likelihood of droplet flocculation. The formation of calcium bridges between peptides present on two different emulsion droplets would also enhance flocculation.

In these emulsions, it is observed that some very large droplets, apparently formed by coalescence, are also formed in the presence of calcium. This is likely to be due to the binding of calcium ions to the negatively charged peptides, causing aggregation of larger, more surface-active peptides. This situation would reduce the effective concentrations of emulsifying peptides available during emulsion formation.

Heat treatment of emulsions stabilized by highly hydrolyzed whey proteins at 121°C for 16 min results in destabilization of the emulsions, which appears to occur mainly via a coalescence mechanism (Agboola *et al.*, 1998b). As the adsorbed peptide layers in these emulsions lack the cohesiveness of the parent proteins and have poor ability to provide steric or charge stabilization, increased collisions between the droplets during heating would cause droplet aggregation, leading to coalescence. It is also possible that desorption of some loosely adsorbed peptides occurs during heating, as indicated by the decrease in the amount of peptides associated with the oil surface after heating, which would also enhance coalescence.

It is well known that the addition of polysaccharide such as xanthan gum or guar gum (in the concentration range 0.01–0.4%) to protein-stabilized emulsions promotes droplet flocculation, through a depletion mechanism, which enhances the creaming rate (Singh *et al.*, 2003). Similar effects are observed in emulsions stabilized with hydrolyzed whey proteins but, interestingly, this depletion flocculation also promotes the coalescence of droplets during the storage of emulsions (Ye *et al.*, 2004a, 2004b).

The rate of coalescence is enhanced considerably with increasing concentration of polysaccharide in the emulsion up to a certain critical concentration. At a given concentration, the rate of coalescence was highest in the emulsions containing guar gum, whereas it was lowest in the emulsions containing κ-carrageenan (Figure 11.4), which

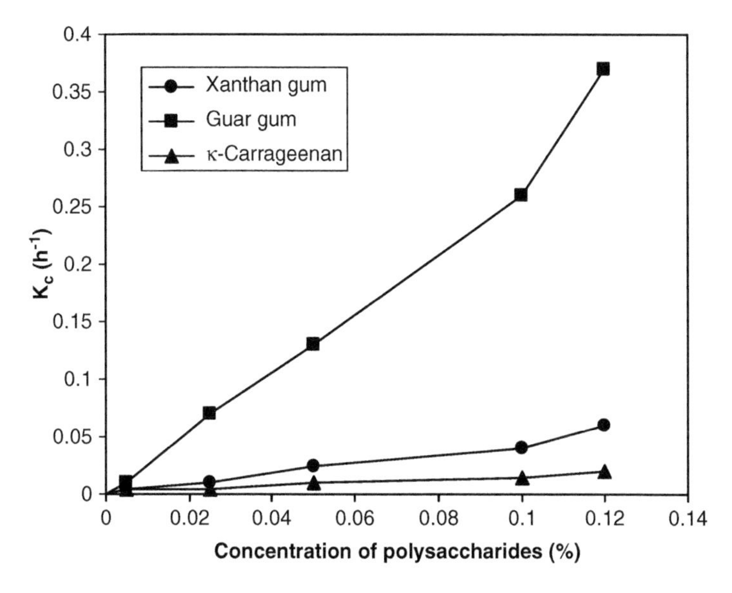

Figure 11.4 Apparent rate of coalescence (K_c) of droplets in emulsions containing xanthan gum (●), guar gum (■) or κ-carrageenan (▲) as a function of polysaccharide concentration (from Ye, Singh and Hemar 2004b, reproduced with the permission of ACS Publications).

could be explained on the basis of the relationship between the strength of the depletion potential and the molecular weight and radius of gyration of the polysaccharide.

Whey protein peptides adsorbed at the droplet surface in these emulsions would almost certainly have a reduced surface viscosity compared with intact proteins and this could lead to reduced stability to drainage and film rupture. It can be concluded that flocculation of droplets, through various mechanisms (e.g. depletion flocculation, calcium-induced aggregation, heat treatment), in these types of emulsions, where the interface is rather weak, would lead to coalescence during storage.

Lactoferrin-based oil-in-water emulsions

Bovine milk contains low levels of lactoferrin, an iron-binding glycoprotein with about 700 amino acid residues and a molecular weight of about 80 000 Da (Baker and Baker, 2005). The polypeptide is folded into two globular lobes, representing its N- and C-terminal halves, commonly referred to as the N-lobe and the C-lobe. The surface of the lactoferrin molecule has several regions with high concentrations of positive charge, giving it a high isoelectric point (pI $\approx$ 9). This positive charge is one of the features that distinguishes lactoferrin from other milk proteins, such as β-lactoglobulin, which have isoelectric points in the range 4.5–5.5 and are negatively charged at neutral pH. This unique difference could allow the formation of oil-in-water emulsions containing cationic emulsion droplets, through adsorption of lactoferrin, over a wider pH range.

Recent work (Ye and Singh, 2006b) has shown that, similar to other milk proteins (e.g. caseinate and β-lactoglobulin), lactoferrin adsorbs on to the interface of oil-in-water emulsion droplets and forms stable emulsions, but emulsion droplets with an overall positive surface charge are produced. In contrast to caseinate- and whey-protein-stabilized emulsions, the cationic emulsion droplets formed by lactoferrin are stable against a change in the pH from 7.0 to 3.0. For emulsions prepared under the same conditions (concentrations of oil and protein, pH, homogenization pressure), the droplet sizes in the lactoferrin emulsions are similar to those in β-lactoglobulin-stabilized emulsions, but the surface protein coverage (mg/m^2) of the emulsions made at pH 7.0 is higher in lactoferrin emulsions, possibly because of its higher molecular weight.

The formation of a positively charged adsorbed layer in lactoferrin-stabilized emulsions over a wide pH range provides an opportunity for electrostatic interactions with other milk proteins that are mostly negatively charged around neutral pH. In aqueous solutions, lactoferrin tends to form a complex with β-lactoglobulin via electrostatic interactions (Wahlgren *et al.*, 1993). Adsorption of such a complex on to the droplet surface during emulsion formation results in greater amounts of protein at the droplet surface and the formation of thick interfacial layers. It is interesting to note that oil-in-water emulsions formed using a binary mixture of β-lactoglobulin and lactoferrin are very stable, even though the overall charge (ζ-potential) of the emulsion droplets is close to zero. This suggests that steric repulsion plays an important role in this binary protein-stabilized emulsion.

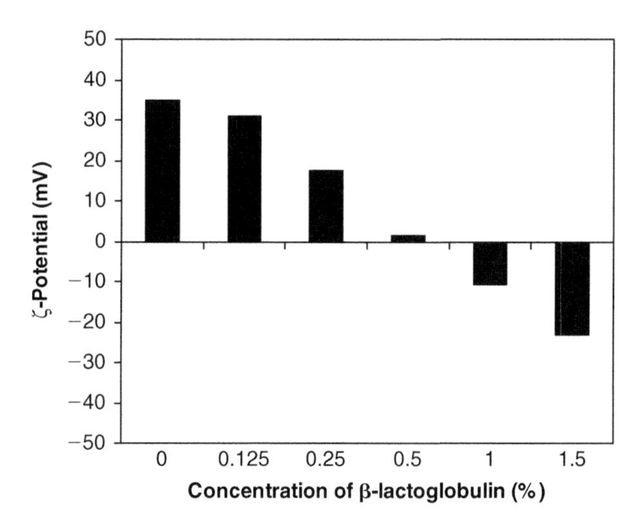

Figure 11.5 Influence of addition of β-lactoglobulin into emulsions formed with 1 wt% lactoferrin (30 wt% soya oil, pH 7.0) on the ζ-potential of the emulsion droplets (from Ye and Singh, 2007, reproduced with the permission of Springer).

Multilayered emulsions can be produced by interactions of oppositely charged milk proteins, i.e. lactoferrin and β-lactoglobulin or caseinate at neutral pH (Ye and Singh, 2007). A primary emulsion, containing either anionic droplets coated with β-lactoglobulin or cationic droplets coated with lactoferrin, can be produced. A secondary emulsion can then be made by mixing either β-lactoglobulin solution or lactoferrin solution with the primary emulsion (Ye and Singh, 2007).

For example, when the emulsions formed with lactoferrin (1 wt%, pH 7.0) were diluted with aqueous phase containing a range of β-lactoglobulin concentrations, the adsorption of β-lactoglobulin increased considerably with an increase in the β-lactoglobulin concentration up to 0.42 wt%, with very little change above this concentration. This increase in β-lactoglobulin on the surface of emulsions formed with lactoferrin was further confirmed by the change in the ζ-potential. In the absence of β-lactoglobulin, the ζ-potential of the emulsion droplets was around +50 mV, because the lactoferrin used to stabilize the droplets has a net positive charge at pH 7.0. The ζ-potential became less positive, and eventually changed from positive to negative, as the β-lactoglobulin concentration in the emulsion was increased (Figure 11.5).

Conclusions

Milk proteins in soluble and dispersed forms have excellent surface-active and emulsion-stabilizing properties. Differences in the emulsifying abilities of milk proteins arise largely from the differences in structure, flexibility, state of aggregation and composition of the proteins. These attributes of milk proteins (and hence their emulsifying abilities) are modified through various interactions occurring during the processing of milk that is required to isolate the protein components as well as during the manufacture of prepared foods. Emulsions with different surface compositions and structures can be made using different kinds of milk proteins and these emulsions exhibit different sensitivities to solution conditions, such as pH and ionic strength, and processing conditions, such as heat and high-pressure treatments. This could offer possibilities for the formation of emulsions with a range of functionalities for different food applications.

Most of the research during the last 20 years has been performed on oil-in-water emulsions using purified or simple mixtures of caseins and whey proteins, with a great deal of information now being available on the conformation of proteins at oil–water interfaces, competitive exchange reactions between adsorbed and unadsorbed proteins, protein–polysaccharide interactions and factors controlling the rheology and stability of emulsions. In addition, some understanding of how processing conditions (heat treatments, high-pressure treatments) influence interfacial structures and emulsion properties has been achieved.

There is much less understanding of the behavior of more complex mixtures of proteins in emulsions and the stability behavior of emulsions under processing environments commonly encountered in the food industry. In addition, there is an almost complete lack of understanding of the behavior of emulsions during oral processing in the mouth as well as during digestion processes. It is critical to understand the oral behavior of emulsions, as common sensorial attributes (e.g. creaminess, smoothness), and that the release of fat-soluble flavors are based on interfacial structures and rheological parameters. There is some evidence to show that the behavior of emulsions in the gastrointestinal tract is affected by their physico-chemical properties and that the properties of the interface modulate fat digestion and consequently influence the bioavailability of lipid nutrients. This area of research needs to be further developed before the knowledge can be used to develop novel products with health and sensory attributes.

References

Agboola, S. O., Singh, H., Munro, P. A., Dalgleish, D. G. and Singh, A. M. (1998a). Destabilization of oil-in-water emulsions formed using highly hydrolyzed whey proteins. *Journal of Agricultural and Food Chemistry*, **46**, 84–90.

Agboola, S. O., Singh, H., Munro, P. A., Dalgleish, D. G. and Singh, A. M. (1998b). Stability of emulsions formed using whey protein hydrolysate: effects of lecithin addition and retorting. *Journal of Agricultural and Food Chemistry*, **46**, 1814–19.

Anema, S. G., Stockmann, R. and Lowe, E. K. (2005). Denaturation of β-lactoglobulin in pressure treated skim milk. *Journal of Agricultural and Food Chemistry*, **53**, 7783–91.

Baker, E. N. and Baker, H. M. (2005). Molecular structure, binding properties and dynamics of lactoferrin. *Cellular and Molecular Life Sciences*, **62**, 2531–39.

Baumy, J. J. and Brule, G. (1988). Binding of bivalent cations to α-lactalbumin and β-lactoglobulin: effect of pH and ionic strength. *Lait*, **68**, 33–48.

Britten, M. and Giroux, H. J. (1991). Emulsifying properties of whey protein and casein composite blends. *Journal of Dairy Science*, **74**, 3318–25.

Brun, J. M. and Dalgleish, D. G. (1999). Some effects of heat on the competitive adsorption of caseins and whey proteins in oil-in-water emulsions. *International Dairy Journal*, **9**, 323–27.

Dalgleish, D. G. (1990). The conformation of proteins on solid/water interfaces—caseins and phosvitin on polystyrene latices. *Colloids and Surfaces*, **46**, 141–55.

Dalgleish, D. G. (1995). Structures and properties of adsorbed layers in emulsions containing milk proteins. In *Food Macromolecules and Colloids* (E. Dickinson and D. Lorient, eds) pp. 23–33. Cambridge: The Royal Society of Chemistry.

Dalgleish, D. G. (1996). Conformations and structures of milk proteins adsorbed to oil-water interfaces. *Food Research International*, **29**, 541–47.

Dalgleish, D. G. (1997). Adsorption of protein and the stability of emulsions. *Trends in Food Science and Technology*, **8**, 1–6.

Dalgleish, D. G. (1999). Interfacial structures and colloidal interactions in protein-stabilised emulsions. In *Food Emulsions and Foams: Interfaces, Interactions and Stability* (E. Dickinson and J. M. Rodriguez-Patino, eds) pp. 1–16. Cambridge: The Royal Society of Chemistry.

Dalgleish, D. G. and Leaver, J. (1993). Dimensions and possible structures of milk proteins at oil-water interfaces. In *Food Colloids and Polymers: Stability and Mechanical Properties* (E. Dickinson and P. Walstra, eds) pp. 113–22. Cambridge: The Royal Society of Chemistry.

Dalgleish, D. G., Goff, H. D., Brun, J. M. and Luan, B. B. (2002). Exchange reactions between whey proteins and caseins in heated soya oil-in-water emulsion systems – overall aspects of the reaction. *Food Hydrocolloids*, **16**, 303–11.

Dam, B. V., Watts, K., Campbell, I. and Lips, A. (1995). On the stability of milk protein-stabilized concentrated oil-in-water food emulsions. In *Foods Macromolecules and Colloids* (E. Dickinson and D. Lorient, eds) pp. 215–22. Cambridge: The Royal Society of Chemistry.

Damodaran, S. (1997). Food proteins: an overview. In *Food Proteins and their Applications* (S. Damodaran, ed.) pp. 1–24. London: CRC Press.

Das, K. P. and Kinsella, J. E. (1993). Droplet size and coalescence stability of whey protein stabilized milkfat peanut oil emulsions. *Journal of Food Science*, **58**, 439–44.

de Hek, H. and Vrij, A. (1981). Interactions in mixtures of colloidal silica spheres and polystyrene molecules in cyclohexane. *Journal of Colloid and Interface Science*, **84**, 409–22.

Demetriades, K. and McClements, D. J. (1998). Influence of pH and heating on physicochemical properties of whey protein-stabilized emulsions containing a nonionic surfactant. *Journal of Agricultural and Food Chemistry*, **46**, 3936–42.

Dickinson, E. (1992). Structure and composition of adsorbed protein layers and the relationship to emulsion stability. *Journal of the Chemical Society – Faraday Transactions*, **88**, 2973–83.

Dickinson, E. (1998a). Proteins at interfaces and in emulsions: stability, rheology and interactions. *Journal of the Chemical Society – Faraday Transactions*, **94**, 1657–69.

Dickinson, E. (1998b). Stability and rheological implications of electrostatic milk protein-polysaccharide interactions. *Trends in Food Science and Technology*, **9**, 347–54.

Dickinson, E. (1999). Adsorbed protein layers at fluid interfaces: interactions, structure and surface rheology. *Colloids and Surfaces B: Biointerfaces*, **15**, 161–76.

Dickinson, E. (2003). Hydrocolloids at interfaces and the influence on the properties of dispersed systems. *Food Hydrocolloids*, **17**, 25–39.

Dickinson, E. and Golding, M. (1997). Depletion flocculation of emulsions containing unadsorbed sodium caseinate. *Food Hydrocolloids*, **11**, 13–18.

Dickinson, E. and Golding, M. (1998). Influence of alcohol on stability of oil-in-water emulsions containing sodium caseinate. *Journal of Colloid and Interface Science*, **197**, 133–41.

Dickinson, E. and James, J. D. (1998). Rheology and flocculation of high pressure-treated β-lactoglobulin-stabilized emulsions: comparison with thermal treatment. *Journal of Agricultural and Food Chemistry*, **46**, 2565–71.

Dickinson, E. and Matsumura, Y. (1991). Time-dependent polymerization of β-lactoglobulin through disulphide bonds at the oil-water interface in emulsions. *International Journal of Biological Macromolecules*, **13**, 26–30.

Dickinson, E. and McClements, D. J. (eds) (1995). *Advances in Food Colloids*. London: Blackie Academic and Professional.

Dickinson, E. and Parkinson, E. L. (2004). Heat-induced aggregation of milk protein-stabilized emulsions: sensitivity to processing and composition. *International Dairy Journal*, **14**, 635–45.

Dickinson, E. and Pawlowsky, K. (1998). Influence of κ-carrageenan on the properties of a protein-stabilized emulsion. *Food Hydrocolloids*, **12**, 417–23.

Dickinson, E. and Stainsby, G. (1988). Emulsion stability. In *Advances in Food Emulsions and Foams* (E. Dickinson and G. Stainsby, eds) pp. 1–44. London: Elsevier Applied Science.

Dickinson, E., Murray, B. S. and Stainsby, G. (1988a). Protein adsorption at air-water and oil-water interfaces. In *Advances in Food Emulsions and Foams* (E. Dickinson and G. Stainsby, eds) pp. 123–62. London: Elsevier Applied Science.

Dickinson, E., Rolfe, S. and Dalgleish, D. G. (1988b). Competitive adsorption of α_{s1}-casein and β-casein in oil-in water emulsions. *Food Hydrocolloids*, **2**, 193–203.

Dickinson, E., Semenova, M. G., Belyakova, L. E., Antipova, A. S., Il'in, M. M., Tsapkina, E. N. and Ritzoulis, C. (2001). Analysis of light scattering data on the calcium ion sensitivity of caseinate solution thermodynamics: relationship to emulsion flocculation. *Journal of Colloid and Interface Science*, **239**, 87–97.

Dumay, E., Lambert, C., Funtenberger, S. and Cheftel, J. C. (1996). Effects of high pressure on the physico-chemical characteristics of dairy creams and model oil/water emulsions. *Lebensmittel-Wissenschaft und -Technologie*, **29**, 606–25.

Euston, S. R. and Hirst, R. L. (1999). Comparison of the concentration-dependent emulsifying properties of protein products containing aggregated and non-aggregated milk protein. *International Dairy Journal*, **9**, 693–701.

Euston, S. E., Singh, H., Munro, P. A. and Dalgleish, D. G. (1995). Competitive adsorption between sodium caseinate and oil-soluble and water-soluble surfactants in oil-in-water emulsions. *Journal of Food Science*, **60**, 1124–31.

Euston, S. E., Singh, H., Munro, P. A. and Dalgleish, D. G. (1996). Oil-in-water emulsions stabilized by sodium caseinate or whey protein isolate as influenced by glycerol monostearate. *Journal of Food Science*, **61**, 916–20.

Euston, S. R., Finnigan, S. R. and Hirst, R. L. (2000). Aggregation kinetics of heated whey protein-stabilized emulsions. *Food Hydrocolloids*, **14**, 155–61.

Fang, Y. and Dalgleish, D. G. (1998). The conformation of α-lactalbumin as a function of pH, heat treatment and adsorption at hydrophobic surfaces studied by FTIR. *Food Hydrocolloids*, **12**, 121–26.

Fellows, C. M. and Doherty, W. O. S. (2006). Insights into bridging flocculation. *Macromolecular Symposia*, **231**, 1–10.

Galazka, V. B., Ledward, D. A., Dickinson, E. and Langley, K. R. (1995). High pressure effects on emulsifying behavior of whey protein concentrate. *Journal of Food Science*, **60**, 1341–43.

Gaucheron, F., Famelart, M. H., Mariette, F., Raulok, K., Michel, F. and Le Graet, Y. (1997). Combined effects of temperature and high-pressure treatments on physico-chemical characteristics of skim milk. *Food Chemistry*, **59**, 439–47.

Hunt, J. A. and Dalgleish, D. G. (1994). Adsorption behavior of whey protein isolate and caseinate in soya oil-water emulsions. *Food Hydrocolloids*, **8**, 175–87.

Hunt, J. A. and Dalgleish, D. G. (1995). Heat stability of oil-in-water emulsions containing milk proteins: effect of ionic strength and pH. *Journal of Food Science*, **60**, 1120–23.

Hunter, R. J. (ed.) (1986). *Foundations of Colloid Science*, Volume 1, p. 674. Oxford: Oxford University Press.

Huppertz, T., Fox, P. F. and Kelly, A. L. (2004). Dissociation of caseins in high-pressure treated bovine milk. *International Dairy Journal*, **14**, 675–80.

López-Fandinó, R., Carrascosa, A. V. and Olano, A. (1996). The effects of high pressure on whey protein denaturation and cheese-making properties of raw milk. *Journal of Dairy Science*, **79**, 929–36.

Lucey, J. A., Srinivasan, M., Singh, H. and Munro, P. (2000). Characterization of commercial and experimental sodium caseinates by multiangle laser light scattering and size-exclusion chromatography. *Journal of Agricultural and Food Chemistry*, **48**, 1610–16.

Mackie, A. R., Mingins, J. and Dann, R. (1993). Preliminary studies of β-lactoglobulin adsorbed on polystyrene latex. In *Food Colloids and Polymers: Stability and Mechanical Properties* (E. Dickinson and P. Walstra, eds) pp. 96–112. Cambridge: The Royal Society of Chemistry.

McClements, D. J. (ed.). (1999) *Food Emulsions: Principles, Practice, and Techniques*. London: CRC Press

McClements, D. J. (2005). Theoretical analysis of factors affecting the formation and stability of multilayered colloidal dispersions. *Langmuir*, **21**, 9777–85.

McClements, D. J. and Decker, E. A. (2000). Lipid oxidation in oil-in-water emulsions: impact of molecular environment on chemical reactions in heterogeneous food systems. *Journal of Food Science*, **65**, 1270–82.

McClements, D. J., Monahan, F. J. and Kinsella, J. E. (1993). Disulfide bond formation affects stability of whey protein isolate emulsions. *Journal of Food Science*, **58**, 1036–39.

Monahan, F. J., McClements, D. J. and German, J. B. (1996). Disulfide-mediated polymerization reactions and physical properties of heated WPI-stabilized emulsions. *Journal of Food Science*, **61**, 504–9.

Mulvihill, D. M. and Fox, P. F. (1989). Physico-chemical and functional properties of milk proteins. In *Developments in Dairy Chemistry – 4. Functional Milk Proteins* (P. F. Fox, ed.) pp. 131–72. London: Elsevier Applied Science.

Pappas, C. P. and Rothwell, J. (1991). The effect of heating, alone or in the presence of calcium or lactose, on calcium binding to milk proteins. *Food Chemistry*, **42**, 183–201.

Parkinson, E. L. and Dickinson, E. (2004). Inhibition of heat-induced aggregation of a α-lactoglobulin-stabilized emulsion by very small additions of casein. *Colloids and Surfaces B: Biointerfaces*, **39**, 23–30.

Patel, H. A., Singh, H., Havea, P., Considine, T. and Creamer, L. K. (2005). Pressure-induced unfolding and aggregation of the protein in whey protein concentrate solutions. *Journal of Agricultural and Food Chemistry*, **53**, 9590–601.

Patel, H. A., Singh, H., Anema, S. G. and Creamer, L. K. (2006). Effects of heat and high hydrostatic pressure treatments on disulfide bonding interchanges among the proteins in skim milk. *Journal of Agricultural and Food Chemistry*, **54**, 3409–20.

Phillips, M. C. (1981). Protein conformation at liquid interfaces and its role in stabilizing emulsions and foams. *Food Technology*, **35**, 424–27.

Ramkumar, C., Singh, H., Munro, P. A., Dalgleish, D. G. and Singh, A. M. (2000). Influence of calcium, magnesium, or potassium ions on the formation and stability of emulsions prepared using highly hydrolyzed whey proteins. *Journal of Agricultural and Food Chemistry*, **48**, 1598–604.

Rollema, H. S. (1992). Casein association and micelle formation. In *Advanced Dairy Chemistry – 1: Proteins* (P. F. Fox, ed.) pp. 111–40. London: Elsevier Applied Science.

Singh, A. M. and Dalgleish, D. G. (1998). The emulsifying properties of hydrolyzates of whey proteins. *Journal of Dairy Science*, **81**, 918–24.

Singh, H. (2005). Milk protein functionality in food colloids. In *Food Colloids: Interactions, Microstructure and Processing* (E. Dickinson, ed.) pp. 179–93. Cambridge: The Royal Society of Chemistry.

Singh, H., Tamehana, M., Hemar, Y. and Munro, P. A. (2003). Interfacial compositions, microstructure and stability of oil-in-water emulsions formed with mixtures of milk proteins and κ-carrageenan: 2. Whey protein isolate (WPI). *Food Hydrocolloids*, **17**, 549–61.

Sliwinski, E. L., Roubos, P. J., Zoet, F. D., van Boekel, M. A. J. S. and Wouters, J. T. M. (2003). Effects of heat on physicochemical properties of whey protein-stabilised emulsions. *Colloids and Surfaces B: Biointerfaces*, **31**, 231–42.

Srinivasan, M., Singh, H. and Munro, P. A. (1996). Sodium caseinate-stabilized emulsions: factors affecting coverage and composition of surface proteins. *Journal of Agricultural and Food Chemistry*, **44**, 3807–11.

Srinivasan, M., Singh, H. and Munro, P. A. (1999). Adsorption behavior of sodium and calcium caseinates in oil-in-water emulsions. *International Dairy Journal*, **9**, 337–41.

Srinivasan, M., Singh, H. and Munro, P. A. (2000). The effect of sodium chloride on the formation and stability of sodium caseinate emulsions. *Food Hydrocolloids*, **14**, 497–507.

Srinivasan, M., Singh, H. and Munro, P. A. (2001). Creaming stability of oil-in-water emulsions formed with sodium and calcium caseinates. *Journal of Food Science*, **66**, 441–46.

Srinivasan, M., Singh, H. and Munro, P. A. (2002). Formation and stability of sodium caseinate emulsions: influence of retorting (121°C for 15 min) before or after emulsification. *Food Hydrocolloids*, **16**, 153–60.

Tuinier, R. and de Kruif, C. G. (1999). Phase behavior of casein micelles/exocellular polysaccharide mixtures: experiment and theory. *Journal of Colloid and Interface Science*, **110**, 9296–304.

Wahlgren, M. C., Arnebrant, T. and Paulsson, M. A. (1993). The adsorption from solutions of β-lactoglobulin mixed with lactoferrin or lysozyme onto silica and methylated silica surfaces. *Journal of Colloid and Interface Science*, **158**, 46–53.

Walstra, P. (1993). Introduction to aggregation phenomena in food colloids. In *Food Colloids and Polymers: Stability and Mechanical Properties* (E. Dickinson and P. Walstra, eds) pp. 3–15. Cambridge: The Royal Society of Chemistry.

Ye, A. and Singh, H. (2000). Influence of calcium chloride addition on the properties of emulsions stabilized by whey protein concentrate. *Food Hydrocolloids*, **14**, 337–46.

Ye, A. and Singh, H. (2001). Interfacial composition and stability of sodium caseinate emulsions as influenced by calcium ions. *Food Hydrocolloids*, **15**, 195–207.

Ye, A. and Singh, H. (2006a). Heat stability of oil-in-water emulsions formed with intact or hydrolysed whey proteins: influence of polysaccharides. *Food Hydrocolloids*, **20**, 269–76.

Ye, A. and Singh, H. (2006b). Adsorption behavior of lactoferrin in oil-in-water emulsions as influenced by interaction with β-lactoglobulin. *Journal of Colloid and Interface Science*, **295**, 249–54.

Ye, A. and Singh, H. (2007). Formation of multilayers at the interface of oil-in-water emulsion via interactions between lactoferrin and β-lactoglobulin. *Food Biophysics*, **2**, 125–32.

Ye, A., Singh, H. and Hemar, Y. (2004a). Enhancement of coalescence by xanthan addition in oil-in-water emulsions formed with highly hydrolysed whey proteins. *Food Hydrocolloids*, **18**, 737–46.

Ye, A., Singh, H. and Hemar, Y. (2004b). Influence of polysaccharides on the rate of coalescence in oil-in-water emulsions formed with highly hydrolyzed whey proteins. *Journal of Agricultural and Food Chemistry*, **52**, 5491–98.

12

Milk protein–polysaccharide interactions

Kelvin K. T. Goh, Anwesha Sarkar and Harjinder Singh

Abstract

Proteins and polysaccharides are common ingredients in many food formulations. They are generally responsible for imparting key sensory attributes (e.g. textural attributes, controlled flavor release) and are capable of modifying phase stability in colloidal food systems. Their physicochemical properties depend not only on the molecular parameters of the individual biopolymers but also on the nature of interactions between the protein and polysaccharide molecules.

This chapter provides a basic overview of the possible types and natures of the interactions that can occur between protein and polysaccharide molecules in aqueous solutions. A description of the phase diagram that is commonly used to estimate phase stability is given. Extensive research carried out in this field over the last few decades, outlining different milk protein–polysaccharide interactions, is summarized in tables. The last sections attempt to categorize the different types of interactions and their impact on the microstructures and the rheological properties of the systems. The chapter concludes by stressing the importance of understanding these interactions, which potentially provide food scientists with the opportunity to modify or create novel food structures and functionalities.

Introduction

Proteins and polysaccharides are broadly classified as biopolymers because of their large molecular structures. These macromolecules are known to play important

Milk Proteins: From Expression to Food
ISBN: 978-0-12-374039-7

physico-chemical roles, such as imparting thickening, stabilizing, gelling, emulsifying properties etc., in food products (Hemar *et al.*, 2001a, 2001b; Dickinson, 2003; Dickinson *et al.*, 2003). The physico-chemical properties of proteins and polysaccharides have individually been studied extensively over the last several decades. It is well established that the factors influencing the physico-chemical properties of these macromolecules in solution include molar mass, molecular conformation, polydispersity, charge density, concentration, pH, ionic strength, temperature, solvent quality and nature of molecular (intra-/inter-) interactions (Tolstoguzov, 1997a; Doublier *et al.*, 2000; de Kruif and Tuinier, 2001).

In many food systems, their physical properties become more complex as both proteins and polysaccharides are present (either naturally or added as ingredients) in the complex multi-component mixtures. The overall stability and the microstructure of the food systems depend not only on the physico-chemical properties of proteins or polysaccharides alone, but also on the nature and strength of the interactions between proteins and polysaccharides (Dickinson, 1998). This chapter reviews a number of recent studies on protein–polysaccharide interactions, with a particular focus on milk proteins and a diverse range of polysaccharides in aqueous systems.

Mixing behavior of biopolymers

When aqueous solutions of proteins and polysaccharides are mixed, one of four phenomena can arise: (a) co-solubility; (b) thermodynamic incompatibility; (c) depletion interaction (or flocculation); (d) complex coacervation (Figure 12.1) (Tolstoguzov, 1991, 1997a; Schmitt *et al.*, 1998; Syrbe *et al.*, 1998; de Kruif and Tuinier, 2001; Benichou *et al.*, 2002; Dickinson, 2003; Tolstoguzov, 2003; de Kruif *et al.*, 2004; Martinez *et al.*, 2005). These phenomena can be explained as follows.

Co-solubility refers to the creation of a stable homogeneous solution, i.e. the generation of one phase in which the two macromolecular species either do not interact or exist as soluble complexes in the aqueous medium. When intermolecular attraction is absent, macromolecules are co-soluble only in dilute solution, where the entropy of mixing favors more randomness in the system (Tolstoguzov, 2003).

To achieve co-solubility from a thermodynamic viewpoint, the Gibbs' free energy of mixing (ΔG_{mixing}), given in Equation (12.1), must be negative. This means that the entropy of mixing should favorably exceed the enthalpy term (note: the highest level of entropy is achieved when the different kinds of molecules are randomly distributed throughout the system) (McClements, 2005). The expression for the Gibbs' free energy accompanying mixing under standard conditions is given by

$$\Delta G_{\text{mixing}} = \Delta H_{\text{mixing}} - T\Delta S_{\text{mixing}} \quad \textbf{(12.1)}$$

where ΔG_{mixing}, ΔH_{mixing} and $T\Delta S_{\text{mixing}}$ are the free energy, enthalpy (interaction energy) and entropy changes between the mixed and unmixed states respectively.

When the size of the molecules is small, as it is in the case of monomer sugars and hydrophilic amino acids, mixing the two species results in a co-soluble system.

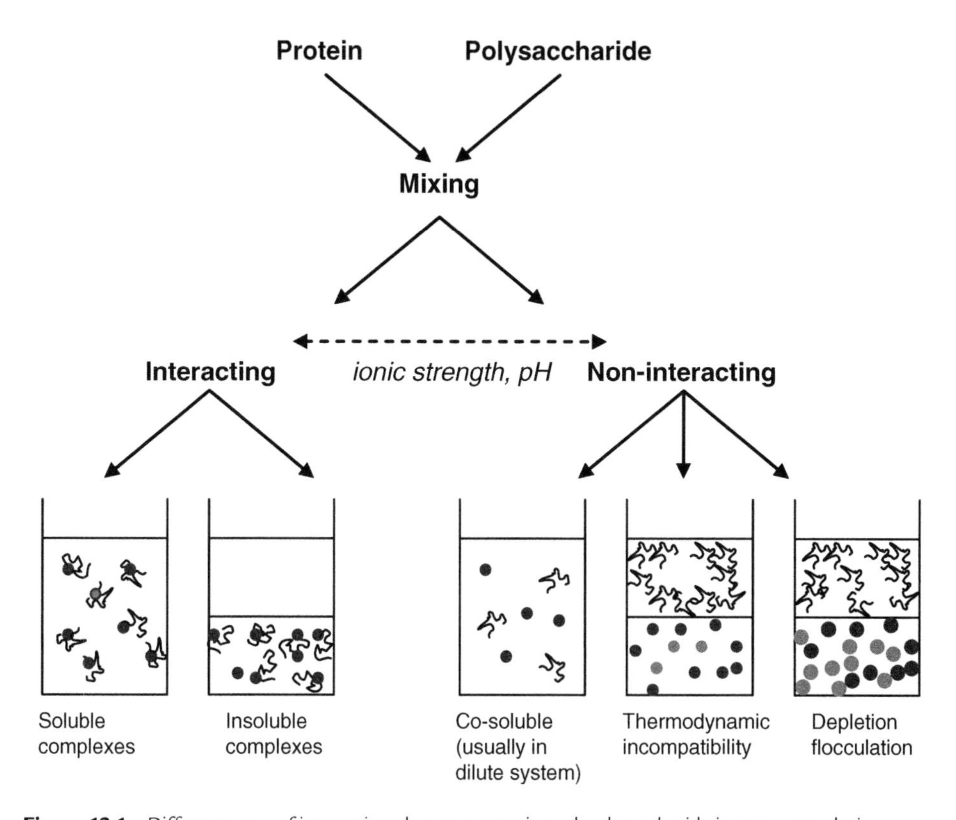

Figure 12.1 Different types of interactions between protein and polysaccharide in aqueous solutions.

However, with increasing molecular weight and concentration of the polymers, the system tends to become less co-soluble as a result of thermodynamic incompatibility (Tolstoguzov, 1991, 1997a), because the entropy of mixing of biopolymers is significantly lower than that of monomers. The bulky size and the rigid structure of biopolymer molecules decrease the entropy of mixing, resulting in a higher free energy. For a mixed biopolymer solution, the enthalpy–entropy balance generally results in mutual exclusion of one biopolymer from the local vicinity of the other biopolymer. This means that biopolymers in mixed solution show a preference to be surrounded by their own type; otherwise, their mixtures separate into liquid phases (Grinberg and Tolstoguzov, 1972; Tolstoguzov *et al.*, 1985; Tolstoguzov, 1988, 1991; Grinberg and Tolstoguzov, 1997; Polyakov *et al.*, 1997).

Thermodynamic incompatibility occurs when the two non-interacting macromolecular species separate into two different phases, as the enthalpy of mixing exceeds the entropy difference (Grinberg and Tolstoguzov, 1997; Schmitt *et al.*, 1998; Benichou *et al.*, 2002; Tolstoguzov, 2002). Each of the two distinct immiscible aqueous phases formed is loaded mainly with only one biopolymer species, i.e. a protein-rich phase and a polysaccharide-rich phase. Phase separation as a result of incompatibility can also occur if each biopolymer shows varying affinity towards the solvent (Tolstoguzov, 1991; Piculell and Lindman, 1992). In this case, solvent–protein (or solvent–polysaccharide) interactions are favored over protein–polysaccharide interactions and solvent–solvent interactions, leading to two phases—one enriched in protein and the other enriched in polysaccharide (Doublier et al., 2000).

Thermodynamic incompatibility can also arise within a mixture of polysaccharides or proteins. Some examples include: polysaccharides with different structures; proteins of different classes, such as water-soluble albumins with salt-soluble globulins; native and denatured forms of the same protein as well as aggregated and non-aggregated forms of the same protein (Tolstoguzov, 2002).

Thermodynamic incompatibility is highly dependent on pH and ionic strength and is prevalent when protein and neutral polysaccharide are present or when both protein and polysaccharide carry the same negative charge at neutral pH (Doublier *et al.*, 2000). Although thermodynamic incompatibility is prevalent in mixed polymer systems, some of these systems do not achieve thermodynamic equilibrium within a limited timescale because of the presence of kinetic energy barriers. When the kinetic energy exceeds the thermal energy of the system, the molecules become "trapped" in a metastable state (McClements, 2005). Some examples of kinetic energy barriers include the formation of a gel network within an incompatible system or a highly viscous continuous phase that slows down the phase separation process. The choice of which phase to gel and the component used to promote gelation depends on the type of biopolymers used in the system (Bryant and McClements, 2000a, 2000b; Norton and Frith, 2001; Kim et al., 2006).

Depletion interaction (or flocculation) usually involves spherical particles in the presence of macromolecules (Asakura and Oosawa, 1954, 1958; Bourriot *et al.*, 1999a). Phase separation of particulate suspensions is enhanced by the addition of a polymer. This phenomenon usually occurs in a colloidal dispersion in the presence of non-interacting polymers (e.g. polysaccharides in an emulsion, polysaccharides and colloidal casein micelles). The higher osmotic pressure of the polymer molecules surrounding the colloidal particles (as compared with the inter-particle region) causes an additional attractive force between the particles, leading to their flocculation. The attractive force depends on the size, shape and concentration of the polymer molecules and the colloidal particles (Hemar *et al.*, 2001b).

When colloidal particles approach each other, the excluded (or depleted) layer starts to overlap, allowing more space for the polymer molecules. The increase in volume causes the total entropy of the system to increase (i.e. the free energy to decrease), which in turn encourages attraction interaction between the colloidal particles (de Bont *et al.*, 2002). In a mixed protein–polysaccharide system where the protein species is casein micelles, phase separation is often attributed to a depletion flocculation phenomenon (Bourriot *et al.*, 1999a; Tuinier and De Kruif, 1999; Tuinier *et al.*, 2000), because of the large colloidal particle size of the casein micelles and because increasing the concentration of polysaccharides results in greater attraction between the casein micelles (Doublier *et al.*, 2000).

Complex coacervation is the formation of electrostatic complexes between protein and polysaccharide molecules, leading to a two-phase system. One phase has both biopolymers in a complex matrix and the other phase contains mainly the solvent water and is depleted in both biopolymers. Complex coacervation commonly occurs between oppositely charged biopolymers. Complex coacervation between oppositely charged proteins and polysaccharides was first reported with the mixing of gelatin and gum arabic in acetic acid solution (Tiebackx, 1911).

The term "coacervation" was first introduced in 1929 to describe a process in which aqueous colloidal solutions separate into two liquid phases: one rich in colloid, i.e. the coacervate, and the other containing little colloid (Bungenberg de Jong and Kruyt, 1929). If the two biopolymers are present in equal proportions by weight at a pH such that they carry net equal opposite charges, the yield of coacervate will be at its maximum (Schmitt *et al.*, 1998). The size and the morphology of these structures may be exploited to bring about new functionalities and textural changes in processed foods.

Phase diagram

Mixing two aqueous solutions of protein and polysaccharide will give rise to a one-phase system or a two-phase system, depending on the solution composition and the environmental conditions, as depicted in Figure 12.1 (Tolstoguzov, 1991, 1997a; Schmitt *et al.*, 1998; Syrbe *et al.*, 1998; de Kruif and Tuinier, 2001; Benichou *et al.*, 2002; Tolstoguzov, 2002; Dickinson, 2003; de Kruif *et al.*, 2004; Martinez *et al.*, 2005).

In a one-phase system, protein and polysaccharide can exist either as individual molecules or as soluble complexes that are uniformly dispersed throughout the entire system. However, with increasing molecular weight and concentration of the biopolymers, the system tends to become less co-soluble and gives rise to a two-phase system, i.e. the system separates into two distinct phases that have different biopolymer concentrations.

For a system with relatively strong net repulsion between protein and polysaccharide in aqueous solution, the two biopolymers move into two different phases as a result of thermodynamic incompatibility. Two distinct immiscible aqueous phases are formed and each is loaded mainly with only one biopolymer species, i.e. one phase is protein rich and the other phase is polysaccharide rich. A typical phase diagram for segregating a biopolymer system is shown in Figure 12.2 and has been explained by many researchers (Grinberg and Tolstoguzov, 1972; Polyakov *et al.*, 1980; Antonov *et al.*, 1982; Tolstoguzov *et al.*, 1985; Grinberg and Tolstoguzov, 1997; Bourriot *et al.*, 1999a; Closs *et al.*, 1999; Clark, 2000; Lundin *et al.*, 2003; Thaiudom and Goff, 2003; Tolstoguzov, 2003; Ercelebi and Ibanoglu, 2007).

The phase diagram consists of a typical binodal curve (the solid line curve), which divides the single-phase miscible region (below the curve) from the two-phase immiscible region (the shaded region). The binodal branches show the points of limited co-solubility. The points of the binodal curve connected by tie lines represent the compositions of the co-existing equilibrium phases. From the phase diagram, it is possible to determine the effective concentrations of biopolymers in the two phases and the concentrations at which maximal co-solubility of the biopolymers is achieved. In addition, it helps to establish which of the two biopolymers forms the continuous phase.

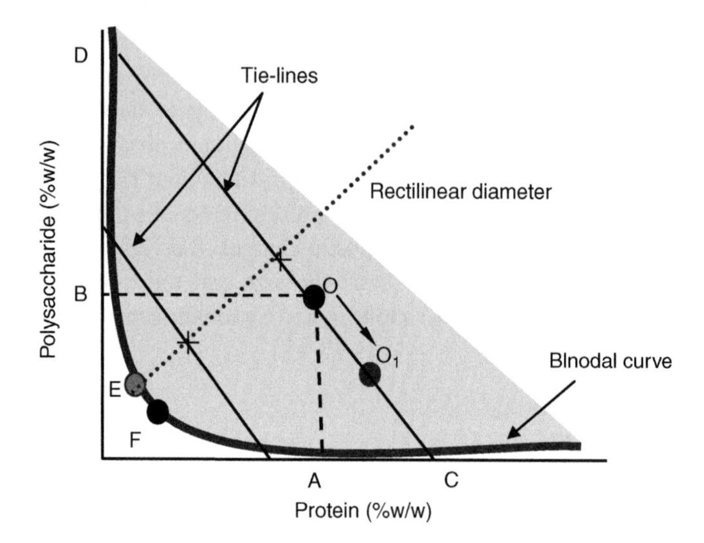

Figure 12.2 A typical phase diagram showing a protein–polysaccharide solution with water as the solvent at a particular pH, temperature and ionic strength. A sample of composition O (which was initially made with A% protein and B% polysaccharide) separates out into two bulk polymer-rich phases. The protein-enriched phase will have composition C% protein whereas the polysaccharide-enriched phase will have composition D% polysaccharide. The binodal (solid) curve separates the single-phase region from the two-phase domain (obtained by direct observation of the phase separation in test tubes). The % protein in the polysaccharide phase will be negligible and vice versa. The tie line is obtained by joining C and D. The ratio DO/OC represents the volume ratio of the protein-rich phase C and the polysaccharide-rich phase D by the inverse-lever rule. If O is shifted along the tie line to O_1, the new phase volume ratio will be DO_1/O_1C. Although any composition lying on the same tie line results in the same effective concentration in the enriched phases, the phase volume varies. The line obtained by joining the mid points (+) of two or more tie lines gives the rectilinear diameter. The co-ordinates of the critical point E (obtained from the intersection of the binodal curve to the rectilinear diameter) show the composition of a system separating into two phases of the same volume and composition, which means that the separated-phase systems will have 50% protein and 50% polysaccharide in the same phase volume ratio. Point F is the separation threshold, which is the minimum critical concentration required for the biopolymers to separate into two phases.

Nature of interactions in protein–polysaccharide systems

The interactions responsible for complex formation between biopolymers can be classified as: weak or strong, specific or non-specific, attractive or repulsive (Dickinson, 1998). The overall interaction between protein and polysaccharide is the sum of the following different intermolecular forces arising between the various segments and chains of the two biopolymers (Dickinson, 1998; Schmitt *et al.*, 1998).

Repulsive interactions

Repulsive interactions are always non-specific and of transient duration. They usually arise from excluded volume effects and/or electrostatic interactions and tend to be weak, except at very close range or at very low ionic strength.

The excluded volume or steric exclusion effects are the non-specific and transient interactions that can be found in non-ionic, non-penetrable polysaccharides and polypeptides that cannot occupy the same solution volume (Tolstoguzov, 1991; Polyakov *et al.*, 1997; Schmitt *et al.*, 1998; Tolstoguzov, 2002, 2003). Excluded volume effects exhibit mutual spatial restrictions and competition between the biopolymers for solution space, i.e. there is a reduction in the mixing entropy of the system because of the reduction in the volume available for the biopolymer molecules to occupy.

Net repulsive interactions, because of electrostatic effects, depend largely on the pH and the ionic strength of the background electrolyte concentration. Electrostatic repulsive interactions are commonly found in mixtures of proteins and anionic polysaccharides under conditions where both biopolymers carry the same net charge, e.g. pH above the isoelectric point (p*I*) of the protein.

Attractive interactions

Attractive interactions between proteins and polysaccharides may be weak or strong and either specific or non-specific. Non-specific attractive interactions arise as a result of a multitude of weak interactions between groups on the biopolymers, such as electrostatic, van der Waals', hydrogen bonding and hydrophobic interactions. Hydrogen bonding and hydrophobic interactions are actually collective interactions (e.g. electrostatic, van der Waals' and steric overlap) including some entropy effects (McClements, 2005).

Electrostatic interactions are the most important forces involved in complex formation between proteins and ionic polysaccharides. These interactions between charged biopolymers lead to a decrease in the electrostatic free energy of the system. Moreover, the enthalpy contribution, because of interactions of oppositely charged biopolymers and liberation of counter ions along with water molecules, often compensates for the loss of configurational entropy of mixing rigid biopolymers (Piculell and Lindman, 1992; Tolstoguzov, 1997a).

Strong electrostatic attractive interactions between positively charged proteins ($pH < pI$) and anionic polysaccharides occur, especially at low ionic strength. Generally, two types of complexes are formed by electrostatic interactions (Tolstoguzov, 1997b; Schmitt *et al.*, 1998; Tolstoguzov, 2002, 2003). Soluble complexes are obtained when the opposite charges carried by the two biopolymers are not equal in number, whereas insoluble complexes result when the net charge on the complex is close to zero.

van der Waals' forces are extremely weak electrical attractions that arise because of temporary dipole interactions (Sherony and Kintner, 1971; Stainsby, 1980; Dickinson, 1998). Basically, every atom has an electron cloud that can yield a temporary electric dipole. The dipole in one atom can induce a corresponding dipole in another atom. This is possible only if the atoms are close. However, if they are too close, repulsive forces between the adjacent negatively charged electron clouds may not allow these van der Waals' attractions. Although these transient electrical attractive forces are very weak, they influence macromolecular interactions together with other non-covalent forces described above (Damodaran, 1997).

Hydrophobic bonding is an entropy-driven long-range interaction between non-polar groups and is promoted by conformational and structural modifications of biopolymers, mostly by unfolding of polymeric chains exposing hydrophobic groups. These kinds of interactions are promoted by an increase in temperature (Stainsby, 1980; Piculell and Lindman, 1992; Samant *et al.*, 1993; Antonov *et al.*, 1996; Tolstoguzov, 1997a).

Hydrogen bonding is a moderately strong bond —O—$H^{\delta+}$...$^{\delta-}$O═, which becomes relatively insignificant at high temperature. These bonds are ionic in nature and refer to the interaction between hydrogen atoms attached to an electronegative atom (oxygen, sulfur) and another electronegative atom (e.g. the sulfur of a sulfate group). A classical example of hydrogen bonding has been shown in the complex coacervation of gelatin and pectin (Braudo and Antonov, 1993), which is obtained over a wide range of pH including the isoelectric pH (4.8) of gelatin. Protein–polysaccharide hydrogen bonding in gelatin–pectin, gelatin–alginate and chitosan–collagen over a wide range of pH has been well established by various studies (Taravel and Domard, 1995; Antonov *et al.*, 1996).

Covalent bonds

Covalent bonds are very strong, specific, non-electrostatic and permanent linkages. There are two principal methods that can be used to generate a covalent linkage between proteins and polysaccharides. The most commonly used method utilizes the chemical reaction between amino groups of proteins and carboxylic groups of polysaccharides (the Maillard reaction) to give an amide covalent bond (Stainsby, 1980). Covalent bonds can also be generated enzymatically using the oxidoreductase family of enzymes (E.C. 1.XXX), which catalyze the oxidation of the phenolic group of tyrosine residues with carbohydrate groups containing phenolic residues, such as cereal arabinoxylans (Boeriu *et al.*, 2004). Tyrosine-containing peptides have also been conjugated with ferulic acid (Oudgenoeg *et al.*, 2001) and with whey proteins through the use of three different oxidoreductases (Faergemand *et al.*, 1998).

Recently, cross-linking of proteins and polysaccharides using transglutaminase (E.C. 2.3.2.13) has been suggested (Flanagan and Singh, 2006). Many polysaccharides contain residual protein; for example, gum arabic, guar gum and locust bean gum all contain low levels of protein. Gum arabic (approximately 2% protein, depending on the source) consists of, among other sub-units, a glycoprotein and an arabinogalactan protein. Provided the residual protein in these polysaccharides contains lysine and/or glutamine residues, theoretically, the treatment of protein and polysaccharide mixtures with transglutaminase could lead to the formation of heteropolymers (i.e. protein–polysaccharide conjugates) in addition to homopolymers (cross-linked protein or cross-linked polysaccharide). Flanagan and Singh (2006) demonstrated that sodium caseinate–gum arabic conjugates catalyzed by transglutaminase can be produced.

These kinds of interactions are generally very stable to pH and ionic strength. Because of their stability properties, this type of bonding has been used intentionally to produce conjugated emulsifiers (Shepherd *et al.*, 2000; Song *et al.*, 2002; Akhtar and Dickinson, 2003, 2007; Neirynck *et al.*, 2004; Dunlap and Côté, 2005; Benichou *et al.*, 2007). In most of these studies, the covalent conjugation between protein and polysaccharides has been studied through use of the Maillard reaction.

Apart from these major interactions, ion bridging involving binding of cations such as Ca^{2+} may also contribute to some extent in protein–polysaccharide interactions although they do not have the predominant influence (Stainsby, 1980; Antonov *et al.*, 1996; Dickinson, 1998). For example, firm sodium caseinate gels ($G' > 100\,Pa$) were formed using pectin concentrations $\geq 0.6\%$ at one particular degree of methylation ($\approx 31\%$) and amidation ($\approx 17\%$) in the presence of Ca^{2+} ions (1.8 mM) at pH $\approx$ 3.6 (Matia-Merino *et al.*, 2004).

Milk protein–polysaccharide interactions

Milk proteins together with polysaccharides dissolved in an aqueous phase form a pseudo-ternary system of milk protein–polysaccharide–water. Various interactions in these systems could lead to complex formation or bulk phase separation. Extensive studies on protein–polysaccharide interactions, particularly using well-studied milk proteins and commercially available polysaccharides, have been carried out (Dickinson, 1998). Table 12.1 and Table 12.2 show a compilation (non-exhaustive) of various milk protein (casein and/or whey protein) and polysaccharide mixtures in aqueous systems and the conditions under which different kinds of interactions occur.

Rheological properties and microstructures of protein–polysaccharide systems

The rheological properties of a solution containing only protein are expected to be different from those of a pure polysaccharide solution. Polysaccharide molecules generally have a greater effect than proteins in causing a significant increase in solution viscosity, because polysaccharide molecules are usually much larger and more extended ($\approx 5.0 \times 10^5$ to 2.0×10^6 Da) than globular proteins ($\approx 1.0 \times 10^4$ to 1.0×10^5 Da). Hence, polysaccharide molecules generally occupy larger hydrodynamic volumes, which give rise to higher solution viscosity. The above assumes that intermolecular interactions are absent or negligible (e.g. in dilute solution). When intermolecular interactions among neighboring polymer molecules (i.e. polysaccharide–polysaccharide or protein–protein interactions) are present, the rheological properties of the systems can be expected to change significantly.

Changes in rheological properties may arise as a result of an increase in the size of the particles (e.g. protein–polysaccharide complexes), or when depletion interactions occur in the mixed system or if one or more polymer species form continuous network structures. The overall effect results in the formation of different microstructures. Schematic illustrations of some possible microstructures formed from mixtures of proteins and polysaccharides under some specific conditions (e.g. pH, ionic strength, heat treatment, etc.) are shown in Figures 12.3a and 12.3b.

Various rheological techniques have been employed to characterize the physicochemical properties of protein–polysaccharide systems. Generally, if the mixtures are liquid like, viscosity measurements using rotational viscometers are commonly

Table 12.1 Casein–polysaccharide interactions in aqueous systems

No.	Casein–polysaccharide aqueous systems	Conditions	Interactions	References
	Milk proteins (Casein micelles + Whey proteins) + **Pectin** (High methoxyl - 62.7% methylated)			
1.	**Milk proteins** (Casein micelles + Whey proteins) + **Gum arabic**	20°C, pH 6.0–10.5, 0–0.5 M NaCl	Thermodynamic incompatibility	(Antonov *et al.*, 1982)
	Milk proteins (Casein micelles + Whey proteins) + **Arabinogalactan**			
2.	**Casein micelles** + **Alginate**	pH 7.2, 25°C	Thermodynamic incompatibility	(Suchkov *et al.*, 1981, 1988)
3.	**Casein micelles** (2.5%) + **Pectin** (Low methoxyl - 35%, High methoxyl - 73%, Low methoxyl amidated - 35% methylated and 20% amidated) (0.1–0.2%)	pH 6.7/5.3, 60°C	pH 6.7: Depletion interaction. Methoxylation affects interaction	(Maroziene and de Kruif, 2000)
4.	**Casein micelles** (0.8–4%) + **Galactomannans** (Guar gum, Locust bean gum) (0.09–0.3%)	5/20°C, pH 6.8/7.0, 0.08/0.25 M NaCl, sucrose (10–40 w/w%)	Depletion interaction. Sucrose affects interaction	(Bourriot *et al.*, 1999a; Schorsch *et al.*, 1999)
5.	**Casein micelles** (1.0%) + **Carrageenan** (ι-, κ-, λ-forms) (0.12%)	pH 6.7/7, 60/50/20°C, 0.25 M NaCl/0.05 M NaCl-0.01 M KCl	Depletion interaction	(Dalgleish and Morris, 1988; Langendorff *et al.*, 1997, 1999, 2000; Bourriot *et al.*, 1999b)
6.	**Sodium caseinate** (0.1–0.5%) + **Gum arabic** (0.01–5%)	0.5 M NaCl, pH 2.0–7.0, slow acidification with glucono-δ-lactone	Soluble electrostatic complexation	(Ye *et al.*, 2006)
7.	**Casein micelles** (0.1%) + **Exopolysaccharide** (5.0%) (*Lactococcus lactis* subsp. *cremoris* B40)	pH 6.6, 25°C	Depletion interaction	(Tuinier & De Kruif, 1999; Tuinier *et al.*, 1999)
8.	**Sodium caseinate** + **Maltodextrin** (2:1, 1:1 and 1:4)	60°C, 2–4 days	Covalent conjugate via Maillard reaction. No phase separation	(Shepherd *et al.*, 2000; Morris *et al.*, 2004)
9.	**Casein** (β-Casein, α_s-Casein) + **Polysaccharide** (Dextran, Galactomannan) (1:1)	60°C, 24 h	Covalent conjugate via Maillard reaction. No phase separation	(Dickinson and Semenova, 1992; Kato *et al.*, 1992)
10.	**Sodium caseinate** (6.0%) + **Sodium alginate** (1%)	pH 7.0, 23°C	Thermodynamic incompatibility	(Guido *et al.*, 2002; Simeone *et al.*, 2002)

Table 12.2 Whey protein–polysaccharide interactions in aqueous systems

No.	Whey protein–polysaccharide aqueous systems	Conditions	Interactions	References
1.	**β-Lactoglobulin (β-Lg)** (0.5%) + **Chitosan** (Degree of deacetylation: 85%) (0–0.1%)	pH 3.0–7.0, 5 mM phosphate buffer	pH-dependent β-Lg–chitosan Soluble/insoluble complex coacervation	(Guzey and McClements, 2006)
2.	**Heat-denatured whey protein isolate (HD-WPI)** (8.0%) + **Pectin** (28, 35, 40, 47 and 65% methylation) (0.1–1.5%)	pH 6.0/7.0, 80°C/85°C, 5.0/10.0 mM $CaCl_2$	Thermodynamic incompatibility	(Beaulieu *et al.*, 2001; Kim *et al.*, 2006)
3.	**β-Lg** (12.0%) + **Alginate** (0.1–1.0%)	pH 7.0/(3.0–7.0), 87°C/30°C, high pressure	pH-dependent β-Lg–chitosan Soluble/insoluble complex coacervation	(Dumay *et al.*, 1999; Harnsilawat *et al.*, 2006)
4.	**β-Lg** (0.05%) + **Pectin** (Low methoxyl - 28.3/42.6%, High methoxyl - 71.3/73.4%) (0.0125%)	4–40°C/25°C, pH 4.0–7.5/6.5, 0.11/0.1–1.0 M NaCl/ 87°C/ high pressure	pH, ionic strength and temp.: Complex coacervation. Precipitation for modified pectin. Methylation affects complexation	(Dumay *et al.*, 1999; Wang and Qvist, 2000; Girard *et al.*, 2002, 2003a, 2003b, 2004; Kazmiersi *et al.*, 2003)
5.	**Whey protein isolate (WPI)** (5.0%) + **Galactomannans** (Locust bean gum) (0–0.4%)	pH 5–7	pH and concentration: Biphasic gel	(Tavares and Lopes da Silva, 2003)
6.	**HD-WPI** (8.5%) + **Xanthan gum** (0–0.2%)	25°C–90°C/75–80°C, pH 7.0/5.4, high-pressure treatment, 0.2 M NaCl	Native WPI: Co-solubility. HD-WPI: Thermodynamic incompatibility	(Bryant and McClements, 2000b; Li *et al.*, 2006)
7.	**WPI** (4–12.5%) + **Xanthan gum** (0.01–1.0%)	pH 5.5/6.0/6.5/7.0, 0.1/0.5 M NaCl, high-pressure treatment	Depletion interaction, pH-dependent electrostatic complexation	(Zasypkin *et al.*, 1996; Laneuville *et al.*, 2000; Hemar *et al.*, 2001b; Benichou *et al.*, 2007; Bertrand and Turgeon, 2007)
8.	**Bovine serum albumin (BSA)** + **Sulfated polysaccharides** (ι-, κ-carrageenan, Dextran sulfate) (2.5:1 and 5:1)	pH 6.5–8, high-pressure treatment	Complex coacervation	(Galazka *et al.*, 1996, 1997, 1999)
9.	**HD-WPI** (10.0%) + **κ-Carrageenan** (0.5%)	80°C, pH 1–12	Complex coacervation	(Mleko *et al.*, 1997)
10.	**β-Lg** (0.5–10.0%) + **κ-Carrageenan** (1.0%) (1:2, 5:1 and 10:1)	pH 7, 45–80°C, 0.1 M NaCl/ 0.01 M $CaCl_2$	Temp., pH and concentration dependent. Phase-separated bicontinuous gel formation	(Capron *et al.*, 1999; Ould Eleya and Turgeon, 2000)

(*Continued*)

Table 12.2 (*Continued*)

No.	Whey protein–polysaccharide aqueous systems	Conditions	Interactions	References
11.	β-Lg + **Gum arabic** (2:1)	pH 3.6–5.0, 0.005–10.7 mM NaCl	Complex coacervation	(Schmitt *et al.*, 1998, 1999, 2000, 2001; Sanchez & Renard, 2002; Sanchez *et al.*, 2002, 2006)
12.	**WPI** + **λ-Carrageenan** (1:1 to 150:1)	pH: Wide range, 0–0.1 M ($NaCl/CaCl_2$)	Electrostatic complexation. Precipitation	(Weinbreck *et al.*, 2004a)
13.	**WPI** + **Gum arabic** (2:1)	pH 4.0–7.0, 0–0.1 M NaCl	Complex coacervation. Glassy state	(Weinbreck *et al.*, 2003a, 2004b, 2004c)
14.	**β-Lg** + **Carboxymethyl dextran** (1:1 and 7:2)	pH 5.5/4.75, 4°C/25°C	β-Lg–carboxymethyl dextran covalent conjugate. No phase separation	(Hattori *et al.*, 1994)
15.	**WPI** + **Carboxymethyl potato starch** (2:1)	pH 7.0, 24°C	WPI–carboxymethyl starch covalent conjugate	(Hattori *et al.*, 1995)
16.	**WPI** + **Exopolysaccharide** (*Lactococcus lactis* subsp. *cremoris* B40) (2:1)	pH: Wide range, 25°C, 0–0.1 M ($NaCl/CaCl_2$), heat treatment of WPI	Electrostatic complexation. Precipitation HD-WPI: Depletion interaction	(de Kruif and Tuinier, 1999; Tuinier & De Kruif, 1999; Weinbreck *et al.*, 2003b)
17.	**β-Lg** + **Pullulan**	0.01 M NaCl, 4°C	Depletion interaction	(Wang et al., 2001)
18.	**β-Lg** + **Carboxymethyl cellulose**	60°C, 0.05–0.2 M, pH 2.5–7.0	Insoluble electrostatic complex, sedimentation	(Hidalgo and Hansen, 1969; Hansen *et al.*, 1974)
19.	**WPI** + **Maltodextrin** (1:2 and 1:3)	80°C, 2 h, 79% RH	Covalent conjugation. No phase separation	(Akhtar and Dickinson, 2007)
20.	**WPI/Whey Protein Concentrate (WPC)** + **Pectin** (4:1, 2:1, 1:1 and 1:2)	60°C, 14 days, pH 7.0	Covalent conjugation. No phase separation	(Mishra *et al.*, 2001; Neirynck *et al.*, 2004)

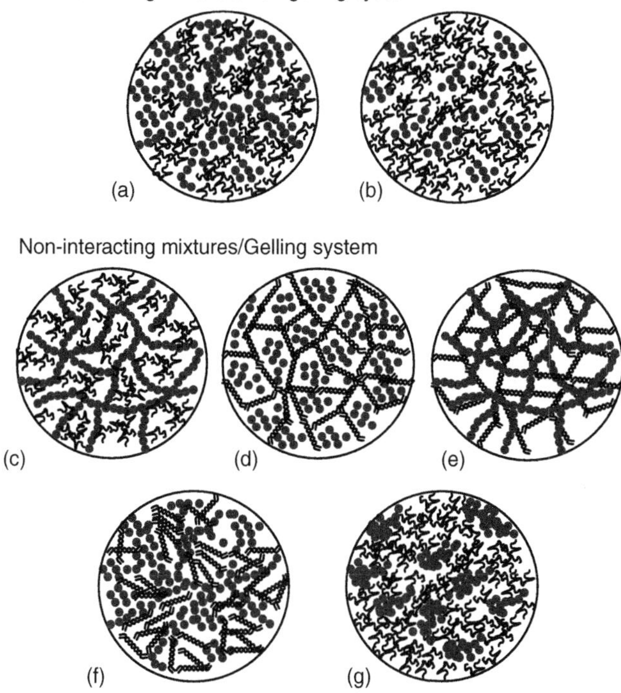

Figure 12.3a Schematic diagrams of some possible microstructures formed between non-interacting protein–polysaccharide mixtures. Circle (•) represents protein; coil structure represents polysaccharide molecules. (a) Flocculated protein network formed with polysaccharide filling the space in the network; (b) polysaccharide molecules overlap and form continuous "network" with protein filling the space; (c) particulate protein gel network formed with polysaccharide filling the space; (d) polysaccharide gel network formed with protein filling the space; (e) bicontinuous network formed from protein and polysaccharide); (f) polysaccharide gels dispersed among a weakly flocculated protein network; (g) protein gels dispersed among entangled polysaccharide molecules.

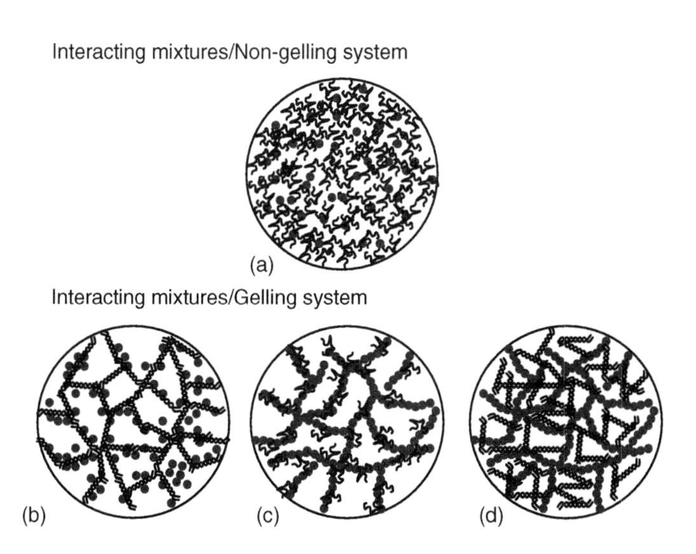

Figure 12.3b Schematic diagrams of some possible microstructures formed between interacting protein-polysaccharide mixtures. Circle (•) represents protein; coil structure represents polysaccharide molecules. (a) Protein–polysaccharide complexes formed; (b) protein interacting with gelling polysaccharide helices; (c) polysaccharide interacting with protein particulate gel network; (d) polysaccharide gel helices interacting with protein particulate gel network.

used to obtain steady-state viscosity curves, yield stress, etc. Other simpler methods include the use of a kinematic viscometer (e.g. an Ubbelodhe capillary viscometer) to obtain a single point relative viscosity measurement. If the samples are viscoelastic (e.g. gels), rheometers are widely used to obtain rheological data (e.g. loss and storage moduli obtained within the linear viscoelastic region), by performing small deformation oscillatory measurements.

The rheological data yield information on the viscosity and viscoelastic properties of the mixed systems. Knowledge of the rheological properties of mixed protein–polysaccharide systems is essential to gain insights into the nature of the interactions and the resulting microstructure of the system. Fundamental understanding of the interactions at the molecular and colloidal levels will provide a strong foundation to exploit the physical functionality of such complex systems in different applications (e.g. microencapsulation technology, imparting specific sensory characteristics, time/temperature/pH/ionic control-release, emulsion stability, etc.).

In the following sections, various examples of mixed systems involving different milk proteins and polysaccharides are provided. An attempt is made to classify these mixed systems into two broad categories (i.e. interacting and non-interacting). Under each of these headings, the systems are further grouped according to whether they form or do not form gels (i.e. gelling or non-gelling). The discussion focuses mainly on the techniques used and the rheological properties of the systems.

Non-interacting protein–polysaccharide mixtures

Non-interacting protein–polysaccharide mixtures existing as one phase are rare but may occur when the two different molecular species have good chemical resemblance of their hydrophilic surfaces (Tolstoguzov, 1991, 2006). However, many polymer mixtures are thermodynamically incompatible and segregative interactions often occur in the absence of electrostatic interaction or in the presence of electrostatic repulsion (Neiser *et al.*, 1998). Protein–polysaccharide mixtures that commonly exist as two separate phases are the result of either thermodynamic incompatibility or depletion flocculation (Doublier *et al.*, 2000).

Non-gelling phase-separated systems

The following are examples of non-interacting, non-gelling protein and polysaccharide mixtures. The proteins and polysaccharides were mixed under conditions where the mixtures did not form gels. The rheological properties of these systems are discussed in relation to their interactions and the microstructures formed.

Casein micelles and galactomannans

A non-interacting protein–polysaccharide mixture where phase separation occurred was reported for a mixed system consisting of micellar casein (3%) and guar gum (0.2%) at pH 7 (Bourriot *et al.*, 1999c). There was a significant change in the flow and viscoelastic properties compared with the individual biopolymer systems. There was an increase in the apparent viscosity in the mixed system. Furthermore, the mechanical spectra (elastic modulus G', viscous modulus G'') of the frequency sweeps showed slightly higher values of the moduli, which were less frequency dependent. The results suggested the formation of a weak network structure within the system because of flocculation of the casein micelles as the polysaccharide molecules were excluded from the protein phase.

The appearance of a slightly thixotropic behavior indicated that the network could be easily broken under shear because the network formed by the micellar casein was weakly flocculated and reversible, presumably attributable to a depletion flocculation mechanism. Also the lower the intrinsic viscosity of the polysaccharide, the higher was the concentration of polysaccharide required before phase separation occurred (Bourriot *et al.*, 1999c). An increase in the concentration of the polysaccharide resulted in stronger flocculation of the casein micelles because the volume occupied by and the osmotic pressure from the surrounding polysaccharides increased.

Similar thixotropic behavior was reported for a ternary solution consisting of micellar casein, locust bean gum (LBG) and sucrose (Schorsch *et al.*, 1999). At pH $\approx$ 6.8, the casein micelles and the LBG were thermodynamically incompatible, behaving as a water-in-water emulsion. The presence of sucrose, even at high concentration (40%), did not significantly improve the compatibility of the biopolymers (Schorsch *et al.*, 1999).

Milk proteins and xanthan gum

Another study investigated the interaction between xanthan gum (0–1% w/w, a polysaccharide with known "weak gel" properties) and different types of milk proteins (5% w/w, sodium caseinate [Na-CN], skim milk powder [SMP], whey protein isolate [WPI] and milk protein concentrate [MPC]) in aqueous solution at neutral pH (Hemar *et al.*, 2001b). The microstructures of the mixtures were different depending on the xanthan gum concentration and the protein type. In the case of xanthan gum mixtures with either MPC or SMP, depletion flocculation of the casein micelles took place. The size of the depleted protein aggregates decreased with increasing xanthan gum concentration (the microstructure resembled a particulate network).

In the case of xanthan gum mixtures with either Na-CN or ultracentrifuged WPI, no phase separation occurred within the timescale of the experiment. This was attributed to the larger size of the casein micelles (average diameter $\approx$ 0.2 μm) compared with the nanometre size scale of WPI and Na-CN (0.05 μm) (Lucey *et al.*, 2000). However, the rheological behavior of the mixtures was very similar to the rheological behavior of xanthan gum. The differences in the microstructures of the mixtures that were observed by confocal laser scanning microscope (CLSM) were not detected by viscosity measurements, probably because the weakly flocculated proteins were easily redispersed by the shearing action of the viscometer during measurement.

Gelling phase-separated systems

In a system where two biopolymer species (e.g. proteins and polysaccharides) do not interact, gelation of one or more of the components in a thermodynamically incompatible system will cause competition between phase separation and gelation (Neiser *et al.*, 1998). Gelation basically means the formation of a three-dimensional aggregated network structure, which is generally induced by heating, cooling, acidification, enzymatic treatments, high-pressure processing, etc.

Generally, heating enhances hydrophobic and covalent interactions. Unfolded proteins interact to give rise to aggregates in the case of whey protein (Kinsella, 1984; Boye *et al.*, 1997). In mixed systems, the microstructure will depend on the rates of phase separation and gel formation (Tavares *et al.*, 2005). The gel may appear to be homogeneous at a macroscopic level, but heterogeneous at a microscopic level. However, the rheological properties of such gels depend on the concentration and arrangement of each species in the different phases. If the gelling species is in the continuous phase, the gel strength is higher than if the gelling species is in the dispersed phase where the network is disrupted (Neiser *et al.*, 1998).

Whey protein and galactomannans

One study was based on a mixture of LBG (a non-gelling neutral polysaccharide) and whey protein at neutral pH and pH 5 (close to the p*I* of whey proteins) (Tavares and Lopes da Silva, 2003). At neutral pH, it is known that whey protein forms clear fine-stranded gels (protein aggregation is hindered by electrostatic repulsion), whereas, at lower pH (e.g. pH 5), an opaque coarse particulate gel is formed (Langton and Hermansson, 1992; Aguilera, 1995). Rheological measurements showed that a WPI gel (13% w/w) had a stronger and more elastic character at pH 5 than at pH 7 because of the thick particulate network formed (Stading *et al.*, 1993; Bertrand and Turgeon, 2007). For the protein gels at pH 7, increasing LBG concentration (>0.25%) decreased the onset temperature for gelation and decreased the gelation time. The presence of LBG was also found to increase the gel rigidity.

The authors attributed this to a decrease in macromolecular mobility within the network in the presence of LBG because of segregative interactions and the "local" concentration of each polymer species. The LBG molecules acted as fillers in the continuous protein network. At pH 5, the elastic character of the particulate gel network was shown to decrease in the presence of LBG, especially at low protein concentration (5%). It was suggested that LBG chains hampered protein–protein interactions and were detrimental to the development of a protein gel. However, at a higher protein concentration (13%), where sufficient particulate gel network was formed, LBG acted as fillers within the network, hence improving the gel strength.

In a subsequent study carried out using WPI and guar gum at pH 7, an increase in protein gel strength with decreasing degree of branching of the galactomannans was found (Tavares *et al.*, 2005). Like LBG, guar gum was dispersed as droplets among the whey protein network at low concentration (0.2%). However, at higher gum concentration (0.6%), the dispersed droplets joined to form a continuous polysaccharide-rich phase. Despite the different microstructures observed, the linear viscoelastic profiles were rather similar, indicating that viscoelasticity was fairly insensitive to microstructural changes of this nature.

WPI and xanthan gum

A very similar trend was observed for whey protein and xanthan gum mixtures after heat treatment (Bertrand and Turgeon, 2007). The microstructures and rheological properties of the gels were highly dependent on pH and salt. At pH 6.5, the presence of xanthan gum improved the elastic modulus of the WPI gel. This was attributed to segregative phase separation, where xanthan gum was dispersed among the protein gel network. However, lowering the pH decreased the elastic character of the gel. At pH 5.5 (close to the p*I* of WPI), the addition of xanthan gum decreased the elastic modulus of the gel. It was suggested that possible WPI–xanthan gum complexes formed decreased protein–protein interactions, producing a weaker gel network.

β-Lactoglobulin and pectin

A different type of network was formed in mixtures of β-lactoglobulin (8% w/w) and low methoxyl pectin (0.85% w/w) after thermal treatment at pH 6.8. The storage modulus of the mixed gel system was significantly lower than that of the protein gel alone. The microstructure observed by CLSM revealed phase separation, with β-lactoglobulin appearing as spherical colloidal particles distributed in a continuous pectin network (Donato *et al.*, 2005). A similar type of protein-depletion-induced phase separation was reported for a mixed system containing aggregated whey protein and an exopolysaccharide (EPS) from lactic acid bacteria (Tuinier *et al.*, 2000).

β-Lactoglobulin and κ-carrageenan

If two gelling species are present in a binary system, the mixed gels may form interpenetrating, coupled or phase-separated networks (Morris, 1986). Interpenetrating networks are the result of two independent continuous networks formed throughout the gel and only topological interactions exist between the networks. Coupled networks (ordered into junction zones, like those of a polysaccharide gel) are formed when favorable interactions between the two molecular species exist. However, such systems involving protein–polysaccharide interactions are uncommon (Rao, 1999).

Phase-separated networks are formed when one polymer species is incompatible with the other polymer species, forming phase-separated regions within the gel network (Piculell and Lindman, 1992; Turgeon and Beaulieu, 2001). An example of a phase-separated gel is that for κ-carrageenan and β-lactoglobulin (Capron *et al.*, 1999). The mixed polymer formed a gel that was weaker than the carrageenan gel alone when the protein was in its native state. On heating the mixture to 90°C, holding for 30 min and then cooling to 20°C, the gel rheology indicated the melting of κ-carrageenan and the gelation of β-lactoglobulin above 65°C. There was no aggregation of κ-carrageenan with β-lactoglobulin on heating. The gelation time of β-lactoglobulin was reduced in the presence of κ-carrageenan, which was attributed to microphase separation, which caused an increase in the local concentration of β-lactoglobulin (Capron *et al.*, 1999). On cooling, the mixed gel system formed a phase-separated bicontinuous network (Ould Eleya and Turgeon, 2000).

Interacting protein–polysaccharide mixtures

Another phase separation phenomenon is associative phase separation, where associative interactions are present. Associative interactions between protein and polysaccharide can occur as a result of electrostatic interactions, hydrogen bonding, hydrophobic interactions or poor solvent conditions (Antonov *et al.*, 1996; Gao and Dubin, 1999; Doublier *et al.*, 2000; de Kruif *et al.*, 2004). In some cases, complexes are formed via electrostatic interactions (known as coacervates). Coacervates of protein–polysaccharide can occur when the pH of the mixture is lower than the isoelectric point of the protein. At this pH, the protein possesses a net positive charge whereas the polysaccharide still possesses a negative charge. The result of the complexation is the formation of a solvent-rich phase and a coacervate-rich phase (Doublier *et al.*, 2000; Ould Eleya and Turgeon, 2000).

The rheological properties of milk protein–polysaccharide complexes are related to the interaction between the complexes and the water molecules, which forms soluble (or liquid coacervate phase) or insoluble (or precipitate) complexes. The solubility of the complexes is based on the energetic difference between biopolymer–biopolymer and biopolymer–solvent interactions (Damodaran, 1997). The main parameters affecting the solubility of biopolymer complexes are charge density, pH, ionic strength and protein:polysaccharide (PP:PS) ratio (Schmitt *et al.*, 1998). It has been suggested that a complex involving a strong polyelectrolyte will form a precipitate rather than a liquid coacervate. A number of examples of protein–polysaccharide systems with complex coacervations have been reviewed (Schmitt *et al.*, 1998; Turgeon *et al.*, 2003; de Kruif *et al.*, 2004). Some examples of interacting polymers in mixed systems and the effect on their rheological properties are given below.

Non-gelling phase-separated systems

β-Lactoglobulin and chitosan

It has been reported that the solubility of a protein increases below its isoelectric pH when it complexes with an anionic polysaccharide (Tolstoguzov *et al.*, 1985; Tolstogusov, 1986). A recent study of a β-lactoglobulin–chitosan complex showed that the complex was either soluble or insoluble, depending on the pH (Guzey and McClements, 2006). The interaction of soluble chitosan (M_W = 15 000 Da, degree of deacetylation = 85%, 0–0.1 wt%, 5 mM phosphate buffer) with β-lactoglobulin (0.5 wt% β-lactoglobulin, 5 mM phosphate buffer) in aqueous solutions studied at pH 3–7 showed that, at pH 3, 4 and 5, the majority of the β-lactoglobulin–chitosan complex in the solutions was soluble, but that at pH 6 and 7, a significant fraction of the two biopolymers was insoluble.

Whey proteins and EPS

"Soluble complexes" formed via electrostatic interactions were reported for EPS B40 (an EPS from *Lactococcus lactis* subsp. *cremoris* NIZO B40) and whey protein (PP:PS = 2:1) under specific pH and ionic conditions (with no macroscopic phase separation) (Weinbreck *et al.*, 2003b). Decreasing the pH of the mixtures increased further aggregation of the complexes, which led to phase separation. In addition, increasing the ionic strength of the solution caused a shift to a lower pH value for the onset of complexation. In this study, complexation in this system led to a decrease in solution viscosity, as intramolecular repulsion of the EPS was reduced in the presence of whey proteins. The decrease in viscosity was attributed to a reduction in the quantity of dispersed phase, i.e. water present within the complexes. Consequently, it was suggested that dilute solution viscosity measurement (which is related to the size of complexes) could be used to determine the optimum conditions for complexation (Weinbreck *et al.*, 2003b). A potential benefit of this complexation is that it protects the protein from loss of solubility as a result of aggregation during thermal or high-pressure treatments (Imeson, 1977; Galazka *et al.*, 1997).

Whey proteins and gum arabic

Viscosity curves were obtained to evaluate the "strength" of electrostatic interactions of whey protein–gum arabic coacervates (Weinbreck and Wientjes, 2004). This study showed that the stronger the interaction, the greater was the shear-thinning behavior and the slower was the reformation of the complexes after shearing. The highly viscous coacervate (at pH 4) was attributed to electrostatic interactions. At pH above the isoelectric point (without electrostatic interactions), the mixtures appeared to be more elastic than viscous.

Sodium caseinate and gum arabic

In contrast to whey proteins, sodium caseinate and gum arabic mixtures showed some peculiar behavior (Ye *et al.*, 2006) as no coacervation was observed in these systems. Below a certain pH (pH 5.4), electrostatic interactions between sodium caseinate and gum arabic led to the formation of stable composite nanoparticles in the size range 100–200 nm.These complexes remained constant in particle size and were stable and soluble over a defined pH range (pH 3.2–5.4). This pH range was dependent on the ratio of sodium caseinate to gum arabic in the mixtures and also on the ionic strength.

The sodium caseinate–gum arabic particles associated to form large particles, which resulted in phase separation when the pH was lower than 3.0. A mechanism for the formation of these nanoparticles, based around self-aggregation of the casein and electrostatic interaction between the aggregated particles of casein and gum arabic, was proposed. As the pH of the mixture decreased below pH 5.4, the caseinate molecules tended towards small-scale aggregation prior to large-scale aggregation and precipitation at pH values closer to their p*I* (pH 4.6).

In this case, the gum arabic molecules may have attached to the outside of these small-scale aggregates in the early stages of aggregation through electrostatic interactions between negatively charged gum arabic and exposed positive patches on the surface of the caseinate aggregates. The presence of hydrophilic gum arabic molecules on the outside of the caseinate aggregates may have been enough to sterically stabilize these nanoparticles and consequently prevent self-aggregation. As the charge on the nanoparticles was quite low, e.g. $\approx$15 mV at pH 4.0, steric stabilization was probably important.

Casein micelles and pectin

Protein–polysaccharide interactions were shown to be pH dependent in the case of pectin and casein micelles (Ambjerg and Jørgensen, 1991; Maroziene and de Kruif, 2000).

At pH 6.7, pectin did not adsorb on to the casein micelles. With sufficient pectin present (0.1–0.2%), phase separation occurred because of depletion interactions of the casein micelles (≈0.1%). However, adsorption of pectin on to the casein micelles occurred at pH 5.3. Viscosity measurements were employed to study the changes that occurred at different polymer concentrations. At low pectin concentrations (≈0.1%) and at pH 5.3, bridging flocculation occurred. A maximum viscosity at this pectin concentration was attributed to bridging flocculation, as bridging among the casein particles was interpreted as having a larger effective volume. As the pectin concentration increased (>0.1%), the casein micelles became fully covered and interactions between the casein particles were reduced, as, typically, in acidified milk.

When the protein was fully covered by the pectin, the viscosity decreased to a certain extent but remained higher than for the pure milk samples (without pectin). The amount of pectin required for full coverage of the casein micelles differed depending on the type of pectin: high methoxyl (HM) < low methoxyl amidated (LMA) < low methoxyl (LM) pectin. Adding more pectin beyond the concentration for full coverage led to phase separation because of depletion interactions. A further increase in pectin reduced the thickness of the casein-depleted layer as the viscosity of the continuous phase became very high and formed gelled polymer networks (Maroziene and de Kruif, 2000). When the pH of the mixture was increased from 5.3 back to 6.7, desorption of pectin from the casein occurred, but over a much longer time scale (≈10–15 min) than for the adsorption process (Maroziene and de Kruif, 2000).

Gelling phase-separated systems

Sodium caseinate and pectin

The dynamic rheological properties of glucono-δ-lactone (GDL)-acidified protein gels (2% w/v Na-CN) were studied in the presence of LMA pectin (0.01–1% w/v) at pH ≈ 4 (Matia-Merino *et al.*, 2004). The presence of pectin (0.01–0.05% w/v) was found to decrease the storage modulus and to increase the gelation time, because pectin adsorbed on to the casein particles. At pectin concentration >0.08% w/v, acid-induced gelation appeared to be completely inhibited over a time period of ≈9 h at 25°C.

Casein micelles and ι-carrageenan

For casein–carrageenan mixed systems, the attractive interactions involved the negatively charged sulfated groups of the polysaccharides and the positive "patches" between residues 97 and 112 of κ-caseins (Snoeren, 1975), despite a pH above the isoelectric point and an overall net negative charge of the casein micelles. The interaction between ι-carrageenan (0.5%) and skim milk (based on 3.3% protein) mixtures was studied above and below the coil–helix transition temperature of carrageenan (Langendorff *et al.*, 1999). At temperatures above the coil–helix transition temperature, carrageenan did not adsorb to the casein micelles, resulting in depletion flocculation.

In contrast, at temperatures below the coil–helix transition temperature, attractive interactions between carrageenan and casein micelles occurred. The higher charge density of the double-helix form, as compared with the coil conformation of carrageenan, probably explained the stronger attractive interaction between casein micelles and carrageenan. The presence of casein micelles increased the gel strength (indicated by higher G' and G'') and the gelation temperature (from 39–47°C) when the mixtures were heated to 65°C and cooled to 25°C.

Depending on the concentration of carrageenan, different types of gel network were deduced from the frequency sweep. At low carrageenan concentrations (<0.2%), one type of network was formed on cooling. This was probably due to the bridging of casein micelles by the adsorbed carrageenan helical chains. The network was much more thermally stable than the pure carrageenan gels. At above 0.2% concentration, as well as the formation of a network as described above, a second network was formed, similar to that of a carrageenan gel in the absence of proteins. This was attributed to interactions between carrageenan chains (Langendorff *et al.*, 1999). Among the different types of carrageenans, the amount required for full coverage increased from $\kappa < \iota < \lambda$ (Langendorff *et al.*, 1997) because the charge density of the polymer determined the strength of adsorption (Pereyra *et al.*, 1997; Maroziene and de Kruif, 2000).

It is clear from the above examples that mixed protein and polysaccharide systems can produce very different rheological properties. The rheological properties of these systems are the results of cumulative effects from the molecular parameters (e.g. size, conformation, charge density, concentration, PP:PS ratio) of the macromolecules, the conditions (e.g. pH, temperature, ionic strength) to which the mixed systems are subjected and the resulting interactions (e.g. type of interaction, strength of interactions, gels or aggregates) among the macromolecules. Understanding the rheological properties of these systems may help in the development of novel food structures with unique sensory properties and of functionalities such as in microencapsulation and controlled-release applications.

Conclusions

Protein and polysaccharide are the two main structural entities in foods and a great deal of work on the interactions between proteins and polysaccharides has been published over the last few decades. In recent years, excellent progress has been made on understanding the key variables and interactions that control the physical stability, rheology and microstructure of protein–polysaccharide mixtures. Although milk proteins, particularly whey proteins, have been most widely used in studies of protein–polysaccharide systems, most deal with the relatively simple binary combination of one protein and one polysaccharide. More complex systems, including ternary mixtures, still remain to be investigated in detail. At a practical level, it seems to be possible to manipulate these interactions and produce different microstructures by controlling internal (pH, ionic strength, biopolymer ratio, molecular weight and charge of the biopolymer) and external (temperature, pressure and shear rate) factors. However, there is a considerable challenge in understanding how different microstructures relate to the sensory properties of food products, such as mouthfeel and flavor release.

The formation of complexes through electrostatic and covalent interactions has been the subject of intensive studies, mainly because of potentially better functionality, e.g. rheology, gelation and interfacial properties, of composites compared with the protein or polysaccharide alone. Protein–polysaccharide complexes can serve as texturizing agents, encapsulating agents, fat replacers and stabilizers of emulsions and other dispersed systems. However, information on the detailed molecular structures of protein–polysaccharide complexes is still lacking and describing the experimental observations within the known theoretical frameworks remains a challenge.

It is now becoming apparent that the modification of food structure through modulation of macromolecular interactions can also be used to control the release of nutrients and bioactive components during digestion and to target where and how such components are released. The basic science underpinning these functions is largely unknown. New knowledge in this area will enable the development of composite food systems and ingredients that are superior in nutritional value and textural characteristics.

References

Aguilera, J. M. (1995). Gelation of whey proteins. *Food Technology*, **49**(10), 83–89.

Akhtar, M. and Dickinson, E. (2003). Emulsifying properties of whey protein–dextran conjugates at low pH and different salt concentrations. *Colloids and Surfaces B: Biointerfaces*, **31**, 125–32.

Akhtar, M. and Dickinson, E. (2007). Whey protein–maltodextrin conjugates as emulsifying agents: an alternative to gum arabic. *Food Hydrocolloids*, **21**, 607–16.

Ambjerg, P. H. C. and Jørgensen, B. B. (1991). Influence of pectin on the stability of casein solutions studied in dependence of varying pH and salt concentration. *Food Hydrocolloids*, **5**, 323–28.

Antonov, Y. A., Grinberg, V. Y., Zhuravskaya, N. A. and Tolstoguzov, V. B. (1982). Concentration of the proteins of skimmed milk by membraneless, isobaric osmosis. *Carbohydrate Polymers*, **2**, 81–90.

Antonov, Y. A., Lashko, N. P., Glotova, Y. A., Malovikova, A. and Markovich, O. (1996). Effect of the structural features of pectins and alginates on their thermodynamic compatibility with gelatin in aqueous media. *Food Hydrocolloids*, **10**, 1–9.

Asakura, S. and Oosawa, F. (1954). On interaction between 2 bodies immersed in a solution of macromolecules. *Journal of Chemical Physics*, **22**, 1255–56.

Asakura, S. and Oosawa, F. (1958). Interaction between particles suspended in solutions of macromolecules. *Journal of Polymer Science*, **33**, 183–92.

Beaulieu, M., Turgeon, S. L. and Doublier, J-L. (2001). Rheology, texture and microstructure of whey proteins/low methoxyl pectins mixed gels with added calcium. *International Dairy Journal*, **11**, 961–67.

Benichou, A., Aserin, A. and Garti, N. (2002). Protein–polysaccharide interactions for stabilization of food emulsions. *Journal of Dispersion Science and Technology*, **23**, 93–123.

Benichou, A., Aserin, A., Lutz, R. and Garti, N. (2007). Formation and characterization of amphiphilic conjugates of whey protein isolate (WPI)/xanthan to improve surface activity. *Food Hydrocolloids*, **21**, 379–91.

Bertrand, M-E. and Turgeon, S. L. (2007). Improved gelling properties of whey protein isolate by addition of xanthan gum. *Food Hydrocolloids*, **21**, 159–66.

Boeriu, C. G., Oudgenoeg, G., Spekking, W. T. J., Berendsen, L., Vancon, L., Boumans, H., Gruppen, H., Van Berkel, W. J. H., Laane, C. and Voragen, A. G. J. (2004). Horseradish peroxidase-catalyzed cross-linking of feruloylated arabinoxylans with beta-casein. *Journal of Agricultural and Food Chemistry*, **52**, 6633–39.

Bourriot, S., Garnier, C. and Doublier, J.-L. (1999a). Phase separation, rheology and microstructure of micellar casein-guar gum mixtures. *Food Hydrocolloids*, **13**, 43–49.

Bourriot, S., Garnier, C. and Doublier, J. L. (1999b). Micellar-casein-κ-carrageenan mixtures. I. Phase separation and ultrastructure. *Carbohydrate Polymers*, **40**, 145–57.

Bourriot, S., Garnier, C. and Doublier, J. L. (1999c). Phase separation, rheology and structure of micellar casein-galactomannan mixtures. *International Dairy Journal*, **9**, 353–57.

Boye, J. I., Ma, C. Y. and Harwalkar, V. R. (1997). Thermal denaturation and coagulation of proteins. In *Food Proteins and their Applications* (S. Damodaran and A. Paraf, eds) pp. 25–56. New York: Marcel Dekker.

Braudo, E. E. and Antonov, Y. A. (1993). Non-coulombic complex formation of protein as a structure forming factor in food systems. In *Food Proteins : Structure and Functionality* (K. D. Schwenke and R. Mothes, eds) pp. 210–15. Weinheim: VCH.

Bryant, C. M. and McClements, D. J. (2000a). Optimizing preparation conditions for heat-denatured whey protein solutions to be used as cold-gelling ingredients. *Journal of Food Science*, **65**, 259–63.

Bryant, C. M. and McClements, D. J. (2000b). Influence of xanthan gum on physical characteristics of heat-denatured whey protein solutions and gels. *Food Hydrocolloids*, **14**, 383–90.

Bungenberg de Jong, H. G. and Kruyt, H. R. (1929). Coacervation (partial miscibility in colloid systems). *Proceedings of the Royal Dutch Academy of Science*, **32**, 849–56.

Capron, I., Nicolai, T. and Durand, D. (1999). Heat induced aggregation and gelation of β-lactoglobulin in the presence of κ-carrageenan. *Food Hydrocolloids*, **13**, 1–5.

Clark, A. H. (2000). Direct analysis of experimental tie line data (two polymer-one solvent systems) using Flory-Huggins theory. *Carbohydrate Polymers*, **42**, 337–51.

Closs, C. B., Conde-Petit, B., Roberts, I. D., Tolstoguzov, V. B. and Escher, F. (1999). Phase separation and rheology of aqueous starch/galactomannan systems. *Carbohydrate Polymers*, **39**, 67–77.

Dalgleish, D. G. and Morris, E. R. (1988). Interactions between carrageenans and casein micelles: electrophoretic and hydrodynamic properties of the particles. *Food Hydrocolloids*, **2**, 311–20.

Damodaran, S. (1997). Food proteins: an overview. In *Food Proteins and their Applications* (S. Damodaran and A. Paraf, eds) pp. 1–24. New York: Marcel Dekker.

de Bont, P. W., van Kempen, G. M. P. and Vreeker, R. (2002). Phase separation in milk protein and amylopectin mixtures. *Food Hydrocolloids*, **16**, 127–38.

de Kruif, C. G. and Tuinier, R. (1999). Whey protein aggregates and their interaction with exo-polysaccharides. *International Journal of Food Science and Technology*, **34**, 487–92.

de Kruif, C. G. and Tuinier, R. (2001). Polysaccharide protein interactions. *Food Hydrocolloids*, **15**, 555–63.

de Kruif, C. G., Weinbreck, F. and de Vries, R. (2004). Complex coacervation of proteins and anionic polysaccharides. *Current Opinion in Colloid and Interface Science*, **9**, 340–49.

Dickinson, E. (1998). Stability and rheological implications of electrostatic milk protein–polysaccharide interactions. *Trends in Food Science and Technology*, **9**, 347–54.

Dickinson, E. (2003). Hydrocolloids at interfaces and the influence on the properties of dispersed systems. *Food Hydrocolloids*, **17**, 25–39.

Dickinson, E. and Semenova, M. G. (1992). Emulsifying properties of covalent protein–dextran hybrids. *Colloids and Surfaces*, **64**, 299–310.

Dickinson, E., Radford, S. J. and Golding, M. (2003). Stability and rheology of emulsions containing sodium caseinate: combined effects of ionic calcium and non-ionic surfactant. *Food Hydrocolloids*, **17**, 211–20.

Donato, L., Garnier, C., Novales, B. and Doublier, J. L. (2005). Gelation of globular protein in presence of low methoxyl pectin: effect of Na^+ and/or Ca^{2+} ions on rheology and microstructure of the systems. *Food Hydrocolloids*, **19**, 549–56.

Doublier, J. L., Garnier, C., Renard, D. and Sanchez, C. (2000). Protein–polysaccharide interactions. *Current Opinion in Colloid and Interface Science*, **5**, 202–14.

Dumay, E., Laligant, A., Zasypkin, D. and Cheftel, J. C. (1999). Pressure- and heat-induced gelation of mixed β-lactoglobulin/polysaccharide solutions: scanning electron microscopy of gels. *Food Hydrocolloids*, **13**, 339–51.

Dunlap, C. A. and Côté, G. L. (2005). β-Lactoglobulin–dextran conjugates: effect of polysaccharide size on emulsion stability. *Journal of Agricultural and Food Chemistry*, **53**, 419–23.

Ercelebi, E. A. and Ibanoglu, E. (2007). Influence of hydrocolloids on phase separation and emulsion properties of whey protein isolate. *Journal of Food Engineering*, **80**, 454–59.

Faergemand, M., Otte, J. and Qvist, K. B. (1998). Cross-linking of whey proteins by enzymatic oxidation. *Journal of Agricultural and Food Chemistry*, **46**, 1326–33.

Flanagan, J. and Singh, H. (2006). Conjugation of sodium caseinate and gum arabic catalyzed by transglutaminase. *Journal of Agricultural and Food Chemistry*, **54**, 7305–10.

Galazka, V. B., Sumner, I. G. and Ledward, D. A. (1996). Changes in protein–protein and protein–polysaccharide interactions induced by high pressure. *Food Chemistry*, **57**, 393–98.

Galazka, V. B., Ledward, D. A., Sumner, I. G. and Dickinson, E. (1997). Influence of high pressure on bovine serum albumin and its complex with dextran sulfate. *Journal of Agricultural and Food Chemistry*, **45**, 3465–71.

Galazka, V. B., Smith, D., Ledward, D. A. and Dickinson, E. (1999). Complexes of bovine serum albumin with sulphated polysaccharides: effects of pH, ionic strength and high pressure treatment. *Food Chemistry*, **64**, 303–10.

Gao, J. Y. and Dubin, P. L. (1999). Binding of proteins to copolymers of varying hydrophobicity. *Biopolymers*, **49**, 185–93.

Girard, M., Turgeon, S. L. and Gauthier, S. F. (2002). Interbiopolymer complexing between β-lactoglobulin and low- and high-methylated pectin measured by potentiometric titration and ultrafiltration. *Food Hydrocolloids*, **16**, 585–91.

Girard, M., Turgeon, S. L. and Gauthier, S. F. (2003a). Thermodynamic parameters of β-lactoglobulin-pectin complexes assessed by isothermal titration calorimetry. *Journal of Agricultural and Food Chemistry*, **51**, 4450–55.

Girard, M., Turgeon, S. L. and Gauthier, S. F. (2003b). Quantification of the interactions between β-lactoglobulin and pectin through capillary electrophoresis analysis. *Journal of Agricultural and Food Chemistry*, **51**, 6043–49.

Girard, M., Sanchez, C., Laneuville, S. I., Turgeon, S. L. and Gauthier, S. F. (2004). Associative phase separation of β-lactoglobulin/pectin solutions: a kinetic study by small angle static light scattering. *Colloids and Surfaces B: Biointerfaces*, **35**, 15–22.

Grinberg, V. Y. and Tolstoguzov, V. B. (1972). Thermodynamic compatibility of gelatin and some β_D-glucans in aqueous media. *Carbohydrate Research*, **25**, 313–20.

Grinberg, V. Y. and Tolstoguzov, V. B. (1997). Thermodynamic incompatibility of proteins and polysaccharides in solutions. *Food Hydrocolloids*, **11**, 145–58.

Guido, S., Simeone, M. and Alfani, A. (2002). Interfacial tension of aqueous mixtures of Na-caseinate and Na-alginate by drop deformation in shear flow. *Carbohydrate Polymers*, **48**, 143–52.

Guzey, D. and McClements, D. J. (2006). Characterization of β-lactoglobulin-chitosan interactions in aqueous solutions: a calorimetry, light scattering, electrophoretic mobility and solubility study. *Food Hydrocolloids*, **20**, 124–31.

Hansen, P. M. T., Hidalgo, J. and Gould, I. A. (1974). Reclamation of whey protein with carboxymethylcellulose. *Journal of Dairy Science*, **54**, 830–34.

Harnsilawat, T., Pongsawatmanit, R. and McClements, D. J. (2006). Characterization of β-lactoglobulin-sodium alginate interactions in aqueous solutions: a calorimetry, light scattering, electrophoretic mobility and solubility study. *Food Hydrocolloids*, **20**, 577–85.

Hattori, M., Nagasawa, K., Amenati, A., Kaminogawa, S. and Takahashi, K. (1994). Functional changes in β-lactoglobulin by conjugation with carboxymethyl dextran. *Journal of Agricultural and Food Chemistry*, **42**, 2120–25.

Hattori, M., Yang, W. and Takahashi, K. (1995). Functional changes of carboxymethyl potato starch by conjugation with whey proteins. *Journal of Agricultural and Food Chemistry*, **43**, 2007–11.

Hemar, Y., Tamehana, M., Munro, P. A. and Singh, H. (2001a). Influence of xanthan gum on the formation and stability of sodium caseinate oil-in-water emulsions. *Food Hydrocolloids*, **15**, 513–19.

Hemar, Y., Tamehana, M., Munro, P. A. and Singh, H. (2001b). Viscosity, microstructure and phase behavior of aqueous mixtures of commercial milk protein products and xanthan gum. *Food Hydrocolloids*, **15**, 565–74.

Hidalgo, J. and Hansen, P. M. T. (1969). Interactions between food stabilizers and β-lactoglobulin. *Journal of Agricultural and Food Chemistry*, **17**, 1089–92.

Imeson, A. P. (1977). On the nature of interactions between some anionic polysaccharides and proteins. *Journal of the Science of Food and Agriculture*, **28**, 661–67.

Kato, A., Mifuru, R., Matsudomi, N. and Kobayashi, K. (1992). Functional casein–polysaccharide conjugates prepared by controlled dry heating. *Bioscience, Biotechnology and Biochemistry*, **56**, 567–71.

Kazmiersi, M., Wicker, L. and Corredig, M. (2003). Interactions of β-lactoglobulin and high-methoxyl pectins in acidified systems. *Journal of Food Science*, **68**, 1673–79.

Kim, H. J., Decker, E. A. and McClements, D. J. (2006). Preparation of multiple emulsions based on thermodynamic incompatibility of heat-denatured whey protein and pectin solutions. *Food Hydrocolloids*, **20**, 586–95.

Kinsella, J. E. (1984). Milk proteins: physical and functional properties. *Critical Reviews in Food Science and Nutrition*, **21**, 197–262.

Laneuville, S. I., Paquin, P. and Turgeon, S. L. (2000). Effect of preparation conditions on the characteristics of whey protein–xanthan gum complexes. *Food Hydrocolloids*, **14**, 305–14.

Langendorff, V., Cuvelier, G., Launay, B. and Parker, A. (1997). Gelation and flocculation of casein-micelle–carrageenan mixtures. *Food Hydrocolloids*, **11**, 35–40.

Langendorff, V., Cuvelier, G., Launay, B., Michon, C., Parker, A. and De Kruif, C. G. (1999). Casein micelle/iota carrageenan interactions in milk: influence of temperature. *Food Hydrocolloids*, **13**, 211–18.

Langendorff, V., Cuvelier, G., Michon, C., Launay, B., Parker, A. and De Kruif, C. G. (2000). Effects of carrageenan type on the behavior of carrageenan/milk mixtures. *Food Hydrocolloids*, **14**, 273–80.

Langton, M. and Hermansson, A. M. (1992). Fine-stranded and particulate gels of beta-lactoglobulin and whey-protein at varying pH. *Food Hydrocolloids*, **5**, 523–39.

Li, J., Ould Eleya, M. M. and Gunasekaran, S. (2006). Gelation of whey protein and xanthan mixture: effect of heating rate on rheological properties. *Food Hydrocolloids*, **20**, 678–86.

Lucey, J. A., Srinivasan, M., Singh, H. and Munro, P. A. (2000). Characterization of commercial and experimental sodium caseinates by multiangle laser light scattering and size-exclusion chromatography. *Journal of Agricultural and Food Chemistry*, **48**, 1610–16.

Lundin, L., Williams, M. A. K. and Foster, T. J. (2003). Phase separation in foods. In *Texture in Foods*, Volume 1, *Semi-solid Foods*, (B. M. McKenna, ed.), Chapter 3, pp. 63–85. Cambridge: Woodhead Publishing Ltd.

Maroziene, A. and de Kruif, C. G. (2000). Interaction of pectin and casein micelles. *Food Hydrocolloids*, **14**, 391–94.

Martinez, K. D., Baeza, R. I., Millan, F. and Pilosof, A. M. R. (2005). Effect of limited hydrolysis of sunflower protein on the interactions with polysaccharides in foams. *Food Hydrocolloids*, **19**, 361–69.

Matia-Merino, L., Lau, K. and Dickinson, E. (2004). Effects of low-methoxyl amidated pectin and ionic calcium on rheology and microstructure of acid-induced sodium caseinate gels. *Food Hydrocolloids*, **18**, 271–81.

McClements, D. J. (2005). Food emulsions in practice. In *Food Emulsions: Principles, Practices and Techniques* (D. J. McClements, ed.) pp. 515–44. Boca Raton, Florida: CRC Press.

Mishra, S., Mann, B. and Joshi, V. K. (2001). Functional improvement of whey protein concentrate on interaction with pectin. *Food Hydrocolloids*, **15**, 9–15.

Mleko, S., Li-Chan, E. C. Y. and Pikus, S. (1997). Interactions of κ-carrageenan with whey proteins in gels formed at different pH. *Food Research International*, **30**, 427–33.

Morris, V. J. (1986). Multicomponent gels. In *Gums and Stabilisers for the Food Industry*, Volume 3 (G. O. Phillips, D. J. Wedlock and P. A. Williams, eds) pp. 87–99. London: Elsevier.

Morris, G. A., Sims, I. M., Robertson, A. J. and Furneaux, R. H. (2004). Investigation into the physical and chemical properties of sodium caseinate-maltodextrin glyco-conjugates. *Food Hydrocolloids*, **18**, 1007–14.

Neirynck, N., Van der Meeren, P., Bayarri Gorbe, S., Dierckx, S. and Dewettinck, K. (2004). Improved emulsion stabilizing properties of whey protein isolate by conjugation with pectins. *Food Hydrocolloids*, **18**, 949–57.

Neiser, S., Draget, K. I. and Smidsrod, O. (1998). Gel formation in heat-treated bovine serum albumin-sodium alginate systems. *Food Hydrocolloids*, **12**, 127–32.

Norton, I. T. and Frith, W. J. (2001). Microstructure design in mixed biopolymer composites. *Food Hydrocolloids*, **15**, 543–53.

Oudgenoeg, G., Hilhorst, R., Piersma, S. R., Boeriu, C. G., Gruppen, H., Hessing, M., Voragen, A. G. J. and Laane, C. (2001). Peroxidase-mediated cross-linking of a tyrosine-containing peptide with ferulic acid. *Journal of Agricultural and Food Chemistry*, **49**, 2503–10.

Ould Eleya, M. M. and Turgeon, S. L. (2000). Rheology of κ-carrageenan and β-lactoglobulin mixed gels. *Food Hydrocolloids*, **14**, 29–40.

Pereyra, R., Schmidt, K. A. and Wicker, L. (1997). Interaction and stabilization of acidified casein dispersions with low and high methoxyl pectins. *Journal of Agricultural and Food Chemistry*, **45**, 3448–51.

Piculell, L. and Lindman, B. (1992). Association and segregation in aqueous polymer/polymer, polymer/surfactant and surfactant/surfactant mixtures: similarities and differences. *Advances in Colloid and Interface Science*, **41**, 149–78.

Polyakov, V. I., Grinberg, V. Y. and Tolstoguzov, V. B. (1980). Application of phase–volume–ratio method for determining the phase diagram of water–casein–soybean globulins system. *Polymer Bulletin*, **2**, 757–60.

Polyakov, V. I., Grinberg, V. Y. and Tolstoguzov, V. B. (1997). Thermodynamic incompatibility of proteins. *Food Hydrocolloids*, **11**, 171–80.

Rao, M. A. (1999). Rheological behavior of food gel systems. In *Rheology of Fluid and Semisolid Foods: Principles and Applications* (M. A. Rao, ed.) p. 357. New York: Aspen Publishers, Inc..

Samant, S. K., Singhal, R. S., Kulkarni, P. R. and Rege, D. V. (1993). Protein–polysaccharide interactions: a new approach in food formulation. *International Journal of Food Science and Technology*, **28**, 547–62.

Sanchez, C. and Renard, D. (2002). Stability and structure of protein–polysaccharide coacervates in the presence of protein aggregates. *International Journal of Pharmaceutics*, **242**, 319–24.

Sanchez, C., Mekhloufi, G., Schmitt, C., Renard, D., Robert, P., Lehr, C.-M., Lamprecht, A. and Hardy, J. (2002). Self-assembly of β-lactoglobulin and acacia gum in aqueous solvent: Structure and phase-ordering kinetics. *Langmuir*, **18**, 10323–33.

Sanchez, C., Mekhloufi, G. and Renard, D. (2006). Complex coacervation between β-lactoglobulin and acacia gum: a nucleation and growth mechanism. *Journal of Colloid and Interface Science*, **299**, 867–73.

Schmitt, C., Sanchez, C., Desobry-Banon, S. and Hardy, J. (1998). Structure and techno-functional properties of protein–polysaccharide complexes: a review. *Critical Reviews in Food Science and Nutrition*, **38**, 689–753.

Schmitt, C., Sanchez, C., Thomas, F. and Hardy, J. (1999). Complex coacervation between β-lactoglobulin and acacia gum in aqueous medium. *Food Hydrocolloids*, **13**, 483–96.

Schmitt, C., Sanchez, C., Despond, S., Renard, D., Thomas, F. and Hardy, J. (2000). Effect of protein aggregates on the complex coacervation between β-lactoglobulin and acacia gum at pH 4.2. *Food Hydrocolloids*, **14**, 403–13.

Schmitt, C., Sanchez, C., Lamprecht, A., Renard, D., Lehr, C.-M., de Kruif, C. G. and Hardy, J. (2001). Study of β-lactoglobulin/acacia gum complex coacervation by diffusing-wave spectroscopy and confocal scanning laser microscopy. *Colloids and Surfaces B: Biointerfaces*, **20**, 267–80.

Schorsch, C., Jones, M. G. and Norton, I. T. (1999). Thermodynamic incompatibility and microstructure of milk protein/locust bean gum/sucrose systems. *Food Hydrocolloids*, **13**, 89–99.

Shepherd, R., Robertson, A. and Ofman, D. (2000). Dairy glycoconjugate emulsifiers: casein-maltodextrins. *Food Hydrocolloids*, **14**, 281–86.

Sherony, D. F. and Kintner, R. C. (1971). Van der Waals forces in a three-phase system. *American Institute of Chemical Engineers*, **17**, 291–94.

Simeone, M., Mole, F. and Guido, S. (2002). Measurement of average drop size in aqueous mixtures of Na-alginate and Na-caseinate by linear oscillatory tests. *Food Hydrocolloids*, **16**, 449–59.

Snoeren, T. H. M. (1975). Electrostatic interaction between kappa-carrageenan and κ-casein. *Milchwissenchaft*, **30**, 393–95.

Song, Y., Babiker, E. E., Usui, M., Saito, A. and Kato, A. (2002). Emulsifying properties and bactericidal action of chitosan-lysozyme conjugates. *Food Research International*, **35**, 459–66.

Stading, M., Langton, M. and Hermansson, A. M. (1993). Microstructure and rheological behavior of particulate β-lactoglobulin gels. *Food Hydrocolloids*, **7**, 195–212.

Stainsby, G. (1980). Proteinaceous gelling systems and their complexes with polysaccharides. *Food Chemistry*, **6**, 3–14.

Suchkov, V. V., Grinberg, V. Y. and Tolstogusov, V. B. (1981). Steady-state viscosity of the liquid two-phase disperse system water–casein–sodium alginate. *Carbohydrate Polymers*, **1**, 39–53.

Suchkov, V. V., Grinberg, V. Y., Muschiolik, G., Schmandke, H. and Tolstoguzov, V. B. (1988). Mechanical and functional properties of anisotropic gel fibers obtained from the 2-phase system of water casein sodium alginate. *Nahrung*, **32**, 661–68.

Syrbe, A., Bauer, W. J. and Klostermeyer, H. (1998). Polymer science concepts in dairy systems – an overview of milk protein and food hydrocolloid interaction. *International Dairy Journal*, **8**, 179–93.

Taravel, M. N. and Domard, A. (1995). Collagen and its interaction with chitosan:II. Influence of the physicochemical characteristics of collagen. *Biomaterials*, **16**, 865–71.

Tavares, C. and Lopes da Silva, J. A. (2003). Rheology of galactomannan-whey protein mixed systems. *International Dairy Journal*, **13**, 699–706.

Tavares, C., Monteiro, S. R., Moreno, N. and da Silva, J. A. L. (2005). Does the branching degree of galactomannans influence their effect on whey protein gelation? *Colloids and Surfaces A: Physicochemical and Engineering Aspects*, **270**, 213–19.

Thaiudom, S. and Goff, H. D. (2003). Effect of κ-carrageenan on milk protein polysaccharide mixtures. *International Dairy Journal*, **13**, 763–71.

Tiebackx, F. W. Z. (1911). Gleichzeitige ausflockung zweier kolloide. *Zeitschrift fur Chemie und Industrie der Kolloide*, **8**, 198–201.

Tolstogusov, V. B. (1986). Functional properties of protein–polysaccharide mixtures. In *Functional Properties of Macromolecules* (J. R. Mitchell and D. A. Ledward, eds) pp. 385–415. London: Elsevier Applied Science Publishers.

Tolstoguzov, V. B. (1988). Some physico-chemical aspects of protein processing into foodstuffs. *Food Hydrocolloids*, **2**, 339–70.

Tolstoguzov, V. B. (1991). Functional properties of food proteins and role of protein–polysaccharide interaction. *Food Hydrocolloids*, **4**, 429–68.

Tolstoguzov, V. B. (1997a). Protein–polysaccharide interactions. In *Food Proteins and their Applications* (S. Damodaran and A. Paraf, eds) pp. 171–98. New York: Marcel Dekker.

Tolstoguzov, V. (1997b). Thermodynamic aspects of dough formation and functionality. *Food Hydrocolloids*, **11**, 181–93.

Tolstoguzov, V. B. (2002). Thermodynamic aspects of biopolymer functionality in biological systems, foods, and beverages. *Critical Reviews in Biotechnology*, **22**, 89–174.

Tolstoguzov, V. (2003). Some thermodynamic considerations in food formulation. *Food Hydrocolloids*, **17**, 1–23.

Tolstoguzov, V. (2006). Phase behavior in mixed polysaccharide systems. In *Food Polysaccharides and their Applications* (A. M. Stephen, G. O. Phillips and P. A. Williams, eds) pp. 590–620. Boca Raton, Florida: CRC Press.

Tolstoguzov, V. B., Grinberg, V. Y. and Gurov, A. N. (1985). Some physicochemical approaches to the problem of protein texturization. *Journal of Agricultural and Food Chemistry*, **33**, 151–59.

Tuinier, R. and De Kruif, C. G. (1999). Phase behavior of casein micelles/exocellular polysaccharide mixtures: experiment and theory. *Journal of Chemical Physics*, **110**, 9296–304.

Tuinier, R., ten Grotenhuis, E., Holt, C., Timmins, P. A. and de Kruif, C. G. (1999). Depletion interaction of casein micelles and an exocellular polysaccharide. *Physical Review E*, **60**, 848–56.

Tuinier, R., Dhont, J. K. G. and De Kruif, C. G. (2000). Depletion-induced phase separation of aggregated whey protein colloids by an exocellular polysaccharide. *Langmuir*, **16**, 1497–507.

Turgeon, S. L. and Beaulieu, M. (2001). Improvement and modification of whey protein gel texture using polysaccharides. *Food Hydrocolloids*, **15**, 583–91.

Turgeon, S. L., Beaulieu, M., Schmitt, C. and Sanchez, C. (2003). Protein–polysaccharide interactions: phase-ordering kinetics, thermodynamic and structural aspects. *Current Opinion in Colloid and Interface Science*, **8**, 401–14.

Wang, Q.a nd Qvist, K. B.(2000). Investigationo ft hec omposites ystemo f β-lactoglobulin and pectin in aqueous solutions. *Food Research International*, **33**, 683–90.

Wang, S., van Dijk, J. A. P. P., Odijk, T. and Smit, J. A. M. (2001). Depletion-induced demixing in aqueous protein-polysaccharide solutions. *Biomacromolecules*, **2**, 1080–88.

Weinbreck, F. and Wientjes, R. H. W. (2004). Rheological properties of whey protein/gum arabic coacervates. *Journal of Rheology*, **48**, 1215–28.

Weinbreck, F., de Vries, R., Schrooyen, P. and de Kruif, C. G. (2003a). Complex coacervation of whey proteins and gum arabic. *Biomacromolecules*, **4**, 293–303.

Weinbreck, F., Nieuwenhuijse, H., Robjin, G. W. and De Kruif, C. G. (2003b). Complex formation of whey proteins–exocellular polysaccharide EPS B40. *Langmuir*, **19**, 9404–10.

Weinbreck, F., Nieuwenhuijse, H., Robijn, G. W. and de Kruif, C. G. (2004a). Complexation of whey proteins with carrageenan. *Journal of Agricultural and Food Chemistry*, **52**, 3550–55.

Weinbreck, F., Tromph, R. H. and de Kruif, C. G. (2004b). Composition and structure of whey protein/gum arabic coacervates. *Biomacromolecules*, **5**, 1437–45.

Weinbreck, F., Rollema, H. S., Tromp, R. H. and de Kruif, C. G. (2004c). Diffusivity of whey protein and gum arabic in their coacervates. *Langmuir*, **20**, 6389–95.

Ye, A. Q., Flanagan, J. and Singh, H. (2006). Formation of stable nanoparticles via electrostatic complexation between sodium caseinate and gum arabic. *Biopolymers*, **82**, 121–33.

Zasypkin, D. V., Dumay, E. and Cheftel, J. C. (1996). Pressure- and heat-induced gelation of mixed β-lactoglobulin/xanthan solutions. *Food Hydrocolloids*, **10**, 203–11.

Interaction between milk proteins and micronutrients

13

T. Considine and J. Flanagan

Abstract

Milk proteins can interact with micronutrients via a variety of mechanisms, with hydrophobic interactions being of particular importance. This chapter focuses on the interaction of individual milk proteins as well as mixtures of milk proteins with a range of micronutrients including vitamins, fatty acids, sugars and minerals. Thus, milk proteins can be used as a carrier of micronutrients and thereby increase the nutritional benefit of milk and milk-based products.

It is widely known that the processing of milk proteins via heat or high pressure can result in modification to protein structure, resulting in altered interactions between the proteins and the micronutrient. Interestingly, the presence of some micronutrients can retard the denaturation of some milk proteins. The addition of specific micronutrients may be used as a processing tool to prevent denaturation of milk proteins under physical conditions which normally result in denaturation.

Introduction

The existence of a three-dimensional, folded protein structure is dependent on several forces. These include hydrogen bonding, hydrophobic interactions, van der Waals' forces and electrostatic interactions. Some amino acid residues may exhibit a hydrophobic character, while electrostatic forces are based on interactions between charged residues. Some proteins contain two or more polypeptide chains. The result of interactions between these components is the quaternary structure, which under

Milk Proteins: From Expression to Food
ISBN: 978-0-12-374039-7

normal physiological conditions is known as the "native state". Thus, the conformation of a protein is extremely dependent upon the presence of amino acids and the variation of residues within the primary structure. Although proteins may be in the native state, interactions, through hydrophobic, electrostatic, van der Waals' and other forces, are possible through exposed regions on the surface of the protein. It is through these mechanisms that interactions between milk proteins and various micronutrients such as retinol, fatty acids, minerals and surfactants, for example, can occur.

Protein structures can be readily destabilized from their native state by relatively minor changes in the environmental conditions. Variations in pH, temperature and pressure, for example, can all induce structural transitions in proteins. In some cases, the objective of processing is to induce changes in protein structure, e.g. the heating of whey proteins to form gels. In other cases, however, changes in the environmental conditions can elicit changes in protein structure that result in undesirable functional properties, e.g. loss of solubility or biological activity.

In addition to pH-, temperature- and pressure-induced changes in protein structure, the presence of micronutrients can affect how the protein structure reacts to variations in pH, temperature or pressure. By interacting with specific sites within the protein's three-dimensional structure, micronutrients can render a protein more, or less, susceptible to denaturation.

Interaction between milk proteins and micronutrients

Micronutrients, such as retinol, sugars, vitamins, fatty acids and minerals, among others, may interact with milk proteins through a variety of mechanisms. The main mechanism is through hydrophobic interaction. In this respect, the majority of studies involving interactions between milk proteins and micronutrients focus on globular whey proteins with their hydrophobic cavities and extensive secondary and tertiary structure. Interactions between caseins and micronutrients are mostly based on electrostatic interactions.

Retinol

Most lipocalin molecules such as β-LG have clear biological roles as ligand carriers. Different researchers have used a variety of methods to determine binding constants, thus making comparison between studies difficult. For example, Muresan *et al.* (2001) compared fluorometry with equilibrium dialysis. The former yielded higher binding affinities than the latter. The pH, the genetic variant and the source of the protein all contribute to the discrepancies in the literature.

Papiz *et al.* (1986) identified that the structure of β-LG was remarkably similar to the structure of retinol-binding protein (RBP). Nonetheless, no definite biological function has been attributed to β-LG. In spite of this, various ligand-binding sites have been defined (Qin *et al.*, 1998; Wu *et al.*, 1999) and the structural changes induced in different environments have been determined by X-ray crystallography (Qin *et al.*, 1998).

However, little endogenous retinol is found bound to β-LG when it is first purified and the ligand most closely associated with the protein is palmitate (Pérez *et al.*, 1989). Vitamin A in the unesterified form is readily oxidized by atmospheric oxygen. The acetate and palmitate esters of vitamin A are somewhat more stable towards oxidation than the free alcohol. *Cis–trans* isomerization is directly promoted by light containing wavelengths of less than 500 nm.

In foods, the retinyl esters and carotenoids are dissolved in the fat matrix, where they are protected from the oxidizing action of atmospheric oxygen by vitamin E and other antioxidants (Ball, 1988). Free retinol is a rather unstable compound, especially in an aqueous environment, but its stability is greatly improved when bound to a RBP (Futterman and Heller, 1972). With the addition of the double bond, free all-*trans*-3-dehydroretinol is much more sensitive to degradation by oxygen than free retinol (Schwieter *et al.*, 1962). When all-*trans*-3-dehydroretinol is bound to RBP, however, it is found to be very stable.

In 1972, Futterman and Heller, using fluorescence measurements, first reported that bovine β-LG, like RBP, formed water-soluble complexes with retinol. It has been suggested by Fugate and Song (1980) that the β-LG binding site makes the retinol more rigid than in free solution. Using fluorescence titration and circular dichroism (CD), β-LG was found to display two high-affinity binding sites for retinol per protein dimer, each with an association constant of 2×10^{-8} M (Fugate and Song, 1980). Fluorescence studies exclude the possibility that both pockets accommodate retinol simultaneously (Futterman and Heller, 1972; Fugate and Song, 1980; Dufour and Haertle, 1990a).

There has been some debate regarding the binding site of retinol. Papiz *et al.* (1986) suggested that it bound inside the main hydrophobic binding pocket, whereas Monaco *et al.* (1987) proposed that a retinol-binding site for β-LG was an external, solvent-accessible hydrophobic cleft located between the three-turn α-helix that is packed against the outer surface of the β-barrel and the β-barrel itself. Site mutation experiments, F136A and K141M, did not support this idea and suggested that retinol binds to an evolutionary conserved interior cavity rather than the surface pocket (Cho *et al.*, 1994).

Several reports have indicated that Trp19 of β-LG is essential for the binding of retinol (Papiz *et al.*, 1986). Katakura *et al.* (1994) used site-directed mutagenesis to investigate whether this completely conserved residue would be indispensable for forming the characteristic structure. Substituting Trp19 with tyrosine was shown not to be critical for the binding of retinol but was important for maintaining stability.

Dufour and Haertle (1991) monitored the binding of retinol, retinyl acetate, retinoic acid and β-carotene to native, esterified and alkylated β-LG by quenching of tryptophan fluorescence. The retinoids bound to native or modified β-LG in a 1:1 molar ratio with apparent dissociation constants in the range of 10^{-8} M, whereas the molar ratio was 1:2 for β-carotene–protein. Chemical modification of β-LG by methods such as methylation, ethylation (Dufour and Haertle, 1990a, 1990b) or alkylation (Dufour and Haertle, 1991) have been shown to enhance the binding affinity for retinol, by opening up a second binding site. It may therefore be assumed that the partial change of β-LG secondary structure produced by these treatments does not destroy the structure of the retinol-binding pocket.

Very few studies have been carried out with ligand binding to α-LA, in comparison with the vast range of studies with β-LG. However, there is the potential of ligands binding to α-LA. Puyol *et al.* (1991) studied the binding of retinol and palmitic acid in a whey protein mixture. From this study, α-LA was shown to bind retinol more strongly than β-LG, but a much lower percentage of palmitic acid bound to α-LA in comparison with to β-LG.

Futterman and Heller (1972) showed that, as with β-LG, BSA formed a strong fluorescent water-soluble complex with retinol. They also postulated that, although no detectable retinol is bound to serum albumin *in vivo*, the possibility exists that this protein could serve as an auxiliary carrier if excess free retinol were introduced into the circulation.

Raica *et al.* (1959) reported that a liposoluble substance such as retinol can also be bound to casein. Modification of casein micelle from its natural state (e.g. through acidification or rennet treatment of milk) affects the nutritional activity of retinol (Adrian *et al.*, 1984).

Vitamin D

The affinity of β-LG for vitamin D_2 is about 10-fold greater than that for vitamin A and other retinoids (Dufour and Haertle, 1991; Cho *et al.*, 1994; Wang *et al.*, 1997a). Further work on the binding of vitamin D and cholesterol to β-LG has been explored by Wang *et al.* (1997a) and Kontopidis *et al.* (2004).

The binding of retinal, vitamin D_2 and retinyl palmitate by β-LG was studied by Wang *et al.* (1999). Analysis of competitive binding experiments with palmitate indicated that retinal and palmitate did not compete for the same site; however, vitamin D_2 appeared to displace palmitate at higher concentrations. Retinoids and vitamin D_2 were bound more tightly than palmitate.

Recently, Forrest *et al.* (2005) reported on the interactions of vitamin D_3 with β-LG A under a range of environmental conditions (i.e. pH and ionic strength). At pH 4.6, β-LG A occurs as octamers (Verheul *et al.*, 1999), whereas dimers predominate at pH 6.6 and pH 8.0. Fogolari *et al.* (2000) demonstrated the importance of pH when binding ligands to β-LG. The results of Forrest *et al.* (2005) indicated that binding depended greatly on the solution conditions, e.g. at low pH, 2.5 ($I = 0.15\,M$), the EF loop (gate) is closed and thus vitamin D_3 was probably weakly bound in the external hydrophobic surface. Upon lowering the ionic strength to 0.08 M, binding increased. It was suggested (Arymard *et al.*, 1996) that lowering the salt concentration allowed more surface binding. A dissociation constant of 0.02–0.29 μM was reported for β-LG A, with apparent mole ratios of vitamin D_3 bound per mole of β-LG A ranging from 0.51 to 2.04 (Forrest *et al.*, 2005).

Two studies have discussed the stability of vitamin D_3 in cheese (Banville *et al.*, 2000; Upreti *et al.*, 2002) and there is one recent study of the binding of vitamin D_3 to β-casein. Forrest *et al.* (2005) reported on the interactions of vitamin D_3 with β-casein under a range of environmental conditions. The binding constants of vitamin D_3 to β-casein were dependent on pH and ionic strength. In agreement with the study of Lietaer *et al.* (1991), an increase in binding as a function of ionic strength was apparent at pH 6.6. This was attributed to reduced solubility of the protein and enhanced hydrophobic interactions, creating more surface area for binding (Lietaer *et al.*, 1991). Increased binding was associated with a weaker affinity, compared with lower ionic strength where binding was stronger. Although stronger interactions at low ionic strength were attributed to fewer protein interactions, the authors could not identify a reason for decreased binding at pH 8.

A dissociation constant of 0.06–0.26 μM was reported for β-casein, with apparent mole ratios of vitamin D_3 bound per mole of β-casein ranging from 1.16 to 2.05. It was suggested by Forrest *et al.* (2005) that the rheomorphic nature of β-casein allowed the hydrophobic area to bind strongly with vitamin D_3, in the most thermodynamically stable conformation. The hydrophobic interactions were aligned with the perturbation of phenylalanine and the quenching of tryptophan, both of which are located in the hydrophobic core.

Vitamin C

Few studies have explored the interactions between vitamin C and milk proteins. Binding of ascorbic acid to BSA was recorded by Tukamoto *et al.* (1974). Oelrichs *et al.* (1984) investigated the interactions between ascorbate and BSA. They suggested an intrinsic association constant of 2600 M^{-1} at 20°C. Dai-Dong *et al.* (1990) observed an increased stability of ascorbic acid in the presence of β-LG compared with in pure water, but also found that vitamin C was more thermoresistant when heated in the presence of β-LG.

In contrast to these studies, Puyol *et al.* (1994) reported the lack of interaction of ascorbic acid with β-LG or indeed any of the other whey proteins. Puyol *et al.* (1994) suggested that the discrepancy between their work and that of Dai-Dong *et al.* (1990) may have been related to the methods used. Monitoring the reducing ability of ascorbic acid may not reflect sufficient allowance for the effects of ascorbate losses through autoxidation. Puyol *et al.* (1994) also suggested that the antioxidant effect of reductive thiols in β-LG and serum albumin may have a protective effect.

Other vitamins

Milk also contains an array of vitamin-binding proteins, including vitamin-B_{12}-binding protein, folate-binding protein, vitamin-D-binding protein and riboflavin-binding protein. These proteins occur at low concentrations, but may play a significant role in the uptake of vital vitamins from the diet (Anderson and von der Lippe, 1979; Salter and Mowlem, 1983). Folate-binding proteins (FBPs) are specifically involved in the uptake of folate from the intestine. *In-vivo* studies on rats have shown that protein-bound folate is absorbed at a lower rate than free folate, resulting in increased retention time of folate, allowing it to reach its target tissues. FBPs also reduce the availability of folate to bacteria in the gut and hence may have antibacterial properties (Ford, 1974).

Raw bovine milk contains a riboflavin-binding protein (Kanno and Kanehara, 1985) and riboflavin bound to this milk protein has been shown to have similar antioxidant activities to riboflavin bound to egg white riboflavin-binding protein (Toyosaki and Mineshita, 1988). More recently, Nixon *et al.* (2004) investigated the source of the co-operativity between FBP and folate and their results suggested stoichiometric interactions. This area, including the binding of trace elements, has been reviewed in detail by Vegarud *et al.* (2000).

Fatty acids

Most of the fatty acids present in milk are found as triglycerides, which form the fat globule (Walstra and Jenness, 1984). The presence of β-LG increases the activity of ruminant pharyngeal lipase, which is deemed to be important during the neonatal period because levels of pancreatic lipase and bile salts are low at that age (Hamosh *et al.*, 1981). The ability of β-LG to remove the released fatty acids, which would otherwise inhibit lipase activity (Calvo *et al.*, 1990), is thus of great importance. Under these conditions, it could be possible that β-LG would bind large amounts of fatty acids, thus displacing retinol (Puyol *et al.*, 1991). Thus, Pérez *et al.* (1992) proposed that ruminant β-LG, because of its activity to bind fatty acids, might play a role in the activity of pregastric lipases.

Diaz de Villegas *et al.* (1987) observed that bovine milk has several long-chain fatty acids that bind to β-LG. The predominant fatty acids bound to β-LG were myristic, palmitic and oleic acids, which together accounted for approximately 83% of the fatty acids from total lipids and 70% of the free fatty acids. These fatty acids also predominate in milk lipids.

Pérez *et al.* (1989) demonstrated that two types of lipids, namely free fatty acids and triglycerides, bound to β-LG. The total amount of fatty acids extracted from β-LG was 0.71 mol per mol of monomer protein. The predominant fatty acids were palmitic (31–35%), oleic (22–23%) and myristic (14–17%) acids, which combined account for 66–75% of the total fatty acids bound to β-LG. The unsaturated fatty acids extracted from β-LG were less than 31% of the total fatty acids and mainly oleic (22–23%) and palmitoleic (4–5%) acids.

As with retinol, there also seems to be controversy regarding the binding location of fatty acids. Narayan and Berliner (1998) suggested that fatty acids bind at the "external site" of β-LG. However, this conflicts with earlier studies by Puyol *et al.* (1991), which suggested competitive binding, and by Creamer (1995), which suggested an internal location as the primary binding site for fatty acids. Since then, several studies have shown ligands to bind in the internal cavity. Qin *et al.* (1998), using X-ray crystallography, showed 12-bromododecanoic acid binding inside the calyx and Wu *et al.* (1999) revealed that palmitate binds in the central cavity (Figure 13.1) in a manner similar to the binding of retinol to the related lipocalin, serum RBP. Ragono *et al.* (2000) provided further evidence for cavity binding of β-LG and palmitic acid, as did Zsila *et al.* (2002) using CD, electronic absorption spectroscopy and electrospray ionization mass spectrometry (ESI-MS) with *cis*-parinaric acid.

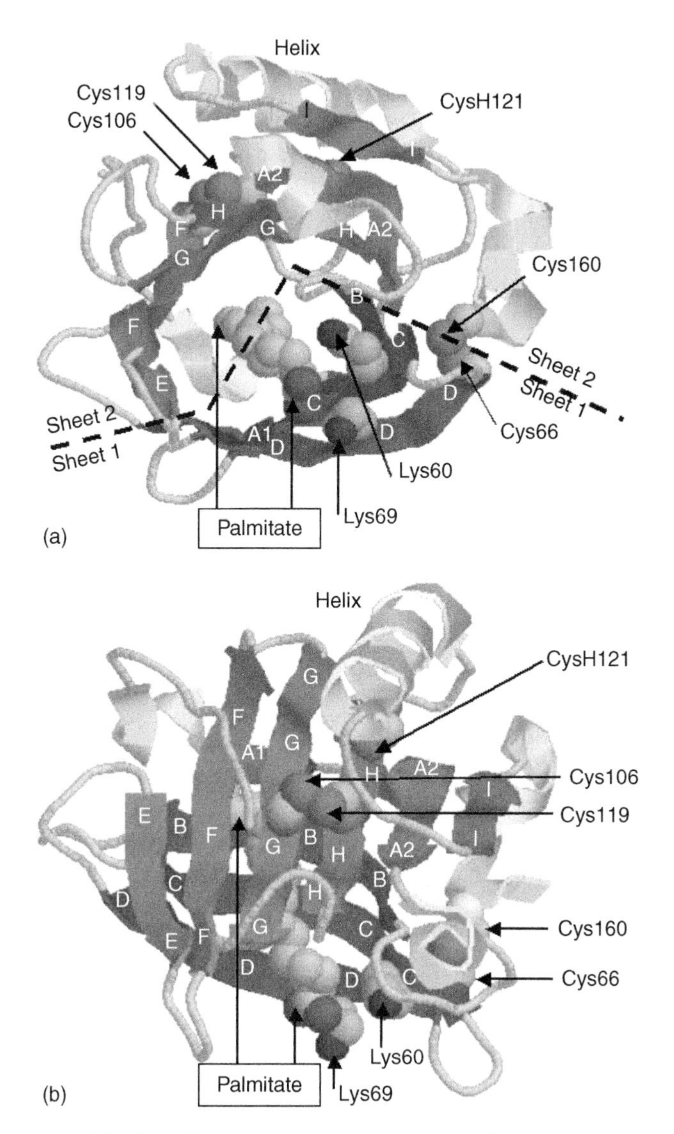

Figure 13.1 Diagram of the three-dimensional structure of β-LG that shows the relative positions of the five Cys residues, Lys60, Lys69 and the bound palmitate (Wu *et al.*, 1999). The helix and the strands that constitute Sheets 1 and 2 are also labeled. The diagram was drawn from the PDB file 1GXA using RASMOL Ver 2.6. Reproduced with the permission of Considine *et al.* (2005a). Copyright 2005 *Journal of Agricultural and Food Chemistry*, American Chemical Society.

Konuma *et al.* (2007) examined palmitic acid binding to a dimeric β-LG mutant A34C using heteronuclear nuclear magnetic resonance (NMR) spectroscopy. Their results suggested a 1:1 binding stoichiometry. They indicated that the protein conformation should be complementary, at least in part, to the ligand's structure, if tight binding (dissociation constant of $<10^{-7}$ M) is to occur. They further highlighted the role of the highly flexible loops above the barrel in ligand binding, which supports the work of Zidek *et al.* (1999) and Stone (2001). Konuma *et al.* (2007) hypothesized that the barrel's entrance accommodates a variety of ligands, because of its plasticity, whereas the bottom of the cavity shows rigid and somewhat selective binding.

Thus, it has been established that β-LG strongly binds one mole of long-chain fatty acids (myristic, palmitic, stearic acid, etc.) per mole of monomeric protein (Spector and Fletcher, 1970; Frapin *et al.*, 1993; Dufour *et al.*, 1994). Frapin *et al.* (1993) explored the binding of a variety of fatty acids to β-LG and strength of binding was associated with chain length. As β-LG has fatty acids physiologically bound, the amount present probably depends on the isolation technique or whether the protein is delipidated. Thus, this consideration is necessary when determining apparent association constants.

Fatty acid binding to β-LG is sensitive to changes in pH. Changes in binding constants are observed over the pH range 5.5–8.5 (Pérez and Calvo, 1995). This may be due to the electrostatic interactions; for example, as the pH increases, β-LG becomes negatively charged, thus making it less electrostatically inviting for a negatively charged fatty acid. The two lysine residues at the opening of β-LG's ligand-binding cavity—Lys60 and Lys69—are likely to play a significant role in ligand affinity. The inability of porcine β-LG to bind fatty acids may be due to the substitution of Lys69 by glutamate, as suggested by Frapin *et al.* (1993) and Pérez *et al.* (1993). Creamer (1995) also hypothesized that lysine was involved in the binding process, whereby, at neutral pH, the carboxylate group of the fatty acid salt bridged to the positively charged ε-amino group.

Puyol *et al.* (1991) studied the competition between the binding of retinol and free fatty acids to β-LG. They observed that, when the ratio between the concentrations of the total fatty acids (as palmitic acid) and retinol is similar to that found in milk, the fatty acids compete with retinol for binding to β-LG. Using intrinsic fluorescence studies, Frapin *et al.* (1993) and Dufour *et al.* (1994) suggested that an external, independent fatty-acid-binding site on the β-LG–retinol complex was in the groove between the α-helix and the β-sheets of the protein. Narayan and Berliner (1997) supported simultaneous binding of retinoids and fatty acids to β-LG. However, binding is more difficult to determine when several ligands are present.

The organic-anion-binding sites of albumin are composed of two parts: a pocket lined with non-polar amino acid chains and a cationic group located at or near the surface of the pocket (Swaney and Klotz, 1970). Most of the information available on the mechanism of binding has been obtained using organic dyes, anionic detergents and fluorinated or spin-labeled derivatives. Free fatty acid binding involves hydrophobic interactions with the hydrocarbon chain and electrostatic interactions with the carboxylate anion of BSA (Spector *et al.*, 1969).

Andersson *et al.* (1971) suggested that the fatty-acid-binding sites are located in clefts between the globular regions of the albumin polypeptide. One tryptophan is located deep inside the globular structure whereas the other is superficially located fairly accessible to solvent. Several of the strong fatty-acid-binding sites are located within 10 Å of the buried tryptophan residue (Spector, 1975). Spector *et al.* (1969) reported that palmitate and palmitoleate were bound more tightly than oleate, linoleate, stearate or myristate and much more tightly than laurate. When a long-chain hydrocarbon did not contain a free carboxyl group (methyl palmitate, cetyl alcohol and hexane), they were bound to a limited extent.

The amount of fatty acids found bound to albumin was 4.8 mol per mol and the predominant acids were oleic, palmitic and stearic acids (Pérez *et al.*, 1989). They suggested that the amount of fatty acids bound to milk albumin could be attributed to the equilibrium between the albumin that carries the fatty acids from lipolysis and the albumin that has been delivered to the tissues. However, milk albumin is not subject to this effect and the fatty acids bound are not taken up by the tissues. Although the number of high-affinity binding sites and the values of apparent association constants for fatty acids to β-LG are lower than those for albumin (Anel *et al.*, 1989), the molar concentration of β-LG in milk is much higher than that of albumin, and therefore β-LG is considered to be the main fatty-acid-binding protein in ruminant whey (Pérez *et al.*, 1990).

Recently, Barbana *et al.* (2006) reported that bovine holo-α-LA neither contains bound fatty acids *in vivo* nor has the ability to bind them *in vitro*. Cawthern *et al.* (1997) observed the lack of binding of stearic acid with bovine holo-α-LA, the fluorescent indicator acrylodated intestinal fatty-acid-binding (ADIFAB) protein. However, these results are in contrast to their spin resonance and intrinsic protein fluorescence results (Cawthern *et al.*, 1997), which showed that stearic acid was bound to holo-α-LA with a dissociation constant of $10–100 \times 10^{-6}$ M. On the other hand, interactions of apo-α-LA with fatty acids have been reported by Barbana *et al.* (2006). Bovine apo-α-LA displayed apparent affinity binding constants of 4.6×10^6 and $5.4 \times 10^5 M^{-1}$ for oleic acid and palmitic acid respectively using partition equilibrium and fluorescence spectroscopy showed a binding constant of $3.3 \times 10^6 M^{-1}$ for oleic acid. The small fluorescence changes observed for palmitic acid made it difficult to obtain a binding constant.

Sugars and polyols

The effect of sugars on the unfolding and denaturation of proteins has often been described by the steric exclusion effect (McClements, 2002; Baier and McClements, 2005). This effect occurs only if solvent and cosolvent molecules have different sizes. As micronutrients are larger than water molecules, there is a region surrounding the protein molecule from which sugars are excluded and this preferential exclusion effect has been verified in a number of studies (Hammou *et al.*, 1998).

As water is able to get into the layer surrounding the protein, a concentration gradient of the sugar molecules between the inner layer and the outer solution arises (Figure 13.2). This is a thermodynamically unfavorable situation because of the free energy that is required to maintain this concentration gradient. Subsequent movement of water molecules from the area surrounding the protein to outer parts leads to a dehydration of the protein molecule. This dehydration can result in tighter folding of the protein molecules. If the transfer free energy of the protein is greater for the native state than for the denatured state, such as in solutions with sugars, the protein is stabilized in its native state. Furthermore, McClements (2002) reported different molecular mechanisms of stabilization. Sugars may be more preferentially excluded from the denatured state than from the native state or they may be more preferentially accumulated by the native state than by the denatured state.

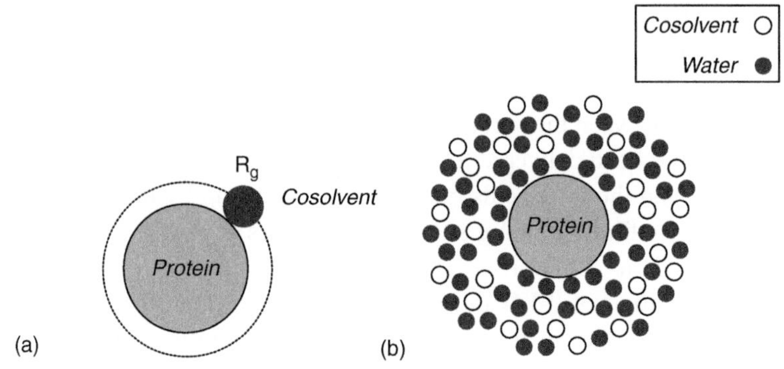

Figure 13.2 Protein-cosolvent-solvent interactions as a result of (a) steric interaction or (b) differential interaction effects. In (b), the exclusion of the cosolvent from the region surrounding the protein is clearly shown. Taken from McClements (2002). Critical Reviews in Food Science and Nutrition, Taylor and Francis Publishers (www.informaworld.com).

An alternative explanation of the effect of non-interacting species on the unfolding and denaturation of proteins has been put forward by Semenova *et al.* (2002), who proposed a direct hydrogen bonding between sugars and proteins, which results in additional hydration; however, the exclusion of sugars from the protein domain is not fully explained by this hypothesis.

The research group of Timasheff has been dominant in research into the interactions of proteins and sugars, or cosolvents as they describe them (Timasheff, 1993). Xie and Timasheff (1997) reported on the exclusion of trehalose from the domain of ribonuclease A at low temperatures. However, at 52°C, where ribonuclease is in the unfolded state, stabilization was brought about by preferential binding of the trehalose to the protein. Sugars showed similar binding affinities to the native and unfolded proteins, but had a lower affinity than water for the exposed peptide groups in the denatured protein. The authors concluded that, in the unfolded state, sugars are more preferentially excluded from the protein domain. The same group conducted a lot of earlier research showing the exclusion of water from the domains of a range of globular proteins, in the presence of sucrose (Lee and Timasheff, 1981), lactose and glucose (Arakawa and Timasheff, 1982). In all cases, they argued that the exclusion of sugars from the protein domain made unfolding of the protein less thermodynamically favorable.

The exclusion of non-interacting species from the protein domain was further shown by Lehmann and Zaccai (1984) who observed, with the aid of neutron small-angle scattering, that glycerol was excluded from the exterior of ribonuclease A at room temperature. Ebel *et al.* (2000), in a hydration study of rabbit muscle aldolase using a variety of sugars, claimed that the hydration parameter increased as the sugar size increased, indicating moderate exclusion volume effects contributions of the sugars. The effect of sugars on the structure of water may also be linked to the ability of the sugars to bind water.

A recent study comparing the effects of trehalose, maltose and sucrose on the structure of water found that trehalose binds to a larger number of water molecules than do maltose and sucrose, thus affecting the structure of water to a greater extent (Lerbret *et al.*, 2005). In a rare study involving sugars and casein, Mora-Gutierrez and Farrell (2000) also proposed preferential exclusion of sugar molecules from the casein domain, resulting in preferential hydration of the caseins. The ability of sugars to alter the heat- and pressure-induced denaturation of milk proteins is discussed further in this chapter.

Flavors

The interaction of milk proteins and volatile flavor has been reviewed in detail by Kühn *et al.* (2006) and the reader should refer to this recent review for more in-depth discussion of protein–flavor interactions. However, this section covers the area briefly. A number of flavor compounds are known to bind to milk proteins. Despite this wide knowledge, there are large discrepancies in the binding data because of the use of different methodologies, which appears to be a common feature of determining binding constants.

β-LG is known to interact with a variety of flavor compounds including ionones (Dufour and Haertle, 1990b; Jouenne and Crouzet, 2000; Jung and Ebeler, 2003), lactones (Sostmann and Guichard, 1998; Guth and Fritzler, 2004), alkanes (Mohammadzadeh *et al.*, 1967, 1969a, 1969b), aldehydes (van Ruth *et al.*, 2002), esters and ketones (Guichard and Langourieux, 2000; Jouenne and Crouzet, 2000). In contrast to β-LG, very few studies have explored flavor binding to α-LA. The binding of aldehydes and methyl ketones (Franzen and Kinsella, 1974) and 2-nonanone and 2-nonanal (Jasinski and Kilara, 1985) to α-LA has been explored. BSA has been shown to bind alkanes (Mohammadzadeh *et al.*, 1967, 1969a, 1969b) and Damodaran and Kinsella (1980a, 1980b, 1981) studied interactions of 2-nonanone and BSA. Jasinski and Kilara (1985) compared the binding of 2-nonanone and nonanal to BSA.

The binding of flavors to caseins or sodium caseinate has also received some attention, including the binding of diacetyl (Reineccius and Coulter, 1969), vanillin (McNeill and Schmidt, 1993), β-ionone, n-hexanol, ethylhexanoate and isoamyl acetate (Voilley *et al.*, 1991) to sodium caseinate.

The most extensively studied flavor compound is 2-nonanone. Thus the binding strengths of this flavor to the whey proteins can be compared. Although different authors have reported different affinity constants, the trend BSA $>$ β-LG $>$ α-LA occurs. Table 13.1 illustrates the interactions between 2-nonanone and various milk proteins (Kühn *et al.*, 2006).

Minerals

The abilities of certain milk proteins, in particular the caseins, to bind calcium are extremely well known. The extent of binding of the caseins is directly related to the number of phosphoserine residues and thus follows the order α_{s2}- $>$ α_{s1}- $>$ β- $>$ κ-casein (Rollema, 1992). Increased binding of calcium to the caseins results in reduced negative charges on the casein molecule, resulting in diminished electrostatic repulsion and consequently inducing precipitation.

Table 13.1 Binding data for the interactions between 2-nonanone and milk proteins (25°C): *n*, number of binding sites per monomer; *K*, intrinsic binding constant; reproduced with the permission of Kühn *et al.* (2006); copyright 2006 *Journal of Food Science*, Institute of Food Technologists Wiley-Blackwell Publishing Ltd

	n	K (M^{-1})	Method	Reference
WPC[a]	61	1 920 000	Equilibrium dialysis	Jasinski and Kilara (1985)
	0.2	53 000 000	Fluorescence spectroscopy	Liu *et al.* (2005)
WPI[b]	1	2059	Headspace SPME[c]	Zhu (2003)
Sodium caseinate	0.3	1858	Headspace SPME	Zhu (2003)
β-LG	1	2439	Equilibrium dialysis	O'Neill and Kinsella (1987)
	0.2	6250 (≤40 ppm)	Static headspace analysis	Charles *et al.* (1996)
	0.5	1667 (≥45 ppm)		
	14	122	Equilibrium dialysis	Jasinski and Kilara (1985)
α-LA	33	11	Equilibrium dialysis	Jasinski and Kilara (1985)
BSA	5–6	1800	Liquid–liquid partitioning	Damodaran and Kinsella (1980a)
	15	14 100	Equilibrium dialysis	Jasinski and Kilara (1985)
	7	833	PFG-NMR[d] spectroscopy	Jung *et al.* (2002)

[a] WPC, whey protein concentrate
[b] WPI, whey protein isolate
[c] SPME, solid phase microextraction
[d] PFG-NMR, pulsed-field gradient NMR

Caseins with low numbers of phosphoserine residues, such as α_{s1}-casein B, α_{s1}-casein C and the α_{s2}-caseins, are insoluble in Ca^{2+} concentrations above about 4 mM. However, β-casein is soluble at high concentrations of Ca^{2+} (0.4 M) at temperatures below 18°C, but β-casein is very insoluble above 18°C, even in the presence of low concentrations of Ca^{2+} (4 mM). κ-Casein, with only one phosphoserine, binds little calcium and remains soluble in Ca^{2+} at all concentrations. Although κ-casein does not bind calcium to any great extent, its ability to stabilize α_{s1}-, α_{s2}- and β-caseins against precipitation by Ca^{2+} is well known and plays a large part in the stabilization of the casein micelle. This is discussed in more detail in Chapter 5 in this volume.

Whereas intact casein has been shown to bind zinc and calcium, tryptic hydrolysates of α_{s1}-, α_{s2}-, β- and κ-caseins also display mineral-binding properties. Termed caseinophosphopeptides (CPPs), these peptides can bind and solubilize high concentrations of calcium because of their highly polar acidic domain. Consumption of high concentrations of calcium in early life contributes to the development of maximal bone density, which in turn can prevent osteoporosis in later life (FitzGerald and Meisel, 2003). In addition, calcium-binding CPPs can have an anti-cariogenic effect in that they inhibit caries lesions through recalcification of the dental enamel (FitzGerald, 1998). CPPs have also been reported to improve the intestinal absorption of zinc, as studied using an isolated perfused rat intestinal loop system (Peres *et al.*, 1998).

Lactoferrin has the ability to bind iron very strongly. *In vivo*, the ferric III form of iron is bound to lactoferrin (Anderson *et al.*, 1989). Considerable interest has been expressed in supplementing bovine-milk-based infant formulas with lactoferrin, as

Table 13.2 Interaction of milk proteins and trace minerals

Milk protein	Mineral	Reference
β-LG A	Chromium	Divsalar *et al.* (2006a)
β-LG A and B	Lead	Divsalar *et al.* (2005)
Caseins and β-LG	Mercury	Mata *et al.* (1997)
α-LA	Copper	Permyakov *et al.* (1988)

bovine milk contains much lower levels of lactoferrin than human milk and lactoferrin, isolated from human milk, can bind two moles of iron per mole of protein (Bezwoda and Mansoor, 1986). The biological importance of lactoferrin has been reviewed recently by Lönnerdal (2003). Nagasko *et al.* (1993) reported that lactoferrin can bind iron at sites other than its chelate-binding sites, probably on the surface of the molecule.

Wieczorek *et al.* (1992) studied the interaction of milk proteins with fluoride ions over a range of pH values. They found that fluoride does not interact with α-, β- or κ-casein. However, interaction between α-LA and fluoride ions was observed at pH 3.9. Wieczorek *et al.* (1994) subsequently showed that both α-LA and apo-α-LA failed to bind fluoride ions at pH 4.6, but did at pH 3.7. Other studies involving interactions of minerals/ions and milk proteins are listed in Table 13.2.

Surfactants

Some interesting non-food-grade ligands, e.g. 1-anilino-8-naphthalenensulfonate (ANS), sodium dodecyl sulfate (SDS) and alkysulfonate (AL) ligands have been widely used to study ligand binding to β-LG. These ligands provide useful information regarding the structure of the molecule. SDS is an amphiphilic ligand that binds strongly to a small number of sites on β-LG at low SDS concentration (Ray and Chatterjee, 1967; Jones and Wilkinson, 1976; Lamiot *et al.*, 1994).

Seibles (1969) studied the interaction of dodecyl sulfate with three of the variants of β-LG. Two moles of dodecyl sulfate were bound to all of the genetic variants of β-LG studied. Creamer (1995) demonstrated that SDS had a profound effect on the equilibrium unfolding of bovine β-LG by maintaining β-LG in the native confirmation despite high concentrations of urea.

Busti *et al.* (1999) examined the interaction of AL with β-LG and demonstrated one binding site per molecule. Their results suggested that the protective action of AL was exerted by the fraction of the AL bound to the monomer. Also, the efficiency of AL stabilizing native β-LG was related to the length of the hydrocarbon tail. Protein–surfactant interactions are greatly influenced by the surfactant's hydrophilic group and aliphatic chain length (Ananthapadmanabhan, 1993).

Waninge *et al.* (1998) showed a substantial increase in unfolding temperature of β-LG–SDS complex at a molar ratio of 1:1. In contrast, a decrease in the unfolding temperature was observed with the addition of an anionic surfactant dodecyl trimethyl ammonium chloride (DOTAC) at a similar ratio. Whereas DOTAC was easily removed via dialysis, it was impossible to remove SDS by dialysis.

Lu *et al.* (2006) showed that the anionic surfactant sodium perfluorooctanoate was a strong denaturant of β-LG. However, the denaturing ability of sodium perfluorooctanoate could be tempered with cationic surfactants, such as alkyl trimethyl ammonium bromide. Sodium n-alkyl sulfates (sodium octyl sulfate, decyl sulfate and dodecyl sulfate) were shown to induce non-native α-helical intermediates of thiol-modified β-LG (Chamani, 2006). Maulik *et al.* (1996) observed the binding of cetyl trimethyl ammonium bromide with β-LG and reported a two-stage interaction by first-order kinetics. Tetradecyl trimethyl ammonium bromide (TTAB) was also shown to interact with α-LA and cause protein unfolding below TTAB concentrations of 2 mM (Housaindokht *et al.*, 2001).

The interactions of sucrose esters with the casein micelle (Fontecha and Swaisgood, 1995) and β-casein (Clark *et al.*, 1992) have also been studied. Creamer (1980) examined the effect of SDS on the β-casein self-association. The results indicated that SDS binds on an external site of β-casein, such that the hydrophobic tail of SDS becomes involved in the casein self-association. This is supported by the lack of displacement of ANS by SDS. It was also postulated that SDS binds to sites in or on the protein such that the amino acid residues involved in the self-association reaction can interact more favorably with one another. At low concentrations, SDS is thought to bind to a limited number of sites. Despite the increase in the negative charge of the protein, the normal monomer–polymer equilibrium moves predominantly to the polymer in solution whereas, at high concentrations, only protein monomers are present.

Phospholipids

Most of the work involving interactions of phospholipids and milk proteins has arisen because of interest in molecular assembly to replicate cell walls. Bos and Nylander (1996) concluded from their studies that both electrostatic and hydrophobic interactions are important for the incorporation of β-LG into phospholipid monolayers. Contrasting effects of β-LG on two zwitterionic phospholipids was reported by Lefevre and Subirade (1999). Whereas dipalmitoyl phosphatidylcholine was unaffected by the presence of β-LG, the conformational disorder of milk sphingomyelin was increased due to hydrophobic interactions with β-LG. Lefevre and Subirade (2001) also studied the effect of phospholipids on the structure of β-LG.

Montich and Marsh (1995) showed, using electron spin resonance, that binding of α-LA to dimyristoyl phosphatidylglycerol bilayers at pH 4.0 caused a motional restriction throughout the full length of the lipid acyl chain. The authors concluded that this restriction in motion would be of direct relevance to the insertion and translocation of protein in the molten globule state across lipid membranes. Earlier work had shown interaction between the milk proteins α-LA and β-LG and phospholipids in monolayer or vesicle formation (Cornell and Patterson, 1989; Kim and Kim, 1989; Cornell *et al.*, 1990).

Other micronutrients

Some of the milk proteins, most particularly the whey proteins, have been used as model proteins in studies involving a range of other micronutrients. The interaction of small heat-shock proteins, such as alpha-crystallin, prevents the precipitation of α-LA when in the molten globule state (Lindner *et al.*, 1997). This finding was confirmed by Sreelakshmi and Sharma (2001) who found that the active site of alpha-cystallin by itself can maintain a significantly denatured and unfolded protein in soluble form. Zhang *et al.* (2005) reported on the chaperone-like activity of β- and α-caseins. β-Casein was found to be able to suppress the thermal and chemical aggregation of insulin, lysozyme and catalase.

The use of milk proteins as chaperones for drugs has also been studied. The interaction of chlorpromazine with β-LG and α_s-casein affected the proteins in different ways. Far UV CD studies revealed that chlorpromazine increased the secondary structure of β-LG, whereas the structure of casein became further disordered (Bhattacharyya and Das, 2001). Divsalar *et al.* (2006b) also reported on the interaction between genetic variants of β-LG and an anti-cancer component.

Effect of processing on milk protein structure

Heat has been used extensively in food processing for centuries and is a widely applied treatment in food production, primarily for the control of microbial populations. Fields of application are pasteurization under mild temperatures and sterilization under higher temperatures. However, heating may also affect texture as well as taste development and may result in flavor and color changes. The latter effects are often described as disadvantages of heat treatment. Changes in the organoleptic properties are generally as a result of structural changes occurring within the constituents of the food, namely the proteins, polysaccharides or fats.

Another technology that is similar in its control of the microbial population of food products is high-pressure treatment. Foods are preserved with minor changes in texture, flavor or color, in contrast to heat processing, and high pressure is usually called cold preservation technology. High pressure is a long-used technique in Japan and has also become increasingly popular worldwide. However, high-pressure treatment may cause some conformational and structural changes to the individual constituents of the food, possibly resulting in altered functional and organoleptic properties.

Heat and high-pressure treatment may cause the denaturation of globular whey proteins such as β-LG; although there may be differences in the mechanisms behind the denaturation process, the general process appears to be similar. These processes have been examined in detail in Chapter 7.

The denaturation of whey proteins during the heat treatment of milk, the interactions of the denatured whey proteins with other milk components and the effect of these reactions on the physical and functional properties of milk products have been extensively reported and reviewed in great detail (Singh and Creamer, 1992; O'Connell and Fox, 2003; Singh and Havea, 2003). Recent studies have shown that heat-induced aggregation and gelation occur along detailed pathways and are influenced by the types of proteins and forces (disulfide bonding and hydrophobic interactions) present (Havea *et al.*, 1998, 2000, 2001; Schokker *et al.*, 1999, 2000; Abbasi and Dickinson, 2002).

There is an increasing number of studies investigating the effect of high pressure on whole milk and individual constituents; they have been reviewed by Huppertz *et al.* (2002) and Trujillo (2002). Much of the research interest has focused on the effect of pressure treatment on the physical and functional properties of milk products; however, some studies have dealt with the changes to the individual proteins during the denaturation process. There have been numerous articles citing the mechanisms of heat and pressure denaturation of milk proteins and interested readers are referred to the recent reviews of Huppertz *et al.* (2006) and Considine *et al.* (2007a), which cite most of these articles.

Protein denaturation by thermal and pressure treatments and effect of micronutrients

As a consequence of their lack of defined secondary and tertiary structure, the caseins have not been suitable candidates for observing changes in protein denaturation. In contrast, the whey proteins have been studied widely as a model globular protein because of their well-defined secondary and tertiary structures, as outlined above.

Interactions between whey proteins and other species induced by either heat treatment or pressure treatment may be divided into two separate classes: covalent interactions and non-covalent interactions. The most important covalent interaction involving whey proteins upon storage is their reaction with reducing sugars via the Maillard reaction to form discolored protein powders, which also have reduced solubilities and nutritional properties. Non-covalent interactions can also occur; these, too, may lead to a loss of protein solubility after association of the proteins with polysaccharides and these non-covalent interactions are primarily driven by reversible electrostatic interactions.

In the present work, the effects of non-interacting species on the unfolding and structural transitions of whey proteins are of specific interest. The marked increase in the thermal and conformational stability of globular proteins in aqueous media in the presence of sugars is well known and has been extensively studied.

Processing treatments involving ligands

Whereas the majority of studies have explored ligand binding after heating or pressure treatment of β-LG, only a few studies have looked at the effect of various ligands during the processing treatment (Dufour *et al.*, 1994; Stapelfeldt and Skibsted, 1999; Considine *et al.*, 2005a, 2005b, 2007b). The studies of Considine *et al.* (2005a, 2005b, 2007b) showed that ligands can retard the heat or pressure denaturation of β-LG, with the type of ligand having an impact on this process. For example, during the heat denaturation of β-LG, both SDS and palmitate stabilized the native structure of β-LG against heat-induced structural flexibility, subsequent unfolding and denaturation up to approximately 70°C, whereas both retinol and ANS provided very little stabilization. When a similar range of ligands was used during pressure denaturation, a similar effect was noted, i.e. higher pressures were required to cause unfolding of β-LG when a ligand was present (Figure 13.3).

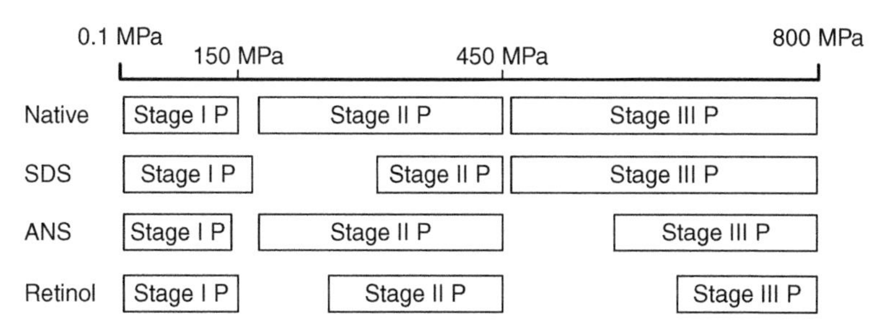

Figure 13.3 Proposed three-stage model of the pressure denaturation of β-LG B and β-LG B with added ANS, retinol or SDS. Reproduced with the permission of Considine *et al.* (2005b). Copyright 2005 *Journal of Agricultural and Food Chemistry*.

It was noted in these studies, and in the comparison study of heat and pressure using myristate and conjugated linoleic acid as ligands, that β-LG unfolds slightly differently with respect to the type of treatment (Figure 13.3). This mechanism is discussed in detail elsewhere (Considine *et al.*, 2005a, 2005b, 2007b).

Celej *et al.* (2005) compared the effects of the binding of two ANS derivatives, namely 1,8-ANS and 6-anilinonaphthalene-2-sulfonic acid (2,6-ANS), on BSA thermostability. 1,8-ANS had a stronger effect on BSA thermal stability and they also indicated that the binding parameters are quite different and suggested that stereochemistry is also an important factor in determining protein–ligand interactions. Thus electrostatic interactions should also be considered along with hydrophobic interactions. The authors emphasized the importance of free ligand concentration rather than ligand to protein mole ratio when determining protein stability.

As discussed earlier, the binding of retinol to casein is through hydrophobic interactions (Poiffait and Adrian, 1991). β-Casein is the most hydrophobic casein and has a highly charged N-terminal domain, containing an anionic phosphoserine cluster, that is clearly distinct from a very hydrophobic C-terminal domain (Swaisgood, 2003). There has been little work on the ability of the caseins to bind retinol. However, Poiffait and Adrian (1991) reported that casein plays an important role in stabilizing retinol over time or during heat treatment. However, information in this area is limited.

Processing treatments involving sugars

In an extensive study, Garrett *et al.* (1988) examined the thermal denaturation of the individual whey proteins β-LG and α-LA and unfractionated whey protein in the presence and absence of sucrose using UV absorption and light scattering methods. The authors found that, in the presence of sucrose, there was increased exposure of the tryptophan residues of both β-LG and whey protein after heating to 90°C. It must be noted that the whey protein results were based on soluble protein only—insoluble protein that formed after heating had been removed centrifugally. The authors postulated that the hydrophobic groups in the protein's interior had paired up with sucrose after the protein had undergone a transitionary conformational change following heat treatment. In this way, aggregation of the protein was inhibited, although the sugars were reported to have promoted conformational changes.

The conformational-promoting properties of sugars as reported by Garrett *et al.* (1988) were contrary to the conformational-retaining properties of sugars as found by Boye and Alli (2000), who reported on the thermal denaturation (by differential scanning calorimetry [DSC]) of 1:1 mixtures of α-LA and β-LG in the presence of a range of sugars. They found that sugars protected against heat-induced denaturation and the protection offered was in the order galactose = glucose > fructose = lactose > sucrose > absence of sugars for increases in the thermal transition temperature of β-LG. No significant effects of sugar were observed with apo-α-LA. Interestingly, an earlier study by the same authors solely on α-LA found an increase in the thermal transition temperature of both the apo and holo forms of α-LA when either 50% sucrose or 50% glucose was added; this increase was fully reversible in the holo form, but only partly reversible in the apo form (Boye *et al.*, 1997). The thermal transition temperature of β-LG was found to be increased in the presence of sucrose, lactose and glucose at 10–50% (Boye *et al.*, 1996b).

Jou and Harper (1996), using the DSC technique, found an increase in the thermal transition temperature of whey protein concentrates following the addition of sugars, with the protection offered by the sugars in the order maltose > trehalose > sucrose. Lactose was also found to provide some protection against heat-induced denaturation. A similar increased thermal stability effect on the heat denaturation temperature of β-LG was observed for sorbitol (Harwalkar and Ma, 1992).

Dierckx and Huyghebaert (2002) followed the heat-induced gelation of a whey protein isolate solution using DSC and dynamic rheology. They found that, by adding increasing concentrations of sucrose or sorbitol, both the thermal transition temperature of the protein denaturation process and the gelation temperature were increased, with a linear relationship existing between the transition and gelation temperatures. They suggested that, because of the differences in the gelation mechanisms observed at different pH values, sucrose and sorbitol affected protein–protein interactions in gels through enhancement of hydrophobic interactions.

Kulmyrzaev *et al.* (2000) had conducted an earlier study on the effect of sucrose on the thermal denaturation, gelation and emulsion stabilization of whey protein isolate and, although they also observed increases in the thermal transition temperatures on the addition of increasing concentrations of sucrose and improved gel formation and enhanced emulsification flocculation, they postulated that sucrose played different roles in a pre-denatured (improved heat stability) and a post-denatured (enhanced protein–protein interactions) whey protein solution system.

In a study on the effects of different lactose concentrations (within a naturally occurring range) on the formation of whey protein microparticulates, Spiegel (1999) put forward a two-stage process in the aggregation of whey proteins. Up to approximately 85°C, the aggregation of whey proteins is limited by the slow unfolding of the individual proteins; above 100°C, however, aggregation is the rate-limiting step as the rate of unfolding is high. Lactose (at 500 mM) was also found to increase the temperature of the denaturation of whey protein isolate at pH 9.0 by approximately 3°C. However, the authors realized the effect that the Maillard reaction was having in these systems, a factor that some reports seem to ignore.

Baier and McClements (2001) found that increased concentrations of sucrose (up to 40%) could increase the thermal stability of BSA; gels formed from these systems had a higher gelation temperature and a lower complex shear modulus. Similar effects were found in a more recent study (Baier and McClements, 2003). A further study by the same group (Baier *et al.*, 2004) showed that 40% glycerol increased the temperature of gelation of BSA, but no change in the temperature of denaturation of BSA with increasing concentration of glycerol was detected.

Only some studies have proposed that heat-induced protein unfolding is promoted in the presence of sugars, with the denatured state formed being less susceptible to aggregation than the denatured state formed under normal heating conditions. It appears that this issue has yet to be fully resolved.

Some early DSC work (Dumay *et al.*, 1994) showed that the presence of 5% sucrose was enough to reduce the extent of β-LG unfolding by 22% following high-pressure treatment at 450 MPa for 15 min.In a subsequent study, Dumay *et al.* (1998) found that adding sucrose to β-LG solutions prior to pressure-induced gelation resulted in gels with decreased pore size and strand thickness. They attributed this to a reduction in the number of protein–protein interactions occurring under the influence of pressure.

Keenan *et al.* (2001) found that low concentrations of sucrose aided in the pressure-induced gel formation of a range of milk-protein-containing systems, but that gel formation was reduced at higher sucrose concentrations. In another group of studies, the pressure-induced gelation properties of skim milk powder were found to be improved by adding low concentrations of sucrose, glucose or fructose, whereas high (45–50%) sugar concentrations inhibited gel formation (Abbasi and Dickinson, 2001).

Boye *et al.* (1996a) found that lactose, sucrose and glucose increased the temperature of denaturation of BSA, with 50% glucose having a greater stabilizing effect than 50% sucrose. Wendorf *et al.* (2004) studied the ability of different proteins (ribonuclease A, BSA and egg white lysozyme) to adsorb to a liquid–solid interface in the presence of a range of sugars. They found that the ability of sugars to reduce protein adsorption followed the trend trisaccharides > disaccharides > 6-carbon polyols > monosaccharides and this was explained by the stabilization of the protein in the native state in solution.

Other studies have also shown the beneficial effects of sugars in protecting against denaturation induced by freeze drying, spray drying and chemicals. At low temperatures, high concentrations of sugars cause a substantial increase in solution viscosities and can thus affect protein denaturation. Tang and Pikal (2005) showed that, by negating the thermal stabilizing effects of sucrose by adding denaturants, the increased stability of β-LG in the freeze drying process could be directly attributed to a viscosity effect. Murray and Liang (1999) explored the addition of sucrose, trehalose, lactose and lactitol to whey protein concentrate solutions prior to spray drying and found that the foaming properties of the spray-dried powders were dramatically decreased when sugars were absent. Trehalose was particularly successful in retaining the original foaming properties of both whey protein concentrate and β-LG, but did not perform as well in spray-dried BSA powders (Murray and Liang, 1999). Trehalose has also been found to be effective in stabilization against chemical denaturants.

Conclusions

The interaction of milk proteins with various micronutrients is primarily governed by the physico-chemical properties of the protein. The whey proteins, with extensive secondary and tertiary structure and significant hydrophobicity (albeit largely shielded in the native form), tend towards hydrophobic interactions with ligands and fatty acids. Steric exclusion effects govern the interaction of sugars and polyols with proteins, thus affecting their denaturing properties in the presence of pressure or heat. Electrostatic interactions drive the association of minerals and proteins.

In the food industry, an increasing emphasis is being placed on foods that will have a physiologically functional benefit, in addition to the nutritional benefit of the food. This is being driven by consumers who are becoming increasingly more health aware and health responsible. The challenge for the food scientist is now to deliver the required physiologically functional activities into the final food product, while retaining product quality and shelf life. Knowledge of the interactions of these micronutrients with milk proteins, a major component in many food products, is necessary to achieve this aim. A relevant example of this is a recent patent by Swaisgood *et al.* (2001) who described the potential of using β-LG as a protein ingredient for carrying lipophilic nutrients such as a range of vitamins, cholesterol and CLA. The binding constants for linoleic acid, CLA and CLA's methyl ester (CLAME) were determined by fluorescence. The use of these β-LG complexes in foods, especially low-fat foods and low-fat dairy foods, is discussed.

A further use of a β-LG–nutrient complex may be in the personal care/hygiene sector. Topical formulations would contain complexes of β-LG with vitamins A or E or CLA and other essential fatty acids. This carrier could be used in a variety of forms including sprays, emulsions, mousses, liquids, creams, oils, ointments and gels. In addition, protein structure can be tailored, by using processing treatments to induce structural changes that may lead to increased interactions with micronutrients.

References

Abbasi, S. and Dickinson, E. (2001). Influence of sugars on high-pressure induced gelation of skim milk dispersions. *Food Hydrocolloids*, **15**, 315–19.

Abbasi, S. and Dickinson, E. (2002). Influence of high-pressure treatment on gelation of skim milk powder plus low methoxyl pectin dispersions. *High Pressure Research*, **22**, 643–47.

Adrian, J., Frangne, R. and Rabache, M. (1984). Rôle des protéines laitières sur l'efficacité du rétinol. *Sciences des Aliments*, n° Hors série III, 305–08.

Ananthapadmanabhan, K. P. (1993). Protein-surfactant interactions. In *Interactions of Surfactants with Polymers and Proteins* (E. D. Goddard and K. P. Ananthapadmanabhan, eds) pp. 319–66. Boca Raton, Florida: CRC Press.

Anderson, B. F., Baker, H. M., Norris, G. E., Rice, D. W. and Baker, E. N. (1989). Structure of human lactoferrin: crystallographic structure analysis and refinement at 2.8 Å resolution. *Journal of Molecular Biology*, **209**, 711–34.

Anderson, K. J. and von der Lippe, G. (1979). The effect of proteolytic enzymes on the vitamin B12-binding proteins of human gastric juice and saliva. *Scandinavian Journal of Gastroenterology*, **18**, 833–38.

Andersson, L. O., Brandt, J. and Johansson, S. (1971). The use of trinitrobenzenesulfonic acid in studies on the binding of fatty acid anions to bovine serum albumin. *Archives of Biochemistry and Biophysics*, **146**, 428–40.

Anel, A., Calvo, M., Naval, J., Iturralde, M., Alava, M. A. and Pineiro, A. (1989). Interaction of rat α-fetoprotein and albumin with polyunsaturated and other fatty acids: determination of apparent association constants. *FEBS Letters*, **250**, 22–24.

Arakawa, T. and Timasheff, S. N. (1982). Stabilization of protein-structure by sugars. *Biochemistry*, **21**, 6536–44.

Arymard, P., Durand, D. and Nicolai, T. (1996). The effect of temperature and ionic strength on the dimerization of β-lactoglobulin. *International Journal of Biological Macromolecules*, **19**, 213–21.

Baier, S. and McClements, J. (2001). Impact of preferential interactions on thermal stability and gelation of bovine serum albumin in aqueous sucrose solutions. *Journal of Agricultural and Food Chemistry*, **49**, 2600–8.

Baier, S. K. and McClements, D. J. (2003). Combined influence of NaCl and sucrose on heat-induced gelation of bovine serum albumin. *Journal of Agricultural and Food Chemistry*, **51**, 8107–12.

Baier, S. and McClements, D. J. (2005). Influence of cosolvent systems on the gelation mechanism of globular protein: thermodynamic, kinetic, and structural aspects of globular protein gelation. *Comprehensive Reviews in Food Science and Safety*, **4**, 43–53.

Baier, S. K., Decker, E. A. and McClements, D. J. (2004). Impact of glycerol on thermostability and heat-induced gelation of bovine serum albumin. *Food Hydrocolloids*, **18**, 91–100.

Ball, G. F. M. (1988). *Fat-soluble Vitamin Assays in Food Analysis. A Comprehensive Review.* London: Elsevier Applied Sciences.

Banville, C., Vuillemard, J. C. and Lacroix, C. (2000). Comparisons of different methods of fortifying Cheddar cheese with vitamin D. *International Dairy Journal*, **10**, 375–82.

Barbana, C., Pérez, M. D., Sánchez, L., Dalgalarrondo, M., Chobert, J. M., Haertlé, T. and Calvo, M. (2006). Interaction of bovine α-lactalbumin with fatty acids as determined by partition equilibrium and fluorescence spectroscopy. *International Dairy Journal*, **16**, 18–25.

Bezwoda, W. R. and Mansoor, N. (1986). Isolation and characterisation of lactoferrin separated from human whey by absorption chromatography using Cibacron Blue F3 G-A liked affinity adsorbent. *Clinical Chimica Acta*, **157**, 89–94.

Bhattacharyya, J. and Das, K. P. (2001). Interaction of chlorpromazine with milk proteins. *Molecular and Cellular Biochemistry*, **221**, 11–15.

Bos, M. A. and Nylander, T. (1996). Interaction between β-lactoglobulin and phospholipids at the air/water interface. *Langmuir*, **12**, 2791–97.

Boye, J. I. and Alli, I. (2000). Thermal denaturation of mixtures of alpha-lactalbumin and β-lactoglobulin: a differential scanning calorimetric study. *Food Research International*, **33**, 673–82.

Boye, J. I., Alli, I. and Ismail, A. A. (1996a). Interactions involved in the gelation of bovine serum albumin. *Journal of Agricultural and Food Chemistry*, **44**, 996–1004.

Boye, J. I., Ismail, A. A. and Alli, I. (1996b). Effects of physicochemical factors on the secondary structure of β-lactoglobulin. *Journal of Dairy Research*, **63**, 97–109.

Boye, J. I., Alli, I. and Ismail, A. (1997). Use of differential scanning calorimetry and infrared spectroscopy in the study of thermal and structural stability of alpha-lactalbumin. *Journal of Agricultural and Food Chemistry*, **45**, 1116–25.

Busti, P., Scarpeci, S., Gatti, C. A. and Delorenzi, N. J. (1999). Interaction of alkylsulfonate ligands with β-lactoglobulin AB from bovine milk. *Journal of Agricultural and Food Chemistry*, **47**, 3628–31.

Calvo, M., Perez, M. D., Oria, R., Aranda, P., Ena, J. M. and Sanchez, L. (1990). In Abstracts of Papers, 20th Meeting of the Federation of European Biochemical Societies, Budapest, August.

Cawthern, K. M., Narayan, M., Chaudhuri, C., Permyakov, E. A. and Berliner, L. J. (1997). Interactions of α-lactalbumin with fatty acids and spin label analogs. *Journal of Biological Chemistry*, **272**, 30812–16.

Celej, M. S., Dassie, S. A., Freire, E., Bianconi, M. L. and Fidelio, G. D. (2005). Ligand-induced thermostability in proteins: thermodynamic analysis of ANS-albumin interaction. *Biochimica et Biophysica Acta*, **1750**, 122–33.

Chamani, J. (2006). Comparison of the conformational stability of the non-native α-helical intermediate of thiol-modified β-lactoglobulin upon interaction with sodium n-alkyl sulfates at two different pH. *Journal of Colloid and Interface Science*, **299**, 636–46.

Charles, M., Bernal, B. and Guichard, E. (1996). Interaction of β-lactoglobulin with flavor compounds. In *Flavor Science: Recent Developments* (A. J. Taylor and D. S. Mottram, eds) pp. 433–36. Cambridge: Royal Society of Chemistry.

Cho, Y. J., Batt, C. A. and Sawyer, L. (1994). Probing the retinol-binding site of bovine β-lactoglobulin. *Journal of Biological Chemistry*, **269**, 11102–7.

Clark, D. C., Wilde, P. J., Wilson, D. R. and Wustneck, R. (1992). The interaction of sucrose esters with β-lactoglobulin A and β-casein from bovine milk. *Food Hydrocolloids*, **6**, 173–86.

Considine, T., Patel, H. A., Singh, H. and Creamer, L. K. (2005a). Influence of binding of sodium dodecyl sulfate, all-*trans*-retinol, palmitate, and 8-anilino-1-naphthalene-sulfonate on the heat-induced unfolding and aggregation of β-lactoglobulin B. *Journal of Agricultural and Food Chemistry*, **53**, 3197–205.

Considine, T., Singh, H., Patel, H. A. and Creamer, L. K. (2005b). Influence of binding of sodium dodecyl sulfate, all-*trans*-retinol, and 8-anilino-1-naphthalenesulfonate on the high-pressure-induced unfolding and aggregation of β-lactoglobulin B. *Journal of Agricultural and Food Chemistry*, **53**, 8010–18.

Considine, T., Patel, H. A., Anema, S. G., Singh, H. and Creamer, L. K. (2007a). Interactions of milk proteins during heat and high hydrostatic pressure treatments – a review. *Innovative Food Science and Emerging Technologies*, **8**, 1–23.

Considine, T., Patel, H. A., Singh, H. and Creamer, L. K. (2007b). Influence of binding of conjugated linoleic acid and myristic acid on the heat- and pressure- induced unfolding and aggregation of β-lactoglobulin B. *Food Chemistry*, **102**, 1270–80.

Cornell, D. G. and Patterson, D. L. (1989). Interaction of phospholipids in monolayers with β-lactoglobulin adsorbed from solution. *Journal of Agricultural and Food Chemistry*, **37**, 1455–59.

Cornell, D. G., Patterson, D. L. and Hoban, N. (1990). The interaction of phospholipids in monolayers with bovine serum-albumin and α-lactalbumin adsorbed from solution. *Journal of Colloid and Interface Science*, **140**, 428–35.

Creamer, L. (1980). A study of the effect of sodium dodecyl sulfate on bovine β-casein self-association. *Archives of Biochemistry and Biophysics*, **199**, 172–78.

Creamer, L. K. (1995). Effect of sodium dodecyl sulfate and palmitic acid on the equilibrium unfolding of bovine β-lactoglobulin. *Biochemistry*, **34**, 7170–76.

Dai-Dong, J. X., Novak, G. and Hardy, J. (1990). Stabilization of vitamin C by β-lactoglobulin during heat treatment. *Sciences des Aliments*, **10**, 393–401.

Damodaran, S. and Kinsella, J. E. (1980a). Flavor Protein Interactions. Binding of Carbonyls to Bovine Serum Albumin – Thermodynamic and Conformational effects. *Journal of Agricultural and Food Chemistry*, **28**, 567–71.

Damodaran, S. and Kinsella, J. E. (1980b). Stabilization of proteins by solvents – effect of pH and anions on the positive cooperativity of 2-nonanone binding to bovine serum albumin. *Journal of Biological Chemistry*, **255**, 8503–08.

Damodaran, S. and Kinsella, J. E. (1981). The effect of neutral salts on the stability of macromolecules – a new approach using a protein-ligand binding system. *Journal of Biological Chemistry*, **256**, 3394–98.

Diaz de Villegas, C., Oria, R., Sala, F. J. and Calvo, M. (1987). Lipid binding by β-lactoglobulin of cow milk. *Milchwissenschaft*, **42**, 357–58.

Dierckx, S. and Huyghebaert, A. (2002). Effects of sucrose and sorbitol on the gel formation of a whey protein isolate. *Food Hydrocolloids*, **16**, 489–97.

Divsalar, A. and Saboury, A. A. (2005). Comparative structural and conformational studies on two forms of beta lactoglobulin (A and B) upon interaction with lead ion. *FEBS Journal*, **272**, 340.

Divsalar, A., Saboury, A. A. and Moosavi-Movahedi, A. A. (2006a). A study on the interaction between beta lactoglobulin-A and a new antitumor reagent (2,2-bipyridinglycinato Pd (ii) chloride). *FEBS Journal*, **273**, 333.

Divsalar, A., Saboury, A. A., Mansoori-Torshizi, H. and Hemmatinejad, B. (2006b). Comparative and structural analysis of the interaction between β-lactoglobulin type A and B with a new anticancer component (2,2'-bipyridin n-hexyl dithiocarbamato Pd(II) nitrate). *Bulletin of the Korean Chemical Society*, **27**, 1801–8.

Dufour, E. and Haertle, T. (1990a). Alcohol induced changes of β-lactoglobulin-retinol-binding stoichiometry. *Protein Engineering*, **4**, 185–90.

Dufour, E. and Haertle, T. (1990b). Binding affinities of β-ionone and related flavor compounds to β-lactoglobulin – effects of chemical modifications. *Journal of Agricultural and Food Chemistry*, **38**, 1691–95.

Dufour, E. and Haertle, T. (1991). Binding of retinoids and β-carotene to β-lactoglobulin. Influence of protein modifications. *Biochimica et Biophysica Acta*, **1079**, 316–20.

Dufour, E., Hui Bon Hoa, G. and Haertle, T. (1994). High-pressure effects on β-lactoglobulin interactions with ligands studied by fluorescence. *Biochimica et Biophysica Acta*, **1206**, 166–72.

Dumay, E. M., Kalichevsky, M. T. and Cheftel, J. C. (1994). High-pressure unfolding and aggregation of β-lactoglobulin and the baroprotective effects of sucrose. *Journal of Agricultural and Food Chemistry*, **42**, 1861–68.

Dumay, E. M., Kalichevsky, M. T. and Cheftel, J. C. (1998). Characteristics of pressure-induced gels of β-lactoglobulin at various times after pressure release. *Lebensmittel-Wissenschaft und -Technologie*, **31**, 10–19.

Ebel, C., Eisenberg, H. and Ghirlando, R. (2000). Probing protein–sugar interactions. *Biophysical Journal*, **78**, 385–93.

FitzGerald, R. J. (1998). Potential uses of caseinophosphopeptides. *International Dairy Journal*, **8**, 451–57.

FitzGerald, R. J. and Meisel, H. (2003). Milk protein hydrolysates and bioactive peptides. In *Advanced Dairy Chemistry*, Volume 1, *Protein,* 3rd edn *Part B*, (P. F. Fox and P. L. H. McSweeney, eds) pp. 675–98. New York: Kluwer Academic/Plenum Publishers.

Fogolari, F., Ragona, L., Licciardi, S., Romagnoli, S., Michelutti, R., Ugolini, R. and Molinari, H. (2000). Electrostatic properties of bovine β-lactoglobulin. *Proteins – Structure Function and Genetics*, **39**, 317–30.

Fontecha, J. and Swaisgood, H. E. (1995). Interaction of sucrose ester with casein micelles as characterized by size-exclusion chromatography. *Journal of Dairy Science*, **78**, 2660–65.

Ford, J. E. (1974). Some observations on the possible nutritional significance of vitamin B12 and folate-binding protein in milk. *British Journal of Nutrition*, **31**, 243–57.

Forrest, S. A., Yada, R. Y. and Rousseau, D. (2005). Interactions of vitamin D3 with bovine β-lactoglobulin, α- and β-casein. *Journal of Agricultural and Food Chemistry*, **53**, 8003–9.

Franzen, K. L. and Kinsella, J. E. (1974). Parameters affecting binding of volatile flavor compounds in model food systems 1. Proteins. *Journal of Agricultural and Food Chemistry*, **22**, 675–78.

Frapin, D., Dufour, E. and Haertle, T. (1993). Probing the fatty acid binding site of β-lactoglobulins. *Journal of Protein Chemistry*, **12**, 443–49.

Fugate, F. D. and Song, P. S. (1980). Spectroscopic characterization of β-lactoglobulin-retinol complex. *Biochimica et Biophysica Acta*, **625**, 28–42.

Futterman, S. and Heller, J. (1972). The enhancement of fluorescence and the decreased susceptibility to enzymatic oxidation of retinol complexed with bovine serum albumin, β-lactoglobulin, and the retinol-binding protein of human plasma. *Journal of Biological Chemistry*, **247**, 5168–72.

Garrett, J. M., Stairs, R. A. and Annett, R. G. (1988). Thermal-denaturation and coagulation of whey proteins – effect of sugars. *Journal of Dairy Science*, **71**, 10–16.

Guichard, E. and Langourieux, S. (2000). Interactions between β-lactoglobulin and flavor compounds. *Food Chemistry*, **71**, 301–8.

Guth, H. and Fritzler, R. (2004). Binding studies and computer-aided modelling of macromolecule/odorant interactions. *Chemical Biodiversity*, **1**, 2001–23.

Hammou, H. O., del Pino, I. M. P. and Sanchez-Ruiz, J. M. (1998). Hydration changes upon protein unfolding: cosolvent effect analysis. *New Journal of Chemistry*, **22**, 1453–61.

Hamosh, M., Scanlon, J. W., Ganot, K., Likely, M., Scanlon, K. B. and Hamosh, P. (1981). Fat digestion in newborn. Characterisation of lipase in gastric aspirates of premature and term infants. *Journal of Clinical Investigation*, **67**, 838–46.

Harwalkar, V. R. and Ma, C. Y. (1992). Evaluation of interactions of β-lactoglobulin by differential scanning calorimetry. In *Protein Interactions* (H. Visser, ed.) pp. 359–77. New York: VCH.

Havea, P., Singh, H., Creamer, L. K. and Campanella, O. H. (1998). Electrophoretic characterization of the protein products formed during heat treatment of whey protein concentrate solutions. *Journal of Dairy Research*, **65**, 79–91.

Havea, P., Singh, H. and Creamer, L. K. (2000). Formation of new protein structures in heated mixtures of BSA and α-lactalbumin. *Journal of Agricultural and Food Chemistry*, **48**, 1548–56.

Havea, P., Singh, H. and Creamer, L. K. (2001). Characterization of heat-induced aggregates of β-lactoglobulin, α-lactalbumin and bovine serum albumin in a whey protein concentrate environment. *Journal of Dairy Research*, **68**, 483–97.

Housaindokht, M. R., Chamani, J., Saboury, A. A., Moosavi-Movahedi, A. A. and Bahrololoom, M. (2001). Three binding sets analysis of α-lactalbumin by interaction of tetradecyl trimethyl ammonium bromide. *Bulletin of the Korean Chemical Society*, **22**, 145–48.

Huppertz, T., Kelly, A. L. and Fox, P. F. (2002). Effects of high pressure on constituents and properties of milk. *International Dairy Journal*, **12**, 561–72.

Huppertz, T., Fox, P. F., de Kruif, K. G. and Kelly, A. L. (2006). High pressure-induced changes in bovine milk proteins: a review. *Biochimica et Biophysica Acta – Proteins and Proteomics*, **1764**, 593–98.

Jasinski, E. and Kilara, A. (1985). Flavor binding by whey proteins. *Milchwissenschaft*, **40**, 596–99.

Jones, M. N. and Wilkinson, A. (1976). The interaction between β-lactoglobulin and sodium n-dodecylsulphate. *Biochemical Journal*, **153**, 713–18.

Jou, K. D. and Harper, W. J. (1996). Effect of di-saccharides on the thermal properties of whey proteins determined by differential scanning calorimetry (DSC). *Milchwissenschaft*, **51**, 509–12.

Jouenne, E. and Crouzet, J. (2000). Determination of apparent binding constants for aroma compounds with β-lactoglobulin by dynamic coupled column liquid chromatography. *Journal of Agricultural and Food Chemistry*, **48**, 5396–400.

Jung, D. M. and Ebeler, S. E. (2003). Investigation of binding behaviour of α- and β-ionones to β-lactoglobulin at different pH values using a diffusion-based NOE pumping technique. *Journal of Agricultural and Food Chemistry*, **51**, 1988–93.

Jung, D. M., de Ropp, J. S. and Ebeler, S. E. (2002). Application of pulse field gradient NMR techniques for investigating binding of flavor compounds to macromolecules. *Journal of Agricultural and Food Chemistry*, **50**, 4262–69.

Kanno, C. and Kanehara, N. (1985). Influence of riboflavin-binding protein in raw cow milk. In *Proc Ann Meeting Soc Heric Chem Jpn*, p. 718.

Katakura, Y., Totsuka, M., Ametani, M. and Kaminogawa, S. (1994). Tryptophan-19 of β-lactoglobulin superfamily is not essential for binding retinol, relevant to stabilizing bound retinol and maintaining its structure. *Biochimica et Biophysica Acta*, **1207**, 58–67.

Keenan, R. D., Young, D. J., Tier, C. M., Jones, A. D. and Underdown, J. (2001). Mechanism of pressure-induced gelation of milk. *Journal of Agricultural and Food Chemistry*, **49**, 3394–402.

Kim, J. G. and Kim, H. M. (1989). Interaction of α-lactalbumin with phospholipid-vesicles as studied by photoactivated hydrophobic labeling. *Biochimica et Biophysica Acta*, **983**, 1–8.

Kontopidis, G., Holt, C. and Sawyer, L. (2004). Invited Review: β-Lactoglobulin: binding properties, structure, and function. *Journal of Dairy Science*, **87**, 785–96.

Konuma, T., Sakurai, M. and Goto, Y. (2007). Promiscuous binding of ligands by β-lactoglobulin involves hydrophobic interactions and plasticity. *Journal of Molecular Biology*, **368**, 209–18.

Kühn, J., Considine, T. and Singh, H. (2006). Interactions of milk proteins and volatile compounds: implications in the development of protein foods. *Journal of Food Science*, **71**, R72–R82.

Kulmyrzaev, A., Bryant, C. and McClements, D. J. (2000). Influence of sucrose on the thermal denaturation, gelation, and emulsion stabilization of whey proteins. *Journal of Agricultural and Food Chemistry*, **48**, 1593–97.

Lamiot, E., Dufour, E. and Haertle, T. (1994). Insect sex pheromone binding by bovine β-lactoglobulin. *Journal of Agricultural and Food Chemistry*, **42**, 695–99.

Lee, J. C. and Timasheff, S. N. (1981). The stabilization of proteins by sucrose. *Journal of Biological Chemistry*, **256**, 7193–201.

Lefevre, T. and Subirade, M. (1999). Structural and interaction properties of β-lactoglobulin as studied by FTIR spectroscopy. *International Journal of Food Science and Technology*, **34**, 419–28.

Lefevre, T. and Subirade, M. (2001). Molecular structure and interaction of biopolymers as viewed by Fourier transform infrared spectroscopy: model studies on β-lactoglobulin. *Food Hydrocolloids*, **15**, 365–76.

Lehmann, M. S. and Zaccai, G. (1984). Neutron small-angle scattering studies of ribonuclease in mixed aqueous-solutions and determination of the preferentially bound water. *Biochemistry*, **23**, 1939–42.

Lerbret, A., Bordat, P., Affouard, F., Descamps, M. and Migliardo, F. (2005). How homogeneous are the trehalose, maltose, and sucrose water solutions? An insight from molecular dynamics simulations. *Journal of Physical Chemistry B*, **109**, 11046–57.

Lietaer, E., Poiffait, A. and Adrian, J. (1991). Nature et propriétés de la liason vitamine A-caséine. *Lebensmittel-Wissenschaft und -Technologie*, **24**, 39–45.

Lindner, R. A., Kapur, A. and Carver, J. A. (1997). The interaction of the molecular chaperone, α-crystallin, with molten globule states of bovine α-lactalbumin. *Journal of Biological Chemistry*, **272**, 27722–29.

Liu, X., Powers, J. R., Swanson, B. G., Hill, H. H. and Clark, S. (2005). Modification of whey protein concentrate hydrophobicity by high hydrostatic pressure. *Innovative Food Science and Emerging Technologies*, **6**, 310–17.

Lönnerdal, B. (2003). Lactoferrin. In *Advanced Dairy Chemistry*, Volume 1, *Proteins,* 3rd edn *Part A*, (P. F. Fox and P. L. H. McSweeney, eds) pp. 449–66. New York: Kluwer Academic/Plenum Publishers.

Lu, R. C., Cao, A. N., Lai, L. H. and Xiao, J. X. (2006). Effect of cationic surfactants on the interaction between sodium perfluorooctanoate and β-lactoglobulin. *Journal of Colloid and Interface Science*, **293**, 61–68.

Mata, L., Sanchez, L. and Calvo, M. (1997). Interaction of mercury with human and bovine milk. *Bioscience Biotechnology and Biochemistry*, **61**, 1641–45.

Maulik, S., Moulik, S. P. and Chattoraj, D. K. (1996). Biopolymer-surfactant interaction. 4. Kinetics of binding of cetyltrimethyl ammonium bromide with gelatin, hemoglobin, β-lactoglobulin and lysozyme. *Journal of Biomolecular Structure and Dynamics*, **13**, 771–80.

McClements, D. J. (2002). Modulation of globular protein functionality by weakly interacting cosolvents. *Critical Reviews in Food Science and Nutrition*, **42**, 417–71.

McNeill, V. L. and Schmidt, K. A. (1993). Vanillin interaction with milk protein isolates in sweetened drinks. *Journal of Food Science*, **58**, 1142–44, 1147.

Mohammadzadeh, K. A., Feeney, R. E., Samuels, R. B. and Smith, L. M. (1967). Solubility of alkanes in protein solutions. *Biochimica et Biophysica Acta*, **147**, 583–89.

Mohammadzadeh, K. A., Feeney, R. E. and Smith, L. M. (1969a). Hydrophobic binding of hydrocarbons by proteins I. Relationship of hydrocarbon structure. *Biochimica et Biophysica Acta*, **194**, 246–55.

Mohammadzadeh, K. A., Smith, G. M. and Feeney, R. E. (1969b). Hydrophobic binding of hydrocarbons by proteins II. Relationship of protein structure. *Biochimica et Biophysica Acta*, **194**, 256–64.

Monaco, H. L., Zanotti, G., Spadon, P., Bolognesi, M., Sawyer, L. and Eliopoulos, E. E. (1987). Crystal-structure of the trigonal form of bovine β-lactoglobulin and of its complex with retinol at 2.5-Å resolution. *Journal of Molecular Biology*, **197**, 695–706.

Montich, G. G. and Marsh, D. (1995). Interaction of α-lactalbumin with phosphatidylglycerol – influence of protein-binding on the lipid phase-transition and lipid acyl-chain mobility. *Biochemistry*, **34**, 13139–45.

Mora-Gutierrez, A. and Farrell, H. M. (2000). Sugar-casein interaction in deuterated solutions of bovine and caprine casein as determined by oxygen-17 and carbon-13 nuclear magnetic resonance: a case of preferential interactions. *Journal of Agricultural and Food Chemistry*, **48**, 3245–55.

Muresan, S., van der Bent, A. and de Wolf, F. A. (2001). Interaction of β-lactoglobulin with small hydrophobic ligands as monitored by fluorometry and equilibrium dialysis: nonlinear quenching effects related to protein–protein association. *Journal of Agricultural and Food Chemistry*, **49**, 2609–18.

Murray, B. S. and Liang, H. J. (1999). Enhancement of the foaming properties of protein dried in the presence of trehalose. *Journal of Agricultural and Food Chemistry*, **47**, 4984–91.

Nagasako, Y., Saito, H., Tamura, Y., Shimamura, S. and Tomita, M. (1993). Iron-binding properties of bovine lactoferrin in iron-rich solution. *Journal of Dairy Science*, **76**, 1876–81.

Narayan, M. and Berliner, L. J. (1997). Fatty acids and retinoids bind independently and simultaneously to β-lactoglobulin. *Biochemistry*, **36**, 1906–11.

Narayan, M. and Berliner, L. J. (1998). Mapping fatty acid binding to β-lactoglobulin: ligand binding is restricted by modification of Cys 121. *Protein Science*, **7**, 150–57.

Nixon, P. F., Jones, M. and Winzor, D. J. (2004). Quantitative description of the interaction between folate and the folate-binding protein from cow's milk. *Biochemical Journal*, **382**, 215–21.

O'Connell, J. E. and Fox, P. F. (2003). Heat-induced coagulation of milk. In *Advanced Dairy Chemistry*, Volume 1, *Proteins*, 3rd edn (P. F. Fox and P. L. H. McSweeney, eds) pp. 879–945. New York: Kluwer Academic/Plenum Publishers.

O'Neill, T. E. and Kinsella, J. E. (1987). Binding of alkanone flavors to β-lactoglobulin – effects of conformational and chemical modification. *Journal of Agricultural and Food Chemistry*, **35**, 770–74.

Oelrichs, B. A., Kratzing, C. C., Kelly, J. D. and Winzor, D. J. (1984). The binding of ascorbate to bovine serum albumin. *International Journal for Vitamin and Nutrition Research*, **54**, 61–64.

Papiz, M. Z., Sawyer, L., Eliopoulos, E. E., North, A. C. T., Findlay, J. B. C., Sivaprasadarao, R., Jones, T. A., Newcomer, M. E. and Kraulis, P. J. (1986). The structure of β-lactoglobulin and its similarity to plasma retinol-binding protein. *Nature*, **324**, 383–85.

Peres, J. M., Bouhallab, S., Petit, C., Bureau, F., Maubois, J. L., Arhan, P. and Bougle, D. (1998). Improvement of zinc intestinal absorption and reduction of zinc/iron interaction using metal bound to the caseinophosphopeptide 1–25 of β-casein. *Reproduction, Nutrition, Development*, **38**, 465–72.

Pérez, M. D. and Calvo, M. (1995). Interaction of β-lactoglobulin with retinol and fatty acids and its role as a possible biological function for this protein: a review. *Journal of Dairy Science*, **78**, 978–88.

Pérez, M. D., Diaz de Villegas, C., Sanchez, L., Aranda, P., Ena, J. M. and Calvo, M. (1989). Interaction of fatty acids with β-lactoglobulin and albumin from ruminant milk. *Journal of Biochemistry*, **106**, 1094–97.

Pérez, M. D., Sanchez, L., Aranda, P., Ena, J. M. and Calvo, M. (1990). Synthesis and evolution of concentration of β-lactoglobulin and α-lactalbumin from cow and sheep colostrums and milk throughout early lactation. *Cellular and Molecular Biology*, **36**, 205–12.

Pérez, M. D., Sanchez, L., Aranda, P., Ena, J. M., Oria, R. and Calvo, M. (1992). Effect of β-lactoglobulin on the activity of pregastric lipase. A possible role for this protein in ruminant milk. *Biochimica et Biophysica Acta*, **1123**, 151–55.

Pérez, M. D., Puyol, P., Ena, J. M. and Calvo, M. (1993). Comparison of the ability to bind lipids of β-lactoglobulin and serum-albumin of milk from ruminant and non-ruminant species. *Journal of Dairy Research*, **60**, 55–63.

Permyakov, E. A., Morozova, L. A., Kalinichenko, L. P. and Derezhkov, V. Y. (1988). Interaction of α-lactalbumin with Cu^{2+}. *Biophysical Chemistry*, **32**, 37–42.

Poiffait, A. and Adrian, J. (1991). Interaction between casein and vitamin A during food processing. *Advances in Experimental and Medical Biology*, **289**, 61–73.

Puyol, P., Perez, M. D., Peiro, J. M. and Calvo, M. (1991). Interaction of bovine β-lactoglobulin and other bovine and human whey proteins with retinol and fatty acids. *Agricultural and Biological Chemistry*, **55**, 2515–20.

Puyol, P., Perez, M. D., Mata, L. and Calvo, M. (1994). Study on interaction between β-lactoglobulin and other bovine whey proteins with ascorbic acid. *Milchwissenschaft*, **49**, 25–27.

Qin, B. Y., Creamer, L. K., Baker, E. N. and Jameson, G. B. (1998). 12-Bromododecanoic acid binds inside the calyx of bovine β-lactoglobulin. *FEBS Letters*, **438**, 272–78.

Ragona, L., Fogolari, F., Zetta, L., Perez, D. M., Puyol, P., De Kruif, K., Lohr, F., Ruterjans, H. and Molinari, H. (2000). Bovine β-lactogobulin: interaction studies with palmitic acid. *Protein Science*, **9**, 1347–56.

Raica, N., Vavich, M. G. and Kemmerer, A. R. (1959). The effect of several milk components and similar compounds on the utilization of carotene by the rat. *Archives of Biochemistry and Biophysics*, **83**, 376–80.

Ray, A. and Chatterjee, R. (1967). *Interactions of β-Lactoglobulin with Large Organic Ions*. New York: Academic Press.

Reineccius, G. A. and Coulter, S. T. (1969). Flavor retention during drying. *Journal of Dairy Science*, **52**, 1219–23.

Rollema, H. S. (1992). Casein association and micelle formation. In *Advanced Dairy Chemistry*, Volume 1, *Proteins* (P. F. Fox, ed.) pp. 111–40. London: Elsevier Applied Science Publishers.

Salter, D. N. and Mowlem, A. (1983). Neonatal role of milk folate-binding proteins. Studies on the course of digestions of goat's milk folate binder in the 6-d child. *British Journal of Nutrition*, **50**, 589–96.

Schokker, E. P., Singh, H., Pinder, D. N., Norris, G. E. and Creamer, L. K. (1999). Characterization of intermediates formed during heat-induced aggregation of β-lactoglobulin AB at neutral pH. *International Dairy Journal*, **9**, 791–800.

Schokker, E. P., Singh, H. and Creamer, L. K. (2000). Heat-induced aggregation of β-lactoglobulin A and B with α-lactalbumin. *International Dairy Journal*, **10**, 843–53.

Schwieter, U., Von Planta, C. V., Rüegg, R. and Isler, O. (1962). Synthesen in der Vitamin-A2-Reihe. Die Darstelluing von vier sterisch ungehinderten Vitamin-A2-Isomeren. *Helvetica Chimica Acta*, **45**, 528–41.

Seibles, T. S. (1969). Interaction of dodecyl sulfate with native and modified β-lactoglobulin. *Biochemistry*, **8**, 2949–54.

Semenova, M. G., Antipova, A. S. and Belyakova, L. E. (2002). Food protein interactions in sugar solutions. *Current Opinion in Colloid and Interface Science*, **7**, 438–44.

Singh, H. and Creamer, L. K. (1992). Heat stability of milk. In *Advanced Dairy Chemistry*, Volume 1, *Proteins* (P. F. Fox, ed.) pp. 621–56. London: Elsevier Applied Science Publishers.

Singh, H. and Havea, P. (2003). Thermal denaturation, aggregation and gelation of whey proteins. In *Advanced Dairy Chemistry*, Volume 1, *Proteins*, 3rd edn (P. F. Fox and P. L. H. McSweeney, eds) pp. 1261–87. New York: Kluwer Academic/Plenum Publishers.

Sostmann, K. and Guichard, E. (1998). Immobilized β-lactoglobulin on a HPLC-column: a rapid way to determine protein-flavor interactions. *Food Chemistry*, **62**, 509–13.

Spector, A. A. (1975). Fatty acid binding to plasma albumin. *Journal of Lipid Research*, **16**, 165–79.

Spector, A. A. and Fletcher, J. E. (1970). Binding of long-chain fatty acids to β-lactoglobulin. *Lipids*, **5**, 403–11.

Spector, A. A., John, K. and Fletcher, J. E. (1969). Binding of long-chain fatty acids to bovine serum albumin. *Journal of Lipid Research*, **10**, 56–67.

Sreelakshmi, Y. and Sharma, K. K. (2001). Interaction of α-lactalbumin with mini-α A-crystallin. *Journal of Protein Chemistry*, **20**, 123–30.

Spiegel, T. (1999). Whey protein aggregation under shear conditions – effects of lactose and heating temperature on aggregate size and structure. *International Journal of Food Science and Technology*, **34**, 523–31.

Stapelfeldt, H. and Skibsted, L. H. (1999). Pressure denaturation and aggregation of β-lactoglobulin studied by intrinsic fluorescence depolarization, Rayleigh scattering, radiationless energy transfer and hydrophobic fluoroprobing. *Journal of Dairy Research*, **66**, 545–58.

Stone, M. J. (2001). NMR relaxation studies of the role of conformational entropy in protein stability and ligand binding. *Accounts of Chemical Research*, **34**, 379–88.

Swaisgood, H. E., Wang, Q. and Allen, J. C. (2001). Protein ingredient for carrying lipophilic nutrients US Patent 6290974.

Swaisgood, H. E. (2003). Chemistry of the caseins. In *Advanced Dairy Chemistry*, Volume 1, *Proteins*, 3rd edn (P. F. Fox and P. L. H. McSweeney, eds) pp. 139–202. New York: Kluwer Academic/Plenum Publishers.

Swaney, J. B. and Klotz, I. M. (1970). Amino acid sequence adjoining the lone tryptophan of human serum albumin. A binding site of the protein. *Biochemical Journal*, **9**, 2570–74.

Tang, X. L. and Pikal, M. J. (2005). Measurement of the kinetics of protein unfolding in viscous systems and implications for protein stability in freeze-drying. *Pharmaceutical Research*, **22**, 1176–85.

Timasheff, S. N. (1993). The control of protein stability and association by weak interactions with water: how do solvents affect these processes? *Annual Review of Biophysics and Biomolecular Structure*, **22**, 67–97.

Toyosaki, T. and Mineshita, T. (1988). Antioxidant effect of protein-bound riboflavin and free riboflavin. *Journal of Food Science*, **53**, 1851–53.

Trujillo, A. J. (2002). Applications of high-hydrostatic pressure on milk and dairy products. *High Pressure Research*, **22**, 619–26.

Tukamoto, T., Ozeki, S., Hattori, F. and Ishida, T. (1974). Drug interactions. I. Binding of ascorbic acid and fatty acid ascorbyl esters to bovine serum albumin. *Chemical and Pharmaceutical Bulletin*, **22**, 385–89.

Upreti, P., Mistry, V. V. and Warthesen, J. J. (2002). Estimation and fortification of vitamin D_3 in pasteurized processed cheese. *Journal of Dairy Science*, **85**, 3173–81.

van Ruth, S. M., de Vries, G., Geary, M. and Giannouli, P. (2002). Influence of composition and structure of oil-in-water emulsions on retention of aroma compounds. *Journal of the Science of Food and Agriculture*, **82**, 1028–35.

Vegarud, G. E., Langsrud, T. and Svenning, C. (2000). Mineral-binding milk proteins and peptides; occurrence, biochemical and technological characteristics. *British Journal of Nutrition*, **84**(Suppl. 1), S91–S98.

Verheul, M., Pedersen, J. S., Roefs, S. P. M. and de Kruif, K. G. (1999). Association behaviour of native β-lactoglobulin A. *Biopolymers*, **49**, 11–20.

Voilley, A., Beghin, V., Charpentier, C. and Peyron, D. (1991). Interaction between aroma substances and macromolecules in model wine. *Lebensmittel-Wissenschaft und -Technologie*, **24**, 469–72.

Walstra, P. and Jenness, R. (1984). *Dairy Chemistry and Physics.* John Wiley and Sons, New York.

Wang, Q., Allen, J. C. and Swaisgood, H. E. (1997a). Binding of vitamin D and cholesterol to β-lactoglobulin. *Journal of Dairy Science*, **80**, 1054–59.

Wang, Q. W., Allen, J. C. and Swaisgood, H. E. (1997b). Binding of retinoids to β-lactoglobulin isolated by bioselective adsorption. *Journal of Dairy Science*, **80**, 1047–53.

Wang, Q., Allen, J. C. and Swaisgood, H. E. (1999). Binding of lipophilic nutrients to β-lactoglobulin prepared by bioselective adsorption. *Journal of Dairy Science*, **82**, 257–64.

Waninge, R., Paulsson, M., Nylander, T., Ninham, B. and Sellers, P. (1998). Binding of sodium dodecyl sulphate and dodecyl trimethyl ammonium chloride to β-lactoglobulin: a calorimetric study. *International Dairy Journal*, **8**, 141–48.

Wendorf, J. R., Radke, C. J. and Blanch, H. W. (2004). Reduced protein adsorption at solid interfaces by sugar excipients. *Biotechnology and Bioengineering*, **87**, 565–73.

Wieczorek, P., Sumujlo, D., Chlubek, D. and Machoy, Z. (1992). Interaction of fluoride ions with milk-proteins studied by gel-filtration. *Fluoride*, **25**, 171–74.

Wieczorek, P., Samujlo, D. and Machoy, Z. (1994). Effect of calcium-ion content on the interaction of alpha-lactalbumin with fluoride ions. *Fluoride*, **27**, 145–50.

Wu, S. Y., Perez, M. D., Puyol, P. and Sawyer, L. (1999). β-Lactoglobulin binds palmitate within its central cavity. *Journal of Biological Chemistry*, **274**, 170–74.

Xie, G. F. and Timasheff, S. N. (1997). The thermodynamic mechanism of protein stabilization by trehalose. *Biophysical Chemistry*, **64**, 25–43.

Zhang, X. F., Fu, X. M., Zhang, H., Liu, C., Jiao, W. W. and Chang, Z. Y. (2005). Chaperone-like activity of β-casein. *International Journal of Biochemistry and Cell Biology*, **37**, 1232–40.

Zhu, X. Q. (2003). *Interactions between* flavor *compounds and milk proteins*. MSc Thesis. Massey University, Palmerston North, New Zealand.

Zidek, L., Novotny, M. V. and Stone, M. J. (1999). Increased protein backbone conformational entropy upon hydrophobic ligand binding. *Nature Structure Biology*, **6**, 1118–21.

Zsila, F., Imre, T., Szabo, B. Z. and Simonyi, M. (2002). Induced chirality upon binding of cis-parinaric acid to bovine β-lactoglobulin: spectroscopic characterization of the complex. *FEBS Letters*, **520**, 81–87.

14

Model food systems and protein functionality

W. James Harper

Abstract

Fabricated foods generally comprise a mixture of components made up of lipids, proteins, simple and complex carbohydrates, emulsifiers and salts, which are capable of interacting with each other and modifying the final characteristics of the food. Often processing utilized in the manufacture of the food also modifies these interactions. Model food systems were first developed because of the disparity between laboratory functional tests for proteins and the functionality in the food. Harper (1984) stated that "disparity between (laboratory) test results and actual functionality in final food formulations necessitates closer scrutiny by researchers of traditional experimental methods". This disparity results from alteration of functionality (and structure of the protein) through component interactions as well as changes brought about by heat and shear during processing.

Model food systems today find utility for investigating the functionality of many other food components, including starches, gums and emulsifiers, as well as factors affecting areas of continued interest (lipid oxidation, Maillard reaction, etc.). Therefore, model food systems provide a means of determining how the ingredients and the process alter the characteristics of the final product, as well as evaluating the sensitivity of the characteristics of the food to the different ingredients and processing steps.

Model food systems are based on the formulation and processing of real foods, using laboratory and pilot plant facilities. Generally, ingredients that do not have a main effect on

Milk Proteins: From Expression to Food
ISBN: 978-0-12-374039-7

the final characteristics of the product are eliminated. One potential limitation is the use of processing equipment that does not scale up to commercial equipment.

The utilization of carefully selected statistical designs is essential to unravel the multiple interactions that do occur and to optimize food formulation and processing.

A major limitation of model food systems is that they do not provide any information as to the mechanisms by which the ingredients and the process control the final characteristics of the product. Thus they have application to only the food under investigation. They do have a major role in food product development.

Introduction

The utilization of proteins in food for nutritional and functional purposes goes back many centuries, but understanding of the relationship of structure and function has been given close attention only during the past 30–40 years (Owusu-Apenten, 2004). Numerous studies and many reviews have contributed to gaining an understanding of just how proteins act in a complex food system and how the structure and the function are altered by the other ingredients in the food, its intrinsic properties and its processing. These include: Anfinsen (1972), Kinsella (1982), Nakai (1983), Mulvihill and Fox (1987), Mangino *et al.* (1994), Zayas (1996), Li Chan (2004), Luyten *et al.* (2004) and Owusu-Apenten (2004).

There are two broad ways of gaining knowledge of the structure and function of protein systems: (a) study of pure proteins in simple systems and (b) study of commercial proteins in the food systems in which they are used. These are entirely different (Luyten *et al.*, 2004; Owusu-Apenten, 2004) and provide quite different information. Functionality tests can be very useful in obtaining reproducible functional properties, even though such tests cannot be used to predict the final characteristics in a real food system (de Wit, 1984, 1989; Harper, 1984; Owusu-Apenten, 2004). Some differences include the following:

- In pure structure/function studies, pure proteins are generally used and are used at concentrations much lower than those used in food systems (Owusu-Apenten, 2004).
- In food systems, proteins are seldom pure and may actually involve complex mixtures of proteins from a given food source (such as milk proteins, egg proteins, soy proteins, etc.) or proteins from multiple food sources (i.e. meat and soy and milk and gluten) or proteins that have been selectively denatured to provide the desired functionality (Mangino *et al.*, 1994).
- In structure/function studies care is taken to avoid interactions with other components and to avoid modifying the secondary and tertiary structure during the experiments. The proteins are fully hydrated (Kinsella, 1982; Owusu-Apenten, 2004).
- In food systems the proteins are constantly exposed to other ingredients, which can modify the structure and hence function, as well as being modified by processes that often include pH, heat and shear (Lee *et al.*, 1992; Kilara, 1994).

Competition for water can also modify functionality, as can changes in intrinsic properties (Zayas, 1996).
- In structure/function studies outcome is generally measured for a specific and single response (Owusu-Apenten, 2004).
- In food systems the ingredients can influence product functionality at different points in the process or functionality can be expressed in more than one outcome with respect to the characteristics of the food (de Wit, 1984, 1989; Harper, 1984).

There is no question that proteins and other hydrocolloids are important and are required to give the food desirable characteristics. However, our knowledge remains incomplete today, because we still cannot fully predict the characteristics of a formulated food on the basis of our knowledge of the structure and function of pure proteins or hydrocolloids under strictly controlled conditions (Kinsella, 1982; de Wit, 1984, 1989; Harper, 1984; Owusu-Apenten, 2004; Zayas, 1996).

Testing for functionality using simplified systems has been reviewed in depth by numerous investigators, including Kinsella (1982), Kilara (1984), Modler and Jones (1987), Mulvihill and Fox (1987), Patel and Fry (1987), Hall (1996), Zayas (1996) and Owusu-Apenten (2004). The continued need to develop standardized testing for protein solubility, viscosity, water absorption, gelation, emulsification and foaming properties has been emphasized by Mulvihill and Fox (1987), Patel and Fry (1987), German and Phillips (1994), Kilara (1994), Hall (1996), Zayas (1996) and Luyten *et al.* (2004).

Protein functionality in foods

Proteins used in foods include plant proteins (soy, wheat, rice, corn and other plant sources), milk proteins (caseins, caseinates, whey proteins and milk protein concentrates [both caseins and whey proteins]), egg proteins (egg white and egg yolk proteins), wheat proteins, meat proteins and fish proteins. Each type of protein exhibits different functional properties and has application in different types of food products (Inglett and Inglett, 1992; Kinsella, 1982; Lee *et al.*, 1992; Kilara, 1994; Mangino *et al.*, 1994; Owusu-Apenten, 2004).

Major functionalities of food proteins include solubility, emulsification, gelation and foaming, water binding and heat stability. As shown in Table 14.1, different types of foods have different functional requirements and may require multiple functionalities.

Factors that may modify the protein, during processing, and hence its effect on the product characteristics include heat, shear, salts and other hydrocolloids (de Wit, 1984; Mangino *et al.*, 1987; Yada, 2004).

Role of interactions in determining food characteristics

Interactions between ingredients and modifications caused by processing are the primary reasons why the functionality of proteins and other colloids cannot be predicted

Table 14.1 Multiple functionality in selected food products

Food type	Multiple functionalities
Beverages	Solubility, Heat stability, pH stability, Color
Baked goods	Emulsification, Foaming, Gelation
Dairy analogs	Gelation, Foaming, Emulsification
Egg substitutes	Foaming, Gelation
Meat emulsions	Emulsification, Foaming, Gelation, Adhesion/Cohesion
Soups and sauces	Viscosity, Emulsification, Water adsorption
Infant formulas	Emulsification, Heat stability
Toppings	Foaming, Emulsification
Frozen desserts	Foaming, Gelation, Emulsification

Adapted from Kinsella (1982), de Wit (1984), Kilara (1994) and Owusu-Apenten (2004)

in food systems (de Wit, 1984, 1989; Harper, 1984; Zayas, 1996; Owusu-Apenten, 2004; Yada, 2004).

The following diagram provides an overview of the potential interactions that can occur in a food product (adapted from Harper, 1984):

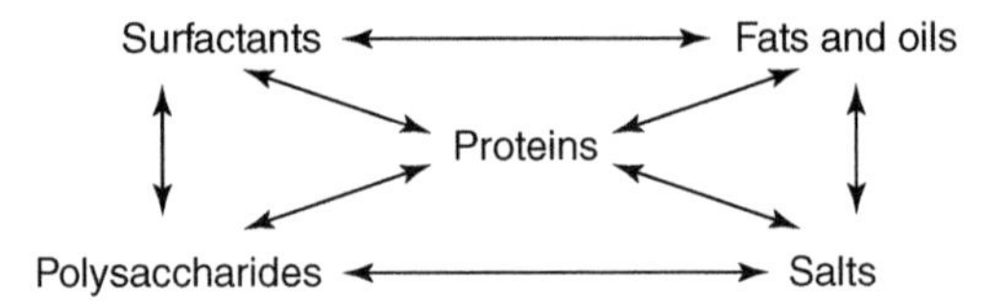

Essentially, almost everything can modify the functionality of everything else. Salts are somewhat unique in that they do not in themselves affect product characteristics, but can act on proteins, surfactants, polysaccharides and, to some extent, polar lipids to modify the functionality of each in the food system. The nature and the extent of these interactions will be modified by pH, ionic strength, ingredient concentrations and process-induced modifications. Some examples include the following:

- Surfactants and proteins can interact competitively at the surface of an oil to modify the characteristics of the emulsion, such as stability, size distribution and light scattering. The extent to which a given component will dominate the characteristics will depend upon the relative concentrations (Figure 14.1) and the chemical natures of the surfactant and the protein.
- Starch is frequently used to provide texture to food products. However, the viscosity during processing and the final viscosity can be greatly altered by interactions with other components. The data in Figure 14.2 show that interactions that are observed with two-component systems do not always predict the effect of three- and four-ingredient interactions. In addition, "who sees who first" can further modify other interactions and change product characteristics.

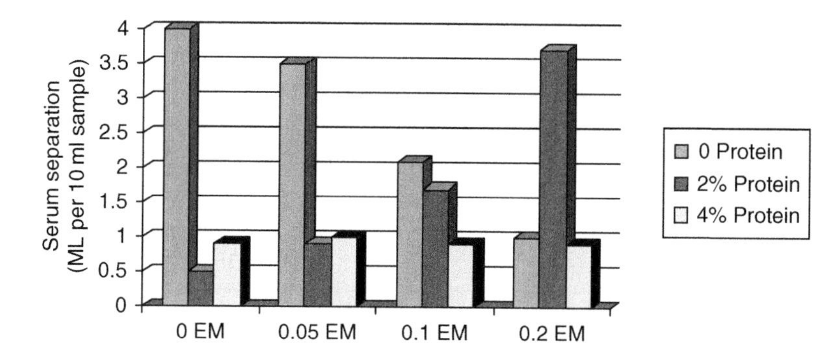

Figure 14.1 Effect of emulsifier (mono- and diglycerides) and protein on the stability of an oil-in-water emulsion (coffee whitener). Adapted from Harper (1984).

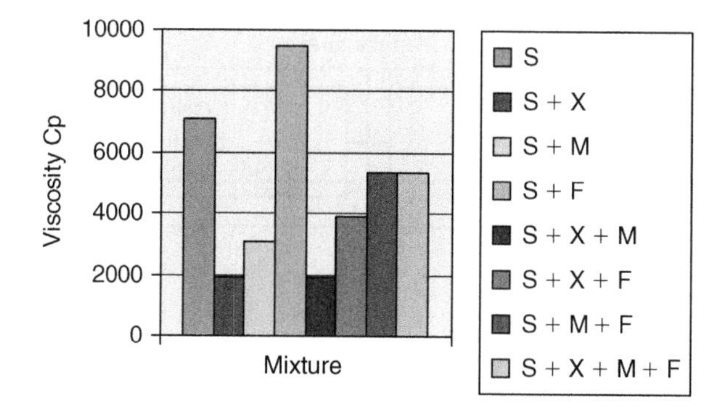

Figure 14.2 Effect of components singly and in combination on the peak viscosity of potato starch (starch (S)): S + X = starch + xanthan gum; S + M = starch + mono- and diglycerides; S + F = starch + fructose; S + X + M = starch + xanthan gum + mono- and diglycerides; S + X + F = starch + xanthan gum + fructose; S + M + F = starch + mono- and diglycerides + fructose; S + X + M + F = starch + xanthan gum + mono- and diglycerides + fructose. Adapted from WJ Harper, unpublished.

Processing effects

The functionality of commercial food proteins, and other hydrocolloids, can be modified both during their production and during the processing of the food product itself. An overview of the conversion of a raw protein source to a functional food ingredient and the subsequent further processing during food manufacture is outlined in Figure 14.3.

During the production of commercial food proteins for use as food ingredients, the proteins may be exposed to a wide range of processing steps that can include thermal processes (pasteurization, sterilization), shear (pumping, mixing, homogenization), pressure (high pressure processing, retorting), concentration (membrane processing, evaporation, drying) and precipitation (heat, acid, salts, solvents). Each of these steps will modify the functional properties of the protein and thus will affect the final characteristics of the food (Kinsella, 1982; de Wit, 1984, 1989; Harper, 1984; Dybing and Smith, 1991; Kilara, 1994; Zayas, 1996; Owusu-Apenten, 2004). Such processes can alter functionality in food through a number of different modifications of the

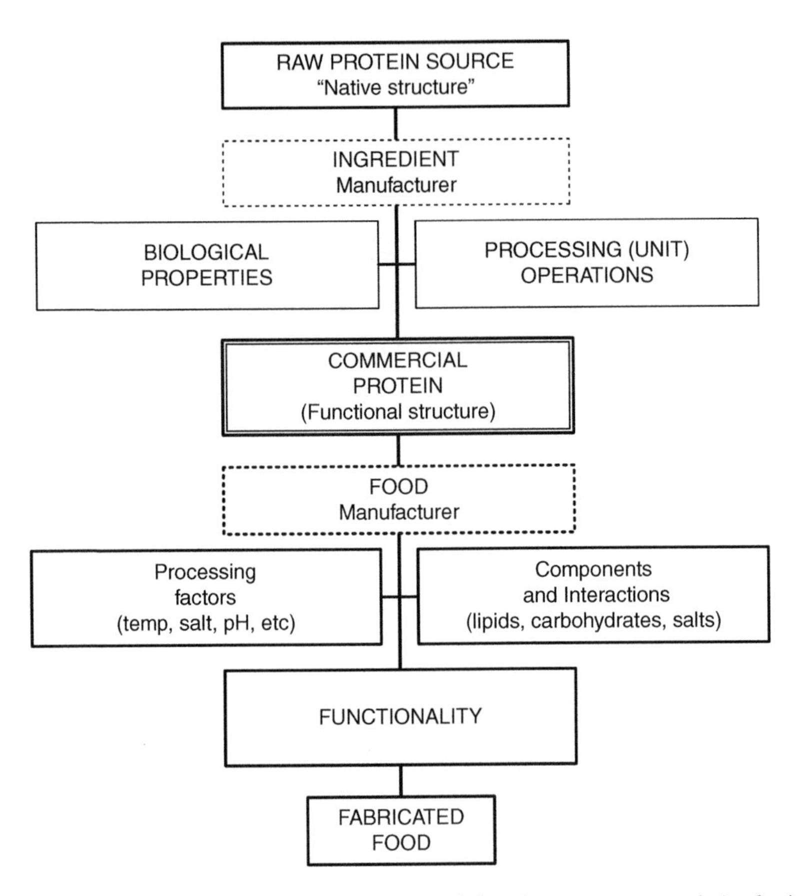

Figure 14.3 Steps in the manufacture of food proteins and the subsequent processes during food manufacture. From Owusu-Apenten (2004).

protein, including changes in sulfydryl interactions, modification of secondary and quaternary structure and shifts in the hydrophilic/lipophilic balance (Kinsella, 1982).

Subsequent processing during use of the protein as a functional ingredient in food will bring further changes in the system, especially those occurring in the presence of other interacting ingredients. Such changes in the characteristics of the food generally cannot be predicted; thus, there is a need for the use of model food systems as an intermediate step in product development (Owusu-Apenten, 2004).

Uses of model food systems

Model food systems can be used in a variety of ways (de Wit, 1984; Harper, 1984; Owusu-Apenten, 2004), including the following:

- determining the relative significance of the main effects of ingredients;
- studying factors in food that affect chemical and physical changes (Maillard reaction, lipid oxidation, etc.);

- evaluating the sensitivity of the food to alterations in formulation and processing;
- defining ingredient interactions;
- optimizing the formulation for robustness;
- determining critical steps in the processing of the product;
- determining interrelationships between ingredients and the process;
- as a means of tailor-making ingredients for a specific food application;
- evaluating and minimizing the sensitivity of product attributes to the formulation and the process.

Owusu-Apenten (2004) stated that the advantages of the use of model food systems over standard functionality tests included: (a) their ease of use, (b) the lack of a need for specialized equipment and methodologies, (c) the ability to aid in product optimization, and (d) the ability to test for multiple factors and interactions with respect to formulation and processing.

Initial steps to developing model food systems

The approach to the development of a model food system will be the same, whether the ingredient being investigated is a protein, lipid, emulsifier, starch or gum.

The development of a model food system begins by reviewing as many formulations as can be found and selecting those that are common to all formulations at a concentration that is at the central point of the various formulas (Harper, 1984). Next, a small-scale process for making the products is developed using processing steps and conditions as close to the commercial process as possible. When more than four or five ingredients are involved, it is often necessary to do a screening experiment to eliminate ingredients that do not have a main effect on important characteristics.

Each different food will have different characteristics, which may include taste, color and texture, that can be modified by the formulation and the process. Key attributes and methods for their evaluation need to be selected. Generally, the methods for evaluation are different from those that are used in research (Owusu-Apenten, 2004).

Statistical design

Statistical design is an essential component in the use of model food systems because of its ability to provide information of ingredient and processing interactions (Dziezak, 1990; Earle *et al.*, 2001; Hanrahan and Lu, 2006). Most fabricated food products have from 5 to 25 variables when both the ingredients and the processing steps are taken into consideration. This makes full factorial designs, which would exceed several hundred experiments, an impractical choice. Thus fractional factorial screening designs are generally required.

For most food products, the experimental design is a step-wise process, starting with screening experiments to minimize the variables that do not have main effects on the characteristics of the products. One of the most common screening designs is the Plackett-Burman, which can be used with up to 36 variables (Mullen and Ennis, 1985; Hanrahan and Lu, 2006). The screening experiments allow determination of the main effects that can be used in further fractional factorial designs to get a better understanding of ingredient and process interactions and to generate response surfaces that give an understanding of the nature of the interactions (Hanrahan and Lu, 2006).

In developing a fractional factorial experimental design in model food systems, it is necessary to know: (a) the critical factors associated with the ingredients and the process, (b) the region of interest where the factor levels influencing the product characteristics are known, (c) that the factors vary continuously throughout the experimental range tested, (d) that a mathematical function relates the variable factors to the measured response, and (e) that the response defined by the function is a smooth curve.

Numerous studies have used statistical design and response surface methodology to determine the effect of interactions on product characteristics and to optimize specific characteristics in a food (Dziezak, 1990).

In developing an experimental design, consultation with a statistician familiar with the factors that affect the outcomes of the specific design is needed to avoid common pitfalls, which include: (a) critical factors may not be correctly defined or specified, (b) the range of factors selected is too narrow or too broad so that the optimum cannot be defined, (c) lack of the use of good statistical practices, (d) too large a variation in the range of the factors utilized, introducing bias and error, (e) over-reliance on computer-generated results and (f) a need to make sure that the results make good sense.

Applications of model food systems

Initially, model food systems were applied to milk proteins to gain a better understanding of what was required to get desired characteristics in complex food products that could not be predicted from standard functionality tests. de Wit (1998) stated "Information obtained from functional characterization tests in model systems is more suitable to explain retroactively protein behaviours in complex food systems than to predict functionality". What has been learned using milk proteins in model food systems has been shown to be equally applicable to other food proteins. In addition to understanding the protein being used, there is a need to know the functionality of other ingredients in the food, the probability of how they will interact and modify the function of the food protein, and the use of statistical design to gain the full potential of the model system approach.

The effect of model food systems to assess their performance in foods has been applied to a large number of different types of foods, as shown in Table 14.2. These include bakery products, dairy products, dairy analogs, meat products, sauces and dressings, fermented foods, wine and infant formulas.

The examples of the model food systems used to illustrate applications in this chapter are primarily from the first generation category. They include bakery products, dairy analogs, meat products, salad dressings and sauces.

Bakery products

Bread represents a system where the methods of evaluation of ingredients have been standardized and covered by AACC-approved methods (AACC methods 10–9, 10–10 and 10–11). Details of the procedures and evaluation techniques have been given

Table 14.2 Model food systems used to assess functionality in foods[a]

First generation model foods[a]	Additional examples[b]
Cakes (angel food, chocolate, yellow, pound)	Low fat spreads
Meringues	Beer batters
Bread	Beef patties
Coffee whitener	Gravies
Ham, restructured meats	Meat emulsions
Infant formula	Cream
Salad dressing	Milk
Sausage	Cheese
Starch pudding	Processed cheese
Whipped topping	Soups and sauces
Ice cream	Surimi
	Wine
	Yoghurt

[a]Adapted from Harper (1984) and de Wit (1984)
[b]Adapted from Owusu-Apenten (2004).

by various investigators (Lindbloom, 1997; Pomerance, *et al.*, 1984; Ranhortra *et al.*, 1992; Fenn *et al.*, 1994; Cauvain and Young, 2006). In general, the substitution or addition of other proteins (milk, whey proteins, etc.) leads to a loss of loaf volume (Harper, *et al.*, 1980; de Wit, 1984). Harper and Zadow (1984) found that heat treatments that prevented loss of loaf volume in bread made with milk powder were ineffective in preventing loss of loaf volume in bread made with whey protein concentrates.

Model food systems have been used widely in cake systems: pound cake (Lee, 1999), Madeira cake (de Wit, 1984), white cake (Harper *et al.*, 1980) and angel food cake (Kissell and Bean, 1978). Of these, angel food cake has received the most attention (Lowe *et al.*, 1969; DeVilbiss *et al.*, 1974; Cunningham, 1976; Regenstein *et al.*, 1978; Johnson and Zabik, 1981a, 1981b; Ball and Winn, 1982; Froning *et al.*, 1987; Froning, 1988; Martinez *et al.*, 1995).

The primary protein evaluated has been egg white, for which the cake height and the cake texture can be related to the individual egg white proteins (Johnson and Zabik, 1981a, 1981b; Ball and Winn, 1982). Attempts to replace egg white with whey proteins have never been completely successful (DeVilbiss *et al.*, 1974; Harper *et al.*, 1980). Arunepanlop *et al.* (1996) were able to replace 25–50% of the egg white with whey protein and could achieve greater replacement by the addition of xanthan gum. Cake volume is essentially the same as with egg white, but the cakes collapse upon baking. This emphasizes the requirement for both foaming and gelation (Owusu-Apenten, 2004). This is due in part to the lower gelation temperature for foams made with egg white (Pernell *et al.*, 2002) and in part to the shear-induced denaturation of egg white with mixing (DeVilbiss *et al.*, 1974).

Other proteins evaluated for angel food cake include blood plasma protein (Kahn *et al.*, 1979; Raeker and Johnson, 1995) and dried beef plasma (Duxbury, 1988).

The factors to consider in developing a model food system for bakery products are outlined briefly in Table 14.3.

Table 14.3 Factors affecting functionality of protein in bakery products

Product type	Functional requirement of protein	Ingredient modifying functionality	Processing factors affecting functionality
Bread	Dough formation, water binding, gelation, elasticity of dough	Protein source, polar lipids, oxidizing and reducing agents, other proteins with sulfydryl groups	Mixing, method of bread making (sponge dough versus mechanical development)
Cakes	Fat binding, foaming, gelation	Protein type and concentration, gums, fat, sugar concentration	Mixing speed and time, pre-emulsification

Dairy analogs

Dairy analogs include coffee whiteners, whipped toppings and processed cheese products.

Coffee whiteners

Coffee whiteners, first developed in the 1950s, generally are protein-stabilized oil-in-water emulsions with vegetable oil as the dispersed phase. A model system, developed by Harper and Raman (1979) and Harper *et al.* (1980), utilized caseinate, soy bean oil, carbohydrate, phosphate, emulsifier and a gum (xanthan gum or carrageenan). The role of the ingredients has been reviewed by Knightly (1969) and Patel *et al.* (1992) and the process has been reviewed by Owusu-Apenten (2004). Factors affecting the properties of coffee whiteners are presented in Table 14.4.

Patented processes include using milk protein retentate (Kosikowski and Jimenez-Florez 1987), reformed casein micelles (McKenna *et al.*, 1992), phosphate-modified milk protein (Melachouris *et al.*, 1994) and soy proteins (Melmychyn, 1973).

Alternative proteins that have been suggested to replace caseinate include milk protein concentrate (Euston and Hirst, 2000), whey protein (Hlavacek *et al.*, 1970; Gruetzmacher and Bradley, 1991; Euston and Hirst, 2000), wheat protein (Golde and Schmidt, 2005; Patil *et al.*, 2006), soy protein (Hlavacek *et al.*, 1970; Golde and Schmidt, 2005), peanut protein (Malundo *et al.*, 1992) and cottonseed protein (Choi *et al.*, 1982).

Coffee whiteners are evaluated to ensure that they do provide an emulsion with a small particle size to maximize whiteness, minimize astringency of the coffee by binding with the coffee tannins, maintain stability in hot coffee under acidic conditions, minimize feathering in the presence of hard water salts and readily disperse in

Table 14.4 Factors affecting functionality of protein in coffee whiteners

Product type	Functional requirement of protein	Ingredient modifying functionality	Processing factors affecting functionality
Coffee whiteners	Emulsification, stability to the pH and temperature of coffee	Emulsifiers, gums, phosphate, calcium	Homogenization, pasteurization time and temperature, temperature and pH of coffee

the coffee (Pearce and Harper, 1982; Tran and Einerson, 1987, Kneifel *et al.*, 1992; Kelly *et al.*, 1999).

Golde and Schmidt (2005) compared coffee whiteners made from sodium caseinate, soy protein isolate and wheat protein isolate and found that they gave similar whiteness (L*) to the coffee. However, the liquid coffee whiteners made with wheat protein tended to separate upon storage and the whiteners made with soy protein isolate tended to show feathering.

Whipped toppings

Most commercial whipped toppings contain sodium caseinate as the protein of choice (Knightly, 1968). Other proteins used for whipped toppings include whey protein concentrate (Peltonen-Shalaby and Mangino, 1986; Liao and Mangino, 1987) and soy protein isolates (Kolar *et al.*, 1979; Lah *et al.*, 1980; Chow *et al.*, 1988; Abdaullah *et al.*, 1993; Shurtleff and Aoyagi, 1994).

Whipped toppings are high-fat, foamed emulsions with about 40% total solids—model food systems generally also contain sugars, gums and small molecular weight emulsifiers (Knightly, 1968; Harper *et al.*, 1980). The model system differs from whipping or foaming tests with respect to both compositions and much lower fat contents (Owusu-Apenten, 2004). Min and Thomas (1977) found that calcium addition to a 15%-fat-containing whipped topping stabilized with sodium caseinate gave improved stability to the system. Peltonen-Shalaby and Mangino (1986) showed that pasteurization also improved the overrun of the topping. Liao and Mangino (1987) used whey proteins to make a model whipped topping and found a positive correlation between exposed hydrophobicity and overrun. Factors affecting the properties of whipped toppings are presented in Table 14.5. Other factors that affect overrun and stability include the hardness of the fat, the type and percentage of emulsifier and the equipment used for mixing (Harper, 1984).

Table 14.5 Factors affecting functionality of protein in whipped toppings

Product type	Functional requirement of protein	Ingredient modifying functionality	Processing factors affecting functionality
Whipped toppings	Emulsification, whipping to ≈200% overrun in the presence of high fat and high solids	Emulsifiers, gums, phosphate, calcium	Homogenization, pasteurization time, equipment used to produce the whipped product

Salad dressings

Salad dressings are high-fat emulsions, frequently stabilized by high shear in the presence of egg yolk as the primary emulsifier (Parker *et al.*, 1995). Mayonnaise—a spoonable dressing—contains 75% oil by definition. Subsequently, starch pastes were used to make a spoonable dressing with about 40% oil. Today, the most common dressings are pourable, with a wide range of oil contents and are stabilized primarily by xanthan gum (Franco *et al.*, 1995).

Model food systems have been used to gain better understanding of both ingredients (Smith, 1977; Paredes *et al.*, 1988) and processing (Parker *et al.*, 1995).

Smith (1977), using a central composite statistical design, found that the coefficients of the regression analysis were larger for the interaction terms than for the main effect terms in pourable salad dressing with 40% oil and containing egg, vinegar, xanthan gum and mustard powder. The order of addition was also found to be important to the viscous properties of the pourable salad dressing.

Meat products

Model meat products, including beef, pork, lamb, poultry or fish, have been utilized for recombined meats (ham, steaks, etc.) and meat macro-emulsions that include bologna, sausages, liver sausages, frankfurters and meat loaves.

Non-meat proteins have been injected into beef and ham, together with water, followed by tumbling to maintain nutritionally equivalent protein levels, increase yield and improve texture (Zayas, 1996; Yada, 2004; Szerman *et al.*, 2007). Szerman *et al.* (2007) found whey protein isolates to be superior to vegetable proteins on the basis of flavor.

Meat emulsions generally have size distributions between 0.1 and 50 μm and many investigators suggest that they are three-dimensional gel networks with entrapped fat (Regenstein, 1978; Krisman and Sharma, 1990; Xiong *et al.*, 1992; Correia and Mittal, 1993; Barbut, 1995). However, most reviewers continue to classify them as meat emulsions (Gordon, 1969; Webb, 1974; Owusu-Apenten, 2004).

The factors that affect the functionality of proteins in these products include: meat extraction temperature, emulsification temperature, shear during emulsification, fat melting point, pH, ionic strength, ratios of ingredients, salt, soluble protein concentration and type of salt (salt, phosphates, citrates, etc.).

Achievement of functionality has been determined by a number of different methods, including:

- emulsification capacity (EC) (Swift *et al.*, 1961; Swift and Sulzbacher, 1963);
- emulsion activity (EA) (Acton and Saffle, 1972);
- emulsion stability (ES) (Carpenter and Saffle, 1964; Townsend *et al.*, 1968; Marshall *et al.*, 1975).

Although the tests for EC and ES for comminuted meat products are widely used, there does not appear to be much collaborative testing of the different methods (Owusu-Apenten, 2004).

The type of protein affects the EC of meat emulsions, with isolated muscle proteins giving different EC values. In general, the EC was in the order of myosin > actomycin > actin for different types of meat (Tsai *et al.*, 1972; Galluzzo and Regenstein, 1978; Li Chan *et al.*, 1984).

Substitution of meat protein by other protein in meat emulsions, as measured by large deformation rheological testing, showed:

- gluten, soy protein isolate or egg white increased the yield after the cooking of meat emulsions (Randall *et al.*, 1976);

- corn germ protein at 2% substitution reduced the shear force and reduced cooking losses (Mittal and Usborne, 1985);
- partial substitution of meat with sodium caseinate, soy protein isolate, whey protein concentrate or wheat germ protein all increased cook yield, increased protein level and decreased fat in frankfurters, without affecting quality (Atughonu *et al.*, 1988);
- addition of bovine blood plasma to meat emulsion products improved emulsion stability and yield, and contents of protein, phenylalanine and valine (Marquez *et al.*, 1997).

Use of model food systems for other food components

In addition to evaluating the performance of proteins in food systems, a wide range of other applications has been utilized. During the past several years, more than 200 papers have been published on other uses of model food systems. A full review of such uses is outside the scope of this chapter. However, selected applications from studies over the past several years are cited both to provide a basis for understanding the scope of the use of model food systems in the food industry and to provide a starting point for obtaining more detailed information.

Applications include:

- factors affecting flavor release in foods (Bylaite *et al.*, 2005; Heineman *et al.*, 2005; Conde-Petit *et al.*, 2006; Nongonierma *et al.*, 2006; Seuvre *et al.*, 2007);
- factors affecting D values in food (Rodriguez *et al.*, 2006);
- lipid oxidation (Jaswir *et al.*, 2004; Sakanaka and Tachibana, 2007; Wijeratne *et al.*, 2006);
- water migration in foods (Guignon *et al.*, 2005; Doona and Moo, 2007);
- Maillard reaction investigations (Severini *et al.*, 2003; Song and Roos, 2006; Acevedo, 2006; Casal, 2006);
- effects of high pressure processing of food (Severini *et al.*, 2003; Sila *et al.*, 2007).

Limitations of model food systems

Model food systems can tell you "what", but cannot tell you the mechanism(s) by which the effects occur. Frequently, the results with a model system cannot be scaled up to full commercial practice because of differences in equipment and processes. However, they do provide insight into directions to take to overcome scale-up problems. Generally, the results are valid only within the parameters that have been established. Optimization of a food system can sometimes be outside the limits of either the processing equipment or the functionality of a specific ingredient.

Conclusions

Historically, model food systems were used first to improve the functionality of milk proteins in food systems. Currently, there are very few publications on the use of milk proteins for these purposes, although it is known that a number of dairy food companies use model food systems in their product development programs. Today, the publications concerning model food systems have a much broader usage, with attention being given to a better understanding of how complex food systems affect such things as oxidation, Maillard reactions, shelf life, etc.

Model food systems can be a valuable tool in product development with respect to developments of both formulations and manufacturing processes and have a role in the development of ingredients for new foods.

Model food systems do not provide information on why interactions occur, but can provide insights as to which interactions need basic study to provide a more robust product.

In the future, it can be expected that model food systems will continue to provide a better understanding of how interactions modify the functionality of proteins in complex food systems and continue to give insight on how to use this information to interface with studies on the basis of protein structure/function.

References

Abdullah, A., Resurreccion, A. V. A. and Beuchat, L. R. (1993). Formulation and evaluation of a peanut milk based whipped topping using response surface methodology. *Lebensmittel-Wissenschaft und -Technologie*, **26**, 167–70.

Acevedo, N., Schebor, C. and Buera, M. P. (2006). Water solids interactions: maxtric, structural properties and the rate of non-enzymatic browning. *Journal of Food Engineering*, **77**(4), 1108–15.

Acton, J. C. and Saffle, R. L. (1972). Emulsifying capacity of muscle protein: phase volumes at emulsion collapse. *Journal of Food Science*, **37**, 904–6.

Anfinsen, C. B. (1972). Studies on the principles that govern the folding of protein chains. Nobel Lecture. Nobel Foundation.

Arunepanlop, B., Morr, C. V., Karleskind, D. and Laye, I. (1996). Partial replacement of egg white proteins with whey proteins in angel food cakes. *Journal of Food Science*, **61**, 1085–93.

Atughonu, A-G., Zayas, J-F., Herald, T-J. and Harbers, L-H. (1988). Thermo-rheology, quality characteristics, and microstructure of frankfurters prepared with selected plant and milk additives. *Journal of Food Quality*, **21**, 223–38.

Ball, H. R. Jr. and Winn, S. E. (1982). Acylation of egg white proteins with acetic anhydride and succinic anhydride. *Poultry Science*, **61**, 1041–46.

Barbut, S. (1995). Importance of fat emulsification and protein matrix characteristics in meat batter stability. *Journal of Muscle Foods*, **6**, 161–77.

Bylaite, E., Adler-Nissen, J. and Meyer, A. S. (2005). Effect of xanthan on flavor release from thickened food model systems. *Journal of Agricultural and Food Chemistry*, **53**, 3577–83.

Carpenter, J. A. and Saffle, R. L. (1964). A simple method of estimating the emulsifying capacity of various sausage meats. *Journal of Food Science*, **29**, 774–81.

Casal, E., Ramirez, P., Ibanez, E., Corzo, N. and Olano, A. (2006). Effect of supercritical carbon dioxide treatment on the Maillard reaction in model food systems. *Food Chemistry*, **97**(2), 272–76.

Cauvain, S. and Young, L. (2006). *Baked Products.* Oxford: Blackwell Publishing Ltd.

Choi, Y. R., Lucas, E. W. and Rhee, K. C. (1982). Formulation of non-dairy coffee whiteners with cottonseed protein isolates. *Journal of the American Oil Chemists' Society*, **59**, 564–67.

Chow, E. T. S., Wei, L. S., DeVor, R. E. and Steinberg, M. P. (1988). Performance of ingredients in a soybean whipped topping: a response surface analysis. *Journal of Food Science*, **53**, 1761–65.

Conde-Petit, B., Escher, F. and Nuessli, J. (2006). Structural features of starch-flavor complexation in food model systems. *Trends in Food Science and Technology*, **17**(5), 227–35.

Correia, L. R. and Mittal, G. S. (1993). Selection of fillers based on z_p values of meat emulsion properties during smokehouse cooking. *International Journal of Food Science and Technology*, **28**(5), 443–451.

Cunningham, F. E. (1976). Properties of egg white foam drainage. *Poultry Science*, **55**, 738–43.

de Wit, J. N. (1984). Functional properties of whey proteins in food systems. *Netherlands Milk and Dairy Journal*, **38**, 71–89.

de Wit, J. N. (1989). The use of whey protein products. In Developments in Dairy Chemistry, Volume 3 (P. F. Fox, ed.) pp. 323–45. London: Elsevier Science Publishers.

de Wit, J. N. (1998). Nutrition and functional characteristics of whey protein in foods. *Journal of Dairy Science*, **81**, 597–608.

DeVilbiss, E. D., Holsinger, V. H., Posati, L. P. and Pallansch, M. J. (1974). Properties of whey protein concentrates. *Food Technology*, **28**(3), 40–42, 46, 48.

Doona, C. J. and Moo, Y. B. (2007). Molecular mobility in model dough systems studied by time-domain nuclear magnetic resonance spectroscopy. *Journal of Cereal Science*, **45**(3), 257–62.

Duxbury, D. D. (1988). Powdered beef plasma replaces eggs in cakes. *Food Processing*, **49**(2), 73–74.

Dybing, S. T. and Smith, D. E. (1991). Relation of chemistry and processing procedures to whey protein functionality. *Cultured Dairy Products Journal*, **26**(1), 4–12.

Dziezak, J. D. (1990). Taking the gamble out of food product development. *Food Technology*, **44**(6), 109–17.

Earle, M. D., Earle, R. L. and Anderson, A. (2001). *Food Product Development.* Cambridge: Woodhead Publishing.

Euston, S. R. and Hirst, R. L. (2000). The emulsifying properties of commercial milk protein products in simple oil-in-water emulsions and in a model system. *Journal of Food Science*, **65**, 934–40.

Fenn, D., Lukow, O. M., Bushuk, W. and Depauw, R. M. (1994). Milling and baking quality of 1BL/1RS translocation wheats. I. Effects of genotype and environment. *Cereal Chemistry*, **71**, 189–95.

Franco, J. M., Guerrero, A. and Gallegos, C. (1995). Rheology and processing of salad dressing emulsions. *Rheologica Acta*, **34**, 513–24.

Froning, G. W. (1988). Nutritional and functional properties of egg proteins. In *Developments in Food Proteins*, Volume 6 (B. J. F. Hudson, ed.) pp. 1–34. London: Elsevier Applied Science Publishers.

Froning, G. W., Wehling, R. L., Ball, H. R. and Hill, R. M. (1987). Effect of ultrafiltration and reverse osmosis on the composition and functional properties of egg white. *Poultry Science*, **66**, 1168–73.

Galluzzo, S. J. and Regenstein, J. M. (1978). Role of chicken breast muscle proteins in meat emulsion formation: myosin, actin and synethetic actomyosin. *Journal of Food Science*, **43**(6), 1761–65.

German, J. B. and Philips, L. (1994). Protein interactions in foams: protein-gas interactions. In *Protein Functionality in Food Systems* (N. S. Hettiarachchy and G. R. Ziegler, eds) pp. 181–208. New York: Marcel Dekker.

Golde, A. E. and Schmidt, K. A. (2005). Quality of coffee creamers as a function of protein sources. *Journal of Food Quality*, **28**(1), 46–61.

Gordon, A. (1969). Emulsification in comminuted meat systems. II. *Food Processing Industry*, **38**(459), 50–52, 54.

Gruetzmacher, T. J. and Bradley, R. L. Jr. (1991). Acid whey as a replacement for sodium caseinate in spray-dried coffee whiteners. *Journal of Dairy Science*, **74**, 2838–49.

Guignon, B., Otero, L., Molina-Garcia, A. D. and Sang, P. D. (2005). Liquid water-ice phase diagrams under high pressure: sodium chloride and suggested model food systems. *Biotechnology Progress*, **21**, 439–45.

Hall, G. M. (ed.) (1996). *Methods of Testing Protein Functionality*. Blackie Academic and Professional, London.

Hanrahan, G. and Lu, K. (2006). Application of factorial and response surface methodology in modern experimental design and optimization. *Critical Reviews in Analytical Chemistry*, **36**, 141–51.

Harper, W. J. (1984). Model food system approaches for evaluating whey protein functionality. *Journal of Dairy Science*, **67**, 2745–56.

Harper, W. J. and Raman, R. V. (1979). Protein interactions in whey protein concentrates and functionality in model food systems. *New Zealand Journal of Dairy Science and Technology*, **14**, 198–99.

Harper, W. J. and Zadow, G. (1984). Heat-induced changes in whey protein concentrates as related to bread manufacture. *New Zealand Journal of Dairy Science and Technology*, **19**, 229–37.

Harper, W. J., Peltonen, R. and Hayes, J. (1980). Model food systems yield clearer utility evaluations of whey protein. *Food Product Development*, **14**(10), 52, 54, 56.

Heinemann, C., Zinsli, M., Renggli, A., Escher, F. and Conde-Petit, B. (2005). Influence of amylose-flavor complexation on build-up and breakdown of starch structures in aqueous food model systems. *Lebensmittel-Wissenschaft und -Technologie*, **38**, 885–94.

Hlavacek, R. G., Robe, K., Stinson, W. S. and Belshaw, F. (1970). Reduce fat at 1/3 in coffee whiteners, emulsifiers-stabilizer also improves flavor. *Food Processing*, **31**(5), 18.

Inglett, M. J. and Inglett, G. E. (1982). *Food Products Formulary*, Volume 4, *Fabricated Foods*. Westport, Connecticut: AVI Publishing Co. Inc.

Jaswir, I., Hassan, T. H. and Said, M. Z. M. (2004). Effect of Malaysian plant extracts in preventing perodidation reactions in model food oil systems. *Journal of Oleo Science*, **53**, 525–29.

Johnson, T. M. and Zabik, M. E. (1981a). Response surface methodology for analysis of protein interactions in angle cake food system. *Journal of Food Science*, **46**, 1226–30.

Johnson, T. M. and Zabik, M. E. (1981b). Egg albumen protein interactions in an angel food system. *Journal of Food Science*, **46**, 1231–36.

Kahn, M. N., Rooney, L. W. and Dill, C. W. (1979). Baking properties of plasma protein isolate. *Journal of Food Science*, **44**, 274–76.

Kelly, P. M., Oldfield, D. J. and O'Kennedy, B. T. (1999). The thermostability of spray dried coffee whiteners. *International Journal of Dairy Technology*, **52**(3), 107–13.

Kilara, A. (1984). Standardization of methodology for evaluating whey proteins. *Journal of Dairy Science*, **67**, 2734–44.

Kilara, A. (1994). Whey protein functionality. In *Protein Functionality in Food Systems* (N. S. Hettiarachchy and G. R. Ziegler, eds) pp. 325–55. New York: Marcel Dekker.

Kinsella, J. N. (1982). Relationship between structure and functional properties of food proteins. In *Food Proteins* (P. F. Fox and J. J. Condon, eds) pp. 51–103. New York: Applied Science Publishers.

Kissell, L. T. and Bean, M. M. (1978). AACC technical committee report: development of a method for angel food cake. *Cereal Foods World*, **23**(3), 136–42.

Kneifel, W., Ulbeth, F. and Shaffer, E. (1992). Evaluation of coffee whitening ability of dairy products and coffee whiteners by means of reflectance colorimetry. *Milchwissenschaft*, **47**, 567–69.

Knightly, W. H. (1968). The role of ingredients in the formulation of whipped toppings. *Food Technology*, **22**(6), 73–74. , 77–78, 81, 85–86.

Knightly, W. H. (1969). The role of ingredients in the formulation of coffee whiteners. *Food Technology*, **32**(2), 37–48.

Kolar, C. W., Cho, I. C. and Watrous, W. L. (1979). Vegetable protein applications in yogurt, coffee creamers and whip toppings. *Journal of the American Oil Chemists' Society*, **56**, 389–91.

Kosikowski, F. and Jimenez-Flores, R. (1987). Low-fat dairy coffee whitener. United States Patent US4689245.

Krishnan, K. R. and Sharma, N. (1990). Studies on emulsion-type buffalo meat sausages incorporating skeletal and offal meat with different levels of pork fat. *Meat Science*, **28**(1), 51–60.

Lah, C. L., Cheryan, M. and DeVor, R. E. (1980). A response surface methodology approach to the optimization of whipping properties of an ultrafiltered soy product. *Journal of Food Science*, **45**, 1720–26, 1731.

Lee, K. M. (1999). *Functionality of 34% whey protein concentrate (WPC) and its application in selected model food systems.* PhD Dissertation. The Ohio State University, Columbus, Ohio.

Lee, S. Y., Morr, C. V. and Ha, E. Y. W. (1992). Structural and functional properties of caseinate and whey protein isolate as affected by temperature and pH. *Journal of Food Science*, **57**, 1210–13.

Li-Chan, E., Nakai, S. and Wood, D. F. (1984). Hydrophobicity and solubility of meat proteins and their relationship to emulsifying properties. *Journal of Food Science*, **49**, 345–50.

Li Chan, E. C. Y. (2004). Properties of proteins in food systems: an introduction. In *Proteins in Food Processing* (R. Y. Yada, ed.) pp. 2–29. Cambridge: Woodhead Publishing.

Liao, S. Y. and Mangino, M. E. (1987). Characterization of the composition, physicochemical and functional properties of acid whey protein concentrates. *Journal of Food Science*, **52**, 1033–37.

Lindblom, M. (1977). Bread baking properties of yeast protein concentrates. *Lebensmittel-Wissenschaft und -Technologie*, **10**, 341–45.

Lowe, E., Durkee, E. L., Mersen, R. L., Ijichi, K. and Cimino, S. L. (1969). Egg white concentrated by reverse osmosis. *Food Technology*, **23**(6), 753, 757–59, 760–62.

Luyten, H., Vereijken, J. and Buecking, M. (2004). Using proteins as additives in foods: an introduction. In *Proteins in Food Processing* (R. Y. Yada, ed.) pp. 421–37. Cambridge: Woodhead Publishing.

Malundo, T. M. M., Resurrección, A. V. A. and Koehler, P. E. (1992). Sensory quality and performance of spray-dried coffee whitener from peanuts. *Journal of Food Science*, **57**, 222–25, 226.

Mangino, M. E., Liao, Y. Y., Harper, N. J., Morr, C. V. and Zadow, J. G. (1987). Effects of heat processing on the functionality of whey protein concentrates. *Journal of Food Science*, **52**, 1522–24.

Mangino, M. E., Hettiarachchy, N. S. and Ziegler, G. R. (1994). *Protein Functionality in Food Systems.* New York: Marcel Dekker.

Marshall, W. H., Dutson, T. R., Carpenter, Z. L. and Smith, G. C. (1975). A simple method for emulsion end-point determinations. *Journal of Food Science*, **40**, 896–97.

Marquez, E., de Barboza, M. Y., Izquierdo, P. and Torres, G. (1997). Studies on the incorporation of bovine plasma in emulsion type of meat product. *Journal of Food Science and Technology*, **34**(4), 337–39.

Martinez, R. M., Dawson, P. L., Ball, H. R. Jr., Swartzel, K. R., Winn, S. E. and Giesbrecht, F. G. (1995). The effects of ultrapasteurization with and without homogenization on the chemical, physical, and functional properties of aseptically packaged liquid whole egg. *Poultry science*, **74**(4), 742–52.

McKenna, R. J., Keller, D. J. and Andersen, D. L. (1992). Process for preparing an alternative protein source for coffee whiteners and other products. United States Patent US5128156, US616909.

Melachouris, N., Moffitt, K. R., Rasilowicz, C. E. and Tonner, G. F. (1994). Powdered coffee whitener containing reformed casein micelles. United States Patent US5318793, US937574.

Melmychyn, P. (1973). Acelated protein for coffee whitener formulations. United States Patent US3764711.

Miao, S. and Roos, Y. H. (2006). Isothermal study of non-enzymatic browning in spray-dried systems at a different relative vapor pressure environment. *Innovative Food Science and Emerging Technologies*, **7**(3), 182–84.

Min, D. B. S. and Thomas, E. L. (1977). A study of physical properties of dairy whipped topping mixtures. *Journal of Food Science*, **42**, 221–24.

Mittal, G. S. and Usborne, W. R. (1985). Meat emulsion extenders. *Food Technology*, **39**(4), 121–30.

Modler, H. and Jones, J. D. (1987). Selected processes to improve the functionality of dairy ingredients. *Food Technology*, **41**(10), 114–17, 129.
Mullen, K. and Ennis, D. (1985). Fractional factorials in food product development. *Food Technology*, **39**(5), 90. , 92, 94, 97–98, 100, 102–3.
Mulvihill, D. M. and Fox, P. F. (1987). Assessment of the functional properties of milk protein products. *Bulletin of the International Dairy Federation*, **209**, 3–11.
Nakai, S. (1983). Structure–function relationships of food proteins: with an emphasis on the importance of protein hydrophobicity. *Journal of Agricultural and Food Chemistry*, **31**, 876–83.
Nongonierma, A. B., Springett, M., Le Quere, J-L., Cavot, P. and Voilley, A. (2006). Flavor release at gastric/matrix interfaces of stirred yogurt models. *International Dairy Journal*, **16**, 102–10.
Owusu-Apenten, R. K. (2004). Testing protein functionality. In *Proteins in Food Processing* (R. Y. Yada, ed.) pp. 217–44. Woodhead Publishing, Cambridge.
Paredes, M. D. C., Rao, M. A. and Bourne, M. C. (1988). Rheological characterization of salad dressings. 1. Steady shear, thixotropy and effect of temperature. *Journal of Texture Studies*, **19**, 247–58.
Parker, A., Gunning, P. A., Ng, K. and Robins, M. M. (1995). How does xanthan stabilize salad dressing. *Food Hydrocolloids*, **9**, 333–42.
Patel, P. D. and Fry, J. C. (1987). The search for standardized methods for assessing protein functionality. In *Developments in Food Proteins*, Volume 5 (B. J. F. Hudson, ed.) pp. 299–333. London: Elsevier Applied Science.
Patel, J. R., Dave, R. I., Joshi, N. S. and Thakar, P. N. (1992). Coffee whiteners. *Journal of Dairy Science Association*, **44**, 18–25.
Patil, S. K., Baczynski, M. and McCurry, T. (2006). Wheat protein isolates – alternative to sodium caseinate. *Cereal Foods World*, **51**(5), 279–81.
Pearce, R. J. and Harper, W. J. (1982). A method for the quantitative evaluation of emulsion stability. *Journal of Food Science*, **47**, 680–81.
Peltonen-Shalaby, R. and Mangino, M. E. (1986). Compositional factors that affect the emulsifying and foaming properties of whey protein concentrate. *Journal of Food Science*, **51**, 91–95.
Pernell, C. W., Luck, P. J., Foegeding, E. A. and Daubert, C. R. (2002). Heat-induced changes in angel food cakes containing egg-white protein or whey protein isolate. *Journal of Food Science*, **67**, 2945–51.
Pomeranz, Y., Bolling, H. and Zwingelberg, H. (1984). Wheat hardness and baking properties of wheat flours. *Journal of Cereal Science*, **2**(3), 137–43.
Raeker, M. O. and Johnson, L. A. (1995). Cake-baking (high-ratio white layer) properties of egg white, bovine blood plasma, and their protein fractions. *Cereal Chemistry*, **72**, 299–303.
Randall, C. J., Raymond, D. P. and Voisey, P. W. (1976). Effects of various animal and vegetable materials on replacing the beef component in a meat emulsion system. *Canadian Institute of Food Science and Technology Journal*, **9**(4), 216–21.
Ranhortra, G. S., Gelroth, J. A. and Eisenbraun, G. J. (1992). Gluten index and breadmaking quality of commercial dry glutens. *Cereal Foods World*, **37**(3), 261–63.
Regenstein, J. M., Grunden, L. P. and Baker, P. C. (1978). Proteolytic enzymes and the functionality of chicken egg albumin. *Journal of Food Science*, **43**, 279–80.

Rodriguez, O., Castell-Perez, M. E. and Moreira, R. G. (2007). Effect of sugar and storage temperature on the survival and recovery or irradiated *Escherichia coli* K-12-MG 1655. *Food Science and Technology*, **40**(4), 690–96.

Sakanaka, S. and Tachibana, Y. (2006). Active oxygen scavenging activity of egg-yolk protein hydrolysates and their effects on lipid oxidation in beef and tuna homogenates. *Food Chemistry*, **95**, 243–49.

Seuvre, A-M., Philippe, E., Rochard, S. and Voilley, A. (2007). Kinetic study of the release of aroma compounds in different model food systems. *Food Research International*, **40**(4), 480–92.

Severini, C., Baiano, A., Rovere, P. and Dall'Aglio, G. (2003). Effect of high pressure on olive oil oxidation and the Maillard reaction in model and food systems. *Italian Food and Beverage Technology*, **31**, 5–10.

Shurtleff, W. and Aoyagi, A. (1994). *Non-dairy Whip Toppings (With or Without Soy Protein) – Bibliography and Source Book: 1944 to 1994.* Lafayette, California: Soy Foods Center.

Sila, D. N., Smout, C., Satara, Y., Vu-Truong, L. and Hendricloc, M. (2007). Combined thermal and high pressure effect on carrot pectinmethylesterase stability and catalytic activity. *Journal of Food Engineering*, **78**(3), 755–64.

Smith, R. A. (1977). *Emulsion formation and stability in salad dressings*. M.S. Thesis. The Ohio State University, Columbus, Ohio.

Stanley, D. W., Goff, H. D. and Smith, A. K. (1996). Texture–structure relationships in foamed dairy emulsions. *Food Research International*, **29**(1), 1–3.

Swift, C. E. and Sulzbacher, W. L. (1963). Comminuted meat emulsions: factors affecting meat proteins as emulsion stabilizers. *Food Technology*, **17**, 224–26.

Swift, C. E., Lockett, C. and Fryar, A. S. (1961). Comminuted meat emulsions: The capacity of meats for emulsifying fat.. *Food Technology,*, **15**(11), 468–73.

Szerman, N., Gonzalez, C. B., Sancho, A. M., Grigioni, G., Carduza, F. and Vaudagna, S. R. (2007). Effect of whey protein concentrate and sodium chloride addition plus tumbling procedures on technological parameters, physical properties and visual appearance of sous vide cooked beef. *Meat Science*, **76**, 463–73.

Townsend, W. E., Witnauer, L. P., Riloff, J. A. and Swift, C. E. (1968). Comminuted meat emulsions: Differential thermal analysis of fat transitions. *Food Technology*, **22**, 319–23.

Tsai, R., Cassens, R. G. and Briskey, E. J. (1972). The emulsifying properties of purified muscle proteins. *Journal of Food Science*, **37**, 286–88.

Tran, K. H. and Einerson, M. A. (1987). A rapid method for the evaluation of emulsion stability of non-dairy creamers. *Journal of Food Science*, **52**, 1109–10.

Webb, N. B. (1974). Emulsion technology. *Proceedings Meat Industry Research Conference* pp. 1–15.

Wijeratne, S. S. K., Amarowicz, R. and Shahidi, F. (2006). Antioxidant activity of almonds and their by-products in food model systems. *Journal of the American Oil Chemists' Society*, **83**, 223–30.

Xiong, Y. L., Blanchard, S. P. and Means, W. J. (1992). Properties of broiler myofibril gels containing emulsified lipids. *Poultry Science*, **71**(9), 1548–55.

Yada, R. Y. (2004). *Proteins in Food Processing.* Cambridge: Woodhead Publishing.

Zayas, J. F. (1996). *Functionality of Proteins in Food.* Berlin: Springer-Verlag.

15

Sensory properties of dairy proteins

M. A. Drake, R. E. Miracle and J. M. Wright

Abstract

Production and applications for dairy proteins are increasing globally. The dairy protein category is a wide one, encompassing milk proteins to whey proteins and many subcategories of these products such as caseins, hydrolysates and serum or native whey proteins. While functionality and nutrition continue to be key aspects of these products, flavor is a critical parameter that should not be overlooked. An array of sensory analysis techniques can be applied to measure flavor intensities and flavor variability and to determine flavor sources when applied in conjunction with analytical chemistry. This chapter addresses the current status and ongoing research on the sensory properties of dairy proteins.

Introduction

Dairy proteins are valuable dried ingredients with a host of functional and nutritional properties (Foegeding *et al.*, 2002; Miller, 2005; O'Connell and Flynn, 2007). Within the category of dairy proteins, there are a variety of ingredients including whey proteins and milk proteins of various protein contents. Dried caseins and caseinates as well as serum proteins ("native" whey proteins or whey proteins separated from milk prior to the cheesemaking process) are also contained within this category. Dairy

Milk Proteins: From Expression to Food
ISBN: 978-0-12-374039-7

proteins (primarily dried) are used in an increasingly wide array of ingredient applications for functionality but, with the current consumer focus on health and nutrition, these ingredients are also used widely to enhance nutrition.

Milk proteins play a crucial role in the flavor of all dairy foods. As part of the sensory experience, proteins provide mouthfeel, viscosity and structure to dairy foods. Amino acids and peptides can elicit basic tastes but can also serve as the starting substrates for numerous volatile aroma-active compounds. Proteolysis and the subsequently released amino acids and peptides are the sources as well as the substrates for many desirable and undesirable flavors in cheeses and other fermented dairy products (Singh *et al.*, 2003, 2005; Carunchia Whetstine *et al.*, 2005a; Drake *et al.*, 2007). Heat processing influences the flavor potential of proteins via denaturation and the release of sulfurous compounds and the typical eggy aroma of scalded milk. Denaturation can also make proteins more susceptible to breakdown and thus influences flavor and flavor development in this manner as well. Theoretically, pure undegraded protein should be flavorless. However, dairy proteins as food ingredients are not 100% protein. Fat, ash, carbohydrate and other components are present in various amounts and also clearly influence the final flavor and flavor stability of dairy proteins.

As with all foods, flavor plays a large role in acceptance and product success. Dried ingredients certainly affect the quality of the final product (Caudle *et al.*, 2005) and the sensory properties of these valuable ingredients should not be overlooked. Dried dairy proteins should ideally be bland or mild and dairy-like in flavor. Recent research has demonstrated that dairy proteins are not flavorless and display a wide array of flavor variability. Understanding and documenting the flavor of these proteins is the key to strategic research and marketing. This chapter addresses and reviews current research on the sensory properties of dairy proteins.

Sensory analysis

Sensory analysis is a scientific discipline that encompasses the depth and breadth of all properties of a food that are perceived by the human senses (Drake, 2007). As such, sensory properties are crucial for product success. Dairy foods continue to enjoy a positive flavor perception by the consumer (Drake and Gerard, 2003; Russell *et al.*, 2006) and this competitive edge should ideally be maximized. This means that a complete understanding of flavor, flavor variability, sources of flavors and consumer perception is mandatory.

A wide array of sensory tests is available to objectively or subjectively measure the sensory properties of foods. These tests and their specific application to dairy products are covered in several recent textbooks and review articles (Lawless and Heymann, 1999; Singh *et al.*, 2003; Drake, 2004, 2007; Meilgaard *et al.*, 2007). Two basic categories exist: analytical tests and affective tests. Several types of tests exist within each category and selection of the specific and appropriate test is dependent on the specific objective in mind. Analytical sensory tests are a group of sensory tests that are objective in nature and use trained or screened panelists. Some examples include descriptive analysis, discrimination tests and threshold tests. Affective tests are subjective tests and comprise tests that use consumers in qualitative or quantitative measurements.

Whey proteins

Whey proteins are recovered from membrane processing and concentration of the liquid whey stream resulting from cheesemaking. Thus, one source of flavor and flavor variability of whey proteins is the flavor of the liquid whey source. The flavor of liquid whey varies, not surprisingly, with the type of cheese (Tomaino *et al.*, 2004; Gallardo Escamilla *et al.*, 2005). The flavor (sensory perception and volatile components) of fresh liquid whey from thermophilic starters (pasta filata cheeses) will differ from that from mesophilic starter cultures (Cheddar cheese). The flavor of whey from direct acid-set curd will deviate further (Table 15.1). The addition of enzymes such as lipases will increase the free fatty acid content of the whey and this will also influence flavor in the form of rancid, waxy and/or animal flavors depending on the lipase and the milk source.

Within a single type of cheese, the flavor of the whey will vary depending on starter culture rotation and/or other variables in the cheesemaking process. Carunchia Whetstine *et al.* (2003) documented tremendous variability in flavor and volatile compound profiles within and between two Cheddar cheese facilities with starter culture rotation. These results were further confirmed by Karagul-Yuceer *et al.* (2003a). Free fatty acid profiles and proteolysis were also distinct. Tomaino *et al.* (2004) documented that cold storage of liquid pasteurized whey increased lipid oxidation products and resulted in cardboard flavors. They speculated that lactic starter culture enzymes accelerated these storage-induced changes because the concentrations of lipid oxidation products were higher in fermented whey than in whey from direct

Table 15.1 Flavor profiles of fluid whey obtained from Mozzarella cheese, Cheddar cheese or acid casein manufacture[a]

Attribute[b]	Cheddar	Mozzarella	Acid casein
Heated milk[c]	41a	20c	27b
Caramelized milk	21a	4b	4b
Natural yoghurt	4c	29a	5c
Stale	9b	9b	22a
Rancid	16c	24b	38a
Oaty	8a	1b	7a
Dirty	2b	7b	35a
Acid	5b	39a	36a
Sweet	37a	8b	12b
Bitter	2b	14a	7ab
Salty	6c	12b	18a

[a]Adapted from Gallardo Escamilla *et al.* (2005)
[b]Attributes were scored on a scale from 0 to 100
[c]Means in a row followed by different letters are different ($p < 0.05$)

acid coagulation. Further, differences in lipid oxidation products were observed in fresh whey from three different single mesophilic starter strains. Clearly, the initial raw product stream in whey protein processing displays tremendous flavor, flavor variability and flavor precursors.

It is not unexpected then that finished dried protein concentrates and isolates also display flavor variability. Liquid whey is subjected to a host of processing techniques to concentrate and separate the whey protein. Pasteurization, membrane filtration, concentration and spray drying are all steps that can induce the formation of flavor compounds. Although there is a general process of whey protein production, each facility is distinct, with facility-specific storage parameters and/or time/temperature profiles that further contribute variability to the finished product. In the case of whey protein from colored Cheddar cheese, a bleaching process with hydrogen or benzoyl peroxide is also involved. The process of oxidizing the whey stream to decolorize it will result in a host of possible flavors and flavor precursors. There is currently no recent published work on the impact of these specific processing steps (other than storage of liquid whey, mentioned in the previous paragraph) on final whey protein flavor.

Application of defined sensory analysis in combination with instrumental volatile analysis has shed light recently on the sources of many dairy flavors and will ultimately aid in the identification of methods to control flavor. This approach recently led to the identification of a method to enhance the nutty flavor in Cheddar cheese (Avsar *et al.*, 2004; Carunchia Whetstine *et al.*, 2006) and the specific identification of a cabbage off-flavor in whey protein isolate (Wright *et al.*, 2006). The reader is referred to three recent reviews on the application of these techniques to control the flavor in dairy products (Singh *et al.*, 2003; Drake, 2004; Drake *et al.*, 2006). A host of defined flavors in whey proteins have been documented (Drake *et al.*, 2003; Carunchia Whetstine *et al.*, 2005b; Drake, 2006; Russell *et al.*, 2006; Wright *et al.*, 2006) (Table 15.2) and volatile compound flavor variability has also been documented (Morr and Ha, 1991; Mills, 1993; Quach *et al.*, 1999; Carunchia Whetstine *et al.*, 2005b; Wright *et al.*, 2006).

The many processing variables listed above undoubtedly contribute to these differences in flavor among fresh products. Figure 15.1 demonstrates the flavor variability in fresh (<1 month old) product collected from different manufacturers. Products 4, 5, 6, 8, 9 and 10 were manufactured from Mozzarella or white Cheddar cheese whey whereas products 1, 2, 3 and 7 were manufactured from colored Cheddar cheese whey. These differences in flavor and volatile compounds also suggest that there are some flavors and volatile components that are specifically formed from whey protein manufacturing bleaching processes and this is certainly an area of research that should be investigated in ongoing efforts to minimize whey protein flavor. The flavor intensities of many whey proteins are comparable with those of soy proteins (Russell *et al.*, 2006) and this is a crucial issue for competitive global marketing.

Storage of whey proteins is another source of flavor variability. The purported shelf life of whey protein concentrate (WPC80) and whey protein isolate (WPI) varies from 12 to 24 months depending on the supplier. There is no published work that demonstrates the flavor stability of these products with storage although recent volatile compound work on WPC80 subjected to accelerated storage conditions has demonstrated key volatile component changes with storage (Javidipour and Qian, 2008).

Table 15.2 Flavors reported in whey proteins (whey protein concentrate [WPC80] and whey protein isolate [WPI]) by sensory analysis[a]

Term	Definition	Reference	Example/Preparation
Overall aroma intensity	The overall orthonasal aroma impact		Evaluated as the lid is removed from the cupped sample
Flavors, tastes, feeling factors (evaluated in the mouth)			
Sweet aromatic	Sweet aroma associated with dairy products		Vanilla cake mix or 20 ppm vanillin in milk
Cooked/milky	Aromatic associated with cooked milk	Cooked milk	Heating skim milk to 85°C for 30 min
Doughy/fatty/Pasta	Aroma associated with canned biscuit dough and cooked pasta	(Z)-4-heptenal	1 ppm (Z)-4-heptenal in water from canned biscuit dough, or cooked pasta water
Fatty/frying oil	Aromatics associated with old frying oil and lipid oxidation products	2,4-Decadienal	Old (stored) vegetable oil
Metallic/meat serum	Aromatics associated with metals or with juices of raw or rare beef	Aroma of fresh raw beef steak or ground beef or juices from seared beef steak	
Cucumber	Aroma associated with freshly sliced cucumber	(E)-2-nonenal	1 ppm (E)-2-nonenal or freshly sliced cucumbers
Brothy	Aromatics associated with broth or boiled potatoes	Methional	1 ppm methional in water or boiled potatoes
Cabbage	Aromatics associated with medium-chain fatty acids and soaps	Dimethyl trisulfide	Boiled fresh cut cabbage, 10 ppb dimethyl trisulfide on filter paper in sniff jar
Cardboard/wet paper	Aroma associated with cardboard	Cardboard paper	Brown cardboard or brown paper bag cut into strips and soaked in water
Animal/wet dog	Aroma associated with wet dog hair	Knox gelatin	One bag of gelatin (28 g) dissolved in two cups of distilled water
Pasta water	Aroma associated with water after pasta has been boiled in it		Pasta boiled in water for 30 min
Soapy	Aroma associated with soap	Lauric acid	1 ppm lauric acid or shaved bar soap
Bitter	Basic taste associated with bitterness	Caffeine	Caffeine, 0.5% in water[b]
Astringency	Drying tongue sensation	Alum	Alum, 1% in water[b]

[a] Adapted from Drake *et al.* (2003), Karagul-Yuceer *et al.* (2003a), Carunchia Whetstine *et al.* (2005b) and Wright *et al.* (2006)
[b] From Meilgaard *et al.* (2007)

Furthermore, these products are often agglomerated to enhance their functional properties. The agglomeration process (rewet or single pass) can include addition of lecithin to further increase wettability.

Both of these processes (agglomeration and agglomeration with lecithin) may impact flavor and decrease shelf life. Figure 15.2 demonstrates the flavor changes with storage time of non-agglomerated, agglomerated and instantized (agglomerated with lecithin) WPC80 from a single supplier, as documented by a trained sensory panel. The results suggest that agglomeration, especially agglomeration with lecithin,

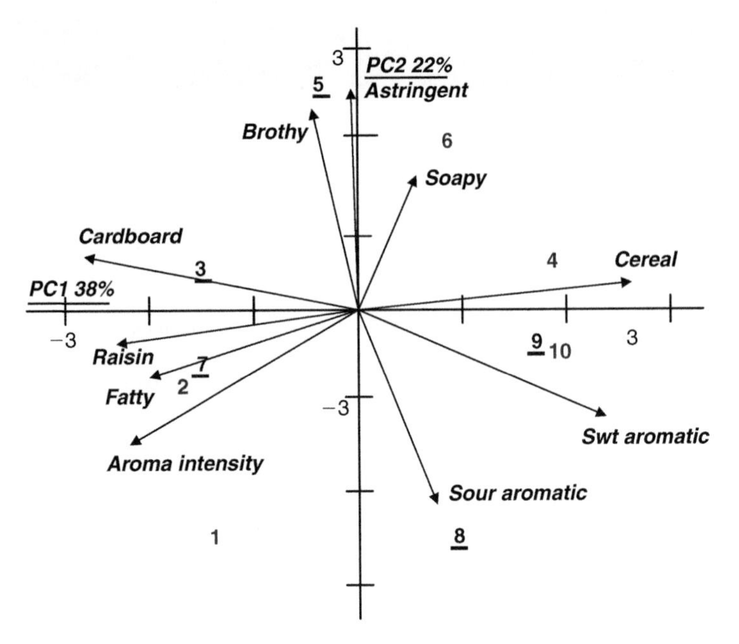

Figure 15.1 Principal component biplot of descriptive sensory analysis of WPC80. The WPC80s are represented by numbers.

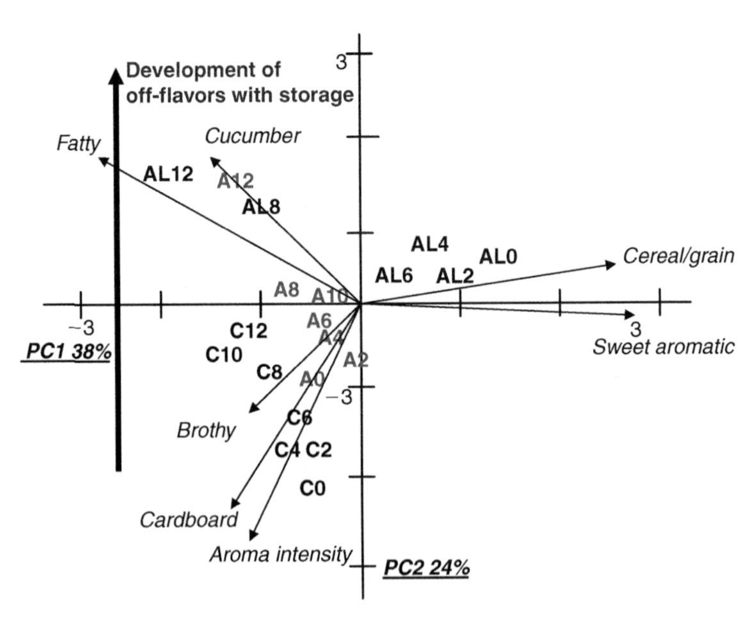

Figure 15.2 Flavor changes during storage for 12 months at 21 °C of non-agglomerated, agglomerated and instantized (agglomerated with lecithin) WPC80s. C, control non-agglomerated product; A, steam-agglomerated product; AL, product agglomerated with lecithin.

affects the storage stability of WPC80, with more rapid development of fatty, cucumber and lipid oxidation types of off-flavors.

Samples of agglomerated and non-agglomerated WPI and WPC80 were collected from suppliers whose products were previously noted to develop a cucumber off-flavor.

The samples were stored at 21°C and were monitored by descriptive sensory analysis (rehydrated to 10% solids w/w) and by instrumental volatile analysis (headspace solid phase microextraction gas chromatography–olfactometry [HS-SPME GC–O] with gas chromatography–mass spectrometry [GC–MS]).

Samples for volatile analysis were prepared as previously described (Wright *et al.*, 2006). Briefly, 20 g of each reconstituted whey protein, a stirring bar and 1 g of NaCl were placed in a 40 mL amber glass SPME vial and sealed air tight with a Teflon™-sided silicon septum (PTFE/silicon) and a plastic cap (Supelco, Bellefonte, Pennsylvania, USA). Samples were heated to 40°C and stirred for 30 min before the SPME fiber (three phase: 2 cm– 50/30 μm DVB/Carboxed™/PDMS Stable Flex, Supelco) was exposed in the headspace at a depth of 3.8 cm for an additional 30 min prior to injection on to the gas chromatograph. The fiber was desorbed at 250°C for 5 min in the injection port fitted with an SPME inlet at a depth of 7.6 cm.

GC–O analysis was performed using an HP 5890 series II gas chromatograph equipped with a flame ionization detector (FID), a sniffing port and a splitless injector. For GC–MS, samples were prepared analogously prior to injection on to the GC–MS system by a CTC Analytics CombiPal Autosampler (Zwingen, Switzerland). The fiber was desorbed at 250°C for 5 min in the injection port fitted with an SPME inlet at a depth of 50 mm. GC–MS analysis was performed using an Agilent 6890N gas chromatograph with a 5973 inert MSD with a DB-5ms (20 m × 0.25 mm internal diameter × 0.25 μm film thickness) column. Each sample was analyzed in triplicate.

The sensory profiles confirm that agglomerated samples, particularly those with lecithin, are more prone to cucumber off-flavor (Figures. 15.3 and 15.4). A number of

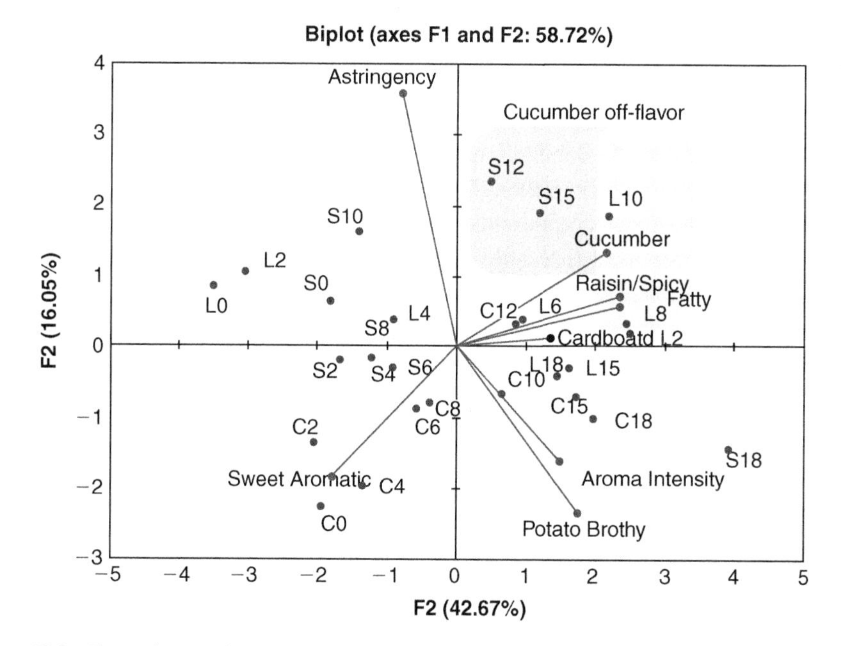

Figure 15.3 Flavor changes during storage for 18 months at 21°C of non-agglomerated, agglomerated and instantized (agglomerated with lecithin) WPC80s. L, lecithinated; C, control; S, steam agglomerated. Number indicates months of storage.

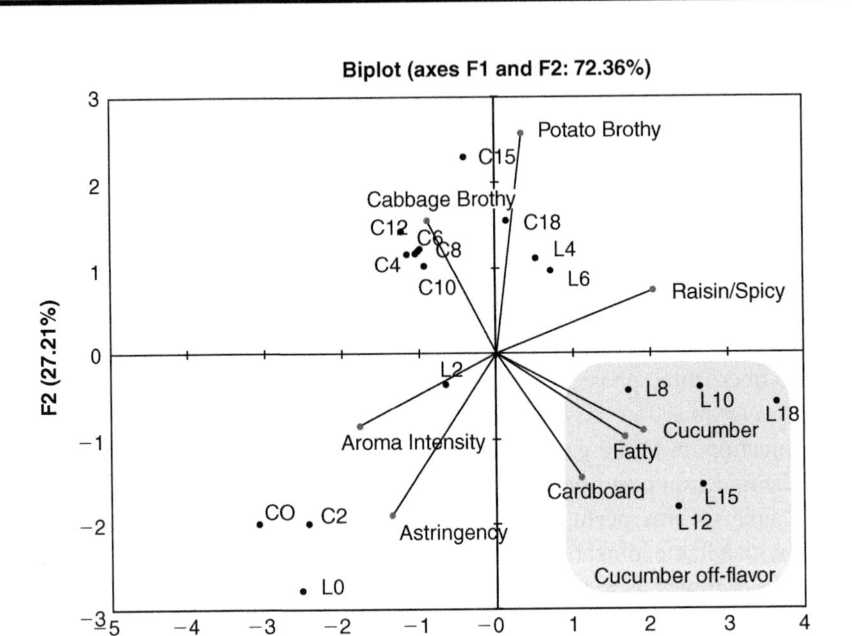

Figure 15.4 Flavor changes during storage for 18 months at 21°C of non-agglomerated, agglomerated and instantized (agglomerated with lecithin) WPIs. L, lecithinated agglomerated; C, control. Number indicates months of storage.

compounds with cucumber or fatty aromas were recorded by GC–O in samples with and without cucumber flavor (Table 15.3). These compounds were unfortunately at or below MS detection limits by the extraction technique used. The aroma of a compound when it is isolated is not necessarily the aroma or flavor that compound elicits when it is in a food matrix, which means that sensory analysis of the compound in the food matrix is recommended (Drake and Civille, 2003). The character of an aroma can also change with compound concentration (Drake and Civille, 2003).

However, when suspect compounds were placed into WPC80 without cucumber flavor within their reported threshold range and were presented to trained panelists ($n = 8$), many of them elicited cucumber flavors (Table 15.4), suggesting that one or a combination of these compounds is responsible for this off-flavor that develops during the storage of whey proteins. In agreement with the previous example (Figure 15.2), these compounds are also lipid oxidation compounds, again indicating that lipid oxidation is a major source of off-flavor development in these protein products.

Recent work has suggested that native whey proteins might provide a product with the functional and nutritional benefits of whey proteins with superior flavor properties. Native whey proteins are simply whey proteins that are removed from fluid milk prior to the initiation of cheesemaking. In fact, serum or whey proteins can be removed from fluid milk and the cheesemaking procedure can subsequently be initiated as normal with few or no effects on cheese yield. As the

Table 15.3 Aroma-active "green" compounds identified by HS-SPME GC–O from stored agglomerated and non-agglomerated WPC80 and WPI. C refers to non-agglomerated product, S to steam agglomerated product and L to product agglomerated with added lecithin. The number following the treatment letter designation indicates storage time at 21 °C in months

WPC80

Compound	Odor[a]	Post Peak Intensity[b]												RI[c]	Method of ID[d]
		C8	C10	C12	C14	S8	S10	S12	S14	L8	L10	L12	L14		
Hexanal	grassy/green	2.50	2.50	2.50	3.00	3.25	2.50	2.50	3.00	2.50	2.75	1.50	3.25	806	RI, odor, MS[e]
E-2-nonenal	carpet/green	ND[f]	2.00	1.50	2.25	3.00	2.00	3.00	ND	ND	2.75	2.00	3.00	1163	RI, odor
E,Z-2,6-nonadienal	cucumber/herb	ND	3.00	2.00	ND	3.50	ND	3.00	ND	2.00	3.00	ND	3.00	1168	RI, odor
Isobutyl-methoxy-pyrazine	green pepper	3.00	ND	ND	ND	3.00	ND	ND	ND	3.00	2.50	ND	ND	1185	RI, odor
E,Z-2,4-nonadienal	carpet/green	ND	ND	ND	ND	ND	ND	ND	2.25	1.75	ND	ND	ND	1189	RI, odor
6-Decenal	cucumber/rosy	2.50	ND	2.50	ND	2.00	ND	1.50	ND	ND	2.00	1.50	ND	1205	RI, odor

Table 15.3 (Continued)

WPI

Compound	Odor[a]	Post Peak Intensity[b]								RI[c]	Method of ID[d]
		C8	C10	C12	C14	L8	L10	L12	L14		
Methyl-2-butenol	green/hay	ND[f]	ND	ND	ND	1.50	ND	ND	ND	777	RI, odor
Hexanal	Grassy/green	ND	2.25	1	2	2.00	2.00	1.5	2.5	806	RI, odor, MS[e]
E-2-nonenal	Carpet/green	ND	ND	ND	ND	ND	1.75	ND	ND	1163	RI, odor
E,Z-2,6-nonadienal	cucumber/herb	ND	ND	ND	ND	1.75	1.75	ND	3.0	1168	RI, odor
Isobutyl-methoxy-pyrazine	green pepper	3.25	2.50	ND	ND	5.00	ND	ND	ND	1185	RI, odor
E,Z-2,4-nonadienal	Carpet/green	ND	ND	ND	ND	2.00	ND	ND	ND	1189	RI, odor
6-Decenal	cucumber/rosy	2.75	ND	ND	ND	3.50	1.50	ND	ND	1205	RI, odor

[a] Odor description at the gas chromatograph sniffing port
[b] Mean post peak intensities as determined by two experienced sniffers at the gas chromatograph sniffing port ([37]van Ruth, 2001)
[c] Retention index was calculated from GC–O data on a DB-5 column
[d] Compounds were identified by comparison with authentic standards on the following criteria: retention index (RI) on a DB-5 column, odor property at the gas chromatograph sniffing port and mass spectra in the electron impact mode. Positive identifications indicate that mass spectral data were compared with authentic standards.
[e] MS = mass spectra
[f] ND = not detected

Table 15.4 Sensory analysis of model whey protein systems for cucumber flavor. WPC80 without discernible cucumber flavor was used as the base. WPC80 was rehydrated to 10% solids (w/w) and then spiked with suspect compounds at their reported threshold concentration

Compound	Concentration in WPC80 model (ppb)	Aroma[a]	Reported threshold (ppb)
E-2-nonenal	0.4	New carpet/cucumber	0.4
E-2-nonenal	4.0	Cucumber	
E,Z-2,6-nonadienal	0.14	Cucumber	0.14
E,Z-2,6-nonadienal	1.4	Cucumber/fatty	
E,Z-2,4-nonadienal	0.158	Fatty/cucumber	Not reported
E,Z-2,4-nonadienal	1.58	Fatty/cucumber	
Hexanal	73.3	Sweet chemical-like	50
2-Pentyl furan	7.97	Cucumber/fatty (very faint)	6
2-Pentyl furan	79.7	Hay/licorice	
2-Ethyl 1 hexanol	300	Chemical/cleaning agent	300

[a]Aroma as perceived by eight trained sensory panelists who were also experienced GC–O sniffers. Panelists were provided with the list of descriptors used for GC–O and were asked to select one or two descriptors that best described the aroma.

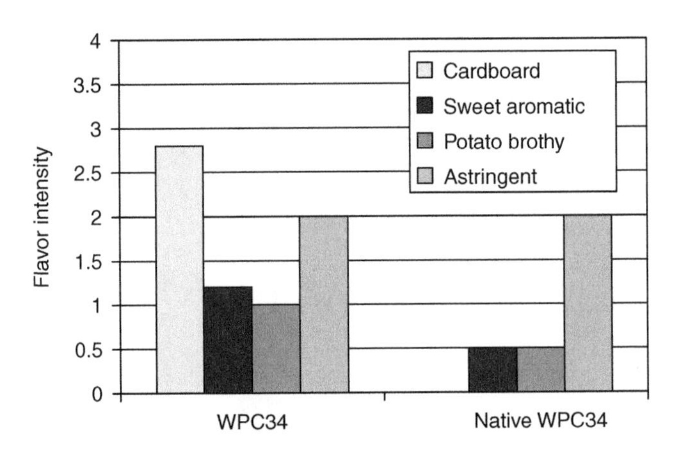

Figure 15.5 Sensory profiles of rehydrated WPC34 from Cheddar whey and rehydrated native WPC34 (serum proteins).

native whey proteins have not been subjected to the normal cheesemaking and whey protein processing procedures, their flavor profiles are remarkably bland and nearly free from flavor (Figure 15.5). Significant economic challenges face the industrial scale-up of these products but, at a minimum, research with these products may reveal key information to minimizing the flavor of traditional whey proteins.

Dairy products and, by association, dairy ingredients continue to enjoy a positive flavor image by consumers (Drake and Gerard, 2003; Russell *et al.*, 2006). However, many consumers are unaware that whey proteins are dairy proteins and this is a

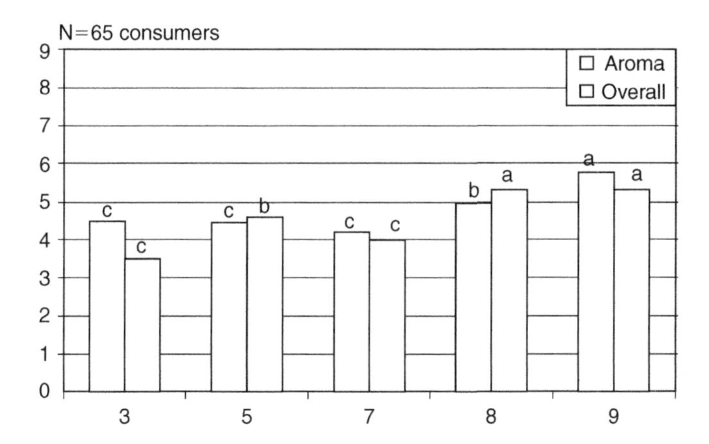

Figure 15.6 Aroma and overall liking of peach beverages with 7% (w/w) WPC80. Numbers represent beverages with fresh WPC80 from different manufacturers (Figure 15.1, underlined samples). Letters above bars within each attribute represent significant differences ($p < 0.05$).

sensory issue because consumer perception influences consumer liking. Drake (2006) documented that consumers were generally less sure of their responses when asked to comment on the properties of specific protein types. The US consumer was generally less informed about whey proteins and was more confident and aware of soy protein than New Zealand consumers (Jones *et al.*, 2008). In a follow-up study (Childs *et al.*, 2007), focus groups with US consumers confirmed that most US consumers were unaware that whey proteins were dairy or milk proteins. Consumer education is a current challenge to the dairy protein industry.

One other issue pertinent to whey proteins and consumer acceptance is whether the flavor variability documented by trained panelists is detected by consumers and/or whether it affects the quality of the finished product. Intuitively, the freshest and highest quality ingredients make the best finished product. However, research also indicates that consumers can discern differences in whey protein flavors and that these flavors carry through into ingredient applications (Drake, 2006). Figure 15.6 demonstrates this concept with protein beverages manufactured from different fresh WPC80s. The nature of the off-flavor and the ingredient application will also influence flavor carry-through (Drake, 2006).

Some ingredient applications will be more tolerant than others of variability in the flavor of the ingredients. Childs *et al.* (2007) recently demonstrated flavor and texture/mouthfeel differences between meal replacement beverages and bars made with whey proteins or soy proteins or mixtures of whey and soy proteins. The ingredient applications were made using standard formulas to allow direct comparison of the influence of the different proteins. Trained panelists documented discernible flavor carry-through of whey and soy proteins in vanilla meal replacement beverages (Figure 15.7). In contrast, no differences in flavor between bars made with whey or soy proteins were noted although several differences in bar texture were impacted by the protein type (Figures. 15.8 and 15.9).

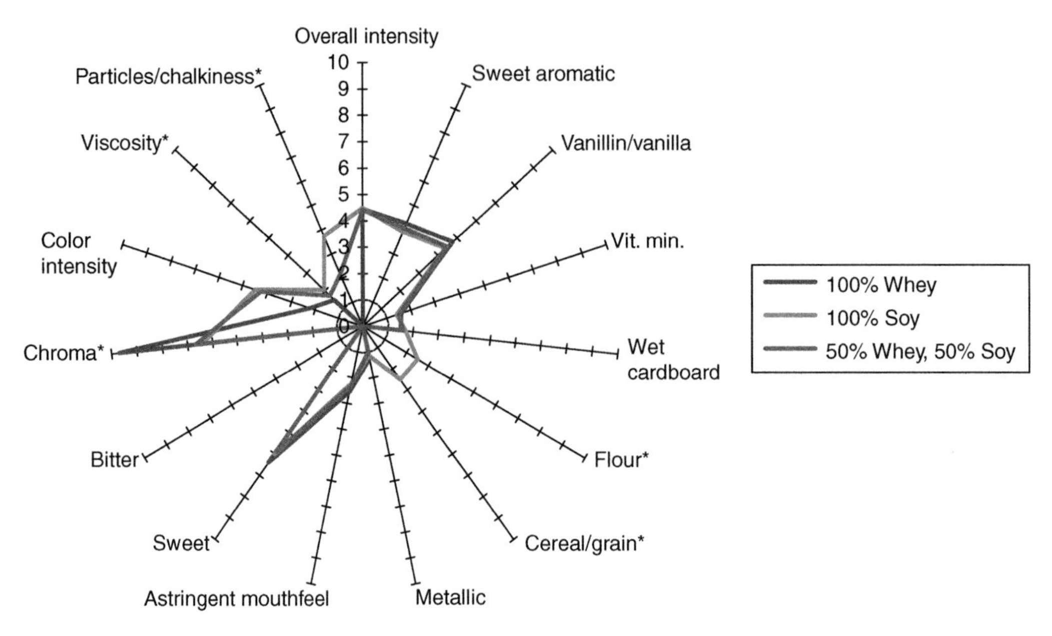

Figure 15.7 Trained panel flavor and mouthfeel profiles of vanilla meal replacement shakes made with whey protein, soy protein or a mixture of whey protein and soy protein. * Indicates significant attributes ($p < 0.05$). Adapted from Childs *et al.* (2007).

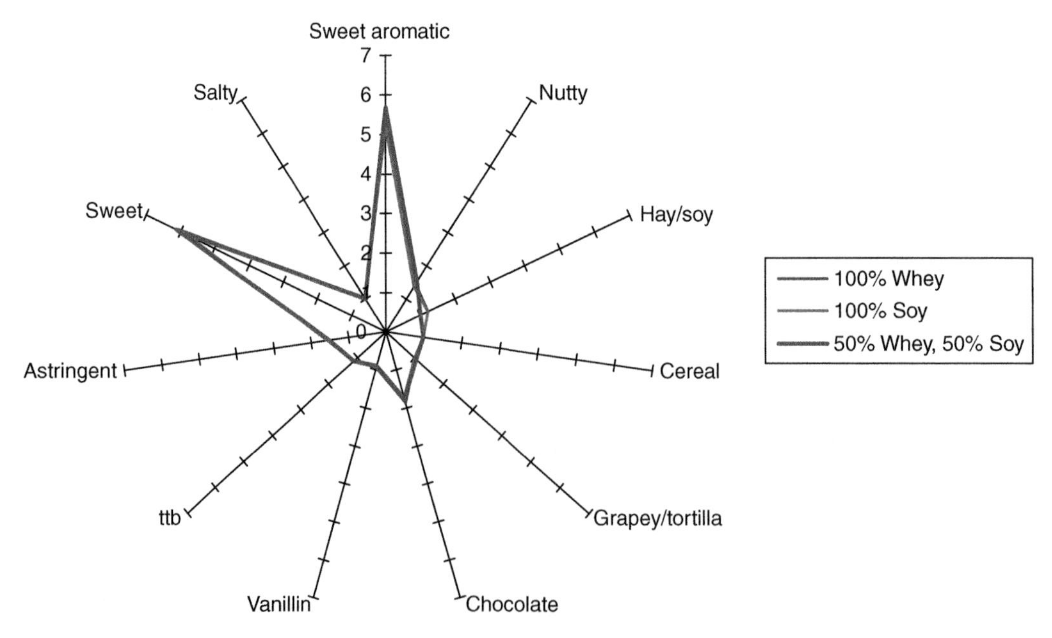

Figure 15.8 Trained panel flavor profiles of peanut-butter-flavored meal replacement bars made with whey protein, soy protein or a mixture of whey protein and soy protein. No attribute differences were noted ($p > 0.05$). Adapted from Childs *et al.* (2007).

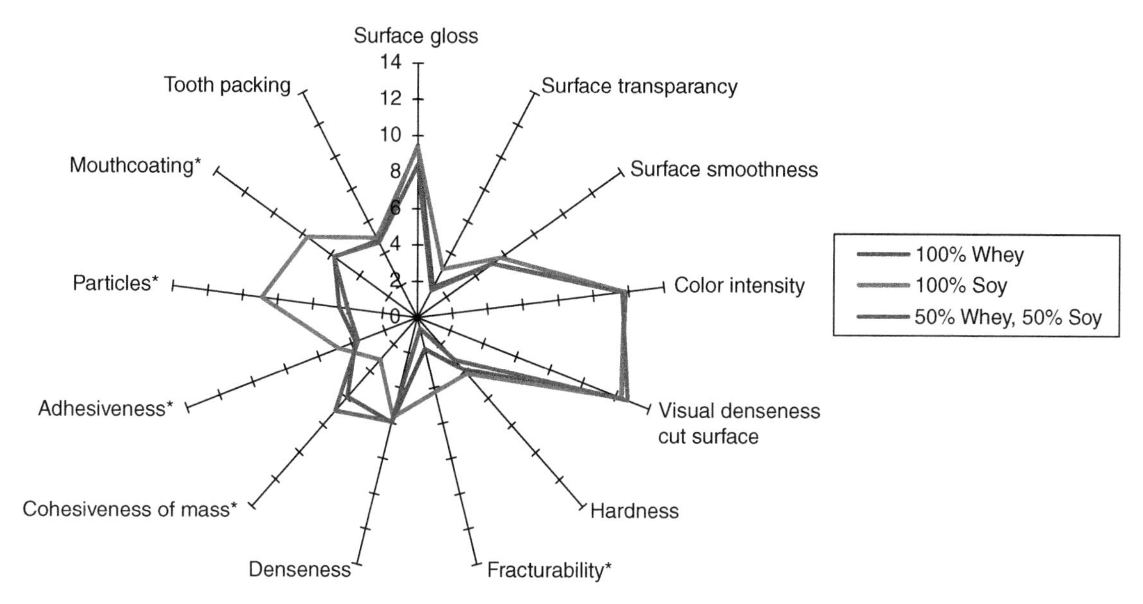

Figure 15.9 Trained panel texture profiles of peanut-butter-flavored meal replacement bars made with whey protein, soy protein or a mixture of whey protein and soy protein. * Indicates significant attributes ($p < 0.05$). Adapted from Childs *et al.* (2007).

Milk proteins

Milk protein concentrates (MPCs) and isolates (MPIs) represent a newer category of dried dairy ingredients that are rapidly gaining in popularity. These products are manufactured by concentrating milk proteins (whey proteins and caseins) from fluid milk by membrane processing followed by spray drying. Recent work in the primary author's laboratory has addressed the sensory properties of milk proteins across increasing protein concentration. MPCs with lower protein contents (56, 70% protein dry weight) are characterized by fluid milk types of flavors: cooked/milky, sweet aromatic, sweet taste and cereal (Figure 15.10). As the protein content is increased, the flavor profiles change and MPC77, MPC80 and MPI are characterized by tortilla, brothy, cardboard and animal flavors and higher astringency.

Changes in flavor with increasing protein content were also observed when the sensory properties of lower protein WPCs were compared with those of WPC80 and WPI. Increases in whey protein content likewise resulted in decreases in sweet aromatic and milky flavors. These changes in flavor are probably directly linked to changes in composition and different concentrations of resulting volatile components. A comparison of aroma-active volatile components isolated from WPC80/WPI and whey powder revealed few differences (Mahajan *et al.*, 2004; Carunchia Whetstine *et al.*, 2005b). Differences in flavor are probably due to differences in the relative abundance of specific compounds. Similarly, volatile compound changes are evident in MPCs and MPIs as the protein content is increased. MPCs with higher final protein content have lower sulfur compound response as well as lower aldehyde

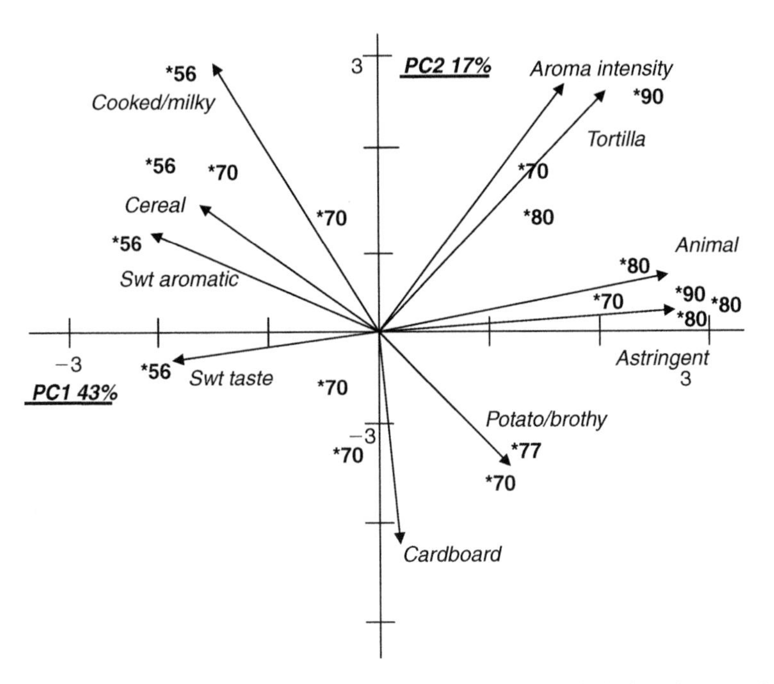

Figure 15.10 Principal component biplot of trained panel flavor profiles of rehydrated MPC and MPI. Number indicates protein content.

levels when analyzed by HS-SPME techniques (Table 15.5). Changes appear to be due to changes in relative abundance rather than the evolution of new compounds.

Caseins and hydrolysates

Caseins represent the primary protein constituent of milk; whey or serum proteins are the other fraction. Just as whey proteins comprise a large group of functional ingredients, so do caseins and caseinates. Caseins are traditionally produced by acid or rennet precipitation of casein followed by spray or roller drying. Caseinate or soluble casein is produced when casein curd (usually acid precipitated) is treated with alkali at pH 6–7 and fully dissolved prior to spray drying (O'Connell and Flynn, 2007). Potassium, sodium and calcium are commonly used counter-ions.

Caseins display a unique set of functional properties including solubility and heat stability and are thus used for a host of ingredient applications. Caseins have relied on functionality for their success because a host of relatively intense and unpleasant flavors, including sulfur, animal, tortilla, musty, cardboard, burnt feathers, glue and bitter taste, have been associated with them (Ramshaw and Dunstone, 1969; Walker and Manning, 1976; Drake *et al.*, 2003; Karagul-Yuceer *et al.*, 2003b) (Figure 15.11). Micellar casein can be manufactured by membrane fractionation and spray drying of fluid milk and may represent a blander option. The net result is a more mildly flavored product that still displays some of the previously reported flavors.

Dairy protein hydrolysates are another promising category of protein-derived ingredients with valuable functional and nutritional properties (Nnanna and Wu, 2007).

Table 15.5 Mean relative concentration (ppb) of selected volatile components extracted from the headspace of rehydrated (10% solids w/w) domestic and international MPCs with various protein contents; all concentrations given as mean relative concentration and (standard deviation)

MPC Sample	Protein (%)	Dimethyl sulfide	Propanal, 2-methyl-	Furan, 2-methyl-	Butanal, 3-methyl-	Butanoic acid, methyl ester	Hexanal	2-Heptanone	Heptanal	Hexanoic acid, methyl ester	Pentanoic acid, 1-methyl ethyl ester	Benzaldehyde	Furan, 2-pentyl-	Octanal	2-Nonanone	Nonanal	Octanoic acid, methyl ester	Decanal
1	56	0.66	2.33	0.22	0.07	0.22	2.40	0.33	0.44	0.15	0.50	0.32	0.56	0.17	0.18	1.12	0.56	0.10
		(0.18)	(0.06)	(0.03)	(0.10)	(0.03)	(0.14)	(0.07)	(0.13)	(0.01)	(0.07)	(0.05)	(0.02)	(0.02)	(0.02)	(0.22)	(0.05)	(0.02)
2	56	0.45	2.73	0.21	0.35	0.43	7.44	0.70	1.72	1.11	0.46	1.49	1.86	0.36	0.33	1.43	0.76	0.13
		(0.04)	(0.42)	(0.02)	(0.03)	(0.01)	(0.13)	(0.01)	(0.11)	(0.18)	(0.01)	(0.10)	(0.01)	(0.00)	(0.01)	(0.04)	(0.01)	(0.02)
3	56	0.53	2.70	0.15	0.20	0.26	3.12	0.74	0.49	0.06	0.49	0.61	0.71	0.18	0.30	1.00	0.08	0.19
		(0.00)	(0.69)	(0.00)	(0.02)	(0.01)	(0.06)	(0.03)	(0.02)	(0.01)	(0.02)	(0.24)	(0.00)	(0.01)	(0.02)	(0.04)	(0.00)	(0.03)
4	56	0.26	3.72	0.11	0.11	0.18	3.96	0.59	0.59	0.07	0.53	0.77	0.80	0.26	0.24	1.07	0.09	0.18
		(0.02)	(0.09)	(0.00)	(0.01)	(0.02)	(0.08)	(0.00)	(0.11)	(0.02)	(0.01)	(0.05)	(0.01)	(0.01)	(0.02)	(0.03)	(0.01)	(0.02)
5	70	0.23	2.56	0.11	0.00	0.24	2.25	0.29	0.40	0.04	0.45	0.95	0.43	0.15	0.17	1.29	0.17	0.10
		(0.01)	(0.44)	(0.01)	(0.00)	(0.02)	(0.04)	(0.04)	(0.04)	(0.01)	(0.02)	(0.01)	(0.01)	(0.00)	(0.00)	(0.02)	(0.01)	(0.01)
6	70	0.39	2.50	0.08	0.00	0.31	0.27	0.36	0.05	0.12	0.52	0.26	0.41	0.08	0.30	0.18	0.38	0.16
		(0.05)	(0.16)	(0.01)	(0.00)	(0.05)	(0.04)	(0.01)	(0.01)	(0.03)	(0.01)	(0.01)	(0.03)	(0.00)	(0.01)	(0.02)	(0.01)	(0.01)
7	70	0.18	2.60	0.09	0.14	0.29	3.27	0.39	0.44	0.30	0.53	1.03	0.94	0.20	0.22	1.37	1.04	0.15
		(0.02)	(0.44)	(0.01)	(0.06)	(0.07)	(0.42)	(0.07)	(0.04)	(0.06)	(0.07)	(0.12)	(0.06)	(0.02)	(0.02)	(0.20)	(0.03)	(0.01)
8	70	0.34	4.13	0.06	0.07	0.30	2.34	0.20	0.58	0.15	0.52	0.23	0.20	0.16	0.16	1.28	0.20	0.18
		(0.07)	(0.45)	(0.02)	(0.01)	(0.13)	(0.31)	(0.01)	(0.19)	(0.01)	(0.00)	(0.00)	(0.03)	(0.06)	(0.02)	(0.21)	(0.01)	(0.02)
9	70	0.00	2.69	0.29	0.14	0.28	0.94	0.15	0.94	0.03	0.58	1.61	0.09	0.18	0.10	0.80	0.01	0.07
		(0.00)	(0.04)	(0.01)	(0.01)	(0.08)	(0.03)	(0.00)	(0.05)	(0.00)	(0.02)	(0.13)	(0.03)	(0.05)	(0.03)	(0.01)	(0.01)	(0.01)
10	70	0.11	3.08	0.04	0.08	0.84	0.82	0.39	0.21	1.21	0.52	0.19	0.12	0.06	0.16	0.17	1.41	0.06
		(0.02)	(0.40)	(0.01)	(0.01)	(0.24)	(0.21)	(0.04)	(0.02)	(0.04)	(0.01)	(0.02)	(0.04)	(0.00)	(0.01)	(0.01)	(0.06)	(0.00)
11	80	0.01	1.62	0.01	0.17	0.15	2.97	0.72	0.46	0.04	0.44	2.08	1.75	0.17	0.75	0.92	0.10	0.24
		(0.02)	(1.00)	(0.01)	(0.09)	(0.14)	(0.29)	(0.04)	(0.08)	(0.00)	(0.05)	(0.20)	(0.08)	(0.03)	(0.04)	(0.08)	(0.00)	(0.03)
12	80	0.00	4.20	0.02	0.03	0.19	1.92	0.30	0.26	0.03	0.45	1.83	0.34	0.11	0.12	0.33	0.00	0.13
		(0.00)	(0.30)	(0.01)	(0.05)	(0.00)	(0.19)	(0.02)	(0.13)	(0.00)	(0.04)	(0.13)	(0.04)	(0.05)	(0.01)	(0.05)	(0.00)	(0.02)
13	80	0.00	0.00	0.03	0.04	0.14	0.49	0.82	0.09	0.19	0.26	0.00	0.00	0.00	0.00	0.00	1.69	2.39
		(0.00)	(0.00)	(0.04)	(0.06)	(0.20)	(0.69)	(0.74)	(0.13)	(0.27)	(0.37)	(0.00)	(0.00)	(0.00)	(0.00)	(0.00)	(1.61)	(3.28)
14	80	0.06	2.35	0.02	0.00	0.20	0.47	0.21	0.32	0.03	0.49	0.46	0.03	0.04	0.09	0.32	0.08	0.08
		(0.00)	(0.04)	(0.02)	(0.00)	(0.00)	(0.00)	(0.04)	(0.05)	(0.00)	(0.06)	(0.04)	(0.01)	(0.02)	(0.00)	(0.05)	(0.02)	(0.00)

Table 15.5 (Continued) Mean relative abundance of volatile compounds averaged across protein levels; MPC volatile compound mean relative concentration (ppb); all concentrations given as mean relative concentration

Protein (%)	Dimethyl sulfide	Propanal, 2-methyl-	Furan, 2-methyl-	Butanal, 3-methyl-	Butanoic acid, methyl ester	Hexanal	2-Heptanone	Heptanal	Hexanoic acid, methyl ester	Pentanoic acid, 1-methyl ethyl ester	Benzaldehyde	Furan, 2-pentyl-	Octanal	2-Nonanone	Nonanal	Octanoic acid, methyl ester	Decanal
56	0.48	2.87	0.17	0.18	0.27	4.23	0.59	0.81	0.35	0.50	0.80	0.98	0.24	0.26	1.16	0.37	0.15
70	0.21	2.93	0.11	0.07	0.38	1.65	0.30	0.44	0.31	0.52	0.71	0.37	0.14	0.19	0.85	0.54	0.12
80	0.02	2.04	0.02	0.06	0.17	1.46	0.51	0.28	0.07	0.41	1.09	0.53	0.08	0.24	0.39	0.47	0.71

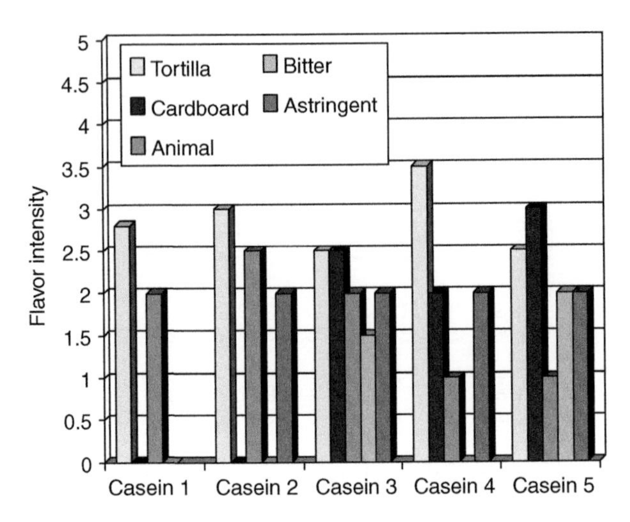

Figure 15.11 Trained panel flavor profiles of rehydrated caseins (10% solids [w/w]). Caseins 1 and 2 are rennet caseins, caseins 3 and 5 are acid caseins and casein 4 is a sodium caseinate.

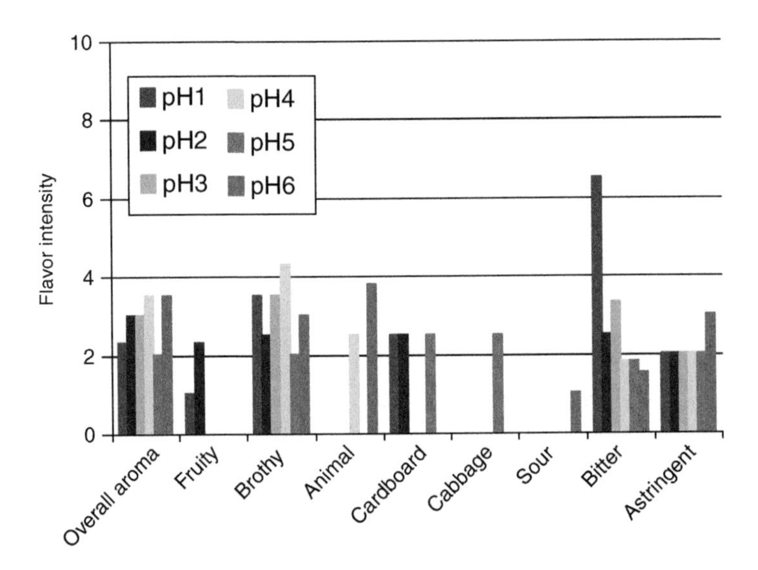

Figure 15.12 Trained panel flavor profiles of commercial rehydrated whey protein hydrolysates (5% solids [w/w]). PH1 has a higher degree of hydrolysis than PH2–PH6. PH2–PH6 represent different enzymatic digestions.

Hydrolysis improves digestibility and hydrolysates are used widely in infant formulas. Recent research has demonstrated that peptides with specific bioactive properties can also be generated and certainly an array of functional properties such as solubility and heat stability can be altered via hydrolysis.

Whey protein hydrolysates are commonly added to meal replacement bars to inhibit bar hardening with storage time (Childs *et al.*, 2007). These products can be manufactured from casein, whey protein or milk protein by enzymatic hydrolysis and are classified based on their degree of protein hydrolysis (molecular weight). Flavor is a significant challenge to increased usage of these products, particularly in beverages. Brothy and free fatty acid flavors and bitter taste are distinct (Figure 15.12) and intensities can vary with the degree of hydrolysis, the specific processing steps, the enzyme used and the protein source. Because of their intense aromas and flavors, these products should be rehydrated to a lower solids concentration prior to sensory analysis (e.g. 5% w/w compared with 10% w/w for all other dairy proteins).

Flavor binding

Although somewhat beyond the scope of this chapter, it is important to note that, in addition to displaying and contributing flavors, dairy proteins can interact and bind with desirable flavors in foods and influence flavor in this manner as well. An excellent review on this subject has been published recently (Kuhn *et al.*, 2006). Most of the research in this arena has been conducted with instrumental analysis (e.g. headspace analysis and calculation of binding constants) and very little research to relate these results directly back to sensory perception has been attempted. Future research should address this issue.

Conclusions

Applications for dairy proteins continue to increase and flavor will remain a crucial aspect. An abundance of research on the functional properties of dairy proteins exists, but there is still a relative dearth of information on the flavor of dairy proteins. Flavor sources, flavor formation during processing and flavor carry-through and stability in ingredient applications are key areas for future research.

The flavor of dairy proteins and their flavor performance in ingredient applications will ultimately influence their widespread usage and competitiveness with other protein sources. Published research has only recently begun to reflect and emphasize the importance of this issue. The positive flavor image of dairy foods, combined with the numerous functional and nutritional benefits of dairy proteins, provides a powerful marketing juggernaut for these products, but specific flavor properties and the flavor variability of these proteins should not be overlooked in ongoing research.

Acknowledgments

The authors gratefully acknowledge Dairy Management, Inc. and the California Dairy Research Foundation for providing financial support.

References

Avsar, Y. K., Karagul-Yuceer, Y., Drake, M. A., Singh, T., Yoon, Y. and Cadwallader, K. R. (2004). Characterization of nutty flavor in Cheddar cheese. *Journal of Dairy Science*, **87**, 1999–2010.

Carunchia Whetstine, M. E., Parker, J. D., Drake, M. A. and Larick, D. K. (2003). Determining flavor and flavor variability in commercially produced liquid Cheddar whey. *Journal of Dairy Science*, **86**, 439–48.

Carunchia Whetstine, M. E., Cadwallader, K. R. and Drake, M. A. (2005a). Characterization of aroma compounds responsible for the rosy/floral flavor in Cheddar cheese. *Journal of Agricultural and Food Chemistry*, **53**, 3126–32.

Carunchia Whetstine, M. E., Croissant, A. E. and Drake, M. A. (2005b). Characterization of WPC80 and WPI flavor. *Journal of Dairy Science*, **88**, 3826–29.

Carunchia Whetstine, M. E., Drake, M. A., Broadbent, J. R. and McMahon, D. J. (2006). Enhanced nutty flavor formation in Cheddar cheese made with a "malty" *Lactococcus lactis* adjunct culture. *Journal of Dairy Science*, **89**, 3277–84.

Caudle, A. D., Yoon, Y. and Drake, M. A. (2005). Influence of flavor variability in skim milk powder on consumer acceptability of ingredient applications. *Journal of Food Science*, **70**, S427–431.

Childs, J. L., Yates, M. D. and Drake, M. A. (2007). Sensory properties of meal replacement bars and beverages made from whey or soy proteins. *Journal of Food Science*, **72**, S425–434.

Drake, M. A. (2004). Defining dairy flavors. *Journal of Dairy Science*, **87**, 777–784.

Drake, M. A. (2006). Flavor and flavor carry-through of whey proteins in beverages. In *The Wonders of Whey … Catch the Power. Proceedings of the 4th International Whey Conference*, pp. 292–300. American Dairy Products Institute, Elmhurst, Illinois.

Drake, M. A. (2007). Sensory analysis of dairy foods. *Journal of Dairy Science*, **90**, 4925–37.

Drake, M. A. and Civille, G. V. (2003). Flavor lexicons. *Comprehensive Reviews in Food Science and Food Safety*, **2**, 33–40.

Drake, M. A. and Gerard, P. D. (2003). Consumer attitudes and acceptability of soy-fortified yogurts. *Journal of Food Science*, **68**, 1118–22.

Drake, M. A., Karagul-Yuceer, Y., Cadwallader, K. R., Civille, G. V. and Tong, P. S. (2003). Determination of the sensory attributes of dried milk powders and dairy ingredients. *Journal of Sensory Studies*, **18**, 199–216.

Drake, M. A., Miracle, R. E., Caudle, A. D. and Cadwallader, K. R. (2006). Relating sensory and instrumental analyses. In *Sensory-directed Flavor Analysis* (R. Marsili, ed.) *Chapter 2*, pp. 23–55. Boca Raton, Florida: CRC Press, Taylor and Francis Publishing.

Drake, S. L., Carunchia Whetstine, M. E., Drake, M. A., Courtney, P., Fligner, K., Jenkins, J. and Pruitt, C. (2007). Sources of umami taste in Cheddar and Swiss cheeses. *Journal of Food Science*, **72**, S360–66.

Foegeding, E. A., Davis, J. P., Doucet, D. and McGuffey, M. (2002). Advances in modifying and understanding whey protein functionally. *Trends in Food Science and Technology*, **13**, 151–159.

Gallardo Escamilla, F. J., Kelly, A. L. and Delahunty, C. M. (2005). Sensory characteristics and related volatile flavor compound profiles of different types of whey. *Journal of Dairy Science*, **88**, 2689–99.

Javidipour, I. and Qian, M. (2008). Volatile component change in whey protein concentrate during storage investigated by headspace solid-phase microextraction gas chromatography. *Lait*, **88**, 95–104.

Jones, V. S., Drake, M. A., Harding, R. and Kuhn-Sherlock, B. (2008). Consumer perception of soy and dairy products: a cross-cultural study. *Journal of Sensory Studies*, **23**, 65–79.

Karagul-Yuceer, Y., Drake, M. A. and Cadwallader, K. R. (2003a). Aroma active components of liquid Cheddar whey. *Journal of Food Science*, **68**, 1215–19.

Karagul-Yuceer, Y., Vlahovich, K. N., Drake, M. A. and Cadwallader, K. R. (2003b). Characteristic aroma components of rennet casein. *Journal of Agricultural and Food Chemistry*, **51**, 6797–801.

Kuhn, J., Considine, T. and Singh, H. (2006). Interactions of milk proteins and volatile flavor compounds: implications in the development of protein foods. *Journal of Food Science*, **71**, R72–R82.

Lawless, H. T. and Heymann, H. (1999). *Sensory Evaluation of Food: Principles and Practices*, 1st edn. New York: Chapman and Hall.

Mahajan, S. S., Goddik, L. and Qian, M. C. (2004). Aroma compounds in sweet whey powder. *Journal of Dairy Science*, **87**, 4057–63.

Meilgaard, M. M., Civille, G. V. and Carr, T. (2007). *Sensory Evaluation Techniques*, 4th edn. New York: CRC Press.

Miller, G. (2005). Healthy growth ahead for wellness drinks. *Food Technology*, **59**, 21–26.

Mills, O. E. (1993). Flavor of whey protein concentrate. In Food Flavors, Ingredients, and Composition. Proceedings of the 7th International Flavor Conference, pp. 139–149. New York: Elsevier.

Morr, C. V. and Ha, E. Y. W. (1991). Off-flavors in whey protein concentrates: a literature review. *International Dairy Journal*, **1**, 1–11.

Nnanna, I. A. and Wu, C. (2007). Dairy protein hydrolysates. In *Handbook of Food Products Manufacturing* (Y. H. Hui, ed.) *Chapter 72,* pp. 537–56. Hoboken, New Jersey: John Wiley and Sons, Inc.

O'Connell, J. E. and Flynn, C. (2007). The manufacture and applications of casein-derived ingredients. In *Handbook of Food Products Manufacturing* (Y. H. Hui, ed.) *Chapter 73,* pp. 557–91. Hoboken, New Jersey: John Wiley and Sons, Inc.

Quach, M. L., Chen, X. G. and Stevenson, R. L. (1999). Headspace samplings of whey protein concentrate solutions using solid phase microextraction. *Food Research International*, **31**, 371–79.

Ramshaw, E. H. and Dunstone, E. A. (1969). Volatile compounds associated with the off-flavor in stored casein. *Journal of Dairy Research*, **36**, 215–23.

Russell, T. A., Drake, M. A. and Gerard, P. D. (2006). Sensory properties of whey and soy proteins. *Journal of Food Science*, **71**, S447–55.

Singh, T., Drake, M. A. and Cadwallader, K. R. (2003). Flavor of Cheddar cheese: a chemical and sensory perspective. *Comprehensive Reviews in Food Science and Food Safety*, **2**, 139–62.

Singh, T. K., Young, N. D., Drake, M. A. and Cadwallader, K. R. (2005). Production and sensory characterization of a bitter peptide from β-casein. *Journal of Agricultural and Food Chemistry*, **53**, 1185–89.

Tomaino, R. M., Turner, L. G. and Larick, D. L. (2004). The effect of *Lactococcus* lactic starter cultures on the oxidative stability of liquid whey. *Journal of Dairy Science*, **87**, 300–7.

Van Ruth, S. (2001). Methods for gas chromatography–olfactometry: a review. *Biomolecular Engineering*, **17**, 121–28.

Walker, N. J. and Manning, D. J. (1976). Components of the musty off-flavor of stored dried lactic casein. *New Zealand Journal of Dairy Science and Technology*, **11**, 1–9.

Wright, J. M., Carunchia Whetstine, M. E., Miracle, R. E. and Drake, M. A. (2006). Characterization of a cabbage off-flavor in whey protein isolate. *Journal of Food Science*, **71**, C91–C96.

16

Milk protein gels

John A. Lucey

Abstract

The formation and the properties of the main types of milk protein gels are described, i.e. casein gels made with rennet or acid, heat-induced whey protein gels and gels made by a combination of approaches. The impact of various factors on these gelation properties is discussed. Recent key advances are highlighted, including the use of high pressure and transglutaminase cross-linking of proteins and new insights into the ubiquitous use of thermal processing to alter the texture of these gels.

Introduction

Gelation of the proteins in milk is the basis for the manufacture of cheese and fermented milk products. Various different approaches can be used to destabilize the milk proteins including heat (whey proteins), use of rennet enzyme (caseins) and acidification (caseins and denatured whey proteins). Combinations of these approaches can also be used to form dairy products, e.g. the use of a low concentration of rennet in cottage cheese (or quarg), which is primarily a cultured product. Yoghurt is a cultured product in which caseins and denatured whey proteins are responsible for the gelation properties.

Milk protein gels are irreversible, in contrast to many polysaccharide gels which are thermoreversible. Milk gels are often classified as particle gels although it is now recognized that they are not simple particle gels, as the internal structure of

Milk Proteins: From Expression to Food
ISBN: 978-0-12-374039-7

the casein particle plays an important role in their rheological properties (Horne, 2001, 2003). The properties of milk protein gels have been reviewed (Green, 1980; de Kruif *et al.*, 1995; Lucey, 2002; van Vliet *et al.*, 2004). The casein particles in rennet gels undergo rearrangement, fusion and syneresis in the process of forming cheese curd; thus they are inherently dynamic in nature and the rearrangement processes involved have been studied (Dejmek and Walstra, 2004).

Rennet-induced gels

Introduction

Coagulation of milk by rennet probably occurred initially by accident, as warm milk was stored in sacks made from the stomachs of ruminant animals that contained some residual proteinase enzymes. Crude extracts, prepared from the fourth stomach of young calves (called rennets, which are a type of aspartic proteinase), have been used for cheesemaking for thousands of years. Pepsin is the predominant proteinase in adult mammals. Naturally produced calf chymosin (EC 3.4.23.4) may contain up to six molecular species, which have slight differences in their amino acid residues (Crabbe, 2004). Chymosin has been cloned into several genetically modified organisms to produce fermentation-derived chymosin, which is widely used in many countries around the world (Crabbe, 2004). The rennet coagulation of milk has been reviewed (Dalgleish, 1987, 1993; Hyslop, 2003; Horne and Banks, 2004).

Primary phase of rennet coagulation

The basic building blocks of rennet-induced gels are the casein micelles. Both α_s- and β-caseins are sensitive to precipitation by the Ca^{2+} in milk and are protected by association with κ-casein, which is one reason for the formation of micelles. κ-Casein molecules are thought to have a predominantly surface position on micelles (although some κ-casein is also present in the interior of the micelle), where the hydrophilic C-terminal apparently acts as a "hairy" layer providing steric stabilization and a barrier against association with other micelles (Walstra, 1990; Chapter 5 in this volume).

The two stages of the rennet coagulation of milk are shown in Figure 16.1. In the primary phase of rennet coagulation, the C-terminal part (residues 106–169) of the κ-casein molecule is hydrolyzed and this hydrophilic peptide diffuses away from the micelle (called *para*-casein) into the serum phase. This macropeptide is called caseinomacropeptide (CMP) or, if it is highly glycosylated, glycomacropeptide (GMP). Most microbial coagulants, including those derived from *Rhizomucor miehei*, hydrolyze the same Phe_{105}–Met_{106} bond as chymosin; however, *Cryphonectria parasitica* hydrolyses the Ser_{104}–Phe_{105} bond (Drøhse and Foltmann, 1989). The proteolysis of other proteins in milk by chymosin occurs at a much slower rate (Crabbe, 2004).

The enzymatic reaction in milk appears to obey first-order kinetics. The proteolysis of κ-casein is usually described by standard Michaelis-Menten kinetics, although Hyslop (2003) questioned whether this was truly appropriate. It should be noted that

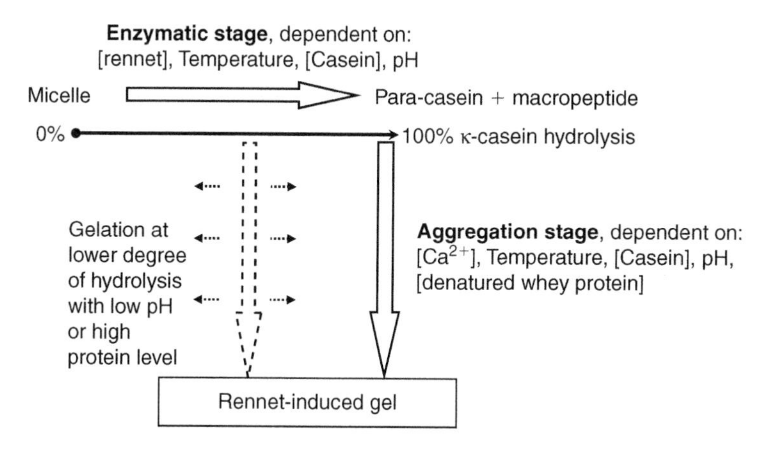

Figure 16.1 The two stages of the rennet coagulation of milk.

the primary phase and the secondary phase of clotting overlap as the aggregation begins before the enzymatic reaction is complete.

Secondary phase of rennet coagulation

The stability of the casein micelles of milk is attributed to their net negative charge and to steric repulsion by the flexible macropeptide region of κ-casein (the so-called hairs that extend out into the solution), calcium-induced interactions between protein molecules, hydrogen bonding and electrostatic and hydrophobic interactions. The release of the CMP (or GMP), which diffuses away from the micelles, leads to a decrease in the zeta potential, by ≈5–7 mV (≈50%), which reduces electrostatic repulsion between rennet-altered micelles. Removal of the "hairs" results in a decrease in the hydrodynamic diameter by ≈5 nm and a loss of steric stabilization, and causes a slight minimum in the viscosity during the initial lag phase of renneting.

Various attempts have been made to model the aggregation reaction (Horne and Banks, 2004). The nature of the attractive forces during the aggregation of casein micelles is still not completely clear, although calcium bridges, van der Waals' forces and hydrophobic interactions may be involved. Destabilized micelles will aggregate only in the presence of free Ca^{2+}. Rennet acts on casein at temperatures as low as 0°C, but milk does not clot at temperatures below 18°C whereas aggregation is very rapid at high temperature (e.g. 55°C).

When milk is clotted under normal conditions of pH and protein content, the viscosity does not increase until the enzymatic phase is mostly complete, i.e. at >60% of the (visual) rennet coagulation time. Coagulation does not occur until the enzymatic phase is at least ≈87% complete. Sandra *et al.* (2007) studied the rennet gelation process using diffusing wave spectroscopy, which allowed gelation to be monitored without the need for dilution. Sandra *et al.* (2007) suggested that partially renneted casein micelles do not begin to approach one another until the extent of breakdown of the κ-casein hairs has reached about 70%; above this point, they interact increasingly strongly with an increase in the extent of proteolysis. This interaction initially restricts the diffusive motion of the particles rather than causing true aggregation. Only after more extensive removal of the protective κ-casein hairs does true aggregation occur, with the appearance of a space-filling gel (as defined by rheology terms, such as having a loss tangent value <1). A micrograph of a rennet-induced skim milk gel is shown in Figure 16.2.

Srinivasan and Lucey (2002) studied the impact of plasmin enzyme on the rennet coagulation of skim milk. They found that hydrolysis of α_s- and β-caseins (as plasmin hardly degrades κ-casein) accelerated the rennet coagulation time. Srinivasan and Lucey (2002) hypothesized that plasmin could have degraded non-κ-casein "hairs" present on the surface of micelles and that this could have reduced the repulsive barrier to aggregation of rennet-altered micelles such that aggregation could take place at a lower degree of κ-casein hydrolysis.

Completely hydrolyzed micelles initially form small linear chains and these continue to aggregate to form clumps, clusters and eventually a system-spanning network that has a fractal-like appearance.

Little aggregation occurs at low temperatures (e.g. $<15°C$), which is usually taken as an indication of the importance of hydrophobic interactions. It is more likely that, with decreasing temperature, the activation free energy for flocculation increases, presumably because of the presence of β-casein chains on the outside of the micelle (Walstra, 1993). There is an increase in the strength of rennet gels at lower temperatures (where hydrophobic interactions are weak), reflecting swelling of casein particles, which results in an increase in the contact area between aggregated particles and strands.

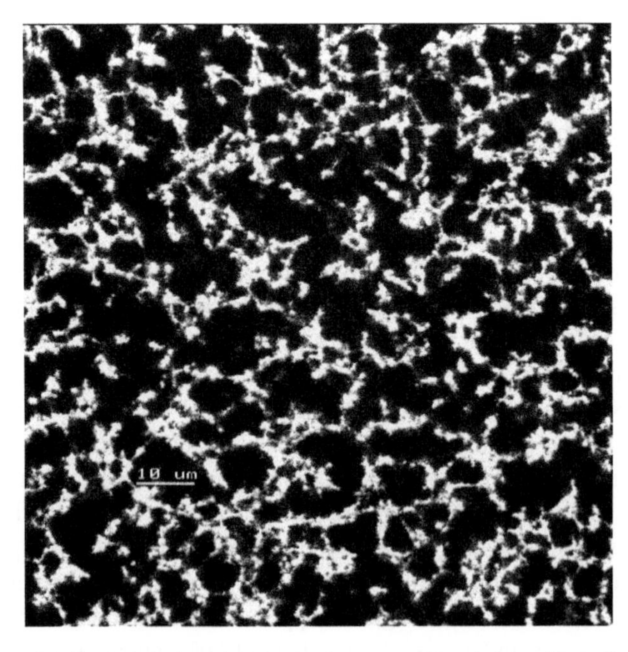

Figure 16.2 A confocal laser scanning micrograph of a rennet gel made from skim milk. Protein is white; dark areas are water. Scale bar = 10 μm.

Monitoring gelation

There have been several recent reviews of techniques to monitor milk gelation (Lucey, 2002; O'Callaghan *et al.*, 2002; Klandar *et al.*, 2007). The interest in monitoring gelation comes from the desire by cheesemakers to know when it is the "optimum" time to initiate cutting as well as from the desire by researchers to better understand this complex process.

Two promising techniques for the study of milk gels are diffusing wave spectroscopy and ultrasonic spectroscopy (Alexander and Dalgleish, 2004; Dalgleish *et al.*, 2006). These techniques could be used to complement existing approaches. For example, Wang *et al.* (2007) used both ultrasonic and (traditional) rheological methods to investigate the effects of milk pretreatment at ultra-high temperatures on the rennet gelation of a whey-protein-free casein solution. Wang *et al.* (2007) found that the ultrasonic velocity was able to measure the enzymatic hydrolysis and aggregation process, but was not as sensitive in detecting gel formation. In contrast, the oscillatory rheological method was not able to detect the enzymatic hydrolysis reaction, but was very suitable for characterizing the formation of a gel network.

Rheological properties of rennet-induced milk gels

Rennet-induced gels are viscoelastic and their rheological properties can be characterized using dynamic low-amplitude oscillatory rheology, which determines both the viscous component and the elastic component. These measurements should be performed in the linear viscoelastic range, where the deformation (strain) is proportional to the applied stress. Often, for rennet and acid gels, that means trying to operate at $\leq$3% strain, which can be difficult during the early stage of gel formation for many (of the popular) controlled stress rheometers (because of the very low torque resulting on the measuring geometry of the rheometer from such a weak gel). Some new techniques/software can be used to reduce this problem in commercial rheometers (e.g. Lauger *et al.*, 2002).

Parameters that can be determined include the elastic or storage modulus (G'), which is a measure of the energy stored per oscillation cycle, the viscous or loss modulus (G''), which is a measure of the energy dissipated as heat per cycle, and the loss tangent (tan δ), which is the ratio of the viscous properties to the elastic properties (loss tangent = G''/G'). The loss tangent is related to the relaxation of bonds in the gel during deformation and is a useful parameter.

During gelation, there is a lag period before a measurable storage modulus value is obtained (this depends on the sensitivity of the rheometer to measure events close to the gelation point). The loss tangent decreases from $>>1$ to <1 at the gelation point and then attains a relatively constant value (about 0.35 for rennet gels). The dynamic moduli initially increase relatively rapidly and then, after a period of several hours, tend to plateau. In commercial practice, rennet-induced gels are cut once they have attained a certain firmness (usually assessed subjectively by the cheesemaker) or, more commonly, at a fixed time after rennet addition. The increase in the moduli after gelation probably reflects ongoing fusion of micelles, which results in an increase in the contact area between aggregated particles, and possibly the incorporation of additional particles into the gel network.

Some micelles that have incomplete hydrolysis of their κ-casein hairs could be trapped within the space-filling network at the point of network formation and they might later become attached to the matrix once their κ-casein hairs get completely hydrolyzed. Mellema *et al.* (2002) reported that their analysis suggested that nearly all casein was incorporated in the rennet gel, at least very soon after network formation. Mellema *et al.* (2002) also considered that changes in the storage modulus and microstructure during aging could be explained in terms of (various types of) rearrangements of the gel network at various length scales.

Typical plateau values for the storage modulus of rennet-induced gels (made from unconcentrated milk) range from 100 to 200 Pa. Both moduli have lower values at low frequencies, reflecting relaxation of more bonds when the timescale of the applied stress is longer. The loss tangent at low frequencies is an important indicator of rearrangements as this is approximately the same timescale as that over which rearrangement processes related to syneresis in rennet gels are estimated to occur (van Vliet *et al.*, 1991).

The development of the complex or shear modulus as a function of time after rennet addition can be replotted against a reduced time t/t_g, where t_g is the gelation time. Various individual renneted milk curves can be normalized against their complex or shear modulus value at a low multiple (two or three) of the t_g. These various curves then collapse into a single or master curve because of the scaling behavior of the dynamics of the gel formation process (Horne, 1995, 1996). Various mathematical, empirical and kinetic models have been applied to predict the development of gel firmness or shear moduli; their effectiveness in performing this function has been reviewed recently by Horne and Banks (2004).

Syneresis of rennet-induced milk gels

The syneresis of rennet-induced gels has been reviewed (Walstra *et al.*, 1985; Pearse and MacKinlay, 1989; Walstra 1993; van Vliet and Walstra, 1994; Dejmek and Walstra, 2004). Rennet-induced gels remain stable for several hours if left undisturbed. They rapidly synerese if disturbed by cutting or by wetting the gel surface. A rennet-induced gel may lose up to two-thirds of its volume (as whey) under quiescent conditions and more than 90% if external pressure is applied (Dejmek and Walstra, 2004).

Cheesemaking can be viewed as a dehydration process and syneresis is the crucial method by which most of the moisture is lost from curd particles. As syneresis is the main method that cheesemakers have to control cheese moisture, it is also the process that is most often manipulated; it also helps to facilitate differentiation between cheese varieties. Most of the water in milk gels is not chemically "bound" to proteins but rather is physically entrapped in the network structure (van Vliet and Walstra, 1994).

Because of the complexity of the syneresis process, researchers have often used thin gel slabs to monitor one-dimensional shrinkage (e.g. van Dijk and Walstra, 1986). The one-dimensional syneresis of rennet gels is related to the flow of liquid (whey) through the network (because liquid flows out of the gel concomitantly with gel shrinkage) and is governed by the equation of Darcy:

$$v = \frac{B}{\eta}\frac{p}{x}$$

where v is the superficial flow velocity of the syneresing liquid, B is the permeability coefficient, η is the viscosity of the liquid, p is the pressure acting on the liquid and x is the distance over which the liquid must flow.

It is believed (Walstra, 1993) that there is an internal (endogenous) pressure or driving force within rennet gels that is responsible for the shrinkage of the gel once the initial gel is disturbed (presumably this overcomes the yield stress of the system). It has not been possible to measure this small endogenous pressure experimentally. Endogenous syneresis pressure (i.e. the pressure within the rennet gel causing the syneresis) is not constant. It increases initially as a function of time after renneting but decreases at longer times, presumably because of fusion of *para*-casein micelles and a reduction in permeability of the contracting network. In practice, syneresis in curd particles occurs in three dimensions simultaneously and is much harder to study than the one-dimensional model.

In rennet-induced milk gels, the mechanism responsible for the strong tendency of the gels to exhibit syneresis is related to the (extensive) rearrangements of the casein network that occur after gel formation. As acid-induced gels undergo much less rearrangement, they synerese less. The rearrangement process is accelerated, and is more extensive, at high temperatures and lower pH values (<6.5 but >5.1) (the loss tangent is also higher under these conditions). Aging of rennet-induced gels results in a coarsening (sometimes called microsyneresis) of the gel (i.e. as a result of more rearrangements) and there is an increase in the permeability and the fractal dimensionality.

Rearrangements of casein particles into a more compact structure would increase the number of bonds and hence decrease the total free energy of the system (Walstra, 1993). However, the casein particles are already part of the gel network, which must be deformed or broken locally to form new junctions. Breakage of the bonds in the strands requires a sufficiently low yield stress if it is to be exceeded. In cheesemaking, conditions such as cutting, stirring, acid production and the increase in the cooking temperature all encourage syneresis and the rearrangement processes that facilitate syneresis of the gel network. If the strands become too thick (e.g. because of a very high casein concentration), syneresis hardly occurs.

One-dimensional syneresis of rennet-induced skim milk gels was studied in gels with different thicknesses and at pH values of 6.4 and 6.0 using a laser displacement sensor (Lodaite *et al.*, 2000). Syneresis was much faster at the lower pH and the initial syneresis rate increased linearly with slab thickness.

Several (mostly empirical) techniques have been used to estimate the syneresis of rennet gels, including shrinkage of gel slabs, determining the volume of whey expelled as a function of time, the dry matter content or density of curd particles and low-resolution nuclear magnetic resonance (NMR) (Dejmek and Walstra, 2004). A recent development has been a light backscatter sensor, with a large field of view relative to curd size, for continuous on-line monitoring of coagulation and syneresis to help cheesemakers improve their control over the moisture content of the curd (Fagan *et al.*, 2006).

Some factors influencing the texture of rennet-induced gels

Many factors influence the milk clotting process and the consistency of rennet gels including pH, temperature, casein content, ionic strength, enzyme concentration, calcium content, presence of homogenized fat globules, concentration of denatured whey proteins and casein hydrolysis by proteinases such as plasmin. These factors have been reviewed many times (Dalgleish, 1987, 1993; Green and Grandison, 1993; Lomholt and Qvist, 1999; Hyslop, 2003; Horne and Banks, 2004) because of the importance of rennet gels for the cheese industry.

The effects of pH (5.19–6.21) and NaCl concentration (0, 1.75 and 3.50%) on the rheological and microstructural properties of rennet-induced casein gels made from ultrafiltered skim milk (19.8%, w/w casein) were recently investigated (Karlsson *et al.*, 2007a). Low pH and high NaCl concentration reduced the rate of development of the gel elasticity after coagulation. Strain at fracture and stress at fracture 48 h after coagulation showed maximum and minimum values at pH 5.8 and 5.29 respectively. The microstructure examined with confocal laser scanning microscopy was unaffected by the changes in pH and the concentrations of NaCl, probably because of the very high volume fraction of caseins in this type of gel (Karlsson *et al.*, 2007a).

Rennet-induced coagulation of ultrafiltered skim milk (19.8%, w/w casein) at pH 5.8 was studied and compared with coagulation of unconcentrated skim milk of the same pH (Karlsson *et al.*, 2007b). At the same rennet concentration, coagulation occurred at a slower rate in ultrafiltered skim milk but started at a lower degree of κ-casein hydrolysis, compared with the unconcentrated skim milk. Confocal laser scanning micrographs revealed that, during storage for up to 60 days (at 13°C), the microstructure and the size of the protein strands of the ultrafiltered gel hardly changed, probably because of the high zero shear viscosity of the concentrated system (Karlsson *et al.*, 2007b).

Plant coagulants obtained from the flowers of *Cynara* sp. have been used to make rennet gels and cheeses (Esteves *et al.*, 2001, 2002, 2003). These coagulants are less sensitive to changes in gelation temperature, they cause more casein rearrangements during gelation and they have higher values for the storage modulus (at least during the initial stages of gelation), compared with gels made with chymosin (probably because of greater proteolysis of the caseins).

Choi *et al.* (2007) demonstrated that the concentration of insoluble calcium phosphate (CCP) associated with the casein micelles had an important influence on the properties of rennet gels. Removal of some CCP from milk prior to gelation using calcium-chelating agents lowered the storage modulus of rennet gels because of the reduction in the amount of CCP cross-linking in the casein micelles. Reduction in the CCP content prior to rennet gelation resulted in gels with higher loss tangent values, indicating greater bond mobility.

Choi *et al.* (2007) also studied the impact of preacidification of milk prior to gelation. They found that gels made at pH 6.4 had higher storage modulus values than gels made at pH 6.7 probably because of the reduction in electrostatic repulsion, whereas the CCP content only slightly decreased at this pH value. The storage modulus values were highest at pH 6.4 and decreased with decreasing pH from 6.4 to 5.4 because of the reduction in CCP cross-linking within the casein micelles (Choi *et al.*, 2007).

Milk heat treatment

It is well known that severe heat treatment of milk at temperatures sufficiently high to denature the whey proteins results in an increased rennet coagulation time as well as weaker gels (Lucey, 1995). There are some reports that the interaction of denatured whey proteins with the κ-casein inhibits the primary phase of rennet action on κ-casein (to some extent). For example, Reddy and Kinsella (1990) reported that very high heat treatments decreased the initial velocity (V_i) and GMP release. However, most studies have concluded that the secondary phase of the coagulation process is the step that is mainly inhibited by the presence of denatured whey proteins on the micelle surface. These denatured whey proteins probably sterically interfere with the (normal) aggregation of rennet-altered micelles (Lucey, 1995).

Vasbinder *et al.* (2003) concluded that whey protein denaturation had only a small effect on rennet activity and that the release of GMP (or the formation of *para*-κ-casein) was similar in heated and unheated milks. Anema *et al.* (2007) adjusted the pH of the milk prior to heat treatment, which allowed them to manipulate the distribution of denatured whey proteins and κ-casein between the serum and micellar phases; they reported that the retardation in rennet gelation as a result of heat treatment was observed regardless of whether the denatured whey proteins were associated with the casein micelles or in the serum phase.

Enzymatic cross-linking of caseins

Transglutaminase (TGase; EC 2.3.2.13) catalyzes covalent intermolecular protein cross-linking through an acyl-transfer reaction, between the γ-carboxyamide group of a peptide-bound glutamine residue (acyl donor) and the primary amino group of an amine (acyl acceptor). The application of TGase in various types of dairy products has been reviewed (Jaros *et al.*, 2006). In a system where caseins and whey proteins are available as substrates for TGase, such as milk, the caseins are preferentially cross-linked over native whey proteins (Han and Damodaran, 1996).

Lorenzen (2000) incubated preheated milk with TGase for various incubation times prior to rennet addition and found that increasing TGase incubation times, as well as an increasing intensity of preheat treatment of the milk, resulted in increasing coagulation times up to the point of a complete absence of coagulation. Lorenzen (2000) attributed the reduced rennetability of preheated milk to a "surface sealing" of the casein micelles with cross-linked β-lactoglobulin, leading to a steric inhibition of the release of the macropeptide from the surface of the casein micelle.

O'Sullivan *et al.* (2002b) also attributed the loss of rennetability to the impact of TGase cross-linking on the primary enzymatic phase, i.e. reduced rate of hydrolysis of κ-casein. Huppertz and de Kruif (2007) criticized the analytical method used by O'Sullivan *et al.* (2002b) to study the hydrolysis reaction because they suggested that this method detects only the products of hydrolysis of non-cross-linked milk; hydrolysis products of cross-linked κ-casein would not be adequately detected because the macropeptide remains attached to the micelle. Huppertz and de Kruif (2007) suggested instead that TGase treatment affects mainly the secondary stage of rennet-induced coagulation. They suggested that this inhibition was due to the progressive cross-linking of the κ-casein located on the surface of the casein micelles, which provided additional steric hindrance to the aggregation of renneted micelles.

High hydrostatic pressure

High hydrostatic pressure influences various properties of milk including a reduction in the size of the casein micelles, denaturation of β-lactoglobulin and a reduction in the CCP content. Huppertz *et al.* (2005) studied the impact of milk heat treatment (90°C for 10 min) and subsequent application of high-pressure treatment at pressures from 0 to 600 MPa. Heated unpressurized milk or heated milk treated for 0 min at 100 MPa was not coagulable by rennet; however, heated milk treated at 250–600 MPa for 0–30 min had a rennet coagulation time equal to, or lower than, that of unheated unpressurized milk; the coagulation time decreased with increasing pressure and treatment time. The strength of the rennet-induced coagulum from heated milk treated at 250–600 MPa for 30 min, or 400 or 600 MPa for 0 min, was considerably greater than that of the rennet-induced coagulum from unheated unpressurized milk. There was also an increase in the yield of cheese curd by ≈15%.

Acid-induced milk gels

Impact of acid on casein micelles

In cultured products, such as yoghurt, as the pH of milk is reduced, CCP is dissolved and the internal casein micelle structure is altered because of the loss of CCP. Little casein dissociation occurs at the high temperatures (>40°C) commonly used for yoghurt manufacture. Aggregation of casein occurs as the isoelectric point (pH ≈ 4.6) is approached (Horne, 1999). Native casein micelles (in milk of normal pH) are stabilized by their negative charge and steric repulsion (Lucey and Singh, 2003). Lucey (2003) distinguished three (arbitrary) pH regions in the acidification of milk from pH 6.7 to 4.6:

(a) *pH from 6.7 to ≈6.0*. The decrease in pH causes a reduction in the net negative charge on the casein micelles, thereby reducing electrostatic repulsion. As only a relatively small amount of CCP is dissolved above pH 6.0, the structural features of the micelles are relatively unchanged.
(b) *pH from ≈6.0 to ≈5.0*. The decrease in pH causes a reduction in the net negative charge on the casein micelles, thereby reducing electrostatic repulsion. As the κ-casein "hairs" on the micelle surface are charged, these charged "hairs" may shrink/collapse as the pH decreases. The net result is a decrease in both electrostatic repulsion and steric stabilization. The CCP within the casein micelles is dissolved completely by pH ≈ 5.0.

(c) *pH* $\leq$ *5.0*. The net negative charge on the casein micelles declines with the approach of the isoelectric point and there are increased +/− charge interactions (and van der Waals' forces). The reduction in electrostatic repulsion allows increased hydrophobic interactions (Horne, 1998, 2001). In unheated milk gels where acidification is the only coagulation method, gelation occurs at around pH 4.9; if acidification is performed at very high temperatures, a higher gelation pH is observed.

On acidification, casein particles aggregate as a result of (mainly) charge neutralization. Acidification eventually leads to the formation of chains and clusters that are linked together to form a three-dimensional network (Kaláb *et al.*, 1983). Acid casein gels can be formed from sodium caseinate (this ingredient is sometimes used as a yoghurt stabilizer). Direct acidification of milk at a low temperature and subsequent warming is another approach to acid gel formation. Glucono-δ-lactone (GDL) is also used to acidify milk but these acid-induced gels have different rheological and structural properties from gels produced by bacterial cultures (Lucey *et al.*, 1998a).

Hydrophobic interactions are unlikely to play a direct role in the strength of acid gels as the storage modulus of acid gels increases with decreasing measurement temperature (Lucey, 2003). Cooling gels results in an increase in the storage modulus, probably as a result of the swelling of casein particles (caused by the weaker hydrophobic interactions) and an increase in the contact area between particles (Lucey, 2003). With increasing ionic strength, the charged groups on casein are screened, thereby weakening interactions between casein particles.

Milk has been reversibly acidified by means of carbonation—injecting pressurized CO_2 as the acidifying agent—in order to reduce the pH (usually done at low temperature). Neutralization is obtained by pressure release followed by degassing under vacuum. Upon CO_2 treatment, the zeta potential and the size of the casein micelles were restored although the total amount of CCP was not restored (Raouche *et al.*, 2007). The rheological properties of acid gels (made using GDL) from CO_2-treated milk were similar to those of acid gels from untreated milk (Raouche *et al.*, 2007).

Some factors influencing the texture of yoghurt gels

It is well established that the way in which the milk is handled or prepared, including the processing conditions used in yoghurt manufacture, greatly influences the gel texture, strength and stability (Lucey and Singh, 1998; Walstra, 1998; Tamime and Robinson, 1999; Jaros and Rohm, 2003a, 2003b), and that these factors include: (a) fortification level and material(s) used in the mix, (b) stabilizer type and usage levels, (c) fat content and homogenization conditions, (d) milk heat treatment conditions, (e) starter culture (type, rate of acid development and production of exopolysaccharides), (f) incubation temperature (influences growth of starter cultures, gel aggregation, bond strength), (g) pH at the breaking of the gel (stirred) and/or the start of cooling (set), (h) cooling conditions (when cooling is started, rate of cooling) and (i) post-manufacture handling of the product, e.g. physical abuse (e.g. vibration) and temperature fluctuations (i.e. if the product is not maintained at $\approx$5°C).

Homogenization and fat globule surface material

The fat globules in milk are surrounded by membrane proteins and, unless homogenized, fat acts as an inert filler in milk gels. Cho *et al.* (1999) prepared fat globules with different surface materials and studied the effects of these surface materials on the rheological properties of acid milk gels. Gels containing fat globules stabilized by non-interacting materials ("structure breaker") (i.e. Tween and unheated whey protein concentrate [WPC]) had low storage moduli compared with interacting surface materials ("structure promoter") (skim milk powder, sodium caseinate and heated WPC).

Milk for the manufacture of yoghurt is normally homogenized (15–20 MPa) to increase the yoghurt consistency and to decrease whey separation during storage (Tamime and Robinson, 1999). High-pressure homogenization has a similar principle to conventional homogenization but works at significantly higher pressures (up to 400 MPa). Milk given a high-pressure (>200 MPa) treatment gave firmer yoghurt gels than milk heat treated (90°C for 90 s) and traditionally homogenized at 15 MPa (Serra *et al.*, 2007); presumably this effect reflects a combination of the creation of very small fat globules, whey protein denaturation and possible modification to the CCP content (Huppertz and de Kruif, 2006; López-Fandiño, 2006).

High hydrostatic pressure

High-hydrostatic-pressure treatment of milk enhances the mechanical properties of yoghurt gels (Needs *et al.*, 2000). The storage moduli of gels made from high-pressure-treated milk were considerably higher than those of gels made from heat-treated milk (85°C for 20 min), although the yield stress and the yield strain were lower in the pressure-treated gel (Needs *et al.*, 2000). The combined use of high thermal treatment and high hydrostatic pressure results in extensive whey protein denaturation and casein micelle disruption respectively (Harte *et al.*, 2003). The net effect of the combined high hydrostatic pressure and thermal treatments was an improvement in the yield stress of the yoghurt and a reduction in syneresis (Harte *et al.*, 2003). High-pressure treatment up to 600 MPa (for 20 min) improved the viscosity of stirred yoghurt, which had similar rheological properties to yoghurt made from milk heated at 90°C for 30 min (Knudsen *et al.*, 2006).

Enzymatic modification of proteins

Acid-induced gelation of TGase-cross-linked casein resulted in increased gel firmness, lower permeability, finer protein networks and improved whey drainage (Færgemand and Qvist, 1997; Færgemand *et al.*, 1999; Schorsch *et al.*, 2000). Lauber *et al.* (2000) reported that even a very small amount of casein cross-linking, due to the action of TGase, is capable of inducing significant changes in yoghurt texture (i.e. a large increase in gel strength). A slightly slower acidification rate by the starter culture was observed in yoghurts made from TGase-treated milk; possibly there was a reduction in availability of the low molecular weight peptides required by *Streptococcus thermophilus* as a result of the cross-linking reaction (Færgemand *et al.*, 1999; Ozer *et al.*, 2007).

Cross-linking of caseins restored the sensory texture profile of a lower protein yoghurt to be comparable with that of a higher protein yoghurt, suggesting that TGase could be used instead of some of the milk solids currently used in yoghurt fortification (Færgemand *et al.*, 1999). Excessive protein cross-linking increased the gel firmness but the yoghurt became grittier than the control samples (Færgemand *et al.*, 1999). TGase is capable of cross-linking caseins even under high pressure (Lauber *et al.*, 2001).

When TGase treatment was performed during high-hydrostatic-pressure treatment, a markedly higher final storage modulus was observed in acid milk gels compared with gels with only pressure treatment or when a separate TGase treatment was performed (Anema *et al.*, 2005). Anema *et al.* (2005) proposed that there is an increase in cross-linking of the whey proteins and an increase in cross-linking between the whey proteins and casein when TGase treatment is performed under high pressure.

Heat treatment

Acid gels formed from unheated milk are very weak and this may arise, at least partly, because the interparticle contact area is still dominated by the presence of the κ-casein hairs (GMP), which have collapsed but are still present (Li and Dalgleish, 2006). The κ-casein hairs are rich in hydroxylated amino acids, some of which are glycosylated, and also acidic and basic residues. Thus, the interface between the aggregating particles will tend to be highly hydrated and attractive interactions will be partly offset by the hydrophilic tendency of the κ-casein hairs (Li and Dalgleish, 2006).

There has been considerable recent research on the topic of how whey proteins influence yoghurt texture. Native whey proteins in unheated milk are inert fillers in yoghurt (Lucey *et al.*, 1999). Added whey proteins alter yoghurt gelation and texture as long as the mix is given a sufficiently high heat treatment to denature the whey proteins and cause them to associate with the casein micelles (Lucey *et al.*, 1999). Commercially, WPC is often used to increase the solids content of yoghurt and to give improved viscosity and lower whey drainage. High-heat treatment causes considerable whey protein denaturation (e.g. 85°C for 15 min results in >80% β-lactoglobulin denaturation). As a result, β-lactoglobulin becomes mostly attached to the κ-casein of the casein micelles or forms soluble complexes (with serum casein), depending on the heating conditions (i.e. pH) (Lucey *et al.*, 1998b).

Denatured whey proteins (DWP) attached to the surface of casein micelles during heating (i.e. bound DWP) are a critical factor in the increased stiffness of yoghurt gels made from heated milk. DWP cause micelles to aggregate at higher pH because of the higher isoelectric pH (≈5.3) of the main whey protein, β-lactoglobulin, than that of caseins (Lucey *et al.*, 1997; Guyomarc'h *et al.*, 2003). An alternative view is that the DWP associated with the micelles alter the hydrophobic interactions between heated micelles, which facilitates gelation at higher pH values (although there is greater electrostatic repulsion at higher pH) (Jean *et al.*, 2006). More cross-linking of gels by bound DWP increases the gel strength. Soluble DWP are not able to increase the gel stiffness of milk in which there are no bound DWP present, i.e. the micelle surface does not contain any "bound" DWP, which can be created experimentally (Lucey *et al.*, 1998b) (Figure 16.3).

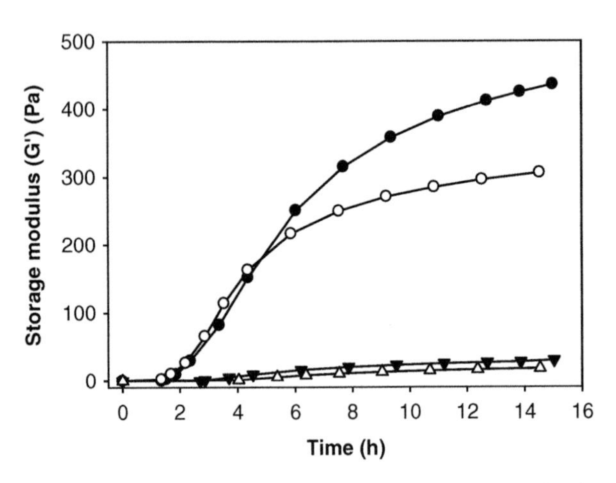

Figure 16.3 Storage modulus as a function of time during the formation of acid-induced milk gels made from heated milk (●), heated milk containing bound DWP (○), heated milk containing soluble DWP (▼) and unheated milk (△). Heat treatment was at 80°C for 30 min and acidification was at 30°C with 1.3% GDL. (Reproduced, with the permission of Cambridge University Press, from Lucey *et al.*, 1998b.)

In industrial practice, heating milk always creates some bound DWP, which allows soluble DWP to become attached to the micelles and to contribute to the gel strength. The pH at heating influences the association of DWP with casein micelles. At pH 6.5, most DWP are associated with micelles (e.g. >70% for milk heated at 90°C for 30 min). At higher pH (e.g. 7.0), fewer DWP are associated with micelles as more κ-casein dissociates from the micelles to interact with β-lactoglobulin during heating.

The gel strength of acid gels made from milk heated at high pH is higher than that of acid gels made from milk heated at the natural pH of milk (Lucey *et al.*, 1998b; Anema *et al.*, 2004); this may not be valid for situations in which there is a lot of added whey protein. At high pH values there is an increase in the concentration of CCP (additional cross-linking) in milk (McCann and Pyne, 1960), which could potentially increase the stiffness of acid gels made from high-pH milk. Increasing the pH of heat treatment of the milk from 6.5 to 7.0 should also alter protein unfolding and disulfide bond formation, involving β-lactoglobulin, as the pK value of its free thiol group is 9.35 (Kella and Kinsella, 1988a). The creation of additional covalent disulfide bonds that involve whey protein and caseins should increase the strength of the yoghurt gel.

Irrespective of the pH of the milk at heating, DWP (i.e. those designated as "soluble" and "bound" at the pH of heat treatment) are insoluble at low pH and should associate with casein at the pH values involved in yoghurt fermentation. As the pH decreases during fermentation virtually all the residual soluble complexes become attached to caseins via the bound DWP. The rate of acidification and the gelation temperature may also influence how these complexes associate with the caseins during acidification. The extent of denaturation of the whey proteins is often determined by their loss of solubility at pH 4.6 (de Wit, 1981), so that all the DWP should precipitate as the pH approaches pH 4.6.

The addition of WPC to milk that was then given a high-heat treatment resulted in an increase in the pH of gelation, an increase in gel stiffness and a reduction in fracture strain compared with gels made from heated milk without added WPC (Lucey *et al.*, 1999). If WPC was added to heated milk and this mixture was not given any further heat treatment, the acid gels formed after acidification were weaker than those made from heated milk without WPC. This suggests that any added whey proteins must be denatured in order to reinforce the network, even when DWP are already present in the milk.

Schorsch *et al.* (2001) examined the effect of heating whey proteins in the presence or absence of casein micelles on the subsequent acid gelation properties of milk. The acid-induced gelation occurred at a higher pH (around pH 6.0) and in a shorter time when the whey proteins (concentration of 1 g whey protein/kg) were denatured separately from the casein micelles than when the whey proteins were heated in the presence of the casein micelles. However, the gels formed were very weak, probably because of the formation of a weak network in which whey proteins entrapped caseins.

Various studies have shown some conflicting results about the relative importance of the soluble and bound DWP fractions to the texture of acid milk gels (Lucey *et al.*, 1998b; Guyomarc'h *et al.*, 2003; Anema *et al.*, 2004). Differences in the proportions of soluble and bound DWP fractions in these studies could have contributed to these conflicting results. Guyomarc'h *et al.* (2003) had only a small proportion (10–15%) of β-lactoglobulin in the bound DWP fraction whereas Lucey *et al.* (1998b) had around 80% β-lactoglobulin in the bound DWP fraction. Guyomarc'h *et al.* (2003) suggested that differences in the quantitative amounts of aggregates (and the total amount of DWP) present in the systems, independently of whether they were soluble or not, could be the reason for some of the conflicting results reported by the different groups.

In gels made from heated milk, because of the high gelation pH, the gel goes through a period of solubilization of the CCP that is present within casein particles that are already part of the gel network (this event is responsible for the maximum in the loss tangent during gelation) (Lucey *et al.*, 1997). This process loosens the interactions between caseins in the gel network, and the higher bond mobility in yoghurt gels during this period has been associated with whey separation (Lucey, 2001).

The rheological changes during the acid-induced gelation (with GDL) of unheated and heated milk at 30°C are shown in Figure 16.4. Note the much shorter gelation time, the large increase in the storage modulus and the maximum in the loss tangent (as indicated by the hatched region between the two arrows, region A) in the heated milk sample. As the low gelation pH (4.8) of the unheated milk gel occurs after most or all of the CCP is already solubilized, there is no maximum in the loss tangent in this type of gel. When acid-induced gelation of heated milk occurs rapidly at high temperature, a plateau in the storage modulus, which corresponds to the region where there is a maximum in the loss tangent, can be observed (Horne, 2001).

Bikker *et al.* (2000) reported that the addition of β-lactoglobulin variant B or variant C to the milk prior to heating and acidification caused a larger increase in the storage modulus of acid gels than the addition of β-lactoglobulin variant A.

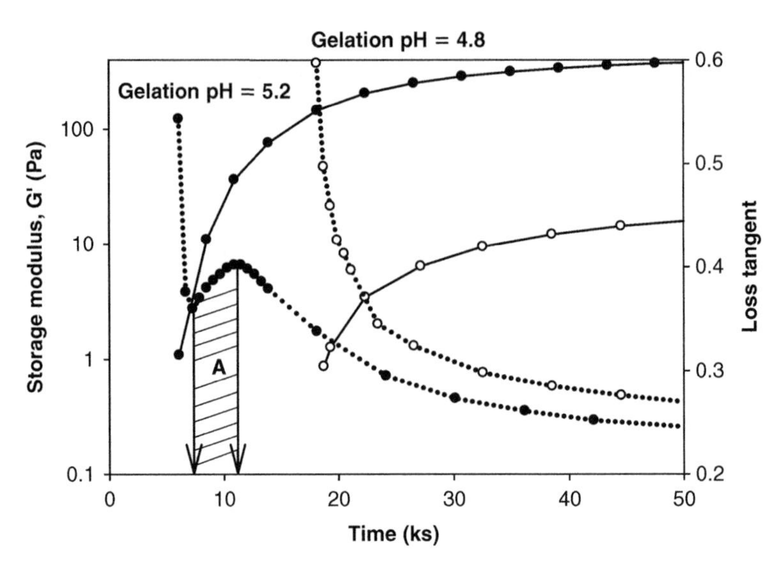

Figure 16.4 Storage modulus (solid lines) and loss tangent (dashed lines) of acid gels made from heated milk (●) and unheated milk (○). Heat treatment was at 80°C for 30 min and acidification was at 30°C with 1.3% GDL. The area marked by the letter A indicates the region in which the loss tangent increases after gelation because of solubilization of CCP in casein particles that are already part of the gel network.

Soluble whey protein polymers have been used as ingredients for yoghurt applications (Britten and Giroux, 2001). The use of whey protein polymers to standardize the protein content of milk increased the yoghurt viscosity to about twice that obtained using skim milk powder at the same protein concentration. The water-holding capacity of yoghurt standardized with whey protein polymers was considerably higher than that of yoghurt standardized with skim milk powder (Britten and Giroux, 2001).

Incubation temperature

Although 42°C is a commonly used fermentation temperature for yoghurt, the use of slightly lower incubation temperatures (e.g. 40°C) leads to slightly longer gelation times, but firmer and more viscous gels that are less prone to whey syneresis are formed (Lee and Lucey, 2004). At a lower incubation temperature, there is an increase in the size of the casein particles because of a reduction in hydrophobic interactions which, in turn, leads to an increased contact area between the casein particles (Lee and Lucey, 2004); a similar trend occurs when the gels are cooled. A high incubation temperature also makes the gel network more prone to rearrangements (more flexible) during gelation and these changes can lead to greater whey separation (Lucey, 2001; Mellema *et al.*, 2002).

Whey protein gels

As whey is usually obtained as a by-product of cheesemaking (although recent developments in membrane technology mean that, in future, "whey" will come not

Table 16.1 Composition of rennet and acid wheys (adapted from Oakenfull *et al.*, 1997)

	Average composition	
	Rennet whey	Acid whey
Total whey protein (g/L)	6.7	5.8
β-lactoglobulin (g/L)	3.5	3.5
α-lactalbumin (g/L)	1.3	1.3
Serum albumin (g/L)	0.1	0.1
Immunoglobulins (g/L)	0.4	0.4
Proteose peptones (g/L)	0.2	0.2
Glycomacropeptide (g/L)	1.0	-
Lactose (g/L)	5.0	4.4
Lipid (g/L)	0.6	0.1
Ash (g/L)	0.5	0.6
Na (mg/100 g)	35	40
K (mg/100 g)	109	133
Ca (mg/100 g)	22	86
Mg (mg/100 g)	6	9
P (mg/100 g)	42	63

Table 16.2 Typical composition of some whey powders (approximate, wet or as is basis)

Whey ingredient	Moisture (%)	Fat (%)	Protein (%)	Lactose (%)	Ash (%)
Sweet whey	3-5	1.1-1.5	11-14.5	75	8-10
Acid whey	3.5	0.5-1.5	11-13.5	70	10-12
WPC35	3-4.5	3-4.5	34-36	48-52	6.5-8
WPC80	3.5-4.5	6-8	80-82	4-8	3-4
WPI	4-5	<1.0	90-92	<1.0	2.5-3.5

necessarily from a cheese vat but as "native" whey directly from milk prior to cheesemaking), its composition depends on the cheesemaking conditions, e.g. acid whey derived from cottage cheese has different mineral (ash), lactic acid and pH values from whey derived from rennet-coagulated cheeses such as Cheddar (Table 16.1).

Whey products are widely used as food ingredients because of their excellent functional and nutritional properties. Various types of whey products are made commercially, ranging from dried whey to WPC (WPC has protein contents ranging from ≈35–80%) to whey protein isolate (WPI) (protein contents ≥90%) (Table 16.2). Membrane filtration, i.e. ultrafiltration (UF) and diafiltration (DF), is used to concentrate the protein fraction before spray drying into WPC. Two different approaches are used to produce WPI: (a) membrane filtration (microfiltration, UF and DF) and (b) ion-exchange chromatography coupled with UF/DF. These two approaches result in WPI with different protein profiles (Table 16.3; Wang and Lucey, 2003). Many serum proteins take part in heat-induced gelation whereas GMP and proteose peptones do not (Walstra *et al.*, 2005).

Whey proteins are globular proteins and heating induces denaturation and aggregation. At sufficiently high protein levels (usually ≥6%, except for purified individual whey

Table 16.3 Approximate protein composition of whey protein isolates made by different technologies (data from several sources including Wang and Lucey, 2003)

Protein type	Membrane filtration	Ion exchange chromatography
β-lactoglobulin	48–55%	60–73%
α-lactalbumin	15–22%	12–25%
Bovine serum albumin and Immunoglobulins	4–7%	6–16%
Glycomacropeptide	17–26%	0.2–1.4%

proteins; this depends on many factors especially pH), gelation occurs during heating or cooling. The formation and the properties of whey protein gels are influenced by many factors:

- pH;
- protein content;
- ionic strength;
- rate and temperature/time of heating;
- types and ratios of the serum proteins;
- concentration of divalent ions (e.g. Ca^{2+});
- concentration of sugars;
- concentration of lipids including phospholipids.

There have been several reviews of the gelation of globular proteins (Oakenfull, 1987; Clark, 1992, 1996, 1998; Doi, 1993; Gosal and Ross-Murphy, 2000), as well as reviews of the thermal denaturation and gelation of whey proteins (Mulvihill and Kinsella, 1987; Mangino, 1992; Aguilera, 1995; Singh and Havea, 2003; Foegeding, 2006).

Thermal denaturation of whey proteins

Many studies on the denaturation of whey proteins have been conducted (see the review by Mulvihill and Donovan, 1987), especially β-lactoglobulin as this is the major whey protein and its behavior dominates the gelation behavior of whey protein products. Denaturation has been used to describe both the loss of native structure (conformational change) and the loss of solubility (e.g. at pH values close to the isoelectric point). At around neutral pH values, denaturation becomes irreversible above about 65°C (Holt and Sawyer, 2003); with a decrease in the pH, the denaturation temperature increases (Kella and Kinsella, 1988b). Disulfide bond formation is favored as the pH is increased towards the pK value of the thiol group on β-lactoglobulin (9.35; Kella and Kinsella, 1988b).

Denaturation, i.e. conformational change, can be reversible and, for whey proteins, the cause of irreversibility is often the formation of new covalent (mostly disulfide) bonds. Various mechanisms for the thermal denaturation/aggregation of β-lactoglobulin (at neutral pH) have been proposed, in which the basic steps are: (a) the dissociation

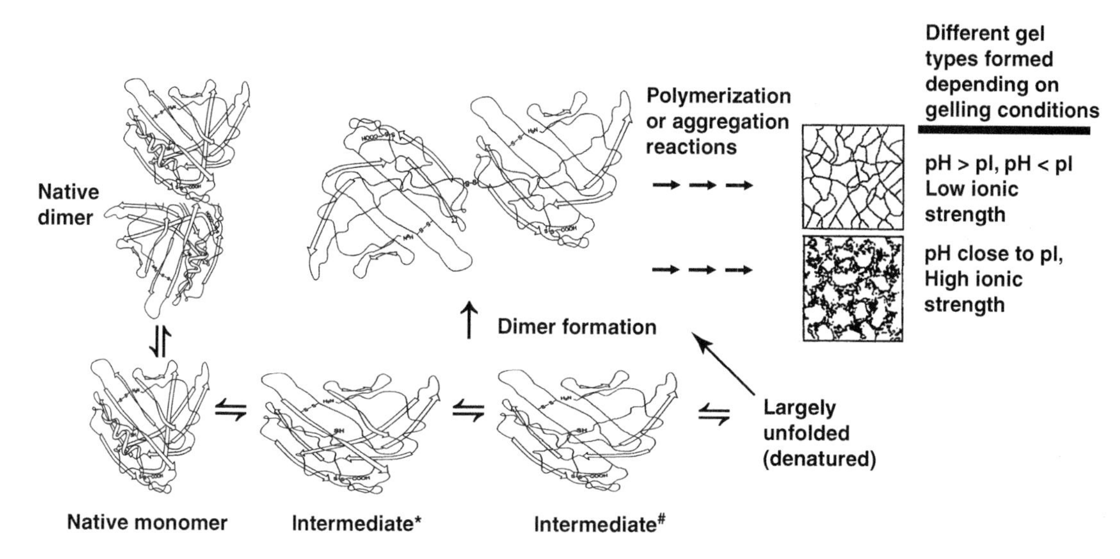

Figure 16.5 Model for the aggregation and formation of heat-induced β-lactoglobulin gels.

of the dimer into monomers and a conformational change leading to the exposure of Cys_{121}, which initiates sulfydryl–disulfide interchange reactions, (b) an endothermic transition to a "molten globule" state, and (c) the unfolding of the protein and a second, high-temperature endothermic transition (Holt and Sawyer, 2003).

The reactive monomers formed during the denaturation process initially form dimers and trimers via the thiol–disulfide exchange reaction and the conversion of dimer to trimer is considered to be the rate-limiting step in the aggregation process (Prabakaran and Damodaran, 1997). Patel *et al.* (2006) proposed that the following reactions occur when milk is heated at ≈85°C; the major whey proteins (β-lactoglobulin and α-lactalbumin) alter their structures and the free cysteine ($CysH_{121}$) of β-lactoglobulin initially reacts reversibly with the adjacent Cys_{106}–Cys_{119} disulfide bond to give a free $CysH_{119}$, which, in turn, reacts with the Cys_{66}–Cys_{160} disulfide bond of the same or another β-lactoglobulin molecule to give a free $CysH_{160}$. $CysH_{160}$ is mobile and free to move because it is so close to the C-terminus of the molecule. Thus, it reacts with disulfide bonds in other proteins, allowing a chain reaction with other β-lactoglobulin or casein molecules to occur (Patel *et al.*, 2006).

A possible model of these reactions during the denaturation and aggregation of β-lactoglobulin is shown in Figure 16.5. In the presence of different types of whey proteins, various heteropolymers (e.g. β-lactoglobulin–α-lactalbumin or β-lactoglobulin–bovine serum albumin) are formed during heating (Havea *et al.*, 2001).

During the heating of β-lactoglobulin, the loss of native structure occurs via both disulfide-linked aggregate formation and non-covalently linked aggregates (e.g. hydrophobic interactions) (McSwiney *et al.*, 1994). When β-lactoglobulin was heated at 75°C, gelation was not observed until most of the protein had aggregated (McSwiney *et al.*, 1994). Pure α-lactalbumin is very heat stable (because it does not have a free thiol group), although it does undergo a reversible transition at

64°C (Ruegg *et al.*, 1977). In the presence of β-lactoglobulin, it undergoes irreversible aggregation via the thiol–disulfide exchange reaction as well as other types of interactions (Elfagm and Wheelock, 1978).

During the heating of β-lactoglobulin, most of the helical conformation is lost by about 65°C; with increasing temperature there is progressive loss of β-sheet structure (Qi *et al.*, 1997). However, in β-lactoglobulin, a considerable amount of secondary structure, particularly β-sheet, still remains intact even at 90°C (Bhattacharjee *et al.*, 2005). Aggregation of globular proteins starts when heat causes some unfolding of the molecule, which exposes reactive groups or sites (e.g. hydrophobic regions) that favor intermolecular interactions (Foegeding, 2006). Gupta *et al.* (1999), using Monte Carlo computer simulations, indicated that protein-like molecules need only partially unfold before they are susceptible to aggregation. Aggregation ultimately results in gelation if the protein concentration and other gelling conditions are favorable. This aggregation process is governed by a balance between attractive hydrophobic and repulsive electrostatic interactions.

Fractal aggregation theory has been applied to the aggregation and formation of whey protein gels (Vreeker *et al.*, 1992; Ikeda *et al.*, 1999). Euston (2004) argued that theories of fractal aggregation are not necessarily a good representation of protein gel structure as they treat the aggregating protein as a rigid particle and ignore any structural changes that occur in the protein during denaturation and aggregation. This criticism could be particularly important for the gelation of globular proteins, such as β-lactoglobulin.

A gel is formed when the extent of aggregation exceeds a critical level for the formation of a self-supporting network that is able to entrap the solvent.

Types and properties of whey protein gels

Different types of gel networks can be formed by globular proteins, such as whey proteins. The network structure in a heat-induced globular protein gel is strongly dependent on the balance between attractive and repulsive forces among (partially) denatured protein molecules during the aggregation process. As whey proteins have isoelectric points (pI) in the vicinity of pH 5, they are negatively charged at neutral pH values. In whey protein solutions, the ionic strength is important as it regulates the amount of ions available for the screening of charged groups on the proteins. At neutral pH values and under low ionic strengths, there is intermolecular repulsion. Aggregation of denatured proteins proceeds via hydrophobic sites and this leads to the formation of fine-stranded gels (with a transparent or translucent appearance and strands that are often 10–20 nm in thickness) (Stading and Hermansson, 1991).

Intermolecular repulsion can be reduced by increasing the ionic strength or by adjusting the pH to be closer to the isoelectric point of the whey proteins ($\approx$5). Under gelation conditions of high ionic strength or pH values close to 5, whey proteins form opaque or particulate or turbid gels. The particles/clusters in this type of gel are in the micron size range. This type of gel structure has a poorer water-holding capacity than fine-stranded gels (Bottcher and Foegeding, 1994). Particulate gels break down rapidly during mastication to yield a homogeneous distribution of small particles, whereas fine-stranded gels break down into large, inhomogeneous particles with irregular shapes (Foegeding, 2006).

A fine-stranded gel formed at neutral pH is rubbery and deformable to a large strain with a small fracture stress (Stading and Hermansson, 1991). At acidic pH values, intermolecular disulfide bonding is unlikely to occur and the fine-stranded networks formed at very low pH values (e.g. 3) are brittle. Particulate gels normally fracture at a small strain, but the stress required to reach the fracture strain is relatively large (Stading and Hermansson, 1991; Bottcher and Foegeding, 1994; Foegeding *et al.*, 1995). After heat-induced gelation, cooling results in strengthening of the network because of hydrogen bond formation.

Heat-induced β-lactoglobulin gels exhibit the characteristics of a "strong gel", i.e. they have a low frequency dependence on the storage modulus (the linear slope, n, of the plot of log frequency versus log storage modulus is <0.06) (Stading and Hermansson, 1990). The slope n is slightly higher for particulate gels than for fine-stranded gels (Stading and Hermansson, 1990).

At pH values around 2 and low ionic strengths, whey protein gels that have some similarities in structure to β-amyloid fibrils are formed (Gosal *et al.*, 2004; Bolder *et al.*, 2006). Fibrils are usually rigid, non-branching and filamentous structures, around 8 nm (or larger) in width (for β-lactoglobulin) and often more than 1 μm long, that arise from linear aggregation of partly unfolded proteins (Gosal *et al.*, 2004). α-Lactalbumin and bovine serum albumin can also form fibrils during heating at pH 2 (Goers *et al.*, 2002; Veerman *et al.*, 2003).

Other factors influencing properties of whey protein gels

pH and ionic strength greatly impact on the type of gel formed and its properties. The strength of whey protein gels increases with protein content. The minimum protein content needed for gelation depends on whether an individual whey protein (e.g. β-lactoglobulin) or a commercial mixture (e.g. WPC) is used as well as the gelation conditions (e.g. pH, heat treatment, ionic strength). Pure solutions of β-lactoglobulin can form a self-supporting gel at 5% protein content when tested at pH 8.0 and a heat treatment of 90°C for 15 min (100 mM Tris-HCl buffer) (Matsudomi *et al.*, 1991).

The protein profile is important for whey gelation, e.g. higher gelling whey products can be made by increasing the proportion of β-lactoglobulin and decreasing the proportion of GMP. As α-lactalbumin is a poorer gelling protein than β-lactoglobulin, increasing the proportion of β-lactoglobulin to α-lactalbumin also increases the gelation properties of whey products (Hines and Foegeding, 1993); commercial whey products with a higher ratio of β-lactoglobulin to α-lactalbumin are available (e.g. WPI made by ion-exchange chromatography compared with WPI made by membrane filtration, or acid whey WPC; both have little or no GMP).

Salts have a major effect on the type of whey protein gel formed as a result of heat treatment and its mechanical/sensory properties. It is generally recognized that the addition of NaCl or $CaCl_2$ to dialyzed samples of WPC or WPI results in an increase in gel strength. Above a level of 10–20 mM $CaCl_2$ and 100–200 mM NaCl, the gel firmness starts to decrease (Schmidt *et al.*, 1979; Kuhn and Foegeding,

1991). Excessive calcium has been speculated to cause rapid protein aggregation, which limits protein unfolding and network formation (Mangino, 1992). Caussin *et al.* (2003) reported that the addition of calcium to whey proteins resulted in the formation of very large protein aggregates during heating. Most commercially available WPC products probably have calcium contents that are greater than that required for optimal gel strength (Mangino, 1992).

There is considerable variability in the thermal aggregation behavior of commercial whey products and some of these differences could be removed by dialysis of these samples to a common ionic strength (McPhail and Holt, 1999). The concentrations of divalent cations are higher in WPC made from cheese whey than in WPC made from acid whey and these cations are not easily removed by dialysis, suggesting some binding with the whey proteins (Havea *et al.*, 2002). Although acid whey starts with a higher calcium content than cheese whey (Table 16.1), it is presumably easier to remove these salts in the manufacture of acid whey WPC than in the manufacture of cheese whey WPC.

Acid whey WPC is known as a superior heat-gelling product compared with cheese whey WPC (Veith and Reynolds, 2004). These differences could be due to the absence of GMP and the low calcium concentration in acid whey WPC. It has been reported that polyphosphates have been added to WPC to improve the gelling properties (Veith and Reynolds, 2004). There are various possible processing approaches to reduce the calcium/mineral content of cheese whey WPC (e.g. electrodialysis, addition of chelating agents, low-pH UF/DF) in order to improve its gelling properties.

The gelling time is also dependent on temperature, with the time required for gelation decreasing with increasing temperature although, at very high temperatures, gelation may occur only during the subsequent cooling stage (Hillier and Cheeseman, 1979). Many reports show that, when all other factors are kept constant, the gel strength increases with increasing temperature (presumably reflecting greater unfolding and reactivity of the proteins) (Mulvihill and Kinsella, 1987). The presence of lipids and lactose impairs the gelation of whey proteins (Mulvihill and Kinsella, 1987). Sugars, such as lactose, are known to protect the protein against loss of solubility during heat treatment and increase the thermal denaturation temperature of whey proteins (de Wit, 1981; Jou and Harper, 1996). Possibly, lipids might interfere with the hydrophobic interactions that play a role in the aggregation of partly unfolded whey proteins during heat treatment.

Cold gelation of whey proteins

Gels can also be produced using a two-step process that involves heat treatment at low ionic strength and/or far from the isoelectric point, followed by an increase in ionic strength and/or an adjustment in pH (Barbut and Foegeding, 1993; Britten and Giroux, 2001). These gels are labelled as cold-set gels, as the initial heat treatment produces a polymerized solution, with gelation occurring during the subsequent cold-set conditions through screening of the repulsive forces. To obtain gels via the cold-set gelation method, it is necessary to prepare a heat-denatured solution, with a protein concentration below the critical gelation concentration. Gelation can then be induced at low temperatures by the addition of mono- or polyvalent cations (e.g. Ca^{2+}).

Britten and Giroux (2001) acidified whey protein polymers to pH 4.6 with GDL and formed opaque particulate gels. The storage modulus and the firmness of the gels were affected by the conditions used to prepare the protein polymers.

Enzymatic modification of whey protein (for gelation purposes)

Extensive hydrolysis of whey protein using proteinases results in gelation mainly via hydrophobic interactions, with hydrogen bonding and electrostatic interactions also playing a minor role (Otte *et al.*, 1996; Doucet *et al.*, 2003).

The casein fractions in milk are more susceptible to TGase cross-linking than the globular whey protein fractions (Jaros *et al.*, 2006). Some unfolding of β-lactoglobulin improves the extent of cross-linking with TGase (Færgemand *et al.*, 1997; O'Sullivan *et al.*, 2002a). Cold-set whey protein gels at low pH have been cross-linked with the TGase enzyme under either low pH or alkaline conditions (Eissa *et al.*, 2004; Eissa and Khan, 2005).

One approach involved two steps, firstly cross-linking whey proteins with TGase at pH 8 and 50°C and secondly cold-set acidifying the resulting solution using GDL (Eissa *et al.*, 2004). During the first step, the whey proteins undergo enzyme-catalyzed ε-(γ-glutamyl)lysine bond formation with a substantial increase in viscosity. Enzyme-cross-linked gels had significantly higher yield/fracture stress and strain than cold-set gels prepared without TGase enzyme or conventional heat-set gels. In addition, the elastic modulus of the enzyme-catalyzed gel was higher than that of its non-enzyme-treated counterpart.

Mixed gels made with rennet and acid

Milk coagulation can be induced by the combined action of acid and enzyme (i.e. mixed gels). The study of mixed milk coagulation has received very little attention when compared with rennet- or acid-induced coagulation, although there have been several recent studies (Roefs *et al.*, 1990; Lucey *et al.*, 2000, 2001; Tranchant *et al.*, 2001).

Cottage cheese is generally manufactured by acid coagulation of pasteurized skim milk and a small concentration of rennet is sometimes added after the starter has been allowed to develop some acidity (i.e. at pH around 5.5) (Castillo *et al.*, 2006). The use of rennet in combination with acid development initiates gelation at a high pH and the gel can undergo a "weakening" stage (as indicated by a decrease/plateau of the storage modulus, a decrease in the light backscatter ratio or an increase in the loss tangent).

This weakening is more pronounced with unheated milk gels and where there have been very high levels of κ-casein hydrolysis prior to acidification (Li and Dalgleish, 2006). This "weakening" stage is related to rearrangements caused by CCP demineralization of the casein particles in the gel network because this CCP solubilization occurs after gelation (gelation is initiated at a high pH in mixed gels) (Lucey *et al.*, 2000). The final storage modulus of mixed gels can be considerably higher than that of acid gels made without rennet. Mixed gels made from heated milk formed firmer gels, as they were cross-linked by denatured whey proteins and underwent fewer large-scale rearrangements (Lucey *et al.*, 2000).

The rheological and microstructural properties of mixed gels are complex and these properties can be adjusted by varying the rennet level or the acidification rate (Tranchant *et al.*, 2001). The use of low rennet levels during the fermentation of milk resulted in a coarser acid gel network and higher syneresis (Aichinger *et al.*, 2003). Micelle fusion was faster in gels with rennet added because of the removal of the κ-casein hairs (Aichinger *et al.*, 2003).

Gastaldi *et al.* (2003) studied the acid-induced gelation of milk samples in which chymosin was used to vary the degree of κ-casein hydrolysis prior to acidification (further chymosin activity during acidification was blocked using an inhibitor). The gelation pH increased and the gelation time decreased with an increasing degree of κ-casein hydrolysis. Gels with much higher storage moduli were formed as a result of partial κ-casein hydrolysis prior to gelation, although the loss tangent and the serum-holding capacity were lower (Gastaldi *et al.*, 2003). Presumably, partial κ-casein hydrolysis prior to acid gelation facilitated greater rearrangements/fusion of casein, which was responsible for the increase in the storage modulus but also increased the serum separation (Lucey *et al.*, 2001).

Conclusions

The formation and the physical properties of milk protein gels have been the subject of intense study because of the great economic impact of these gels for dairy products such as cheese, yoghurt and heat-set whey gels. There is a growing recognition that the internal structure of casein micelles plays an important role in the structural properties of rennet, acid and mixed gels. These gels are dynamic in nature and undergo rearrangements.

Technologists have recently studied the impact of high pressure and enzymatic cross-linking of these proteins to modify their functionality. The interaction between DWP and caseins has received a lot of attention and this interaction has been used to alter the texture of acid gels, although there is disagreement about the exact mechanism(s) involved. DWP polymers have been used for making cold-set gels and they have interesting possible applications in various milk gels/products. Fine-stranded whey proteins made at very low pH values have been shown to be similar in structure to amyloid fibrils. From an industrial viewpoint, these fine-stranded fibril types of gels might have some useful applications because they gel at low protein concentrations.

References

Aichinger, P-A., Michel, M., Servais, C., Dillmann, M-L., Rouvet, M., D'Amico, N., Zink, R., Klostermeyer, H. and Horne, D. S. (2003). Fermentation of a skim milk concentrate with *Streptococcus thermophilus* and chymosin: structure, viscoelasticity and syneresis of gels. *Colloids and Surfaces B: Biointerfaces*, **31**, 243–55.

Aguilera, J. M. (1995). Gelation of whey proteins. *Food Technology*, **49**, 83–89.

Alexander, M. and Dalgleish, D. G. (2004). Application of transmission diffusive wave spectroscopy to the study of gelation of milk by acidification and rennet. *Colloids and Surfaces B: Biointerfaces*, **38**, 83–90.

Anema, S. G., Lee, S. K., Lowe, E. K. and Klostermeyer, H. (2004). Rheological properties of acid gels prepared from heated pH-adjusted skim milk. *Journal of Agricultural and Food Chemistry*, **52**, 337–43.

Anema, S. G., Lauber, S., Lee, S. K., Henle, T. and Klostermeyer, H. (2005). Rheological properties of acid gels prepared from pressure- and transglutaminase-treated skim milk. *Food Hydrocolloids*, **19**, 879–87.

Anema, S. G., Lee, S. K. and Klostermeyer, H. (2007). Effect of pH at heat treatment on the hydrolysis of κ-casein and the gelation of skim milk by chymosin. *Lebensmittel-Wissenschaft und -Technologie*, **40**, 99–106.

Barbut, S. and Foegeding, E. A. (1993). Calcium-induced gelation of preheated whey protein isolate. *Journal of Food Science*, **58**, 867–71.

Bhattacharjee, C., Saha, S., Biswas, A., Kundu, M., Ghosh, L. and Das, K. P. (2005). Structural changes of β-lactoglobulin during thermal unfolding and refolding – an FT-IR and circular dichroism study. *Protein Journal*, **24**, 27–35.

Bikker, J. F., Anema, S. G., Li, Y. and Hill, J. P. (2000). Rheological properties of acid gels prepared from heated milk fortified with whey protein mixture containing the A, B, and C variants of β-lactoglobulin. *International Dairy Journal*, **10**, 723–32.

Bolder, S. G., Hendrickx, H., Sagis, L. M. C. and van der Linden, E. (2006). Fibril assemblies in aqueous whey protein mixtures. *Journal of Agricultural and Food Chemistry*, **54**, 4229–34.

Bottcher, S. R. and Foegeding, E. A. (1994). Whey protein gels – fracture stress and strain and related microstructural properties. *Food Hydrocolloids*, **8**, 113–23.

Britten, M. and Giroux, H. J. (2001). Acid-induced gelation of whey protein polymers: effects of pH and calcium concentration during polymerization. *Food Hydrocolloids*, **15**, 609–17.

Castillo, M., Lucey, J. A. and Payne, F. A. (2006). The effect of temperature and inoculum concentration on rheological and light scatter properties of milk coagulated by a combination of bacterial fermentation and chymosin. Cottage cheese-type gels. *International Dairy Journal*, **16**, 131–46.

Caussin, F., Famelart, M. H., Maubois, J-L. and Bouhallab, S. (2003). Mineral modulation of thermal aggregation and gelation of whey proteins: from β-lactoglobulin model system to whey protein isolate. *Lait*, **83**, 1–12.

Cho, Y. H., Lucey, J. A. and Singh, H. (1999). Rheological properties of acid milk gels as affected by the nature of the fat globule surface material and heat treatment of milk. *International Dairy Journal*, **9**, 537–45.

Choi, J., Horne, D. S. and Lucey, J. A. (2007). Effect of insoluble calcium concentration on rennet coagulation properties of milk. *Journal of Dairy Science*, **90**, 2612–23.

Clark, A. H. (1992). Gels and gelling. In *Physical Chemistry of Foods* (H. G. Schwartzberg and R. W. Hartel, eds) pp. 263–305. New York: Marcel Dekker Inc.

Clark, A. H. (1996). Biopolymer gels. *Current Opinion in Colloid and Interface Science*, **1**, 712–17.

Clark, A. H. (1998). Gelation of globular proteins. In *Functional Properties of Food Macromolecules*, 2nd edn (S. E. Hill, D. A. Ledward and J. R. Mitchell, eds) pp. 77–142. Gaithersburg, Maryland: Aspen Publishers.

Crabbe, M. J. C. (2004). Rennet: general and molecular aspects. In *Cheese: Chemistry, Physics and Microbiology*, Volume 1, *General Aspects*, 3rd edn (P. F. Fox, P. L. H. McSweeney, T. M. Cogan and T. P. Guinee, eds) pp. 19–43. London: Elsevier.

Dalgleish, D. G. (1987). The enzymatic coagulation of milk. In *Cheese: Chemistry, Physics and Microbiology* (P. F. Fox, ed.) pp. 63–95. London: Elsevier Applied Science.

Dalgleish, D. G. (1993). The enzymatic coagulation of milk. In *Cheese: Chemistry, Physics and Microbiology*, Volume 1, *General Aspects*, 2nd edn (P. F. Fox, ed.) pp. 69–100. London: Chapman and Hall.

Dalgleish, D. G., Alexander, M. and Corredig, M. (2006). Studies of the acid gelation of milk using ultrasonic spectroscopy and diffusing wave spectroscopy. *Food Hydrocolloids*, **18**, 747–55.

de Kruif, K. G., Hoffman, M. A. M., van Marle, M. E., van Mil, P. J. J. M., Roefs, S. P. F. M., Verheul, M. and Zoon, N. (1995). Gelation of proteins from milk. *Faraday Discussions*, **101**, 185–200.

de Wit, J. N. (1981). Structure and functional behaviour of whey proteins. *Netherlands Milk and Dairy Journal*, **35**, 47–64.

Dejmek, P. and Walstra, P. (2004). The syneresis of rennet-coagulated curd. In *Cheese: Chemistry, Physics and Microbiology,* Volume 1, *General Aspects*, 3rd edn (P. F. Fox, P. L. H. McSweeney, T. M. Cogan and T. G. Guinee, eds) pp. 71–103. London: Elsevier.

Doi, E. (1993). Gels and gelling of globular proteins. *Trends in Food Science and Technology*, **4**, 1–5.

Doucet, D., Gauthier, S. F., Otter, D. E. and Foegeding, E. A. (2003). Enzyme-induced gelation of extensively hydrolyzed whey proteins by alcalase: comparison with the plastein reaction and characterization of interactions. *Journal of Agricultural and Food Chemistry*, **51**, 6036–42.

Drøhse, H. B. and Foltmann, B. (1989). Specificity of milk-clotting enzymes towards bovine κ-casein. *Biochimica et Biophysica Acta*, **995**, 221–24.

Eissa, A. S. and Khan, S. A. (2005). Acid-induced gelation of enzymatically modified, preheated whey proteins. *Journal of Agricultural and Food Chemistry*, **53**, 5010–17.

Eissa, A. S., Bisram, S. and Khan, S. A. (2004). Polymerization and gelation of whey protein isolates at low pH using transglutaminase enzyme. *Journal of Agricultural and Food Chemistry*, **52**, 4456–64.

Elfagm, A. A. and Wheelock, J. V. (1978). Interaction of bovine α-lactalbumin and β-lactoglobulin during heating. *Journal of Dairy Science*, **61**, 28–32.

Esteves, C. L. C., Lucey, J. A. and Pires, E. M. V. (2001). Mathematical modelling of the formation of rennet-induced gels by plant coagulants and chymosin. *Journal of Dairy Research*, **68**, 499–510.

Esteves, C. L. C., Lucey, J. A. and Pires, E. M. V. (2002). Rheological properties of milk gels made using coagulants of plant origin and chymosin. *International Dairy Journal*, **12**, 427–34.

Esteves, C. L. C., Lucey, J. A., Hyslop, D. B. and Pires, E. M. V. (2003). Effect of gelation temperature on the properties of skim milk gels made from plant coagulants and chymosin. *International Dairy Journal*, **13**, 877–85.

Euston, S. R. (2004). Computer simulation of proteins: adsorption, gelation and self-association. *Current Opinion in Colloid and Interface Science*, **9**, 321–27.

Færgemand, M. and Qvist, K. B. (1997). Transglutaminase: effect on rheological properties, microstructure and permeability of set style acid skim milk gel. *Food Hydrocolloids*, **11**, 287–92.

Færgemand, M., Otte, J. and Qvist, K. B. (1997). Enzymatic crosslinking of whey proteins by a Ca^{2+} independent microbial transglutaminase from *Streptomyces lydicus*. *Food Hydrocolloids*, **11**, 19–25.

Færgemand, M., Sørensen, M. V., Jørgensen, U., Budolfsen, G. and Qvist, K. B. (1999). Transglutaminase: effect on instrumental and sensory texture of set style yoghurt. *Milchwissenschaft*, **54**, 563–66.

Fagan, C. C., Leedy, M., Castillo, M., Payne, F. A., O'Donnell, C. P. and O'Callaghan, D. J. (2006). Development of a light scatter sensor technology for on-line monitoring of milk coagulation and whey separation. *Journal of Food Engineering*, **83**, 61–67.

Foegeding, E. A. (2006). Food biophysics of protein gels: a challenge of nano and macroscopic proportions. *Food Biophysics*, **1**, 41–50.

Foegeding, E. A., Bowland, E. I. and Hardin, C. C. (1995). Factors that determine the fracture properties and microstructure of globular protein gels. *Food Hydrocolloids*, **9**, 237–49.

Gastaldi, E., Trial, N., Guillaume, C., Bourret, E., Gontard, N. and Cuq, J. L. (2003). Effect of controlled κ-casein hydrolysis on rheological properties of acid milk gels. *Journal of Dairy Science*, **86**, 704–11.

Goers, J., Permyakov, S. E., Permyakov, E. A., Uversky, V. N. and Fink, A. L. (2002). Conformational prerequisites for alpha-lactalbumin fibrillation. *Biochemistry*, **41**, 12546–51.

Gosal, W. S. and Ross-Murphy, S. B. (2000). Globular protein gelation. *Current Opinion in Colloid and Interface Science*, **5**, 188–94.

Gosal, W. S., Clark, A. H. and Ross-Murphy, S. B. (2004). Fibrillar β-lactoglobulin gels: Part 1. Fibril formation and structure. *Biomacromolecules*, **5**, 2408–19.

Green, M. L. (1980). The formation and structure of milk protein gels. *Food Chemistry*, **6**, 41–49.

Green, M. L. and Grandison, A. S. (1993). Secondary (non-enzymatic) phase of rennet coagulation and post-coagulation phenomena. In *Cheese: Chemistry, Physics and Microbiology*, Volume 1, *General Aspects*, 2nd edn (P. F. Fox, ed.) pp. 101–40. Gaithersburg, Maryland: Aspen Publishers.

Gupta, P., Hall, C. K. and Voegler, A. (1999). Computer simulation of the competition between protein folding and aggregation. *Fluid Phase Equilibria*, **158–60**, 87–93.

Guyomarc'h, F., Queguiner, C., Law, A. J. R., Horne, D. S. and Dalgleish, D. G. (2003). Role of the soluble and micelle-bound heat-induced protein aggregates on network formation in acid skim milk gels. *Journal of Agricultural and Food Chemistry*, **51**, 7743–50.

Han, X-Q. and Damodaran, S. (1996). Thermodynamic compatability of substrate proteins affects their cross-linking by transglutaminase. *Journal of Agricultural and Food Chemistry*, **44**, 1211–17.

Harte, F., Luedecke, L., Swanson, B. and Barbosas-Cánovas, G. V. (2003). Low-fat set yogurt made from milk subjected to combinations of high hydrostatic pressure and thermal processing. *Journal of Dairy Science*, **86**, 1074–82.

Havea, P., Singh, H. and Creamer, L. K. (2001). Characterization of heat-induced aggregates of β-lactoglobulin, α-lactalbumin and bovine serum albumin in a whey protein concentrate environment. *Journal of Dairy Research*, **68**, 483–97.

Havea, P., Singh, H. and Creamer, L. K. (2002). Heat-induced aggregation of whey proteins: comparison of cheese WPC with acid WPC and relevance of mineral composition. *Journal of Agricultural and Food Chemistry*, **50**, 4674–81.

Hillier, R. M. and Cheeseman, G. C. (1979). Effect of proteose-peptone on the heat gelation of whey protein isolate. *Journal of Dairy Research*, **46**, 113–20.

Hines, M. E. and Foegeding, E. A. (1993). Interactions of α-lactalbumin and bovine serum albumin with β-lactoglobulin in thermally induced gelation. *Journal of Agricultural and Food Chemistry*, **41**, 341–46.

Holt, C. and Sawyer, L. (2003). The principal bovine whey protein β-lactoglobulin: a structure: function analysis. In *Industrial Proteins In Perspective*, Progress in Biotechnology 23 (W. Y. Aalbersberg, R. J. Hamer, P. Jasperse, H. H. J. de Jong, C. G. de Kruif, P. Walstra and F. A. de Wolf, eds) pp. 44–49. Amsterdam: Elsevier Science.

Horne, D. S. (1995). Scaling behaviour of shear moduli during the formation of rennet milk gels. In *Food Macromolecules and Colloids* (E. Dickinson and D. Lorient, eds) pp. 456–61. Cambridge: Royal Society of Chemistry.

Horne, D. S. (1996). Aspects of scaling behaviour in the kinetics of particle formation. *Journal de Chimie Physique et de Physico-Chimie Biologique*, **93**, 977–86.

Horne, D. S. (1998). Casein interactions: casting light on the black boxes, the structure in dairy products. *International Dairy Journal*, **8**, 171–77.

Horne, D. S. (1999). Formation and structure of acidified milk gels. *International Dairy Journal*, **9**, 261–68.

Horne, D. S. (2001). Factors influencing acid-induced gelation of skim milk. In *Food Hydrocolloids: Fundamentals of Formulation* (E. Dickinson and R. Miller, eds) pp. 345–51. Cambridge: Royal Society of Chemistry.

Horne, D. S. (2003). Casein micelles as hard spheres: limitations of the model in acidified milk gels. *Colloids and Surfaces A: Physicochemical Engineering Aspects*, **213**, 255–63.

Horne, D. S. and Banks, J. M. (2004). Rennet-induced coagulation of milk. In *Cheese: Chemistry, Physics and Microbiology*, Volume 1, *General Aspects*, 3rd edn (P. F. Fox, P. L. H. McSweeney, T. Cogan and T. Guinee, eds) pp. 47–70. London: Elsevier.

Huppertz, T. and de Kruif, C. G. (2006). Disruption and reassociation of casein micelles under high pressure: influence of milk serum composition and casein micelle concentration. *Journal of Agricultural and Food Chemistry*, **54**, 5903–9.

Huppertz, T. and de Kruif, C. G. (2007). Rennet-induced coagulation of enzymatically cross-linked casein micelles. *International Dairy Journal*, **17**, 442–47.

Huppertz, T., Hinz, K., Zobrist, M. R., Uniacke, T., Kelly, A. L. and Fox, P. F. (2005). Effects of high pressure treatment on the rennet coagulation and cheese-making properties of heated milk. *Innovative Food Science and Emerging Technologies*, **6**, 279–85.

Hyslop, D. B. (2003). Enzymatic coagulation of milk. In *Advanced Dairy Chemistry*, Volume 2, *Proteins*, 2nd edn (P. F. Fox and P. L. H. McSweeney, eds) pp. 839–78. Gaithersburg, Maryland: Aspen Publishers.

Ikeda, S., Foegeding, E. A. and Hagiwara, T. (1999). Rheological study of the fractal nature of the protein gel structure. *Langmuir*, **15**, 8584–89.

Jaros, D. and Rohm, H. (2003a). The rheology and textural properties of yoghurt. In *Texture in Food*, Volume 1 (B. M. McKenna, ed.) pp. 321–49. Cambridge: Woodhead Publishing.

Jaros, D. and Rohm, H. (2003b). Controlling the texture of fermented dairy products: the case of yoghurt. In *Dairy Processing: Improving Quality* (G. Smit, ed.) pp. 155–84. Cambridge: Woodhead Publishing.

Jaros, D., Partschefeld, C., Henle, T. and Rohm, H. (2006). Transglutaminase in dairy products: chemistry, physics, applications. *Journal of Texture Studies*, **37**, 113–55.

Jean, K., Renan, M., Famelart, M. H. and Guyomarc'h, F. (2006). Structure and surface properties of the serum heat-induced protein aggregates isolated from heated skim milk. *International Dairy Journal*, **16**, 303–15.

Jou, K. D. and Harper, W. J. (1996). Effect of di-saccharides on the thermal properties of whey proteins determined by differential scanning calorimetry (DSC). *Milchwissenschaft*, **51**, 509–12.

Kaláb, M., Allan-Wojtas, P. and Phipps-Todd, B. E. (1983). Development of microstructure in set-style nonfat yoghurt – a review. *Food Microstructure*, **2**, 51–66.

Karlsson, A. O., Ipsen, R. and Ardö, Y. (2007a). Influence of pH and NaCl on rheological properties of rennet-induced casein gels made from UF concentrated skim milk. *International Dairy Journal*, **17**, 1053–62.

Karlsson, A. O., Ipsen, R. and Ardö, Y. (2007b). Rheological properties and microstructure during rennet induced coagulation of UF concentrated skim milk. *International Dairy Journal*, **17**, 674–82.

Kella, N. K. and Kinsella, J. E. (1988a). Structural stability of β-lactoglobulin in the presence of kosmotropic salts. A kinetic and thermodynamic study. *International Journal of Peptide and Protein Research*, **32**, 396–405.

Kella, N. K. and Kinsella, J. E. (1988b). Enhanced thermodynamic stability of β-lactoglobulin at low pH. A possible mechanism. *Biochemical Journal*, **255**, 113–18.

Klandar, A. H., Lagaude, A. and Chevalier-Lucia, D. (2007). Assessment of the rennet coagulation of skim milk: a comparison of methods. *International Dairy Journal*, **17**, 1151–60.

Knudsen, J. C., Karlsson, A. O., Ipsen, R. and Skibsted, L. H. (2006). Rheology of stirred acidified skim milk gels with different particle interactions. *Colloids and Surfaces A: Physicochemical and Engineering Aspects*, **274**, 56–61.

Kuhn, P. R. and Foegeding, E. A. (1991). Mineral salt effects on whey protein gelation. *Journal of Agricultural and Food Chemistry*, **39**, 1013–16.

Lauber, S., Henle, T. and Klostermeyer, H. (2000). Relationship between the crosslinking of caseins by transglutaminase and the gel strength of yoghurt. *European Food Research and Technology Journal*, **210**, 305–9.

Lauber, S., Noack, I., Klostermeyer, H. and Henle, T. (2001). Stability of microbial transglutaminase to high pressure treatment. *European Food Research and Technology Journal*, **213**, 273–76.

Lauger, J., Wollny, K. and Huck, S. (2002). Direct strain oscillation: a new oscillatory method enabling measurements at very small shear stresses and strains. *Rheologica Acta*, **41**, 356–61.

Lee, W. J. and Lucey, J. A. (2004). Structure and physical properties of yogurt gels: effect of inoculation rate and incubation temperature. *Journal of Dairy Science*, **87**, 3153–64.

Li, J. and Dalgleish, D. G. (2006). Controlled proteolysis and the properties of milk gels. *Journal of Agricultural and Food Chemistry*, **54**, 4687–95.

Lodaite, K., Östergren, K., Paulsson, M. and Dejmek, P. (2000). One-dimensional syneresis of rennet-induced gels. *International Dairy Journal*, **10**, 829–34.

Lomholt, S. B. and Qvist, K. B. (1999). The formation of cheese curd. In *Technology of Cheesemaking* (B. A. Law, ed.) pp. 66–98. Sheffield: Sheffield Academic Press.

López-Fandiño, R. (2006). High pressure-induced changes in milk proteins and possible applications in dairy technology. *International Dairy Journal*, **16**, 1119–31.

Lorenzen, P. C. (2000). Renneting properties of transglutaminase-treated milk. *Milchwissenschaft*, **55**, 433–37.

Lucey, J. A. (1995). Effect of heat treatment on the rennet coagulability of milk. In *Heat-induced Changes in Milk*, 2nd edn, IDF Special Issue 9501 (P. F. Fox, ed.) pp. 171–87. Brussels: International Dairy Federation.

Lucey, J. A. (2001). The relationship between rheological parameters and whey separation in milk gels. *Food Hydrocolloids*, **15**, 603–8.

Lucey, J. A. (2002). Formation and physical properties of milk protein gels. *Journal of Dairy Science*, **85**, 281–94.

Lucey, J. A. (2003). Formation, structure, properties and rheology of acid-coagulated milk gels. In *Cheese: Chemistry, Physics and Microbiology,* Volume 1, *General Aspects*, 3rd edn (P. F. Fox, P. L. H. McSweeney, T. M. Cogan and T. P. Guinee, eds) pp. 105–22. London: Elsevier.

Lucey, J. A. and Singh, H. (1998). Formation and physical properties of acid milk gels: a review. *Food Research International*, **30**, 529–42.

Lucey, J. A. and Singh, H. (2003). Acid coagulation of milk. In *Advanced Dairy Chemistry,* Volume 2, *Proteins*, 2nd edn (P. F. Fox and P. L. H. McSweeney, eds) pp. 1001–26. Gaithersburg, Maryland: Aspen Publishers.

Lucey, J. A., Teo, C. T., Munro, P. A. and Singh, H. (1997). Rheological properties at small (dynamic) and large (yield) deformations of acid gels made from heated milk. *Journal of Dairy Research*, **64**, 591–600.

Lucey, J. A., Tamehana, M., Singh, H. and Munro, P. A. (1998a). A comparison of the formation, rheological properties and microstructure of acid skim milk gels made with a bacterial culture or glucono-δ-lactone. *Food Research International*, **31**, 147–55.

Lucey, J. A., Tamehana, M., Singh, H. and Munro, P. A. (1998b). Effect of interactions between denatured whey proteins and casein micelles on the formation and rheological properties of acid skim milk gels. *Journal of Dairy Research*, **65**, 555–67.

Lucey, J. A., Munro, P. A. and Singh, H. (1999). Effects of heat treatment and whey protein addition on the rheological properties and structure of acid skim milk gels. *International Dairy Journal*, **9**, 275–79.

Lucey, J. A., Tamehana, M., Singh, H. and Munro, P. A. (2000). Rheological properties of milk gels formed by a combination of rennet and glucono-δ-lactone. *Journal of Dairy Research*, **67**, 415–27.

Lucey, J. A., Tamehana, M., Singh, H. and Munro, P. A. (2001). Effect of heat treatment on the physical properties of milk gels made with both rennet and acid. *International Dairy Journal*, **11**, 559–65.

Mangino, M. E. (1992). Gelation of whey-protein concentrates. *Food Technology*, **46**, 114–17.

Matsudomi, N., Rector, D. and Kinsella, J. E. (1991). Gelation of bovine serum albumin and β-lactoglobulin; effects of pH, salts and thiol reagents. *Food Chemistry*, **40**, 55–69.

McCann, T. C. A. and Pyne, G. T. (1960). The colloidal phosphate of milk. III. Nature of its association with casein. *Journal of Dairy Research*, **27**, 403–17.

McPhail, D. and Holt, C. (1999). Effect of anions on the denaturation and aggregation of β-lactoglobulin as measured by differential scanning microcalorimetry. *International Journal of Food Science and Technology*, **34**, 477–81.

McSwiney, M., Singh, H. and Campanella, O. H. (1994). Thermal aggregation and gelation of bovine β-lactoglobulin. *Food Hydrocolloids*, **8**, 441–53.

Mellema, M., Walstra, P., van Opheusden, J. H. J. and van Vliet, T. (2002). Effects of structural rearrangements on the rheology of rennet-induced casein particle gels. *Advances in Colloid and Interface Science*, **98**, 25–50.

Mulvihill, D. M. and Donovan, M. (1987). Whey proteins and their thermal denaturation. A review. *Irish Journal of Food Science and Technology*, **11**, 43–77.

Mulvihill, D. M. and Kinsella, J. E. (1987). Gelation characteristics of whey proteins and β-lactoglobulin. *Food Technology*, **41**, 102–11.

Needs, E. C., Capellas, M., Bland, A. P., Manoj, P., Macdougal, D. and Paul, G. (2000). Comparison of heat and pressure treatments of skim milk, fortified with whey protein concentrate, for set yogurt preparation: effects on milk proteins and gel structure. *Journal of Dairy Research*, **67**, 329–48.

O'Callaghan, D. J., O'Donnell, C. P. and Payne, F. A. (2002). Review of systems for monitoring curd setting during cheesemaking. *International Journal of Dairy Technology*, **55**, 65–74.

O'Sullivan, M. M., Kelly, A. L. and Fox, P. F. (2002a). Effect of transglutaminase on the heat stability of milk: a possible mechanism. *Journal of Dairy Science*, **85**, 1–7.

O'Sullivan, M. M., Kelly, A. L. and Fox, P. F. (2002b). Influence of transglutaminase treatment on some physico-chemical properties of milk. *Journal of Dairy Research*, **69**, 433–42.

Oakenfull, D. (1987). Gelling agents. CRC Critical Reviews in *Food Science and Nutrition*, **26**, 1–25.

Oakenfull, D., Pearce, J. and Burley, R. W. (1997). Protein gelation. In *Food Proteins and their Applications* (S. Damodaran and A. Paraf, eds) pp. 111–42. New York: Marcel Dekker.

Otte, J., Ju, Z. Y., Faergemand, M., Lomholt, S. B. and Qvist, K. B. (1996). Protease-induced aggregation and gelation of whey proteins. *Journal of Food Science*, **61**, 911–15, 923.

Ozer, B., Kirmaci, H. A., Oztekin, S., Hayalogluc, A. and Atamer, M. (2007). Incorporation of microbial transglutaminase into non-fat yogurt production. *International Dairy Journal*, **17**, 199–207.

Patel, H. A., Singh, H., Anema, S. G. and Creamer, L. K. (2006). Effects of heat and high hydrostatic pressure treatments on disulfide bonding interchanges among the proteins in skim milk. *Journal of Agricultural and Food Chemistry*, **54**, 3409–20.

Pearse, M. J. and MacKinlay, A. G. (1989). Biochemical aspects of syneresis: a review. *Journal of Dairy Science*, **72**, 1401–7.

Prabakaran, S. and Damodaran, S. (1997). Thermal unfolding of β-lactoglobulin: characterization of initial unfolding events responsible for heat-induced aggregation. *Journal of Agricultural and Food Chemistry*, **45**, 4303–08.

Qi, X. L., Holt, C., McNulty, D., Clarke, D. T., Brownlow, S. and Jones, G. R. (1997). Effect of temperature on the secondary structure of β-lactoglobulin at pH 6.7, as determined by CD and IR spectroscopy: a test of the molten globule hypothesis. *Biochemical Journal*, **324**, 341–46.

Raouche, S., Dobenesque, M., Bot, A., Lagaude, A., Cuq, J. L. and Marchesseau, S. (2007). Stability of casein micelle subjected to reversible CO_2 acidification: impact of holding time and chilled storage. *International Dairy Journal*, **17**, 873–80.

Reddy, I. M. and Kinsella, J. E. (1990). Interaction of β-lactoglobulin with κ-casein in micelles as assessed by chymosin hydrolysis: effect of temperature, heating time, β-lactoglobulin concentration, and pH. *Journal of Agricultural and Food Chemistry*, **38**, 50–58.

Roefs, S. P. F. M., van Vliet, T., van den Bijgaart, H. J. C. M., de Groot-Mostert, A. E. A. and Walstra, P. (1990). Structure of casein gels made by combined acidification and rennet action. *Netherlands Milk and Dairy Journal*, **44**, 158–88.

Ruegg, M., Moor, U. and Blanc, B. (1977). A calorimetric study of the thermal denaturation of whey proteins in simulated milk ultrafiltrate. *Journal of Dairy Research*, **44**, 509–20.

Sandra, S., Alexander, M. and Dalgleish, D. G. (2007). The rennet coagulation mechanism of skim milk as observed by transmission diffusing wave spectroscopy. *Journal of Colloid and Interface Science*, **308**, 364–73.

Schmidt, R. H., Illingworth, B. L., Deng, J. C. and Cornell, J. A. (1979). Multiple regression and response surface analysis of the effects of calcium chloride and cysteine on heat-induced whey protein gelation. *Journal of Agricultural and Food Chemistry*, **27**, 529–32.

Schorsch, C., Carrie, H. and Norton, I. T. (2000). Cross-linking casein micelles by a microbial transglutaminase: influence of cross-links in acid-induced gelation. *International Dairy Journal*, **10**, 529–39.

Schorsch, C., Wilkins, D. K., Jones, M. G. and Norton, I. T. (2001). Gelation of casein-whey mixtures: effects of heating whey proteins alone or in the presence of casein micelles. *Journal of Dairy Research*, **68**, 471–81.

Serra, M., Trujillo, A. J., Quevedo, J. M., Guamis, B. and Ferragut, V. (2007). Acid coagulation properties and suitability for yogurt production of cows' milk treated by high-pressure homogenization. *International Dairy Journal*, **17**, 782–90.

Singh, H. and Havea, P. (2003). Thermal denaturation, aggregation and gelation of whey proteins. In *Advanced Dairy Chemistry,* Volume 2, *Proteins*, 2nd edn (P. F. Fox and P. L. H. McSweeney, eds) pp. 1261–87. Gaithersburg, Maryland: Aspen Publishers.

Srinivasan, M. and Lucey, J. A. (2002). Effects of added plasmin on the formation and rheological properties of rennet-induced skim milk gels. *Journal of Dairy Science*, **85**, 1070–78.

Stading, M. and Hermansson, A-M. (1990). Viscoelastic behaviour of β-lactoglobulin gel structures. *Food Hydrocolloids*, **4**, 121–35.

Stading, M. and Hermansson, A-M. (1991). Large deformation properties of β-lactoglobulin gel structures. *Food Hydrocolloids*, **5**, 339–52.

Tamime, A. Y. and Robinson, R. K. (1999). *Yoghurt – Science and Technology.* Cambridge: Woodhead Publishers.

Tranchant, C. C., Dalgleish, D. G. and Hill, A. R. (2001). Different coagulation behaviour of bacteriologically acidified and renneted milk: the importance of fine-tuning acid production and rennet action. *International Dairy Journal*, **11**, 483–94.

van Dijk, H. J. M. and Walstra, P. (1986). Syneresis of curd. 2. One-dimensional syneresis of rennet curd in constant conditions. *Netherlands Milk and Dairy Journal*, **40**, 3–30.

van Vliet, T. and Walstra, P. (1994). Water in casein gels; how to get it out or keep it in. *Journal of Food Engineering*, **22**, 75–88.

van Vliet, T., van Dijk, H. J. M., Zoon, P. and Walstra, P. (1991). Relation between syneresis and rheological properties of particle gels. *Colloid and Polymer Science*, **269**, 620–27.

van Vliet, T., Lakemond, C. M. M. and Visschers, R. W. (2004). Rheology and structure of milk protein gels. *Current Opinion in Colloid and Interface Science*, **9**, 298–304.

Vasbinder, A. J., Rollema, H. S. and de Kruif, C. G. (2003). Impaired rennetability of heated milk; study of enzymatic hydrolysis and gelation kinetics. *Journal of Dairy Science*, **86**, 1548–55.

Veerman, C., Sagis, L. M. C., Heck, J. and van der Linden, E. (2003). Mesostructure of fibrillar bovine serum albumin gels. *International Journal of Biological Macromolecules*, **31**, 139–46.

Veith, P. D. and Reynolds, E. C. (2004). Production of a high gel strength whey protein concentrate from cheese whey. *Journal of Dairy Science*, **87**, 831–40.

Vreeker, R., Hoekstra, L. L., den Boer, D. C. and Agterof, W. G. M. (1992). Fractal aggregation of whey proteins. *Food Hydrocolloids*, **6**, 423–35.

Walstra, P. (1990). On the stability of casein micelles. *Journal of Dairy Science*, **73**, 1965–79.

Walstra, P. (1993). The syneresis of curd. In *Cheese: Chemistry, Physics and Microbiology,* Volume 1, *General Aspects*, 2nd edn (P. F. Fox, ed.) pp. 141–91. London: Chapman and Hall.

Walstra, P. (1998). Relationship between structure and texture of cultured milk products. *Texture of Fermented Milk Products and Dairy Desserts*, IDF Special Issue 9802, pp. 9–15. Brussels: International Dairy Federation.

Walstra, P., van Dijk, H. J. M. and Geurts, T. J. (1985). The syneresis of curd. 1. General considerations and literature review. *Netherlands Milk and Dairy Journal*, **39**, 209–46.

Walstra, P., Wouters, J. T. M. and Geurts, T. J. (2005). *Dairy Science and Technology*, 2nd edn. New York: CRC Press.

Wang, Q., Bulca, S. and Kulozik, U. (2007). A comparison of low-intensity ultrasound and oscillating rheology to assess the renneting properties of casein solutions after UHT heat pre-treatment. *International Dairy Journal*, **17**, 50–58.

Wang, T. and Lucey, J. A. (2003). Use of multi-angle laser light scattering and size-exclusion chromatography to characterize the molecular weight and types of aggregates present in commercial whey protein products. *Journal of Dairy Science*, **86**, 3090–101.

17

Milk proteins: a cornucopia for developing functional foods

Paul J. Moughan

Abstract

Milk proteins have a central role to play in the development of functional foods—foods that have targeted physiological effects in the body over and above the normal effects of food nutrients. Milk proteins contain high amounts of bioavailable amino acids making them ideal ingredients for the manufacture of nutritionals—foods designed for specific nutritional purposes. Certain amino acids (e.g. tryptophan as a precursor of serotonin or leucine in the regulation of muscle metabolism) have specific physiological roles and some isolated milk proteins have particularly high concentrations of these amino acids, allowing foods to be developed to target physiological end points.

Milk proteins, and especially whey protein and glycomacropeptide, have an application in inducing satiety in humans and the relatively low yield of ATP per unit amino acid in comparison with glucose or fatty acids means that milk proteins are ideal ingredients for weight-loss foods.

Finally, milk proteins are known to be a rich source of bioactive peptides, released in the gut naturally during digestion. These peptides have a plethora of physiological effects and notable effects locally at the gut level. This chapter discusses the multiple nutritional and physiological properties of milk proteins and peptides in the context of functional foods.

Milk Proteins: From Expression to Food
ISBN: 978-0-12-374039-7

Introduction

Over the last century, scientists have gradually come to better understand the complexity of foods. In addition to delivering nutrients, satisfying hunger and providing pleasure, food components are now known to have a role in directly influencing physiological processes in the body. In particular, certain food components may assist in maintaining or promoting health and well-being and preventing the development of disease. These components may be non-nutrients (e.g. the antioxidant effect of polyphenols or the blood-cholesterol-lowering effect of plant sterols) or nutrients (e.g. the effects of short-chain volatile fatty acids in modulating gut development and function). In fact, the often multiple effects of some chemical compounds released from foods during digestion, traditionally considered to have a sole role in nourishment, call into question the very definition of "a nutrient".

It is also now widely appreciated that body growth and maintenance processes are a subtle interaction between nutrient supply and genetically regulated metabolism. The assimilation and metabolism of food compounds is subject to genetic and epigenetic control, which will vary among individuals and diverse populations, leading to important gene–nutrient interactions. In turn, nutrients and food non-nutrients also greatly influence gene expression, with effects once again varying among individuals; the complexity and the subtlety of nutrigenomics and nutrigenetics are really only beginning to be understood.

Over the next 50 years, it is expected that great advances will be made in understanding how food influences gene expression and how genetic regulation influences the assimilation and utilization of nutrients, and how the individual's genome explains differences among individuals in their physiological and nutritional response to different foods under different conditions. Such understanding will pave the path towards personalized nutrition and personalized health foods and dietary/exercise regimens.

Milk is an excellent example of a food having both nutritional and non-nutritional physiological roles in the human diet. Milk and, in particular, milk proteins not only supply the body with amino acids necessary for the maintenance and growth of body protein, but also give rise, during food manufacture and/or food digestion, to a myriad of protein fragments and large and small peptides that have distinct biological functions (Ward and German, 2004).

Also, certain amino acids (e.g. leucine, tryptophan) released during digestion have regulatory functions or act as precursors for the synthesis of key non-protein metabolites. Such compounds are a rich source of bioactive components for the development of functional foods. Although these compounds are undoubtedly of major significance to the suckled infant, whereby milk should be regarded as a biological fluid specifically designed by nature for optimal growth and development, they are also probably of importance in the adult diet, where milk has long been an important constituent.

Functional foods

A functional food has been defined by Diplock *et al.* (1999) as:

> A food can be regarded as functional if it has beneficial effects on target functions in the body beyond nutritional effects in a way that is relevant to health and well-being and/or the reduction of disease.

In a sense, and following this definition strictly, most if not all foods could be described as "functional" and perhaps the definition needs to be expanded to include the notion that the "beneficial effect on target functions" occurs at a meaningful level. Minor beneficial effects on target functions, which would be consequent upon the ingestion of many mainstream foods, may be relevant in the long run, in the context of a balanced diet, to health, well-being and disease prevention, but such foods are not widely regarded as functional foods.

Under the latter broader definition of a functional food, foods may be functional naturally (e.g. oily fish containing high amounts of omega 3 fatty acids) or may be rendered functional usually by adding a bioactive component to a food or by removing some component that is inhibiting bioactivity. Foods may also be enriched with a given bioactive component or components, through conventional animal or plant breeding practices, by genetic engineering or by manipulation of the feeding and nutrition of the plant or animal.

Functional foods is a rapidly growing sector of the international food industry, with development spurred by a number of technical, social and economic drivers. Firstly, there is, today, a high degree of awareness of the link between diet and health established largely through well-publicized epidemiological studies. This knowledge has been expanded by the more cogent "proof of cause and effect", from human intervention studies. Well-educated consumers, aware of the importance of diet to health, are demanding healthy and functional foods and are prepared to pay a price premium for such foods. Escalating healthcare costs, which are a major concern to governments, encourage "disease prevention rather than cure" community health strategies.

Completion of the human genome project with a rapid accumulation of knowledge concerning single and multiple gene effects (heralding pre-symptomatic testing for disposition to particular conditions), coupled with a better understanding of inter-generational nutritional effects (Barker hypothesis) on predisposition to chronic disease states, is likely to lead in the near future to personalized nutrition strategies and further demand for specific functional foods tailored to the requirements of the individual (mass customization).

If the functional foods industry is to achieve its full potential, however, it will be critical that regulatory bodies have a clear, consistent and rigorous approach to safety, labeling and health claims issues and that food manufacturers contract reputable science providers to independently establish "proof of concept" around their products (Roberfroid, 2000). There will also need to be a considerable investment in research and development to clearly establish cause-and-effect relationships between food compounds and targeted physiological end points and recognized disease risk factors.

Consumer confidence can quickly be eroded by conflicting messages received from the scientific community, underlining the critical role of regulatory authorities in ensuring that sufficient adequate information is available and is correctly analyzed using meta-analysis techniques, to make sustainable health claims, both qualified and approved claims. The burden of proof for generic health claims and for claims on the efficacy of specific food products needs to be considerable, such that consumers can have high levels of confidence.

The need for adequate evidence to substantiate claims and the need for regulatory authorities to be cautious in allowing claims is borne out by the recently published conclusion of the American Heart Association (Sacks *et al.*, 2006) that:

> earlier research indicating that soy protein has clinically important favorable effects [decreased blood LDL cholesterol] as compared with other proteins has not been confirmed.

The US Food and Drug Administration (FDA) had previously (October, 1999) approved labeling for foods containing soy protein as protective against heart disease, which in the light of fuller information does not appear to be supported. There has been widespread acceptance in the food industry and among consumers of the cholesterol-lowering properties of soy protein, but it would now appear that such an effect is meagre even at high daily intakes of soy protein. There are many other examples of conflicting information. The food industry must be careful that, while forging ahead with new product development and associated implied or stated health claims, long-term consumer confidence is not eroded.

Foods undoubtedly have a major role to play in preventing disease and ensuring health and vitality. However, if functional foods are to achieve their potential as part of an overall lifestyle stratagem towards healthfulness, then consumers must be guided by the highest quality information and distilled findings that have a strong likelihood of remaining substantiated over time. The food industry, science providers and government bodies all have a responsibility to ensure that the functional foods movement is led by ethical informed decision making. The recent Institute of Food Technologists (IFT, USA) Expert Report on Functional Foods (2005), although recognizing the enormous potential for functional foods, stated:

> But that is not to say that IFT believes that all foods on the market for which claims are being made are being properly represented based on science and proper regulatory policies. IFT does not support some claims on foods marketed today because they are not supported by today's science.

Clearly there is a challenge! The IFT Expert Panel goes on to recommend basing structure/function health claims on broad-based scientific criteria that address the underlying link between health and nutrition and meet the need for sound scientific substantiation supporting the structure/function effect. The panel discusses principles around ensuring the safety of functional foods and has introduced the concept of GRAE (generally recognized as efficacious), analogous to GRAS (generally recognized as safe), to encourage public confidence in the labeling of functional foods.

The dairy industry is in an excellent position to take advantage of the trend towards more healthy diets, more healthy foods and functional foods. Because milk proteins are readily available sources of amino acids and give rise to many bioactive compounds, they have a central place in the development of both specialized nutritionals and functional foods.

Milk proteins as a source of amino acids

Milk and milk proteins have long been regarded as a rich source of readily digestible and nutritionally available amino acids.

Specialized nutritionals

Early *in vivo* determinations of protein digestibility were based on fecal measurement, which is now known to be flawed given the significant degree of colonic microbial metabolism known to occur and that amino acids either are not absorbed as such from the large intestine or are absorbed to only a very limited extent (Moughan, 2003). The preferred accurate method for determining amino acid digestibility is to determine unabsorbed amino acids at the end of the small bowel (terminal ileum). This can be achieved in humans using naso-intestinal intubation or through the co-operation of ileostomates. Alternatively, animal models can be used with rat and pig ileal digestibility assays, both being suitable (Moughan *et al.*, 1994).

Where a protein has undergone structural alteration due to processing or storage (especially at high temperatures), conventional digestibility measures are inappropriate for some amino acids and in particular the often first-limiting amino acid, lysine. A new lysine bioavailability assay (based on the collection of ileal digesta and application of the guanidination reaction) has been developed; it can usefully be applied to damaged proteins (Moughan, 2003). When digestibility determinations are based on the sampling of ileal digesta, it is important to recognize that digesta contain copious quantities of endogenous (of body origin) protein in addition to undigested protein. This endogenous protein component needs to be taken into account (Moughan *et al.*, 1998), to yield "true" rather than "apparent" estimates of digestibility.

There are very few published data on true ileal protein digestibility as determined using human subjects. A comprehensive set of studies in humans (Gaudichon *et al.*, 1994, 1995, 1996; Mahé *et al.*, 1995, 1996) demonstrated a high true ileal digestibility of protein in milk proteins of around 95%. Comparable values, using the same methodology, for soy and pea proteins were 91 and 89% respectively. Sandstrom *et al.* (1986) gave soy- and meat-based diets to ileostomates and reported true ileal digestibility coefficients for total nitrogen in the range 80–85%. The naso-intestinal intubation method with normal volunteers has also been used to determine digestibility coefficients for individual amino acids (Gaudichon *et al.*, 2002).

For cow's milk, true ileal digestibility ranged from 92% for glycine to 99% for tyrosine, whereas, for soy bean, digestibility ranged from 89% for threonine to 97% for tyrosine. In our own laboratory, true ileal amino acid digestibility determined

Table 17.1 Mean true ileal digestibility of selected amino acids in a range of soy and dairy products

Amino acid	Soy protein concentrate	Soy protein isolate	Lactic casein	Sodium caseinate	Whey protein concentrate	α-Lactalbumin	Milk protein concentrate
Lysine	97.3	98.5	98.8	98.0	98.2	94.7	98.3
Methionine	95.3	100.0	100.0	99.6	100.0	99.2	100.0
Cysteine	86.9	95.3	99.2	93.0	99.6	96.1	97.8
Isoleucine	96.4	96.8	94.8	90.6	98.1	95.4	94.9
Leucine	95.7	95.3	99.1	97.6	99.1	96.1	98.9

Source: Adapted from Rutherfurd and Moughan (1997), with permission of the publisher

using ileostomates ranged from 98% for aspartate to 100% for cysteine in sodium caseinate; from 93% for threonine to 99% for cysteine in whey protein concentrate; from 95% for glycine to 99% for arginine in soy protein isolate; and from 91% for cysteine to 100% for arginine in soy protein concentrate (Moughan *et al.*, 2005a).

There are more comprehensive data on the true ileal digestibility of amino acids in various milk proteins, which have been obtained using animal models for digestion in humans (Rutherfurd and Moughan, 1997). Table 17.1 shows ileal digestibility data obtained using the laboratory rat for selected amino acids in soy protein concentrate, soy protein isolate, lactic casein, sodium caseinate, whey protein concentrate, α-lactalbumin and milk protein concentrate. These data confirm the very high digestibility of milk-derived proteins to the end of the ileum in simple-stomached mammals. The dietary amino acids are virtually completely digested.

Almost all dairy proteins have been subjected to some type of processing during their manufacture and, given that milk products often contain the reducing sugar lactose, they are susceptible to damage to the amino acid lysine. A specific assay designed to allow an accurate determination of lysine bioavailability in processed foods (Moughan and Rutherfurd, 1996) has recently been applied to a range of commercially available milk-protein-based products (Table 17.2), once again underscoring the high bioavailability of milk proteins and the limited amount of lysine damage incurred by proteins with modern controlled processing. In contrast, when the same bioassay was applied to grain-based processed foods, including cereals for children, substantial amounts of lysine damage were found (Table 17.3). Milk and milk-based products have an important role in complementing cereal foods and in supplying available lysine.

Figure 17.1 highlights the substantial differences that exist in the amounts of digestible amino acids supplied by plant proteins (e.g. soy) and milk proteins (e.g. α-lactalbumin). The often first-limiting amino acids (lysine and methionine plus cysteine) are found in much higher concentrations in the milk proteins, making them excellent sources of amino acids and very important dietary constituents to afford a balanced dietary protein intake.

Because of their relatively high levels of nutritionally important amino acids, milk proteins are utilized efficiently by humans, when given as a sole protein source. Tomé and Bos (2000) reported net post-prandial protein utilization values of 80% and 72% for milk protein and soy protein respectively, measured over 8 h after the ingestion of standard meals by healthy human subjects.

Table 17.2 True ileal reactive lysine digestibility (bioavailability, %) and digestible total and reactive lysine contents (g/kg air-dry) for 12 dairy protein sources

Product	Digestibility	Digestible lysine	
		Total	Reactive[a]
Whole milk protein	98.3	26.2	24.0
Infant formula A	91.0	8.3	8.6
Infant formula B	92.3	9.1	9.2
Infant formula C	93.1	11.1	11.7
Whey protein concentrate	98.5	79.9	77.5
UHT milk	100.0	31.7	31.4
Evaporated milk	96.7	23.4	20.5
Weight-gain formula	99.0	24.4	24.1
Sports formula	98.0	20.4	19.1
Elderly formula	97.1	11.7	11.8
Hydrolyzed lactose milk powder	98.6	27.2	25.1
High-protein supplement	99.9	14.3	14.3

[a]Bioavailable lysine; minimal difference between total lysine and reactive lysine denotes minimal Maillard damage
Source: Adapted from Rutherfurd and Moughan (2005), with permission of the publisher

Table 17.3 True ileal digestible total and reactive lysine contents (g/kg air-dry) in selected cereal-based foods

Cereal product	Digestible lysine	
	Total	Reactive[a]
Wheat-based (shredded)	1.3	0.8
Corn-based (flaked)	0.4	0.2
Rice-based (puffed)	1.1	0.6
Mixed cereal (rolled)	3.2	1.9

[a]Bioavailable lysine; a difference between total lysine and reactive lysine denotes Maillard damage
Source: Adapted from Rutherfurd *et al.* (2006), with permission of the publisher

Given the high bioavailability of amino acids in milk proteins and their abundant supply, it is hardly surprising that milk proteins are commonly used for the manufacture of so-called nutritionals, i.e. foods designed for a specific nutritional purpose (e.g. infant, sports, elderly formulas).

In the future, with increasing human population growth and greater pressure on food supplies, it is likely that milk proteins will play an ever more important role as protein "balancers" in plant-based processed foods.

Specific physiological roles

Amino acids may have physiological roles that are unrelated to their direct involvement in protein synthesis. These include their roles as neurotransmitters (e.g. glutamate, aspartate and glycine) and as precursors for the synthesis of other molecules

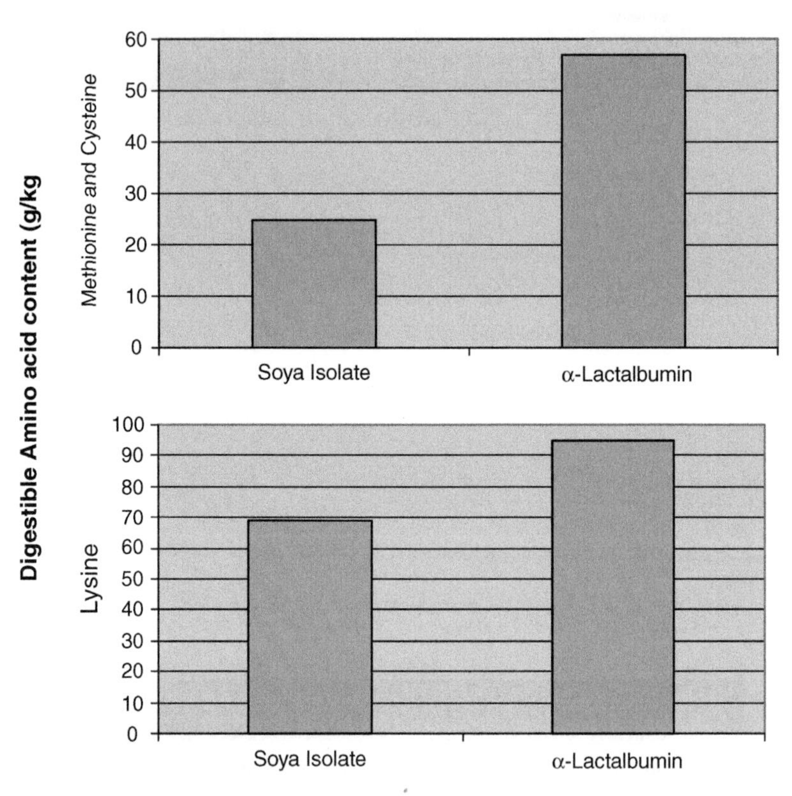

Figure 17.1 Digestible (true ileal) amino acid contents of a plant protein and a milk protein (adapted from Rutherfurd and Moughan, 1997, with permission of the publisher).

involved in neuromuscular function (e.g. creatine and taurine) and in host defences (e.g. glutathione and nitric oxide). Tryptophan is a precursor for the synthesis of serotonin, potentially impacting mood control (van de Poll *et al.*, 2006) and appetite regulation (Fernstrom and Wurtman, 1972; Fernstrom and Fernstrom, 1995), whereas the nitrogen-rich amino acid arginine leads to the production of nitric oxide (Wu and Morris, 1998), which is considered to have a significant role in cell signalling and the control of endothelial tone. Depending on its site of release, nitric oxide exerts several known functions, including stimulation of the pituitary gland, vasodilation, neurotransmission and immune modulation.

Another example of an amino acid with a specific metabolic role is the branched-chain amino acid leucine, which has a unique role in the regulation of muscle protein synthesis (Kimball and Jefferson, 2001). Interestingly, leucine stimulates protein synthesis directly in skeletal muscle but not in liver. The other branched-chain amino acids—isoleucine and valine—are less effective in stimulating muscle protein synthesis compared with leucine. Leucine supplementation has been shown to stimulate recovery of muscle protein synthesis during food restriction and after endurance exercise (Gautsch *et al.*, 1998; Anthony *et al.*, 2000). It has also been suggested (Layman, 2002, 2003) that leucine has a potential regulatory role in glycaemic control.

It has been known for many years that glutamine, glutamate and aspartate are preferred oxidative fuels for the gut—a highly metabolic organ. Consequently, many studies with humans and animals have been undertaken to investigate the effects, especially of dietary glutamine, on intestinal mucosal integrity, glutathione synthesis and immune function. This has led to debate as to whether glutamine should be regarded as a "conditionally essential" dietary amino acid (Grimble, 1993). In the traumatized patient, dietary glutamine may be needed to maintain immune responsiveness and for maintenance of the mucosal barrier against bacterial action and endotoxins.

There are also amino acids not found in proteins (i.e. non-protein amino acids) with specific physiological functions. The classic example is taurine (β-aminoethanesulfonic acid), which is synthesized by the body from cysteine or methionine and is essential for the production of conjugated bile salts (taurocholic acid). Taurine is found in milk, normally in the free form. It is recognized that cow's milk has low concentrations of taurine relative to human milk, raising the question as to whether cow's-milk-based infant formulas should contain added taurine.

The above are examples of specific physiological functions of amino acids; there are many others. It is anticipated that, over the next decade, our understanding of the physiological roles of individual amino acids will increase, leading to opportunities to develop functional foods containing higher or lower amounts of certain amino acids. van de Poll *et al.* (2006) have provided a useful summary of current evidence for proven functional effects (clinical benefits) in humans consequent upon dietary supplementation with specific amino acids.

Arginine has been widely used in supplemental nutrition for surgical patients and patients with burns, to modify the inflammatory response, to enhance organ perfusion and to stimulate wound healing, but the benefits accruing from arginine supplementation are not uniformly proven and accepted. There is some evidence that taurine supplementation improves retinal development in premature babies receiving parenteral nutrition. Taurine is often added to feeding formulas for infants and growing children. Glutamine is one of the more extensively studied amino acids and has been used in the preparation of medical foods. There is evidence that glutamine supplementation may reduce infectious morbidity and the length of hospital stay in surgical patients. Phenylalanine-free preparations have application in phenylketonuria.

Etzel (2004) highlighted an opportunity for the dairy industry, whereby a number of refined high-quality proteins are produced and marketed. The diverse amino acid compositions of these proteins can be exploited. The mixtures of proteins in milk and whey may be fractionated to give isolated proteins (α-lactalbumin, β-lactoglobulin, immunoglobulins, bovine serum albumin, the caseins, lactoferrin, lactoperoxidase and the peptide glycomacropeptide that is cleaved from κ-casein by chymosin) and blends of proteins with unique amino acid patterns. Etzel (2004) compared the amino acid compositions of several milk proteins with the composition of a theoretical "average" protein. The amino acid composition of the theoretical "average" protein was calculated from the frequency of occurrence of each amino acid in 207 unrelated proteins of known sequence.

Table 17.4 provides an abridged version of the Etzel (2004) data set. Some interesting comparisons can be made. Firstly, glycomacropeptide is completely devoid of cysteine, histidine, phenylalanine, tyrosine and tryptophan. Cysteine content is relatively high in α-lactalbumin, whereas glutamine has a relatively high concentration

Table 17.4 The amounts (% air-dry) of selected amino acids in various milk proteins and in an "average" protein

Amino acid	β-Lactoglobulin	α-Lactalbumin	Glycomacropeptide	Whey protein isolate	"Average" protein[a]
Cysteine	2.8	5.8	0	1.7	2.6
Glutamine	6.3	4.5	3.8	3.4	4.6
Glutamic acid	11.3	7.3	15.5	15.4	7.3
Histidine	1.5	2.9	0	1.7	2.6
Isoleucine	6.2	6.4	11.9	4.7	4.8
Leucine	13.6	10.4	1.7	11.8	7.8
Valine	5.4	4.2	8.9	4.7	6.2
Phenylalanine	3.2	4.2	0	3.0	4.7
Tryptophan	2.0	5.3	0	1.3	1.9
Tyrosine	3.6	4.6	0	3.4	5.2
Threonine	4.4	5.0	16.7	4.6	5.5

[a]Based on amino acid compositions of 207 unrelated sequenced proteins
Source: Adapted from Etzel (2004), with permission of the publisher

in β-lactoglobulin and glutamic acid is found at a high concentration in three of the dairy products. Leucine content is some twofold higher in β-lactoglobulin compared with the "average" protein, and the threonine content of glycomacropeptide is extraordinarily high. The branched-chain amino acids as a group are higher in concentration than that found for the "average" protein. It is clear from this type of comparison that milk-based products can be developed with amino acid compositions targeting particular physiological end points.

Role in providing calories and in promoting satiety

In addition to their role as a substrate for body protein synthesis and for the synthesis of various non-protein nitrogenous compounds, amino acids may also be used as a source of dietary energy; and the interaction between dietary protein and energy has long been understood.

However, the fact that different dietary macronutrients give rise, biologically, to quite different amounts of free energy (ATP) per unit gross energy (bomb calorimeter) is often overlooked. It is often argued that a "calorie is a calorie" regardless of the macronutrient giving rise to the energy. This is true but, what often fails to be appreciated, is that usually the numbers of calories in a food deemed to be derived from the respective macronutrients are estimates, not absolute measures. Conversion factors such as the Atwater factors are applied to determined amounts of macronutrients in a food and "available" energy is estimated. The point is that the conversion factors are not completely accurate and thus the "estimated calories" will not be completely accurate. This has particular relevance in the case of amino acids.

Atwater factors attempt to take into account the loss of energy due to incomplete absorption of the amino acid during digestion and the loss of energy in excreted urinary metabolites, post catabolism. However, the net yield of ATP during the catabolism of the amino acid and the ATP cost of synthesizing urea are not accounted for. The capture of net energy as ATP for an amino acid is less efficient than for other nutrients such as glucose and fatty acids, with an accompanying higher dietary-induced thermogenesis. That amino acids are used less efficiently energetically (i.e. have a higher dietary-induced thermogenesis) compared with glucose and fatty acids has important implications for designing weight-loss diets. Foods containing high amounts of protein will provide less "available energy" (i.e. ATP) per unit dry matter or gross energy, compared with foods high in available carbohydrate and/or fat.

Dairy proteins are a highly versatile source of amino acids, for the design of weight-loss foods, and more care should be taken in describing the calorific values especially for functional foods designed for weight loss. Protein has a further advantage for the formulation of weight-loss foods. It is now widely accepted that protein is a satiating nutrient and is effective relative to carbohydrate and fat in suppressing voluntary food intake independent of its calorific value. The role of dietary protein in the regulation of food intake and body weight in humans, and underlying mechanisms, has been the subject of recent reviews (Anderson and Moore, 2004; Westerterp-Plantenga and Lejeune, 2005). There is strong evidence that the protein content of a food is a determinant of short-term satiety and of how much food is eaten. The role of protein in the regulation of long-term food intake and body weight is less clear, because of a paucity of relevant experimental observations.

The role of protein in body weight regulation, in comparison with other macronutrients, is considered to consist of several often-related but different aspects: satiety, thermogenesis, metabolic energy efficiency and body composition. As stated, protein appears to increase satiety and therefore helps to sustain reduced-energy-intake diets.

Firstly, the highly satiating effect of protein has been observed both post-prandially and post-absorptively. Secondly, and also as discussed, high-protein diets are associated with a high dietary-induced thermogenesis, which could be related to the satiety effect of proteins. Thirdly, high-protein diets assist to maintain or increase fat-free body mass and the maintenance of a higher lean mass is costly energetically (i.e. a higher resting energy expenditure), leading to a lower associated metabolic efficiency of energy utilization.

Of particular interest to the dairy industry is the observation that protein source *per se* may be a factor influencing short-term satiety in humans. Whey protein has been identified as a candidate protein that may be highly effective in promoting satiety (Vandewater and Vickers, 1996; Portman *et al.*, 2000; Hall *et al.*, 2003). A basis for differences in satiety related to source of protein may be found in amino acid composition (e.g. a high leucine content stimulating protein synthesis and altering body energetics), in bioactive peptides released from the protein during digestion (refer to the following section), in different kinetics of protein digestion and, in the case of whey, in the presence of glycomacropeptide.

Milk proteins as a source of bioactive peptides

Milk is known to contain proteins (e.g. lactoferrin, lactoperoxidase, immunoglobulins) and free peptides having specific non-nutritional physiological functions. These

Table 17.5 Reported effects of natural food-derived peptides

- Modulation of gastrointestinal motility
- Stimulation of secretory processes
- Mineral binding
- Antibacterial properties
- Immunomodulation
- Antithrombotic activity
- Inhibition of angiotensin-converting enzyme (ACE) in the control of hypertension
- Analgesic (pain relief) and other neuroactive effects

compounds are undoubtedly important in the case of the human infant and may also have a functional role in adults. Of potentially greater significance, however, are the many small (from 3 to 20 amino acids) bioactive peptides encrypted in food protein amino acid sequences and released during digestion. Bioactive peptides are specific protein fragments that influence body function. These peptides are inactive within the sequence of the parent protein and can be released during proteolysis or fermentation. Bioactive peptides may act as physiological modulators locally in the gut and, potentially, systemically. Most, if not all, proteins appear to contain bioactive sequences, although the majority of research to date has been conducted with milk proteins.

An opioid activity of peptides derived from partial enzymatic digestions of milk proteins and wheat gluten was reported in the literature as early as 1979 (Brantl *et al.*, 1979; Zioudrou *et al.*, 1979). Since then, a considerable body of research has been undertaken, many different bioactive amino acid sequences have been discovered and physiological functions have been defined. The potential for bioactive peptides in the development of functional foods is great. It is now appreciated that bioactive peptides have a wide range of physiological effects, some of which are listed in Table 17.5. Specific bioactive peptides and protein hydrolysates can now be produced commercially, allowing for dietary supplementation and protein fortification. Casein-derived peptides are already in commercial use as food supplements (e.g. phosphopeptides) and as pharmaceuticals (Meisel, 1997).

The remainder of this section focuses on the first two functions listed in Table 17.5 (i.e. gut function), as an example of the potential of food-derived peptides as natural bioactive peptides. Several studies have described the involvement of bioactive peptides (exorphins) in regulating stomach emptying rate, gastrointestinal motility and gut secretory activity in mammals (see Rutherfurd-Markwick and Moughan, 2005). A role for bioactive peptides in influencing gut function is not to be unexpected, as the effects may be mediated both directly and hormonally, involving receptor sites in the gut without the need for absorption and systemic uptake of the peptide.

Our own studies within the Riddet Institute at Massey University highlight the potential importance, quantitatively, of the net effect of food-derived peptides on overall gut metabolism. The gut is a highly metabolic organ with changes in the rate of organ metabolism having significant implications for total body energetics, protein dynamics and amino acid and other nutrient requirements. In our series of studies, terminal ileal digesta amino acid or nitrogen flow was determined as an indicator of overall protein dynamics in the upper digestive tract consequent upon the ingestion

Table 17.6 Endogenous lysine loss at the terminal ileum of the growing pig given protein-free (PF-), synthetic amino acid (SAA-), hydrolyzed casein (EHC-) or zein (Zn-) based diets

	Diet				Significance
	PF	SAA[a]	EHC[b]	Zn[c]	
Lysine loss (mg/kg dry matter intake)	252	284	448	389	$P < 0.05$

[a]Devoid of lysine, with intravenous lysine infusion
[b]Digesta were centrifuged and ultrafiltered (10 000 Da molecular weight cut-off)
[c]Naturally devoid of lysine, with intravenous lysine infusion
Adapted from Butts *et al.* (1993), with permission of the publisher

of a meal and is reflective of the various cellular and dietary controls on the protein secretion and amino acid reabsorption processes.

Endogenous amino acid flows (the net result of secretion and reabsorption) at the end of the ileum were determined following the provision of semi-synthetic corn-starch-based diets, differing in the source of dietary nitrogen (protein-free, synthetic amino acids, protein, hydrolyzed protein). A range of methods (Moughan *et al.*, 1998) to determine endogenous (of body origin) as opposed to exogenous (diet origin) nitrogen were applied.

Table 17.6 gives results for endogenous lysine (marker for total protein) flow at the end of the small bowel from a representative study from our series of experiments using the pig as a generalized mammalian model. The results clearly demonstrate that, when amino acids were present in the gut (either directly from the hydrolyzed casein or after being released from the digestion of dietary zein), endogenous protein loss at the end of the small bowel was significantly enhanced. The peptides led to an enhanced secretion of protein into the gut lumen and/or a reduced reabsorption of endogenous amino acids.

In any case, the loss of extra protein into the colon, whereupon the amino acids are not salvageable, represents a considerable loss of amino acids and is associated with a high metabolic energy cost. Further work has demonstrated that the quite dramatic effect (an almost 60% increase in flow for the hydrolyzed casein) of dietary peptides is dose dependent (Hodgkinson *et al.*, 2000; Hodgkinson and Moughan, 2006).

These results combined with those of several other similar studies provide compelling evidence for a significant influence of dietary peptides on gut protein dynamics and overall metabolism. Little is known about how these effects are mediated or how the magnitude of the effect is influenced by the source of dietary protein. Claustre *et al.* (2002) have recently shown that casein and lactalbumin hydrolysates (but not egg or meat hydrolysate) greatly stimulate mucin secretion in rat jejunum. The casein-hydrolysate-mediated effect was blocked by the administration of naloxone (an opioid antagonist), and β-casomorphin-7, an opioid peptide released from β-casein during digestion, also induced mucin secretion. The peptide effect was also inhibited by naloxone. It may be that the effects are more pronounced with milk proteins.

The co-operation of ileostomates has allowed our results obtained from animal studies to be confirmed in humans (Moughan *et al.*, 2005b; Table 17.7).

Table 17.7 Endogenous ileal lysine and total nitrogen losses (μg/g dry matter intake) in adult humans receiving a protein-free (PF-) or hydrolyzed casein (EHC-) based diet

	Diet		Significance
	PF	EHC[a]	
Lysine	383	614	$P < 0.01$
Total nitrogen	2061	4233	$P < 0.001$

[a]Digesta were centrifuged and ultrafiltered (10 000 Da molecular weight cut-off)
Source: Adapted from Moughan *et al.* (2005b), with permission of the publisher

Bioactive peptides, and it would seem particularly those arising from the digestion of milk proteins, have been shown to have multiple physiological effects often at modest dietary intakes. As more is understood about these effects, there will be the possibility to develop novel functional foods. The potential to develop protein hydrolysates, peptide fractions and commercially synthesized peptides, to target physiological end points associated with gut motility, digestion, energetics and satiety, is particularly promising.

Conclusions

In this chapter, a case has been made for the central place of dairy proteins in the development of functional foods and specialized nutritionals. Dairy proteins are a source of highly bioavailable amino acids and offer a diverse range of amino acid patterns and specific amino acid concentration ratios. Certain amino acids, having direct physiological as opposed to nutritional functions in humans, are found in some milk proteins in high concentrations. Milk proteins may have antimicrobial and immunomodulatory functions and induce and maintain satiety.

Additionally, amino acids have a relatively high dietary thermogenesis; thus dairy proteins are ideal for the formulation of specialized weight-loss foods. Almost every week, new information is reported concerning bioactive peptides, which are released in the gut during the natural digestion of milk proteins. Milk is indeed a veritable cornucopia for developing functional foods.

References

Anderson, G. H. and Moore, S. E. (2004). Dietary proteins in the regulation of food intake and body weight in humans. *Journal of Nutrition*, **134**, 974S–979S.

Anthony, J. C., Yoshizawa, F., Gautsch-Anthony, T., Vary, T. C., Jefferson, L. S. and Kimball, S. R. (2000). Leucine stimulates translation initiation in skeletal muscle of postabsorptive rats via a rapamyscin-sensitive pathway. *Journal of Nutrition*, **130**, 2413–19.

Brantl, V. H., Teschemacher, H., Blasig, J., Henschen, A. and Lottspeich, F. (1979). Novel opioid peptides derived from casein (β-casomorphins). I. Isolation from bovine casein peptone. *Hoppe-Seyler's Zeitschrift fur Physiological Chemistry*, **360**, 1211.

Butts, C. A., Moughan, P. J., Smith, W. C. and Carr, D. H. (1993). Endogenous lysine and other amino acid flows at the terminal ileum of the growing pig (20 kg bodyweight): the effect of protein-free, synthetic amino acid, peptide and protein alimentation. *Journal of the Science of Food and Agriculture*, **61**, 31–40.

Claustre, J., Toumi, F., Trompette, A., Jourdan, G., Guignard, H., Chayvialle, J. A. and Plaisancié, P. (2002). Effects of peptides derived from dietary proteins on mucus secretion in rat jejunum. *American Journal of Physiology and Gastrointestinal Liver Physiology*, **283**, G521–G528.

Diplock, A. T., Aggett, P. J., Ashwell, M., Bornet, F., Fern, E. B. and Roberfroid, M. B. (1999). Scientific concepts for functional foods in Europe. Consensus document. *British Journal of Nutrition*, **81**, S1–S27.

Etzel, M. R. (2004). Manufacture and use of dairy protein fractions. *Journal of Nutrition*, **134**, 996S–1002S.

Fernstrom, M. H. and Fernstrom, J. D. (1995). Brain tryptophan concentrations and serotonin synthesis remain responsive to food consumption after the ingestion of sequential meals. *American Journal of Clinical Nutrition*, **61**, 312–19.

Fernstrom, J. D. and Wurtman, R. J. (1972). Brain serotonin content: physiological regulation by plasma neutral amino acids. *Science*, **178**, 414–16.

Gaudichon, C., Roos, N., Mahé, S., Sick, H., Bouley, C. and Tomé, D. (1994). Gastric emptying regulates the kinetics of nitrogen absorption from ^{15}N-labelled milk and ^{15}N-labelled yogurt in miniature pigs. *Journal of Nutrition*, **124**, 1970–77.

Gaudichon, C., Mahé, S., Roos, N., Benamouzig, R., Luengo, C., Huneau, J. F., Sick, H., Bouley, C., Rautureau, J. and Tomé, D. (1995). Exogenous and endogenous nitrogen flow rates and level of protein hydrolysis in the human jejunum after [^{15}N]-labeled milk and [^{15}N]-labeled yogurt ingestion. *British Journal of Nutrition*, **74**, 251–60.

Gaudichon, C., Mahé, S., Luengo, C., Laurent, C., Meaugeais, M., Krempf, M. and Tomé, D. (1996). A ^{15}N-leucine-dilution method to measure endogenous contribution to luminal nitrogen in the human upper jejunum. *European Journal of Clinical Nutrition*, **50**, 261–68.

Gaudichon, C., Bos, C., Morens, C., Petzke, K. J., Mariotti, F., Everwand, J., Benamouzig, R., Daré, S., Tomé, D. and Metges, C. C. (2002). Ileal losses of nitrogen and amino acids in humans and their importance to the assessment of amino acid requirements. *Gastroenterology*, **123**, 50–59.

Gautsch, T. A., Anthony, J. C., Kimball, S. R., Paul, G. L., Layman, D. K. and Jefferson, L. S. (1998). Availability of eIF4E regulates skeletal muscle protein synthesis during recovery from exercise. *American Journal of Physiology*, **274**, C406–C414.

Grimble, G. K. (1993). Essential and conditionally-essential nutrients in clinical nutrition. *Nutrition Research Reviews*, **6**, 97–119.

Hall, W. L., Millward, D. J., Long, S. J. and Morgan, L. M. (2003). Casein and whey exert different effects on plasma amino acid profiles, gastrointestinal hormone secretion and appetite. *British Journal of Nutrition*, **89**, 239–48.

Hodgkinson, S. M. and Moughan, P. J. (2006). An effect of dietary protein content on endogenous ileal lysine flow in the growing rat. *Journal of the Science of Food and Agriculture*, **87**, 233–38.

Hodgkinson, S. M., Moughan, P. J., Reynolds, G. W. and James, K. A. C. (2000). The effect of dietary peptide concentration on endogenous ileal amino acid loss in the growing pig. *British Journal of Nutrition*, **83**, 421–30.

Institute of Food Technologists (IFT, USA) *Expert Report on Functional Foods* (2005). http://www.ift.org/cms/?pid = 1001247

Kimball, S. R. and Jefferson, L. S. (2001). Regulation of protein synthesis by branched-chain amino acids. *Current Opinion in Clinical Nutrition and Metabolic Care*, **4**, 39–43.

Layman, D. K. (2002). Role of leucine in protein metabolism during exercise and recovery. *Canadian Journal of Applied Physiology*, **27**, 592–608.

Layman, D. K. (2003). The role of leucine in weight loss diets and glucose homeostasis. *Journal of Nutrition*, **133**, 261S–267S.

Mahé, S., Benamouzig, R., Gaudichon, C., Huneau, J. F., De Cruz, I. and Tomé, D. (1995). Nitrogen movements in the upper jejunum lumen in humans fed low amounts of caseins or β-lactoglobulin. *Gastroenterology and Clinical Biology*, **19**, 20–26.

Mahé, S., Roos, N., Benamouzig, R., Davin, L., Luengo, C., Gagnon, L., Gausserès, N., Rautureau, J. and Tomé, D. (1996). Gastrojejunal kinetics and the digestion of [^{15}N]beta-lactoglobulin and casein in humans: the influence of the nature and quantity of the protein. *American Journal of Clinical Nutrition*, **63**, 546–52.

Meisel, H. (1997). Biochemical properties of regulatory peptides derived from milk proteins. *Biopolymers*, **43**, 119–28.

Moughan, P. J. (2003). Amino acid availability: aspects of chemical analysis and bioassay methodology. *Nutrition Research Reviews*, **16**, 127–41.

Moughan, P. J. and Rutherfurd, S. M. (1996). A new method for determining digestible reactive lysine in foods. *Journal of Agricultural and Food Chemistry*, **44**, 2202–9.

Moughan, P. J., Cranwell, P. D., Darragh, A. J. and Rowan, A. M. (1994). The domestic pig as a model for studying digestion in humans. In *Digestive Physiology in the Pig* (W. B. Souffrant and H. Hagemeister, eds) Volume II, pp. 389–96. Dummerstorf, Germany: Forschungsinstitut fur die Biologie Landwirtschaftlicher Nutztiere (FBN).

Moughan, P. J., Souffrant, W. B. and Hodgkinson, S. M. (1998). Physiological approaches to determining gut endogenous amino acid flows in the mammal. *Archives of Animal Nutrition*, **51**, 237–52.

Moughan, P. J., Butts, C. A., van Wijk, H., Rowan, A. M. and Reynolds, G. W. (2005a). An acute ileal amino acid digestibility assay is a valid procedure for use in human ileostomates. *Journal of Nutrition*, **135**, 404–9.

Moughan, P. J., Butts, C. A., Rowan, A. M. and Deglaire, A. (2005b). Dietary peptides increase endogenous amino acid losses from the gut in adults. *American Journal of Clinical Nutrition*, **81**, 1359–65.

Portman, R., Bakal, A. and Peikin, S. R. (2000). Ingestion of a premeal beverage designed to release CCK reduces hunger and energy consumption in overweight females. *Obesity Research*, **8**(Suppl 1), PB59 (Abstract).

Roberfroid, M. B. (2000). Concepts and strategy of functional food science: the European perspective. *American Journal of Clinical Nutrition*, **71**(Suppl), 1660S–64S.

Rutherfurd, S. M. and Moughan, P. J. (1997). The digestible amino acid composition of several milk proteins: application of a new bioassay. *Journal of Dairy Science*, **81**, 909–17.

Rutherfurd, S. M. and Moughan, P. J. (2005). Digestible reactive lysine in selected milk-based products. *Journal of Dairy Science*, **88**, 40–48.

Rutherfurd-Markwick, K. J. and Moughan, P. J. (2005). Bioactive peptides derived from food. *Journal of AOAC International*, **88**, 955–66.

Rutherfurd, S. M., Torbatinejad, N. M. and Moughan, P. J. (2006). Available (ileal digestible reactive) lysine in selected cereal-based food products. *Journal of Agricultural and Food Chemistry*, **54**, 9453–57.

Sacks, F. M., Lichtenstein, A., Van Horn, L., Harris, W., Kris-Etherton, P. and Winston, M. (2006). Soy protein, isoflavones, and cardiovascular health. An American Heart Association science advisory for professionals from the Nutrition Committee. *Circulation*, **113**, 1034–44.

Sandstrom, B., Andersson, H., Kivisto, B. and Sandberg, A. S. (1986). Apparent small intestinal absorption of nitrogen and minerals from soy and meat-protein based diets. A study on ileostomy subjects. *Journal of Nutrition*, **116**, 2209–18.

Tomé, D. and Bos, C. (2000). Dietary protein and nitrogen utilization. *Journal of Nutrition*, **130**, 1868S–73S.

van de Poll, M. C. G., Luiking, Y. C., Dejong, C. H. C. and Soeters, P. B. (2006). Amino acids – specific functions. In *Encyclopedia of Human Nutrition*, 2nd edn. (B. Caballero, L. Allen and A. Prentice, eds) pp. 92–100. Amsterdam: Elsevier.

Vandewater, K. and Vickers, Z. (1996). Higher-protein foods produce greater sensory-specific satiety. *Physiology and Behavior*, **59**, 579–83.

Ward, R. E. and German, J. B. (2004). Understanding milk's bioactive components: a goal for the genomics toolbox. *Journal of Nutrition*, **134**, 962S–7S.

Westerterp-Plantenga, M. S. and Lejeune, M. P. G. M. (2005). Protein intake and body-weight regulation. *Appetite*, **45**, 187–90.

Wu, G. and Morris, S. M. (1998). Arginine metabolism: nitric oxide and beyond. *Biochemical Journal*, **336**, 1–17.

Zioudrou, C., Streaty, R. A. and Klee, W. A. (1979). Opioid peptides derived from food proteins, the exorphins. *Journal of Biological Chemistry*, **254**, 2446–49.

18

Milk proteins: the future

Mike Boland

Abstract

This final chapter contemplates future trends and their likely impact on the production and use of milk proteins. We consider first global issues, including energy consumption, the global water economy and specific issues for dairy relating to greenhouse gases. We then review current and emerging trends in consumer demands and how they might impact on the market for milk proteins. Important factors are expected to be food safety and traceability, as well as an increasing concern for the effect of food on health and an increasing importance of nutrigenomics and personalized nutrition. Finally we consider some emerging technologies and how they might affect the future of milk protein production and processing.

Introduction

As a wrap-up of our journey from expression to food, this chapter takes a look at the possible future of food, especially as it relates to milk proteins. Global macro-environmental factors are considered first, then we examine consumer demands and trends and the likely impact of new technologies.

Global issues for food

Global issues that can be expected to have a predominant impact on future food production (and hence production of milk proteins) include: energy—primarily

Milk Proteins: From Expression to Food
ISBN: 978-0-12-374039-7

because of the greenhouse gas implications of energy use, but also because of the rising cost of energy; the effect of the water economy; and the consequence of methane emissions from cows on global warming. Fifty-year predictions for the USA include a 50% loss of available water, as well as land, and a 50% reduction in animal agriculture.

Milk and energy

Milk is energetically very expensive. Milk is an animal product: to produce it requires the cow to eat vegetable material that has already been produced in a nutritional format. However, milk is the most efficiently produced of the animal-produced foods—largely because the animal itself is not consumed.

It has been estimated that production of 50 kg of milk protein in the USA requires 7×10^6 kcal of feed energy (i.e. 585 kJ/kg), an energy efficiency of 30:1 (Pimentel and Pimentel, 1979). In contrast, the total energy input per kilogram for production of corn or soy protein in the USA calculates out to 58 kJ/kg (data calculated from Pimentel and Pimentel, 1979). These figures do not take into account the uptake of direct solar energy through photosynthesis as the crops grow, or the opportunity cost in energy for other use of the land used to grow these products.

These figures mean that dairy has a strong sensitivity to energy and energy-related costs. As energy costs rise, dairy protein products will increasingly be restricted to use in high-value or luxury foods and substitution by vegetable proteins will increase, particularly in the area of nutritional ingredients.

Early signs of energy sensitivity in the market are coming through the use of "food miles"—an inappropriately named measure of the carbon footprint (i.e. the energy cost) expended in producing and distributing foods. These measures can be expected to become more accurate and more stringently applied in future, but also will be potentially misused as non-tariff barriers in some jurisdictions. We note that, because most food products are shipped by sea from remote markets, the greenhouse gas component of shipping is small compared with production costs, even when food is shipped halfway across the world, such as from New Zealand to the UK: the contribution of CO_2 from shipping was estimated at 125 kg CO_2/tonne milk solids out of a total of 1422 kg CO_2/tonne, which in turn compared favorably with the figure of 2921 kg CO_2/tonne milk solids for the locally produced equivalent in the UK (Saunders *et al.*, 2006).

Milk and the water economy

Increasingly, international attention is being paid to the "water economy" as water becomes a limiting resource in many regions. The amount of "virtual water" in a product means the amount of water required to produce it throughout the production chain. The amount of virtual water in a range of products is given in Table 18.1.

Most of the virtual water in these products arises from on-farm activities, with processing water a minor component. Hence, protein product values have been calculated here by simply adjusting for the amount of protein in the parent product, without

Table 18.1 Virtual water content of dairy and related products

Product	Virtual water (m^3/ton)	Reference
Milk	990	Hoekstra and Chapagain (2007)
Milk Powder	4602	Hoekstra and Chapagain (2007)
Milk Protein Powders	18400	Calculated from above
Soy Beans	1789	Hoekstra and Chapagain (2007)
Soy Protein	5400	Calculated from above

adjustment for processing water or credit for the water value of any co-products. The key point is that, as with energy, the cost of water for producing milk-origin products is several-fold higher than for producing similar plant-origin products. This means that only countries that are very water-rich can ever contemplate producing animal-based products. This will impact in future as water distribution changes with climate change, but also threatens production in some parts of the world where existing water use is unsustainable, such as parts of Australia where water offtake has led to saline ingress into soils (Anderies *et al.*, 2006). A full discussion of the implications of climate change on dairy production is beyond the scope of this chapter, but recent droughts in Australia, leading to a downsizing of the dairy herd, may be portents of the future.

Implications of dairy methane production

Methane is worthy of special mention as a greenhouse gas because emissions from cows contribute substantially to the greenhouse gas load as a by-product of rumen digestion. Methane is recognized as a greenhouse gas and is rated as having a global warming potential 21 times that of the equivalent amount of carbon dioxide on a 100-year timescale. It has been estimated that methane is second in effect only to carbon dioxide and is responsible for around 10–15% of the present greenhouse gas effect in the atmosphere. Globally, ruminant livestock produce about 28% of methane emissions from human-related activities. A single adult cow is a relatively minor contributor, emitting only 80–110 kg of methane but, with about 100 million cattle in the USA alone and 1.2 billion large ruminants in the world, ruminants are one of the largest methane sources. In the USA, cattle emit about 5.5 million tonnes of methane per annum into the atmosphere, accounting for 20% of US methane emissions, with dairy cattle producing around one-quarter of the total (see http://www.epa.gov/methane/rlep/faq.html).

Most governments recognize the need to limit greenhouse gases and current international negotiations are expected to impose penalties on greenhouse gas producers. This will have serious implications for the dairy industry and penalties, or costs of compliance, may become prohibitively high in some jurisdictions. Research efforts to specifically target the removal of methanogenic organisms from the rumen are important for the future economic viability of the industry. Because methanogens are believed to have an important role in managing the hydrogen concentration in the rumen, it may be necessary to find or create a micro-organism that can transfer hydrogen into a product other than methane.

Consumer demands and trends for food and ingredients

Food safety and traceability

Throughout the world, awareness of foodborne disease has risen in response to the high level of publicity that such outbreaks receive. The toll exacted in human and economic terms is considerable. Notable dairy outbreaks in recent years include *Salmonella* in ice cream (USA, 1994: 224 000 cases of illness) and staphylococcal enterotoxin in milk (Japan, 2000: 15 000 cases). Contaminated soft cheeses and raw milk are often in the news. Most dairy products, processed to modern standards of hygiene, have an excellent safety record, but consumers are demanding increased surveillance and control of all foods, including dairy. The contamination of animal feed with dioxin in Belgium in 1999 highlighted the importance consumers place on the absence of toxic chemicals in their food. There will be no lessening in the demands on food producers to control risks and deliver assurances of safety. The increased costs from providing this assurance through effective process control will become the norm for dairy businesses in the future.

In recent times, increasing attention has been paid to traceability, so that any food safety issue can be quickly traced to its origin and other food from the same batch can be quickly quarantined. Traceability can also be important because of consumers' desire for products that are sustainably produced or have other connotations of quality (such as organically produced products).

Traceability is usually managed through labeling and tracking of products through manufacture and distribution, usually by means of labels on the packaging. This is usually well handled and food manufacturers and distributors are good at it. There have, however, been attempts at "false-flagging" products in the past and this will no doubt continue. For products containing milk proteins, it is often possible to obtain an internal check on the origin of the product: dairy herds in different countries and regions tend to have a rather unique mixture of breeds and genetics. This is reflected in the distribution of polymorphisms of the proteins, which can be relatively simply analyzed using gel electrophoresis and/or mass spectroscopy. Additional information about processing can be gained from mass spectroscopic analysis of post-production changes in the chemistry of milk proteins (see Chapter 10 earlier in this volume).

Food and health

Consumers are being increasingly sensitized to the effects of diet on health (and appearance). The success of diet clinics attests to this. The occurrence of (and attention being paid to) current high levels of obesity in affluent societies is spurring interest in diet at all levels of society, from individual to government. Food products on supermarket shelves are increasingly differentiated by the presence of (omega-3 fats, antioxidants) or absence of (fat-free, gluten-free) components believed to affect health.

Nutrigenomics

The combination of the availability of individual genetic data on a scale never before seen with a detailed understanding of nutrition has led to the field of "nutrigenomics": the study of the relationship between a person's genetic makeup and their individual nutritional needs. The relationship between the human body and food is now understood to be far more complex than simple nutrition.

Food contains bioactive materials that can interact with the body to stimulate or inhibit the activity of enzymes, the immune system and even the expression of genes. Individuals have differences in the genes that code for synthesis or control the expression of particular enzymes and other proteins called polymorphisms. The simplest and most common of these differ in only a single base pair in the DNA and are called single nucleotide polymorphisms, or SNPs. Individuals with different SNPs will respond differently to some food components and nutrigenomics is the science of how nutrition interacts with different individuals' genomes, with a view to creating a healthier diet for each individual.

Some experts distinguish between nutrigenomics–where the study is based on a whole genome and systems biology approach–and nutrigenetics, which involves hypothesis-driven investigation around specific known genetic variations (for example Mutch *et al.*, 2005). "Nutrigenomics" is the term widely used to cover both aspects of nutritional science and it is used in that sense here. Whole-genome nutrigenomics is still many years away and will not be possible until all the common SNPs are mapped and their interaction with food is understood. Meantime, nutritional advice based around a series of known SNPs that do interact with diet is being provided by several companies, based on specific DNA tests for these SNPs. Such advice is usually general nutritional advice, modified to take into account risk factors associated with "bad" SNPs. This is called personalized nutrition and is dealt with in the following section.

Nutrigenomics is expected to have a wide effect on the food market in affluent countries over the next 10–20 years (Oliver, 2005) and will provide both opportunities and challenges for the food manufacturer using dairy, and most other ingredients. While there are a few known SNPs that are likely to lead to effects from milk consumption (most notably those affecting the ability to metabolize lactose, leading to lactose intolerance), there is little or no specific information relating to milk proteins, apart from the "A2" milk case discussed below.

Personalized nutrition

Individuals can now obtain information about their own genetic profile, with respect to known genetic polymorphisms related to health and metabolism. In the USA, companies such as Sciona and Genelex will provide a mail-order analysis of key genetic polymorphisms together with advice about diet and lifestyle.

Personalized nutrition is a nutritional response to differences between individuals, whether from a nutrigenomics input or through other identified needs and preferences, and attempts to balance an individual's diet to their specific individual and situational needs. Nutrition today is not just about balance of macro- and micronutrients: a plethora of "functional" (bioactive) food components are also known to affect health in ways that extend far beyond the simple supply of nutrients, and they can be modifiers of nutrient uptake and usage, thus modifying the effect of nutritional balance as seen by the body's metabolism. The kinetics of nutrient uptake are just as important as overall absolute uptakes of nutrients. Personalized nutrition attempts to take this into account, to provide optimal customized nutrition for the individual.

In sophisticated markets today, there is increasing acceptance that nutrition has a profound effect on health and wellness and, as individuals become more aware of their specific nutritional needs, the demand for personalized nutrition is set to increase.

The impact of all this on milk proteins has to date been minimal. However, three aspects are notable.

- Allergies to milk, particularly in infants, have been attributed to β-lactoglobulin in cows' milk (although recent work has cast some doubt on this [Brix *et al.*, 2003]). This protein is not produced in human milk and is the dominant whey protein in bovine milk (see Chapter 1 of this volume). Whey proteins are important nutritionally, as they are a valuable source of essential amino acids. β-lactoglobulin is a particularly important source of branched chain amino acids. So-called hypoallergenic products are therefore produced by hydrolyzing milk proteins, more particularly whey proteins, so that fragments are sufficiently small to be non-antigenic.
- There is some literature that suggests a weak correlation between consumption of milk containing the β-casein A1 variant and some diseases, notably type 1 diabetes (Elliott at al., 1997) and ischaemic heart disease (McLachlan, 2001). Further studies on diabetes proved to be inconclusive (Beales *et al.*, 2004) and the heart disease data do not stand up to scrutiny; furthermore, other epidemiological data show that the A2 hypothesis does not hold up (Truswell, 2005). Notwithstanding this, a New Zealand-based company, the A2 Corporation, has been formed to produce and market a niche milk product, called A2 milk, from cows that do not carry the A1 gene. The milk has, to date, been sold mostly in Australia and the company is very careful not to make claims about any specific health benefits after having been prosecuted and fined $15 000 in Queensland in 2004 for making such claims. This product is now being sold in New Zealand and is shortly to be released in the USA—it will be interesting to see how it fares.
- There is increasing evidence that milk proteins and peptides have physiological functionality, in particular effects on cardiovascular health, immune modulation and anti-cancer effects. The validity of these effects remains to be fully proven, but in time may lead to new functional foods based on milk proteins and their products.

New technologies and their possible effect on milk protein ingredients and products

A range of new technologies has the potential to affect dairy production and processing in the near future. They include gene technologies that could lead to new, different milk proteins, new kinds of processing that can produce novel milk protein materials and products containing them, and new analytical techniques that have the potential to improve processing and place ever more stringent requirements on product quality.

Genetic modification

Milk proteins have been genetically modified and expressed in non-bovine animals (e.g. Bleck *et al.*, 1998) and in cows (Brophy *et al.*, 2003). However, it seems unlikely that transgenic modification of milk proteins for functional or nutritional purposes will occur widely in the foreseeable future. There are several reasons for this (Creamer *et al.*, 2002):

- Consumer acceptance of genetically modified (GM) foods is still variable, throughout the world, with most countries now having strict labeling requirements. Because milk is a liquid product handled in large volumes during processing, maintenance of batch identity and keeping GM milk separate are more problematic than with discrete products, although recent efforts with organic milk have proven that this is possible.
- Milk is an animal product that is strongly targeted at the health of babies and young people. This has been identified in consumer surveys as a very sensitive area (compared, for example, with the acceptability of GM fruit and vegetables) and milk will probably be one of the last foods for which genetic modification is acceptable to most consumers.
- The cost of producing herds of GM cows will be very high and developing herds will be very slow unless expensive cloning and embryo transfer methods are used. This is not justified by a small premium for improved nutrition or functionality arising from genetic modification.
- More importantly, a switch to genetic modification will severely limit genetic gain, because the gene pool will be restricted to the genetics of the donor animals for the original GM parents. This segregation from the global bovine gene pool will prevent, or severely limit, participation in the ongoing genetic improvement of the species, currently occurring at about 2% per annum.

Notwithstanding these points, if a strong nutraceutical or pharmaceutical component were to be identified, enhanced expression through genetic modification would not be out of the question. However, much-touted "gene-pharming" in dairy animals has not yet been notably successful commercially.

In contrast to milk proteins, competitive plant-origin proteins are well advanced in improvement using genetic modification. GM soy beans are now predominant in world soy bean crops, covering 54.4 million hectares in 2005, which makes up 60% of worldwide soy bean production (http://www.gmo-compass.org/eng/agri_biotechnology/gmo_planting/194.global_growing_area_gm_crops.html). Soy beans can be genetically modified to remove undesirable proteins such as trypsin inhibitor, soy haemagglutinin and allergens (e.g. Friedman *et al.*, 1991), while at the same time soy proteins can be modified to provide a more favorable nutritional balance of essential amino acids (Mandal and Mandal, 2000). The more efficient production of soy proteins in terms of energy and water, coupled with these improvements from genetic modification, means that soy proteins will increasingly out-compete dairy proteins as bulk nutritional and functional food ingredients.

Novel processing

High-pressure processing was originally developed in the late nineteenth century (Hite, 1899) but did not find application in food processing until the 1990s, when new materials enabled the development of production-scale processing equipment. High-pressure processing has been used commercially as an alternative method for preservation, particularly for acidic foods (Dunne and Kluter, 2001).

When milk is subjected to high pressure, the casein micelle undergoes dramatic non-reversible changes, leading to a smaller micelle that is less opaque (see Chapters 5 and 7 earlier in this volume). It has also been reported that high-pressure processing can alter the functionality of whey proteins (Patel *et al.*, 2005). Whether these novel modified proteins will find application in new foods remains to be seen.

New analytical methods

Recent years have seen a range of new and improved analytical methods that have potential to improve process control and tighten product specification. Particularly important are methods that can control product safety (particularly microbiology), as well as nutritional and functional properties.

One of the weaknesses in product safety in the past has been the need to grow up samples on Petri dishes to test for the presence of undesirable microbial species. This process is time-consuming and laborious and can identify issues with process and product only well after they have occurred. A range of novel microbial detection methods is emerging; they have the potential to allow at-line detection of microbiological problems, or conversely to provide early assurance of food safety. For example, use of flow cytometry was able to reduce times for measuring bacterial numbers from the 3 days required for the traditional plate count to 2 h (Flint *et al.*, 2006). This could be particularly important for proteins manufactured using ultrafiltration, such as whey protein concentrates and milk protein concentrates, because the ultrafiltration step co-concentrates any microbial contaminants that may be present.

Modern electrospray mass spectrometric analysis has enhanced our ability to understand and control processing effects that can alter the nutritional value of milk proteins, particularly the loss of bioavailable lysine due to processing and storage effects (see Chapter 10 earlier in this volume). Similarly, once the relationship between functionality and protein chemistry is well understood, the same techniques will allow better management of functional properties. Novel in-line and at-line methods are becoming possible through a range of techniques such as nuclear magnetic resonance, Fourier transform infrared spectroscopy and surface plasmon resonance analysis, for example, to measure water activities inside packaging and to predict the flavor of cheeses at maturation.

Materials science and nanotechnology

Food structure is important at all dimensional scales for the sensory properties of food (including texture, mouthfeel and flavor release) and may have an important effect on nutrient release and bioavailability (Parada and Aguilera, 2007). Increasingly, attention is being paid to materials science approaches to understanding and potentially managing these effects. An example is the physics of soft materials being applied to food (Mezzenga *et al.*, 2005).

The potential application of nanotechnology and nanoscience to food is likely to become an important area, for the reasons outlined above. Much of the higher dimensional structure of food is a consequence of nanostructures. It is unlikely that nano robotics will be applied to food in the foreseeable future: regardless of the considerable technical challenges, public acceptance can be expected to be a major barrier. Notwithstanding this, nanotechnology is having a considerable impact on food science, in part through the use of new improved instrumentation becoming available to support nanotechnology research (Foegeding, 2006; Weiss *et al.*, 2006).

One of the important features in nanotechnology is the occurrence of self-assembling molecular superstructures (nanostructures). It turns out that foods naturally contain many such systems, examples being the actin–myosin complex in the muscle fibers of meat, starch granules in plant foods and the casein micelle in milk. Whey proteins have also been shown to form self-assembling systems under certain conditions (Bolder *et al.*, 2006; Graveland-Bikker and de Kruif, 2006).

Conclusions

This brief chapter has provided a glimpse of some of the global issues and new technologies that may or may not influence the future development and use of dairy protein products. It is clear that milk proteins are expensive food ingredients and will increasingly be restricted to niche applications, and that global trends such as rising energy costs, scarcity of water and effect of greenhouse gases will increase the cost of production and will restrict land areas where they can be sustainably produced.

References

Anderies, J. M., Ryan, P. and Walker, B. H. (2006). Loss of resilience, crisis, and institutional change: lessons from an intensive agricultural system in Southeastern Australia. *Ecosystems*, **9**, 865–78.

Beales, P., Elliott, R., Flohé, S., Hill, J., Kolb, H., Pozzilli, P., Wang, G-S., Wasmuth, H. and Scott, F. (2004). A multi-center, blinded international trial of the effect of A^1 and A^2 β-casein variants on diabetes incidence in two rodent models of spontaneous Type I diabetes. *Diabetologia*, **45**, 1240–46.

Bleck, G. T., White, B. R., Miller, D. J. and Wheeler, M. B. (1998). Production of bovine alpha-lactalbumin in the milk of transgenic pigs. *Journal of Animal Science*, **76**, 3072–78.

Bolder, S. G., Hendrickx, H., Sagis, L. M. G. and van der Linden, E. (2006). Fibril Assemblies in aqueous whey protein mixtures. *Journal Agricultural Food Chemistry*, **54**, 4229–34.

Brix, S., Bovetto, L., Fritsché, R., Barkholt, V. and Frøkiaer, H. (2003). Immunostimulatory potential of β-lactoglobulin preparations: effects caused by endotoxin contamination. *Journal of Allergy and Clinical Immunology*, **112**, 1216–22.

Brophy, B., Smolenski, G., Wheeler, T., Wells, D., L'Huillier, P. and Laible, G. (2003). Cloned transgenic cattle produce milk with higher β-lactoglobulin and κ-casein. *Nature Biotechnology*, **21**, 157–62.

Creamer, L. K., Pearce, L. E., Hill, J. P. and Boland, M. J. (2002). Milk and Dairy Products in the 21st Century. *Journal of Agricultural and Food Chemistry*, **50**, 7187–93.

Dunne, C. P. and Kluter, R. A. (2001). Emerging non-thermal processing technologies: criteria for success. *Australian Journal of Dairy Technology*, **56**, 109–12.

Elliott, R. B., Wasmuth, H. E., Bibby, N. J. and Hill, J. P. (1997). The role of β-casein variants in the induction of insulin-dependent diabetes in the non-obese diabetic mouse and humans. In *Milk Protein Polymorphism*, (Creamer, L. K. ed.) pp. 445–453. Brussels, Belgium: International Dairy Federation.

Flint, S., Drocourt, J-L., Walker, K., Stevenson, B., Dwyer, M., Clarke, I. and McGill, D. (2006). A rapid, two-hour method for the enumeration of total viable bacteria in samples from commercial milk powder and whey protein concentrate powder manufacturing plants. *International Dairy Journal*, **16**, 379–84.

Foegeding, E. A. (2006). Food Biophysics of Protein Gels: A Challenge of Nano and Macroscopic Proportions. *Food Biophysics*, **1**, 41–50.

Friedman, M., Brandon, D. L., Bates, A. H. and Hymowitz, T. (1991). Comparison of a commercial soybean cultivar and an isoline lacking the Kunitz trypsin inhibitor: composition, nutritional value, and effects of heating. *Journal of Agricultural and Food Chemistry*, **39**, 327–35.

Graveleand-Bikker, J. F. and de Kruif, C. G. (2006). Unique milk protein based nanotubes: Food and nanotechnology meet. *Trends in Food Science and Technology*, **17**, 196–203.

Hite, B. H. (1899). The effect of pressure in the preservation of milk. *Bulletin of the West Virginia University Agricultural Experiment Station*, **58**, 15.

Hoekstra, A. Y. and Chapagain, A. K. (2007). Water footprints of nations: water use by people as a function of their consumption pattern. *Water Resource Management*, **21**, 35–48.

Mandal, S. and Mandal, R. K. (2000). Seed storage proteins and approaches for improvement of their nutritional quality by genetic engineering. *Current Science*, **79**, 576–89.

McLachlan, C. N. (2001). β-casein A1, ischaemic heart disease mortality, and other illnesses. *Medical Hypotheses*, **56**, 262–72.

Mezzenga, R., Schurtenberger, P., Burbridge, A. and Michel, M. (2005). Understanding foods as soft materials. *Nature Materials*, **4**, 729–40.

Mutch, D. M., Wahli, W. and Williamson, G. (2005). Nutrigenomics and nutrigenetics: the emerging faces of nutrition. *FASEB Journal*, **19**, 1602–16.

Oliver, D. (2005). The Future of Nutrigenomics. Institute for the Future Report SR–899, Palo Alto, Ca.:Institute for the Future.

Parada, J. and Aguilera, J. M. (2007). Food microstructure affects the bioavailability of several nutrients. *Journal of Food Science*, **72**, R21–R32.

Patel, H. S., Singh, H., Havea, P., Considine, T. and Creamer, L. K. (2005). Pressure-induced unfolding and aggregation of the proteins in whey protein concentrate solutions. *Journal of Agricultural and Food Chemistry*, **53**, 9590–601.

Pimentel, D. and Pimentel, M. (1979). *Food, Energy, and Society.* New York: Wiley.

Saunders, C., Barber, A., and Taylor, G. (2006). *Food Miles – Comparative Energy/Emissions Performance of New Zealand's Agriculture Industry*. Research Report No. 285. Lincoln University, Lincoln, New Zealand.

Truswell, A. S. (2005). The A2 milk case: a critical review. *European Journal of Clinical Nutrition*, **59**, 623–31.

Weiss, J., Takhistov, P. and McClements, D. J. (2006). Functional materials in food nanotechnology. *Journal of Food Science*, **71**, R107–R116.

Index

Food Science and Technology
International Series

Maynard A. Amerine, Rose Marie Pangborn, and Edward B. Roessler, *Principles of Sensory Evaluation of Food*. 1965.
Martin Glicksman, *Gum Technology in the Food Industry*. 1970.
Maynard A. Joslyn, *Methods in Food Analysis*, second edition. 1970.
C. R. Stumbo, *Thermobacteriology in Food Processing*, second edition. 1973.
Aaron M. Altschul (ed.), *New Protein Foods*: Volume 1, *Technology, Part A*—1974. Volume 2, *Technology, Part B*—1976. Volume 3, *Animal Protein Supplies, Part A*—1978. Volume 4, *Animal Protein Supplies, Part B*—1981. Volume 5, *Seed Storage Proteins*—1985.
S. A. Goldblith, L. Rey, and W. W. Rothmayr, *Freeze Drying and Advanced Food Technology*. 1975.
R. B. Duckworth (ed.), *Water Relations of Food*. 1975.
John A. Troller and J. H. B. Christian, *Water Activity and Food*. 1978.
A. E. Bender, *Food Processing and Nutrition*. 1978.
D. R. Osborne and P. Voogt, *The Analysis of Nutrients in Foods*. 1978.
Marcel Loncin and R. L. Merson, *Food Engineering: Principles and Selected Applications*. 1979.
J. G. Vaughan (ed.), *Food Microscopy*. 1979.
J. R. A. Pollock (ed.), *Brewing Science*, Volume 1—1979. Volume 2—1980. Volume 3—1987.
J. Christopher Bauernfeind (ed.), *Carotenoids as Colorants and Vitamin A Precursors: Technological and Nutritional Applications*. 1981.
Pericles Markakis (ed.), *Anthocyanins as Food Colors*. 1982.
George F. Stewart and Maynard A. Amerine (eds), *Introduction to Food Science and Technology*, second edition. 1982.
Hector A. Iglesias and Jorge Chirife, *Handbook of Food Isotherms: Water Sorption Parameters for Food and Food Components*. 1982.
Colin Dennis (ed.), *Post-Harvest Pathology of Fruits and Vegetables*. 1983.
P. J. Barnes (ed.), *Lipids in Cereal Technology*. 1983.
David Pimentel and Carl W. Hall (eds), *Food and Energy Resources*. 1984.
Joe M. Regenstein and Carrie E. Regenstein, *Food Protein Chemistry: An Introduction for Food Scientists*. 1984.
Maximo C. Gacula, Jr. and Jagbir Singh, *Statistical Methods in Food and Consumer Research*. 1984.
Fergus M. Clydesdale and Kathryn L. Wiemer (eds), *Iron Fortification of Foods*. 1985.
Robert V. Decareau, *Microwaves in the Food Processing Industry*. 1985.
S. M. Herschdoerfer (ed.), *Quality Control in the Food Industry*, second edition. Volume 1—1985. Volume 2—1985. Volume 3—1986. Volume 4—1987.
F. E. Cunningham and N. A. Cox (eds), *Microbiology of Poultry Meat Products*. 1987.
Walter M. Urbain, *Food Irradiation*. 1986.

Peter J. Bechtel, *Muscle as Food*. 1986. H. W.-S. Chan, *Autoxidation of Unsaturated Lipids*. 1986.
Chester O. McCorkle, Jr., *Economics of Food Processing in the United States*. 1987.
Jethro Japtiani, Harvey T. Chan, Jr., and William S. Sakai, *Tropical Fruit Processing*. 1987.
J. Solms, D. A. Booth, R. M. Dangborn, and O. Raunhardt, *Food Acceptance and Nutrition*. 1987.
R. Macrae, *HPLC in Food Analysis*, second edition. 1988.
A. M. Pearson and R. B. Young, *Muscle and Meat Biochemistry*. 1989.
Marjorie P. Penfield and Ada Marie Campbell, *Experimental Food Science*, third edition. 1990.
Leroy C. Blankenship, *Colonization Control of Human Bacterial Enteropathogens in Poultry*. 1991.
Yeshajahu Pomeranz, *Functional Properties of Food Components*, second edition. 1991.
Reginald H. Walter, *The Chemistry and Technology of Pectin*. 1991.
Herbert Stone and Joel L. Sidel, *Sensory Evaluation Practices*, second edition. 1993.
Robert L. Shewfelt and Stanley E. Prussia, *Postharvest Handling: A Systems Approach*. 1993.
Tilak Nagodawithana and Gerald Reed, *Enzymes in Food Processing*, third edition. 1993.
Dallas G. Hoover and Larry R. Steenson, *Bacteriocins*. 1993.
Takayaki Shibamoto and Leonard Bjeldanes, *Introduction to Food Toxicology*. 1993.
John A. Troller, *Sanitation in Food Processing*, second edition. 1993.
Harold D. Hafs and Robert G. Zimbelman, *Low-fat Meats*. 1994.
Lance G. Phillips, Dana M. Whitehead, and John Kinsella, *Structure-Function Properties of Food Proteins*. 1994.
Robert G. Jensen, *Handbook of Milk Composition*. 1995.
Yrjö H. Roos, *Phase Transitions in Foods*. 1995.
Reginald H. Walter, *Polysaccharide Dispersions*. 1997.
Gustavo V. Barbosa-Cánovas, M. Marcela Góngora-Nieto, Usha R. Pothakamury, and Barry G. Swanson, *Preservation of Foods with Pulsed Electric Fields*. 1999.
Ronald S. Jackson, *Wine Tasting: A Professional Handbook*. 2002.
Malcolm C. Bourne, *Food Texture and Viscosity: Concept and Measurement*, second edition. 2002.
Benjamin Caballero and Barry M. Popkin (eds), *The Nutrition Transition: Diet and Disease in the Developing World*. 2002.
Dean O. Cliver and Hans P. Riemann (eds), *Foodborne Diseases*, second edition. 2002. Martin
Kohlmeier, *Nutrient Metabolism*, 2003.
Herbert Stone and Joel L. Sidel, *Sensory Evaluation Practices*, third edition. 2004.
Jung H. Han, *Innovations in Food Packaging*. 2005.
Da-Wen Sun, *Emerging Technologies for Food Processing*. 2005.
Hans Riemann and Dean Cliver (eds) *Foodborne Infections and Intoxications*, third edition. 2006.
Ioannis S. Arvanitoyannis, *Waste Management for the Food Industries*. 2008.
Ronald S. Jackson, *Wine Science: Principles and Applications*, third edition. 2008.
Da-Wen Sun, *Computer Vision Technology for Food Quality Evaluation*. 2008.
Kenneth David and Paul B. Thompson, *What Can Nanotechnology Learn From Biotechnology?* 2008.
Elke K. Arendt and Fabio Dal Bello, *Gluten-Free Cereal Products and Beverages*. 2008.

Debasis Bagchi, *Nutraceutical and Functional Food Regulations in the United States and Around the World*, 2008.
R. Paul Singh and Dennis R. Heldman, *Introduction to Food Engineering*, fourth edition. 2008.
Zeki Berk, *Food Process Engineering and Technology*. 2009.
Abby Thompson, Mike Boland and Harjinder Singh, *Milk Proteins: From Expression to Food*. 2009.
Wojciech J. Florkowski, Stanley E. Prussia, Robert L. Shewfelt and Bernhard Brueckner (eds) *Postharvest Handling*, second edition. 2009.

Printed and bound by CPI Group (UK) Ltd, Croydon, CR0 4YY
08/05/2025
01864821-0001